Student Study Art Notebook

to accompany

AF838499

Anatomy and Physiology

Seventh Edition

Rod R. Seeley
Idaho State University

Trent D. Stephens
Idaho State University

Philip Tate
Phoenix College

McGraw Hill Higher Education

Boston Burr Ridge, IL Dubuque, IA Madison, WI New York San Francisco St. Louis
Bangkok Bogotá Caracas Kuala Lumpur Lisbon London Madrid Mexico City
Milan Montreal New Delhi Santiago Seoul Singapore Sydney Taipei Toronto

The McGraw-Hill Companies

Student Study Art Notebook to accompany
ANATOMY AND PHYSIOLOGY, SEVENTH EDITION
ROD R. SEELEY, TRENT D. STEPHENS, AND PHILIP TATE

Published by McGraw-Hill Higher Education, an imprint of The McGraw-Hill Companies, Inc., 1221 Avenue of the Americas, New York, NY 10020. Copyright © 2006 by The McGraw-Hill Companies, Inc. All rights reserved.

No part of this publication may be reproduced or distributed in any form or by any means, or stored in a database or retrieval system, without the prior written consent of The McGraw-Hill Companies, Inc., including, but not limited to, network or other electronic storage or transmission, or broadcast for distance learning.

This book is printed on recycled, acid-free paper containing 10% postconsumer waste.

2 3 4 5 6 7 8 9 0 QPD/QPD 0 9 8 7 6 5

ISBN 0-07-310788-3

www.mhhe.com

DIRECTORY OF NOTEBOOK FIGURES

TO ACCOMPANY SEELEY/STEPHENS/TATE *ANATOMY AND PHYSIOLOGY 7/e*

1. **Chemical level.** Atoms (*colored balls*) combine to form molecules.
2. **Cell level.** Molecules form organelles, such as the nucleus and mitochondria, which make up cells.
3. **Tissue level.** Similar cells and surrounding materials make up tissues.
4. **Organ level.** Different tissues combine to form organs, such as the urinary bladder.
5. **Organ system level.** Organs such as the urinary bladder and kidneys make up an organ system.
6. **Organism level.** Organ systems make up an organism.

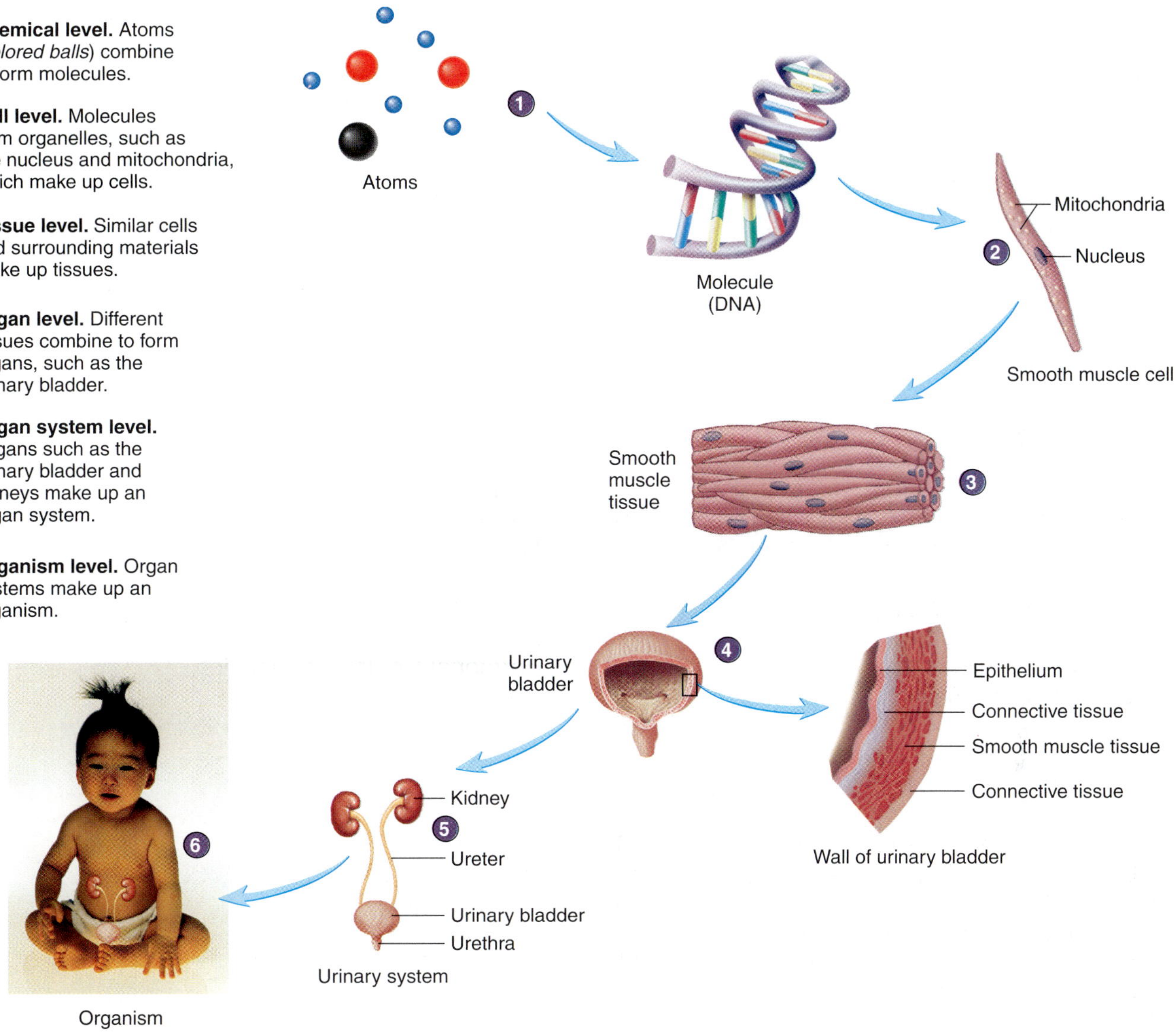

Levels of Organization
Figure 1.1
© Bart Harris/CORBIS

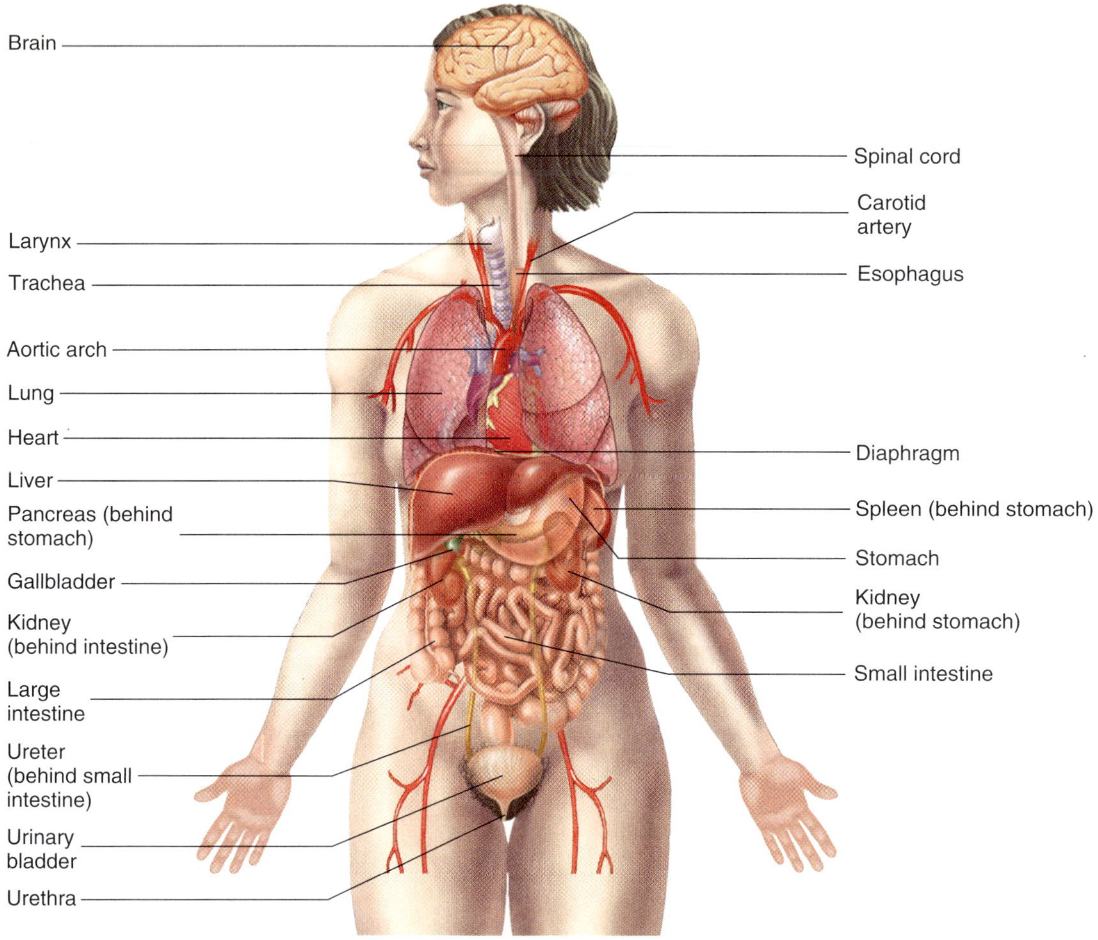
Brain
Spinal cord
Carotid artery
Larynx
Trachea
Esophagus
Aortic arch
Lung
Heart
Diaphragm
Liver
Pancreas (behind stomach)
Spleen (behind stomach)
Stomach
Gallbladder
Kidney (behind intestine)
Kidney (behind stomach)
Small intestine
Large intestine
Ureter (behind small intestine)
Urinary bladder
Urethra

Organs of the Body
Figure 1.2

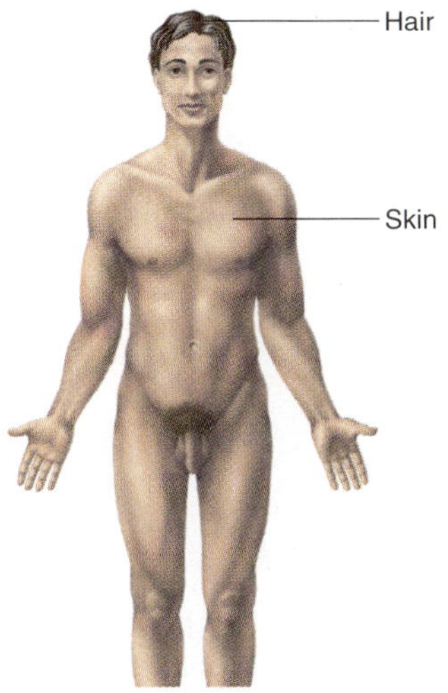

Integumentary System
Provides protection, regulates temperature, prevents water loss, and produces vitamin D precursors. Consists of skin, hair, nails, and sweat glands.

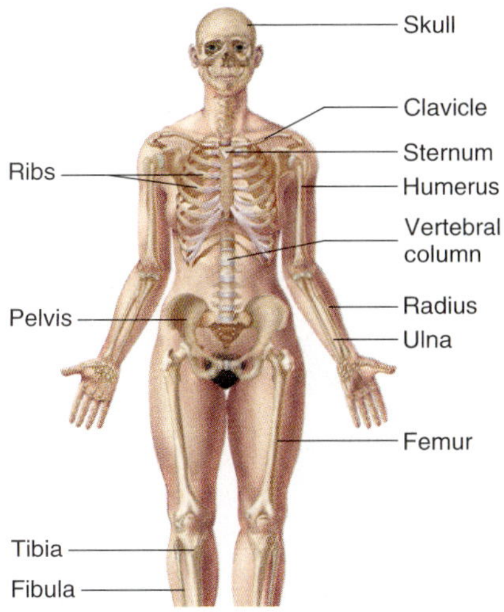

Skeletal System
Provides protection and support, allows body movements, produces blood cells, and stores minerals and fat. Consists of bones, associated cartilages, ligaments, and joints.

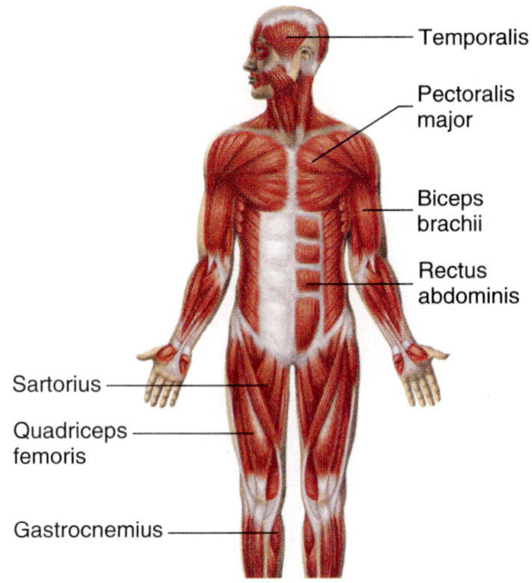

Muscular System
Produces body movements, maintains posture, and produces body heat. Consists of muscles attached to the skeleton by tendons.

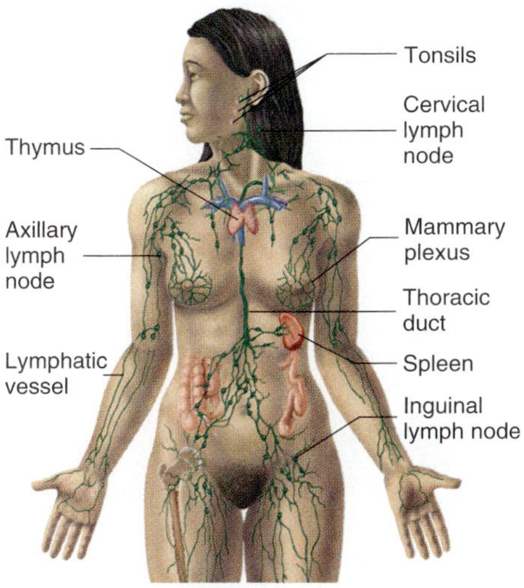

Lymphatic System
Removes foreign substances from the blood and lymph, combats disease, maintains tissue fluid balance, and absorbs fats from the digestive tract. Consists of the lymphatic vessels, lymph nodes, and other lymphatic organs.

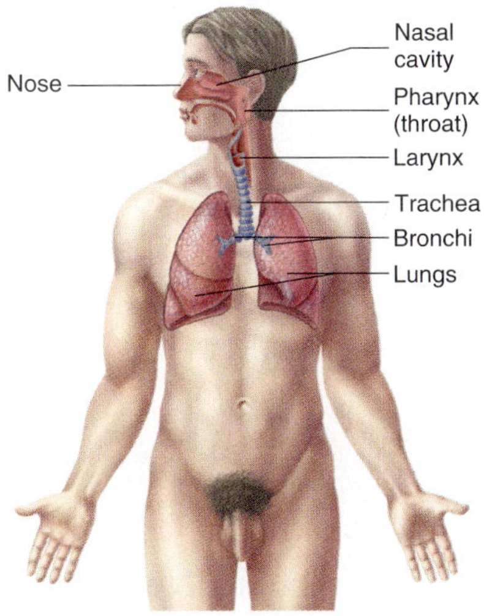

Respiratory System
Exchanges oxygen and carbon dioxide between the blood and air and regulates blood pH. Consists of the lungs and respiratory passages.

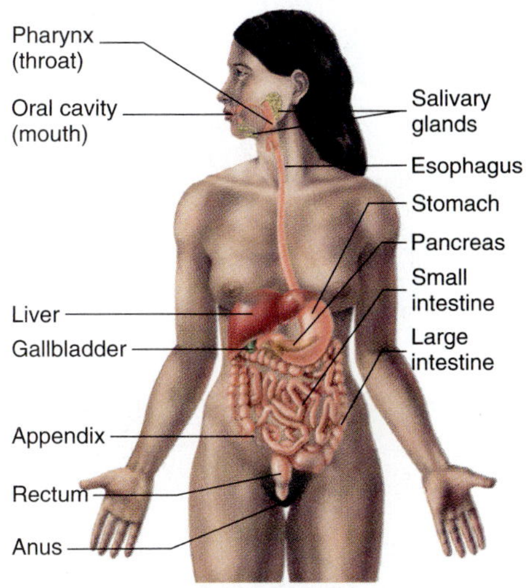

Digestive System
Performs the mechanical and chemical processes of digestion, absorption of nutrients, and elimination of wastes. Consists of the mouth, esophagus, stomach, intestines, and accessory organs.

Organ Systems of the Body
Figure 1.3

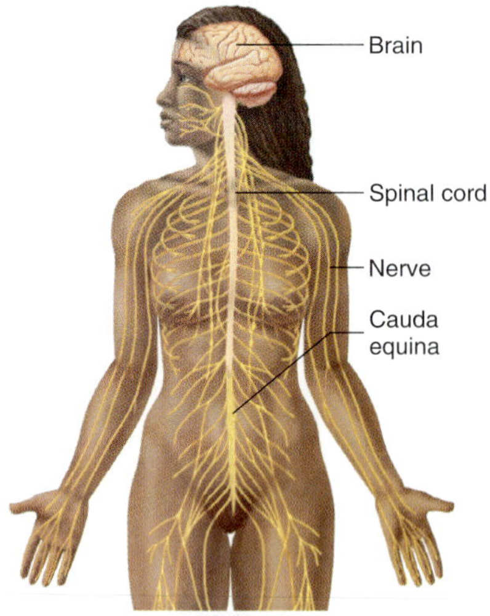

Nervous System

A major regulatory system that detects sensations and controls movements, physiologic processes, and intellectual functions. Consists of the brain, spinal cord, nerves, and sensory receptors.

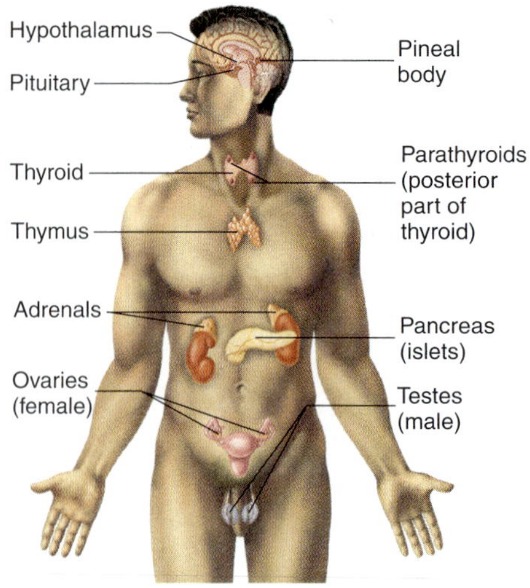

Endocrine System

A major regulatory system that influences metabolism, growth, reproduction, and many other functions. Consists of glands, such as the pituitary, that secrete hormones.

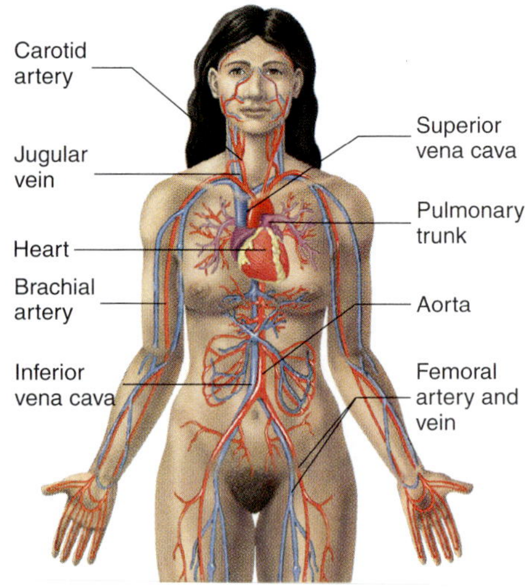

Cardiovascular System

Transports nutrients, waste products, gases, and hormones throughout the body; plays a role in the immune response and the regulation of body temperature. Consists of the heart, blood vessels, and blood.

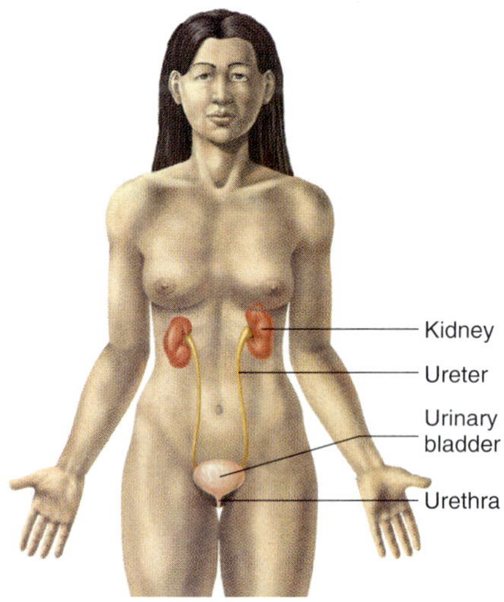

Urinary System

Removes waste products from the blood and regulates blood pH, ion balance, and water balance. Consists of the kidneys, urinary bladder, and ducts that carry urine.

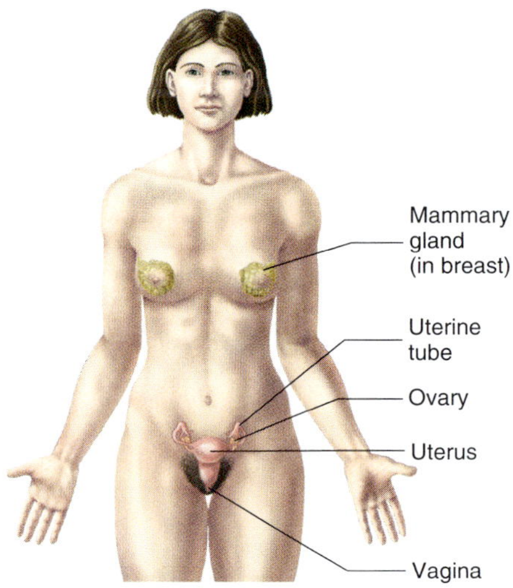

Female Reproductive System

Produces oocytes and is the site of fertilization and fetal development; produces milk for the newborn; produces hormones that influence sexual functions and behaviors. Consists of the ovaries, vagina, uterus, mammary glands, and associated structures.

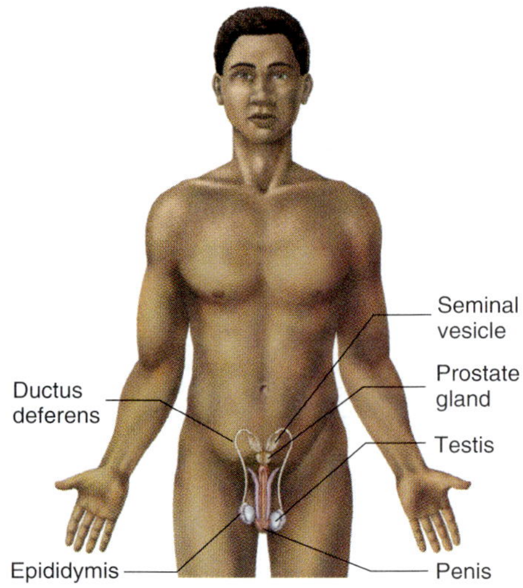

Male Reproductive System

Produces and transfers sperm cells to the female and produces hormones that influence sexual functions and behaviors. Consists of the testes, accessory structures, ducts, and penis.

Figure 1.3—*continued*

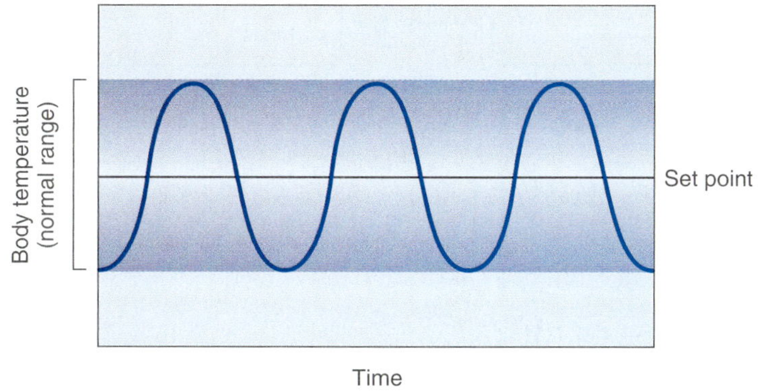

Homeostasis
Figure 1.4

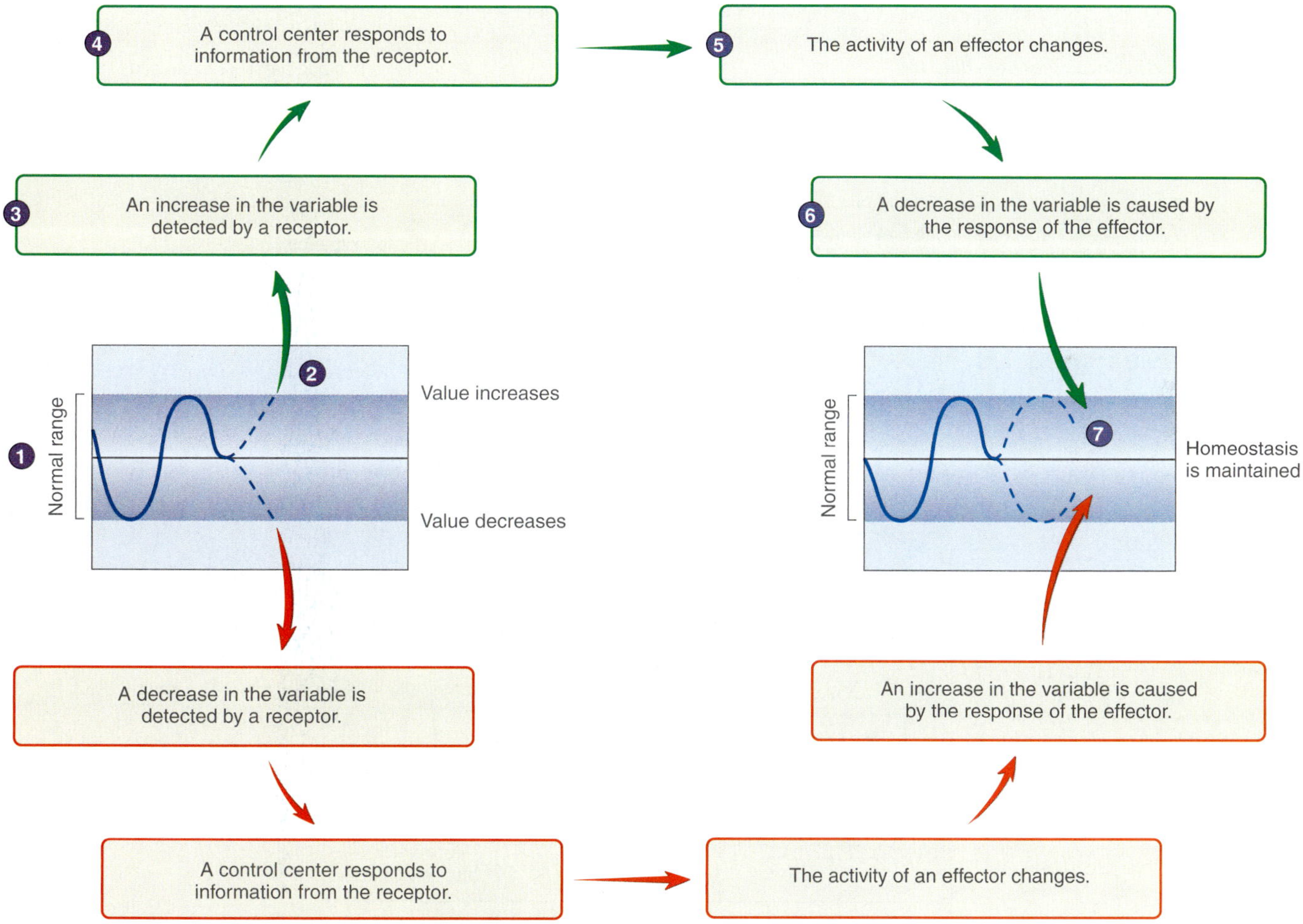

Mechanism of Negative Feedback
Homeostasis Figure 1.5

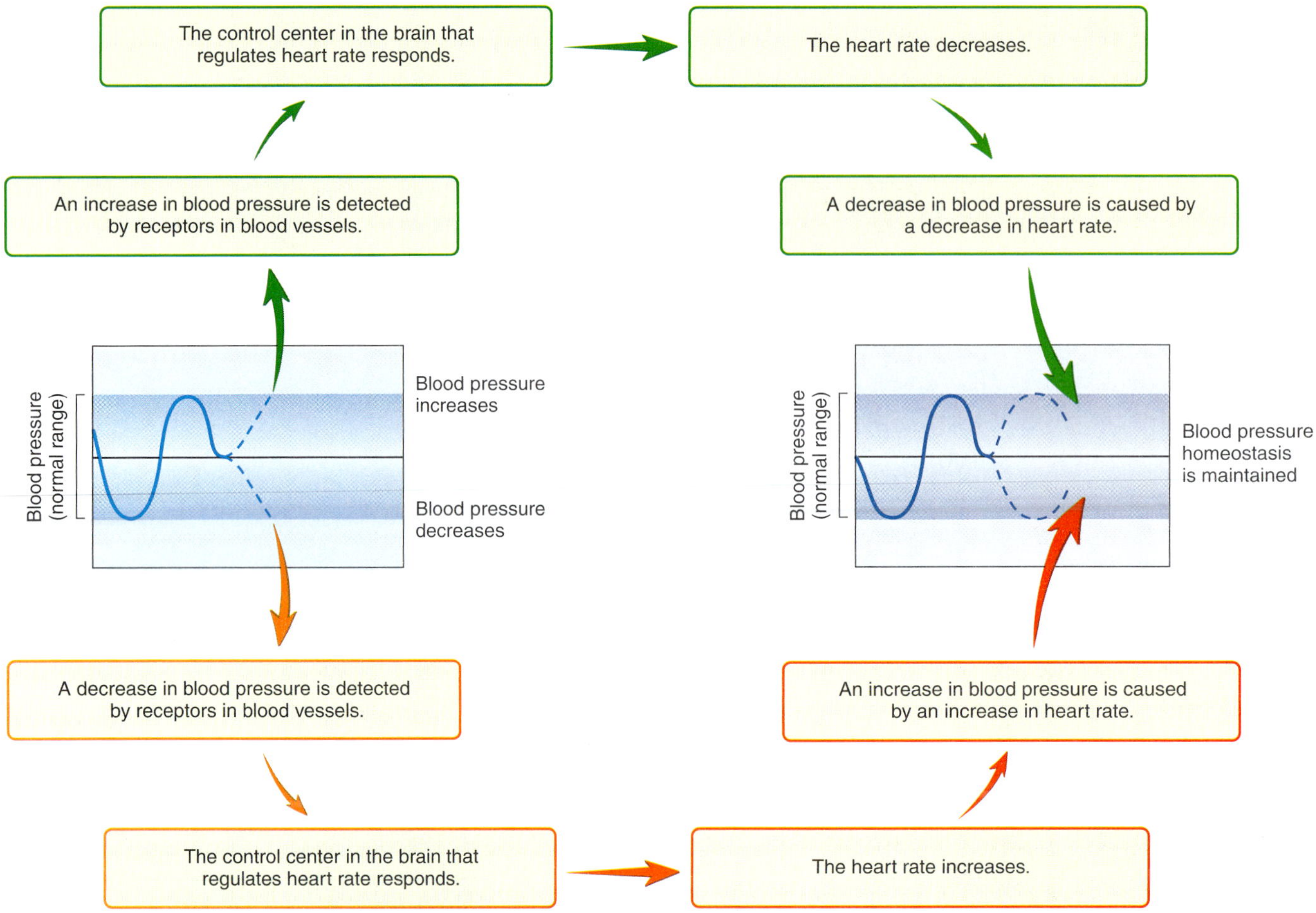

Example of Negative Feedback
Homeostasis Figure 1.6

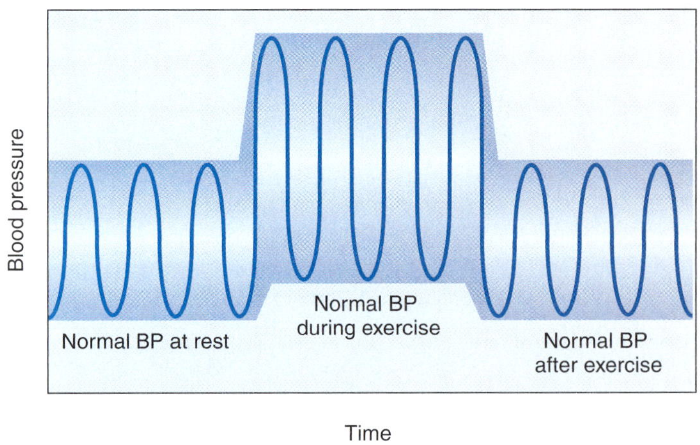

Changes in Blood Pressure During Exercise
Figure 1.7

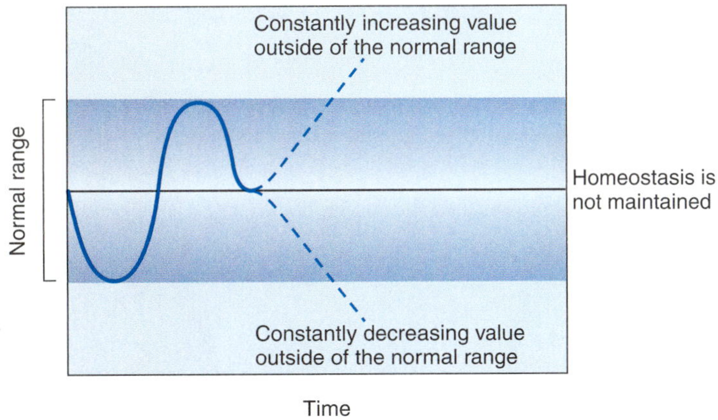

Positive Feedback
Figure 1.8

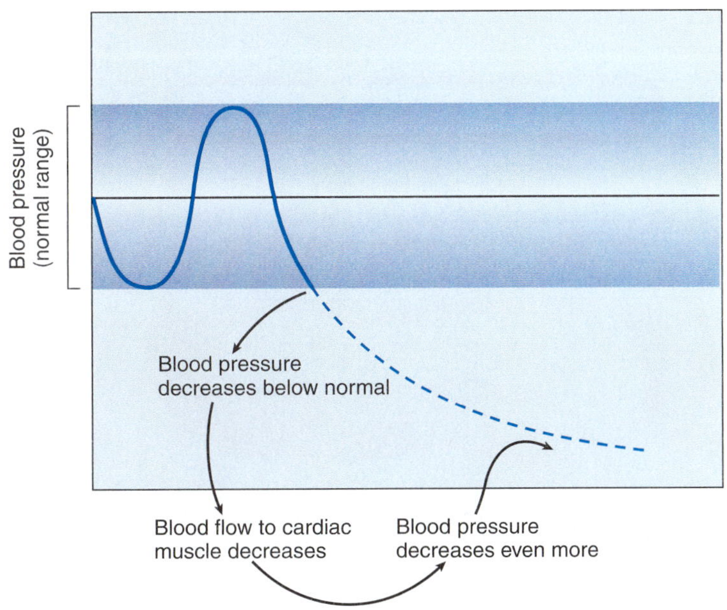

Example of Harmful Positive Feedback
Figure 1.9

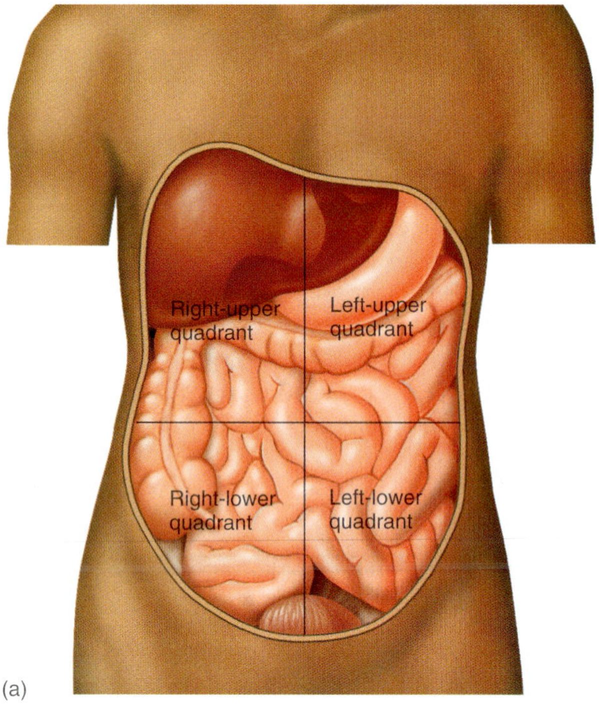

(a)

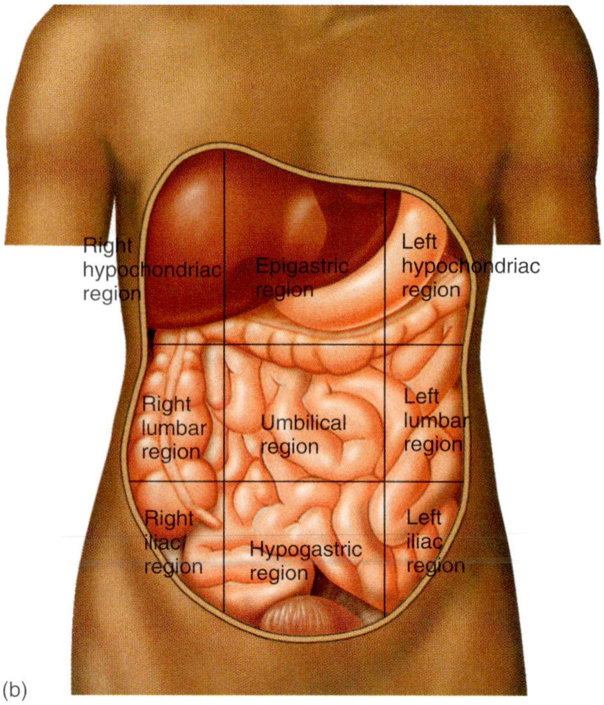

(b)

Subdivisions of the Abdomen
Figure 1.12

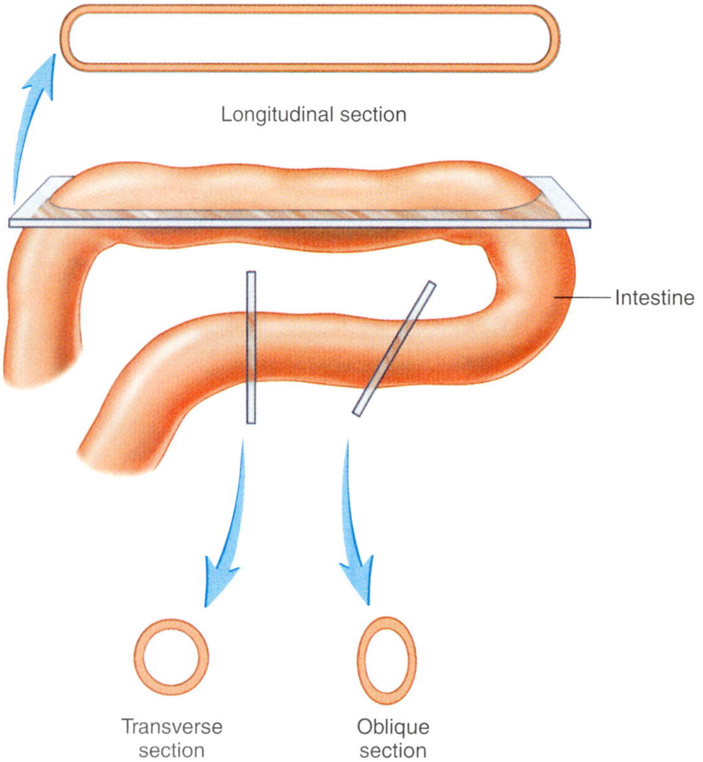

Planes of Section Through an Organ
Figure 1.14

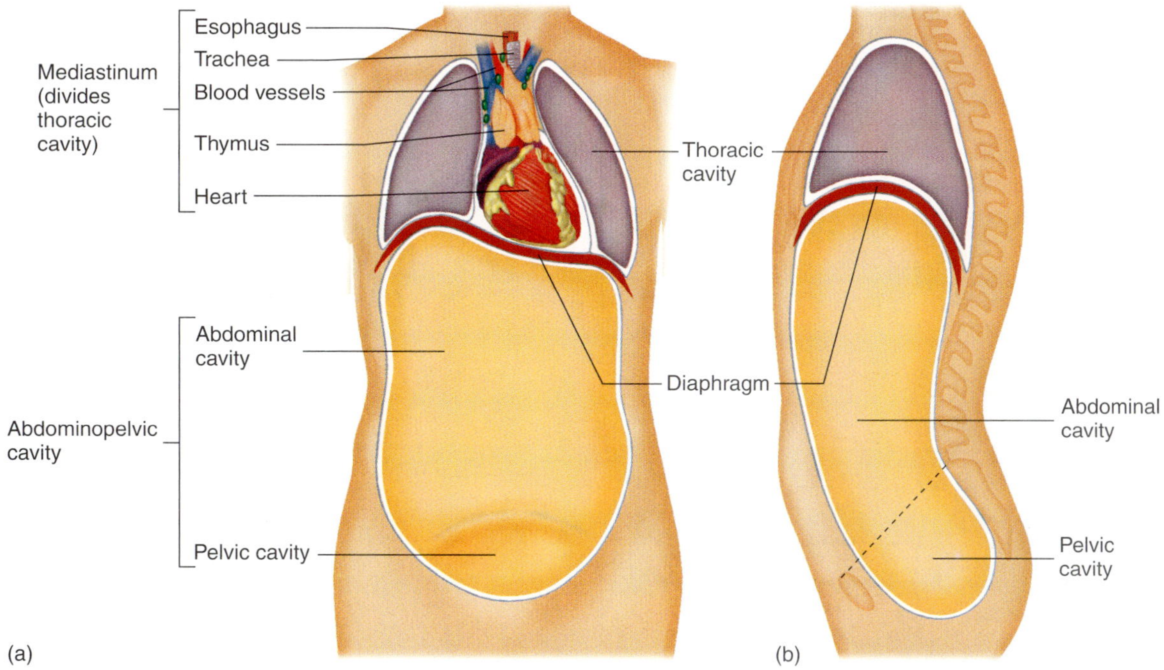

Trunk Cavities
Figure 1.15

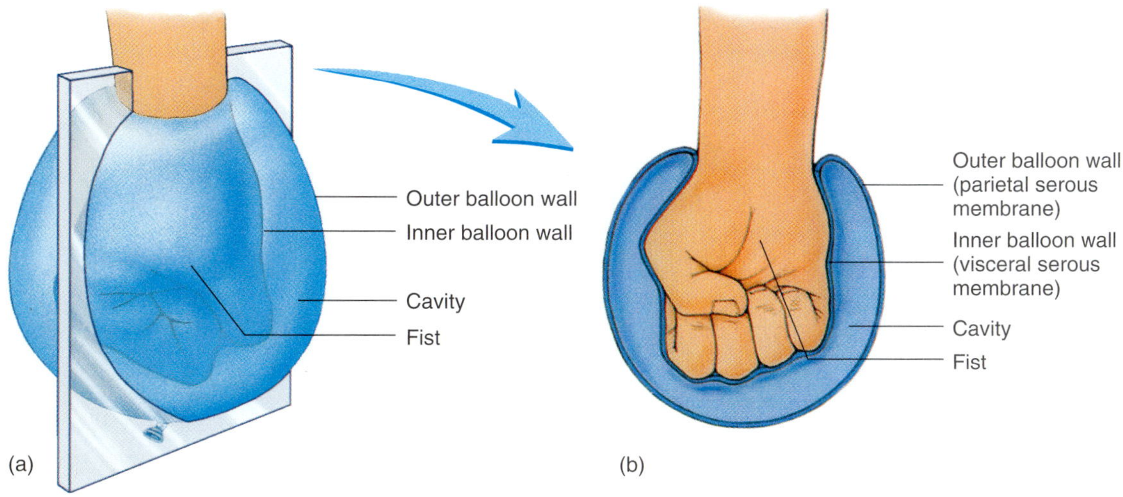

Serous Membranes
Figure 1.16

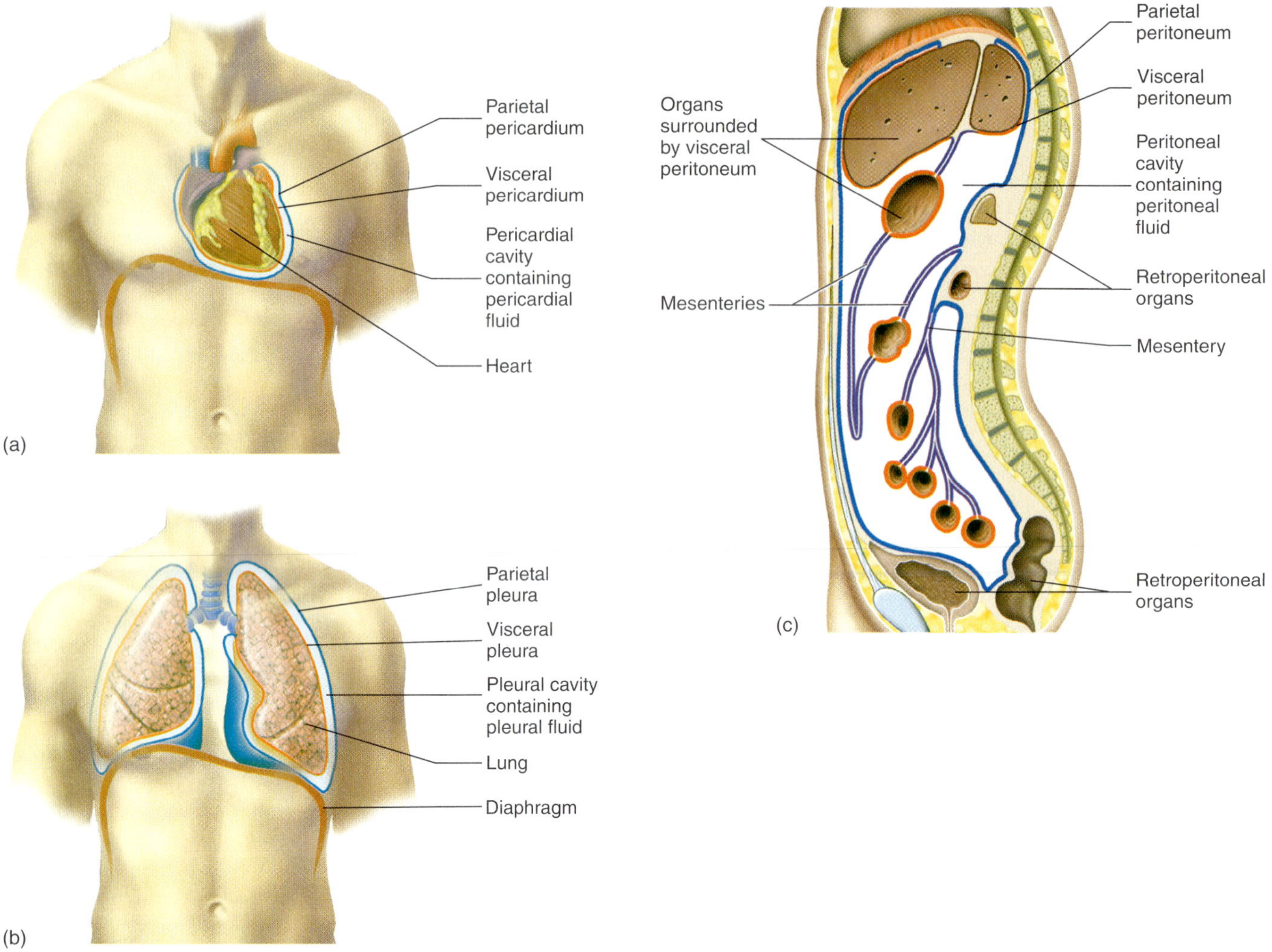

Location of Serous Membranes
Figure 1.17

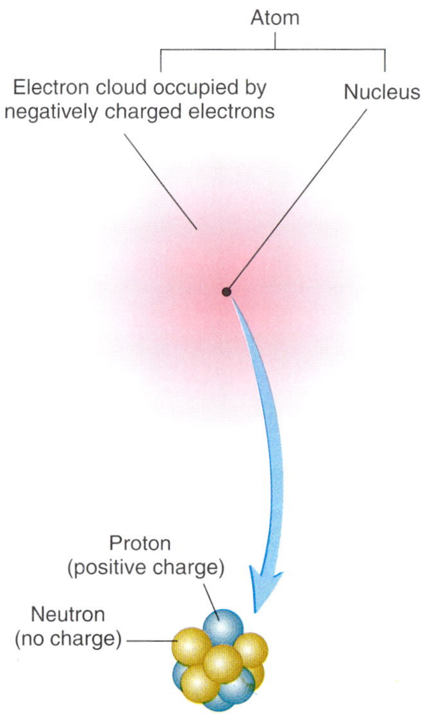

Model of an Atom
Figure 2.1

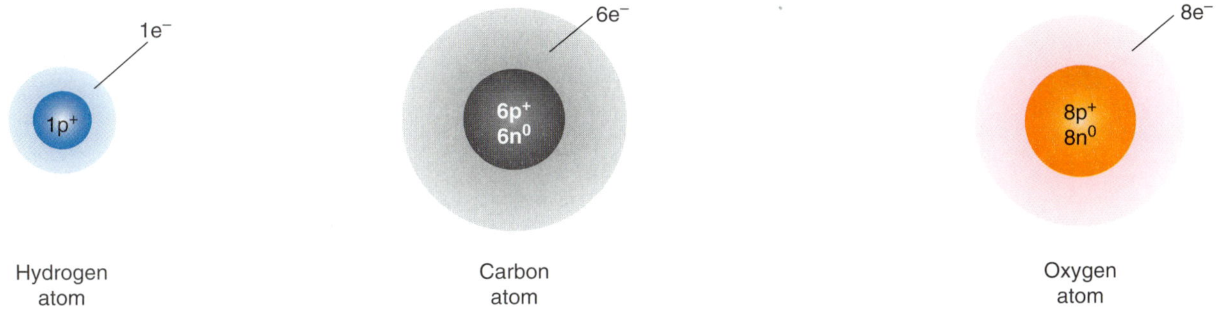

Hydrogen, Carbon, and Oxygen Atoms
Figure 2.2

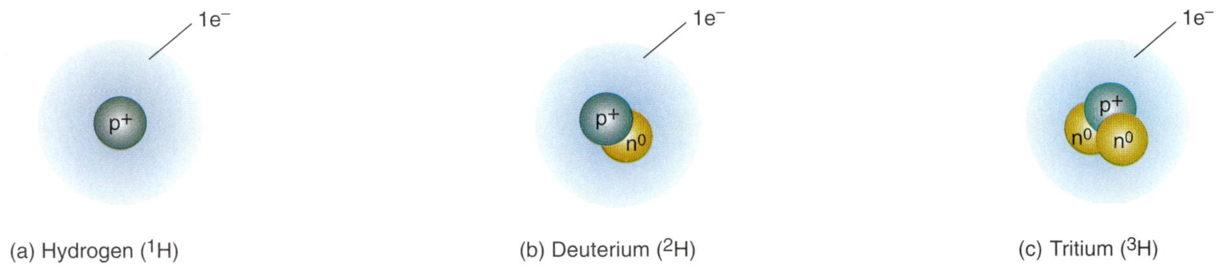

Isotopes of Hydrogen
Figure 2.3

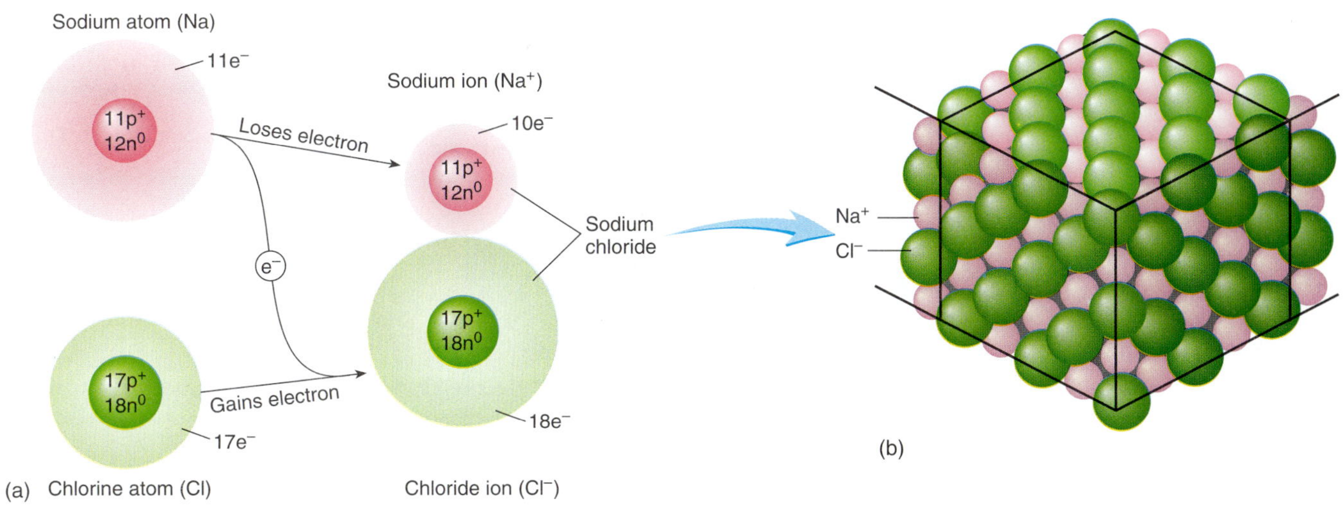

Ionic Bonding
Figure 2.4

No interaction between the two hydrogen atoms because they are too far apart.

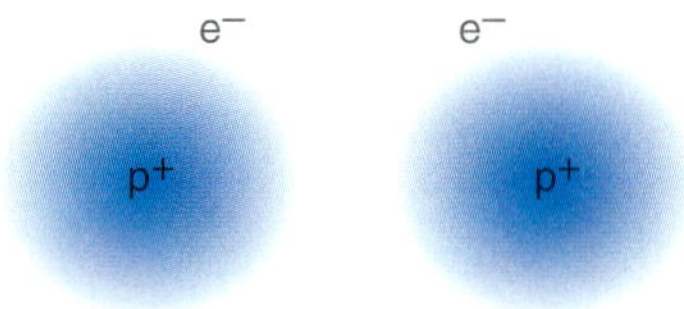

The positively charged nucleus of each hydrogen atom begins to attract the electron of the other.

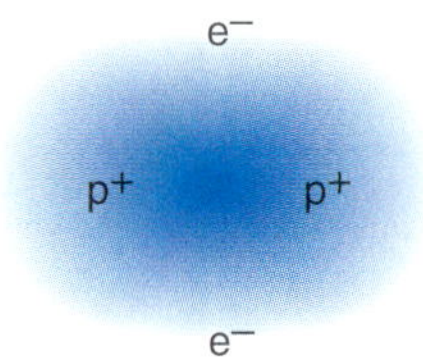

A covalent bond is formed when the electrons are shared between the nuclei because the electrons are equally attracted to each nucleus.

Covalent Bonding
Figure 2.5

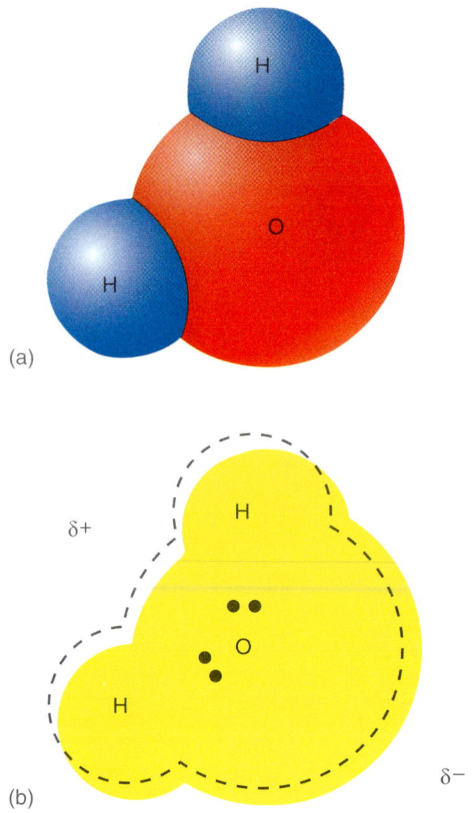

Polar Covalent Bonds
Figure 2.6

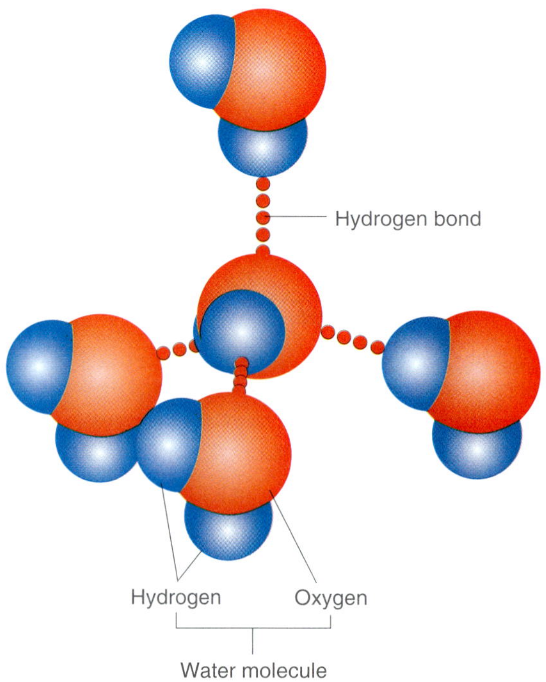

Hydrogen Bonds
Figure 2.7

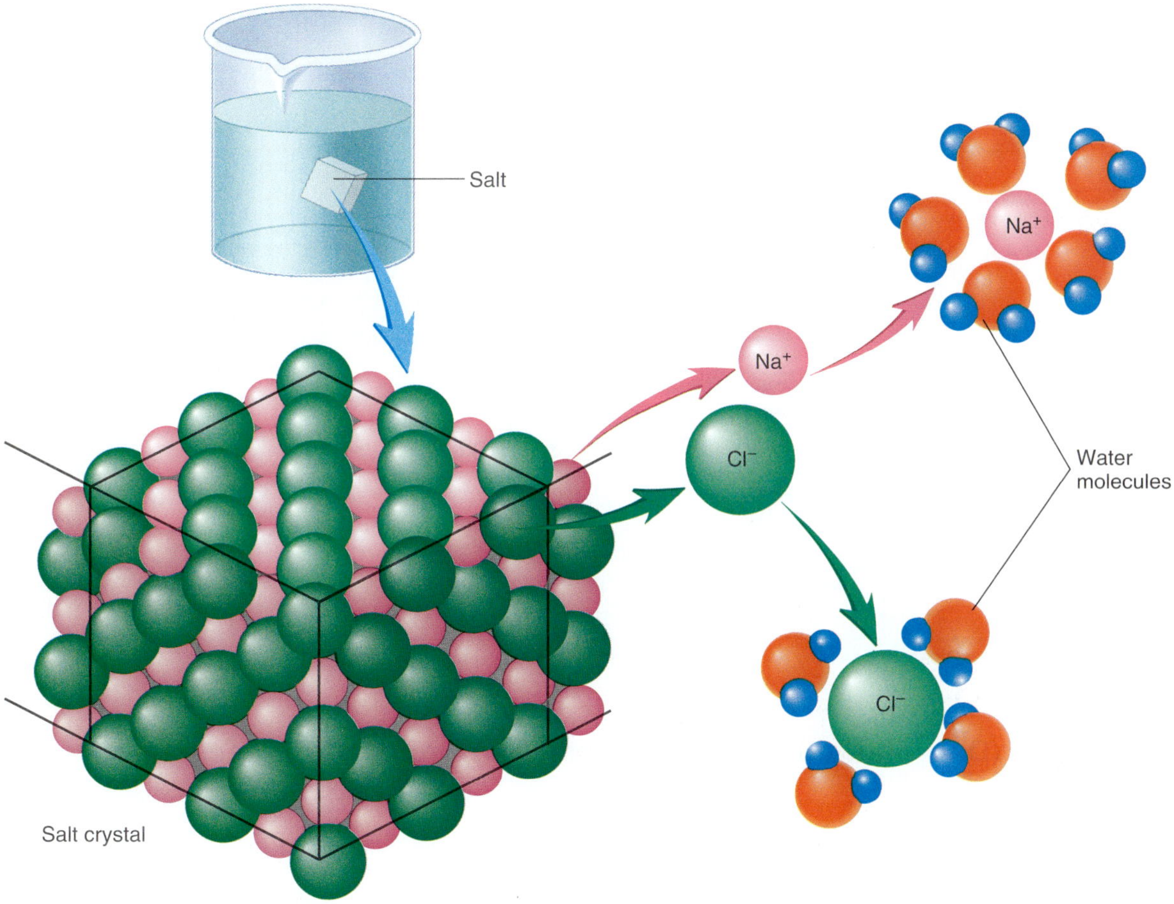

Dissociation
Figure 2.8

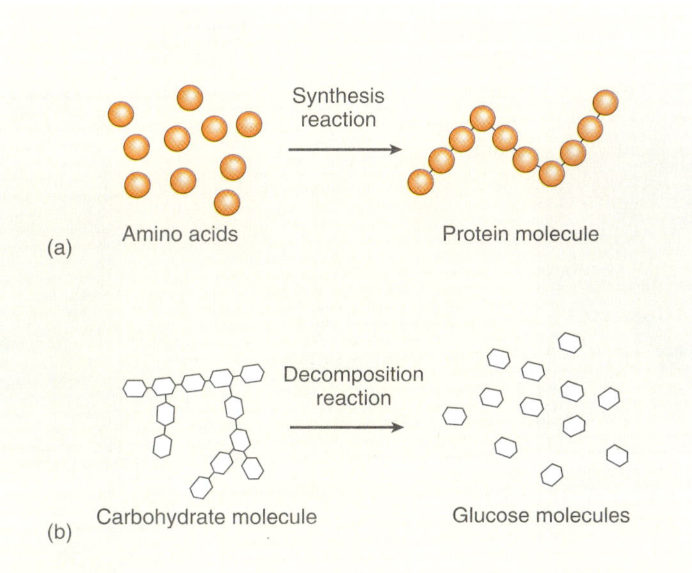

Synthesis and Decomposition Reactions
Figure 2.9

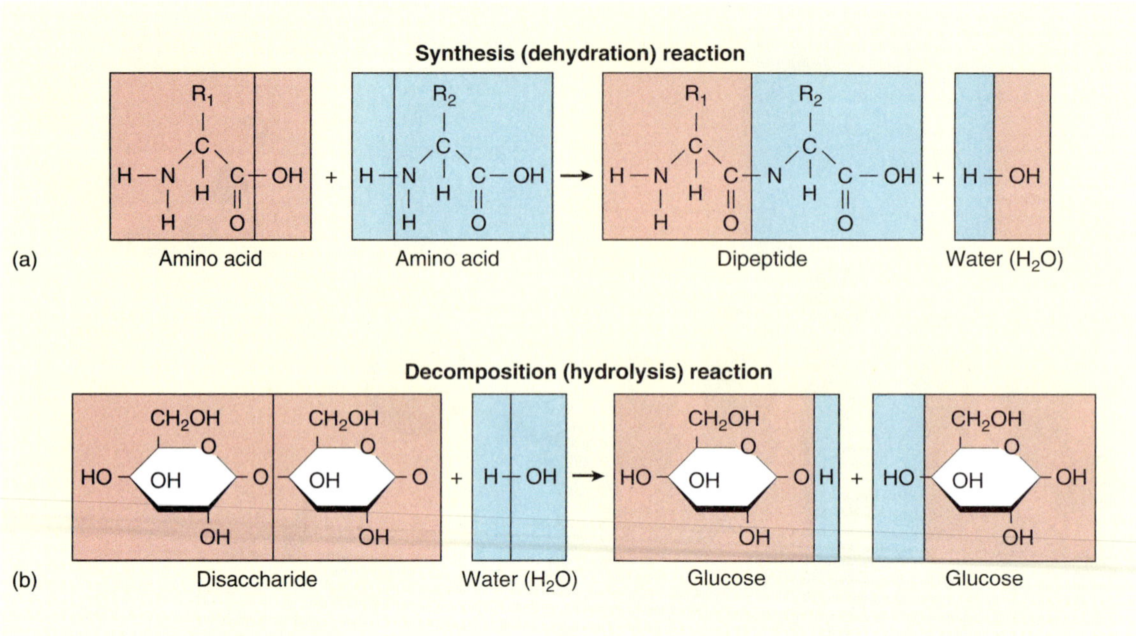

Synthesis (Dehydration) and Decomposition (Hydrolysis) Reactions
Figure 2.10

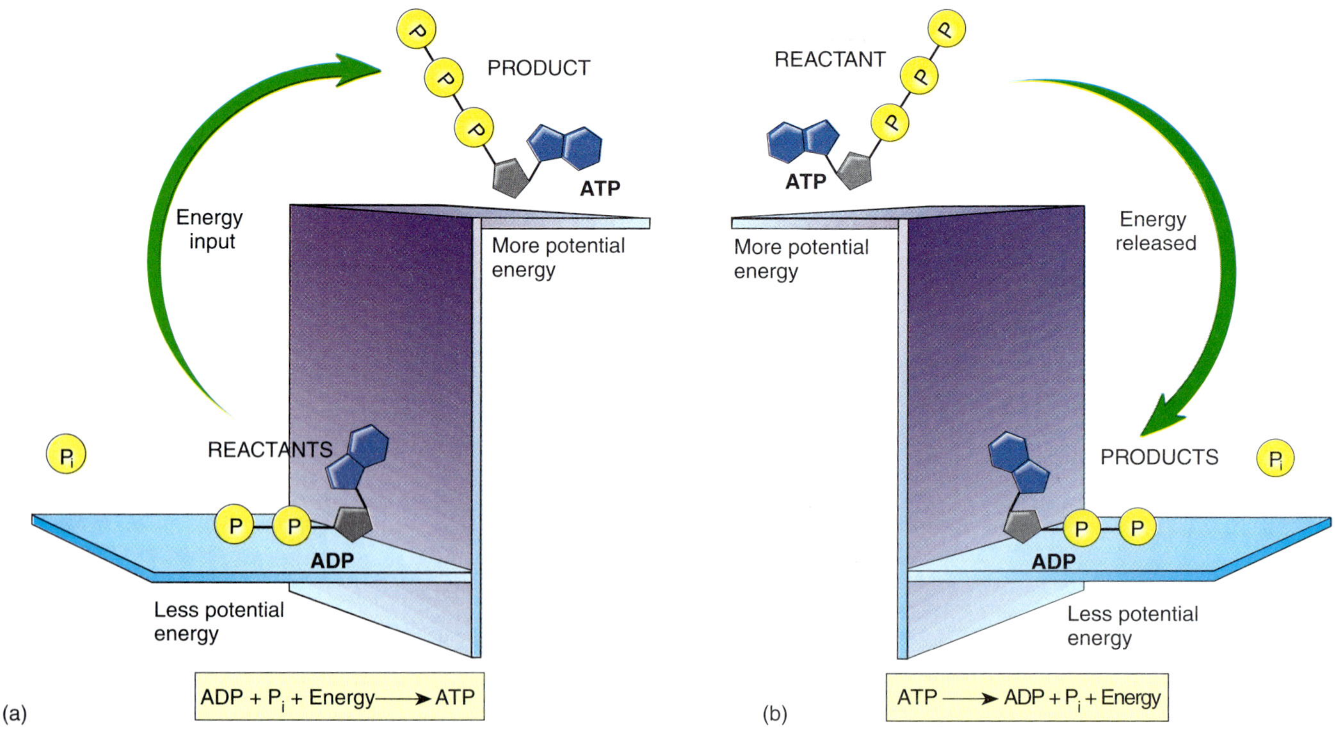

Energy and Chemical Reactions
Figure 2.11

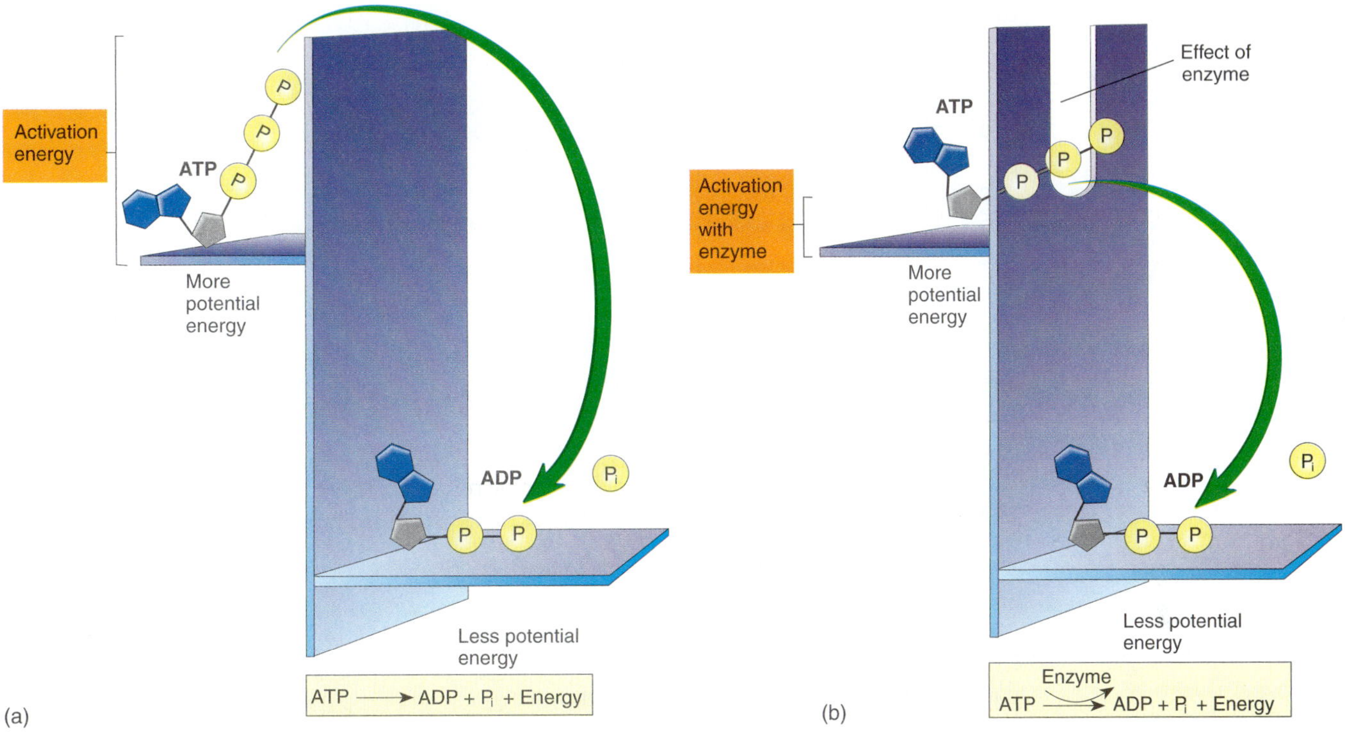
Activation energy
ATP
P
P
P
More potential energy
ADP
P_i
P
P
Less potential energy
ATP ⟶ ADP + P_i + Energy
(a)
Effect of enzyme
ATP
P
P
P
Activation energy with enzyme
More potential energy
ADP
P_i
P
P
Less potential energy
Enzyme
ATP ⟶ ADP + P_i + Energy
(b)

Activation Energy and Enzymes
Figure 2.12

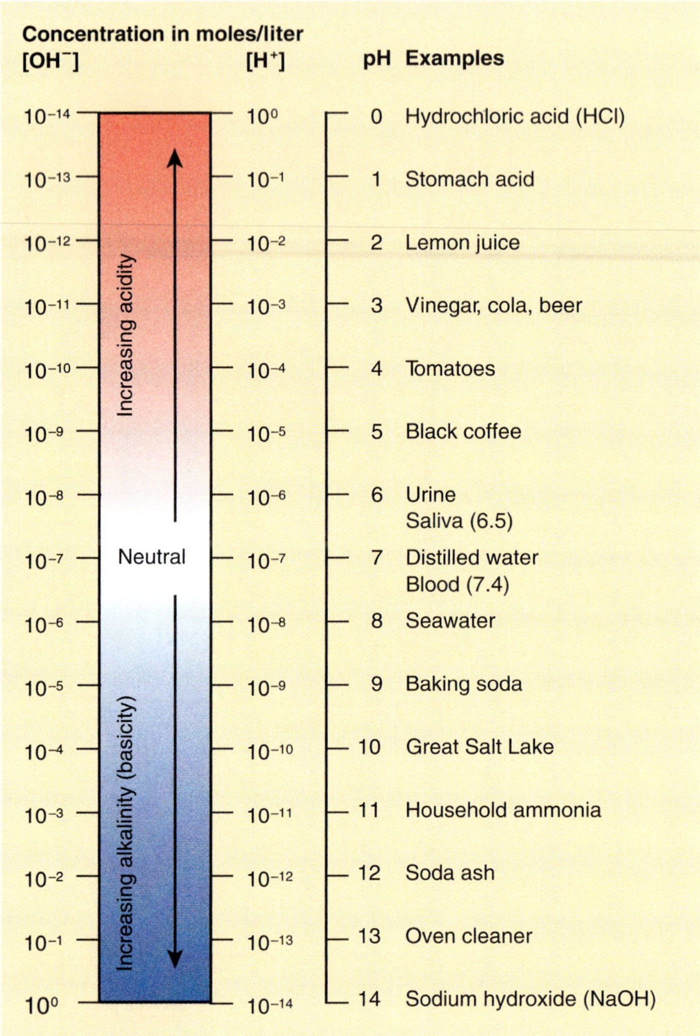
Concentration in moles/liter
[OH⁻]
[H⁺]
pH
Examples
10⁻¹⁴
10⁻¹³
10⁻¹²
10⁻¹¹
10⁻¹⁰
10⁻⁹
10⁻⁸
10⁻⁷
10⁻⁶
10⁻⁵
10⁻⁴
10⁻³
10⁻²
10⁻¹
10⁰
Increasing acidity
Neutral
Increasing alkalinity (basicity)
0 Hydrochloric acid (HCl)
1 Stomach acid
2 Lemon juice
3 Vinegar, cola, beer
4 Tomatoes
5 Black coffee
6 Urine
Saliva (6.5)
7 Distilled water
Blood (7.4)
8 Seawater
9 Baking soda
10 Great Salt Lake
11 Household ammonia
12 Soda ash
13 Oven cleaner
14 Sodium hydroxide (NaOH)

The pH Scale
Figure 2.13

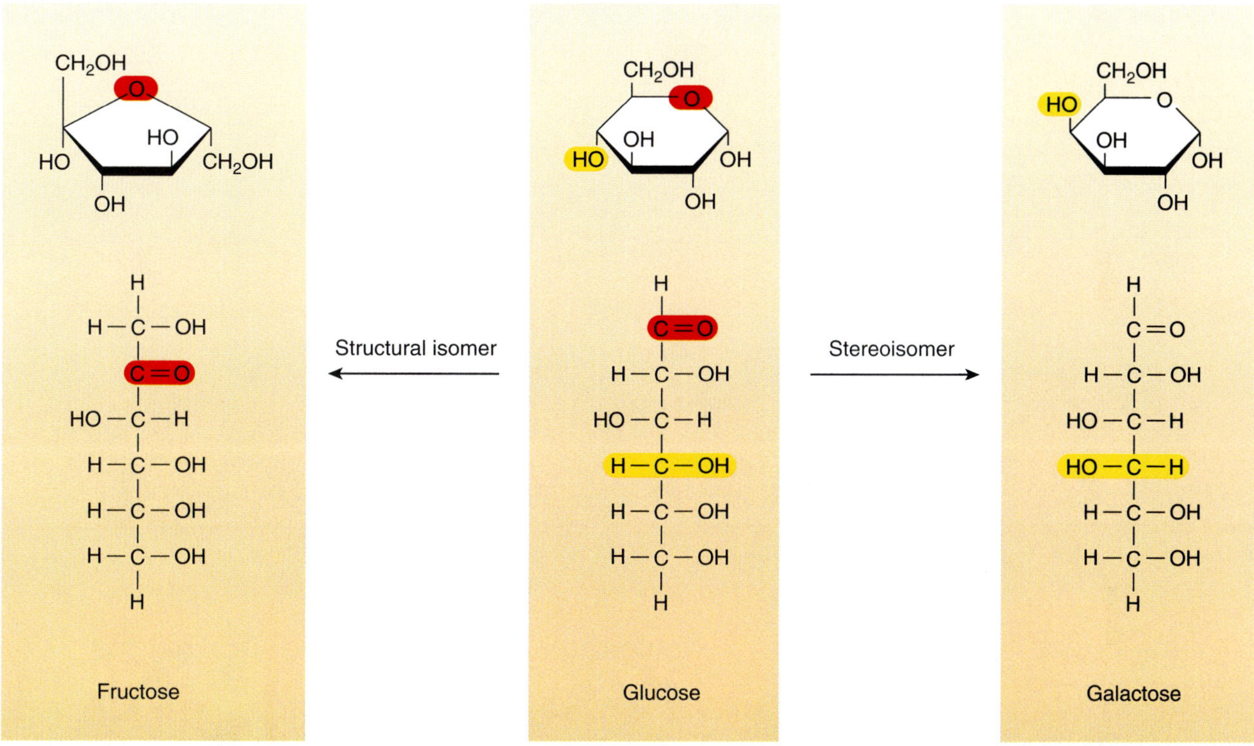
CH2OH
O
HO
HO
CH2OH
OH
H
H—C—OH
C=O
HO—C—H
H—C—OH
H—C—OH
H—C—OH
H
Fructose
Structural isomer
CH2OH
O
OH
HO
OH
OH
H
C=O
H—C—OH
HO—C—H
H—C—OH
H—C—OH
H—C—OH
H
Glucose
Stereoisomer
CH2OH
HO
O
OH
OH
OH
H
C=O
H—C—OH
HO—C—H
HO—C—H
H—C—OH
H—C—OH
H
Galactose

Monosaccharides
Figure 2.14

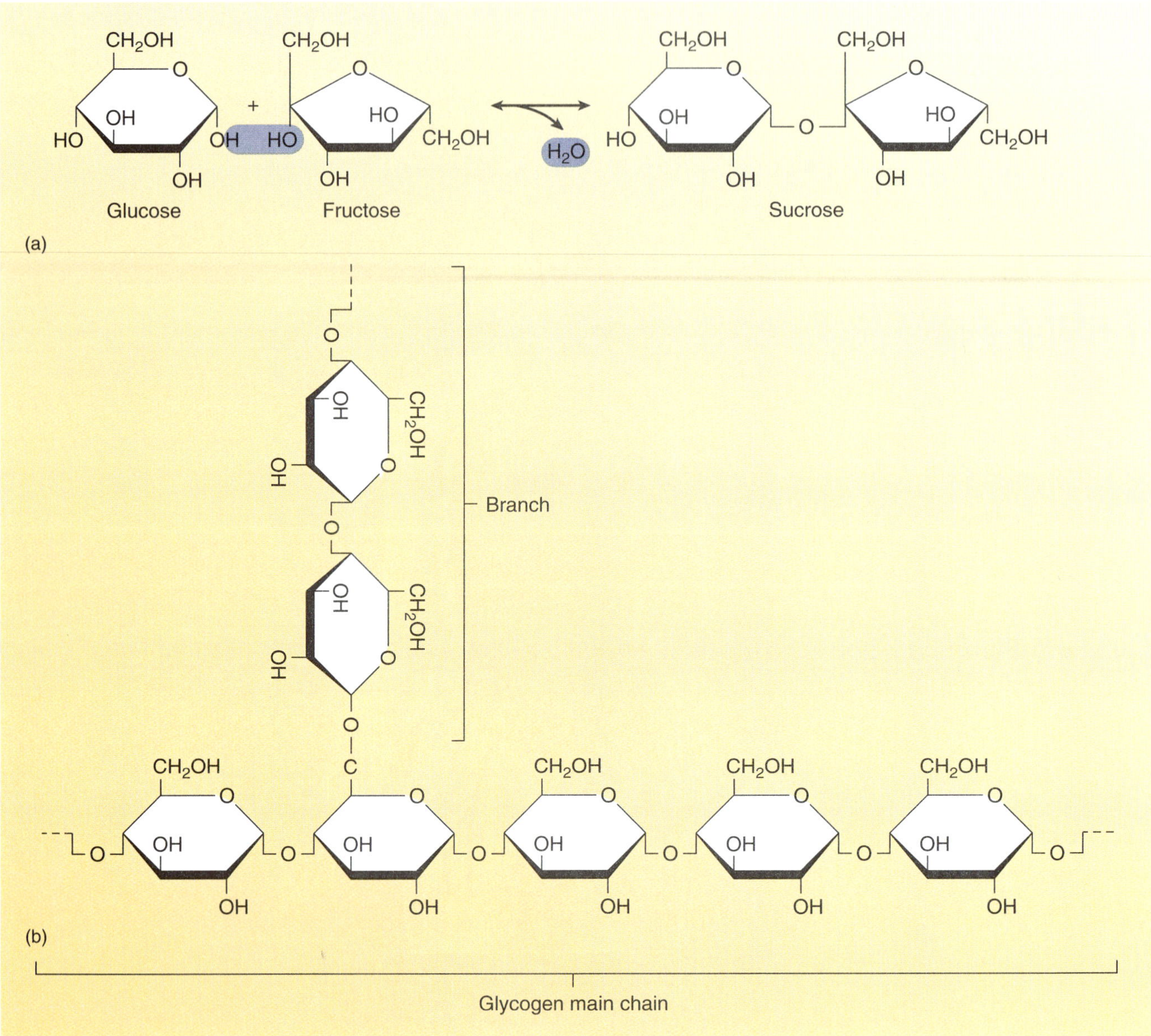
CH₂OH
Glucose
Fructose
Sucrose
(a)
Branch
(b)
Glycogen main chain

Disaccharide and Polysaccharide
Figure 2.15

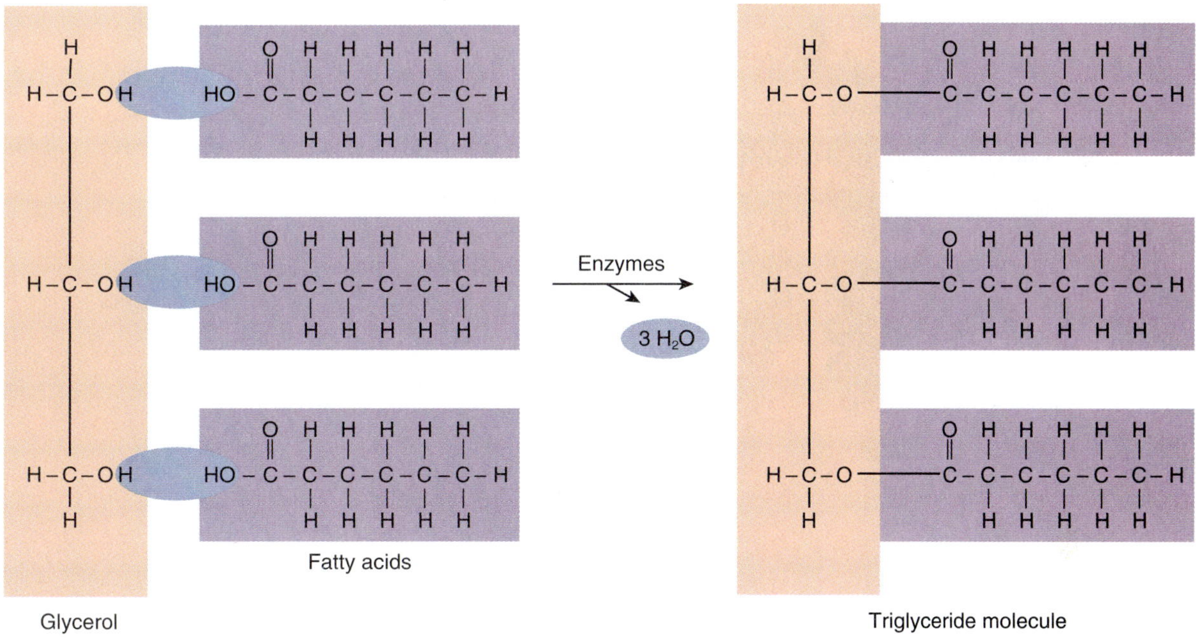
Enzymes
3 H_2O
Fatty acids
Glycerol
Triglyceride molecule

Triglyceride
Figure 2.16

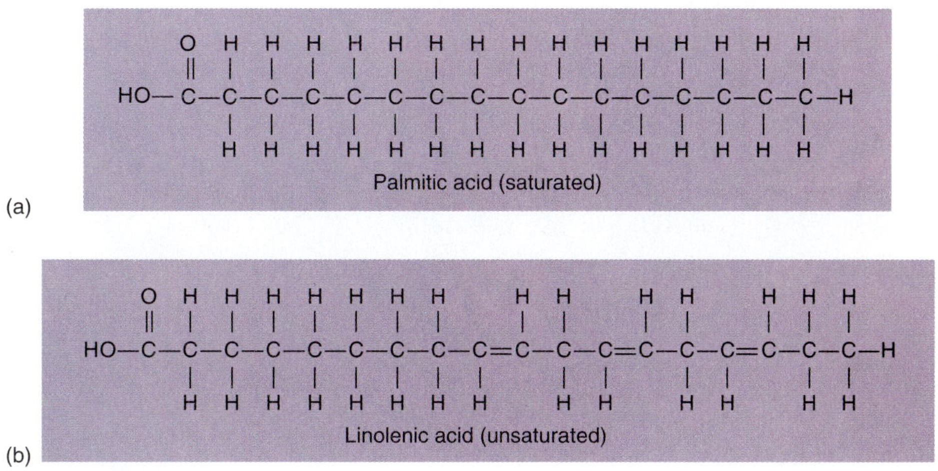
(a)
Palmitic acid (saturated)
(b)
Linolenic acid (unsaturated)

Fatty Acids
Figure 2.17

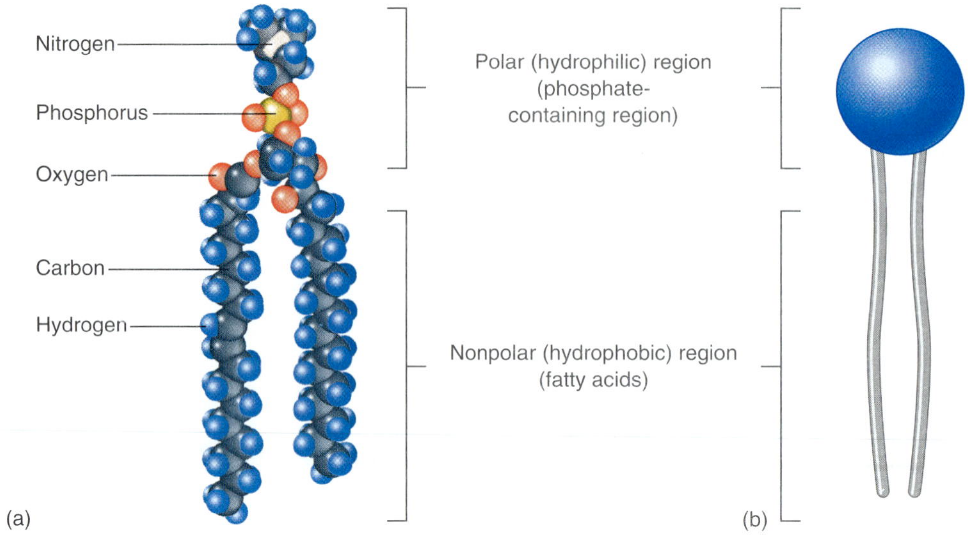

Nitrogen
Phosphorus
Oxygen
Carbon
Hydrogen
Polar (hydrophilic) region (phosphate-containing region)
Nonpolar (hydrophobic) region (fatty acids)
(a)
(b)

Phospholipids
Figure 2.18

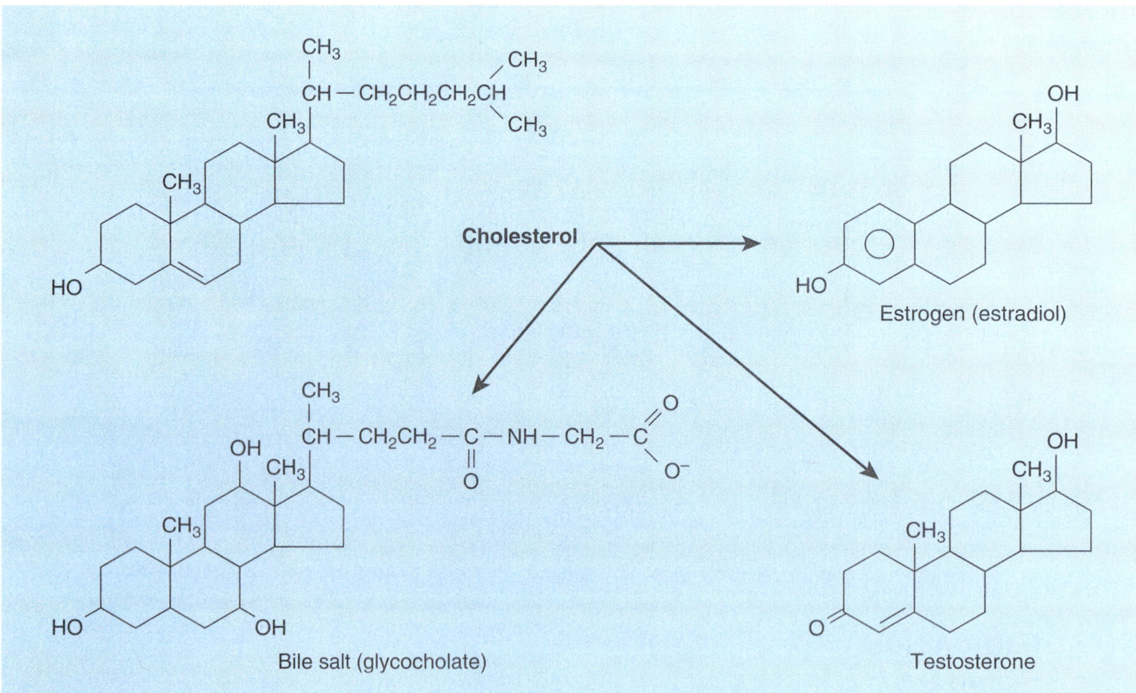

Cholesterol
Estrogen (estradiol)
Bile salt (glycocholate)
Testosterone

Steroids
Figure 2.19

Amino Acids
Figure 2.20

The general structure of an amino acid showing the amine group ($—NH_2$), carboxyl group (—COOH), and hydrogen atom highlighted in yellow. The R side chain is the part of an amino acid that makes it different from other amino acids.

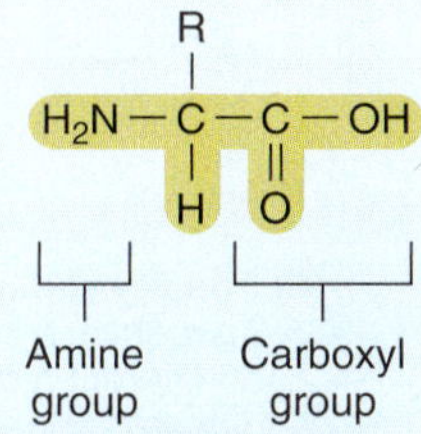

Glycine is the simplest amino acid. The side chain is a hydrogen atom.

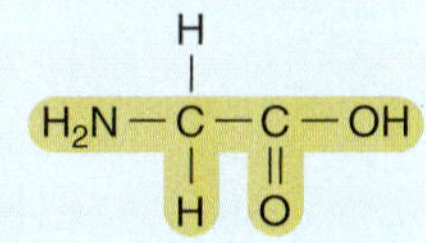

Glycine

Tyrosine, which has a more complicated side chain, is an important component of thyroid hormones.

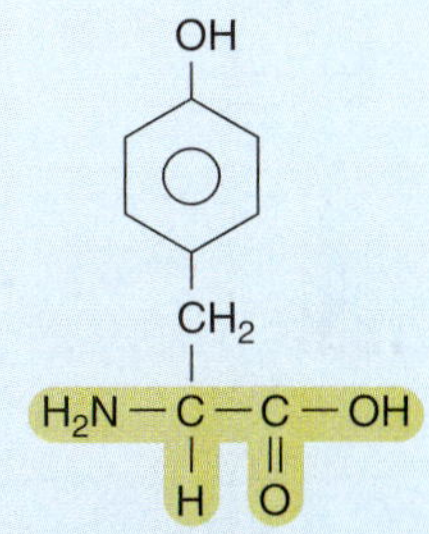

Tyrosine

Improper metabolism of phenylalanine in the genetic disease phenylketonuria (PKU) can cause mental retardation.

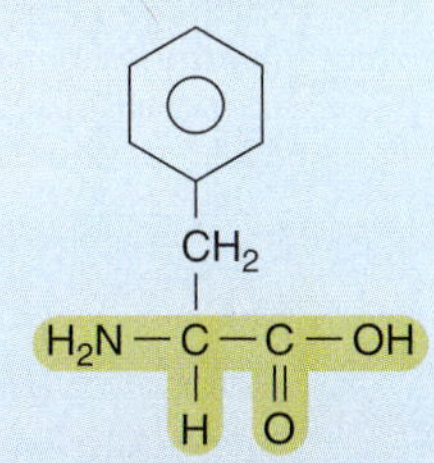

Phenylalanine

Aspartic acid combined with phenylalanine forms the artificial sweetener aspartame (Nutrasweet ™ and Equal ™).

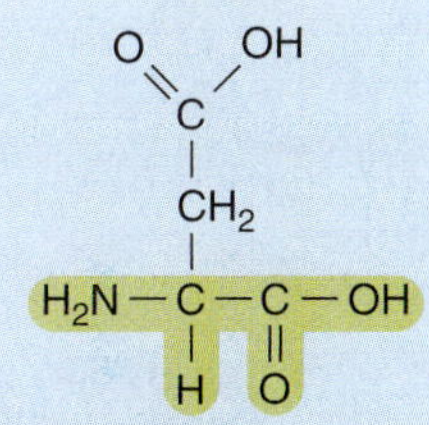

Aspartic acid

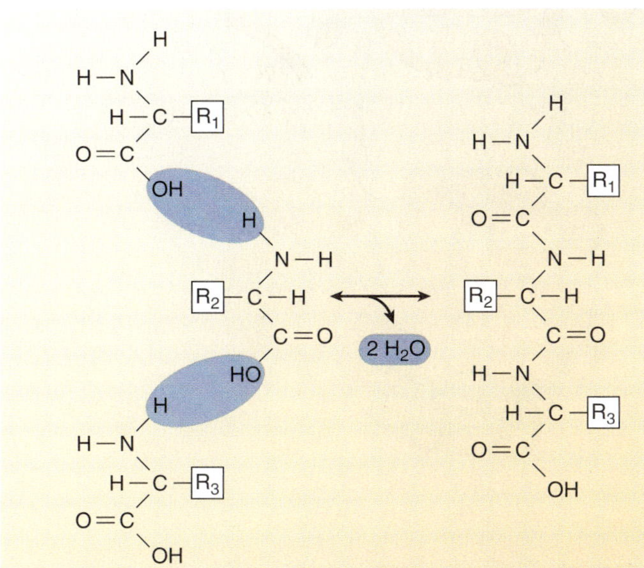

Peptide Bonds
Figure 2.21

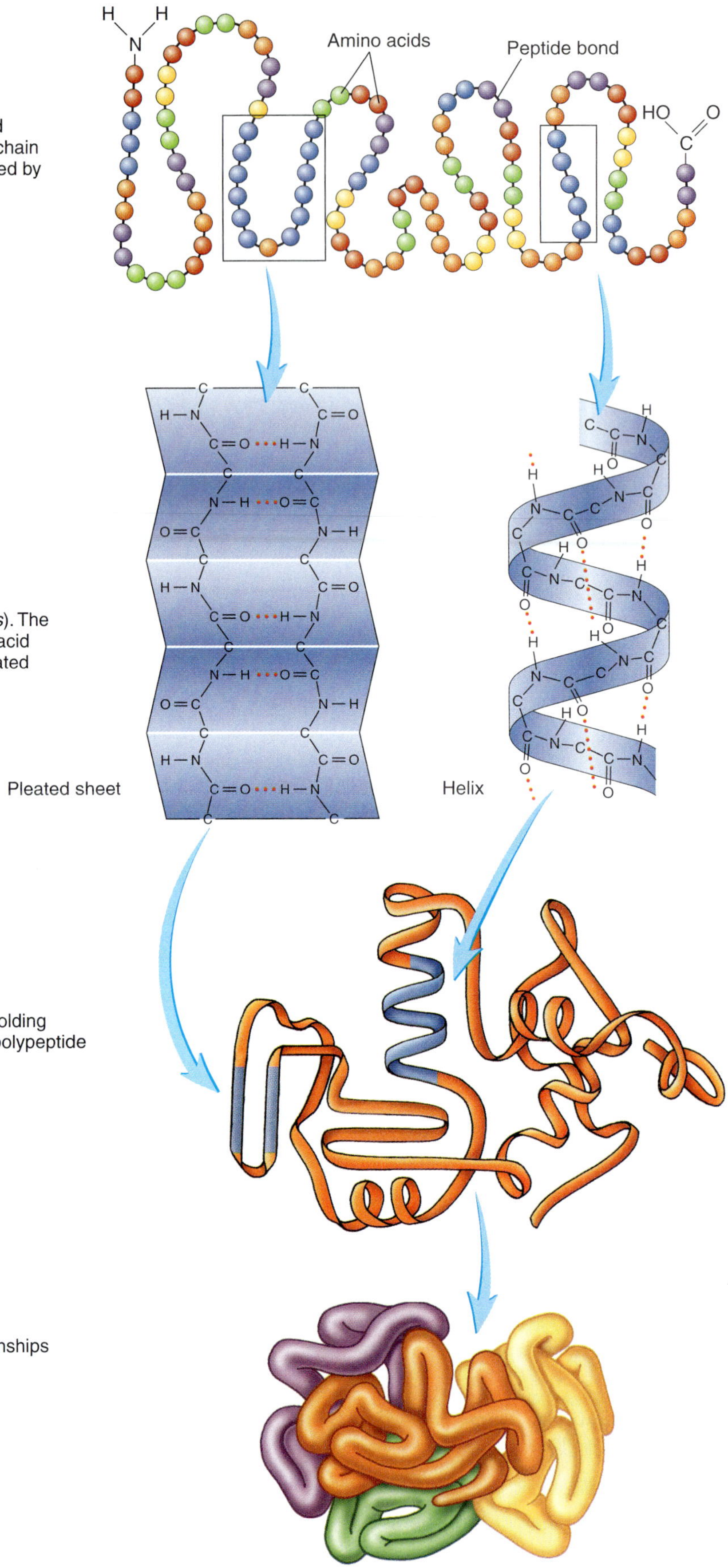

(a) Primary structure—the amino acid sequence. A protein consists of a chain of different amino acids (represented by different colored spheres).

(b) Secondary structure results from hydrogen bonding (*dotted red lines*). The hydrogen bonds cause the amino acid chain to form helices (coils) or pleated (folded) sheets.

(c) Tertiary structure with secondary folding caused by interactions within the polypeptide and its immediate environment

(d) Quaternary structure—the relationships between individual subunits

Protein Structure
Figure 2.22

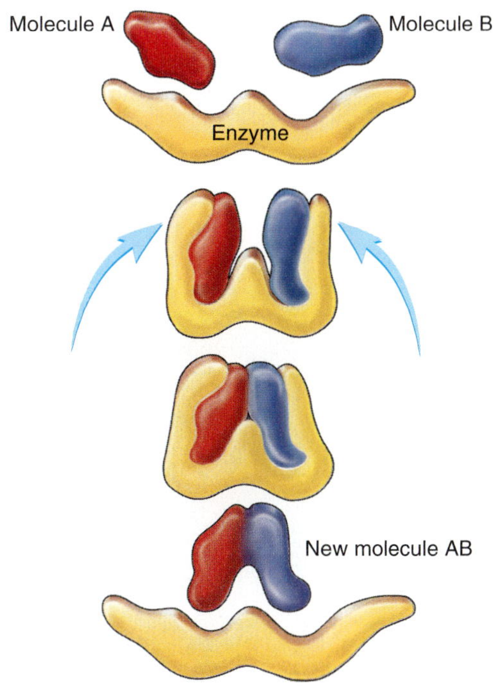

Enzyme Action
Figure 2.23

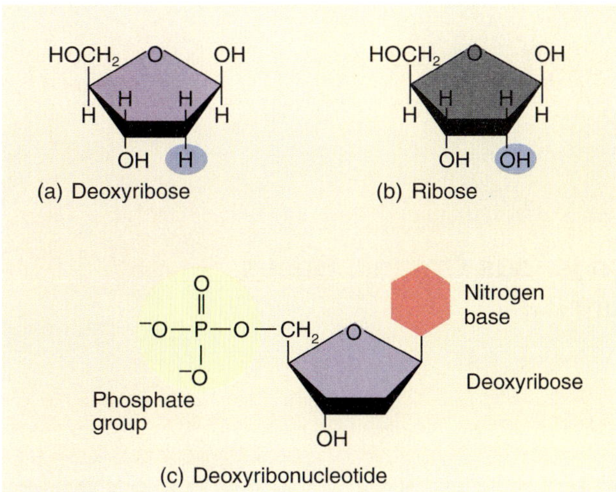

Components of Nucleotides
Figure 2.24

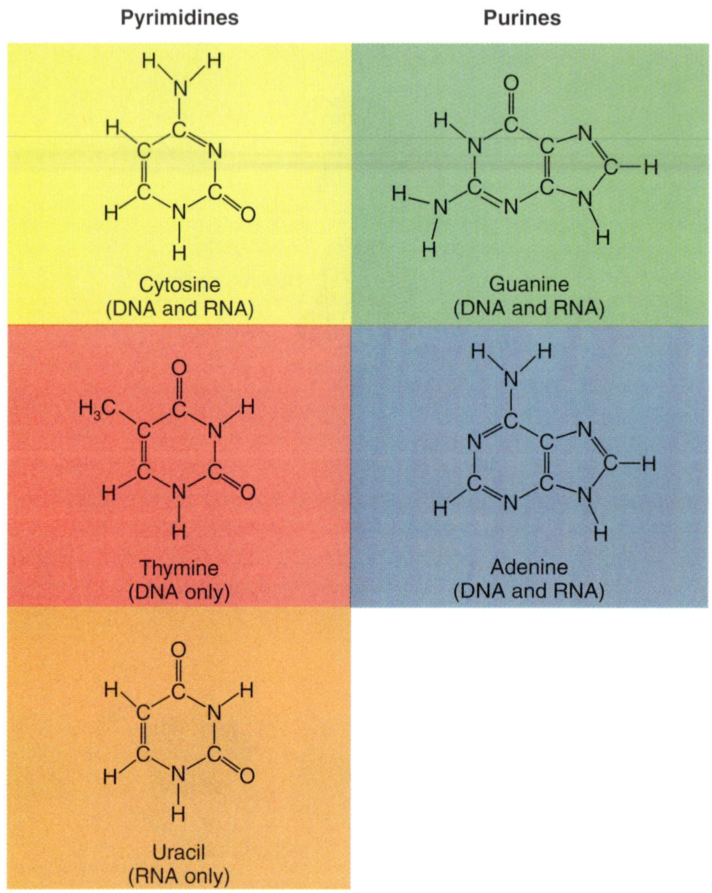
Pyrimidines
Purines
Cytosine
(DNA and RNA)
Guanine
(DNA and RNA)
Thymine
(DNA only)
Adenine
(DNA and RNA)
Uracil
(RNA only)

Nitrogenous Organic Bases
Figure 2.25

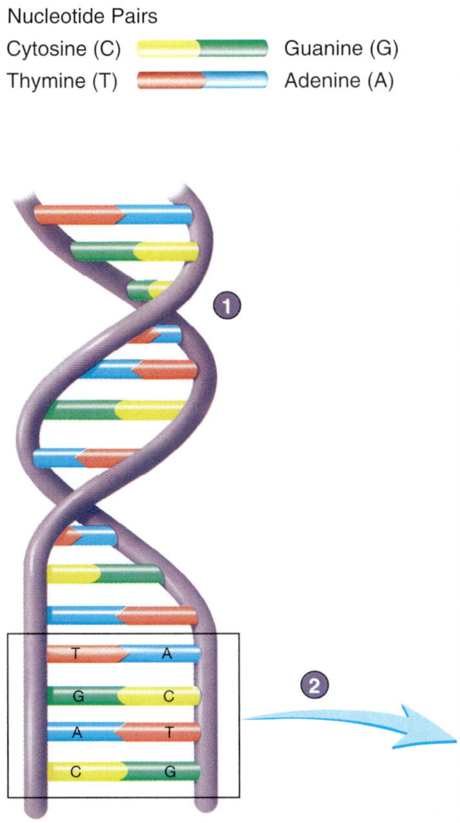

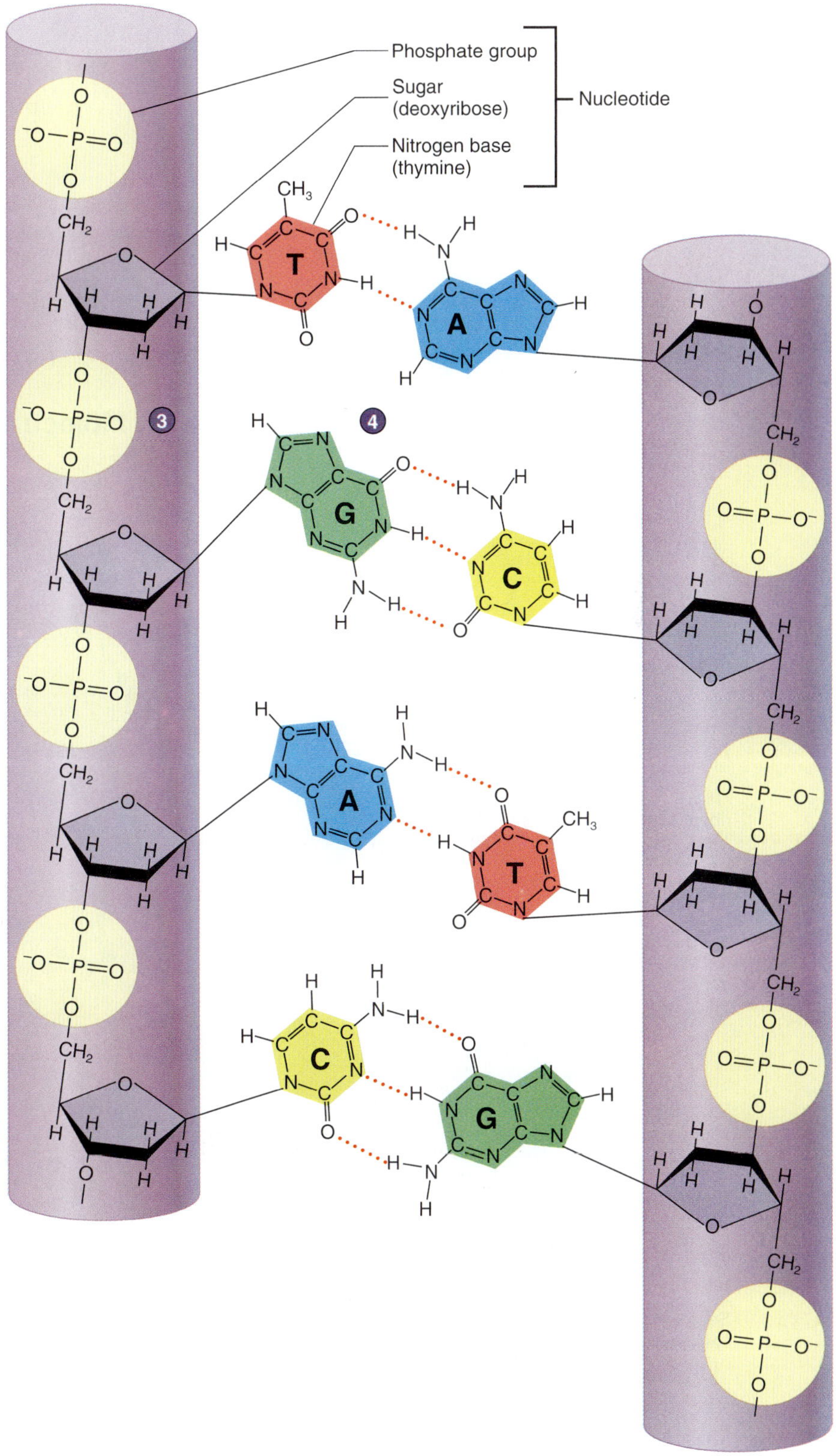

1. A DNA molecule is two strands of nucleotides joined together to form a double-stranded helix.
2. The strands are uncoiled and enlarged.
3. The deoxyribose molecules and phosphate groups of each strand are joined by covalent bonds.
4. The strands are held together by hydrogen bonds (*dotted red lines*) between the bases of the nucleotides.

Structure of DNA
Figure 2.26

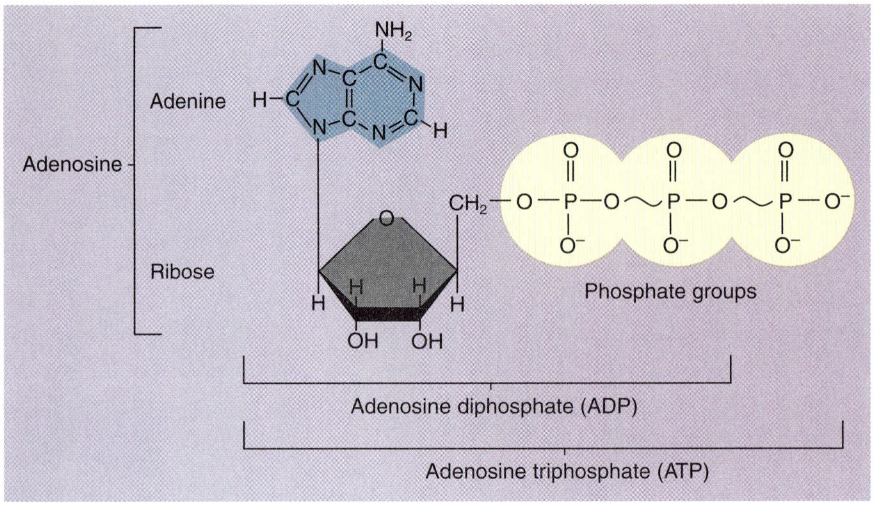

Adenosine Triphosphate (ATP) Molecule
Figure 2.27

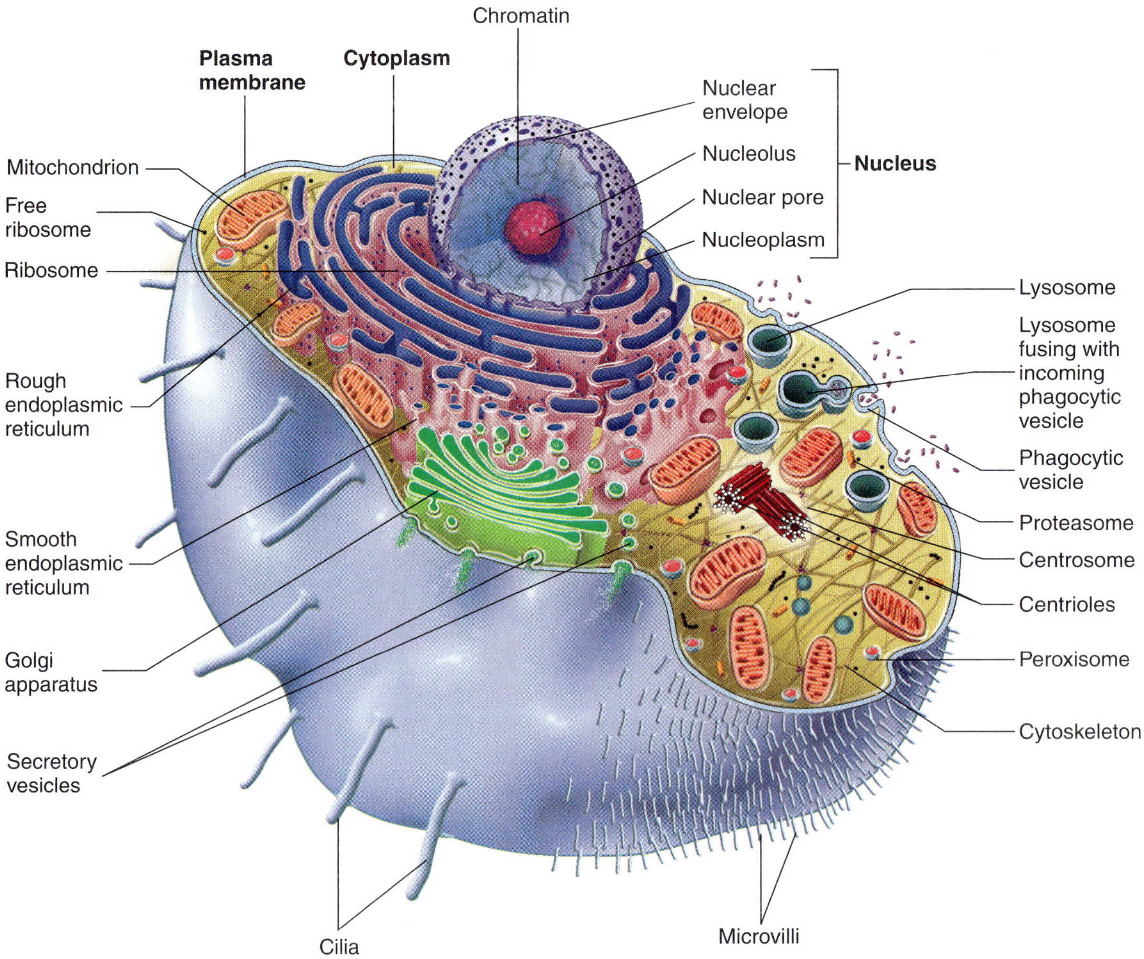

The Cell
Figure 3.1

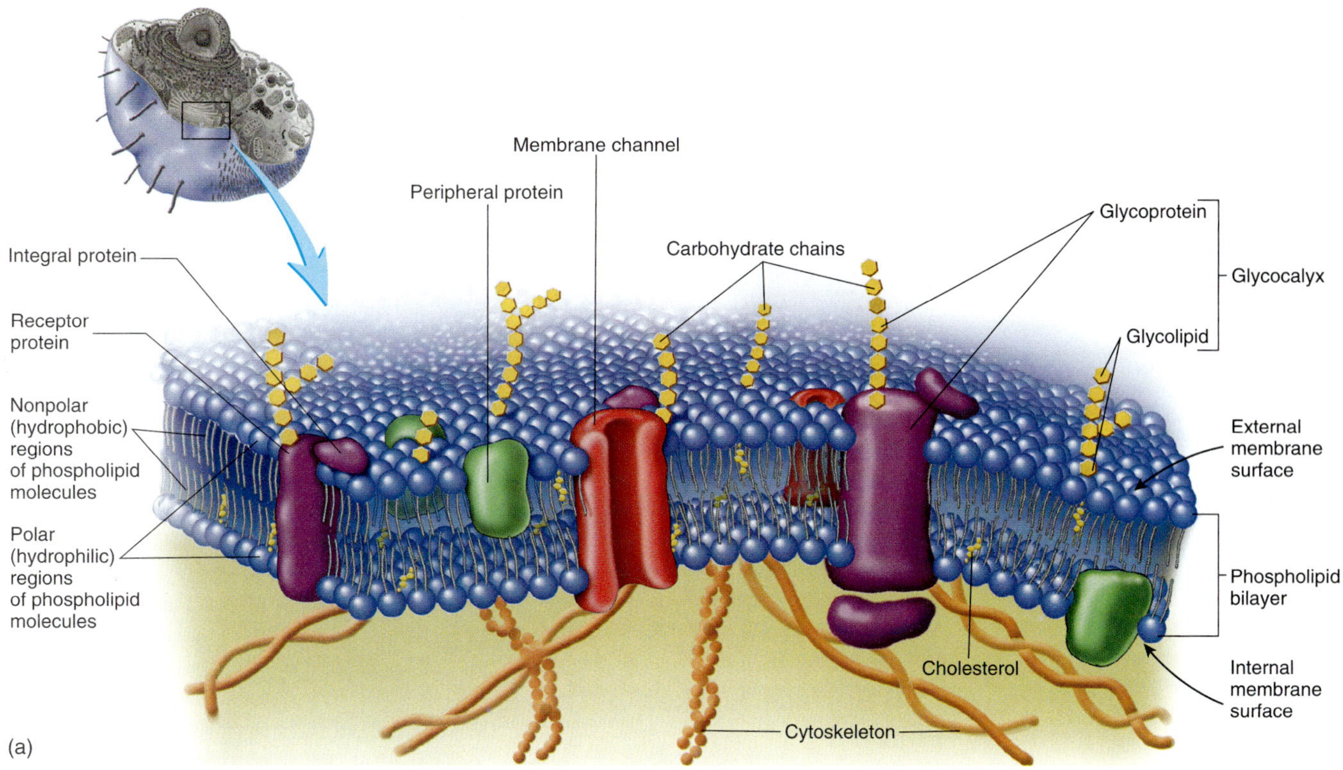

Plasma Membrane
Figure 3.2

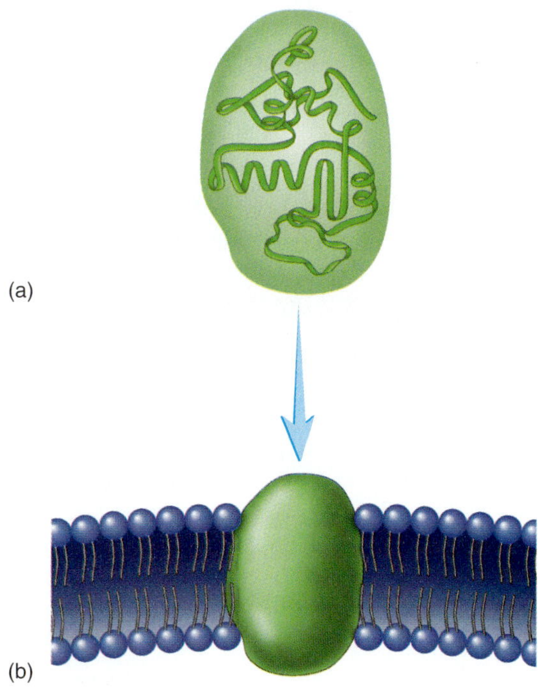

Proteins in the Plasma Membrane
Figure 3.3

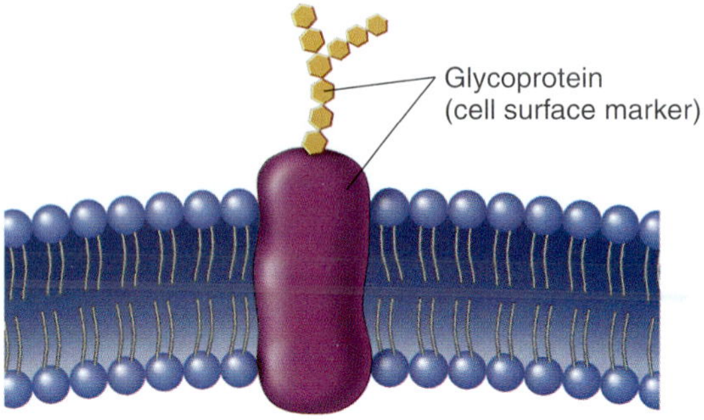

Cell Surface Marker
Figure 3.4

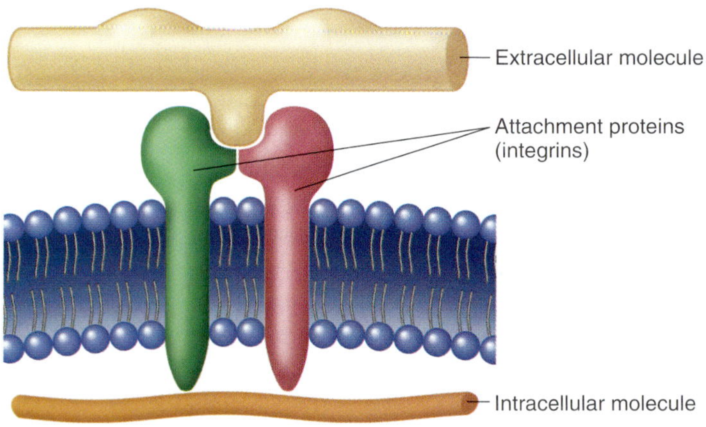

Attachment Sites
Figure 3.5

a. Some regions of a protein are helical. Each helical region can be depicted as a cylinder.

b. In some membrane proteins, the helical regions form a circle with a channel in the center.

c. The ring of cylinders can be depicted as a 3-D globular structure with a channel in the center This is called a channel protein.

d. The channel protein can be depicted cut in half to show the channel.

e. The cut channel protein is depicted within the plasma membrane.

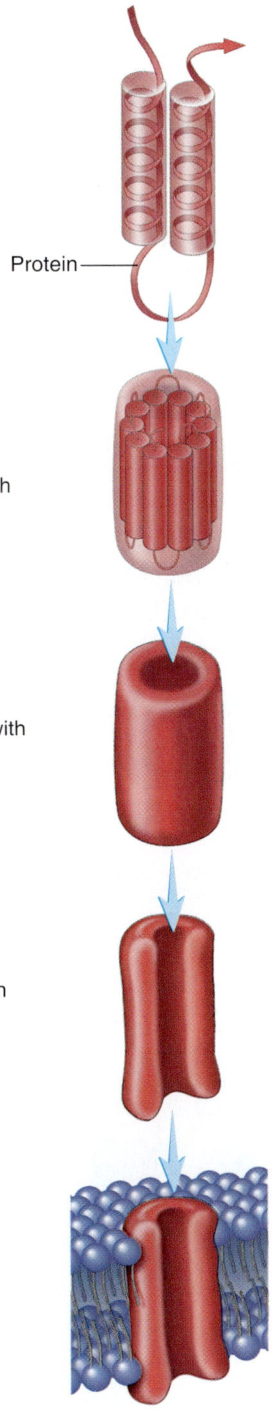

Channel Protein
Figure 3.6

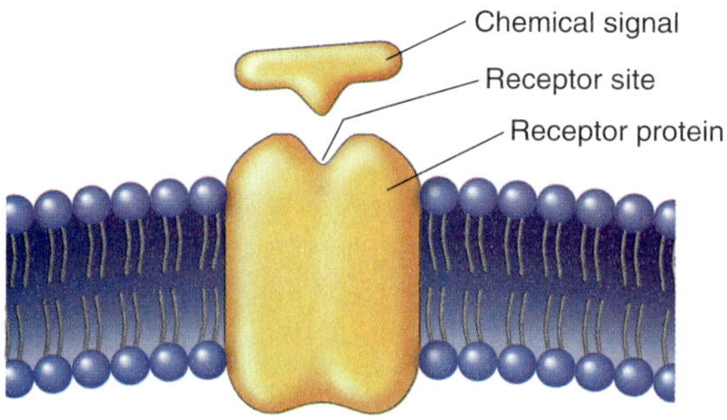

Receptor Protein
Figure 3.7

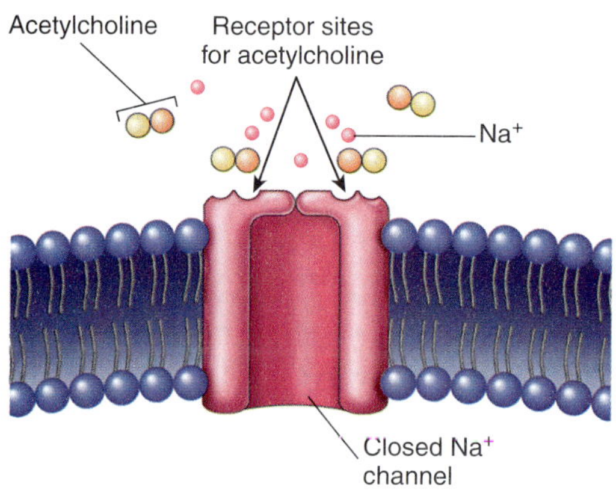

1. The Na^+ channel has receptor sites for the chemical signal, acetylcholine. When the receptor sites are not occupied by acetylcholine, the Na^+ channel remains closed.

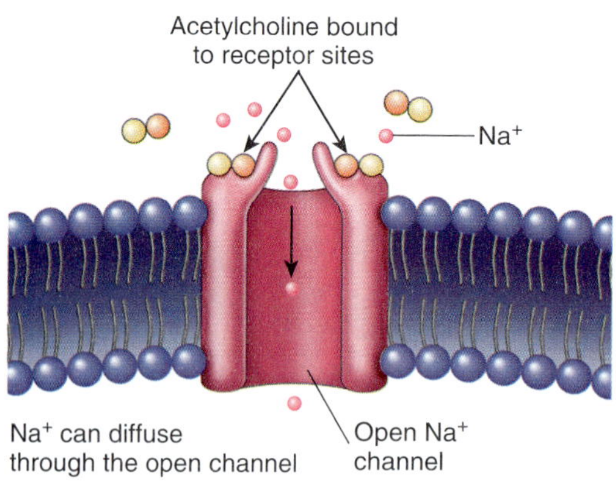

2. When two acetylcholine molecules bind to their receptor sites on the Na^+ channel, the channel opens to allow Na^+ to diffuse through the channel into the cell.

Receptors Linked to a Channel Protein
Process Figure 3.8

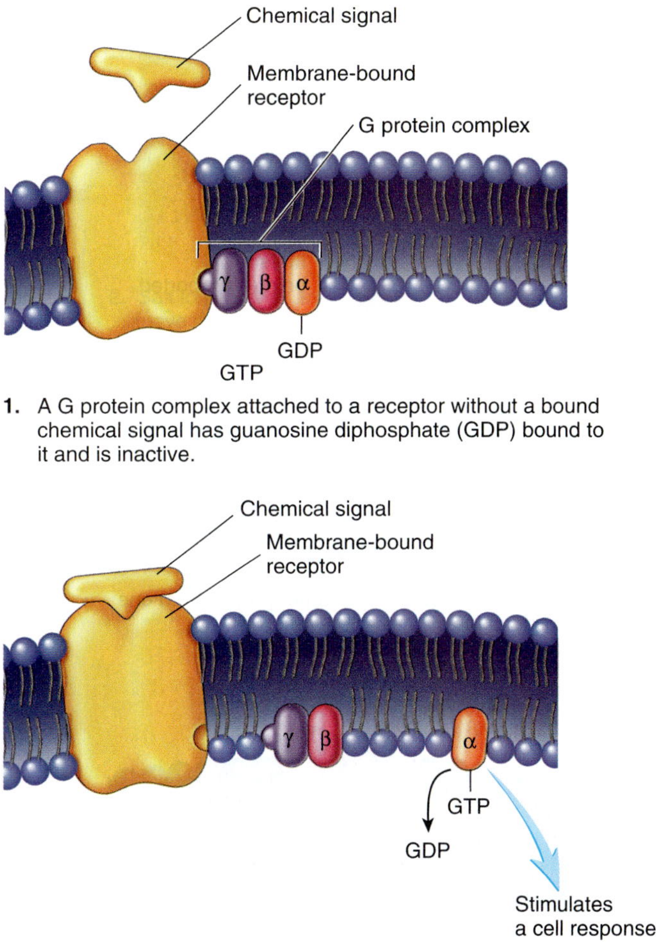

1. A G protein complex attached to a receptor without a bound chemical signal has guanosine diphosphate (GDP) bound to it and is inactive.

2. When a chemical signal attaches to the receptor, guanosine triphosphate (GTP) replaces GDP on the α-subunit of the G protein complex, which separates from the other subunits. The α-subunit, with GTP attached, stimulates a cell response.

A Receptor Linked to a G Protein Complex

Process Figure 3.9

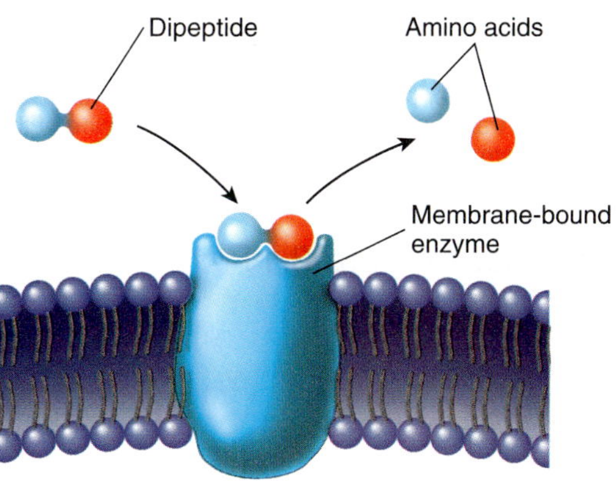

Enzyme in the Plasma Membrane

Figure 3.10

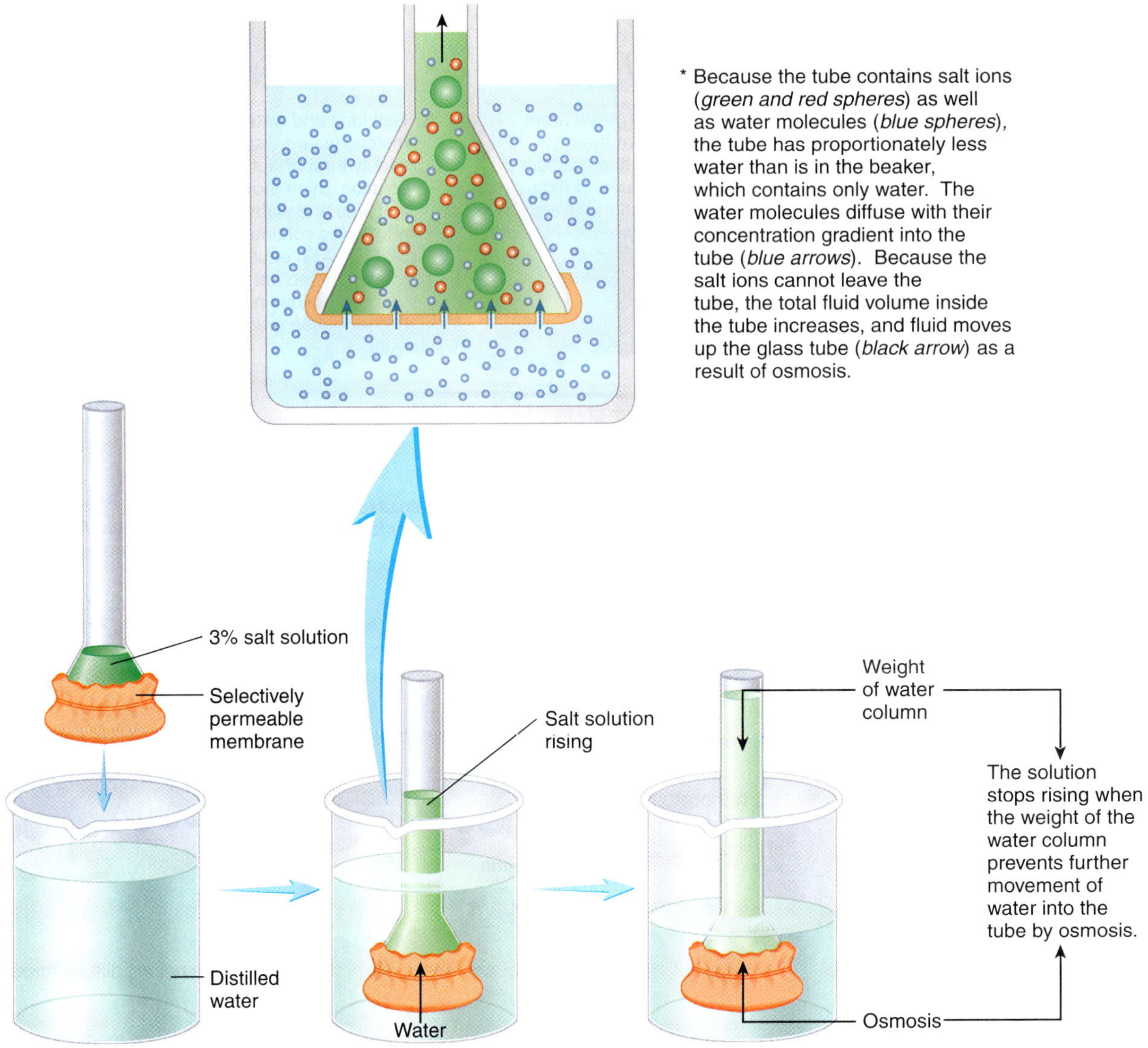

1. The end of a tube containing a 3% salt solution (*green*) is closed at one end with a selectively permeable membrane, which allows water molecules to pass through it but retains the salt ions within the tube.
2. The tube is immersed in distilled water. Water moves into the tube by osmosis (see inset above*). The concentration of salt in the tube decreases as water rises in the tube (*lighter green color*).
3. Water moves by osmosis into the tube until the weight of the column of water in the tube (hydrostatic pressure) prevents further movement of water into the tube. The hydrostatic pressure that prevents net movement of water into the tube is equal to the osmotic pressure of the solution in the tube.

Osmosis
Process Figure 3.15

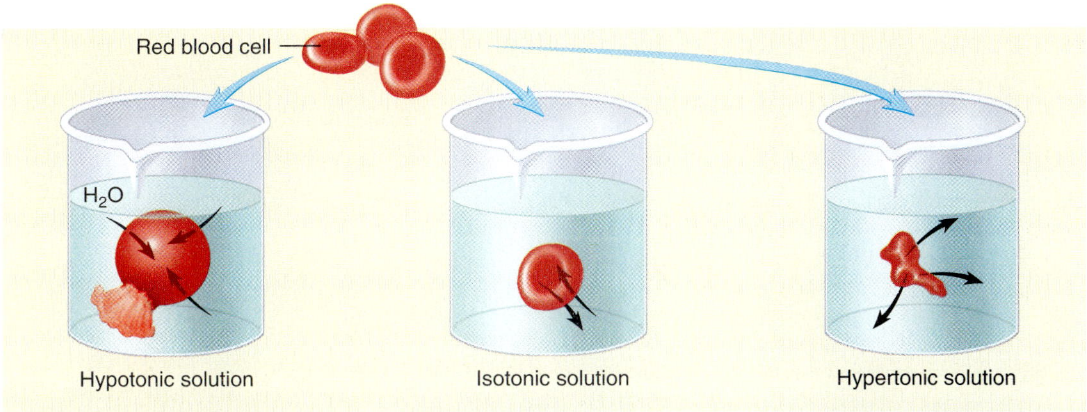

(a) A hypotonic solution with a low solute concentration results in swelling of the red blood cell placed into the solution. Water enters the cell by osmosis *(black arrows),* and the red blood cell lyses (bursts; *puff of red in the lower part of the cell*).

(b) An isotonic solution with a concentration of solutes equal to that inside the cell results in a normally shaped red blood cell. Water moves into and out of the cell at the same rate *(black arrows),* but there is no net water movement.

(c) A hypertonic solution, with a high solute concentration, causes shrinkage (crenation) of the red blood cell as water moves by osmosis out of the cell and into the hypertonic solution *(black arrows).*

Effects of Hypotonic, Isotonic, and Hypertonic Solutions on Red Blood Cells
Figure 3.16

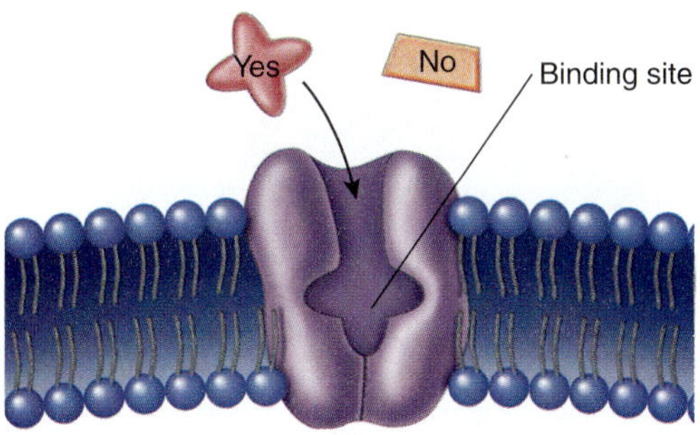

(a) Specificity. Only molecules that are the right shape to bind to the binding site are transported.

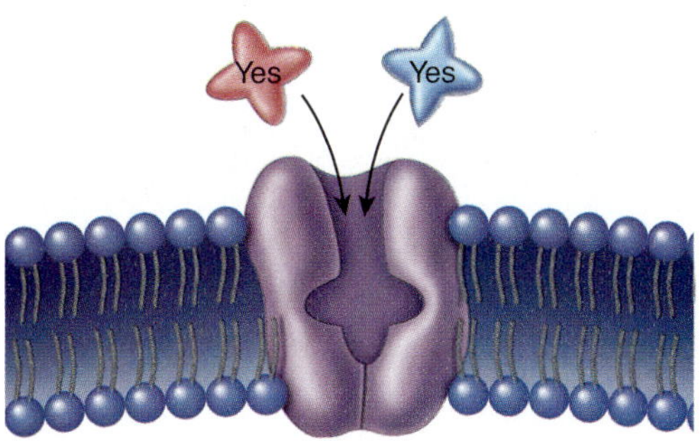

(b) Competition. Similarly shaped molecules can compete for the same binding site.

Mediated Transport: Specificity and Competition
Figure 3.17

The rate of transport of molecules into a cell is plotted against the concentration of those molecules outside the cell minus the concentration of those molecules inside the cell. As the concentration increases, the rate of transport increases and then levels off.

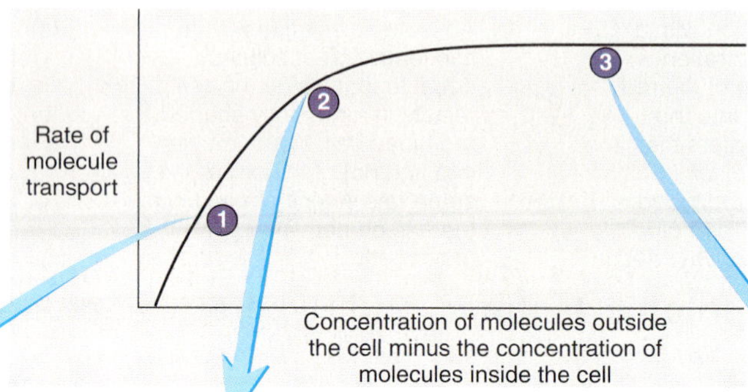

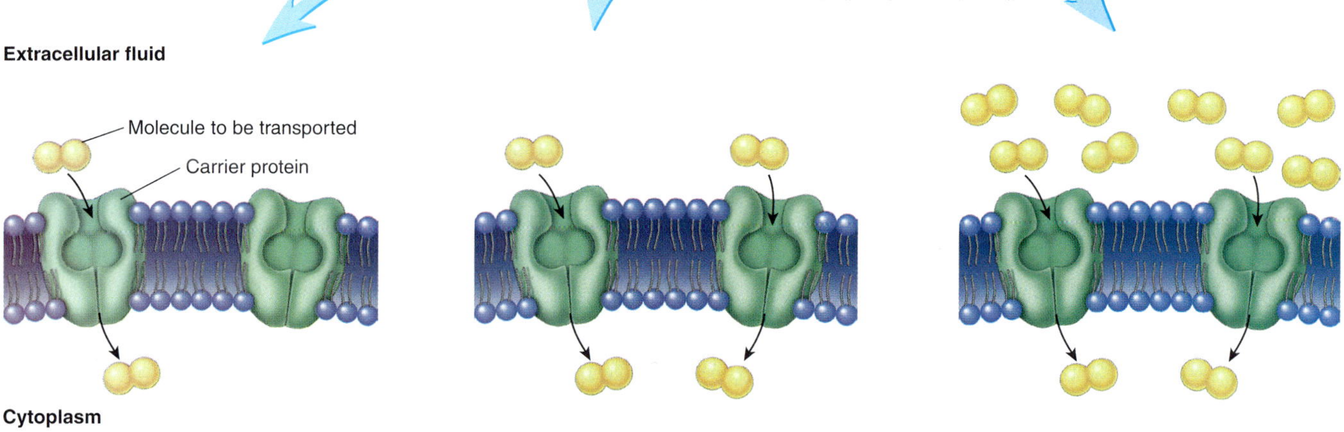

1. When the concentration of molecules outside the cell is low, the transport rate is low because it is limited by the number of molecules available to be transported.

2. When more molecules are present outside the cell, as long as enough carrier proteins are available, more molecules can be transported, and therefore the transport rate increases.

3. The transport rate is limited by the number of carrier proteins and the rate at which each carrier protein can transport solutes. When the number of molecules outside the cell is so large that the carrier proteins are all occupied, the system is saturated and the transport rate cannot increase.

Saturation of a Carrier Protein
Process Figure 3.18

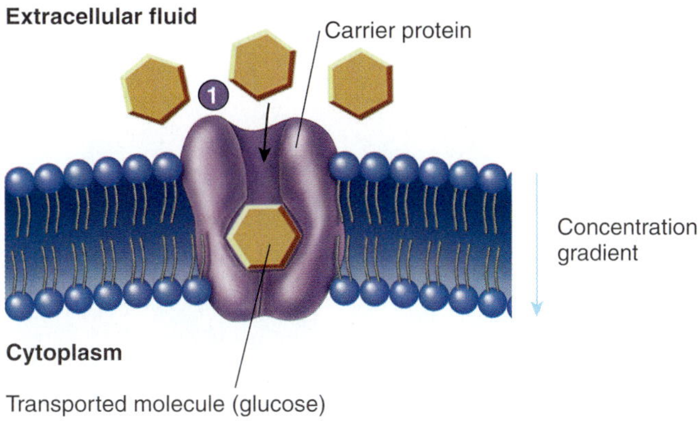

1. The carrier protein binds with a molecule, such as glucose, on the outside of the plasma membrane.

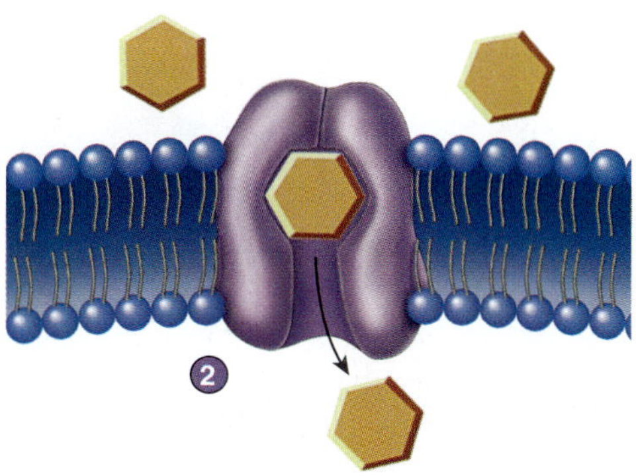

2. The carrier protein changes shape and releases the molecule on the inside of the plasma membrane.

Facilitated Diffusion
Process Figure 3.19

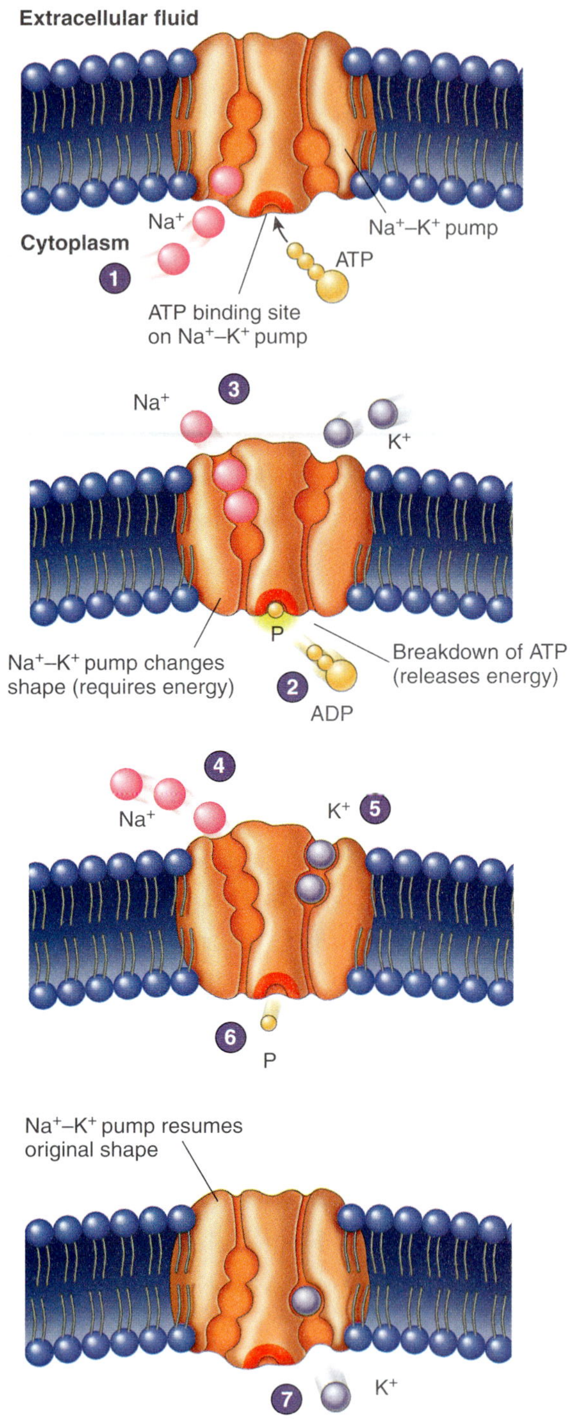

1. Three sodium ions (Na^+) and adenosine triphosphate (ATP) bind to the Na^+–K^+ pump .

2. The ATP breaks down to adenosine diphosphate (ADP) and a phosphate (P) and releases energy. That energy is used to power the shape change in the Na^+–K^+ pump. Phosphate remains bound to the Na^+–K^+–ATP binding site.

3. The Na^+–K^+ pump changes shape, and the Na^+ are transported across the membrane.

4. The Na^+ diffuse away from the Na^+–K^+ pump.

5. Two potassium ions (K^+) bind to the Na^+–K^+ pump.

6. The phosphate is released from the Na^+–K^+ pump binding site.

7. The Na^+–K^+ pump changes shape, transporting K^+ across the membrane, and the K^+ diffuse away from the pump. The Na^+–K^+ pump can again bind to Na^+ and ATP.

Sodium–Potassium Pump
Process Figure 3.20

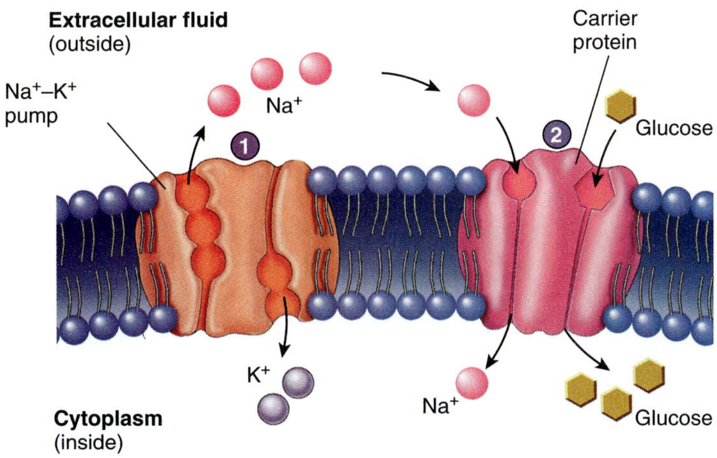

1. A Na^+–K^+ pump maintains a concentration of Na^+ that is higher outside the cell than inside.

2. Na^+ move back into the cell by a carrier protein that also moves glucose. The concentration gradient for Na^+ provides energy required to move glucose against its concentration gradient.

This example shows contransport of Na^+ and glucose. Secondary Active Transport (Cotransport)
Process Figure 3.21

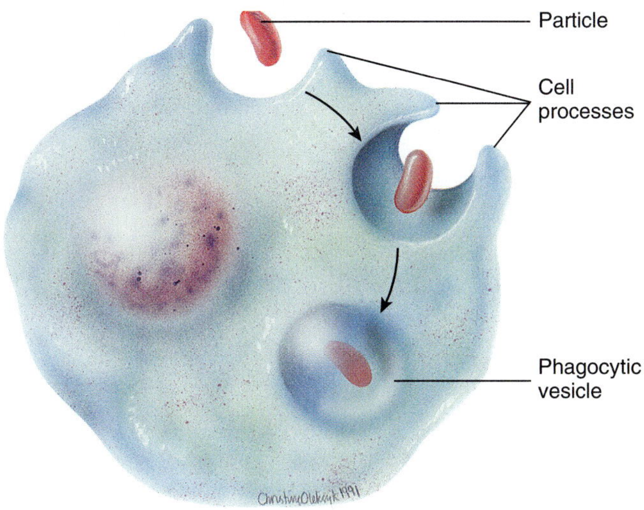

(a)

Endocytosis
Figure 3.22

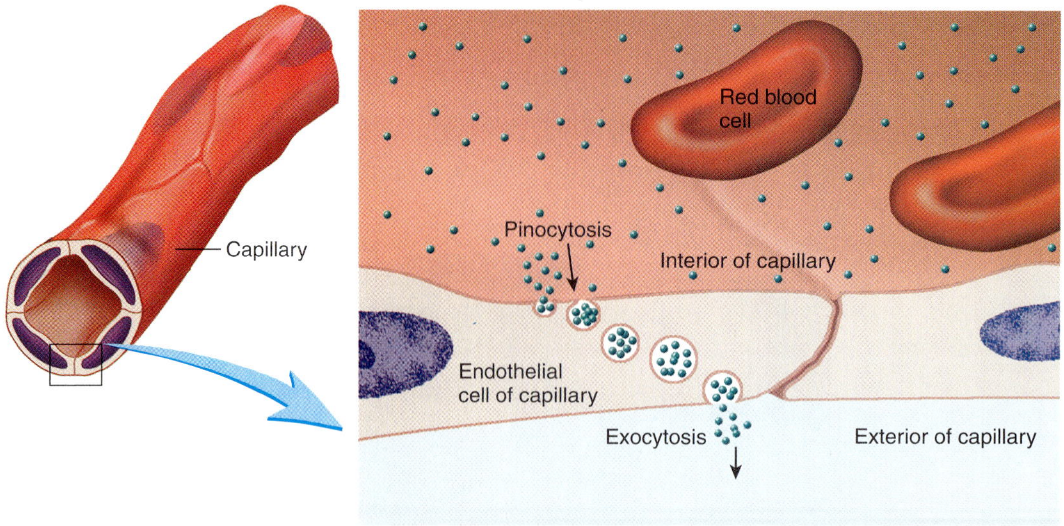

Pinocytosis
Figure 3.23

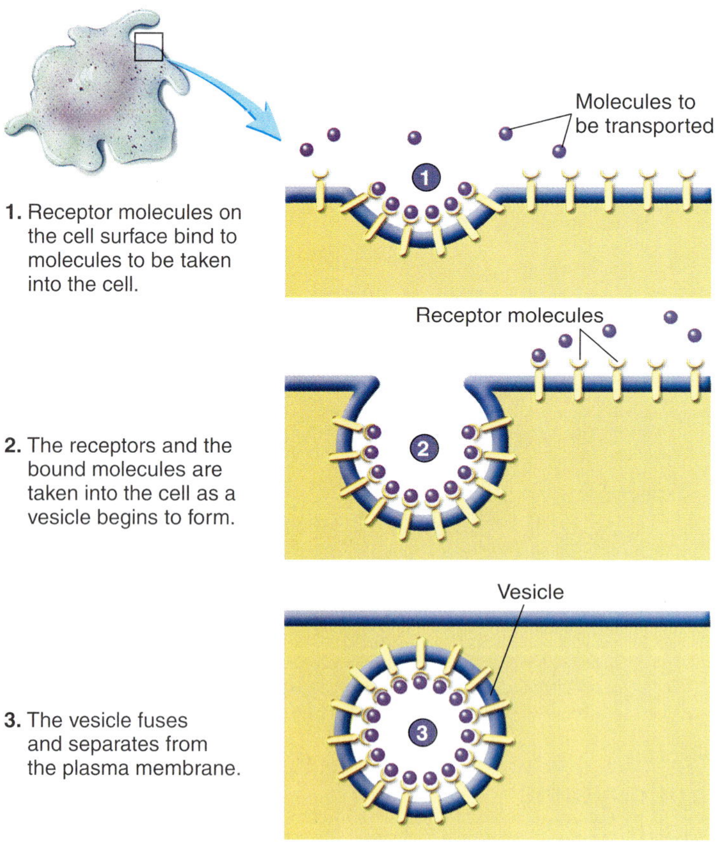

Receptor-Mediated Endocytosis
Process Figure 3.24

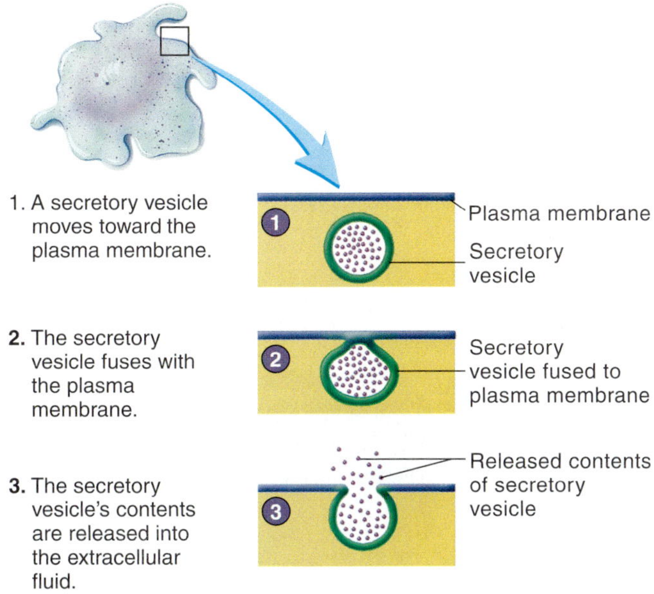

(a)

Exocytosis
Process Figure 3.25

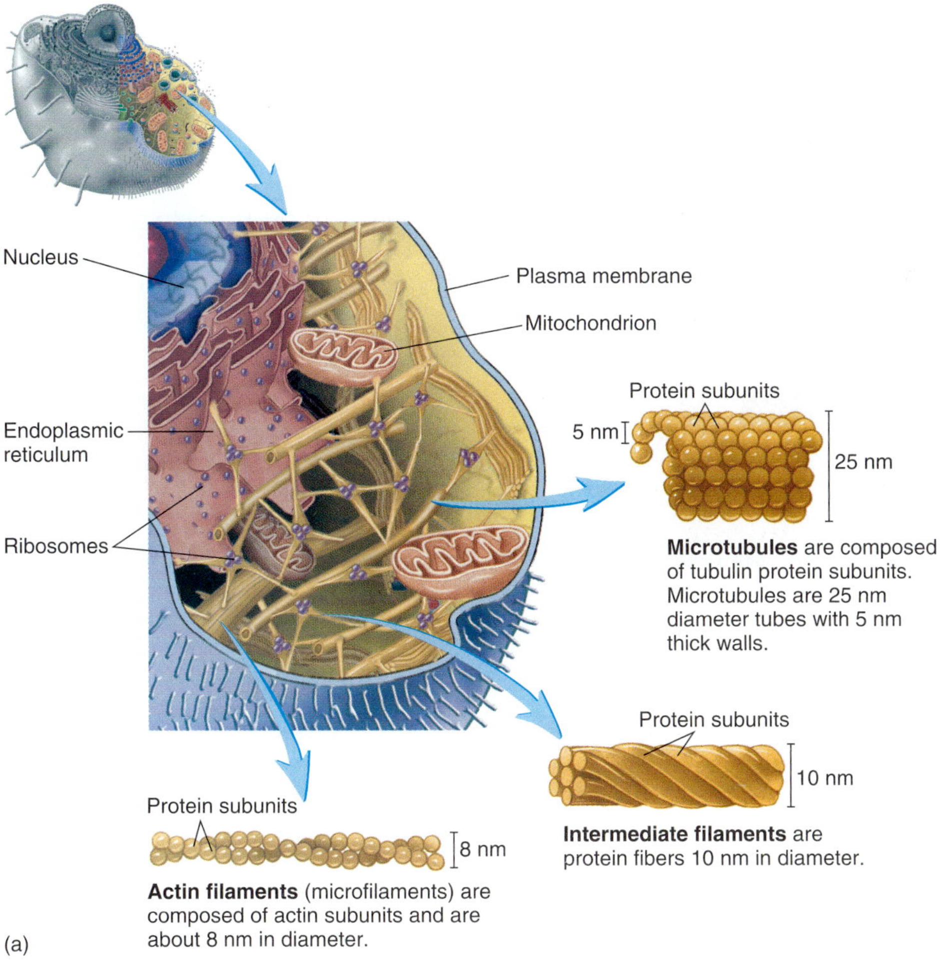

(a)

Cytoskeleton
Figure 3.26

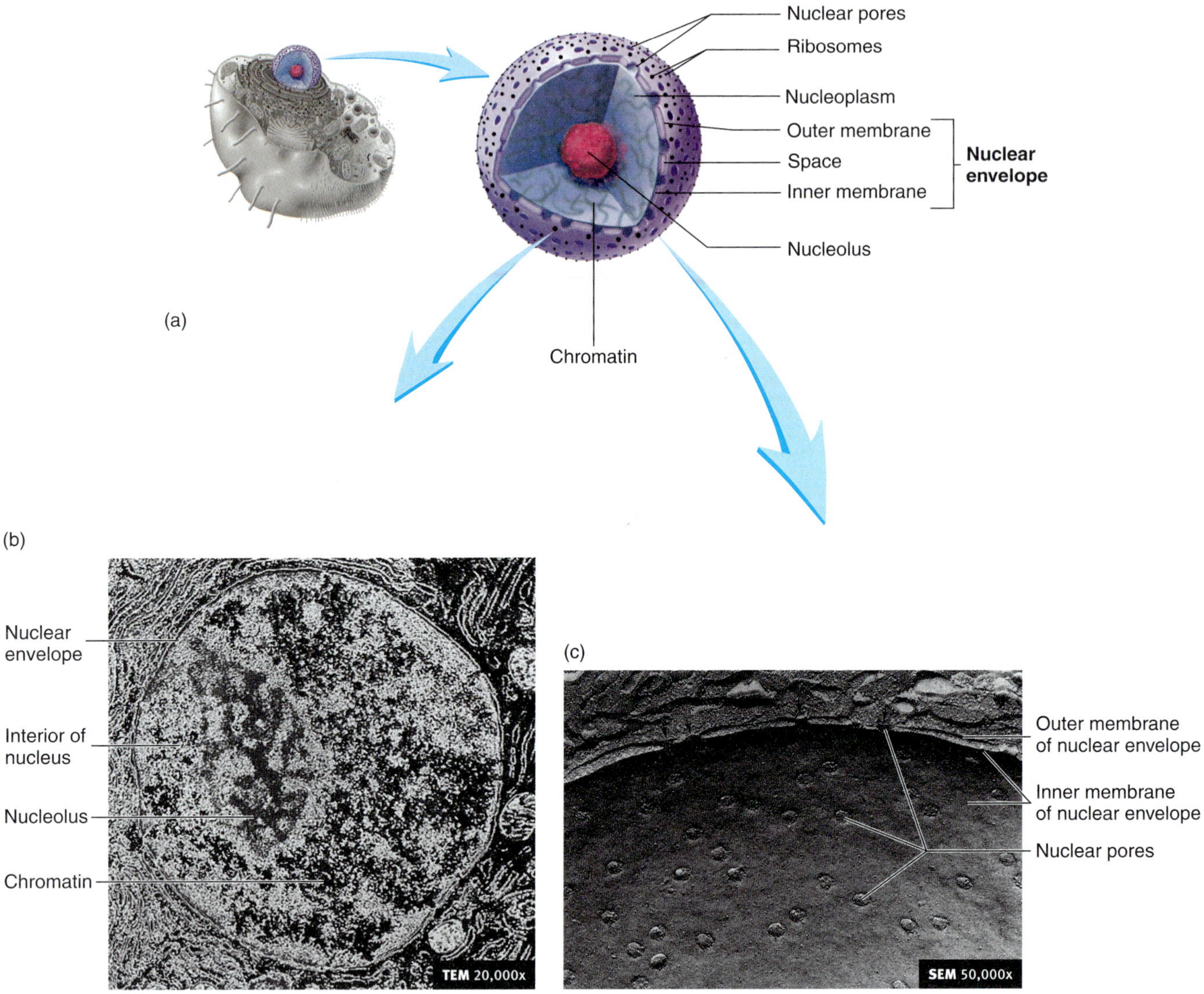

The Nucleus
Figure 3.27

b: © Don Fawcett/Photo Researchers, Inc.; **c:** © Bernard Gilula/Photo Researchers, Inc.

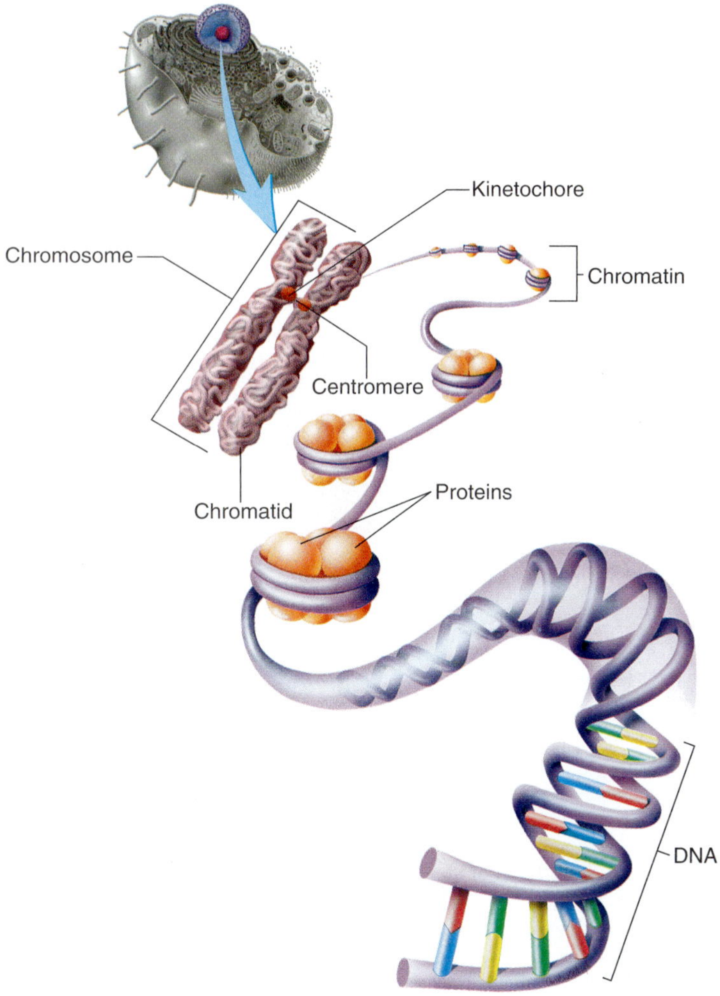

Chromosome Structure
Figure 3.28

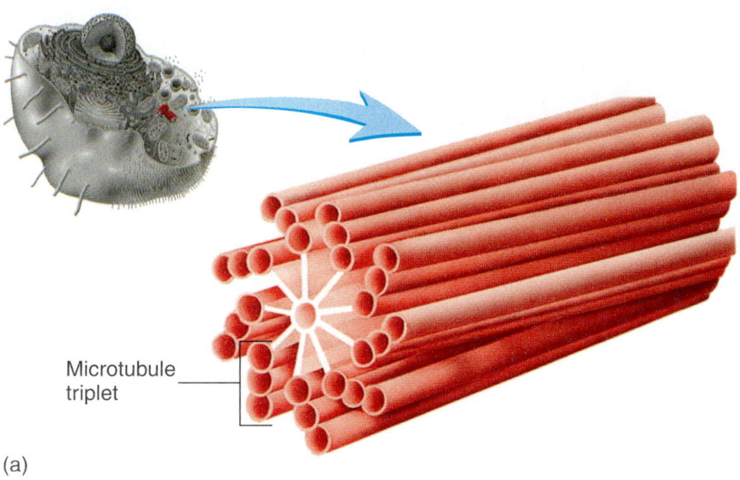

Centriole
Figure 3.29

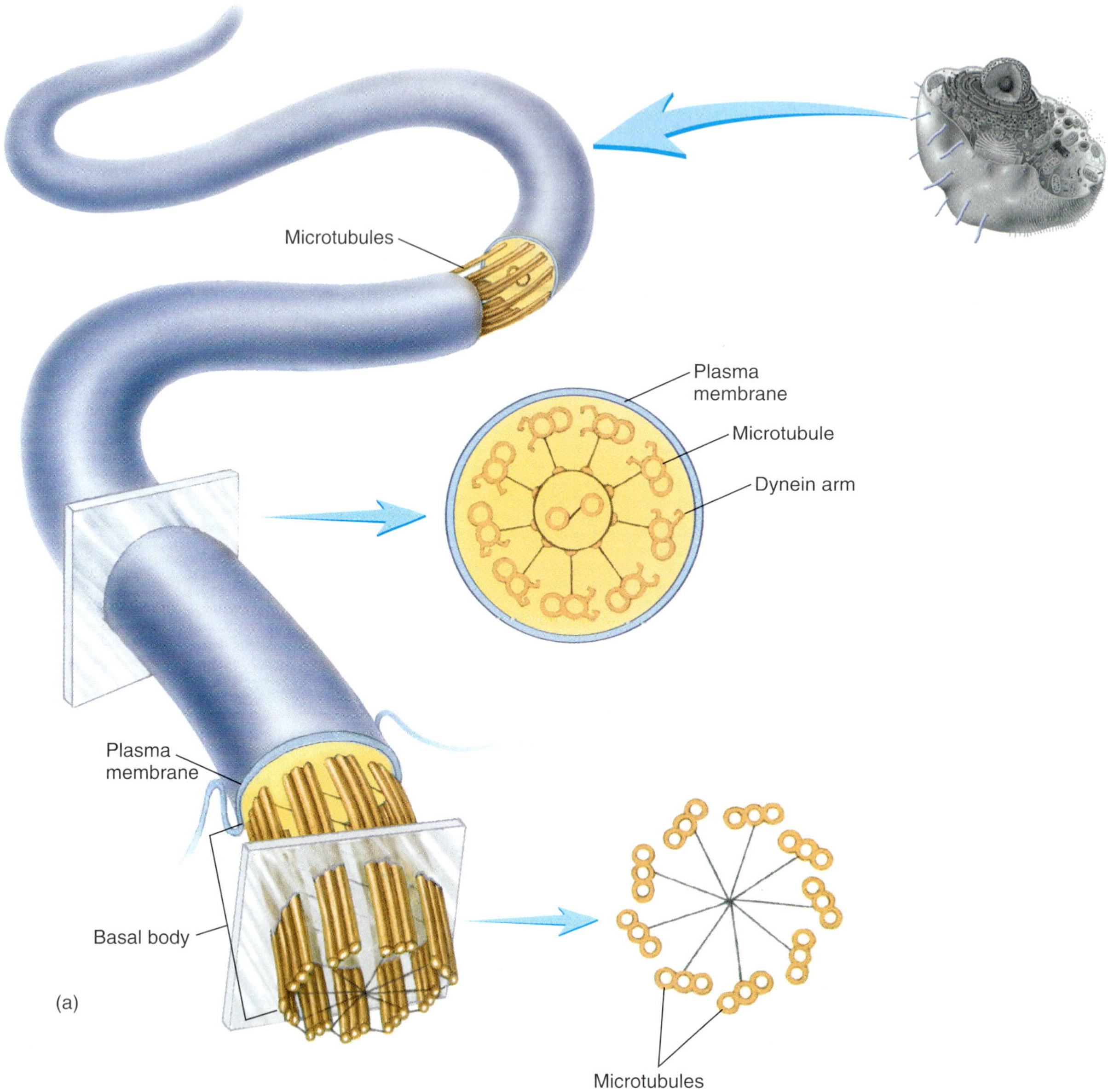
Microtubules
Plasma
membrane
Microtubule
Dynein arm
Plasma
membrane
Basal body
(a)
Microtubules

Cilia and Flagella
Figure 3.30

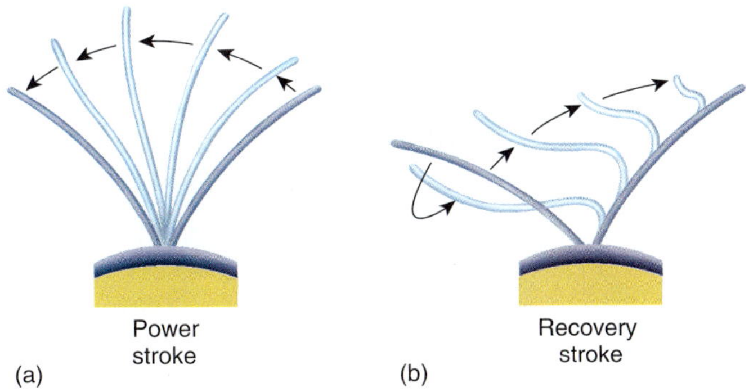

Ciliary Movement
Figure 3.31

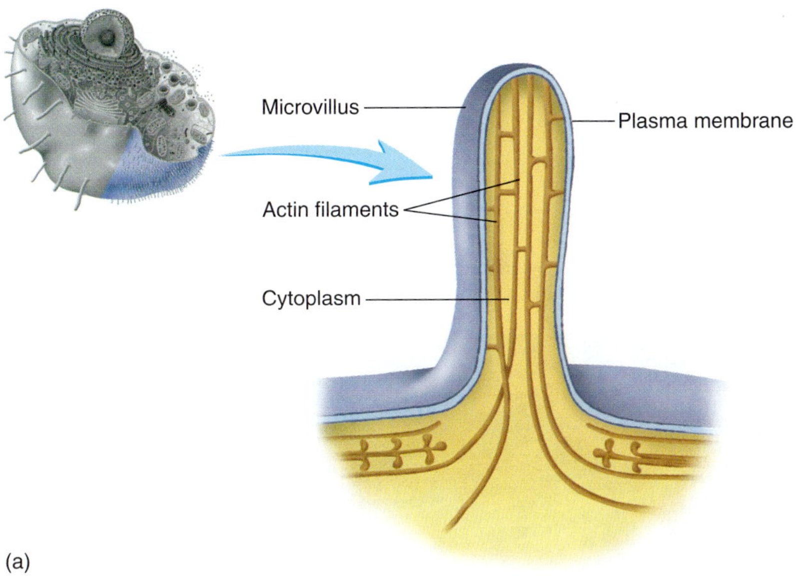

(a)

Microvillus
Figure 3.32

1. Ribosomal proteins, produced in the cytoplasm, are transported through nuclear pores into the nucleolus.
2. rRNA, most of which is produced in the nucleolus, is assembled with ribosomal proteins to form small and large ribosomal subunits.
3. The small and large ribosomal subunits leave the nucleolus and the nucleus through nuclear pores.
4. The small and large subunits, now in the cytoplasm, combine with each other and with mRNA.

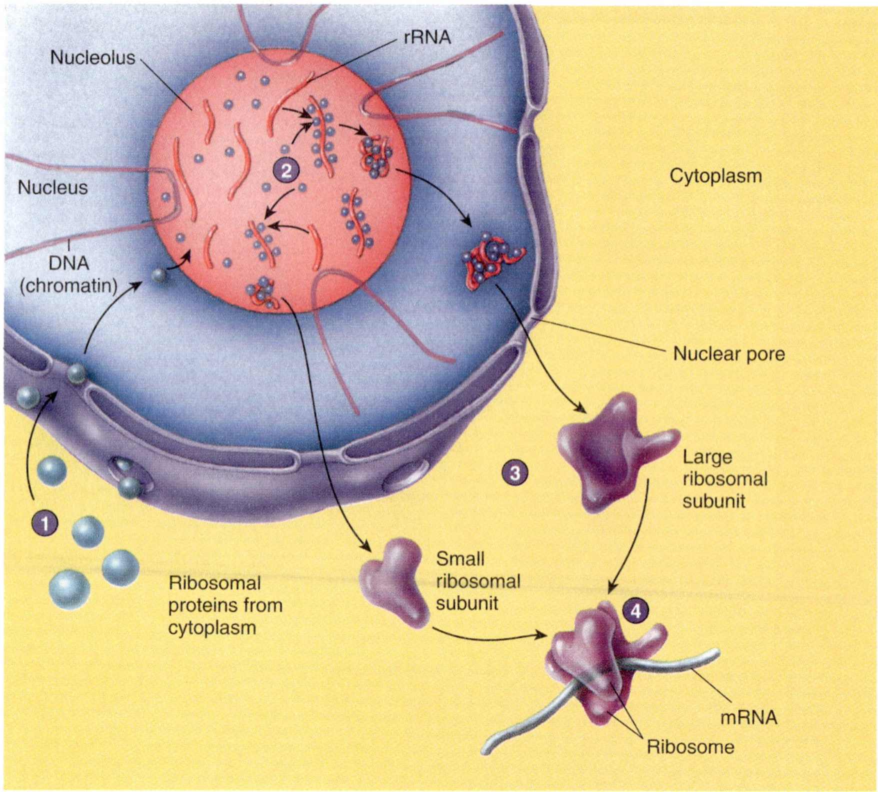

Production of Ribosomes
Process Figure 3.33

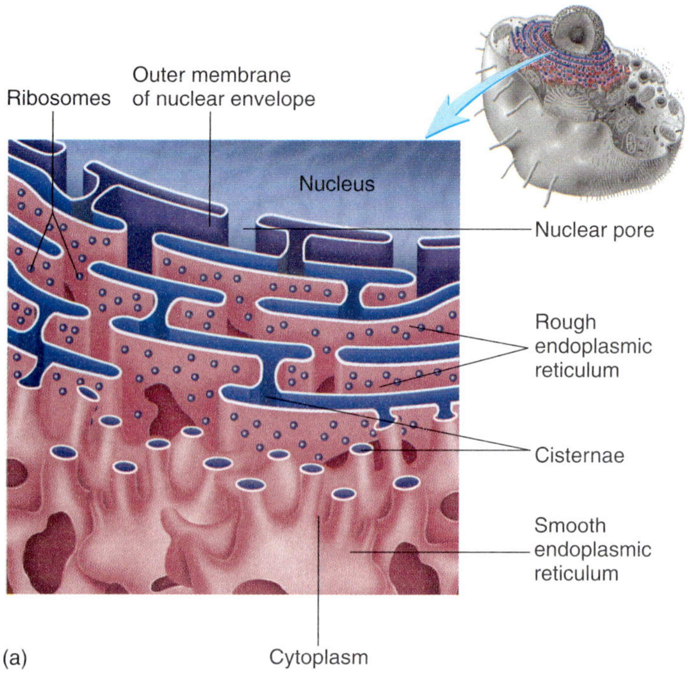

The Endoplasmic Reticulum
Figure 3.34

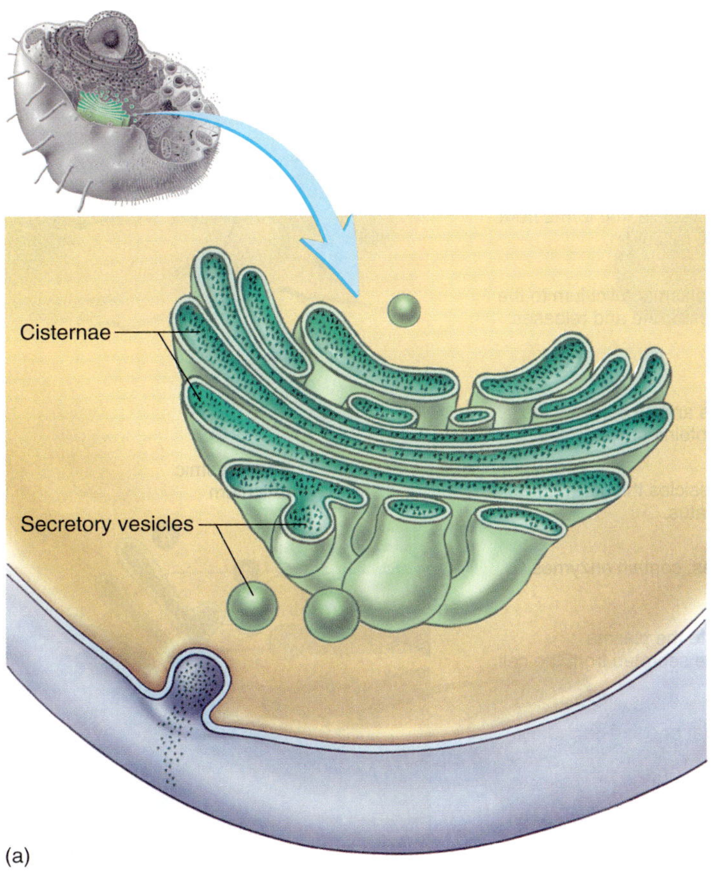
Cisternae
Secretory vesicles

(a)

Golgi Apparatus
Figure 3.35

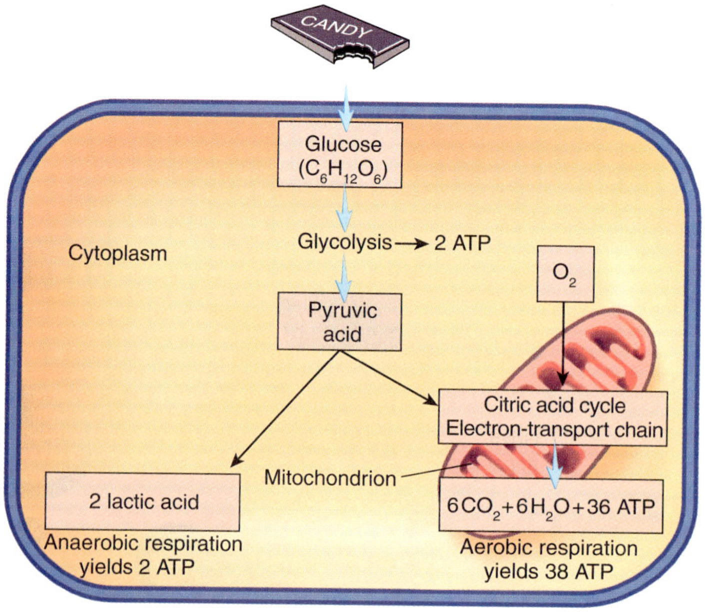

Overview of Cell Metabolism
Figure 3.39

1. DNA contains the information necessary to produce proteins.
2. Transcription of one DNA strand results in mRNA, which is a mirror image copy of the information in the DNA strand needed to make a protein.
3. The mRNA leaves the nucleus and goes to a ribosome.
4. Amino acids, the building blocks of proteins, are carried to the ribosome by tRNAs.
5. In the process of translation, the information contained in mRNA is used to determine the number, kinds, and arrangement of amino acids in the protein.

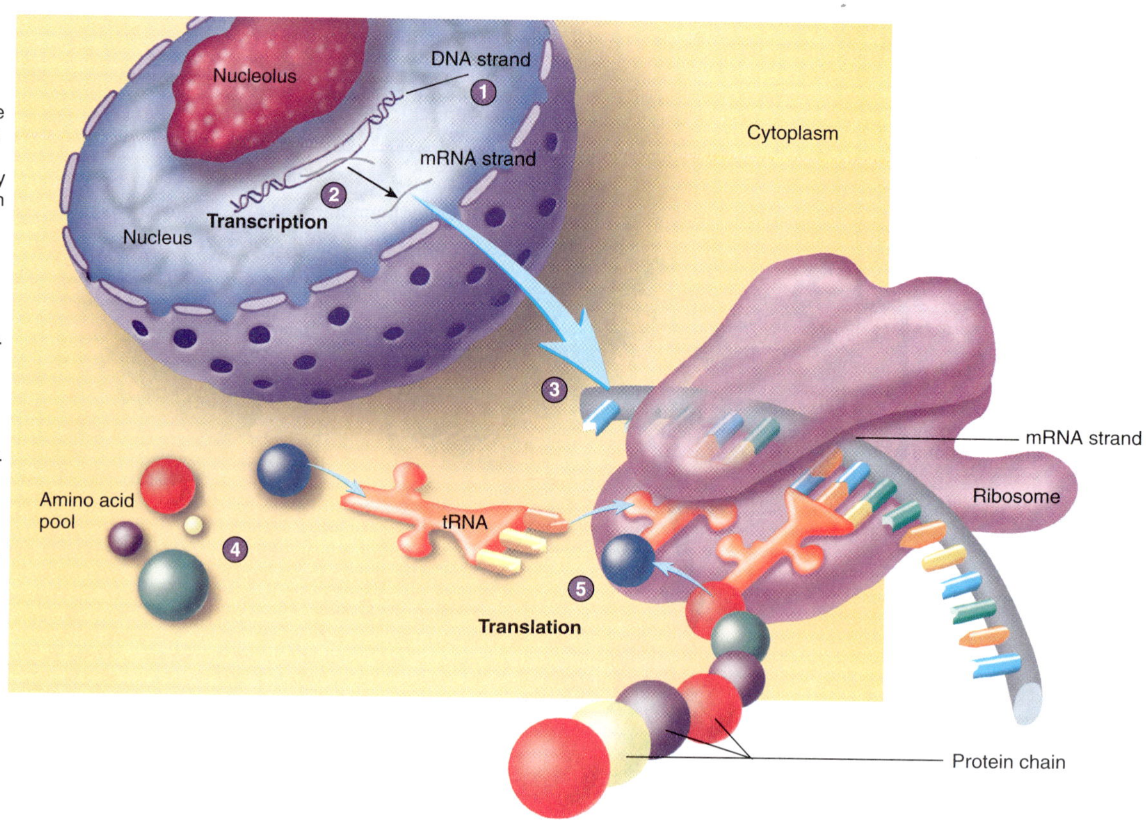

Overview of Protein Synthesis
Process Figure 3.40

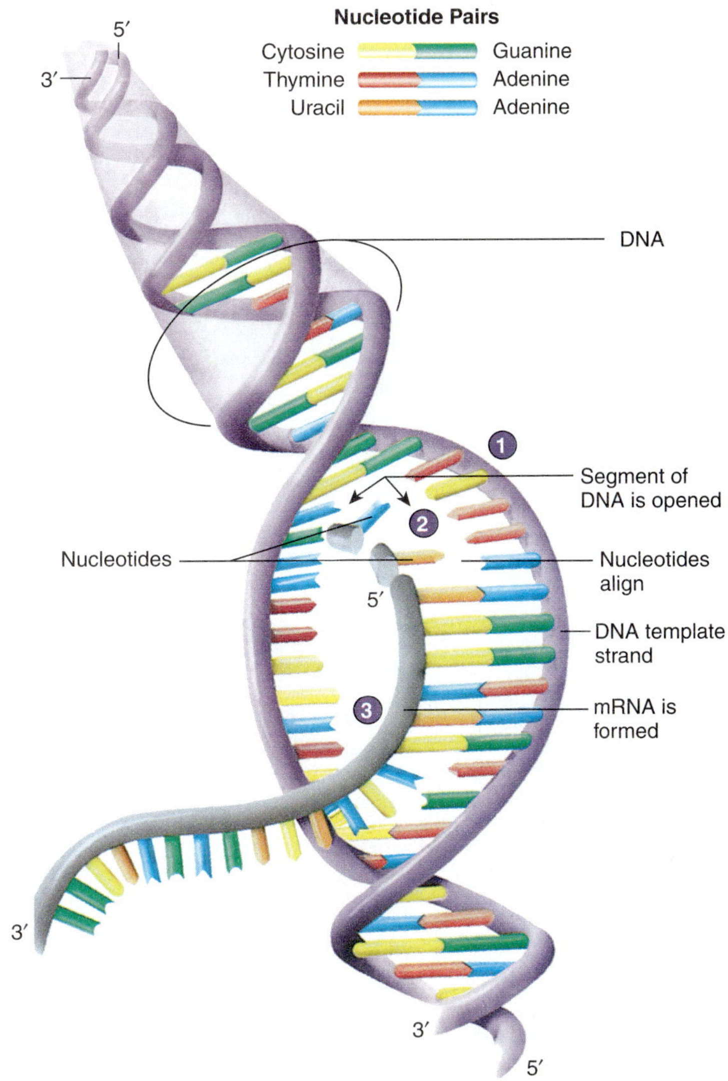

1. The strands of the DNA molecule separate from each other. One DNA strand serves as a template for mRNA synthesis.
2. Nucleotides that will form mRNA pair with DNA nucleotides according to the base-pair combinations shown in the key at the top of the figure. Thus the sequence of nucleotides in the template DNA strand (*purple*) determines the sequence of nucleotides in mRNA (*gray*). RNA polymerase (the enzyme is not shown) joins the nucleotides of mRNA together.
3. As nucleotides are added, an mRNA molecule is formed.

Formation of mRNA by Transcription of DNA
Process Figure 3.41

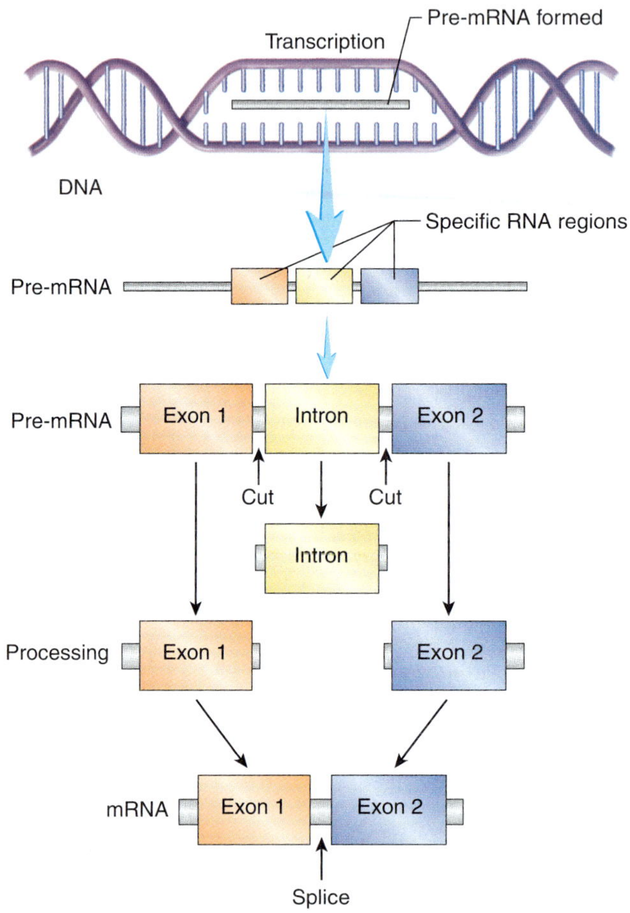
Transcription
Pre-mRNA formed
DNA
Specific RNA regions
Pre-mRNA
Pre-mRNA
Exon 1
Intron
Exon 2
Cut
Cut
Intron
Processing
Exon 1
Exon 2
mRNA
Exon 1
Exon 2
Splice

Posttranscriptional Change in mRNA
Figure 3.42

1. To start protein synthesis a ribosome binds to mRNA. The ribosome also has two binding sites for tRNA, one of which is occupied by a tRNA with its amino acid. Note that the codon of mRNA and the anticodon of tRNA are aligned and joined. The other tRNA binding site is open.

2. By occupying the open tRNA binding site the next tRNA is properly aligned with mRNA and with the other tRNA.

3. An enzyme within the ribosome catalyzes a synthesis reaction to form a peptide bond between the amino acids. Note that the amino acids are now associated with only one of the tRNAs.

4. The ribosome shifts position by three nucleotides. The tRNA without the amino acid is released from the ribosome, and the tRNA with the amino acids takes its position. A tRNA binding site is left open by the shift. Additional amino acids can be added by repeating steps *2* through *4*. Eventually a stop codon in the mRNA ends the addition of amino acids to the protein (polypeptide), which is released from the ribosome.

5. Multiple ribosomes attach to a single mRNA. As the ribosomes move down the mRNA, proteins attached to the ribosomes lengthen and eventually detach from the mRNA.

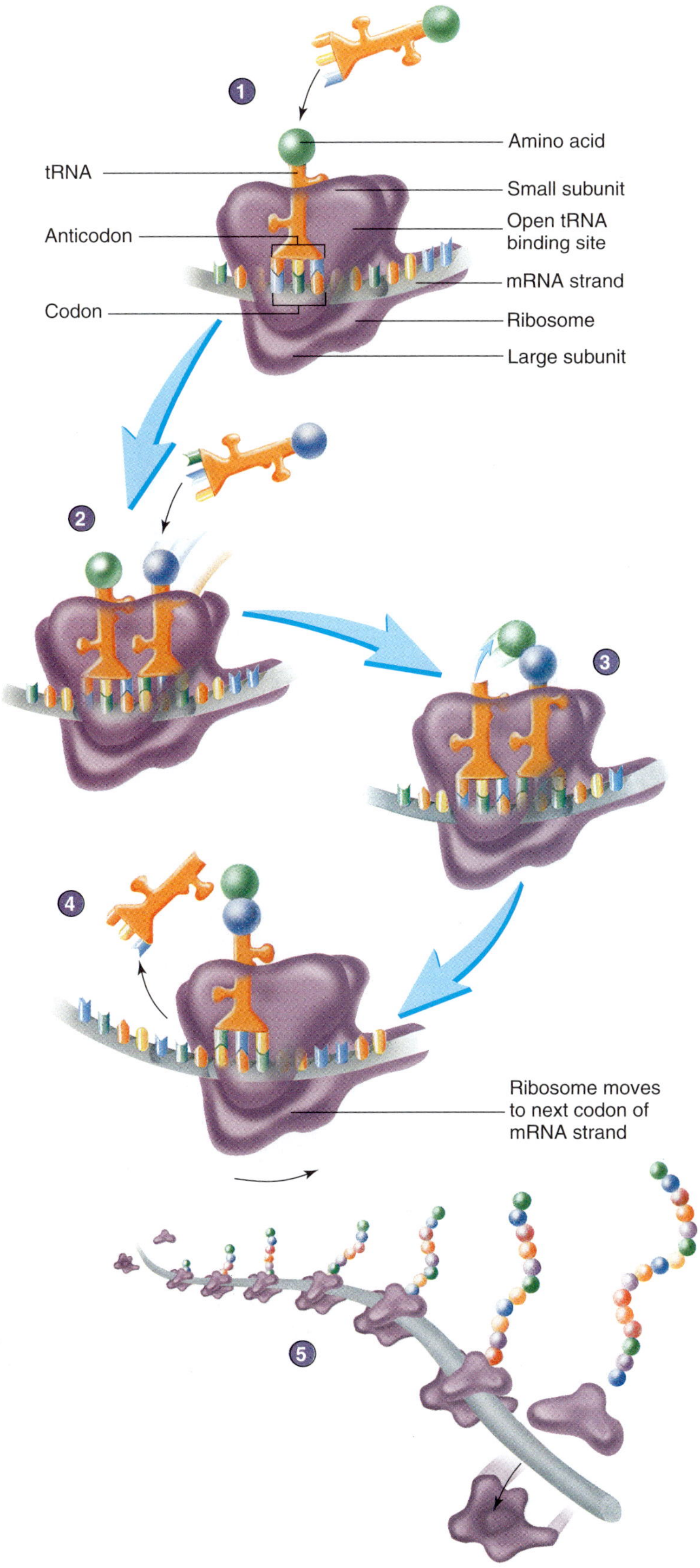

Translation of mRNA to Produce a Protein
Process Figure 3.43

Cell Cycle
Figure 3.44

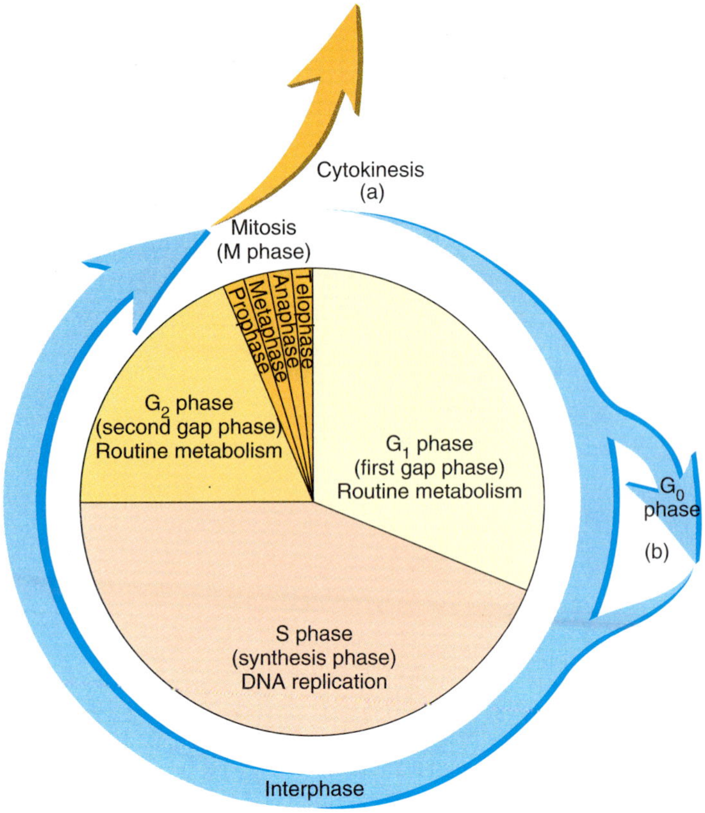

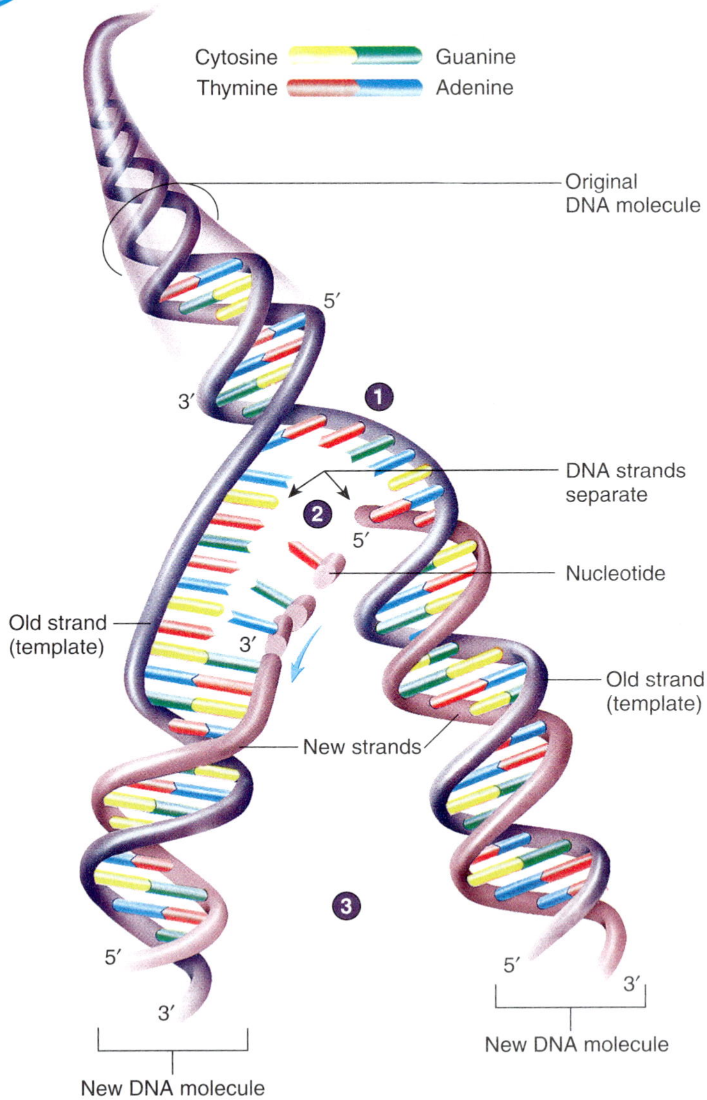

1. The strands of the DNA molecule separate from each other.
2. Each old strand (*dark purple*) functions as a template on which a new strand *(light purple)* is formed. The base-pairing relationship between nucleotides determines the sequence of nucleotides in the newly formed strands.
3. Two identical DNA molecules are produced.

Replication of DNA
Process Figure 3.45

1. **Interphase** is the time between cell divisions. DNA is found as thin threads of chromatin in the nucleus. DNA replication occurs during interphase. Organelles, other than the nucleus, and centrioles duplicate during interphase.

2. In **prophase**, the chromatin condenses into chromosomes. Each chromosome consists of two chromatids joined at the centromere. The centrioles move to the opposite ends of the cell, and the nucleolus and the nuclear envelope disappear. Microtubules form near the centrioles and project in all directions. Some of the microtubules end blindly and are called astral fibers. Others, known as spindle fibers, project toward an invisible line called the equator and overlap with fibers from opposite centrioles.

3. In **metaphase**, the chromosomes align in the center of the cell in association with the spindle fibers. The spindle fibers attach to kinetochores in the centromere of each chromosome.

4. In **anaphase**, the chromatids separate, and each chromatid is then referred to as a chromosome. Thus, when the centromeres divide, the chromosome number is double, and there are two identical sets of chromosomes. The chromosomes, assisted by the spindle fibers, move toward the centrioles at each end of the cell. Separation of the chromatids signals the beginning of anaphase, and by the time anaphase has ended, the chromosomes have reached the poles of the cell. Cytokinesis begins during anaphase.

5. In **telophase**, migration of each set of chromosomes is complete. The chromosomes unravel to become less distinct chromatin threads. The nuclear envelope forms from the endoplasmic reticulum. The nucleoli form, and cytokinesis continues to form two cells.

6. Mitosis is complete, and a new interphase begins. The chromosomes have unraveled to become chromatin. Cell division has produced two daughter cells, each with DNA that is identical to the DNA parent cell.

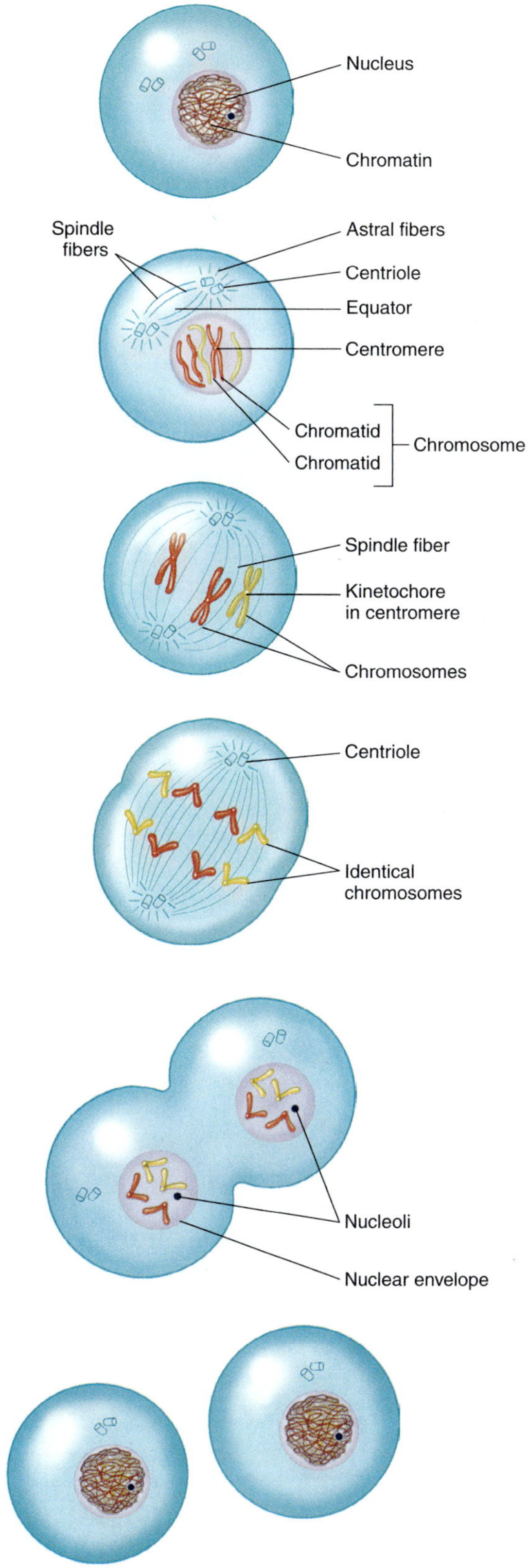

Mitosis and Cytokinesis
Process Figure 3.46

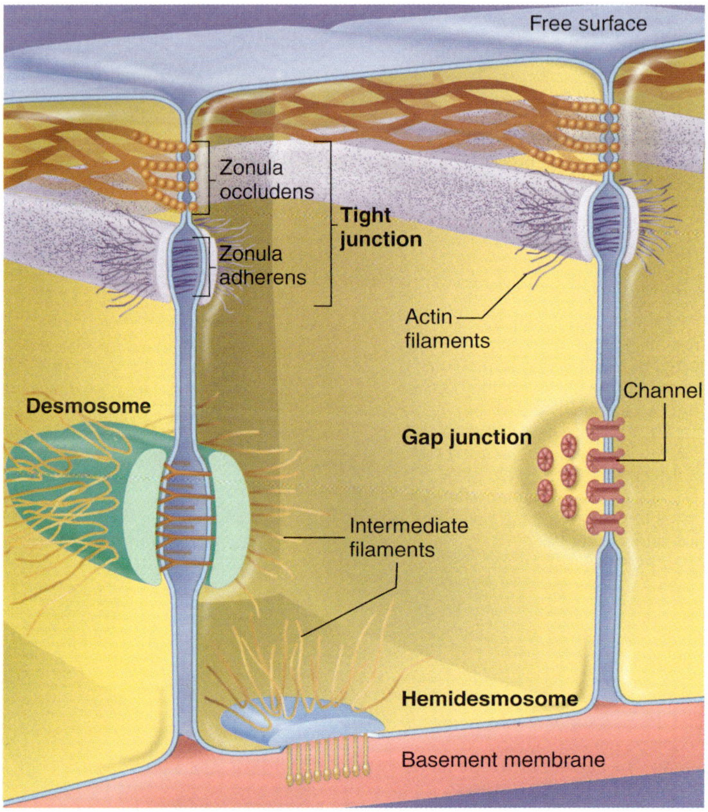
Free surface
Zonula
occludens
Tight
junction
Zonula
adherens
Actin
filaments
Channel
Desmosome
Gap junction
Intermediate
filaments
Hemidesmosome
Basement membrane

Cell Connections
Figure 4.2

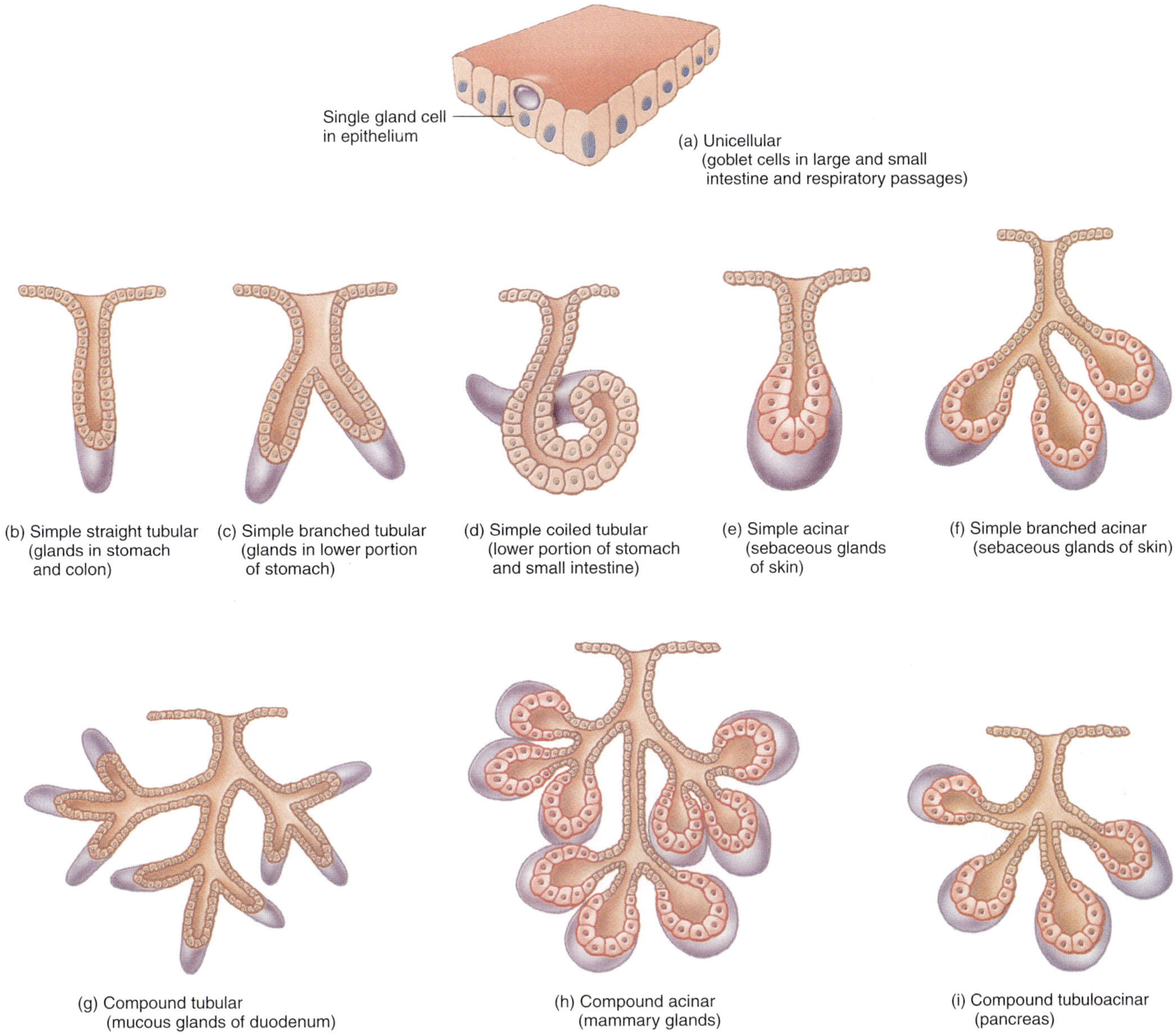
Single gland cell in epithelium
(a) Unicellular (goblet cells in large and small intestine and respiratory passages)
(b) Simple straight tubular (glands in stomach and colon)
(c) Simple branched tubular (glands in lower portion of stomach)
(d) Simple coiled tubular (lower portion of stomach and small intestine)
(e) Simple acinar (sebaceous glands of skin)
(f) Simple branched acinar (sebaceous glands of skin)
(g) Compound tubular (mucous glands of duodenum)
(h) Compound acinar (mammary glands)
(i) Compound tubuloacinar (pancreas)

Structure of Exocrine Glands
Figure 4.3

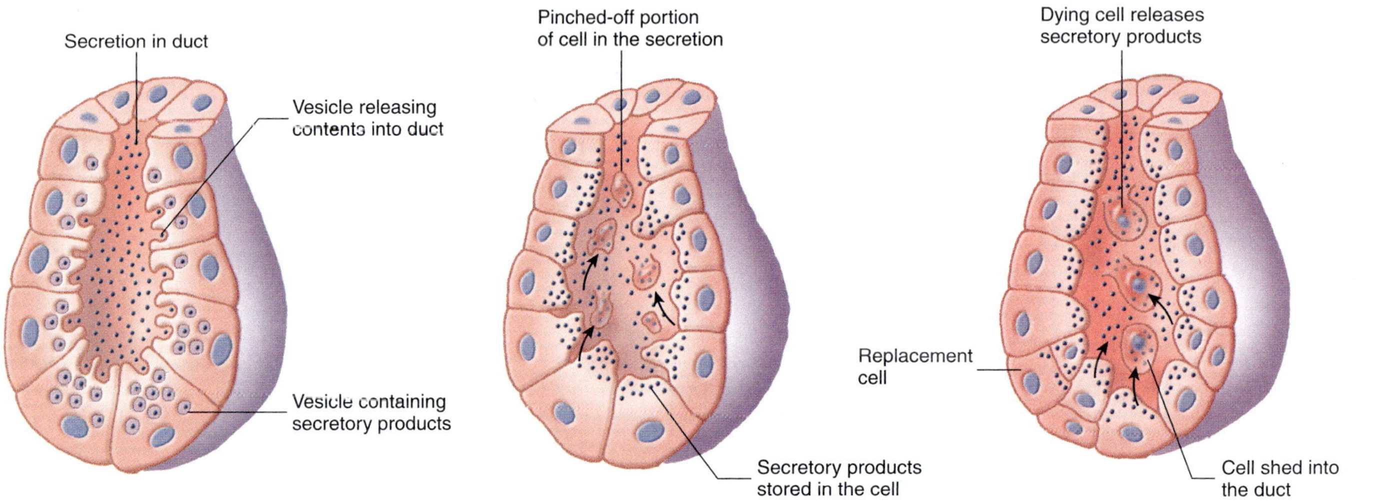

(a) **Merocrine gland**
Cells of the gland produce vesicles that contain secretory products, and the vesicles empty their contents into the duct through exocytosis.

(b) **Apocrine gland**
Secretory products are stored in the cell near the lumen of the duct. A portion of the cell near the lumen containing secretory products is pinched off the cell and joins secretions produced by a merocine process.

(c) **Holocrine gland**
Secretory products are stored in the cells of the gland. Entire cells are shed by the gland and become part of the secretion. The lost cells are replaced by other cells deeper in the gland.

Exocrine Glands and Secretion Types
Figure 4.4

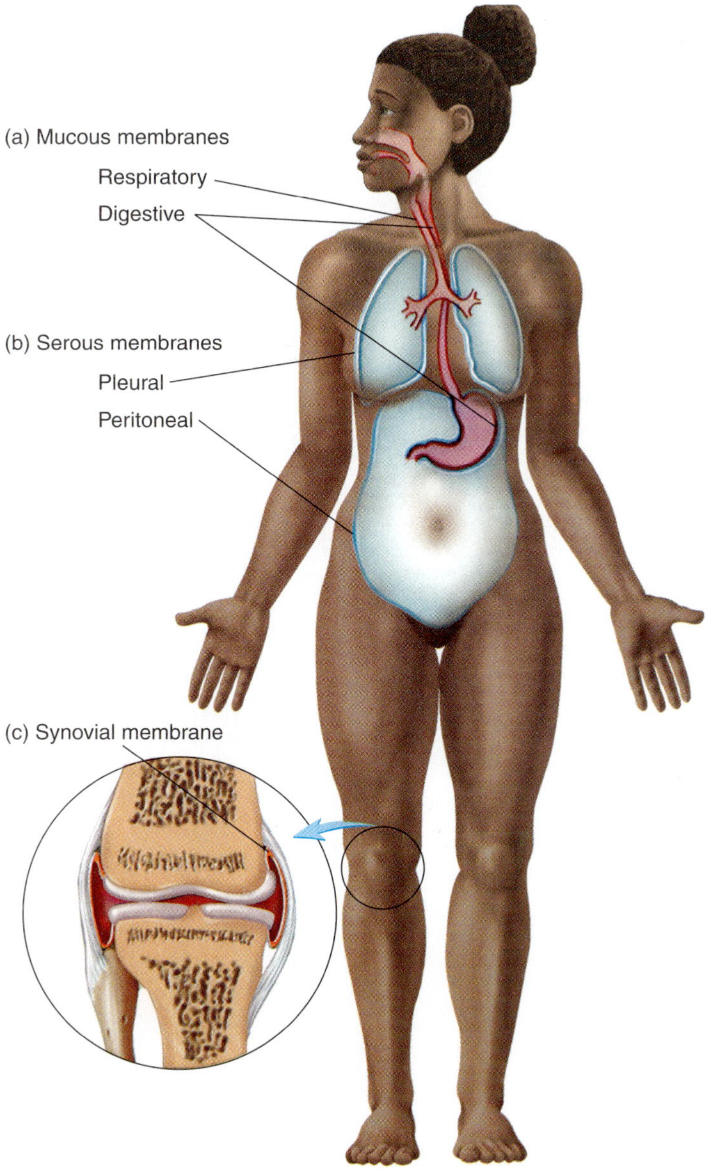
(a) Mucous membranes
Respiratory
Digestive
(b) Serous membranes
Pleural
Peritoneal
(c) Synovial membrane

Membranes
Figure 4.6

1. A splinter in the skin causes damage and introduces bacteria. Mediators of inflammation are released or activated in injured tissues and adjacent blood vessels. Some blood vessels are ruptured causing bleeding.

2. Mediators of inflammation cause capillaries to dilate, causing the skin to become red. Mediators of inflammation also increase capillary permeability, and fluid leaves the capillaries, producing swelling (*arrows*).

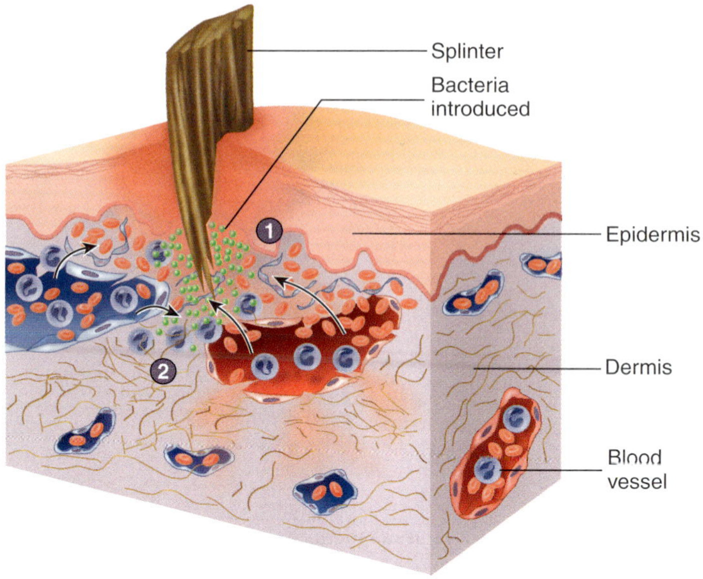

3. White blood cells (e.g., neutrophils and macrophages) leave the dilated blood vessels and move to the site of bacterial infection, where they begin to phagocytize bacteria and other debris.

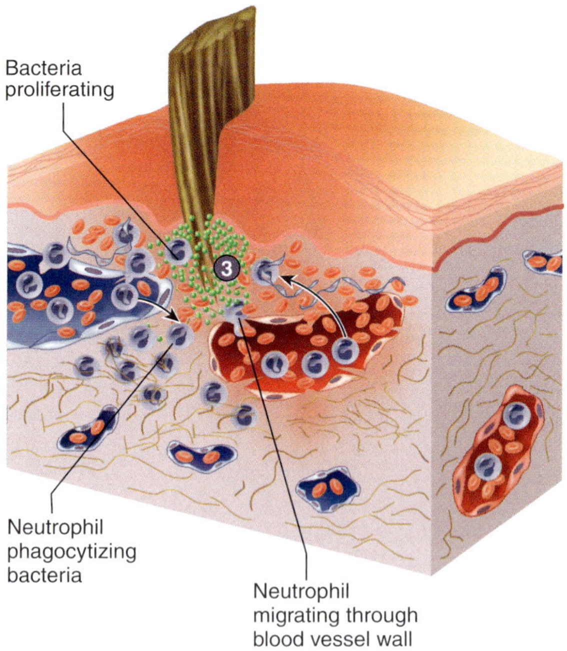

Inflammation
Process Figure 4.7

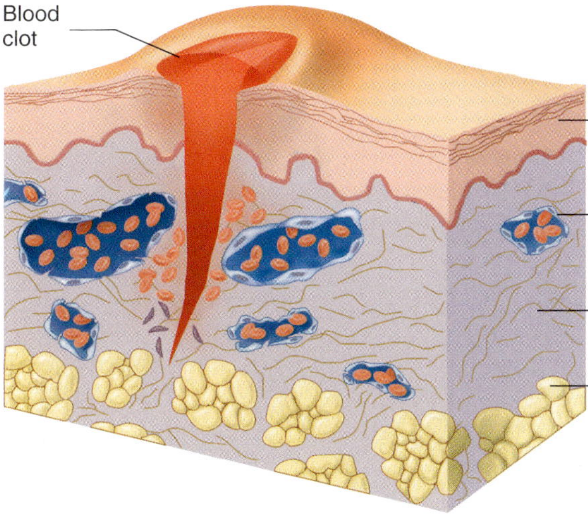

1. Fresh wound cuts through the epithelium (epidermis) and underlying connective tissue (dermis), and a clot forms.

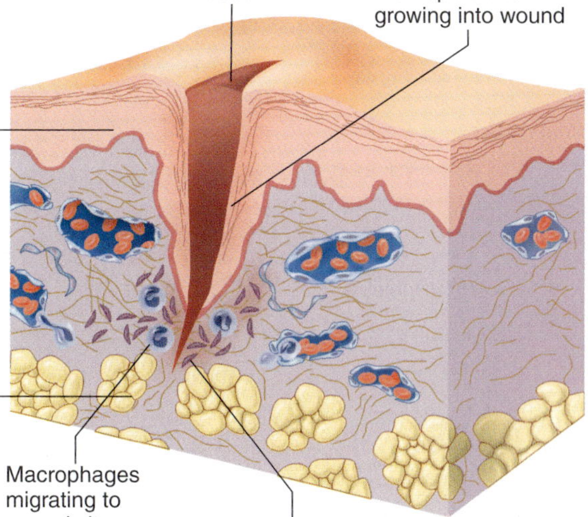

2. Approximately 1 week after the injury, a scab is present, and epithelium (new epidermis) is growing into the wound.

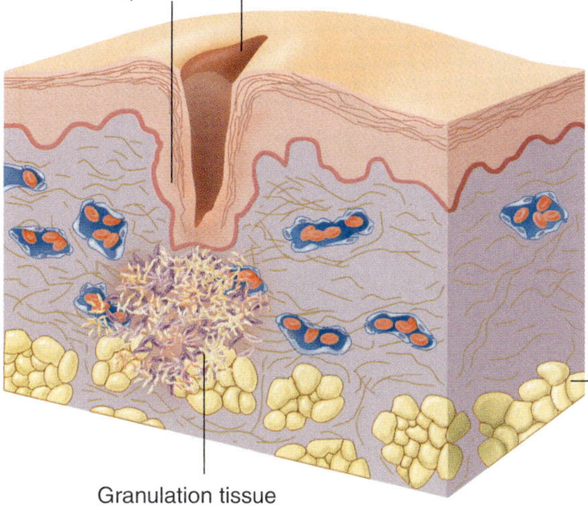

3. Approximately 2 weeks after the injury, the epithelium has grown completely into the wound, and granulation tissue has formed.

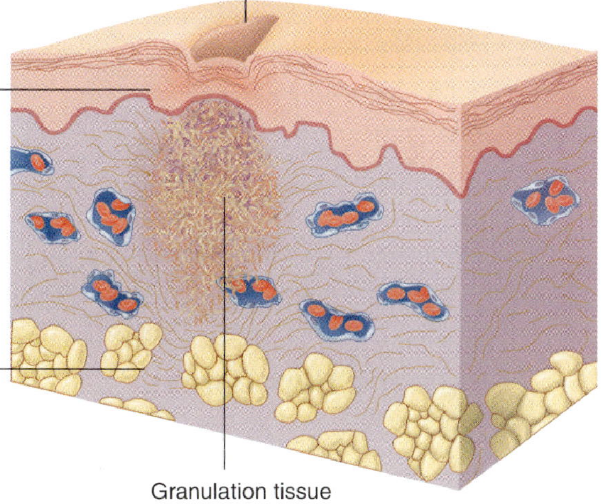

4. Approximately 1 month after the injury, the wound has completely closed, the scab has been sloughed, and the granulation tissue is being replaced with dermis.

Tissue Repair
Process Figure 4.8

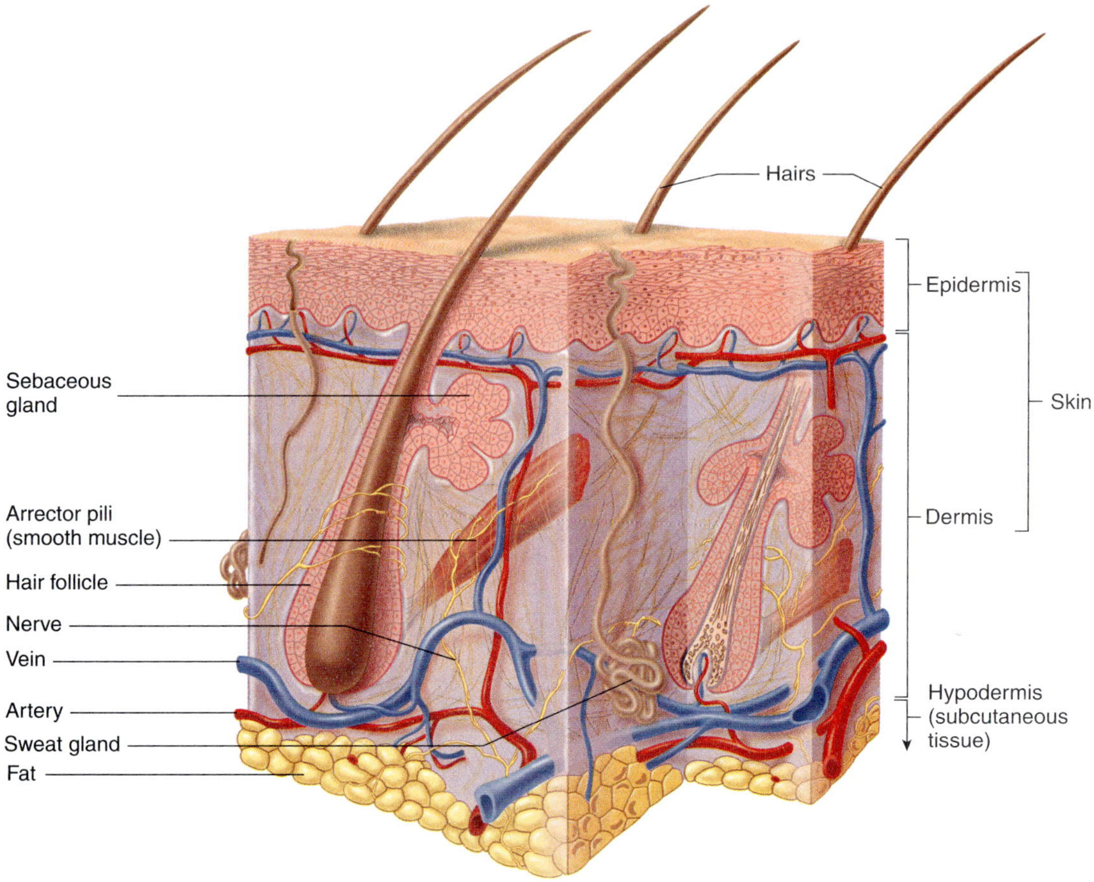
Hairs
Epidermis
Skin
Dermis
Hypodermis
(subcutaneous
tissue)
Sebaceous
gland
Arrector pili
(smooth muscle)
Hair follicle
Nerve
Vein
Artery
Sweat gland
Fat

Skin and Hypodermis
Figure 5.1

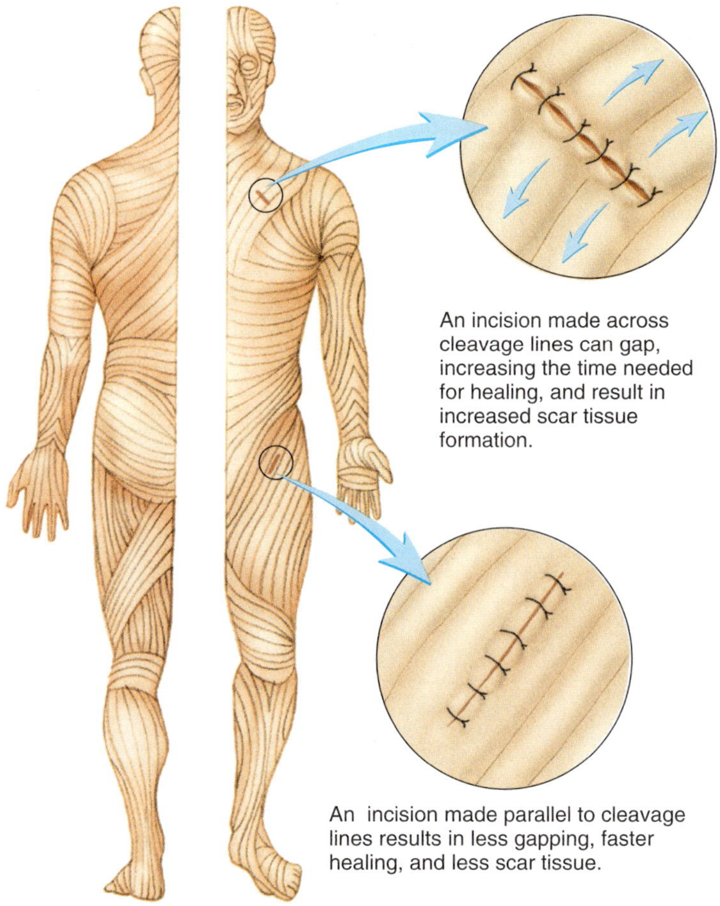

Cleavage Lines
Figure 5.3

1. In the stratum basale, cells divide by mitosis and some of the newly formed cells become the cells of the more superficial strata.
2. In the stratum spinosum, keratin fibers and lamellar bodies accumulate.
3. In the stratum granulosum, keratohyalin granules accumulate and a hard protein envelope forms beneath the plasma membrane; lamellar bodies release lipids; cells die.
4. In the stratum lucidum, the cells are dead and contain dispersed keratohyalin.
5. In the stratum corneum, the dead cells have a hard protein envelope, contain keratin, and are surrounded by lipids.

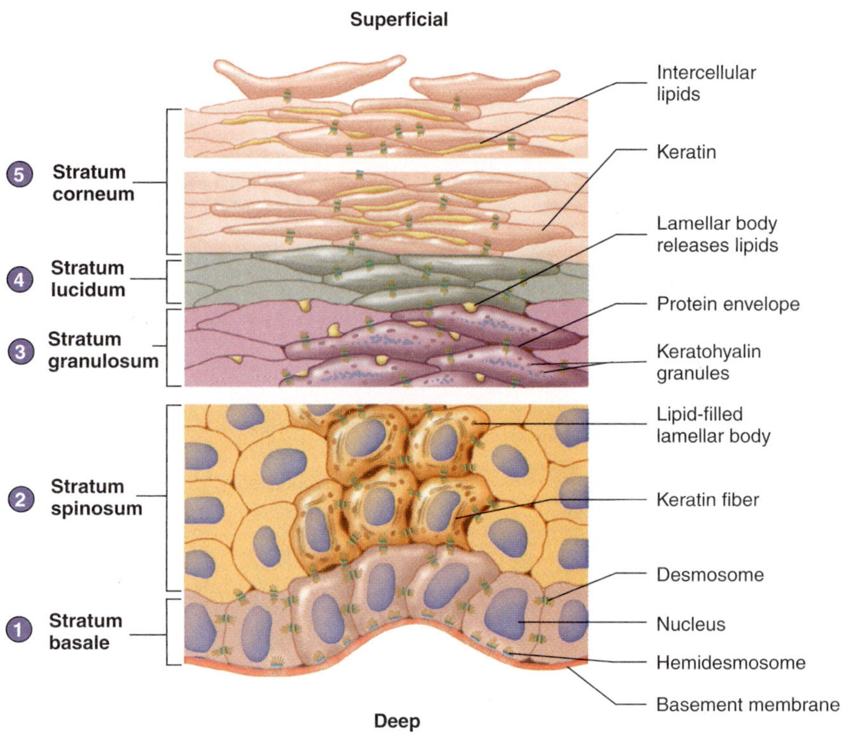

Epidermal Layers and Keratinization
Process Figure 5.4

1. Melanosomes are produced by the Golgi apparatus of the melanocyte.
2. Melanosomes move into melanocyte cell processes.
3. Epithelial cells phagocytize the tips of the melanocyte cell processes.
4. These melanosomes are within epithelial cells.

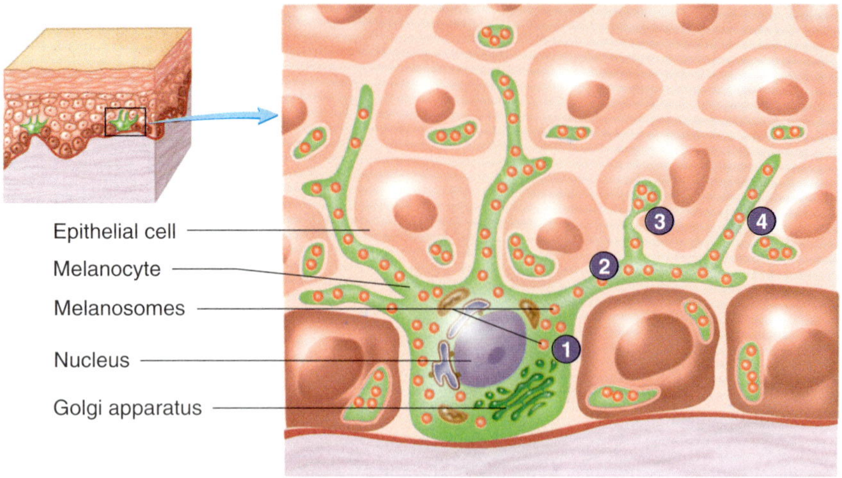

Melanin Transfer from Melanocyte to Keratinocytes
Process Figure 5.5

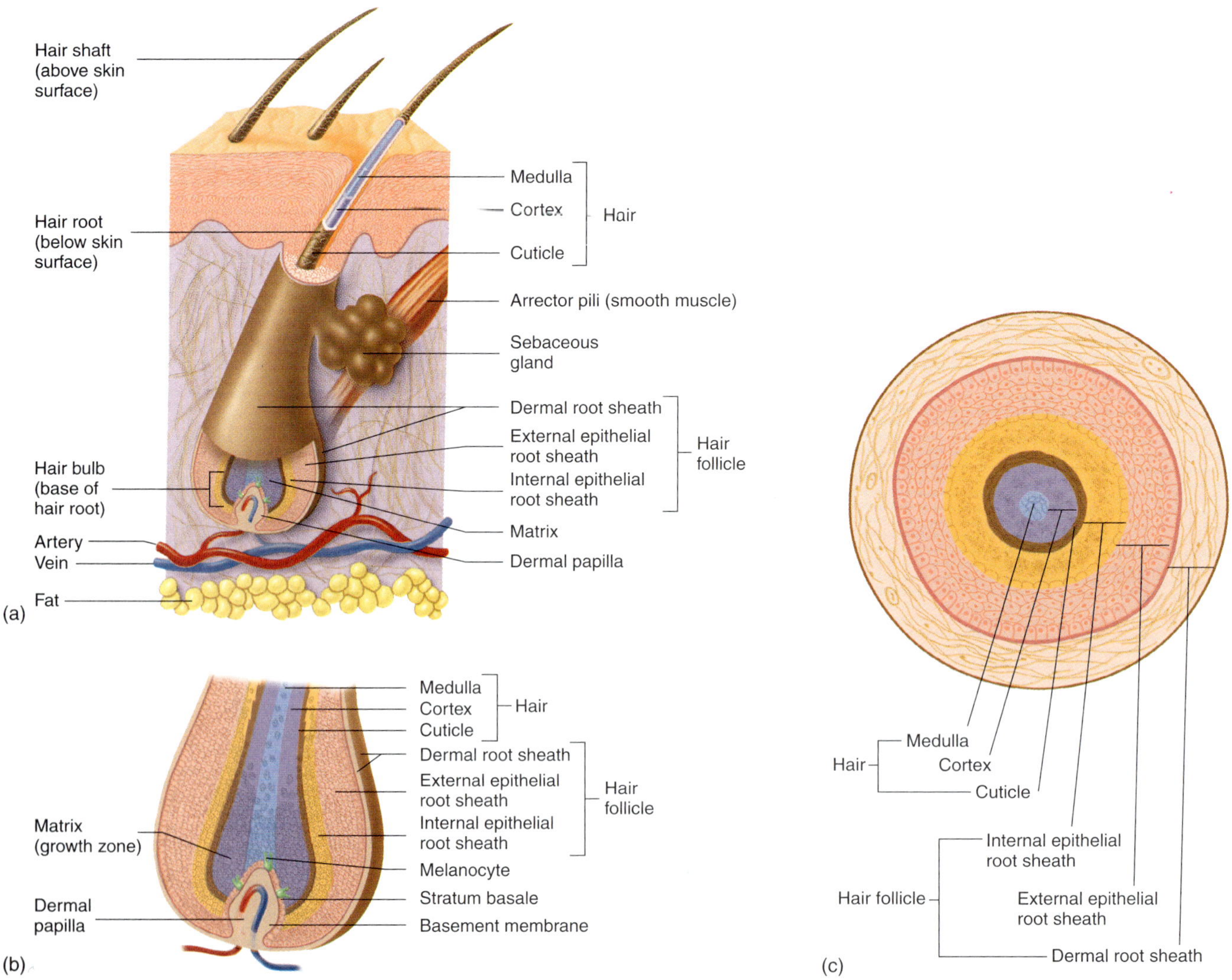

Hair Follicle
Figure 5.6

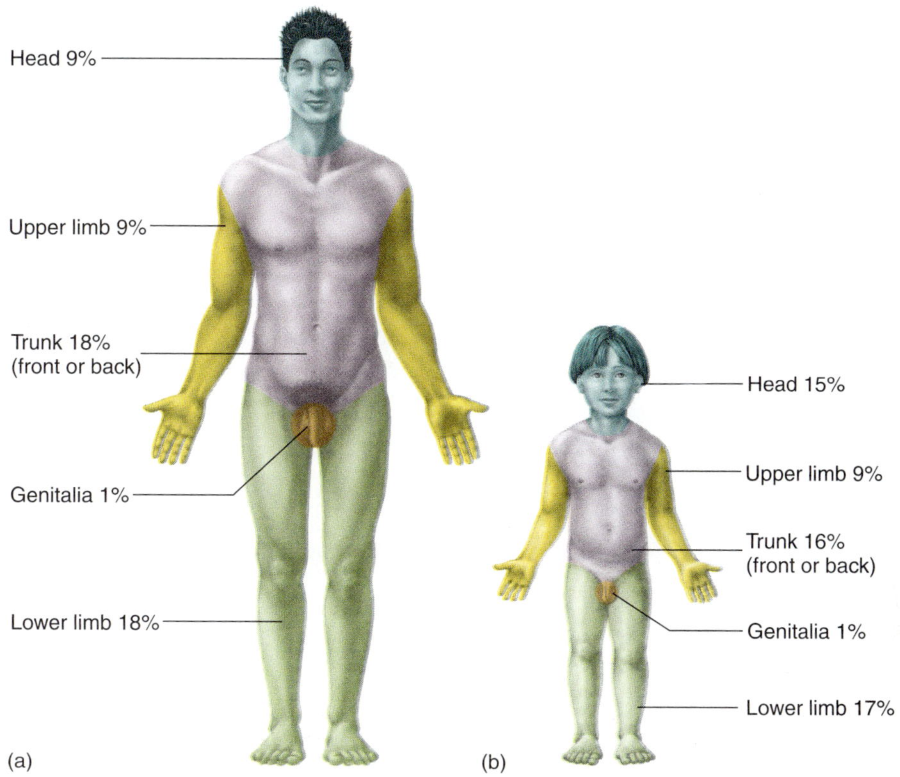
Head 9%
Upper limb 9%
Trunk 18%
(front or back)
Genitalia 1%
Lower limb 18%
(a)
Head 15%
Upper limb 9%
Trunk 16%
(front or back)
Genitalia 1%
Lower limb 17%
(b)

The Rule of Nines
Figure A

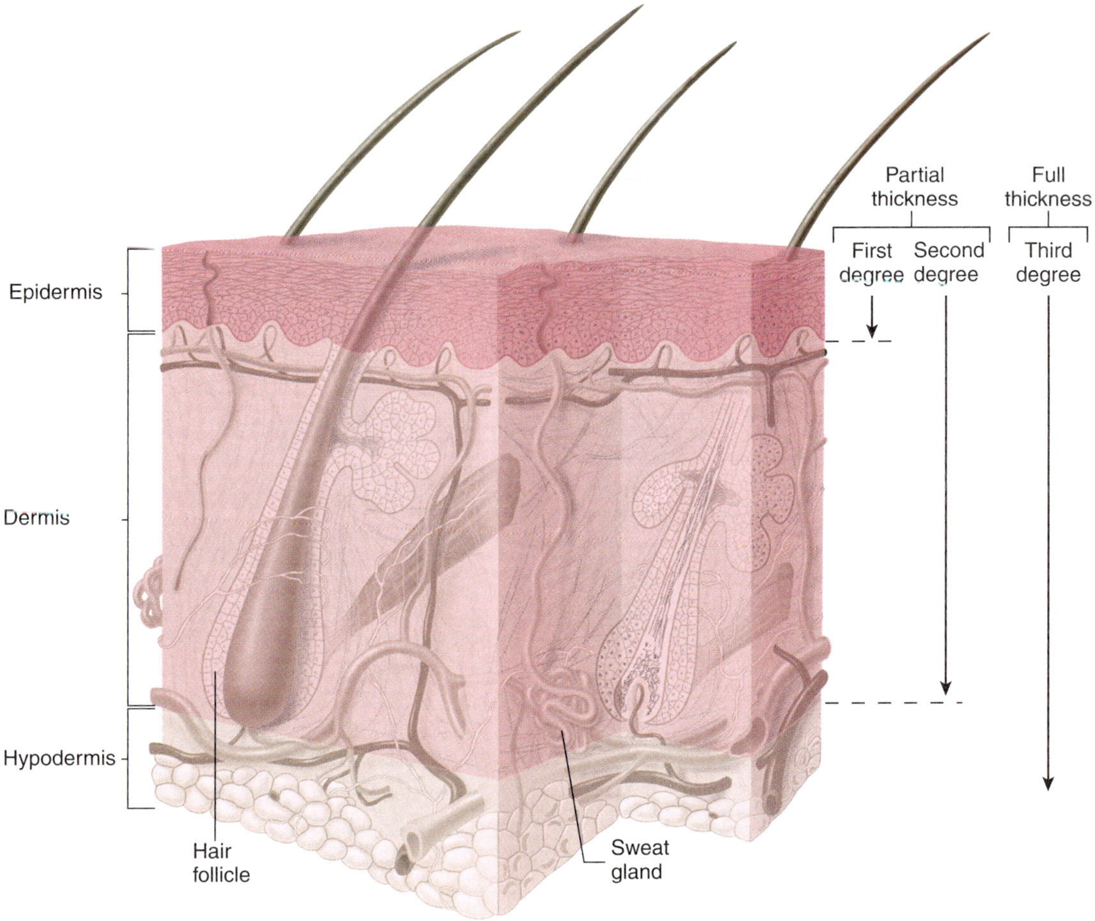
Partial
thickness
Full
thickness
First
degree
Second
degree
Third
degree
Epidermis
Dermis
Hypodermis
Hair
follicle
Sweat
gland

Burns
Figure B

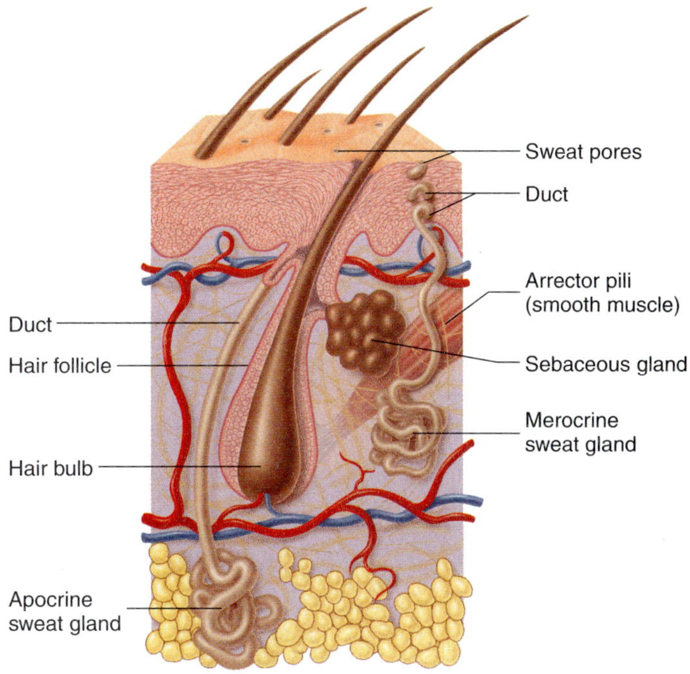
Sweat pores
Duct
Arrector pili
(smooth muscle)
Sebaceous gland
Merocrine
sweat gland
Duct
Hair follicle
Hair bulb
Apocrine
sweat gland

Glands of the Skin
Figure 5.7

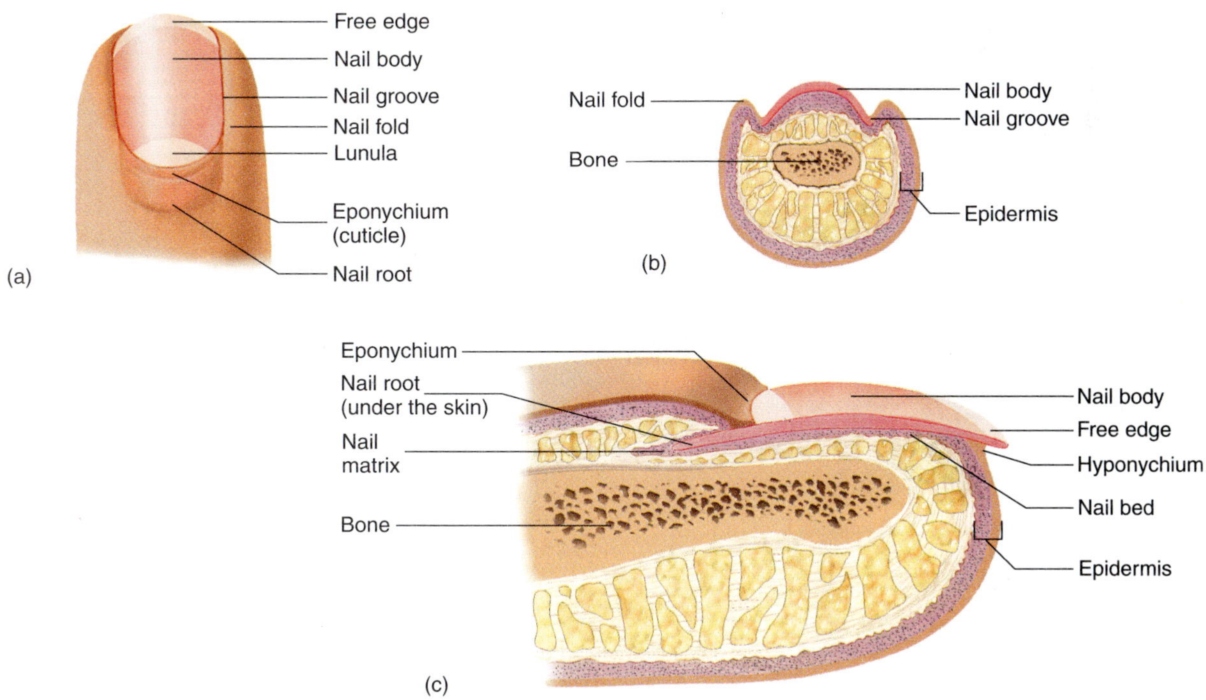
Free edge
Nail body
Nail groove
Nail fold
Lunula
Eponychium
(cuticle)
Nail root
(a)
Nail fold
Bone
Nail body
Nail groove
Epidermis
(b)
Eponychium
Nail root
(under the skin)
Nail
matrix
Bone
Nail body
Free edge
Hyponychium
Nail bed
Epidermis
(c)

Nail
Figure 5.8

1. Blood vessel dilation results in increased blood flow toward the surface of the skin.
2. Increased blood flow beneath the epidermis results in increased heat loss (*gold arrows*).

(a)

3. Blood vessel constriction results in decreased blood flow toward the surface of the skin.
4. Decreased blood flow beneath the epidermis results in decreased heat loss.

(b)

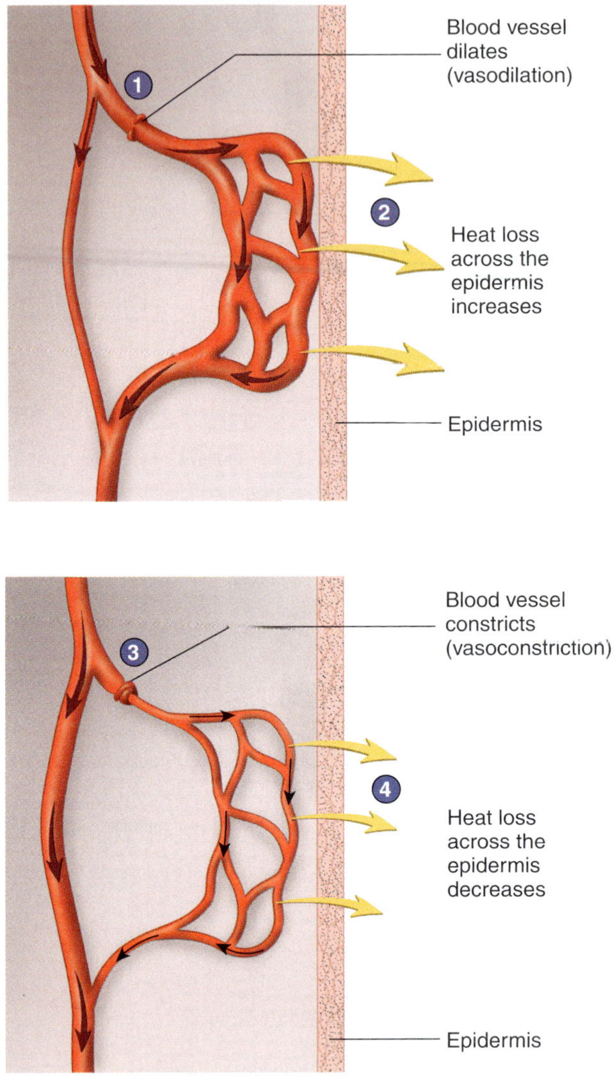

Heat Exchange in the Skin
Process Figure 5.9

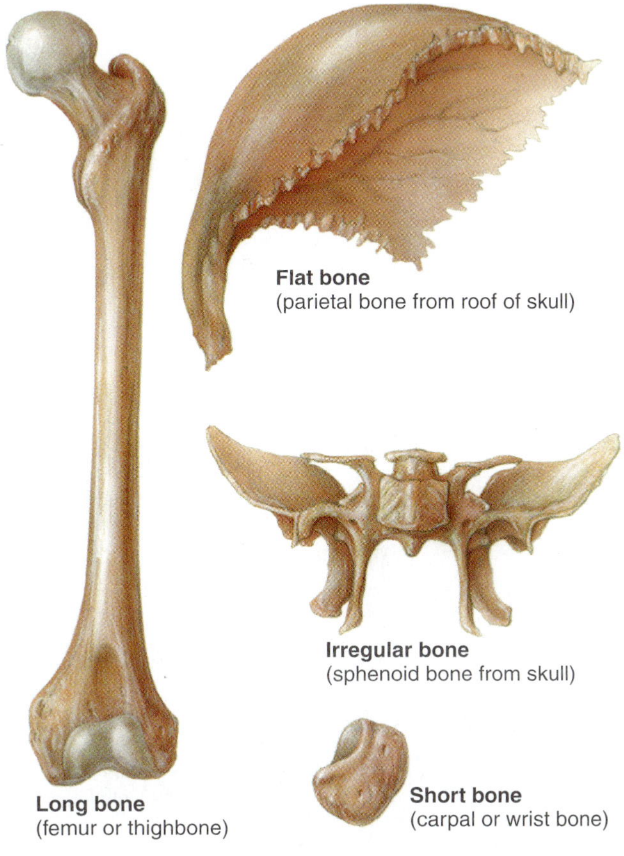

Flat bone
(parietal bone from roof of skull)
Irregular bone
(sphenoid bone from skull)
Long bone
(femur or thighbone)
Short bone
(carpal or wrist bone)

Bone Shapes
Figure 6.2

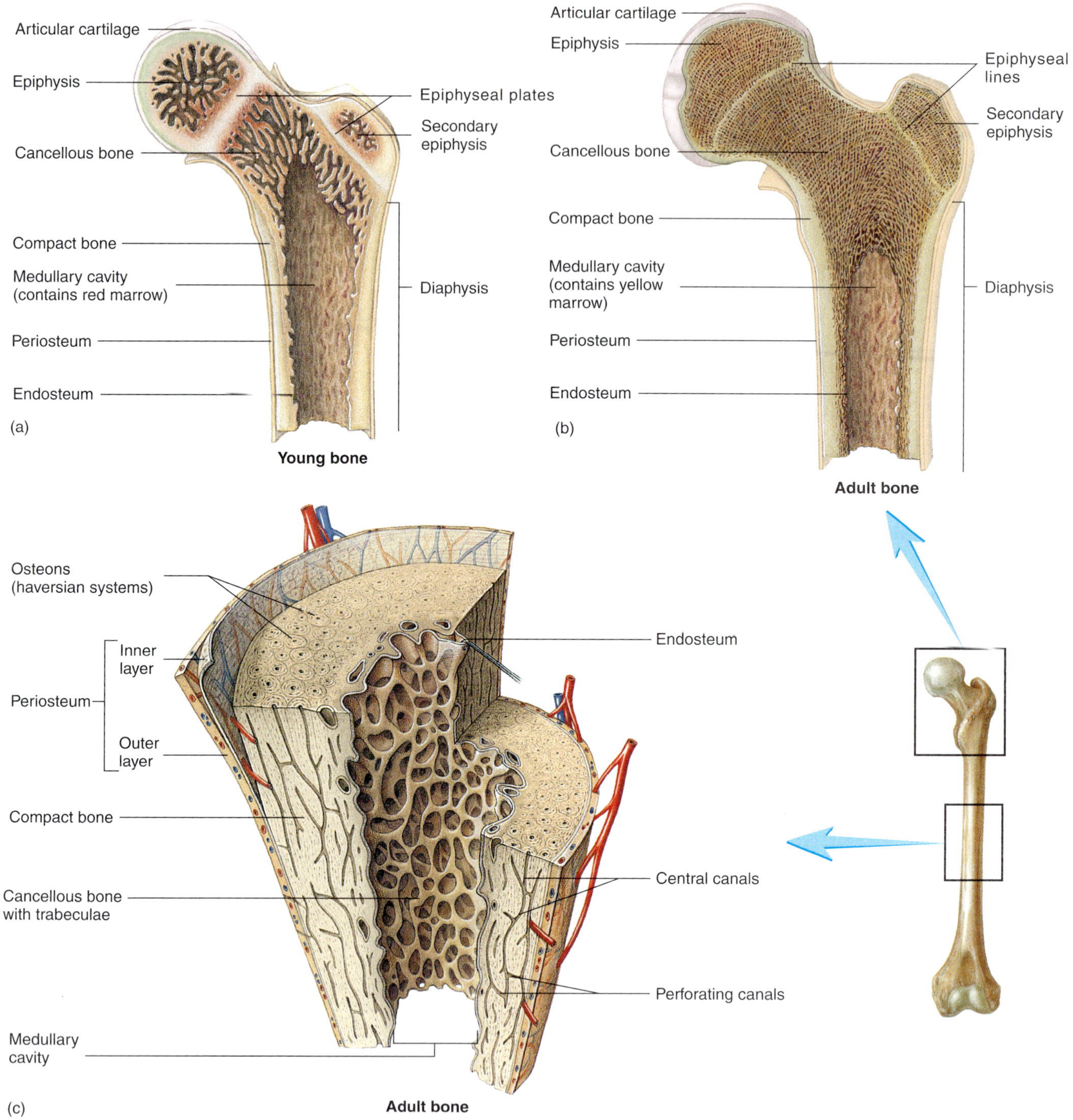

Articular cartilage
Epiphysis
Epiphyseal plates
Secondary epiphysis
Cancellous bone
Compact bone
Medullary cavity (contains red marrow)
Diaphysis
Periosteum
Endosteum
(a)
Young bone
Articular cartilage
Epiphysis
Epiphyseal lines
Secondary epiphysis
Cancellous bone
Compact bone
Medullary cavity (contains yellow marrow)
Diaphysis
Periosteum
Endosteum
(b)
Adult bone
Osteons (haversian systems)
Endosteum
Inner layer
Periosteum
Outer layer
Compact bone
Cancellous bone with trabeculae
Central canals
Perforating canals
Medullary cavity
(c)
Adult bone

Long Bone
Figure 6.3

Bone Marrow
Figure 6.4

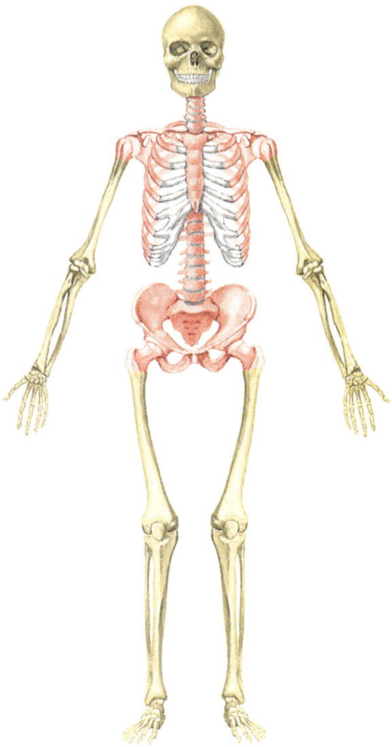

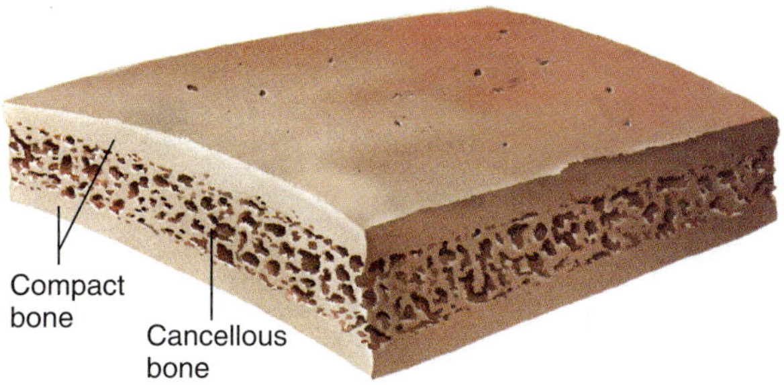

Structure of a Flat Bone
Figure 6.5

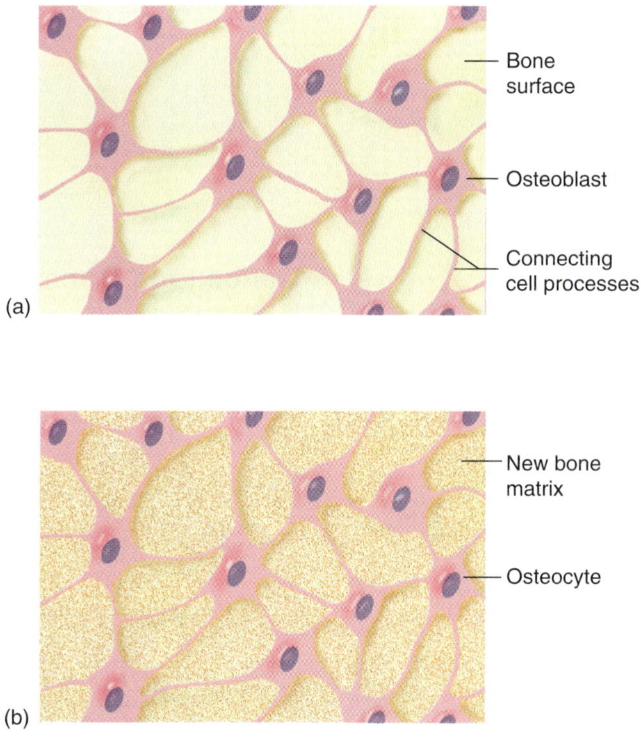

Ossification
Figure 6.7

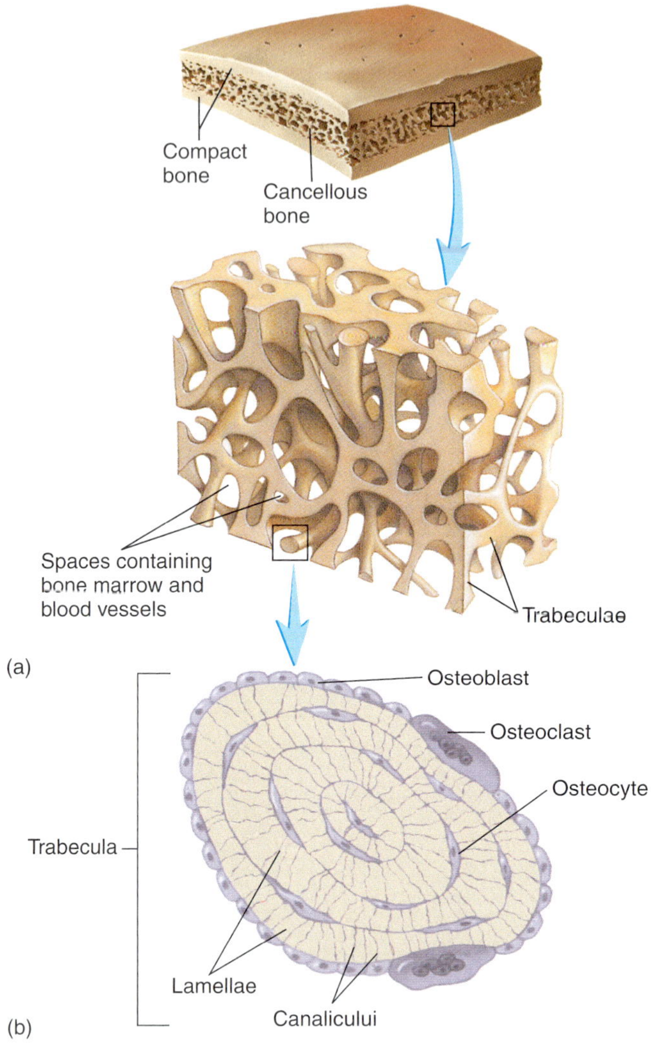
Compact bone
Cancellous bone
Spaces containing bone marrow and blood vessels
Trabeculae
(a)
Osteoblast
Osteoclast
Osteocyte
Trabecula
Lamellae
Canalicului
(b)

Cancellous Bone
Figure 6.8

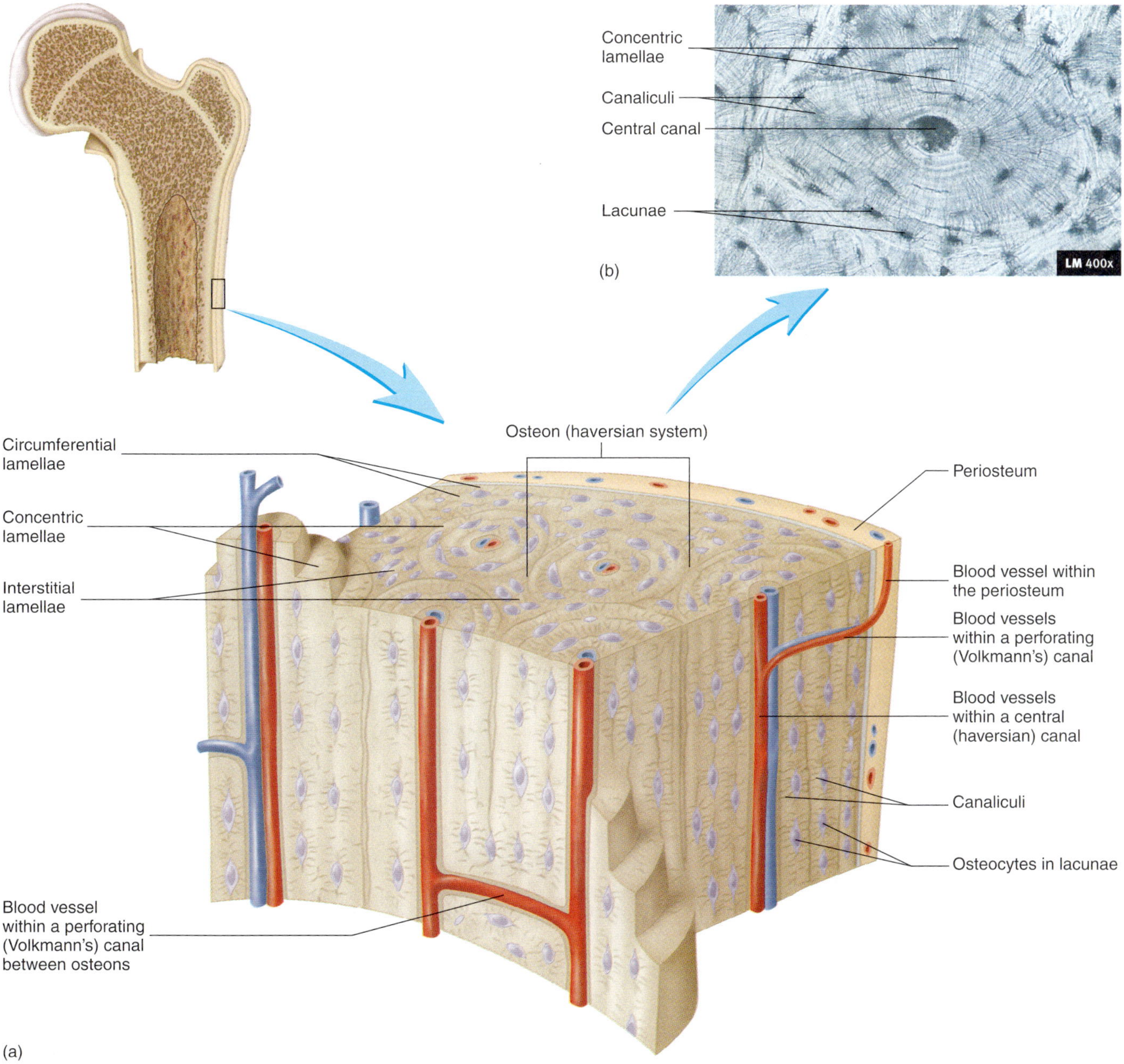
Concentric
lamellae
Canaliculi
Central canal
Lacunae
(b)
LM 400x
Osteon (haversian system)
Circumferential
lamellae
Concentric
lamellae
Interstitial
lamellae
Periosteum
Blood vessel within
the periosteum
Blood vessels
within a perforating
(Volkmann's) canal
Blood vessels
within a central
(haversian) canal
Canaliculi
Osteocytes in lacunae
Blood vessel
within a perforating
(Volkmann's) canal
between osteons
(a)

Compact Bone

Figure 6.10

b: © Trent Stephens

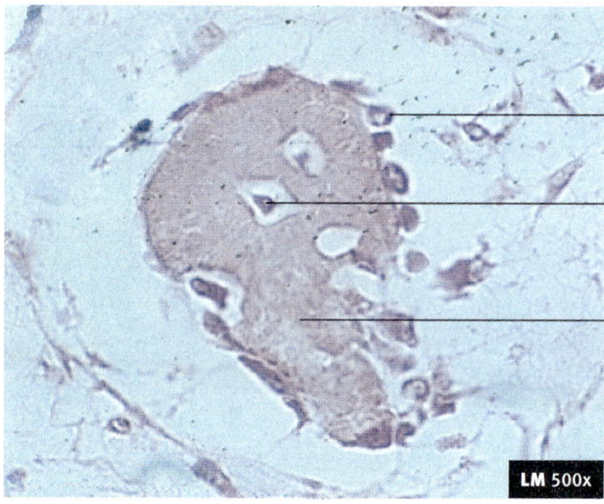

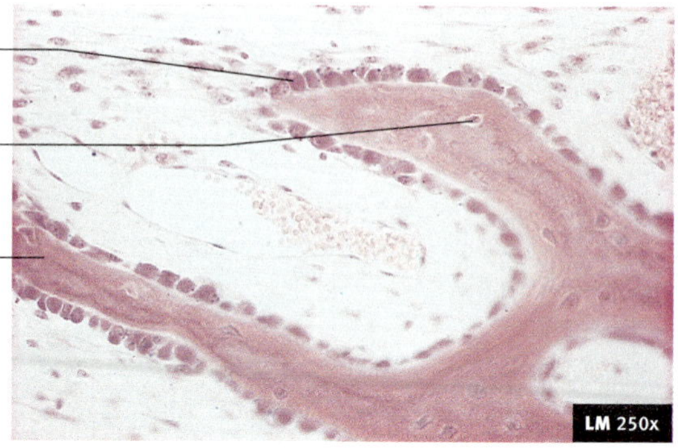

1. A cross section of a newly formed trabecula shows the youngest bone in this series of photomicrographs. Osteocytes are surrounded by bone matrix and osteoblasts are forming a ring on the outer surface of the trabecula. As the osteoblasts lay down bone, the trabeculae increase in size.

2. A lower magnification shows older bone than in step 1. Cancellous bone has formed as a result of the enlargement and interconnections of many trabeculae.

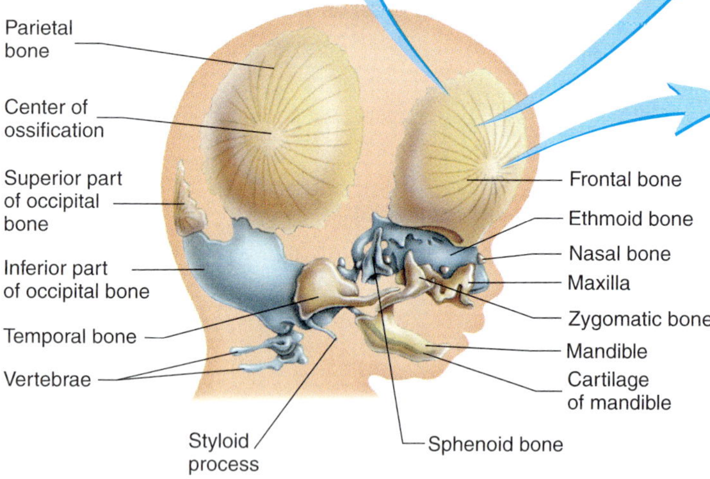

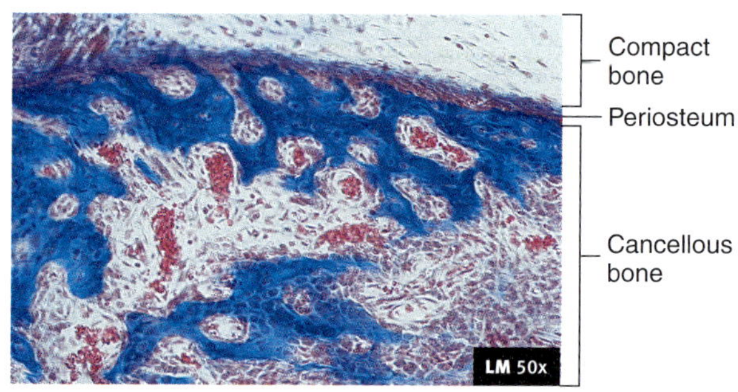

3. A lower magnification than in step 2, with a different stain that makes the bone appear blue, shows the oldest bone in this series. Within the cancellous bone there is trabeculae (*blue*) and developing red bone marrow (*pink*). Beneath the periosteum is an outer layer of developing compact bone.

Intramembranous Ossification
Process Figure 6.11

1: © Victor Eroschenko; **2:** R. Kessel/Visuals Unlimited; **3:** © Victor Eroschenko

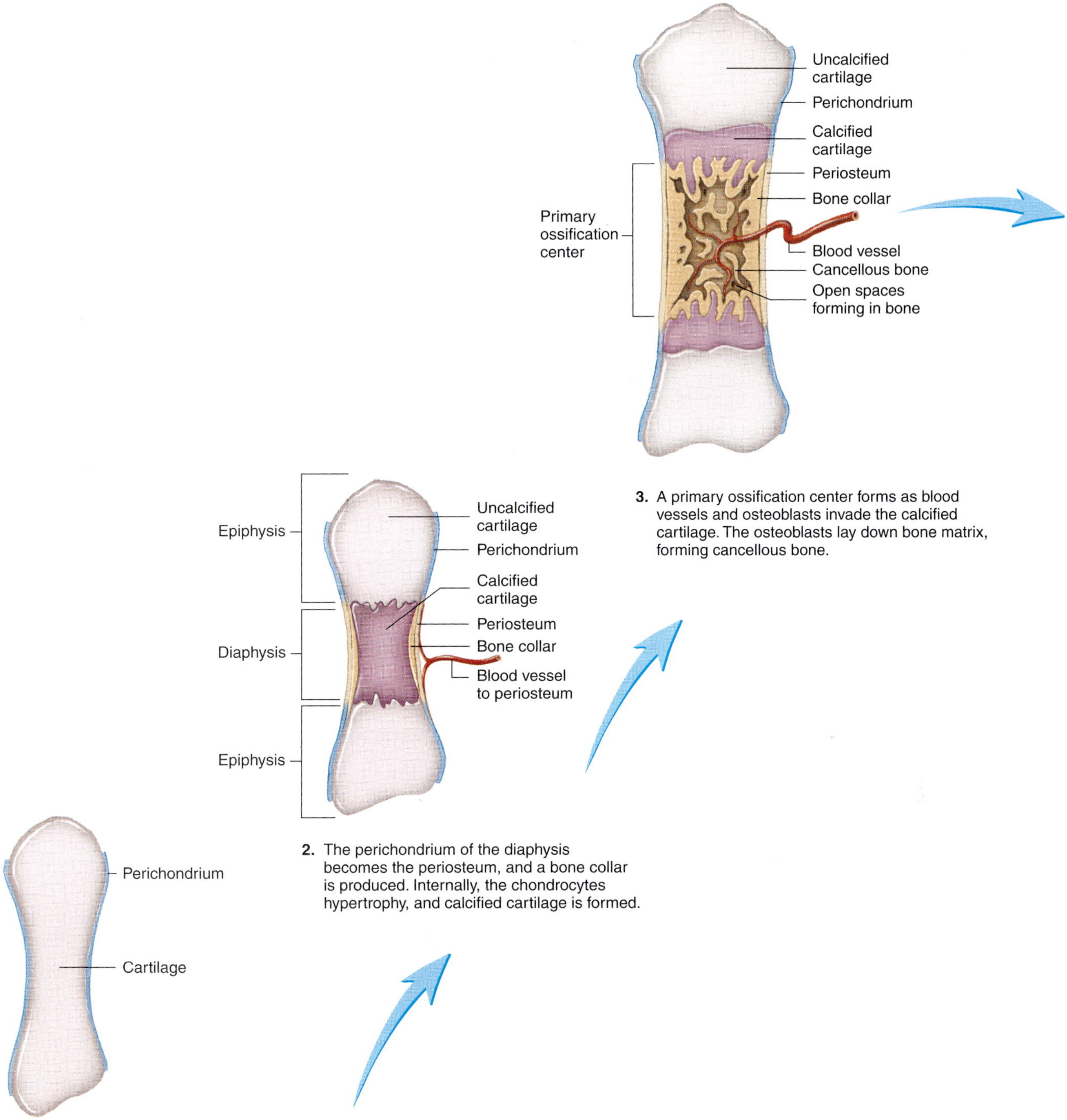

1. A cartilage model is produced by chondroblasts. The cartilage model is surrounded by perichondrium, except where joints will form.

2. The perichondrium of the diaphysis becomes the periosteum, and a bone collar is produced. Internally, the chondrocytes hypertrophy, and calcified cartilage is formed.

3. A primary ossification center forms as blood vessels and osteoblasts invade the calcified cartilage. The osteoblasts lay down bone matrix, forming cancellous bone.

Endochondral Ossification
Process Figure 6.13

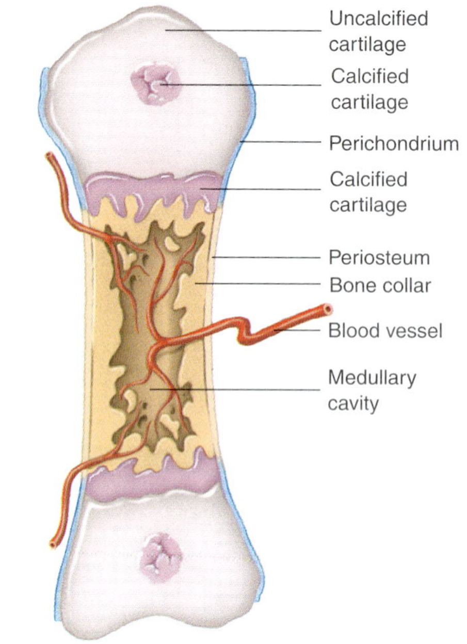

4. The process of bone collar formation, cartilage calcification, and cancellous bone production continues. Calcified cartilage begins to form in the epiphyses. A medullary cavity begins to form in the center of the diaphysis.

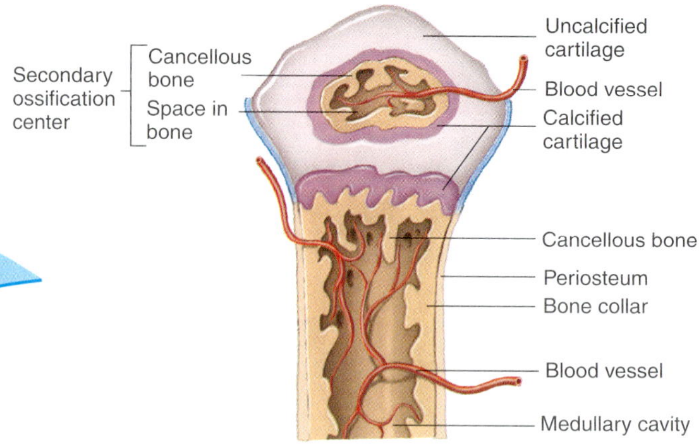

5. Secondary ossification centers form in the epiphyses of long bones.

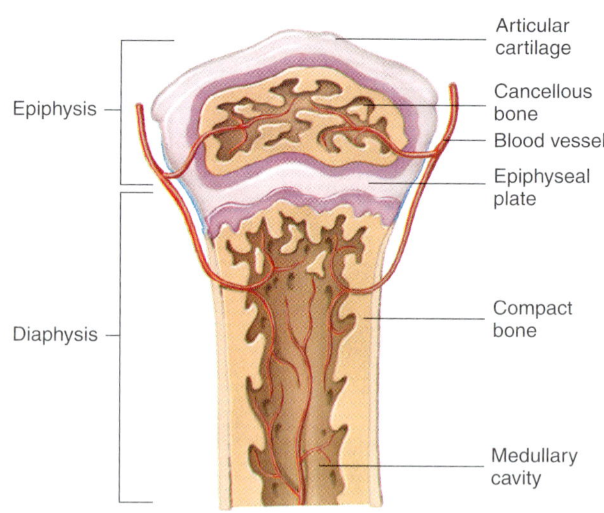

6. The original cartilage model is almost completely ossified. Unossified cartilage becomes the epiphyseal plate and the articular cartilage.

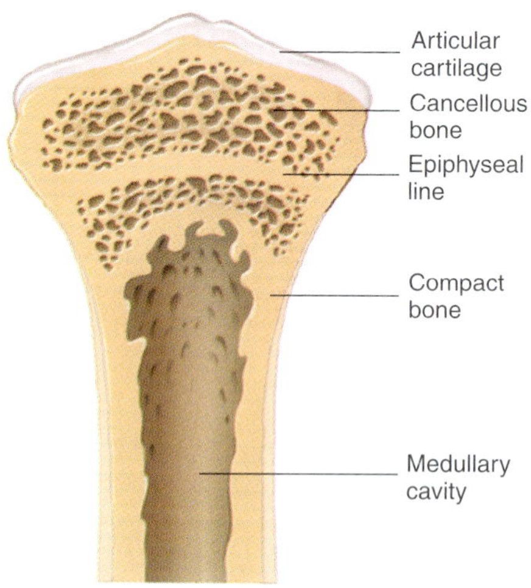

7. In a mature bone, the epiphyseal plate has become the epiphyseal line and all the cartilage in the epiphysis, except the articular cartilage, has become bone.

Process Figure 6.13—*continued*

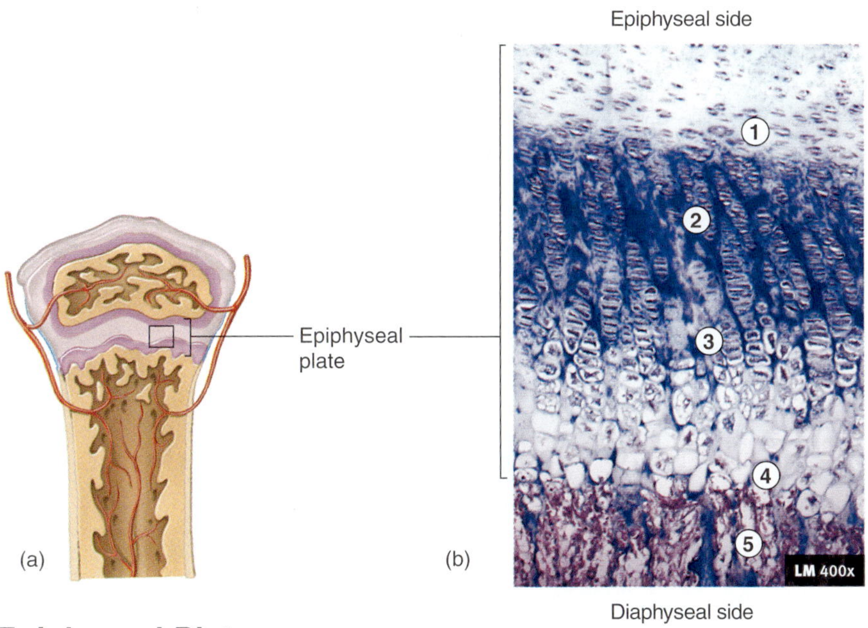

1. **Zone of resting cartilage.** Cartilage attaches to the epiphysis.
2. **Zone of proliferation.** New cartilage is produced on the epiphyseal side of the plate as the chondrocytes divide and form stacks of cells.
3. **Zone of hypertrophy.** Chondrocytes mature and enlarge.
4. **Zone of calcification.** Matrix is calcified, and chondrocytes die.
5. **Ossified bone.** The calcified cartilage on the diaphyseal side of the plate is replaced by bone.

Epiphyseal Plate
Process Figure 6.14
b: © Bio-Photo Assocs/Photo Researchers, Inc.

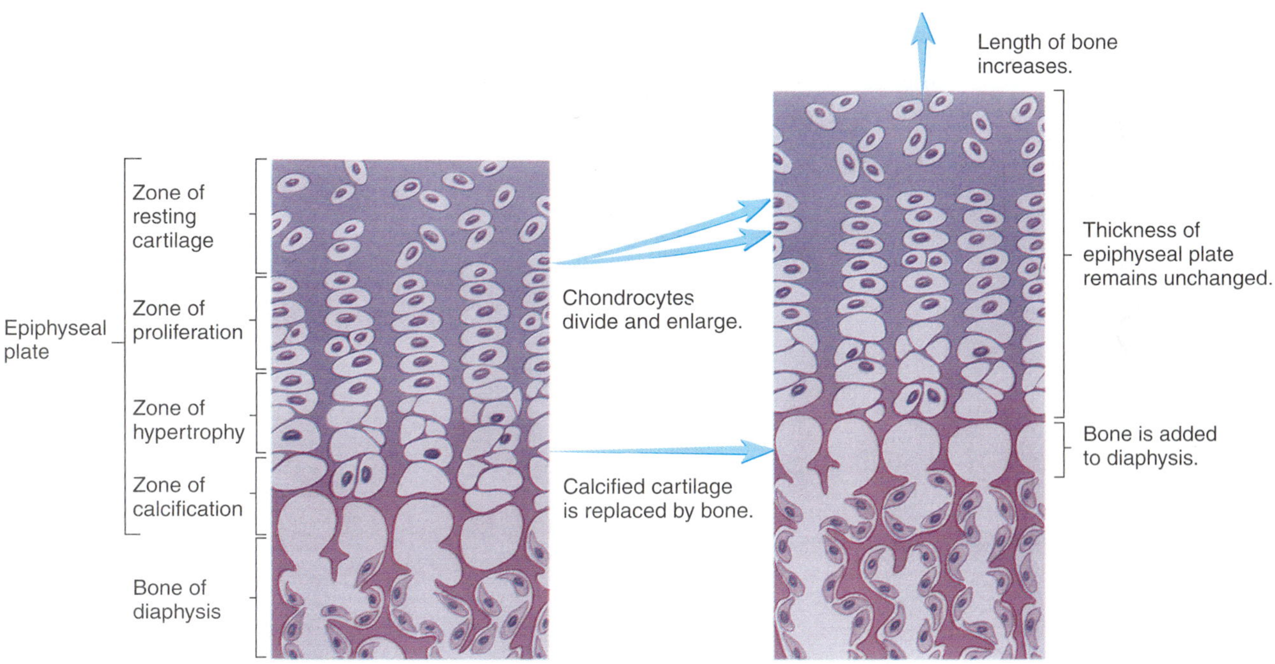

Bone Growth in Length at the Epiphyseal Plate
Figure 6.15

1. Osteoblasts beneath the periosteum lay down bone (*dark brown*) to form ridges separated by grooves. Blood vessels of the periosteum lie in the grooves.

2. The groove is transformed into a tunnel when the bone built on adjacent ridges meets. The periostcum of the groove becomes the endosteum of the tunnel.

3. Appositional growth by osteoblasts from the endosteum results in the formation of a new concentric lamella.

4. The production of additional concentric lamellae fills in the tunnel and completes the formation of the osteon.

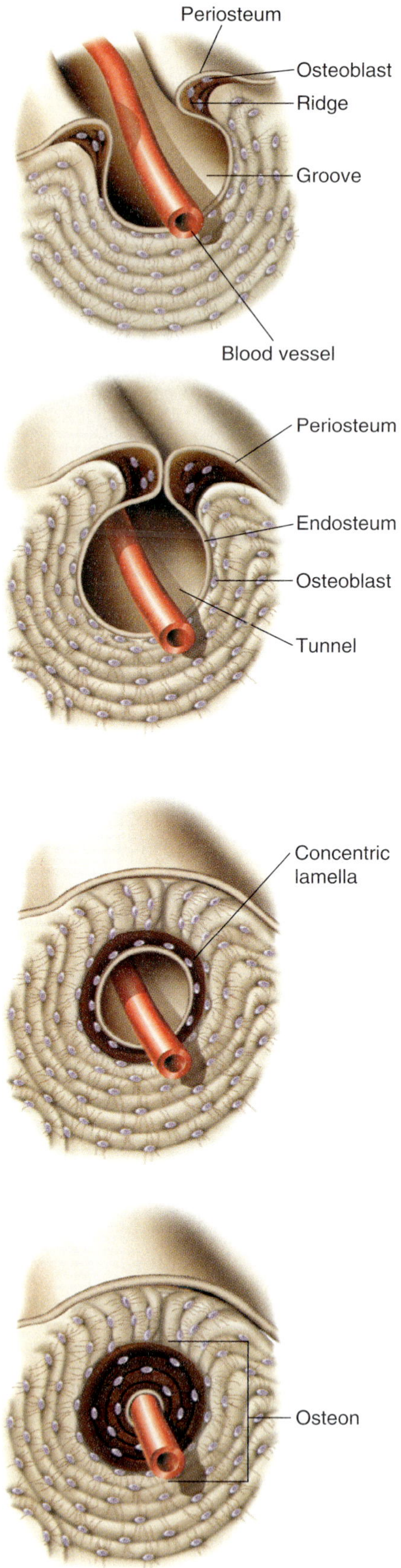

Bone Growth in Width
Process Figure 6.17

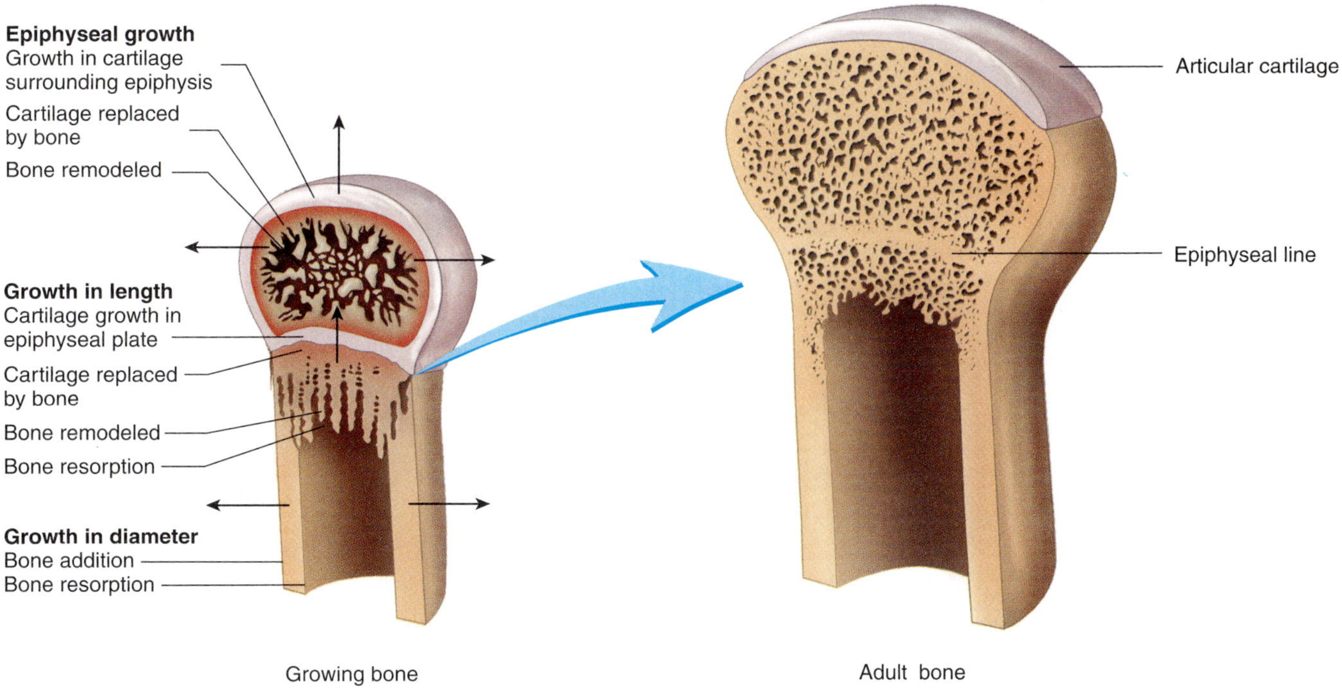

Remodeling of a Long Bone
Figure 6.18

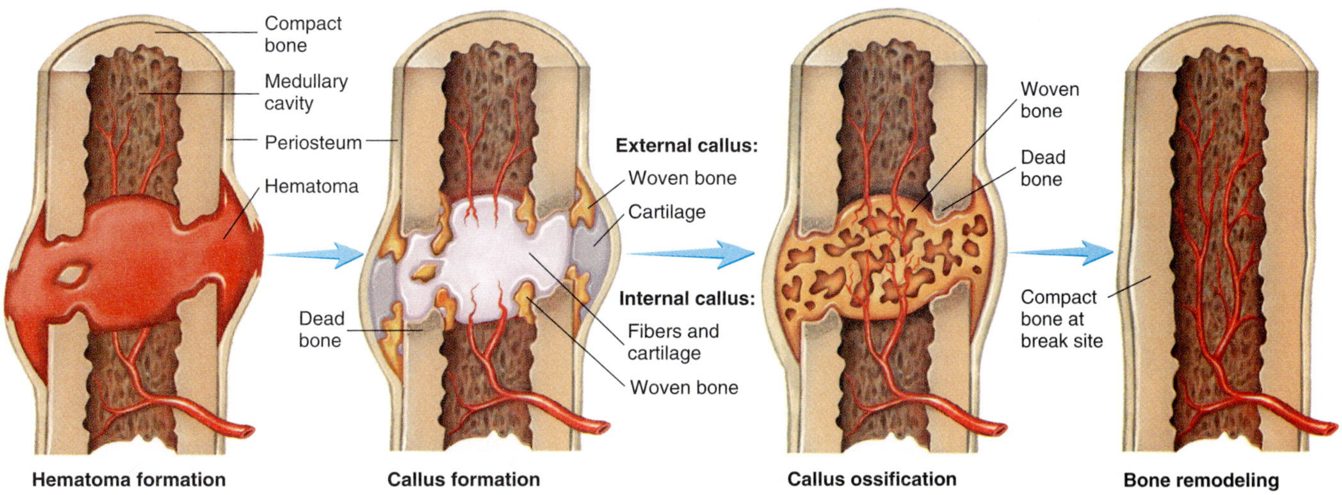

Hematoma formation

1. Blood released from damaged blood vessels forms a hematoma.

Callus formation

2. The internal callus forms between the ends of the bones, and the external callus forms a collar around the break.

Callus ossification

3. Woven, cancellous bone replaces the internal and external calluses.

Bone remodeling

4. Compact bone replaces woven bone and part of the internal callus is removed, restoring the medullary cavity.

Bone Repair
Process Figure 6.19

1. Osteoclasts break down bone and release Ca^{2+} into the blood, and osteoblasts remove Ca^{2+} from the blood to make bone. (*Blue arrows represent the movement of Ca^{2+}.*) Parathyroid hormone (PTH) regulates blood Ca^{2+} levels by indirectly stimulating osteoclast activity, resulting in increased Ca^{2+} release into the blood. Calcitonin plays a minor role in Ca^{2+} maintenance by inhibiting osteoclast activity.
2. In the kidneys, PTH increases Ca^{2+} reabsorption from the urine.
3. In the kidneys, PTH also promotes the formation of active vitamin D (*green arrows*), which increases Ca^{2+} absorption from the small intestine.

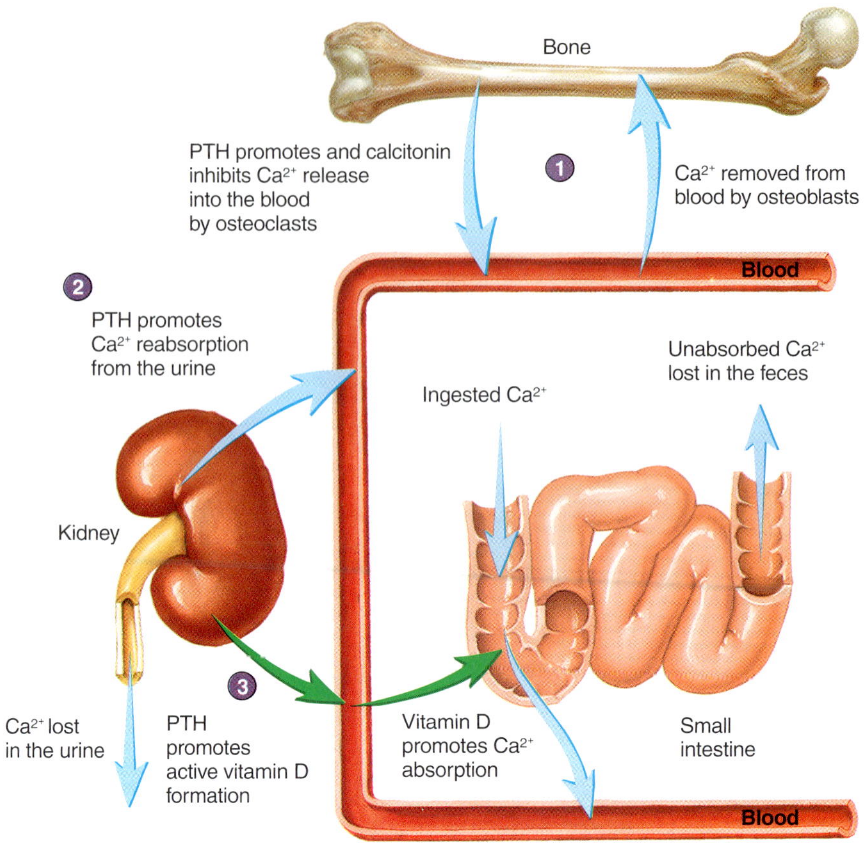

Calcium Homeostasis
Process Figure 6.20

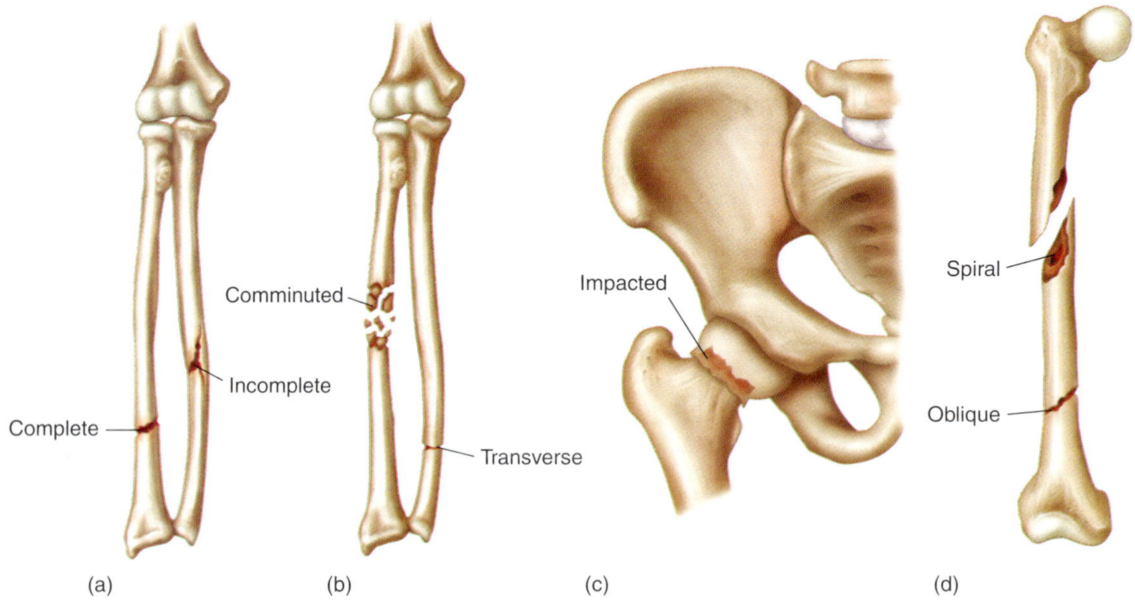

Bone Fractures
Figure B

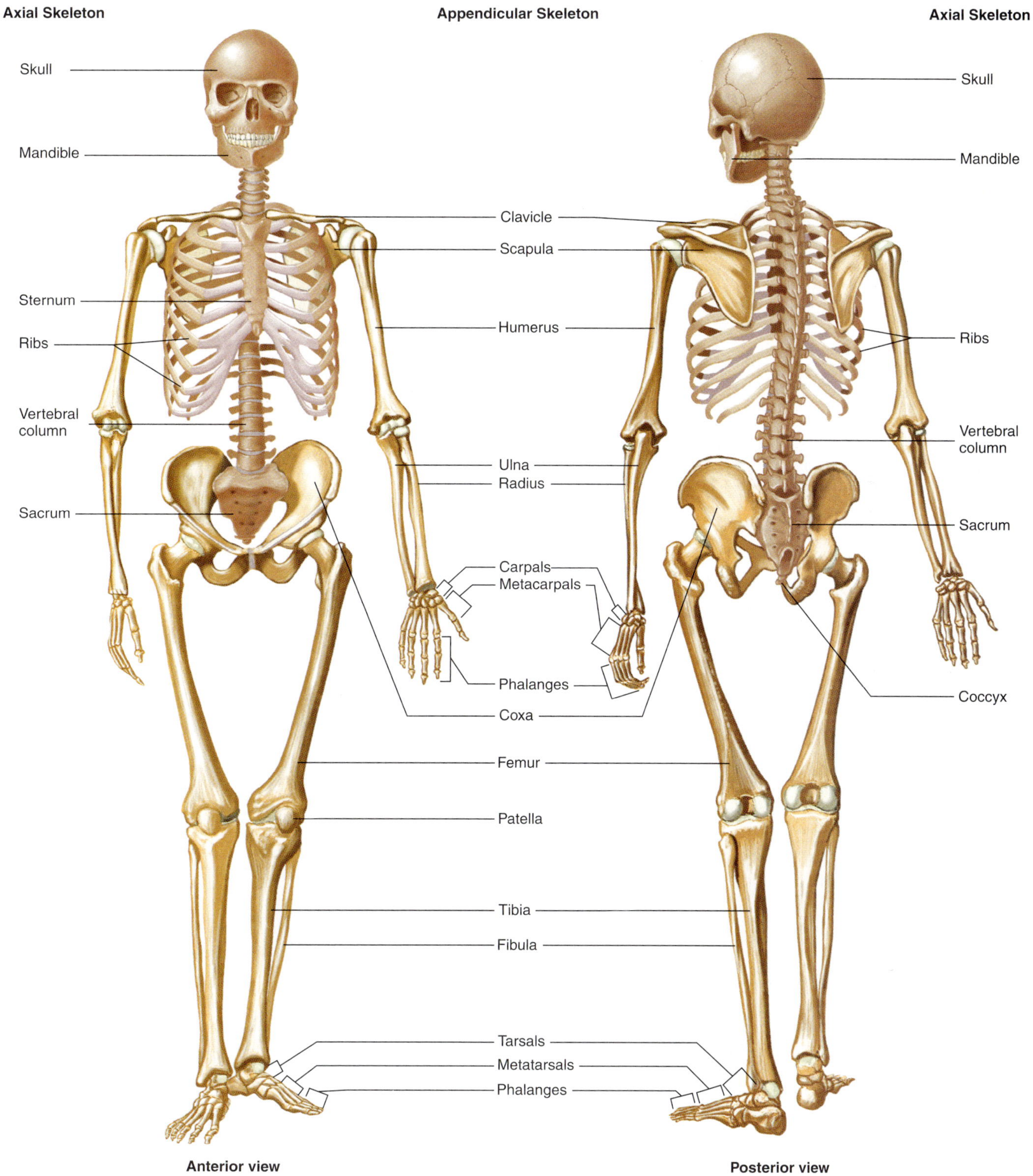
Axial Skeleton
Appendicular Skeleton
Axial Skeleton
Skull
Mandible
Sternum
Ribs
Vertebral column
Sacrum
Clavicle
Scapula
Humerus
Ulna
Radius
Carpals
Metacarpals
Phalanges
Coxa
Femur
Patella
Tibia
Fibula
Tarsals
Metatarsals
Phalanges
Skull
Mandible
Ribs
Vertebral column
Sacrum
Coccyx
Anterior view
Posterior view

The Complete Skeleton
Figure 7.1

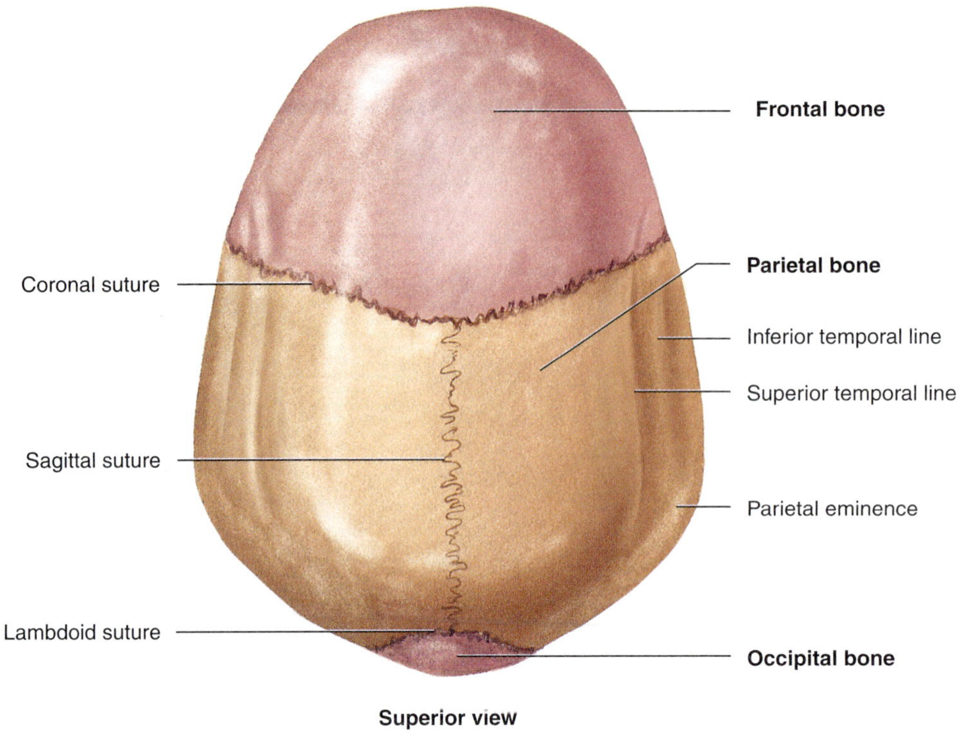

Skull as Seen from the Superior View
Figure 7.2

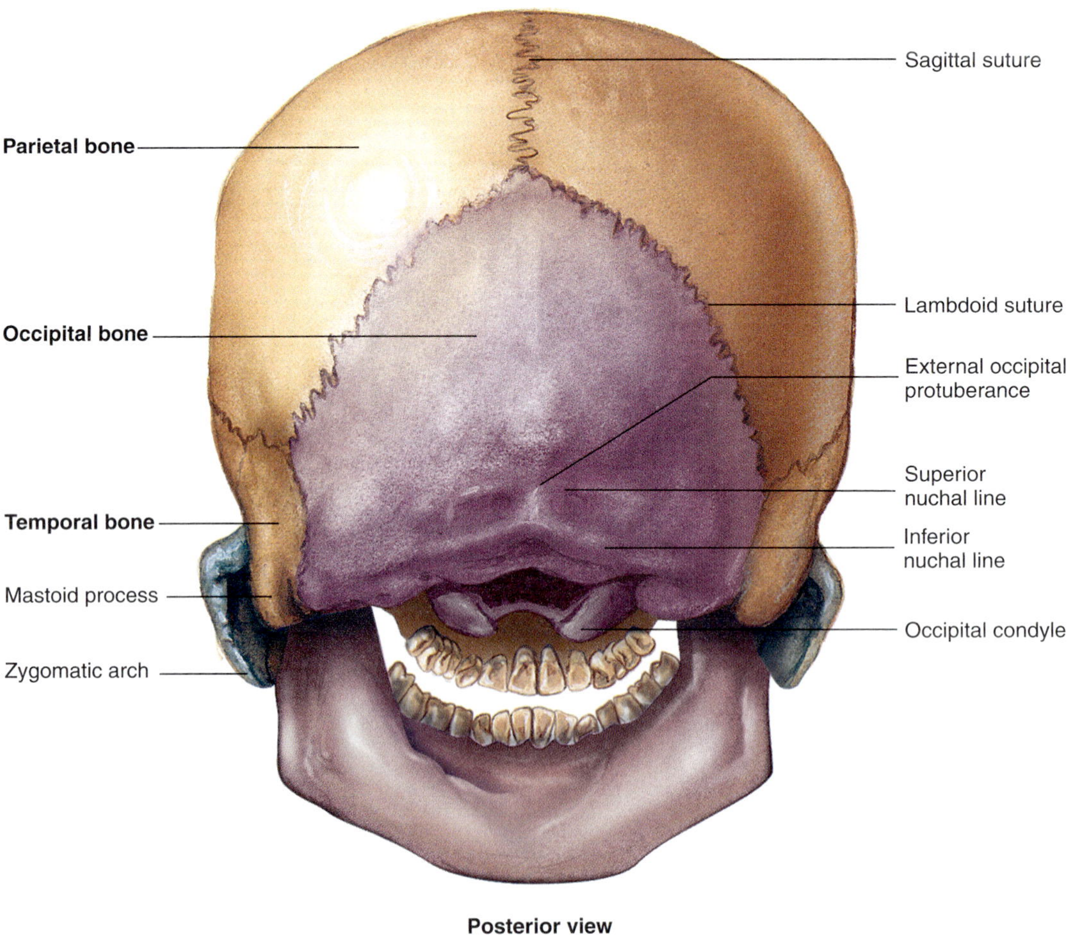

Skull as Seen from the Posterior View
Figure 7.3

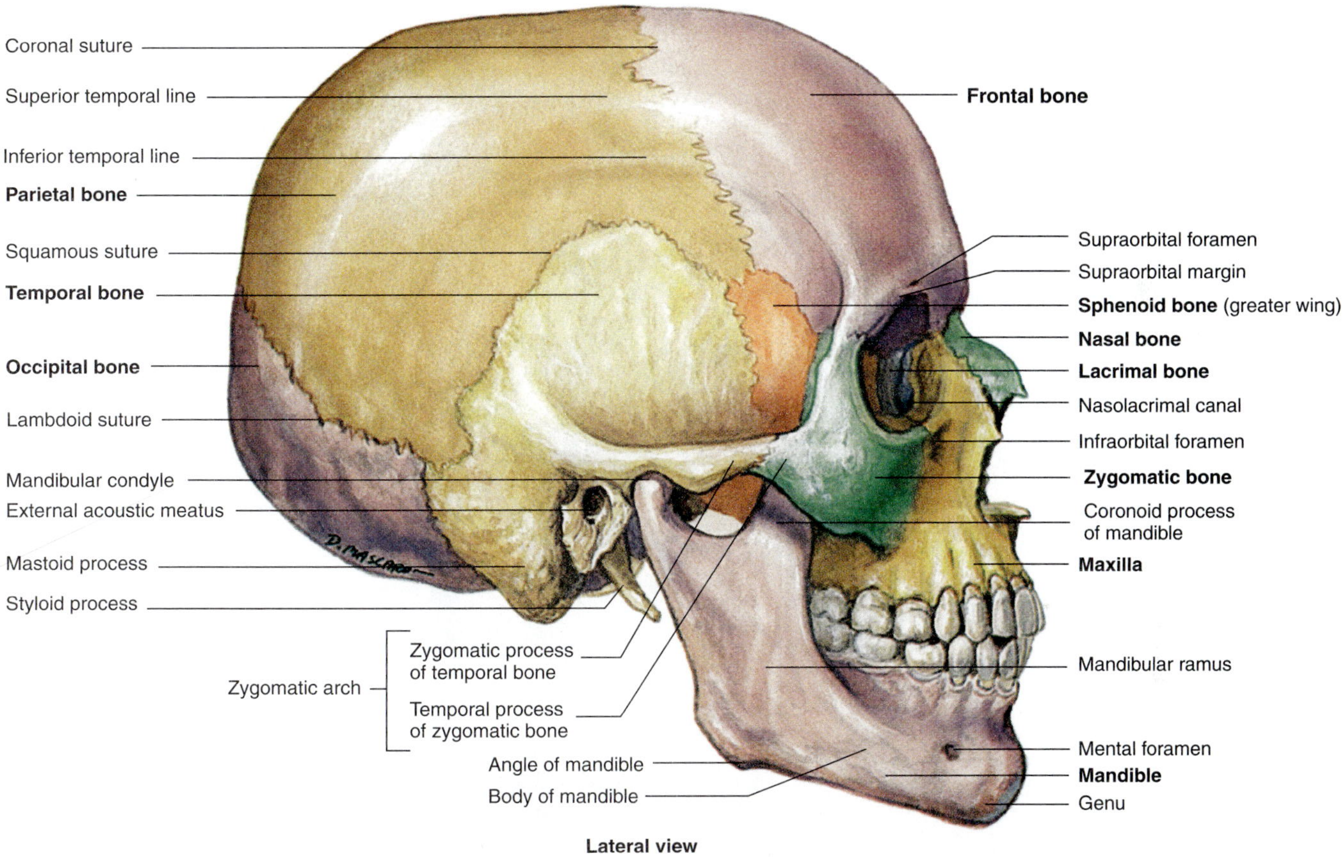

Lateral View of the Skull as Seen from the Right Side
Figure 7.4

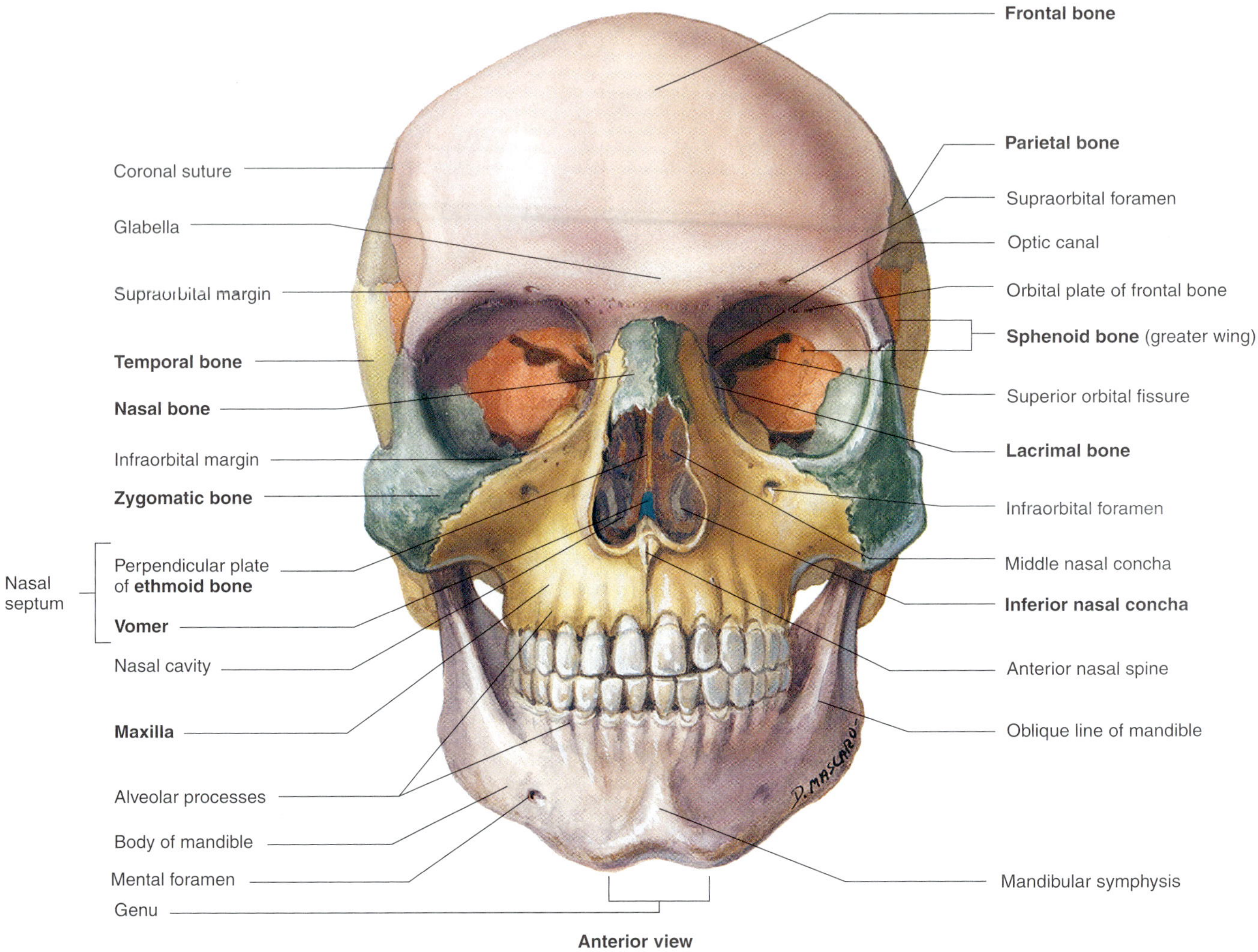
Frontal bone
Parietal bone
Supraorbital foramen
Optic canal
Orbital plate of frontal bone
Sphenoid bone (greater wing)
Superior orbital fissure
Lacrimal bone
Infraorbital foramen
Middle nasal concha
Inferior nasal concha
Anterior nasal spine
Oblique line of mandible
Mandibular symphysis
Coronal suture
Glabella
Supraorbital margin
Temporal bone
Nasal bone
Infraorbital margin
Zygomatic bone
Nasal septum
Perpendicular plate of ethmoid bone
Vomer
Nasal cavity
Maxilla
Alveolar processes
Body of mandible
Mental foramen
Genu
D. MASCARO
Anterior view

Skull as Seen from the Anterior View
Figure 7.6

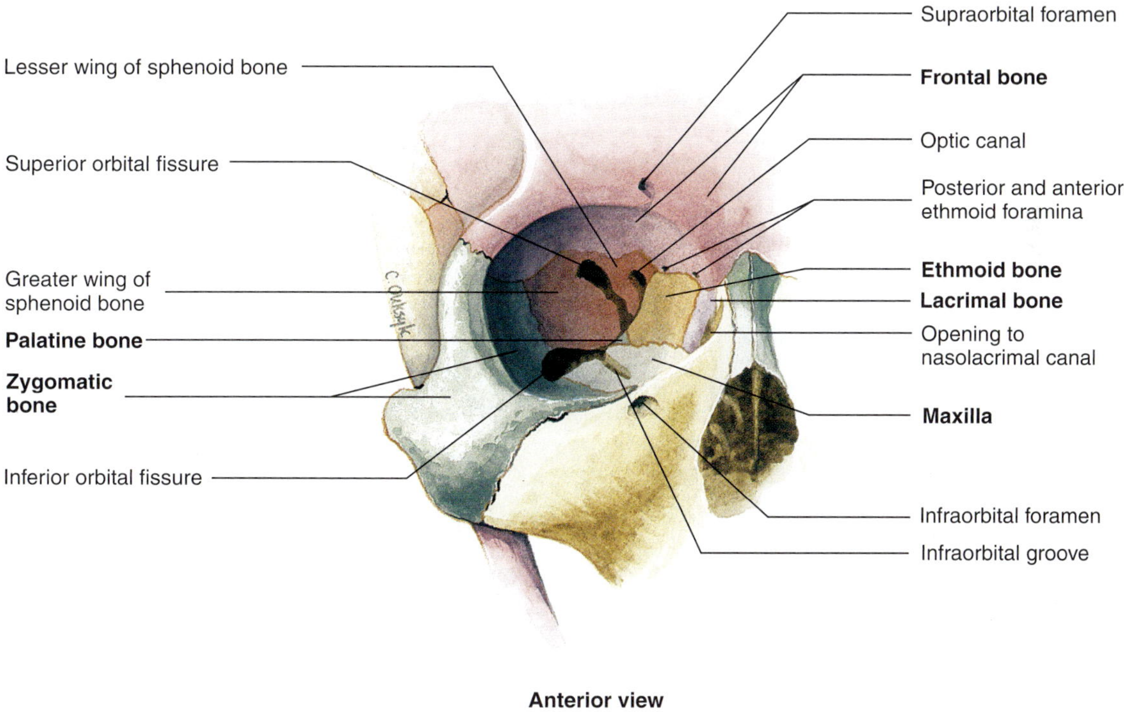
Lesser wing of sphenoid bone
Superior orbital fissure
Greater wing of sphenoid bone
Palatine bone
Zygomatic bone
Inferior orbital fissure
Supraorbital foramen
Frontal bone
Optic canal
Posterior and anterior ethmoid foramina
Ethmoid bone
Lacrimal bone
Opening to nasolacrimal canal
Maxilla
Infraorbital foramen
Infraorbital groove
Anterior view

Bones of the Right Orbit
Figure 7.8

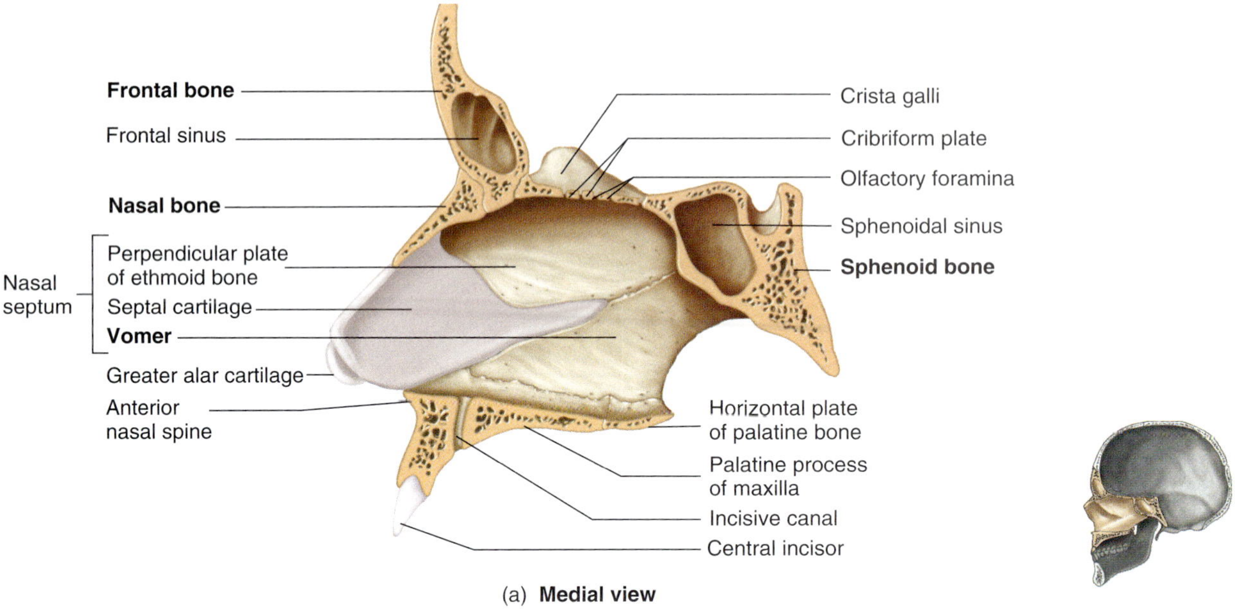

(a) **Medial view**

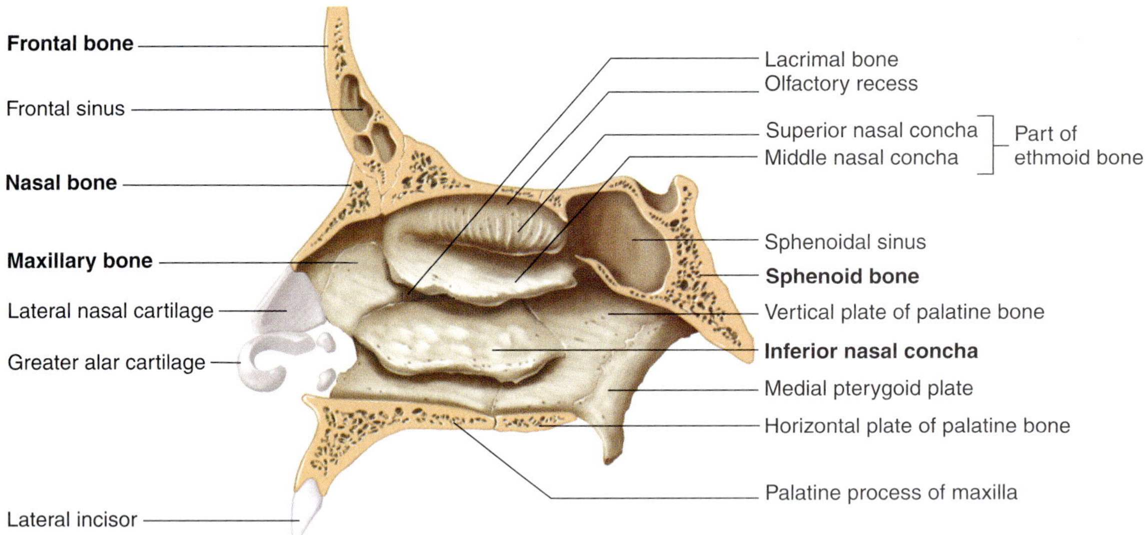

(b) **Medial view**

Bones of the Nasal Cavity
Figure 7.9

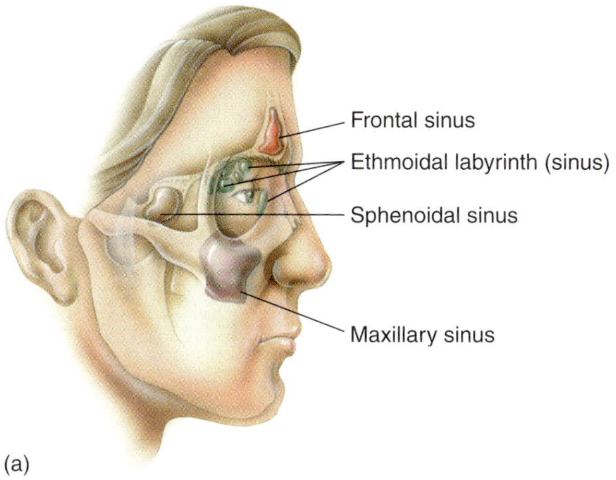

(a)

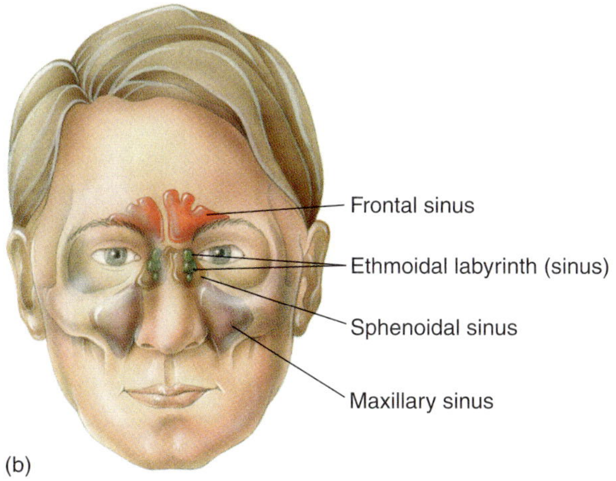

(b)

Paranasal Sinuses
Figure 7.10

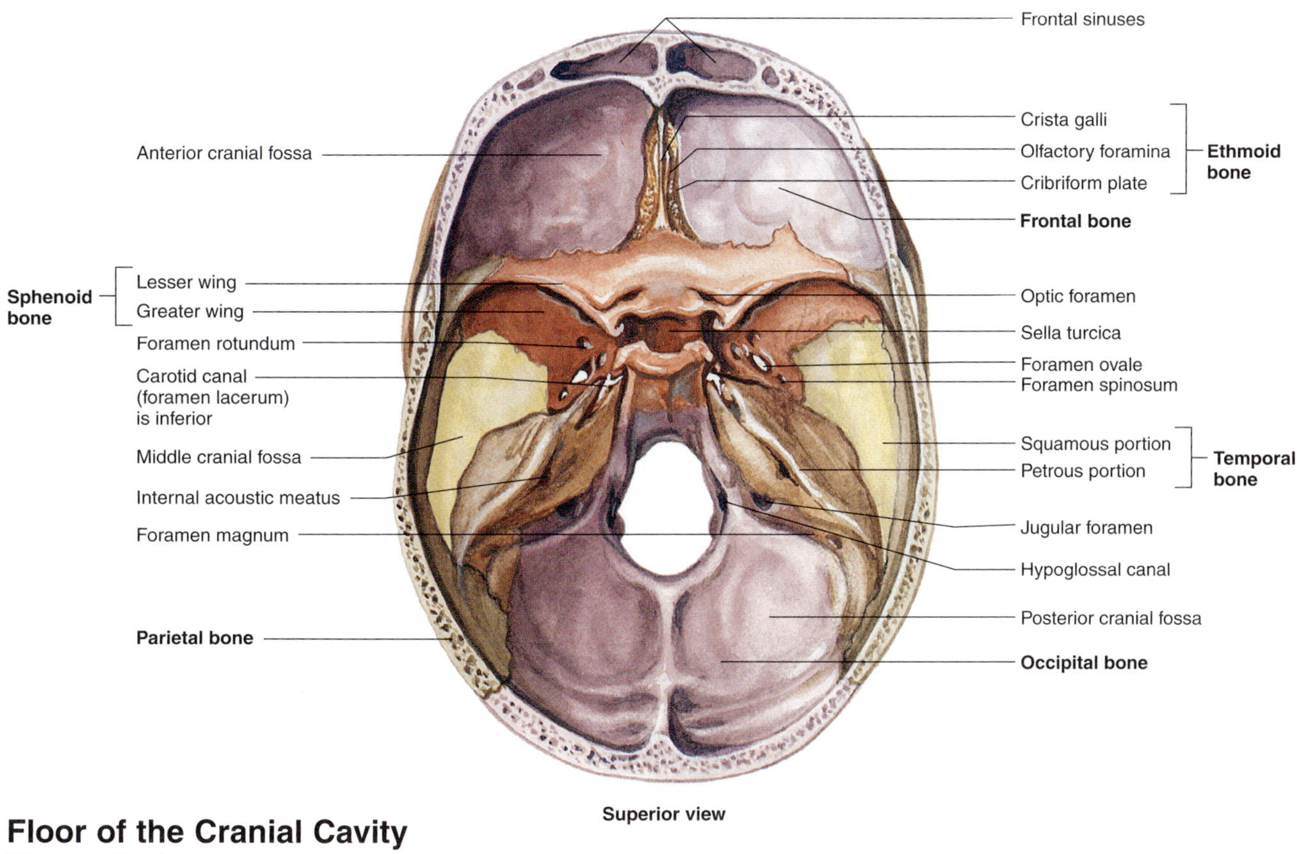

Floor of the Cranial Cavity
Figure 7.11

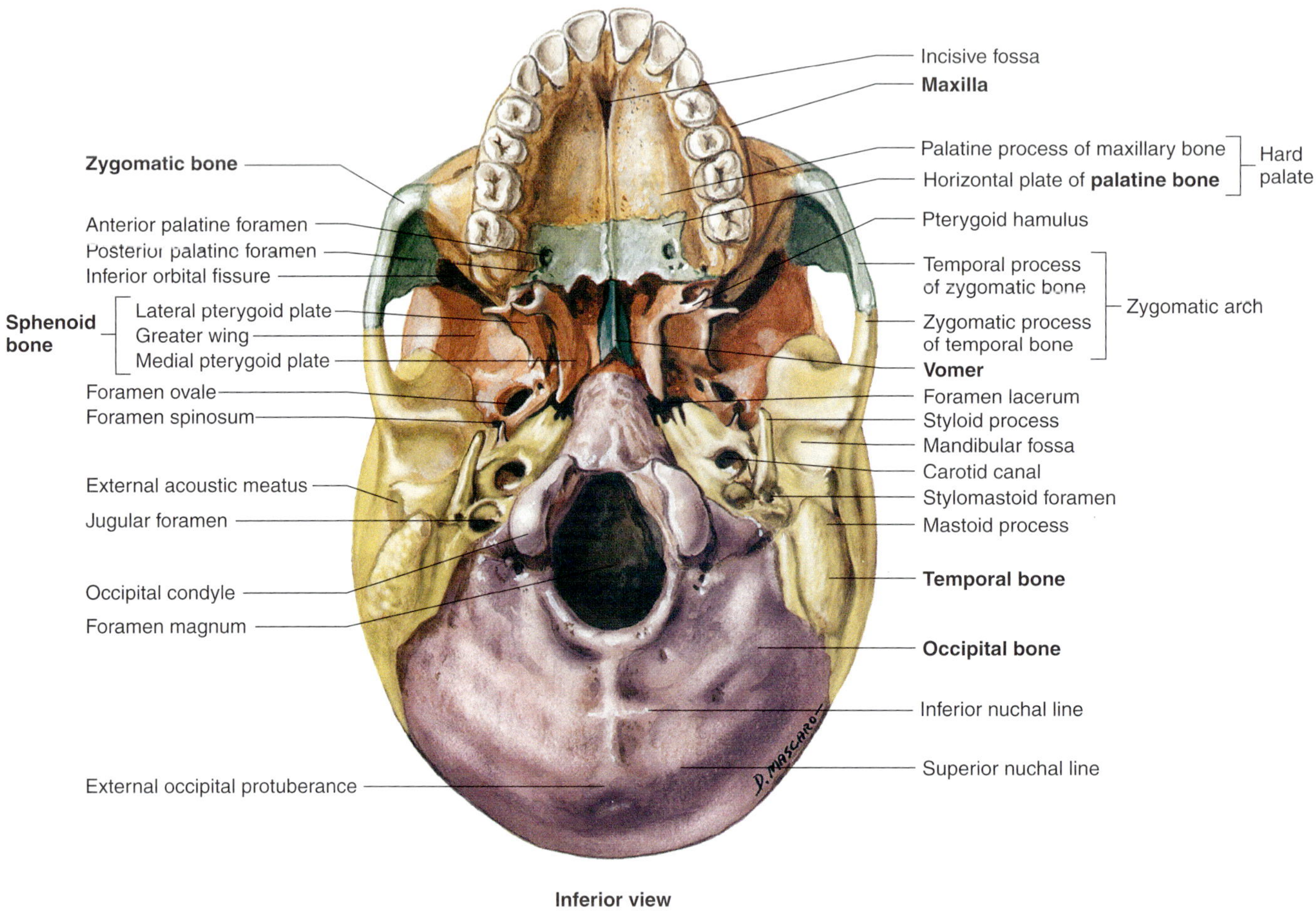
Incisive fossa
Maxilla
Palatine process of maxillary bone
Horizontal plate of palatine bone
Hard palate
Pterygoid hamulus
Temporal process of zygomatic bone
Zygomatic process of temporal bone
Zygomatic arch
Vomer
Foramen lacerum
Styloid process
Mandibular fossa
Carotid canal
Stylomastoid foramen
Mastoid process
Temporal bone
Occipital bone
Inferior nuchal line
Superior nuchal line
Zygomatic bone
Anterior palatine foramen
Posterior palatine foramen
Inferior orbital fissure
Sphenoid bone
Lateral pterygoid plate
Greater wing
Medial pterygoid plate
Foramen ovale
Foramen spinosum
External acoustic meatus
Jugular foramen
Occipital condyle
Foramen magnum
External occipital protuberance
D. MASCARO
Inferior view

Inferior View of the Skull
Figure 7.12

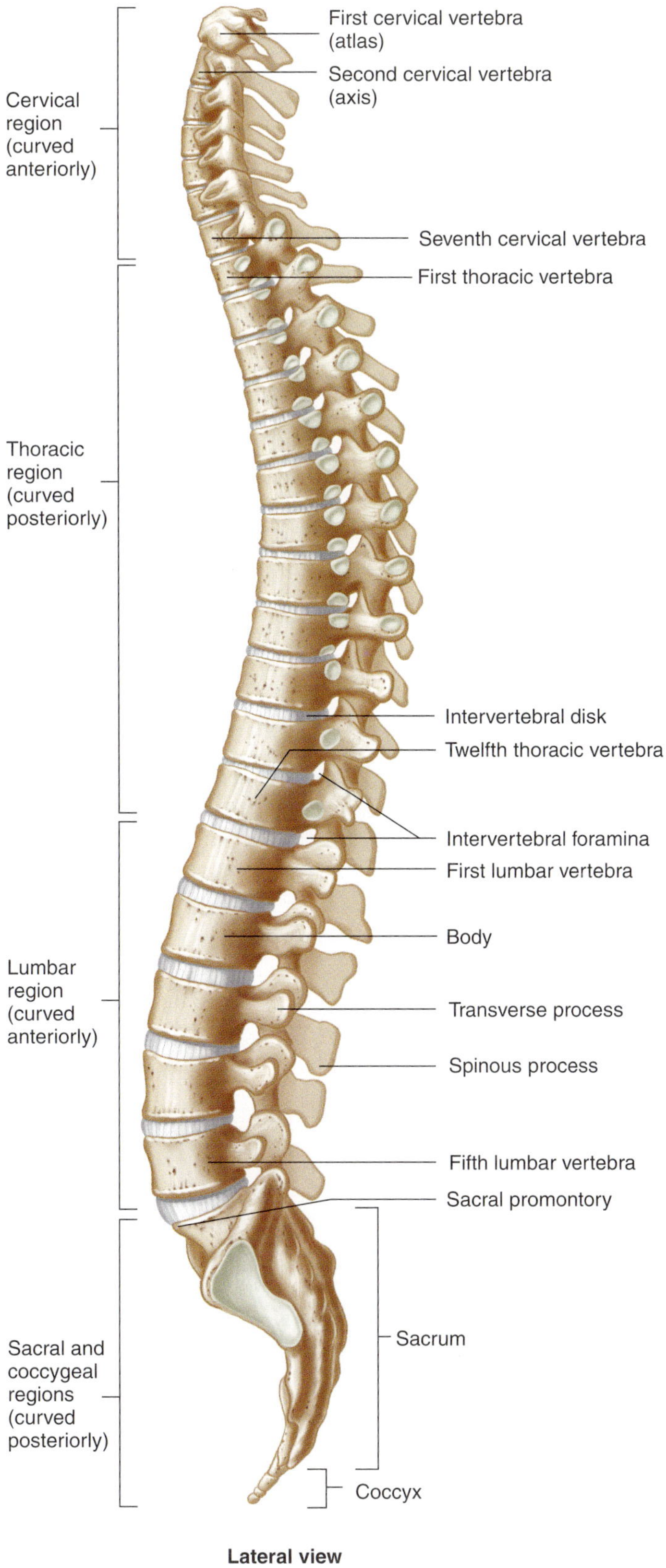
Cervical region (curved anteriorly)
Thoracic region (curved posteriorly)
Lumbar region (curved anteriorly)
Sacral and coccygeal regions (curved posteriorly)
First cervical vertebra (atlas)
Second cervical vertebra (axis)
Seventh cervical vertebra
First thoracic vertebra
Intervertebral disk
Twelfth thoracic vertebra
Intervertebral foramina
First lumbar vertebra
Body
Transverse process
Spinous process
Fifth lumbar vertebra
Sacral promontory
Sacrum
Coccyx
Lateral view

Vertebral Column
Figure 7.13

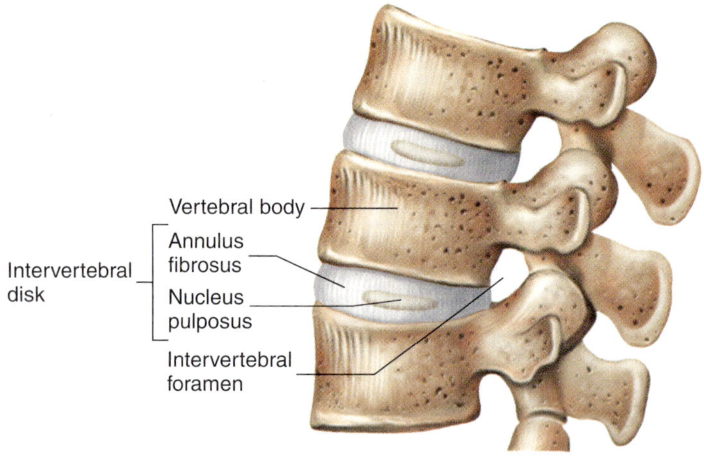

(a) **Lateral view**

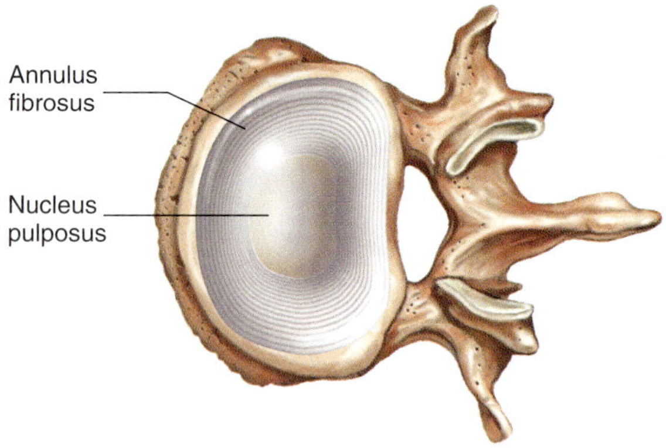

(b) **Superior view**

Intervertebral Disk
Figure 7.14

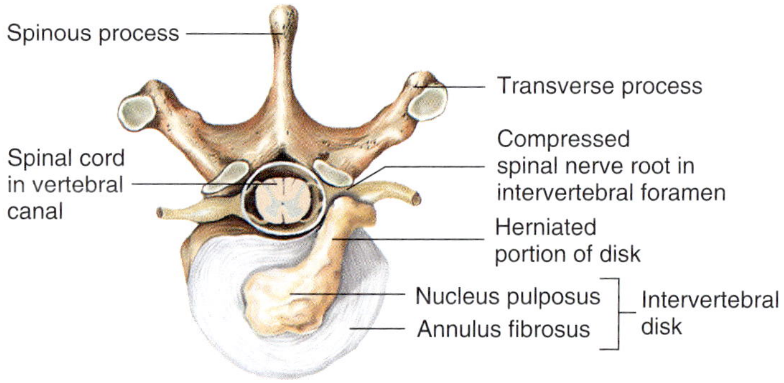

Superior view

Herniated Disk
Figure B

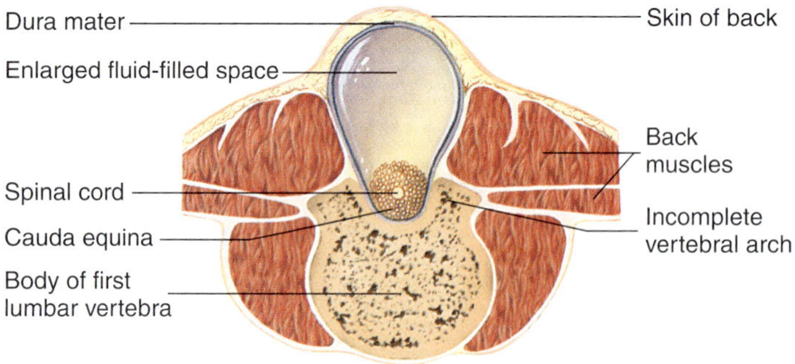

Superior view

Spina Bifida
Figure C

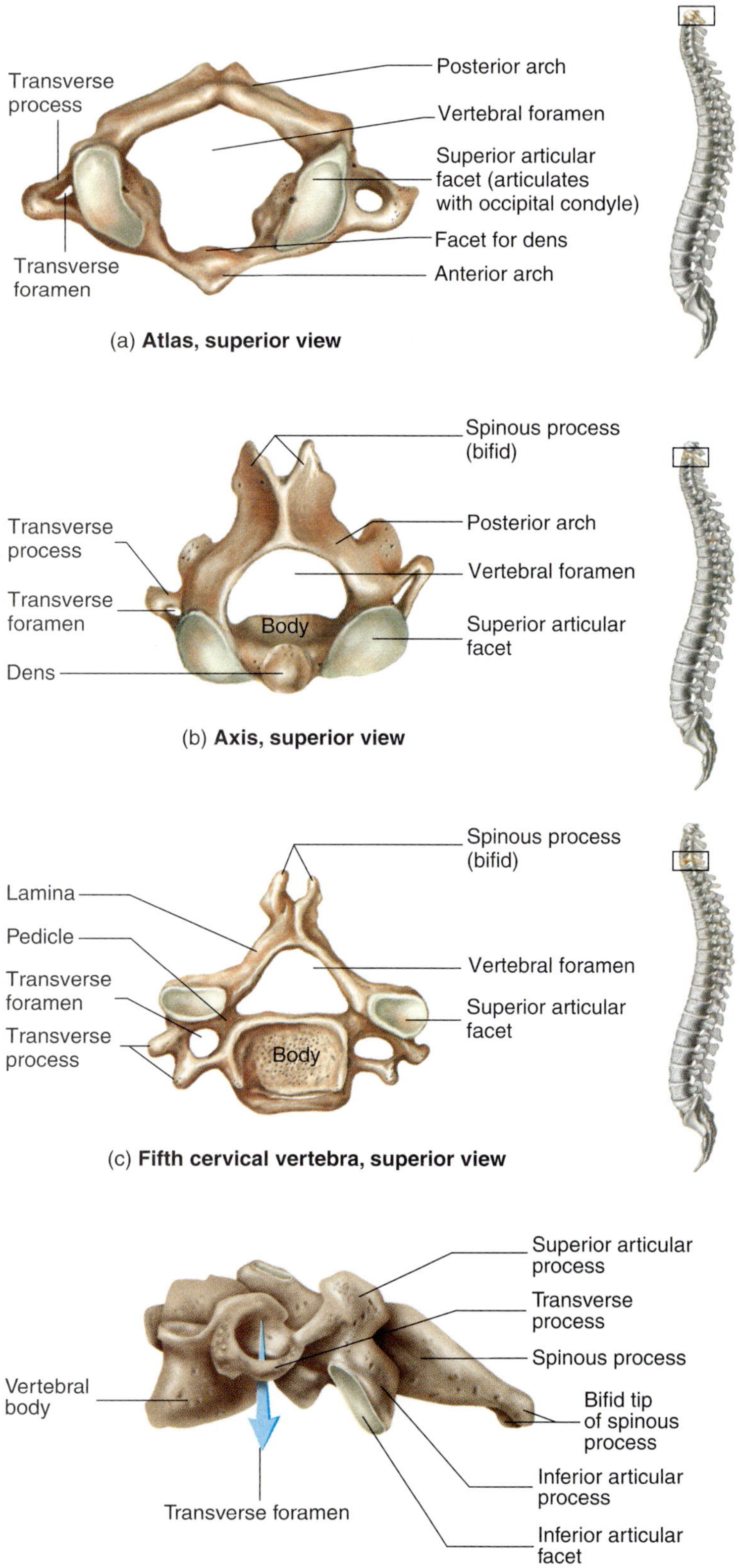

(a) **Atlas, superior view**

(b) **Axis, superior view**

(c) **Fifth cervical vertebra, superior view**

(d) **Lateral view**

Cervical Vertebrae

Figure 7.17

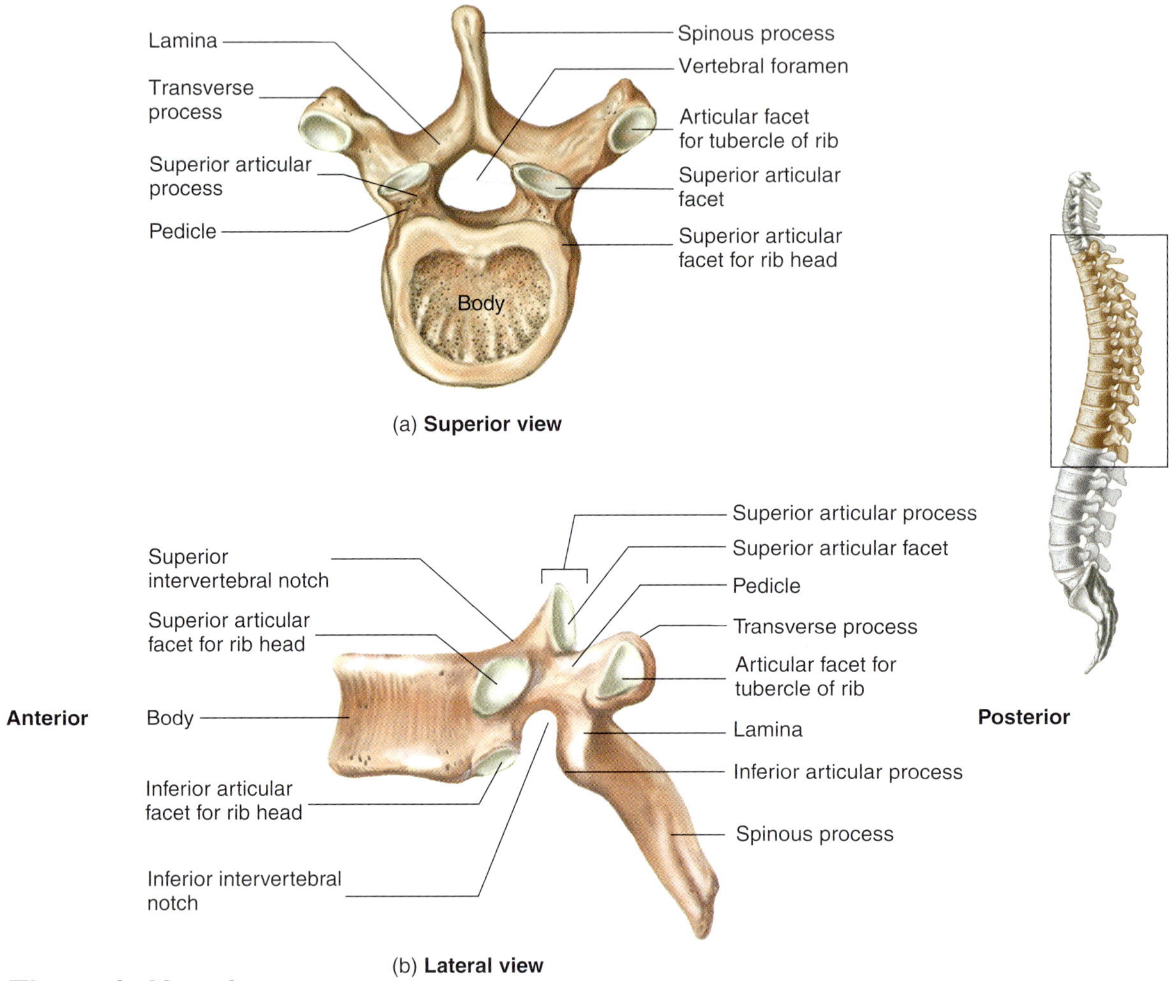

(a) **Superior view**

(b) **Lateral view**

Thoracic Vertebrae
Figure 7.18

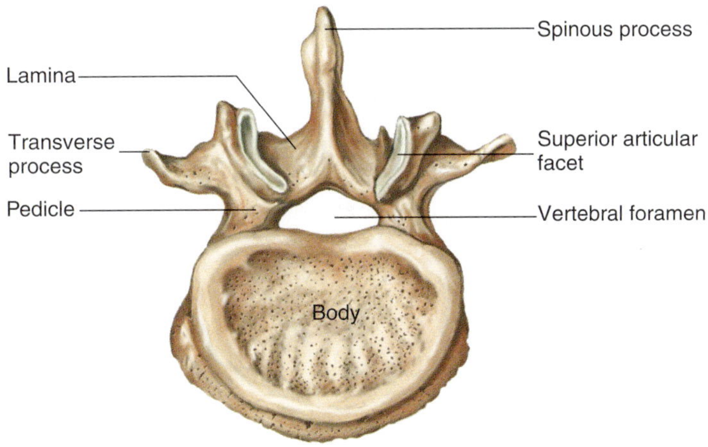

(a) **Superior view**

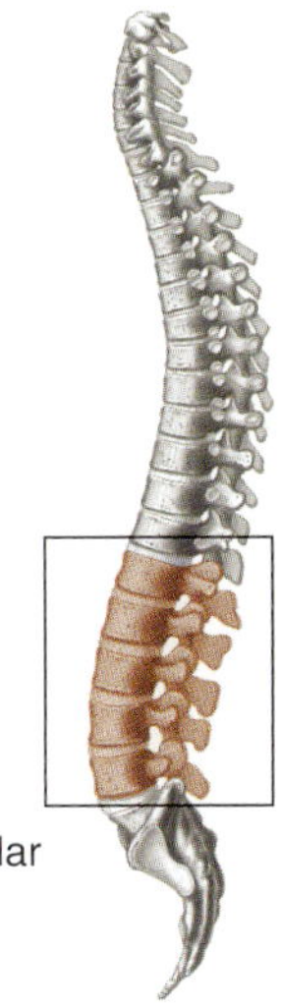

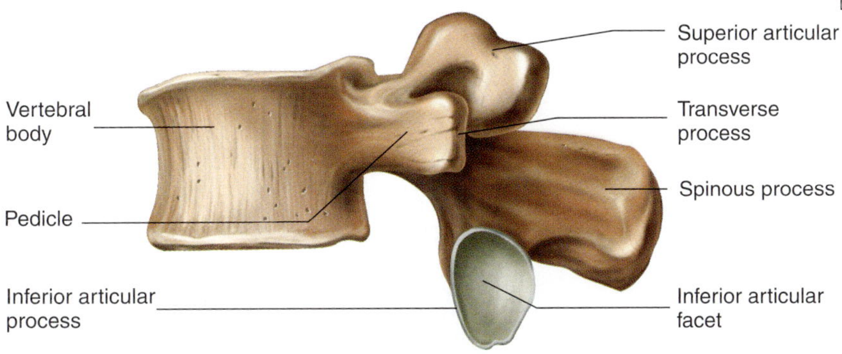

(b) **Lateral view**

Lumbar Vertebrae
Figure 7.19

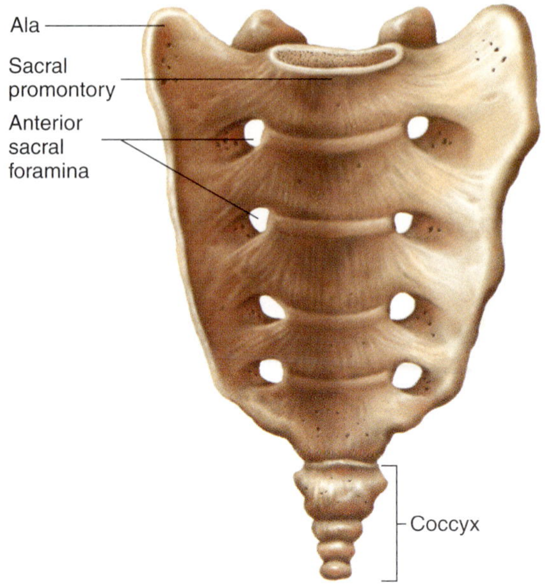

(a) **Anterior view**

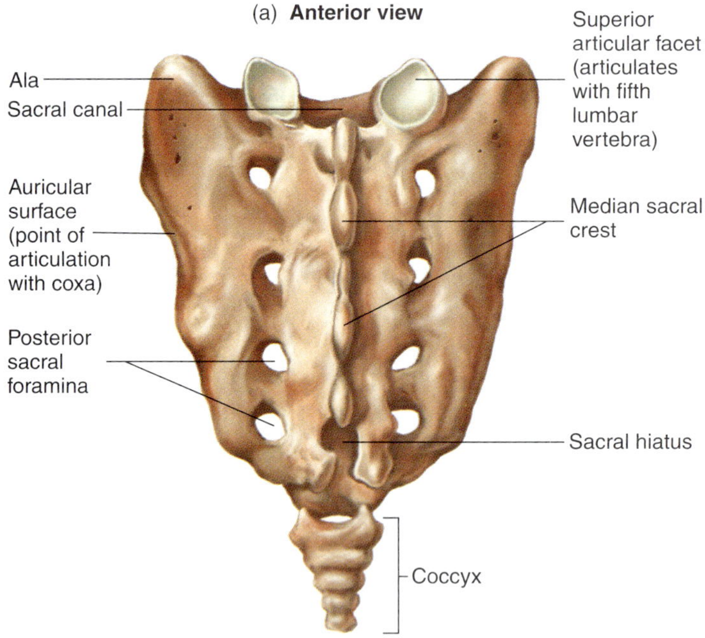

(b) **Posterior view**

Sacrum
Figure 7.20

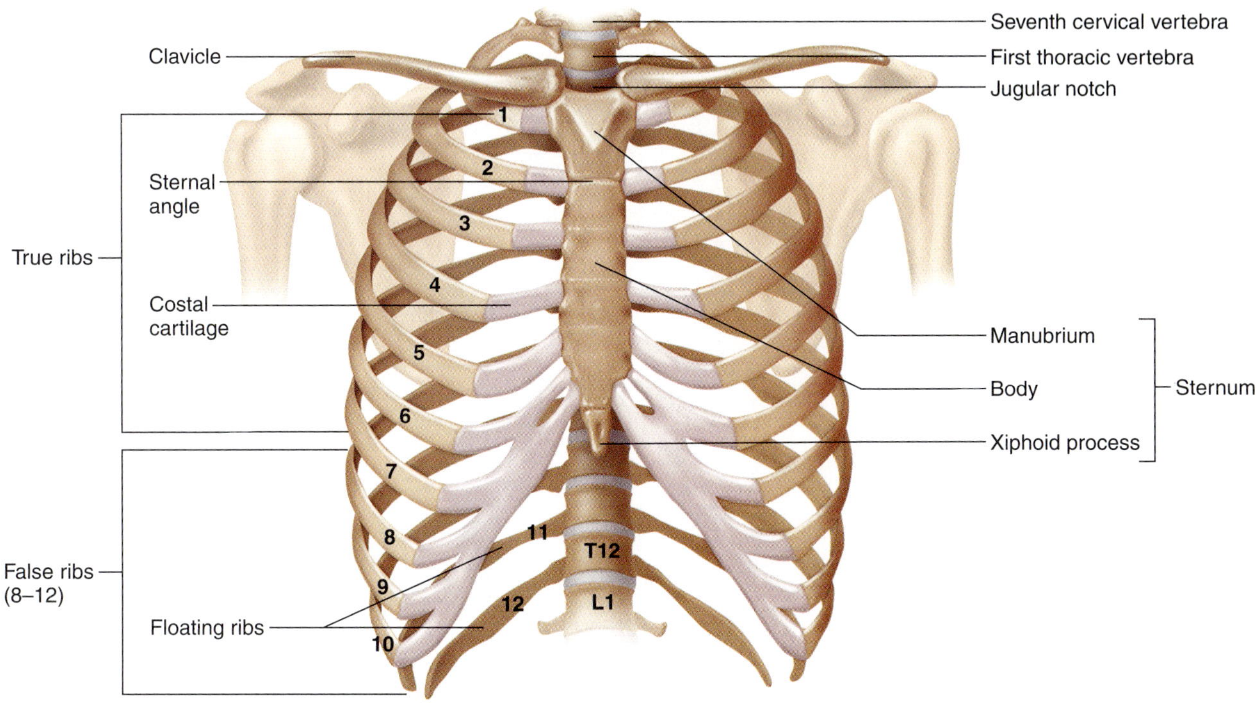

(a) **Anterior view**

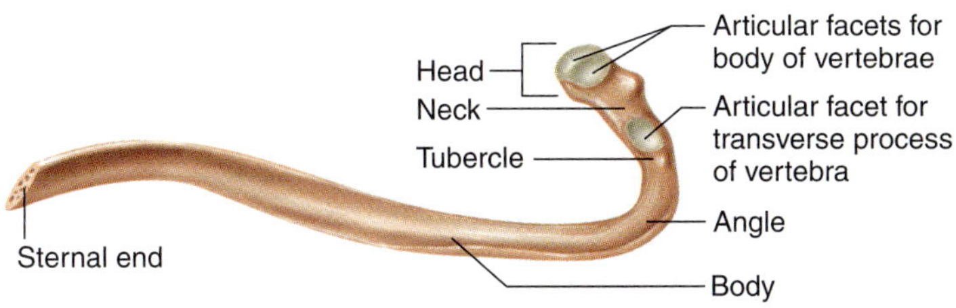

(b) **Inferior view**

Thoracic Cage
Figure 7.21

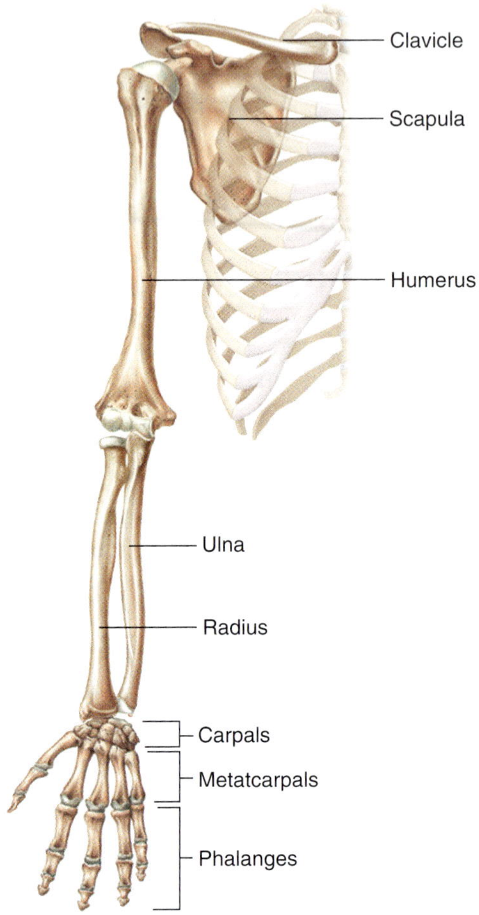

Anterior view

Bones of the Pectoral Girdle and Right Upper Limb
Figure 7.23

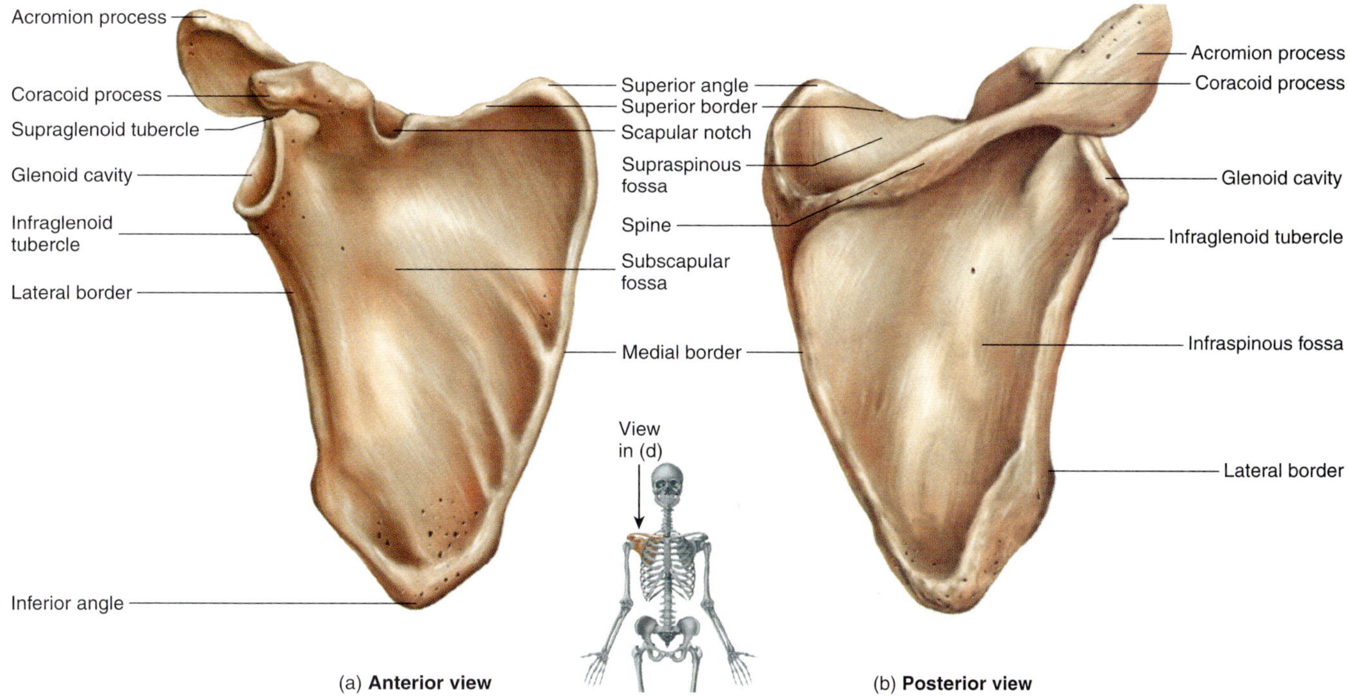

(a) **Anterior view** (b) **Posterior view**

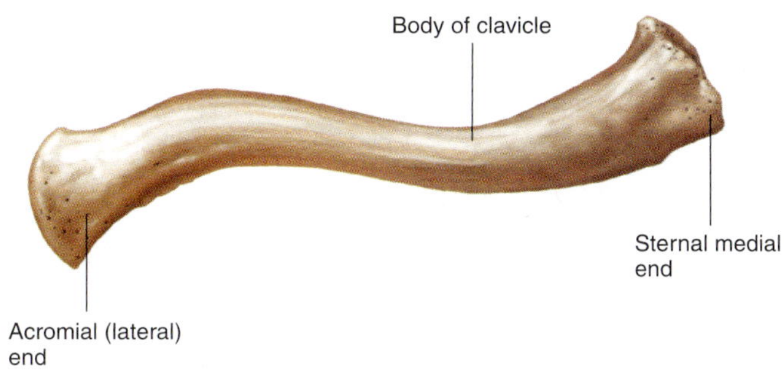

(c) **Superior view**

Right Scapula and Clavicle
Figure 7.24

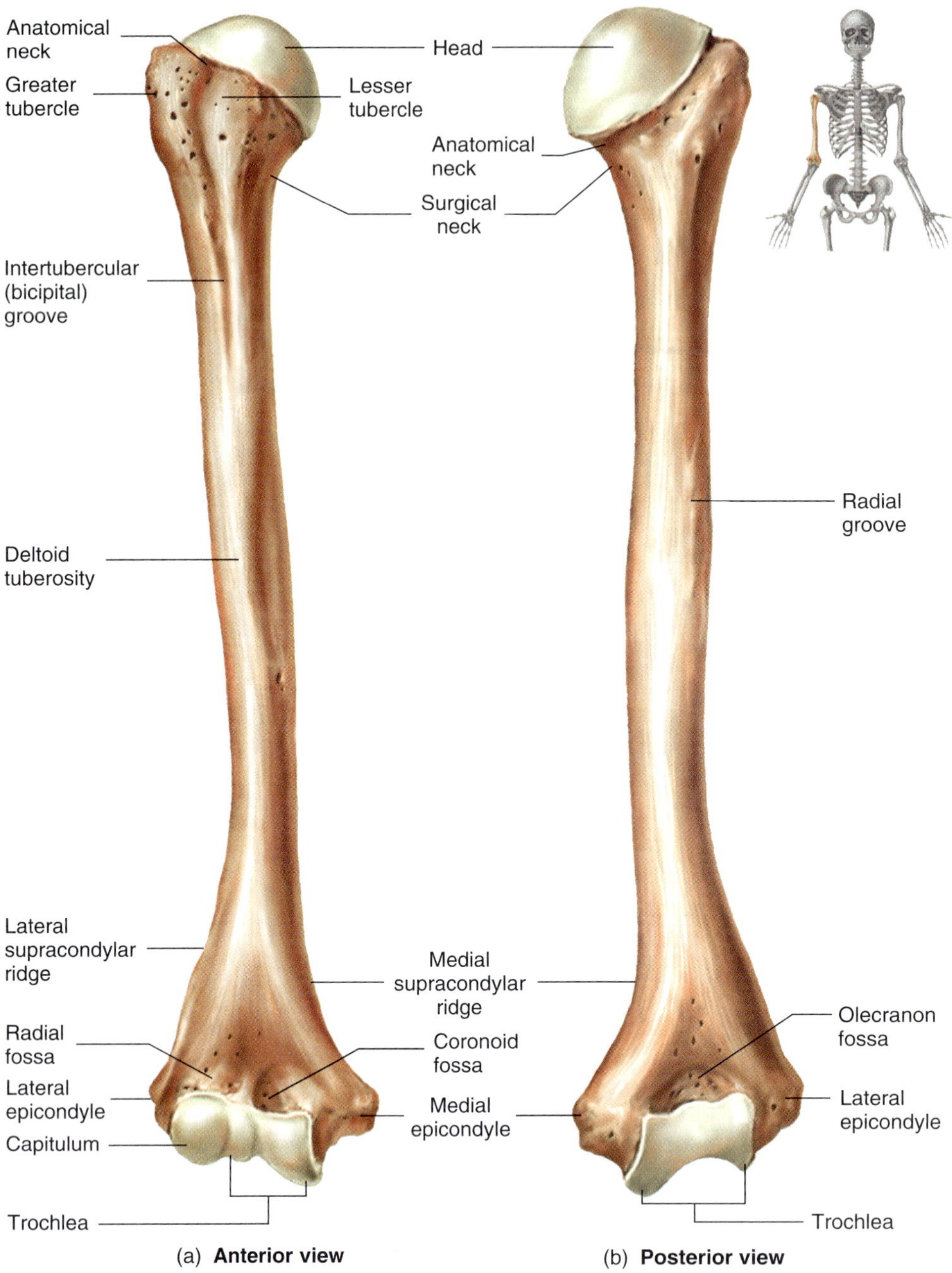
Anatomical neck
Greater tubercle
Head
Lesser tubercle
Head
Anatomical neck
Surgical neck
Intertubercular (bicipital) groove
Radial groove
Deltoid tuberosity
Lateral supracondylar ridge
Medial supracondylar ridge
Radial fossa
Coronoid fossa
Olecranon fossa
Lateral epicondyle
Medial epicondyle
Lateral epicondyle
Capitulum
Trochlea
Trochlea
(a) Anterior view
(b) Posterior view

Right Humerus
Figure 7.25

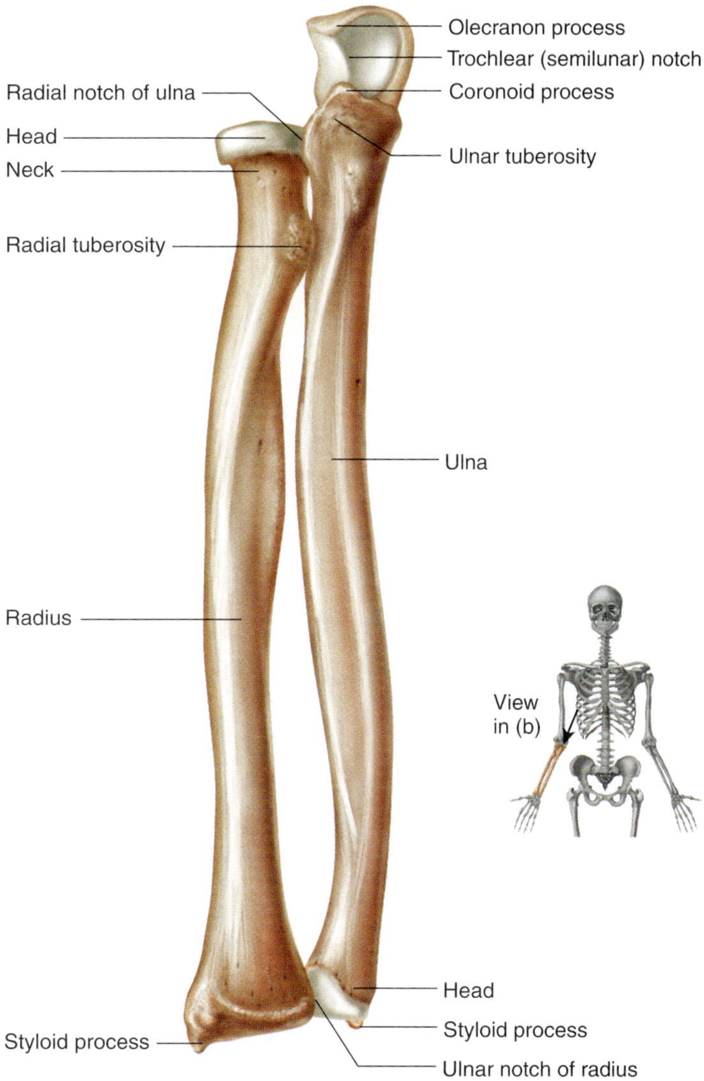

(a) **Anterior view**

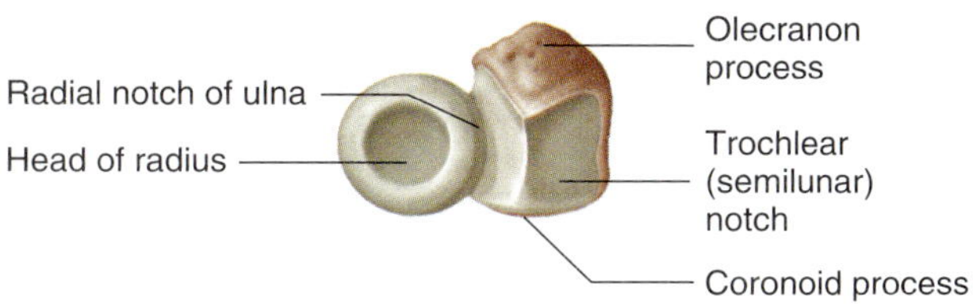

(b) **Proximal view**

Right Ulna and Radius
Figure 7.26

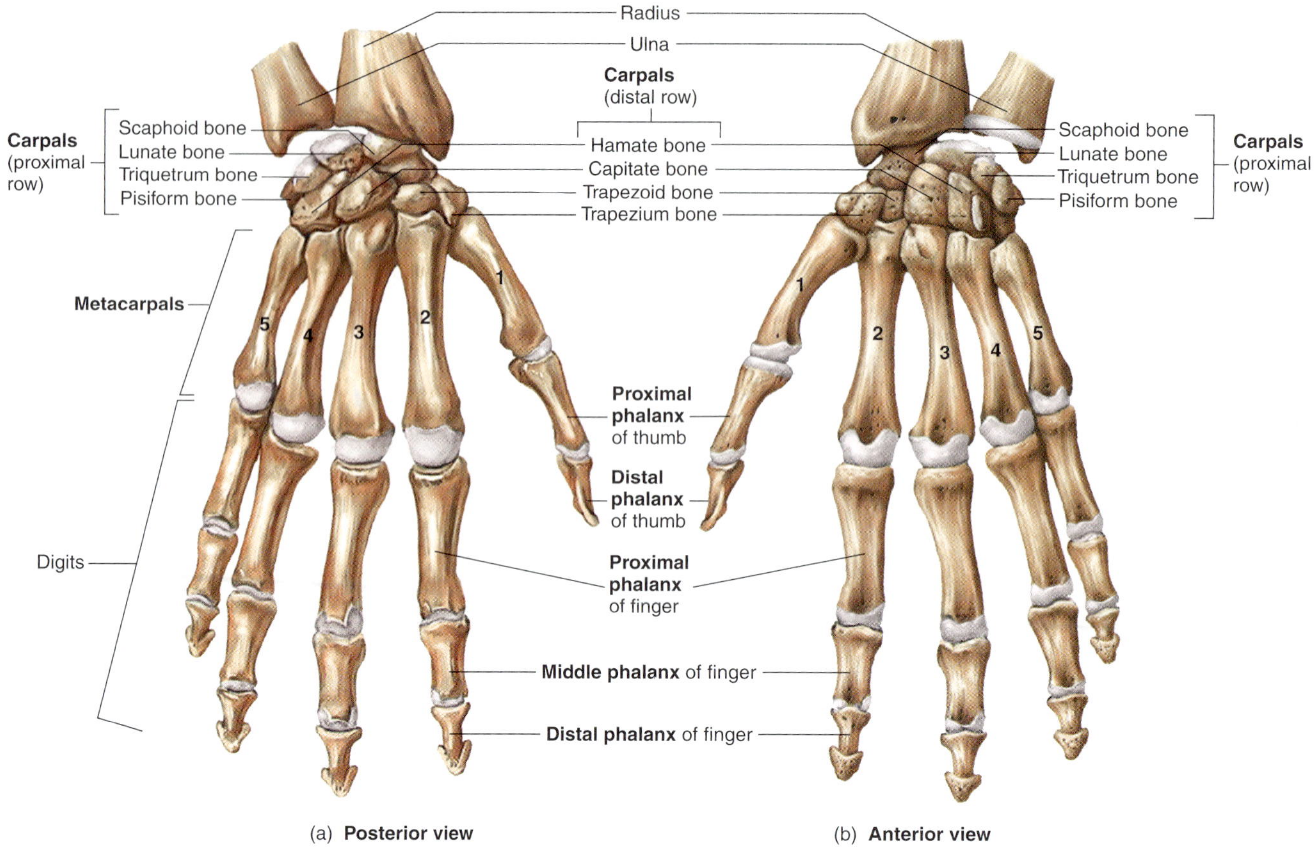

Bones of the Right Wrist and Hand
Figure 7.27

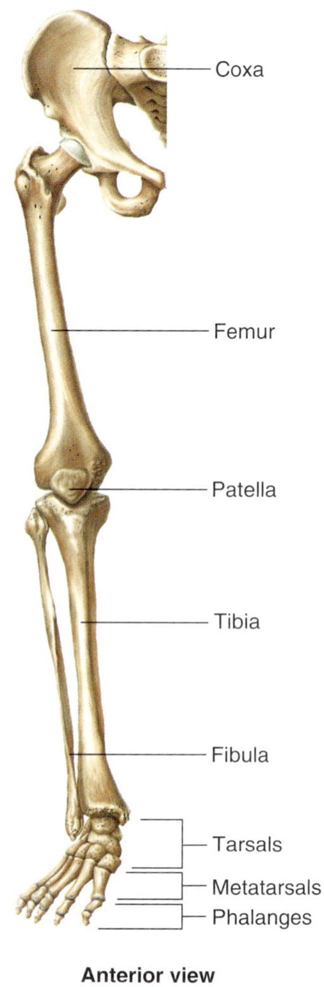

Bones of the Pelvic Girdle and Right Lower Limb
Figure 7.29

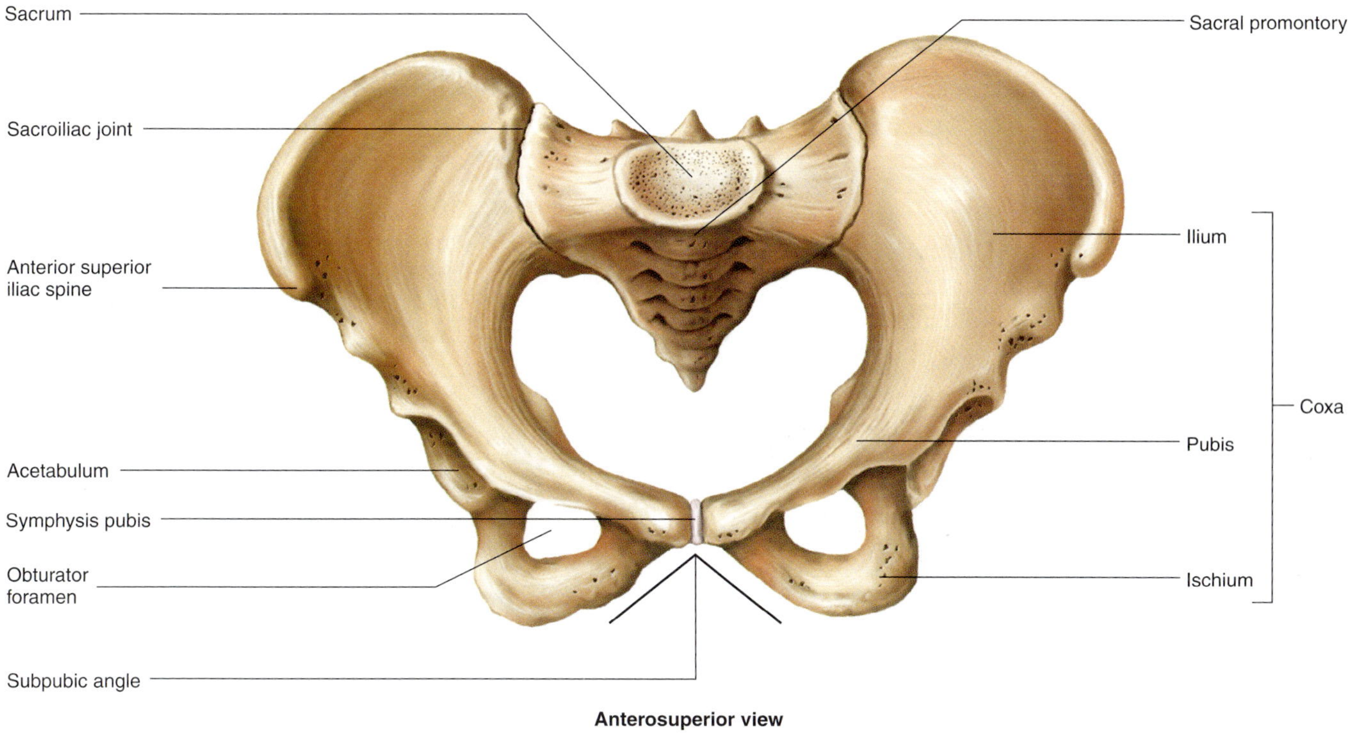

Anterosuperior View of the Pelvis
Figure 7.30

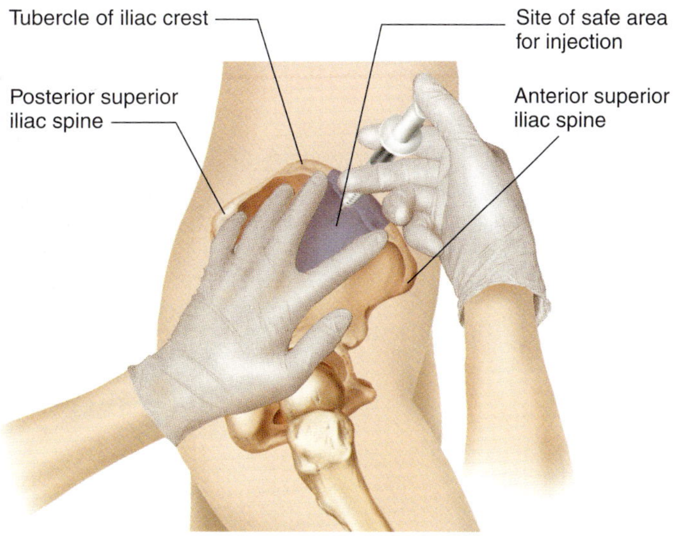

Gluteal Injection Site
Figure D

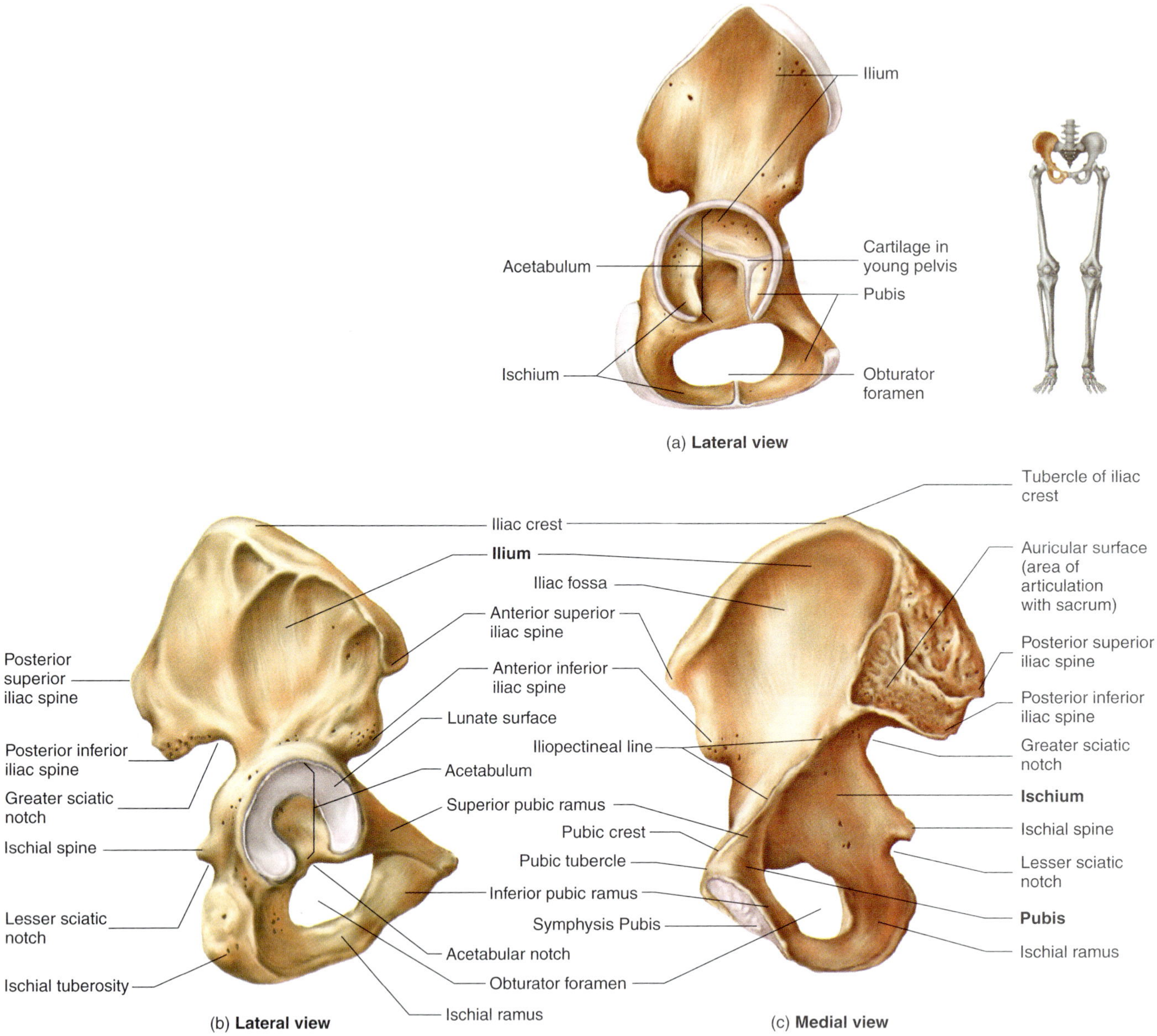
Ilium
Acetabulum
Cartilage in young pelvis
Pubis
Ischium
Obturator foramen
(a) Lateral view
Tubercle of iliac crest
Iliac crest
Ilium
Iliac fossa
Anterior superior iliac spine
Anterior inferior iliac spine
Lunate surface
Iliopectineal line
Acetabulum
Superior pubic ramus
Pubic crest
Pubic tubercle
Inferior pubic ramus
Symphysis Pubis
Acetabular notch
Obturator foramen
Ischial ramus
Posterior superior iliac spine
Posterior inferior iliac spine
Greater sciatic notch
Ischial spine
Lesser sciatic notch
Ischial tuberosity
Auricular surface (area of articulation with sacrum)
Posterior superior iliac spine
Posterior inferior iliac spine
Greater sciatic notch
Ischium
Ischial spine
Lesser sciatic notch
Pubis
Ischial ramus
(b) Lateral view
(c) Medial view

Coxa
Figure 7.31

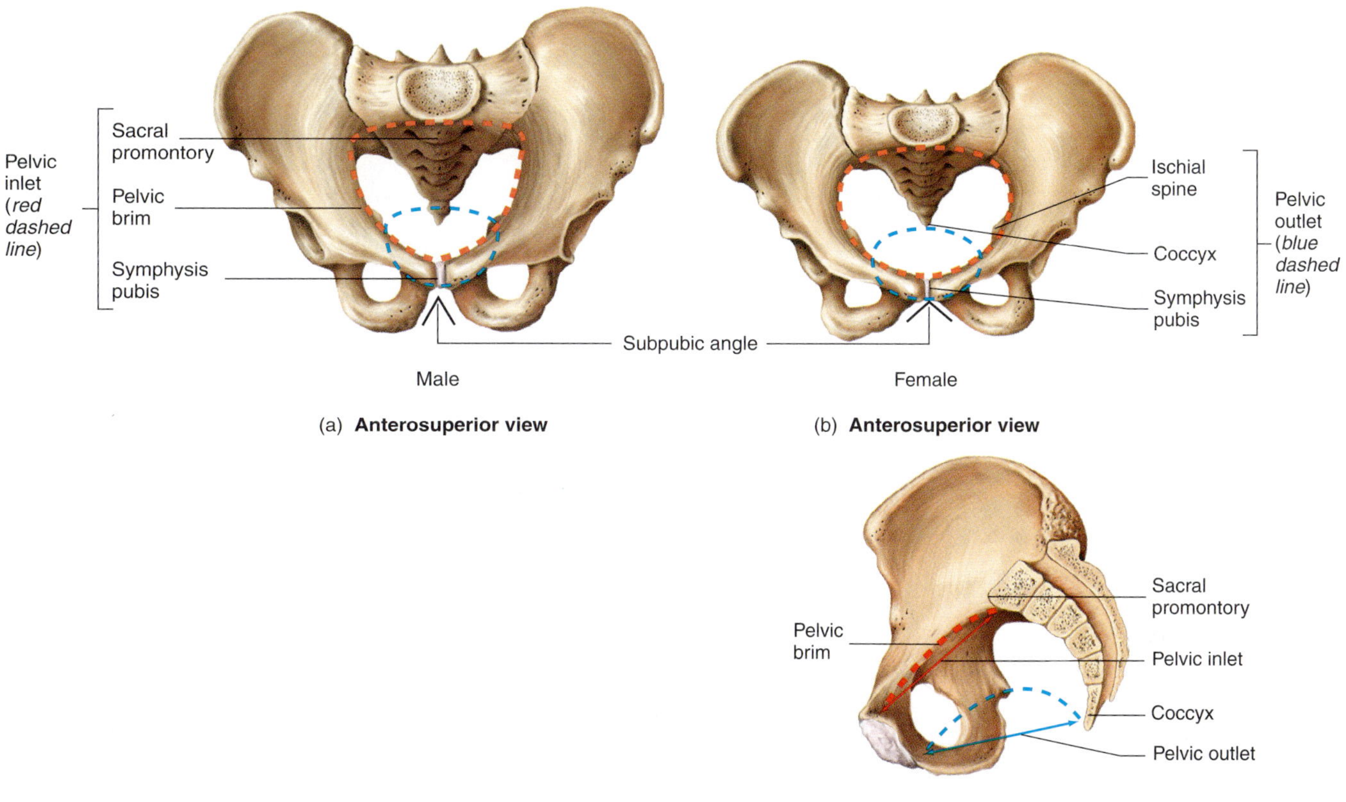

(a) **Anterosuperior view**

(b) **Anterosuperior view**

(c) **Medial view**

Comparison of the Male and Female Pelvis
Figure 7.33

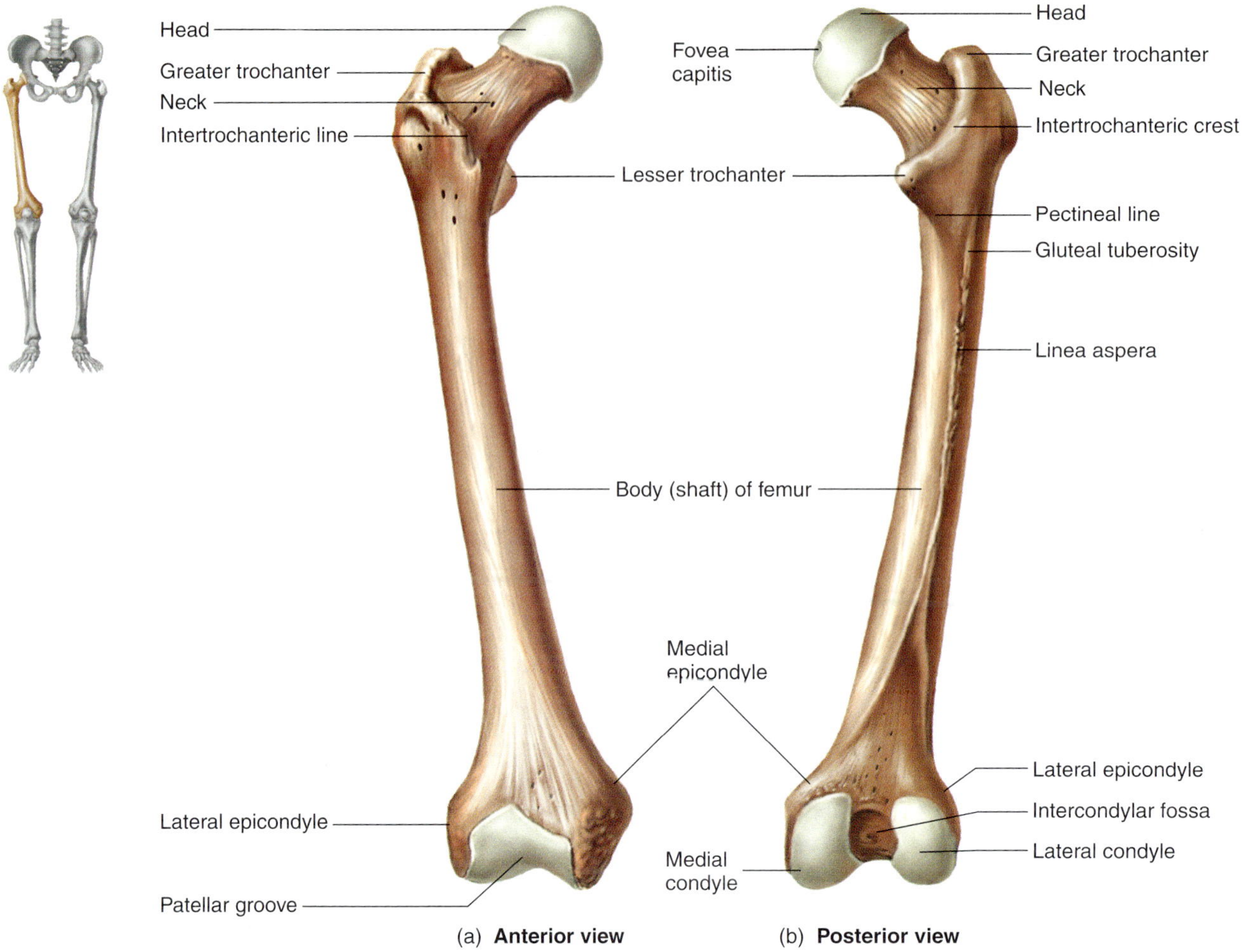
Head
Greater trochanter
Neck
Intertrochanteric line
Lesser trochanter
Body (shaft) of femur
Medial epicondyle
Lateral epicondyle
Patellar groove
(a) Anterior view
Fovea capitis
Head
Greater trochanter
Neck
Intertrochanteric crest
Pectineal line
Gluteal tuberosity
Linea aspera
Lateral epicondyle
Intercondylar fossa
Lateral condyle
Medial condyle
(b) Posterior view

Right Femur
Figure 7.34

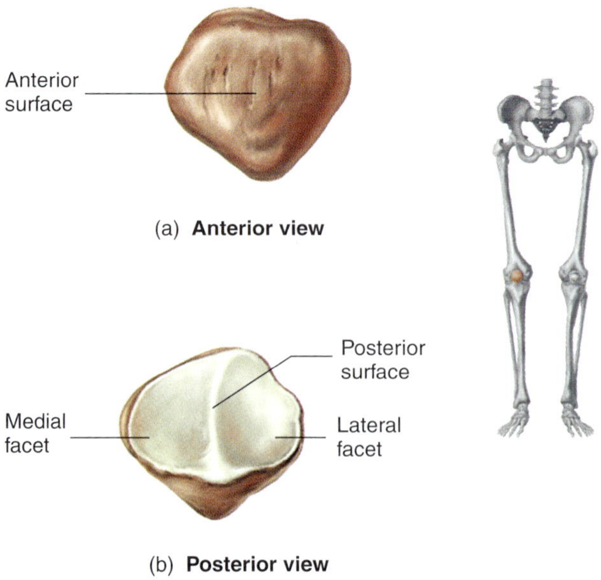
Anterior surface
(a) Anterior view
Medial facet
Posterior surface
Lateral facet
(b) Posterior view

Right Patella
Figure 7.35

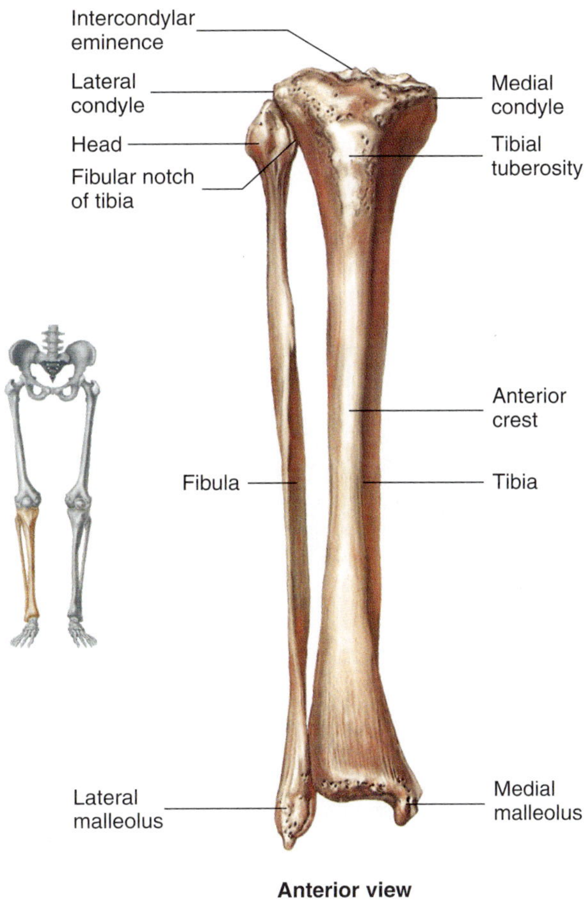

Right Tibia and Fibula, Anterior View
Figure 7.36

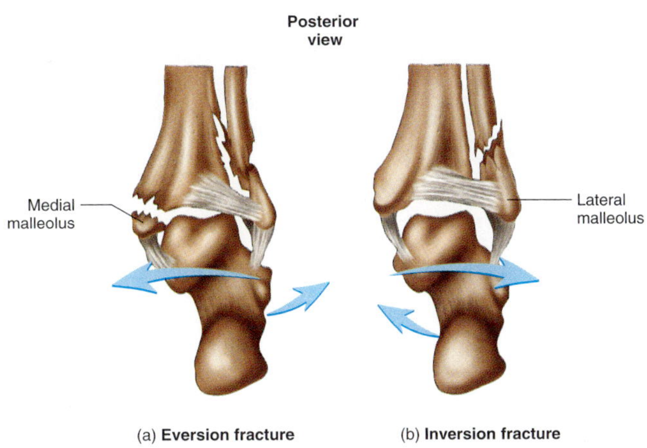

Fracture of the Medial or Lateral Malleolus
Figure E

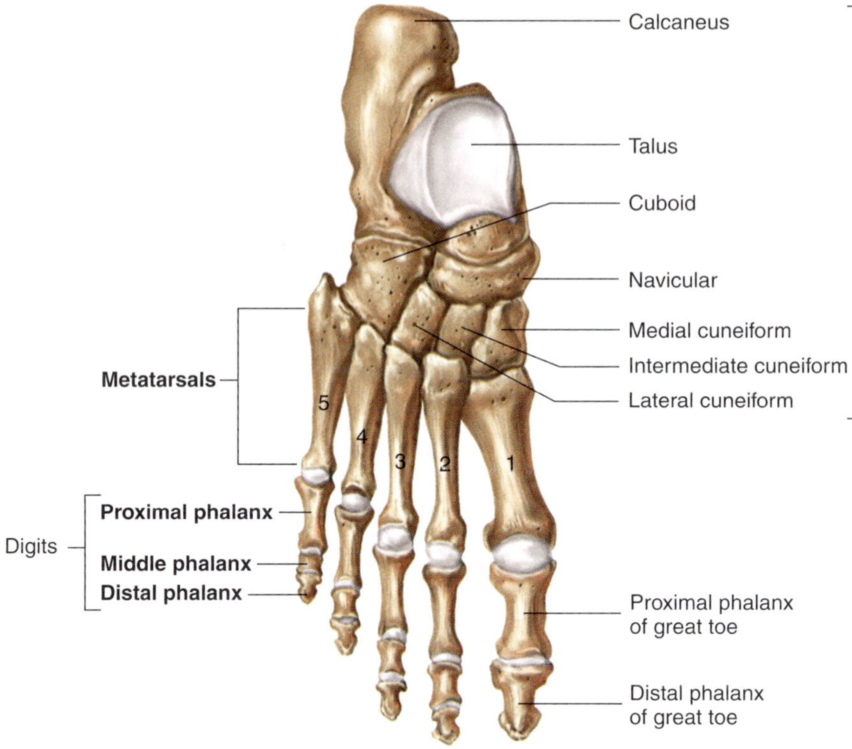

(a) **Superior view**

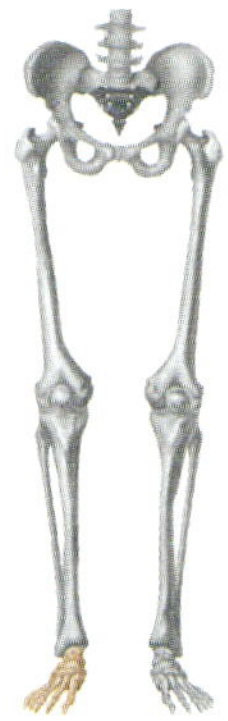

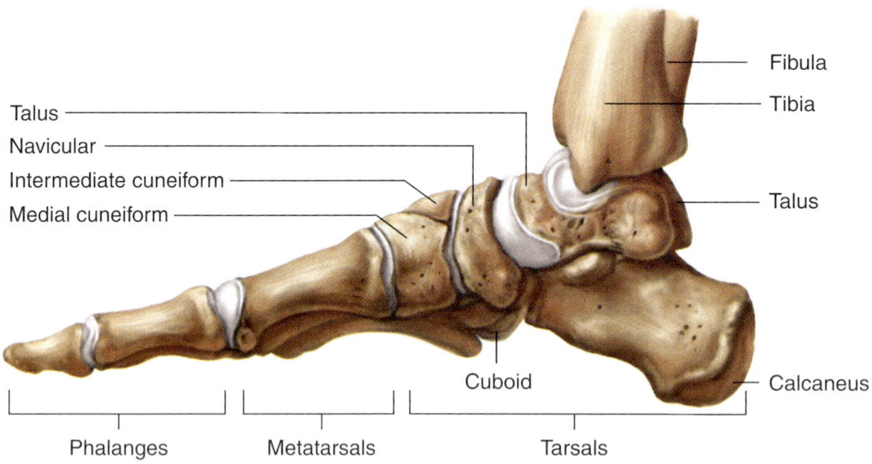

(b) **Medial view**

Bones of the Right Foot
Figure 7.38

Sutures, Superior View
Figure 8.1

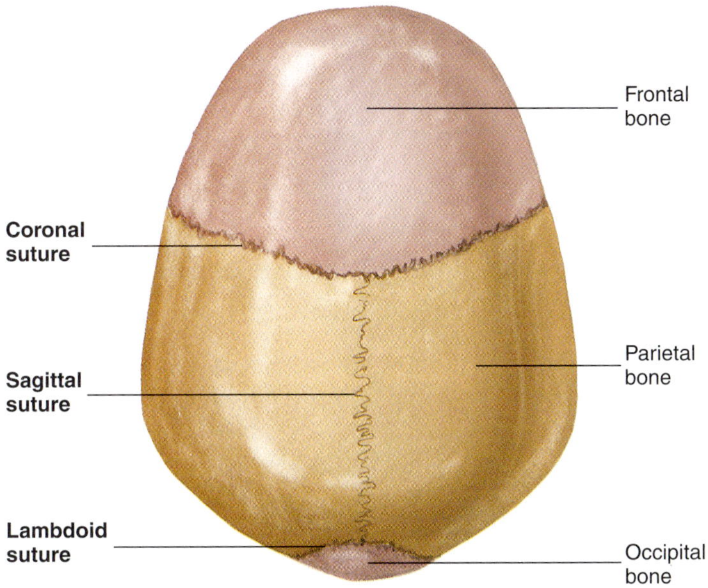

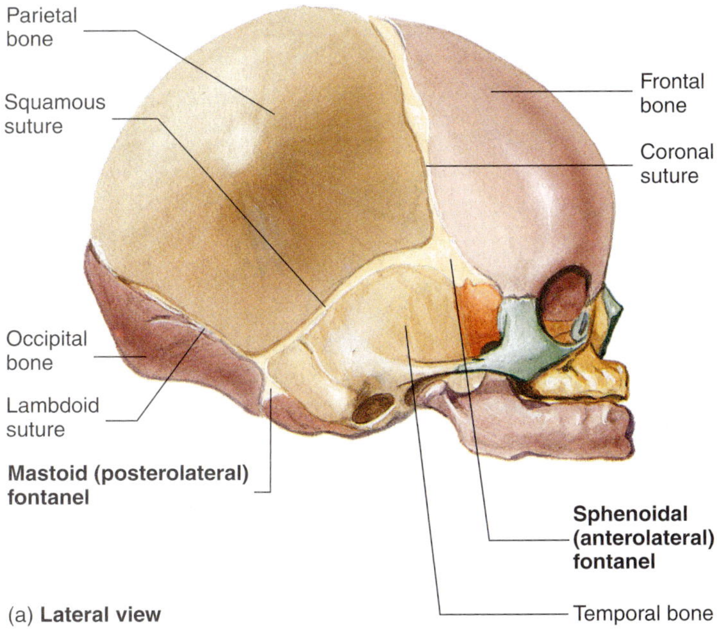

(a) **Lateral view**

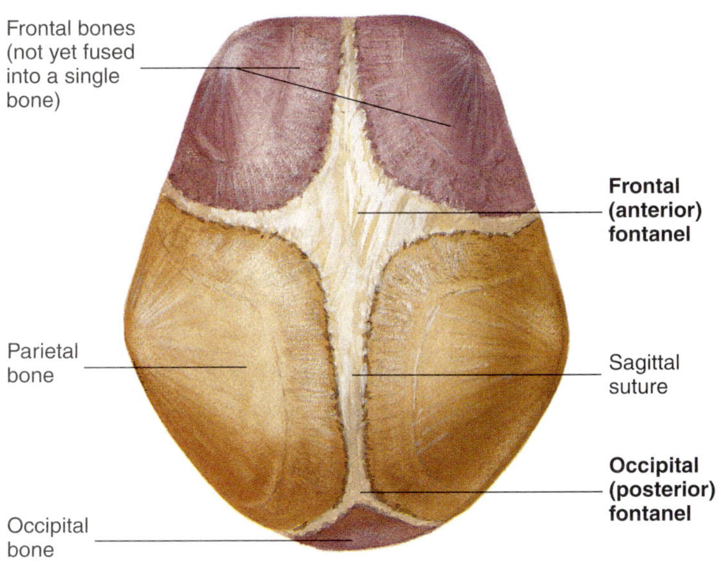

(b) **Superior view**

Fetal Skull Showing Fontanels and Sutures
Figure 8.2

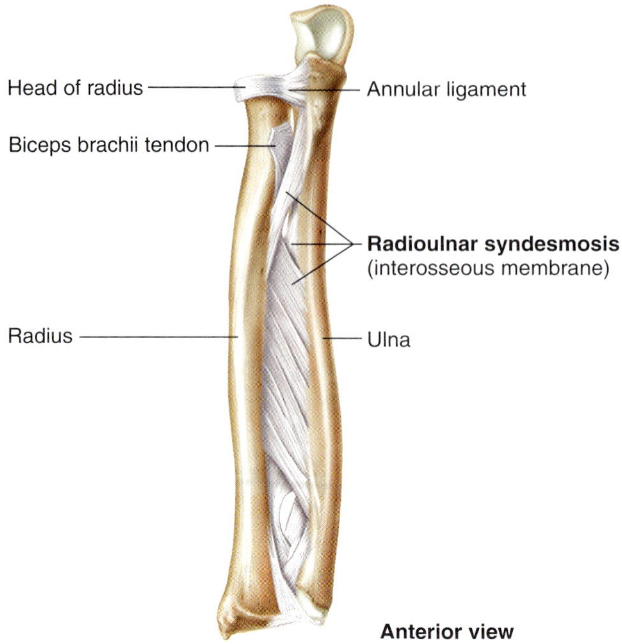

Right Radioulnar Syndesmosis
Figure 8.3

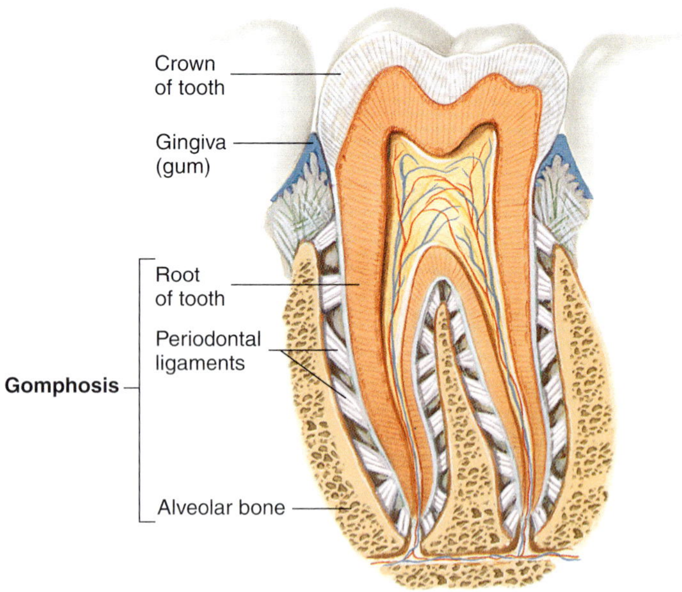

Gomphosis Between a Tooth and Alveolar Bone of the Mandible
Figure 8.4

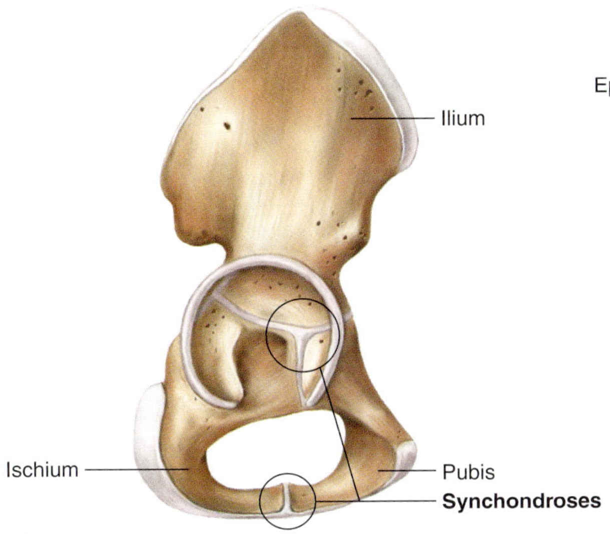

(a) **Lateral view**

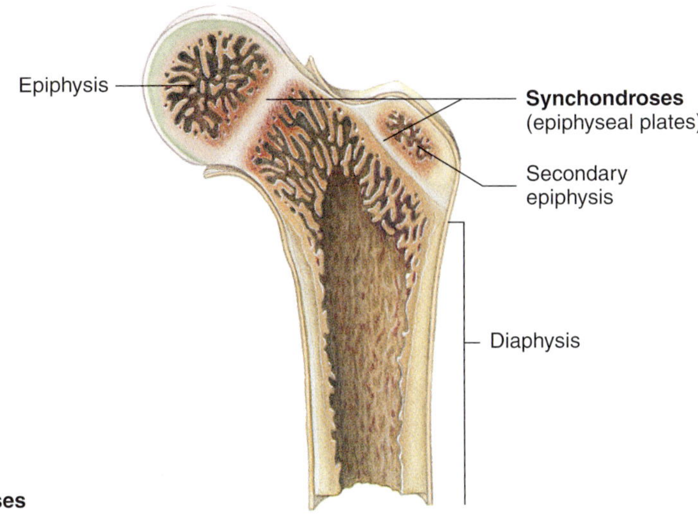

(b)

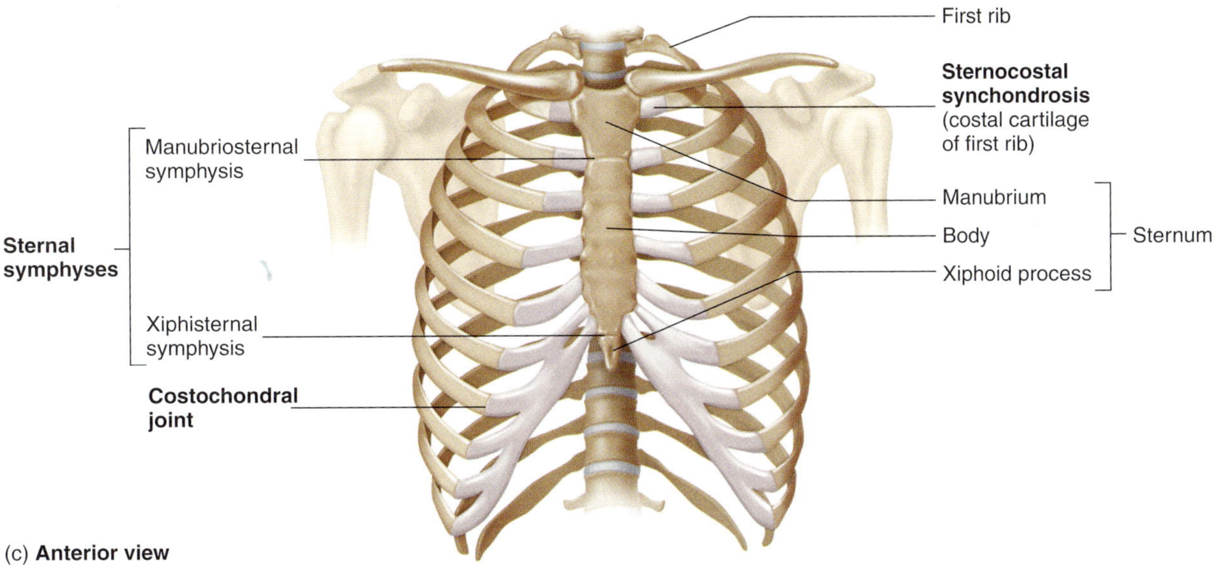

(c) **Anterior view**

Synchondroses
Figure 8.5

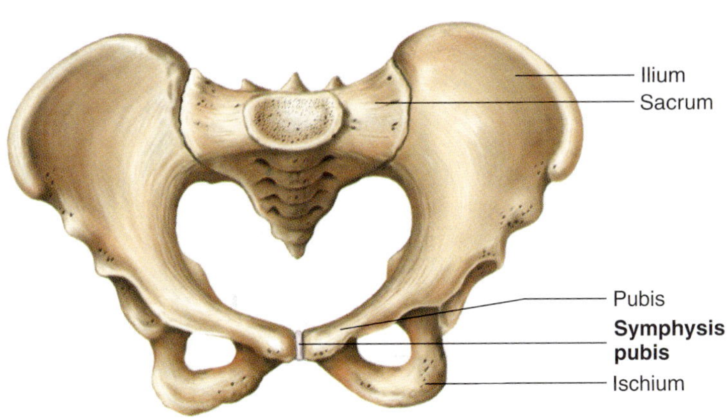

Anterior view

Symphysis Pubis
Figure 8.6

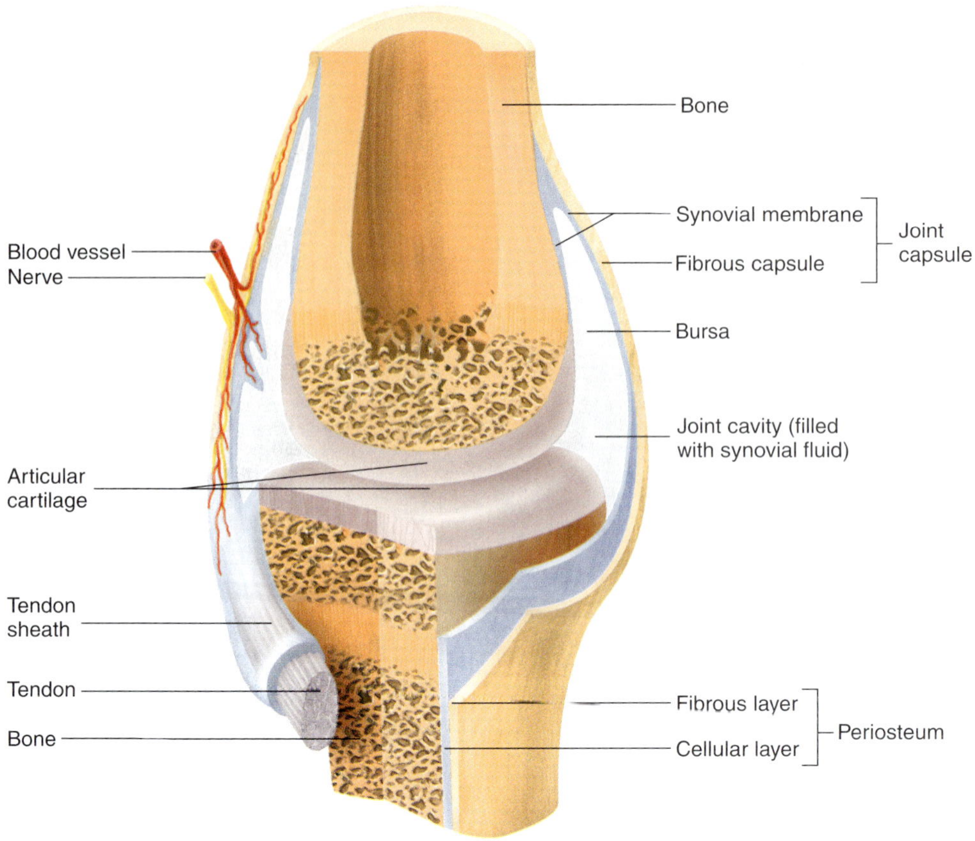

Structure of a Synovial Joint
Figure 8.7

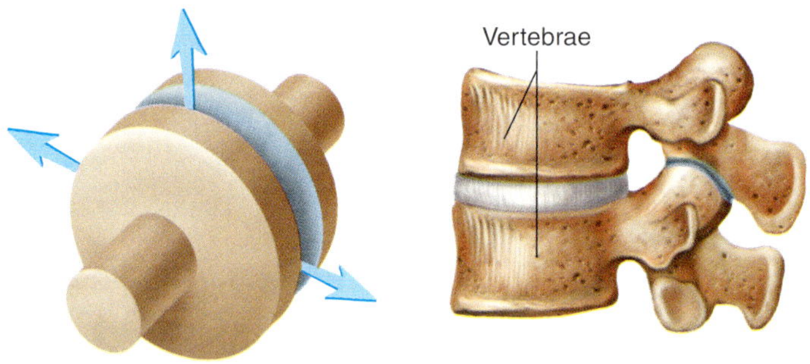

Plane Joint
Figure 8.8

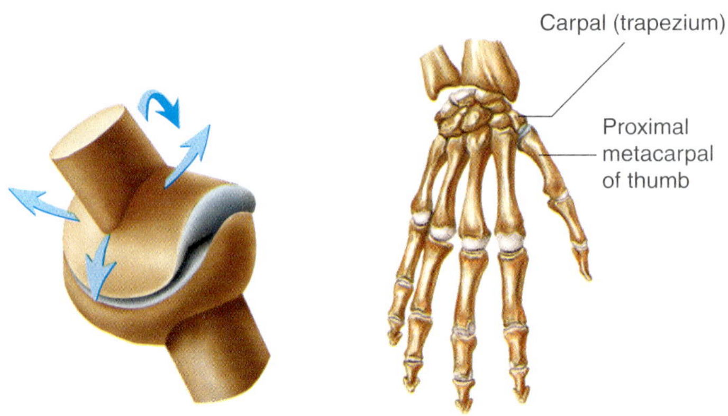

Saddle Joint
Figure 8.9

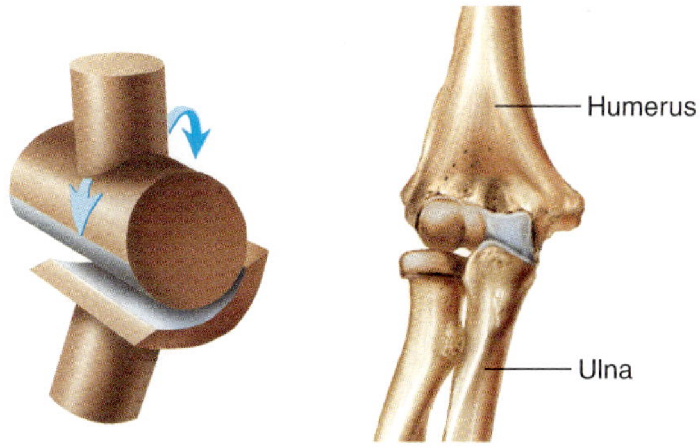

Hinge Joint
Figure 8.10

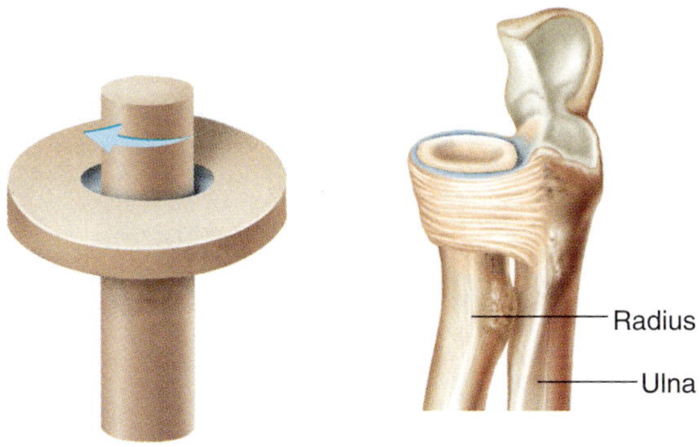

Pivot Joint
Figure 8.11

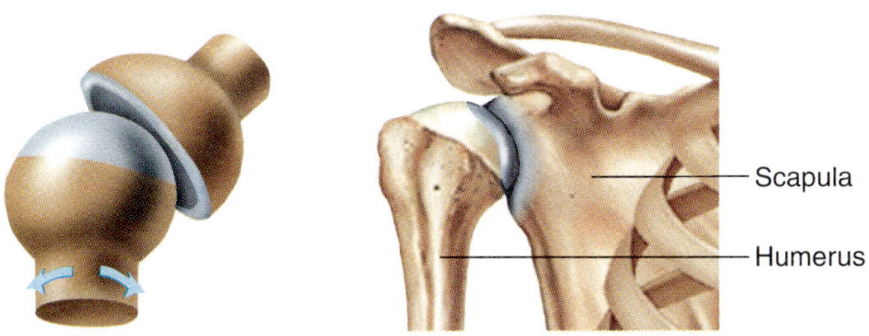

Ball-and-Socket Joint
Figure 8.12

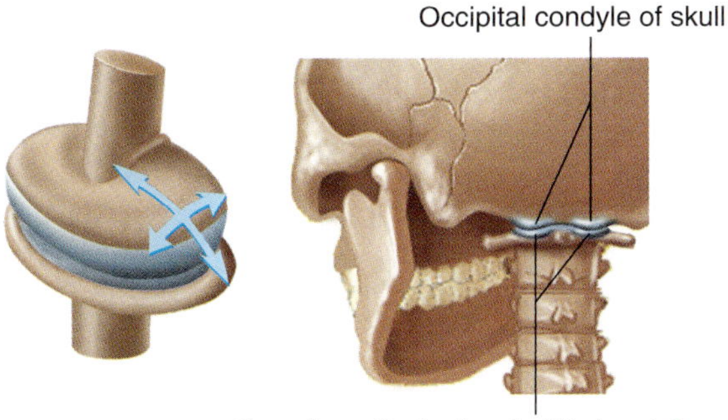

Ellipsoid Joint
Figure 8.13

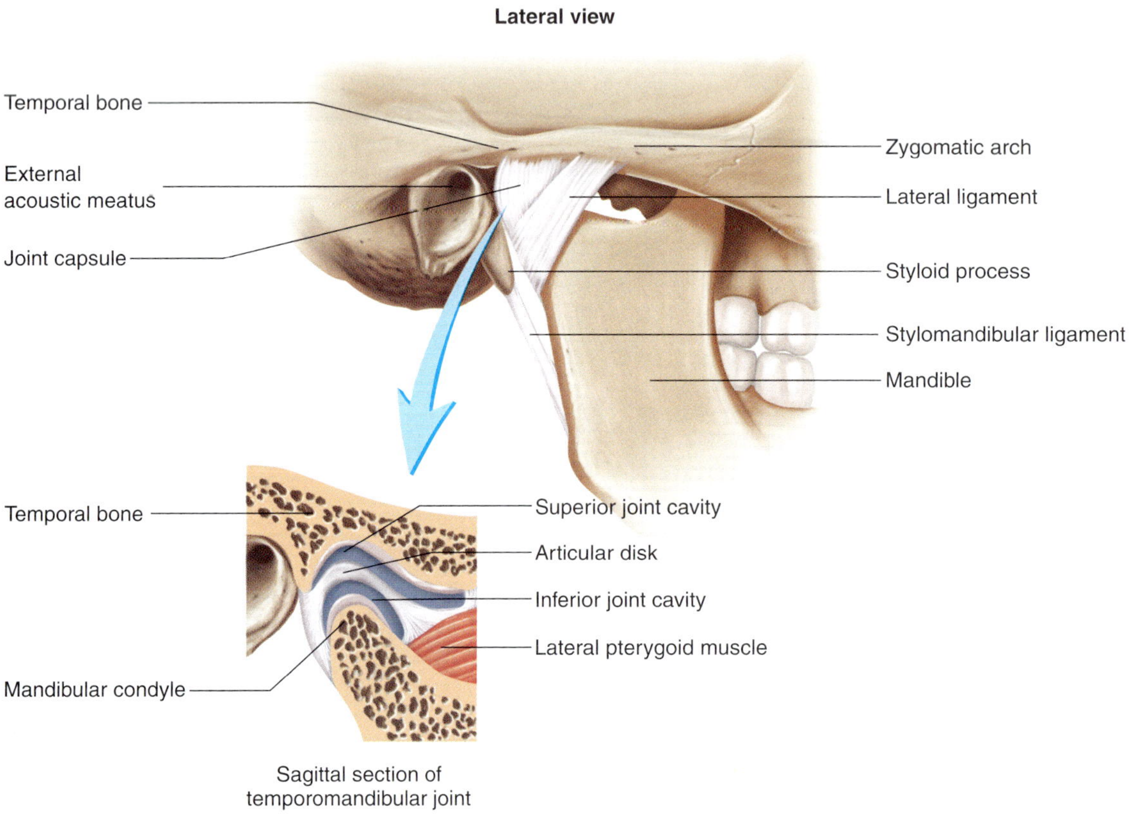

Right Temporomandibular Joint, Lateral View
Figure 8.27

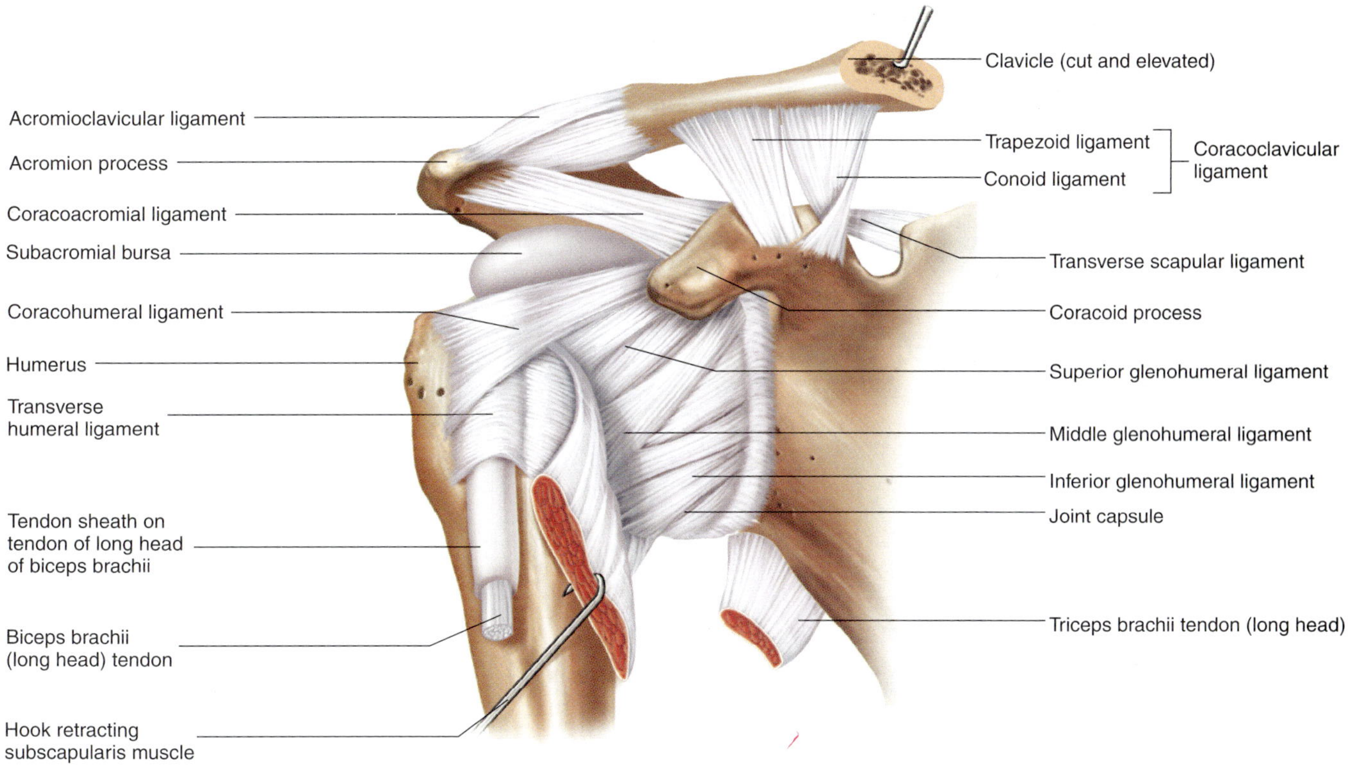

(a) **Anterior view**

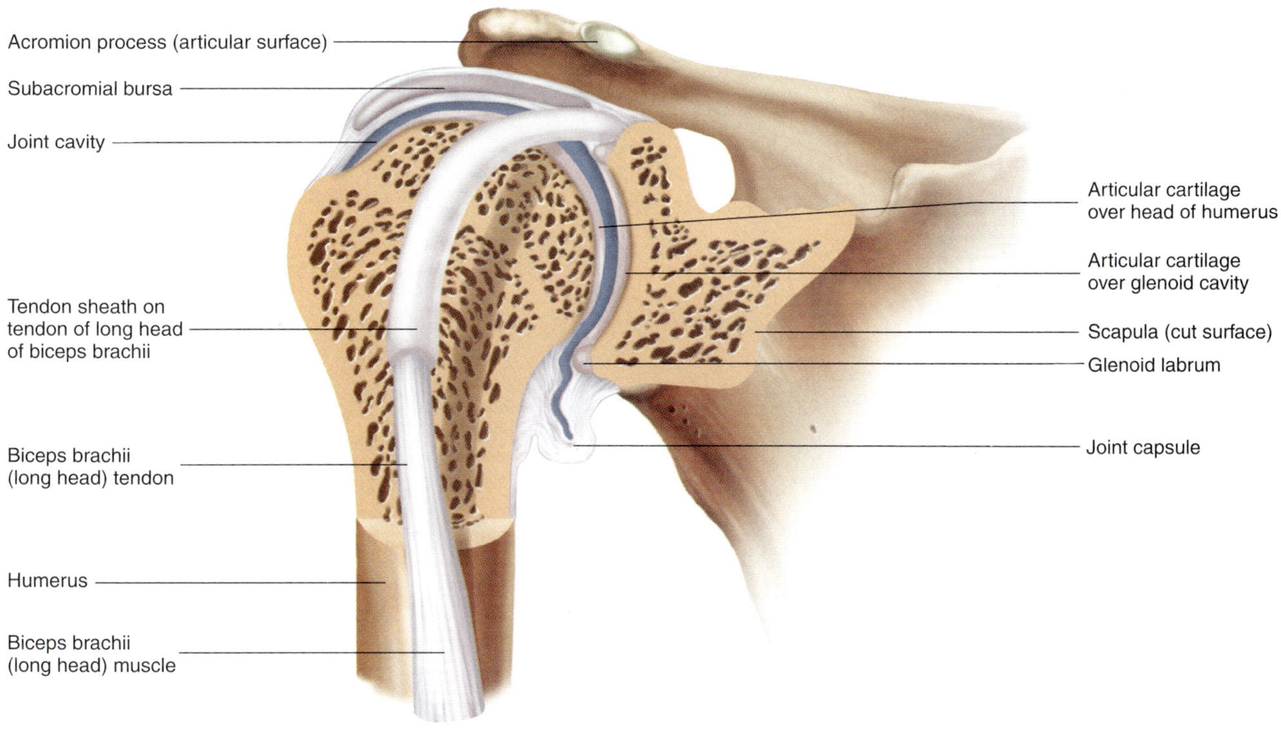

(b) **Frontal section**

Right Shoulder Joint
Figure 8.28

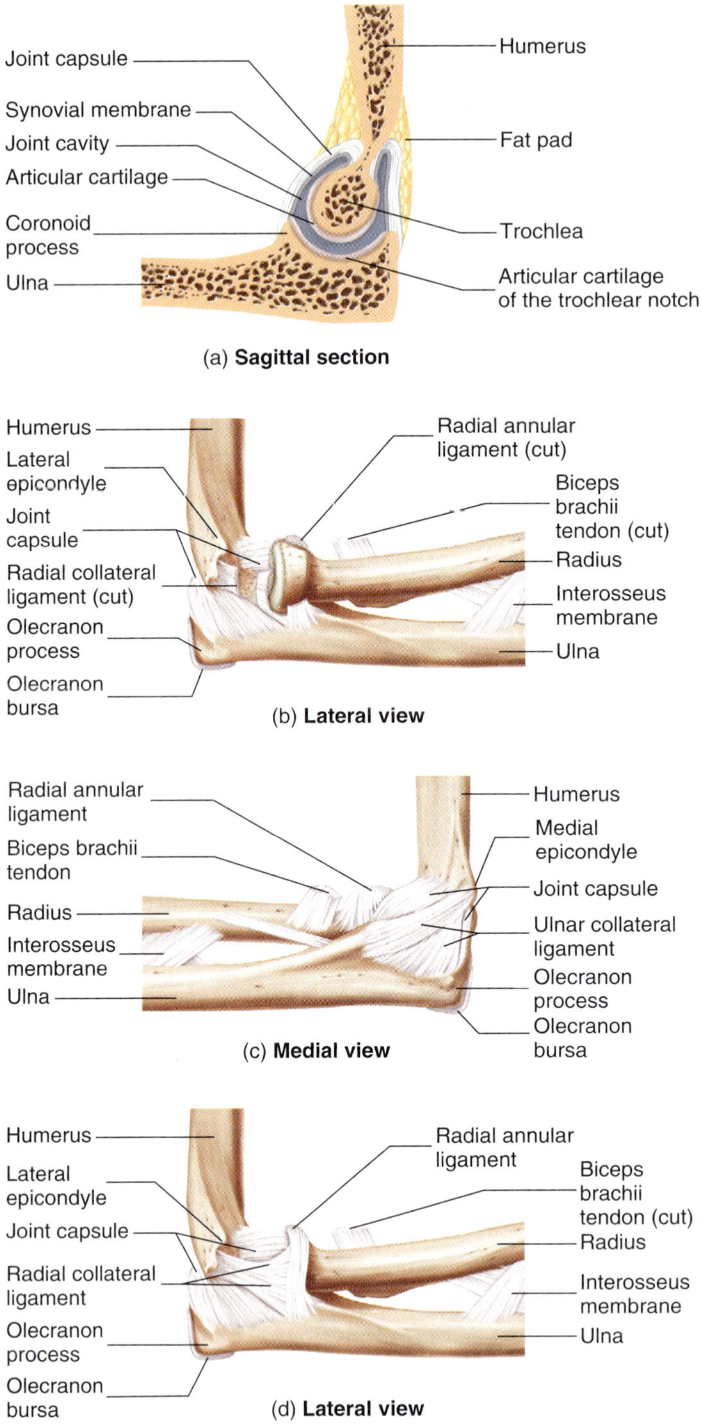
Joint capsule
Synovial membrane
Joint cavity
Articular cartilage
Coronoid process
Ulna
Humerus
Fat pad
Trochlea
Articular cartilage of the trochlear notch
(a) Sagittal section
Humerus
Lateral epicondyle
Joint capsule
Radial collateral ligament (cut)
Olecranon process
Olecranon bursa
Radial annular ligament (cut)
Biceps brachii tendon (cut)
Radius
Interosseus membrane
Ulna
(b) Lateral view
Radial annular ligament
Biceps brachii tendon
Radius
Interosseus membrane
Ulna
Humerus
Medial epicondyle
Joint capsule
Ulnar collateral ligament
Olecranon process
Olecranon bursa
(c) Medial view
Humerus
Lateral epicondyle
Joint capsule
Radial collateral ligament
Olecranon process
Olecranon bursa
Radial annular ligament
Biceps brachii tendon (cut)
Radius
Interosseus membrane
Ulna
(d) Lateral view

Right Elbow Joint
Figure 8.29

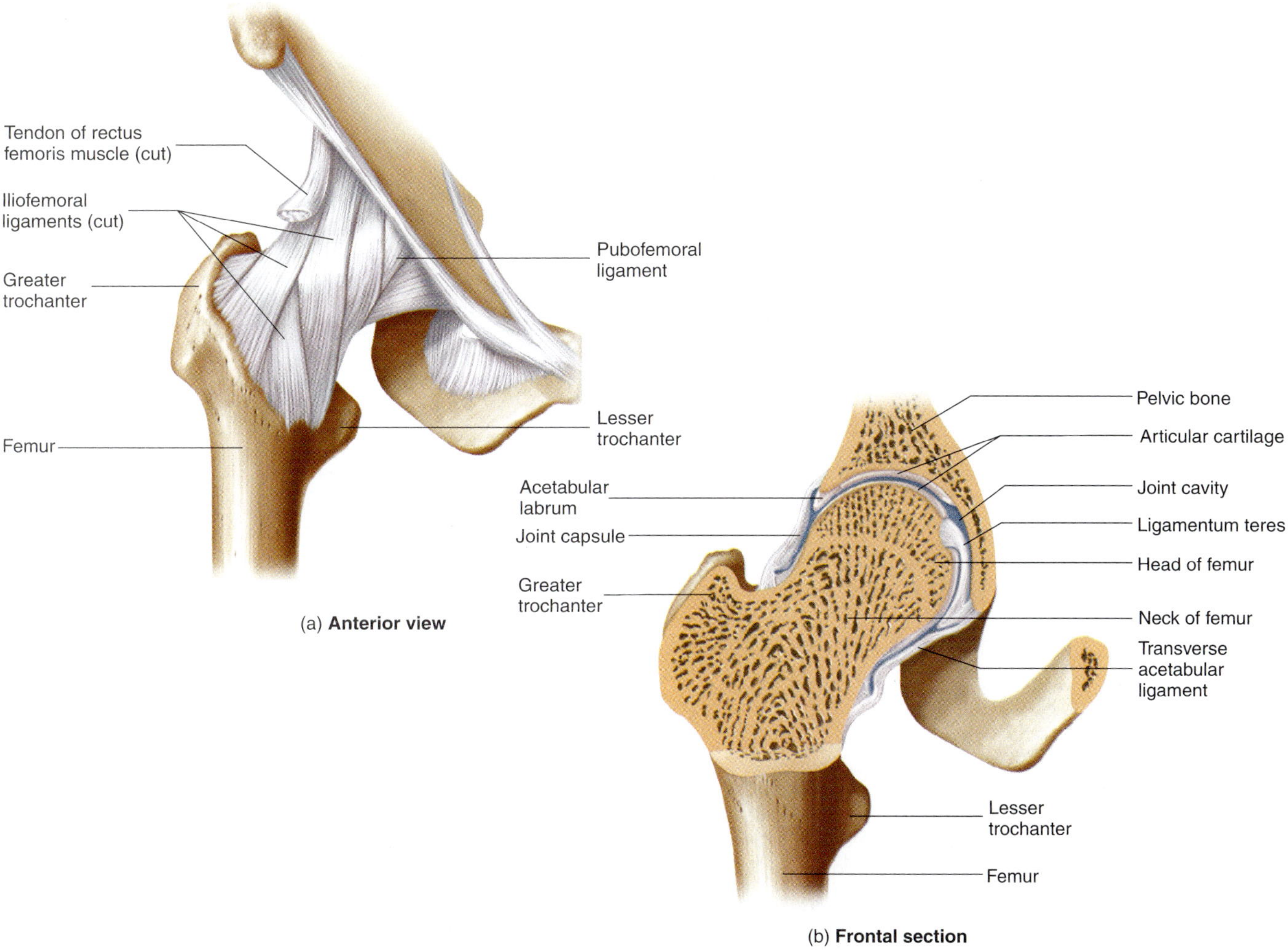
Tendon of rectus femoris muscle (cut)
Iliofemoral ligaments (cut)
Greater trochanter
Femur
Pubofemoral ligament
Lesser trochanter
(a) Anterior view
Pelvic bone
Articular cartilage
Joint cavity
Ligamentum teres
Head of femur
Neck of femur
Transverse acetabular ligament
Acetabular labrum
Joint capsule
Greater trochanter
Lesser trochanter
Femur
(b) Frontal section

Right Hip Joint
Figure 8.30

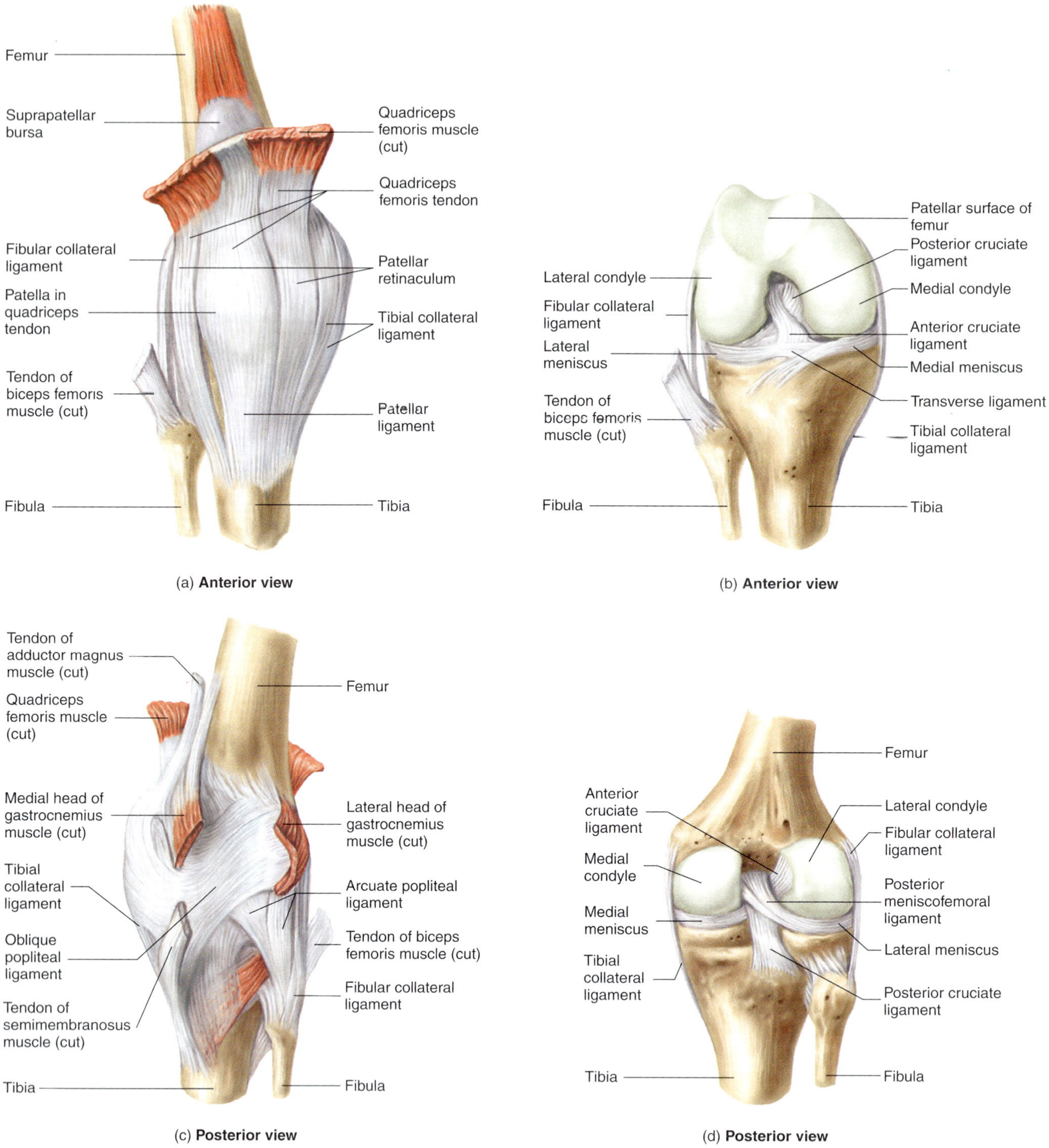
Femur
Suprapatellar bursa
Quadriceps femoris muscle (cut)
Quadriceps femoris tendon
Fibular collateral ligament
Patellar retinaculum
Patella in quadriceps tendon
Tibial collateral ligament
Tendon of biceps femoris muscle (cut)
Patellar ligament
Fibula
Tibia
(a) Anterior view
Patellar surface of femur
Posterior cruciate ligament
Lateral condyle
Medial condyle
Fibular collateral ligament
Anterior cruciate ligament
Lateral meniscus
Medial meniscus
Transverse ligament
Tendon of biceps femoris muscle (cut)
Tibial collateral ligament
Fibula
Tibia
(b) Anterior view
Tendon of adductor magnus muscle (cut)
Femur
Quadriceps femoris muscle (cut)
Medial head of gastrocnemius muscle (cut)
Lateral head of gastrocnemius muscle (cut)
Tibial collateral ligament
Arcuate popliteal ligament
Oblique popliteal ligament
Tendon of biceps femoris muscle (cut)
Fibular collateral ligament
Tendon of semimembranosus muscle (cut)
Tibia
Fibula
(c) Posterior view
Femur
Anterior cruciate ligament
Lateral condyle
Fibular collateral ligament
Medial condyle
Posterior meniscofemoral ligament
Medial meniscus
Lateral meniscus
Tibial collateral ligament
Posterior cruciate ligament
Tibia
Fibula
(d) Posterior view

Right Knee Joint
Figure 8.31

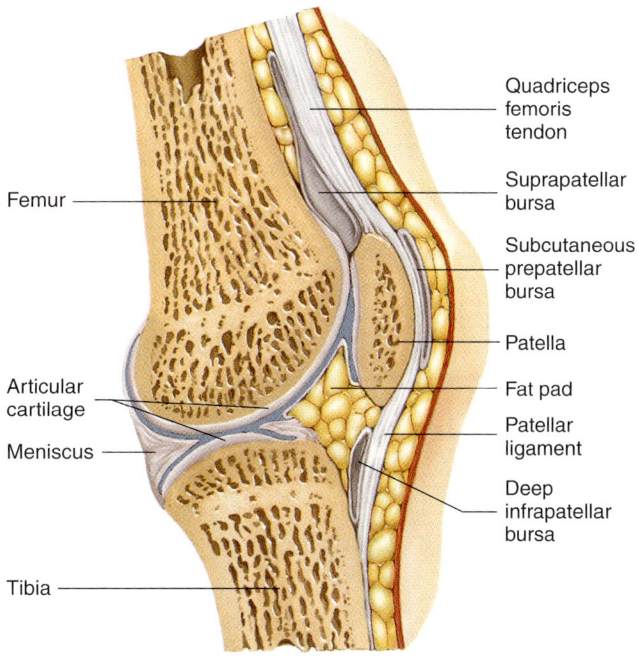

(f) **Sagittal section**

Figure 8.31—*continued*

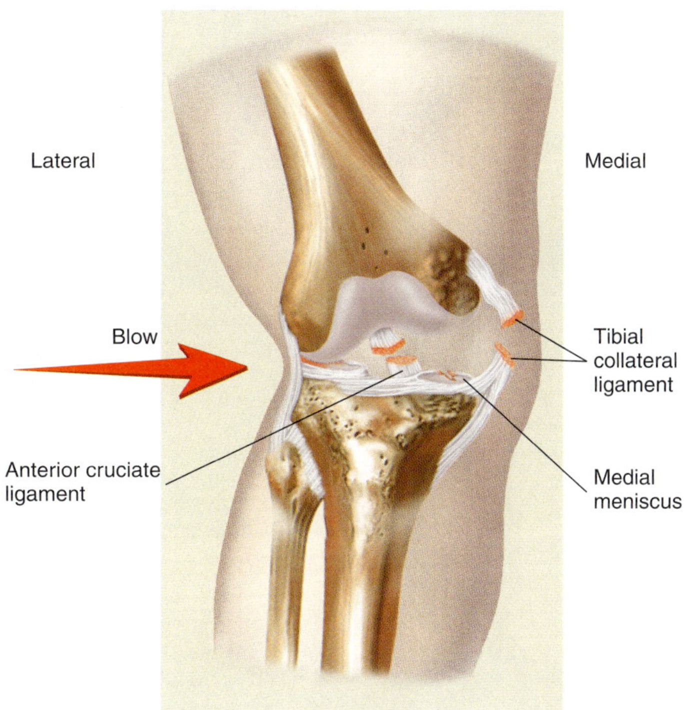

Anterior view

Injury to the Right Knee
Figure A

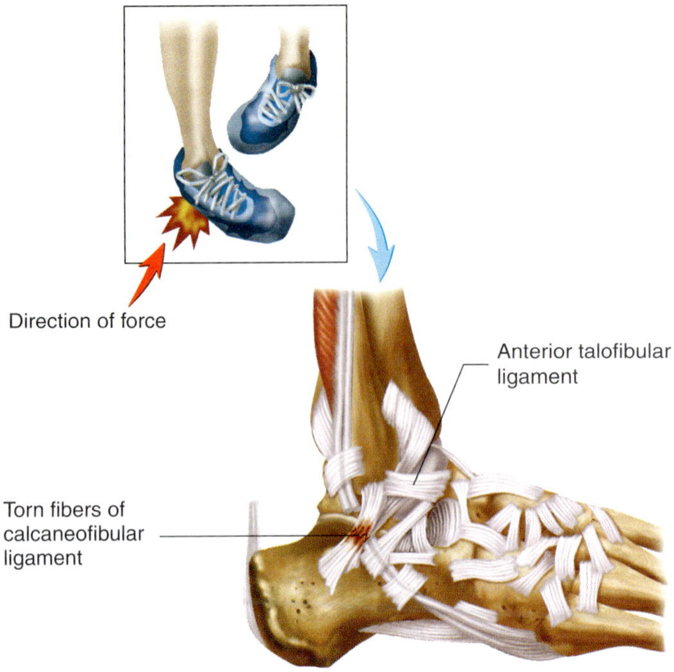

Lateral view

Figure B

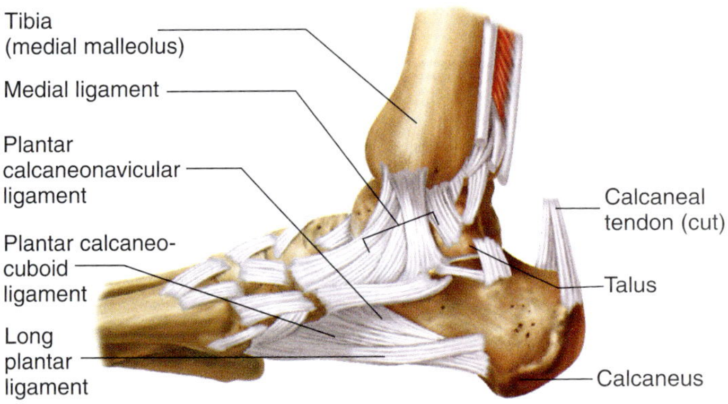

(a) **Medial view**

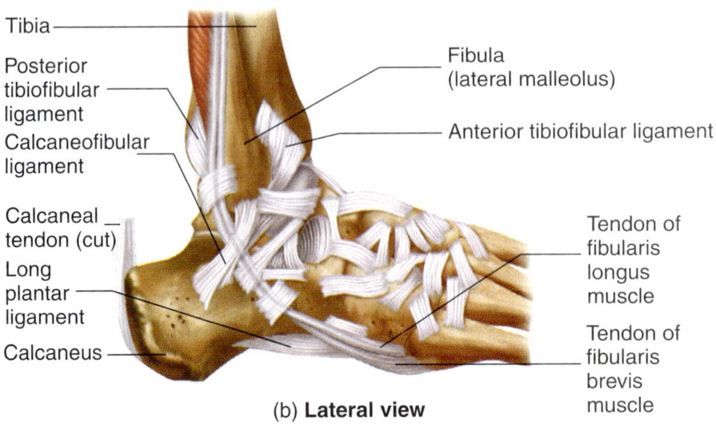

(b) **Lateral view**

Ligaments of the Right Ankle Joint
Figure 8.32

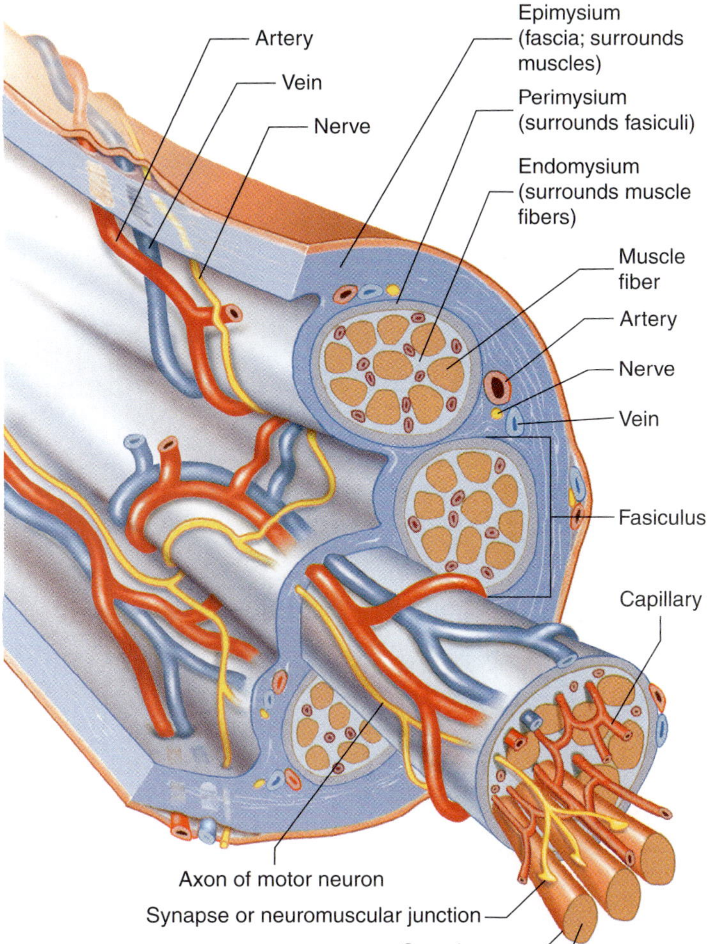

Skeletal Muscle Structure: Connective Tissue, Innervation, and Blood Supply
Figure 9.2

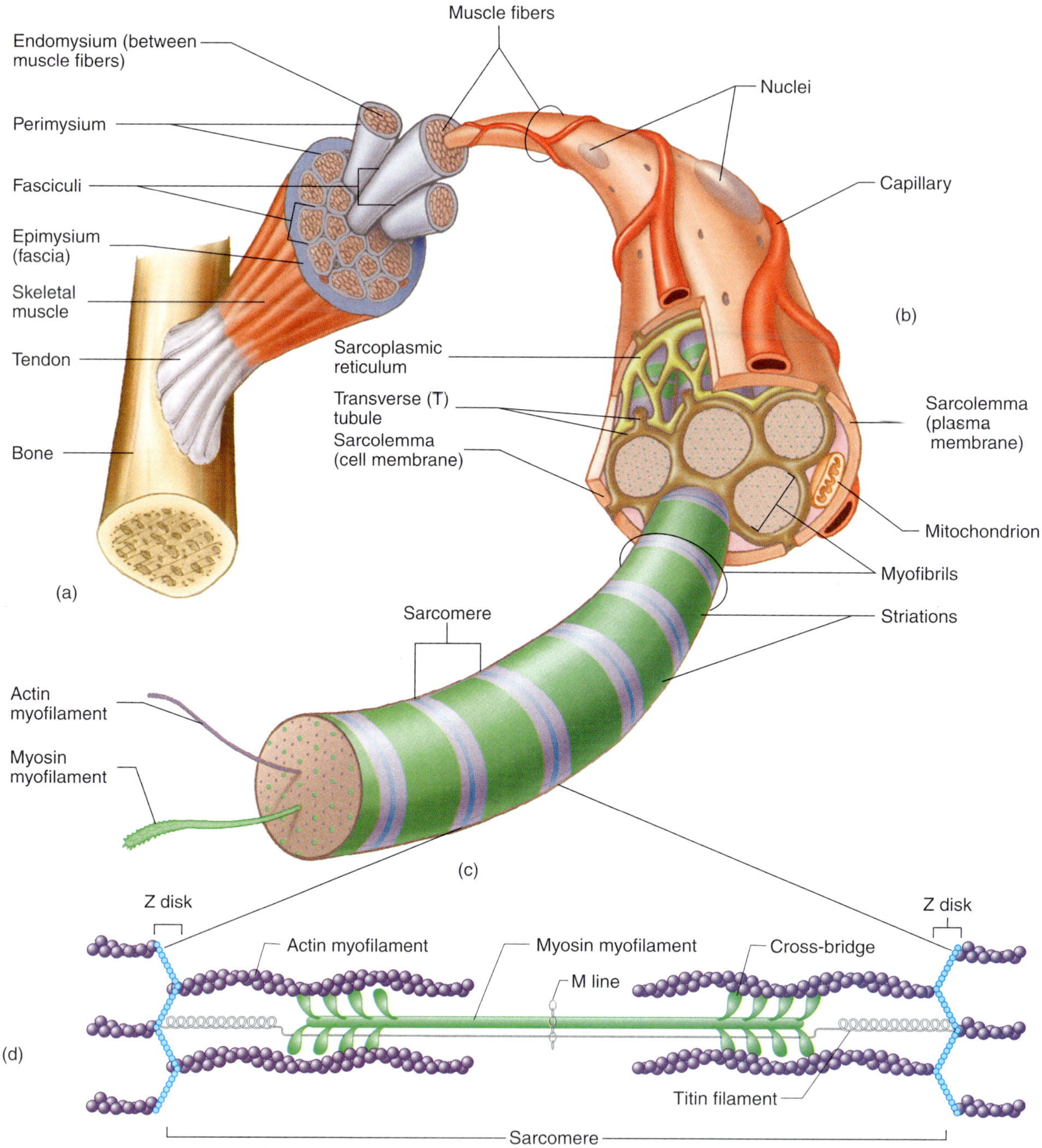

Muscle fibers
Endomysium (between muscle fibers)
Perimysium
Fasciculi
Epimysium (fascia)
Skeletal muscle
Tendon
Bone
(a)
Nuclei
Capillary
(b)
Sarcoplasmic reticulum
Transverse (T) tubule
Sarcolemma (cell membrane)
Sarcolemma (plasma membrane)
Mitochondrion
Myofibrils
Striations
Sarcomere
Actin myofilament
Myosin myofilament
(c)
Z disk
Actin myofilament
Myosin myofilament
Cross-bridge
M line
Z disk
(d)
Titin filament
Sarcomere

Parts of a Muscle
Figure 9.3

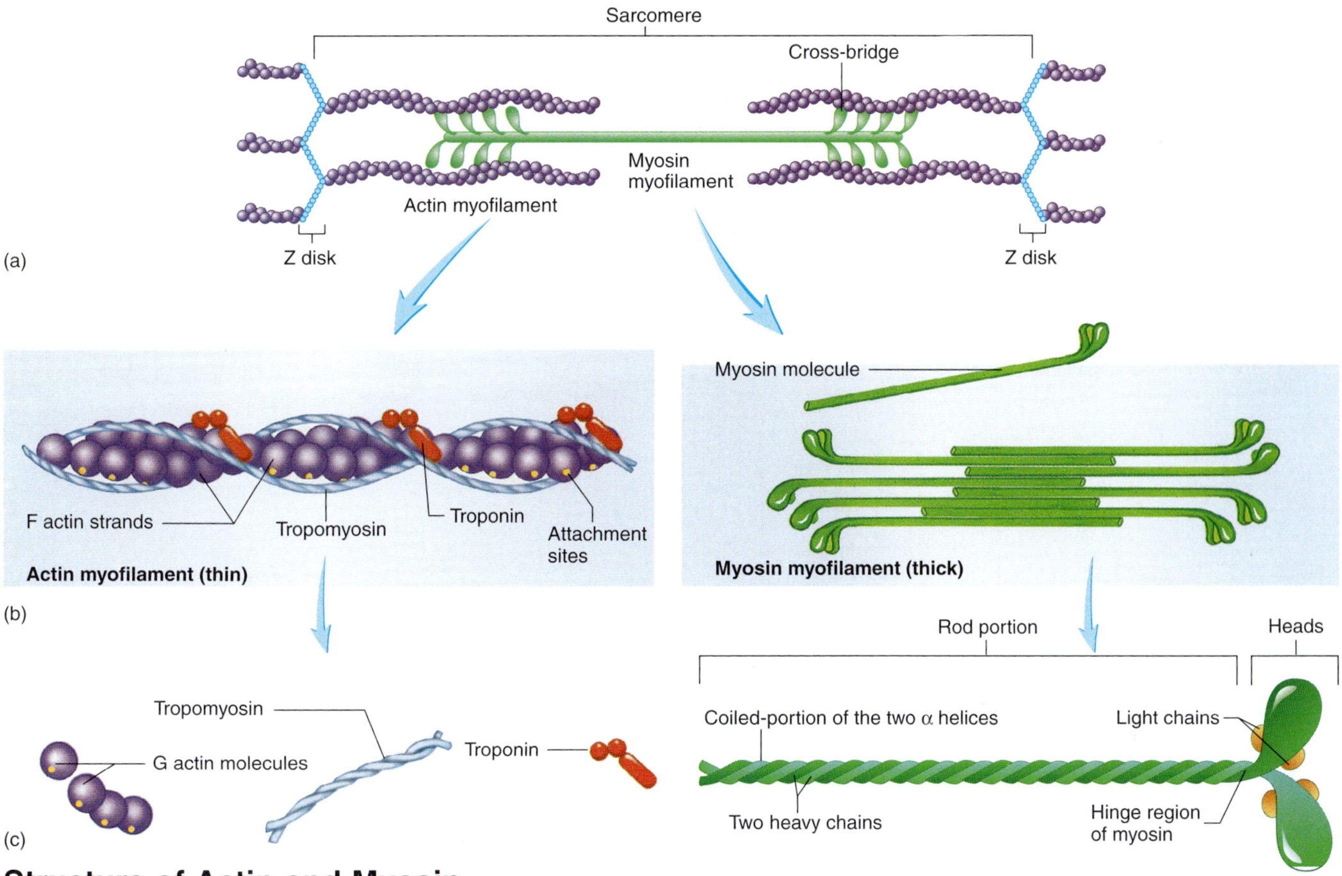

Structure of Actin and Myosin
Figure 9.4

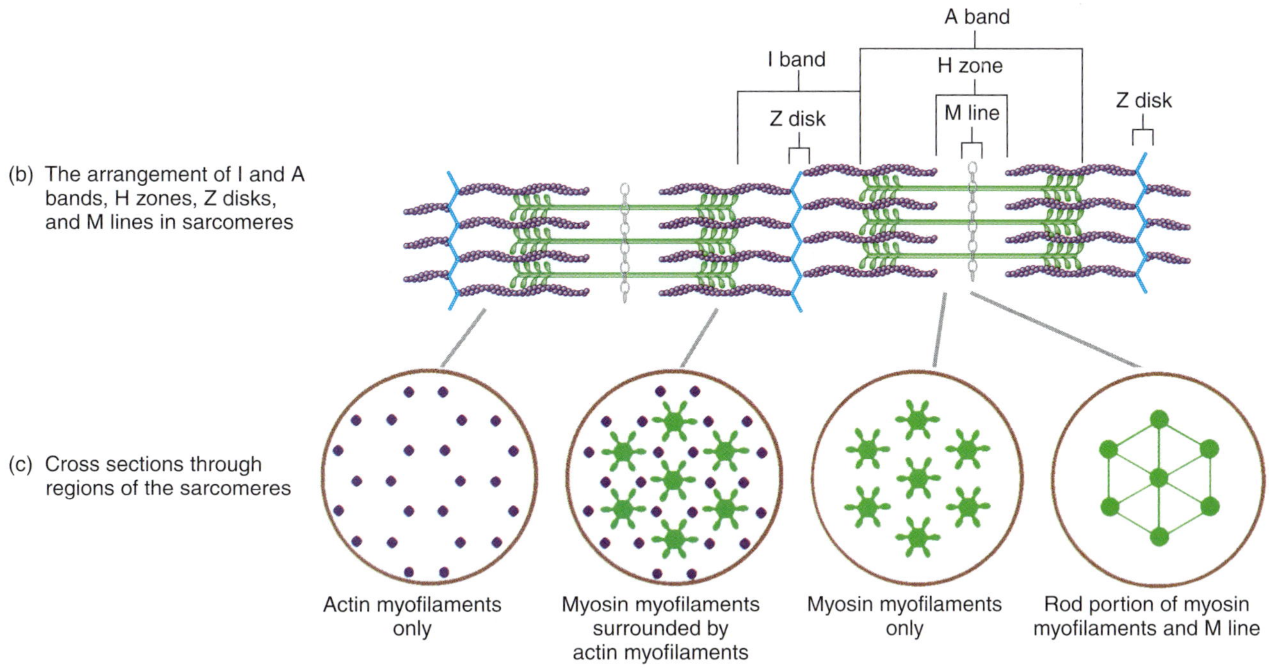

Components of Sarcomeres
Figure 9.5

1. Actin and myosin myofilaments in a relaxed muscle (*right*) and a contracted muscle (*#4 below*) are the same length. Myofilaments do not change length during muscle contraction.

2. During contraction, actin myofilaments at each end of the sarcomere slide past the myosin myofilaments toward each other. As a result, the Z disks are brought closer together, and the sarcomere shortens.

3. As the actin myofilaments slide over the myosin myofilaments, the H zones (*yellow*) and the I bands (*blue*) narrow. The A bands, which are equal to the length of the myosin myofilaments, do not narrow, because the length of the myosin myofilaments does not change.

4. In a fully contracted muscle, the ends of the actin myofilaments overlap and the H zone disappears.

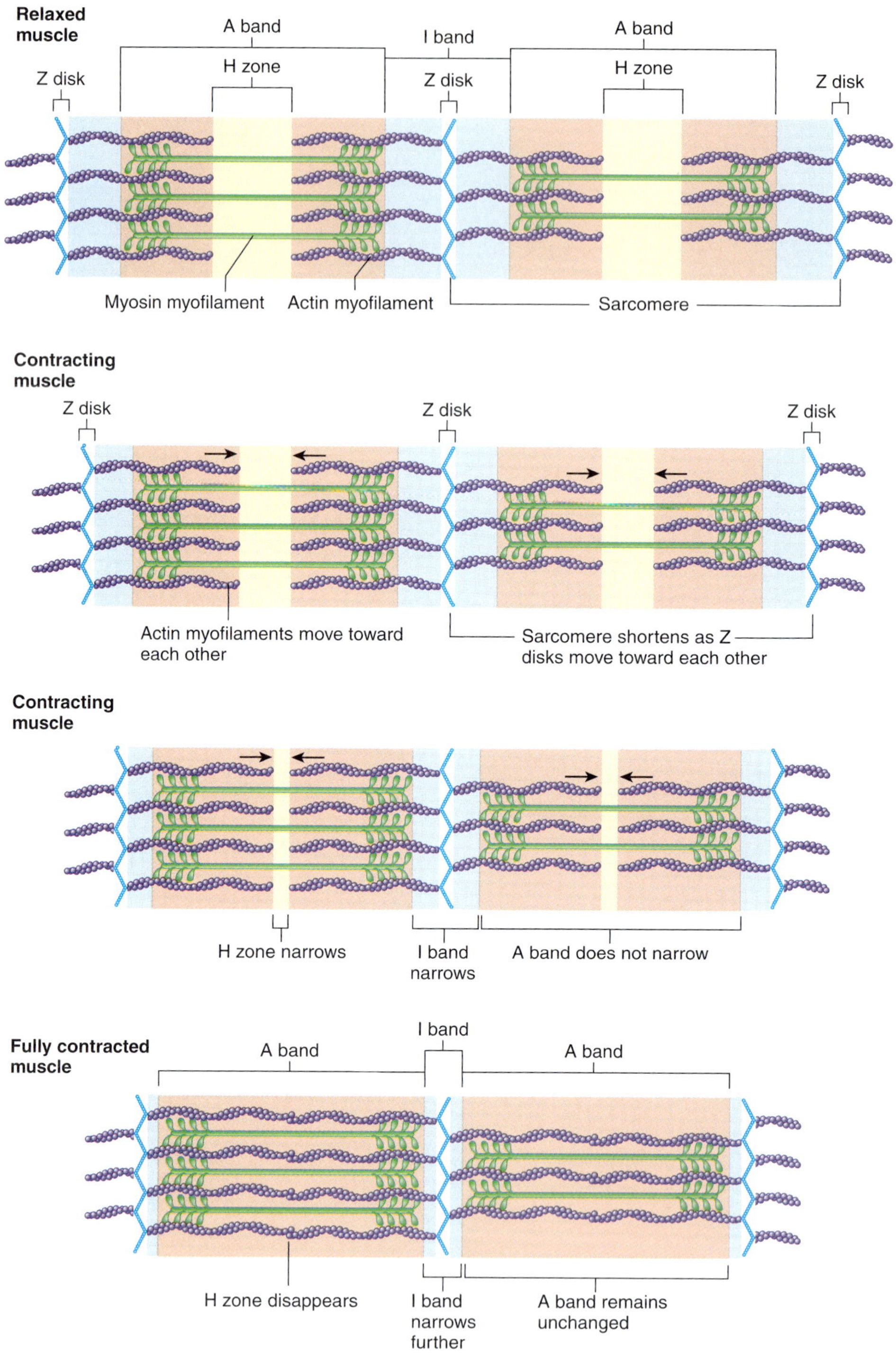

Sarcomere Shortening
Process Figure 9.6

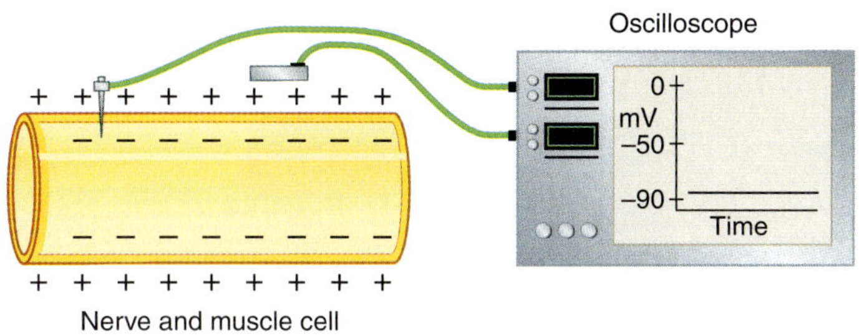

Measuring the Resting Membrane Potential
Figure 9.7

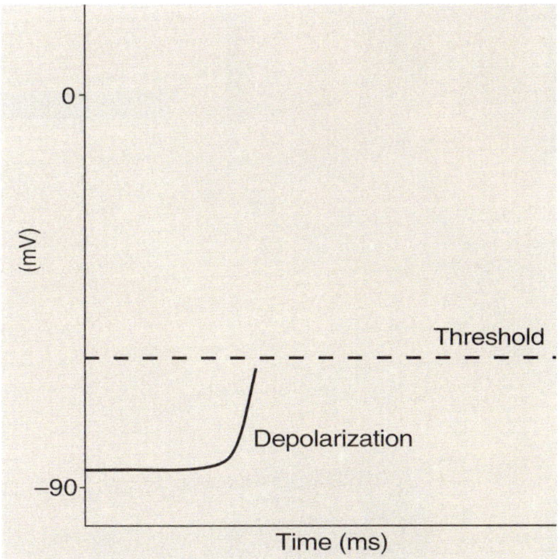

(a) Depolarization is a change of the charge difference across the plasma membrane, making the charge inside of the cell less negative and the outside of the plasma membrane less positive. Once threshold is reached an action potential is produced.

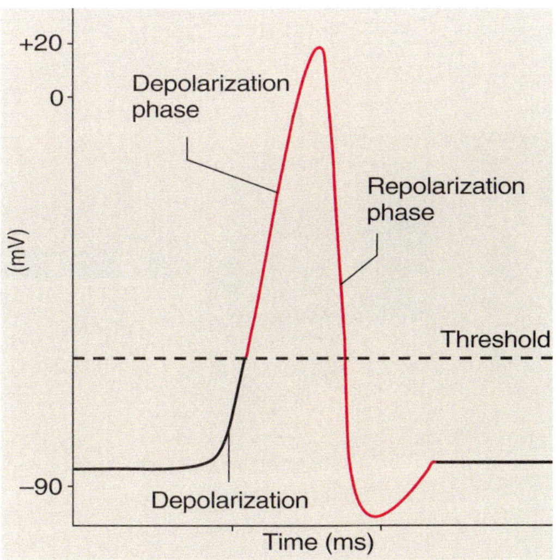

(b) During the depolarization phase of the action potential the membrane potential changes from approximately –85 mV to approximately +20 mV. During the repolarization phase, the inside of the plasma membrane changes from approximately +20 mV to –85 mV.

Depolarization and the Action Potential
Figure 9.8

1. **Resting membrane potential.** In addition to some non-gated K^+ channels (not shown), which are open, voltage-gated Na^+ channels (*pink*) and voltage-gated K^+ channels (*purple*) are closed. The outside of the plasma membrane is positively charged compared to the inside.

2. **Depolarization.** Voltage-gated Na^+ channels are open. Depolarization results because the inward movement of Na^+ makes the inside of the membrane more positive.

3. **Repolarization.** Voltage-gated Na^+ channels are closed and voltage-gated K^+ channels are open. Na^+ movement into the cell stops and K^+ movement out of the cell increases, causing repolarization.

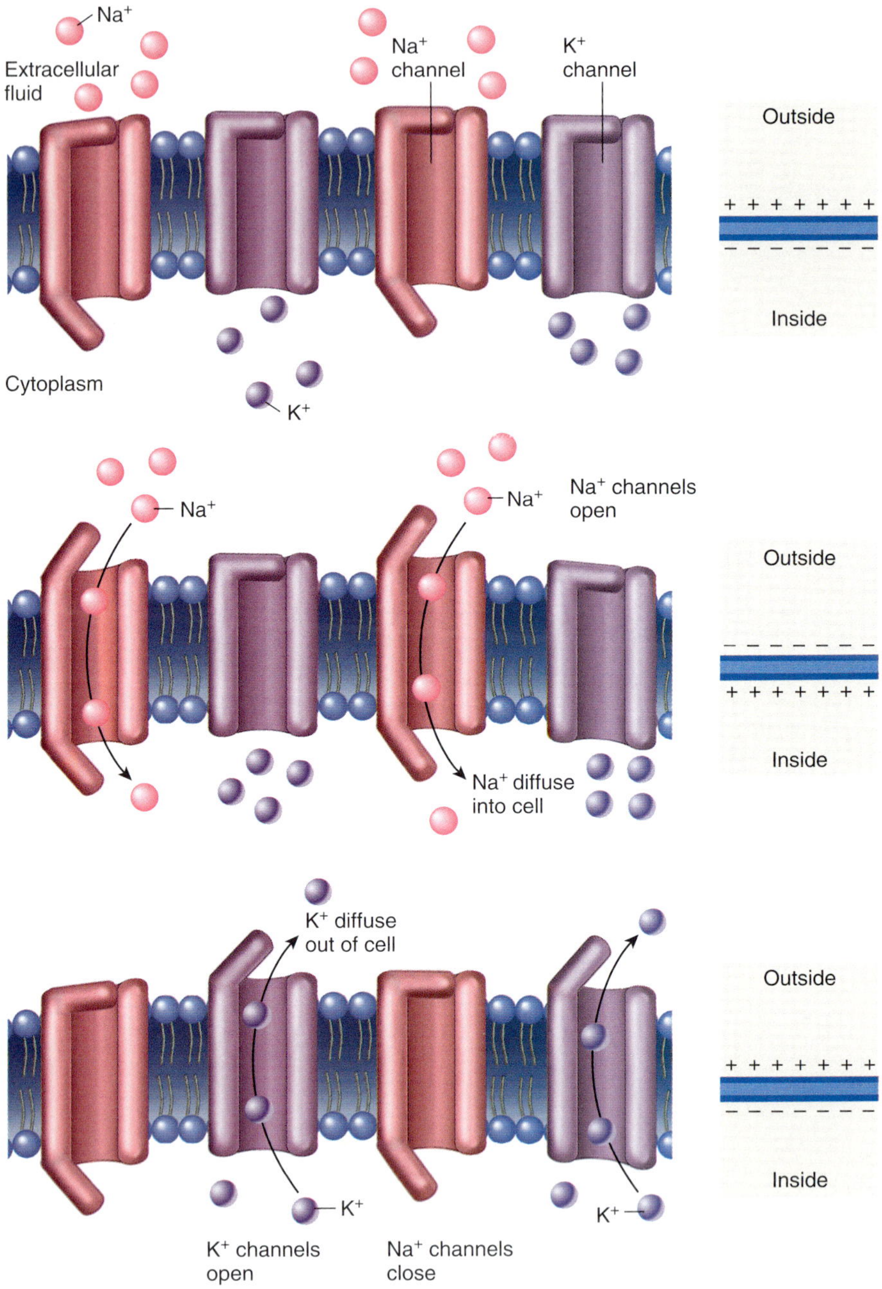

Gated Ion Channels and the Action Potential
Process Figure 9.9

1. An action potential in a local area of the plasma membrane is indicated by the *orange band*. Note the reversal of charge across the membrane.
2. The action potential is a stimulus that causes the production of another action potential in the adjacent plasma membrane.
3. The action potential propagates along the plasma membrane (indicated by *orange arrow*).

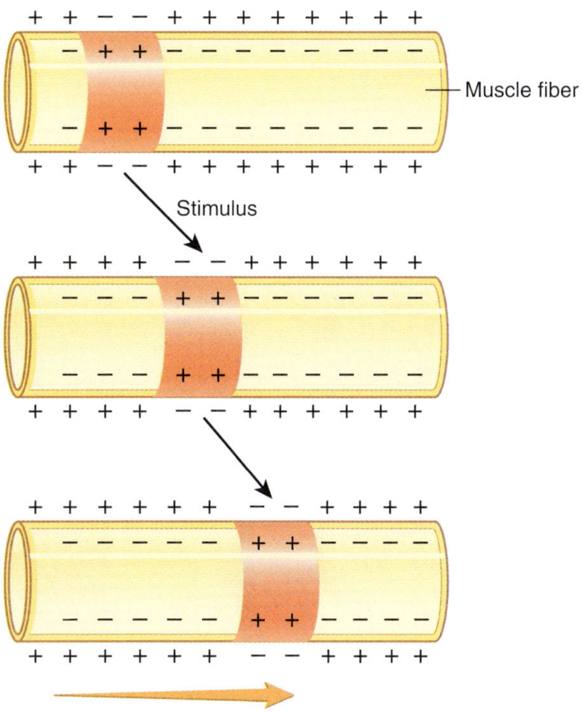

Action Potential Propagation in a Muscle Fiber
Process Figure 9.10

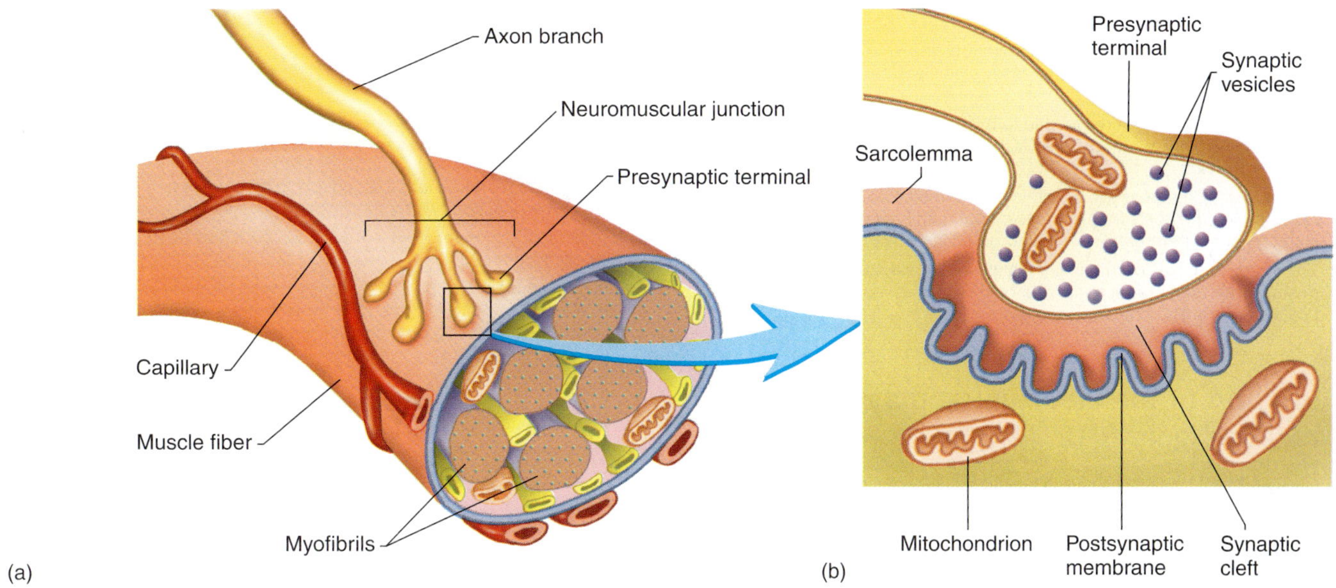

Neuromuscular Junction
Figure 9.11

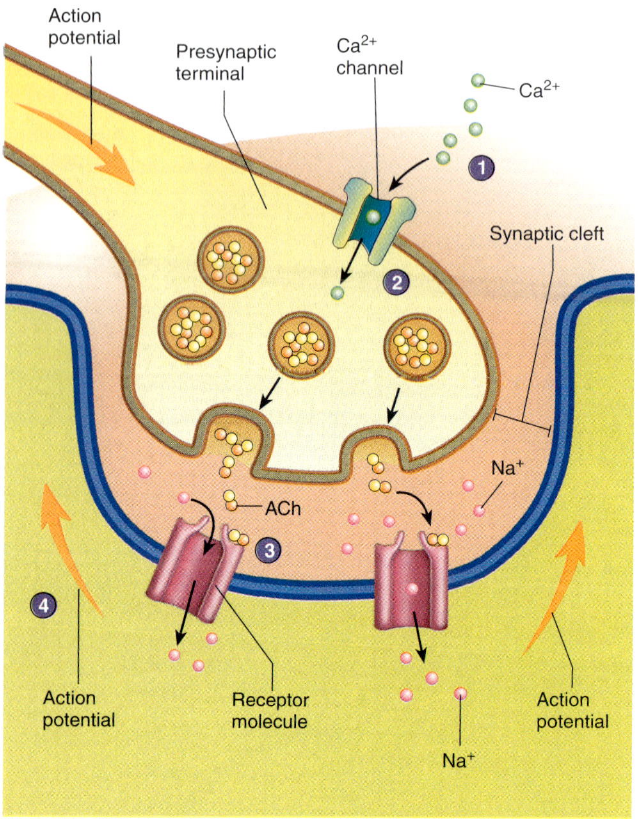

1. An action potential arrives at the presynaptic terminal causing voltage gated Ca^{2+} channels to open, increasing the Ca^{2+} permeability of the presynaptic terminal.
2. Calcium ions enter the presynaptic terminal and initiates the release of a neurotransmitter, acetylcholine (ACh), from synaptic vesicles in the presynaptic cleft.
3. Diffusion of ACh across the synaptic cleft and binding of ACh to ACh receptors on the postsynaptic muscle fiber membrane causes an increase in the permeability of ligand-gated Na^+ channels.
4. The increase in Na^+ permeability results in depolarization of the postsynaptic membrane; once threshold has been reached a postsynaptic action potential results.

(a)

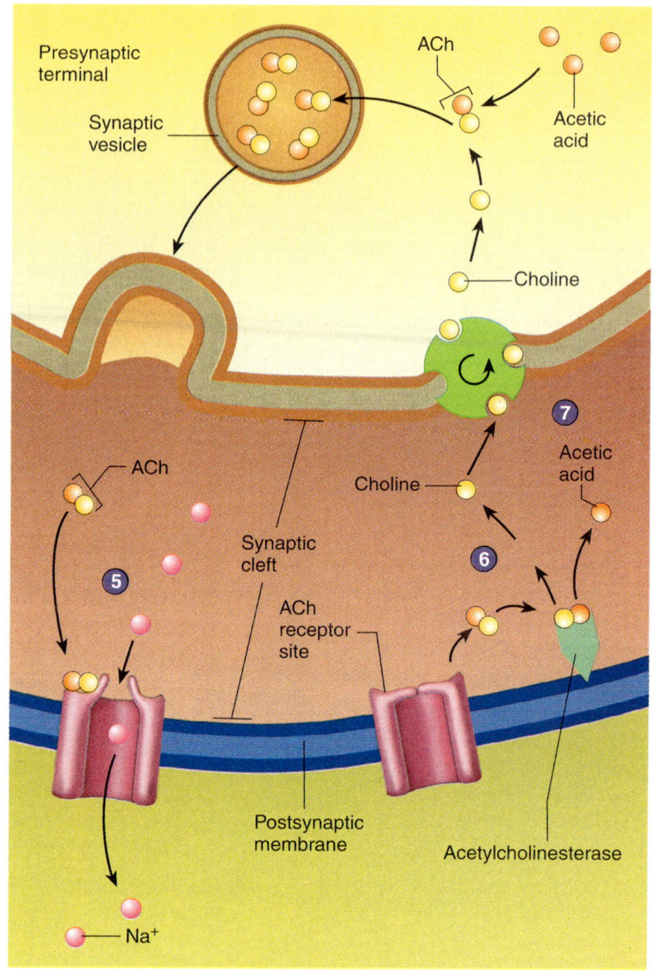

5. Once ACh is released into the synaptic cleft it binds to the receptors for ACh on the postsynaptic membrane and causes Na^+ channels to open.
6. ACh is rapidly broken down in the synaptic cleft by acetylcholinesterase to acetic acid and choline.
7. The choline is reabsorbed by the presynaptic terminal and combined with acetic acid to form more ACh, which enters synaptic vesicles. Acetic acid is taken up by many cell types.

(b)

Function of the Neuromuscular Junction
Process Figure 9.12

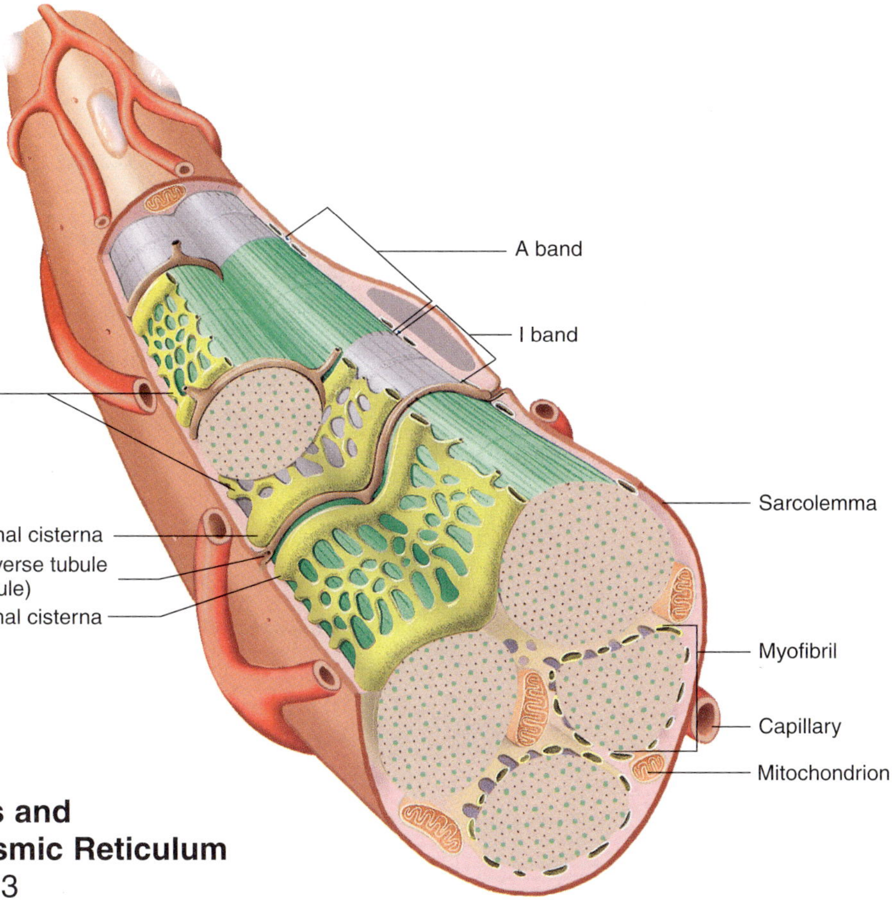

T Tubules and Sarcoplasmic Reticulum
Figure 9.13

1. An action potential produced at the neuromuscular junction is propagated along the sarcolemma of the skeletal muscle. The depolarization also spreads along the membrane of the T tubules.

2. The depolarization of the T tubule causes gated Ca^{2+} channels in the sarcoplasmic reticulum to open, resulting in an increase in the permeability of the sarcoplasmic reticulum to Ca^{2+}, especially in the terminal cisternae. Calcium ions then diffuse from the sarcoplasmic reticulum into the sarcoplasm.

3. Ca^{2+} released from the sarcoplasmic reticulum bind to troponin molecules in the actin myofilament. Consequently, the troponin molecules bound to G actin molecules are released. This causes tropomyosin molecules to move, thereby exposing active sites on the G actin molecules.

4. Once active sites on G actin molecules are exposed, the heads of the myosin myofilaments bind to them to form cross-bridges.

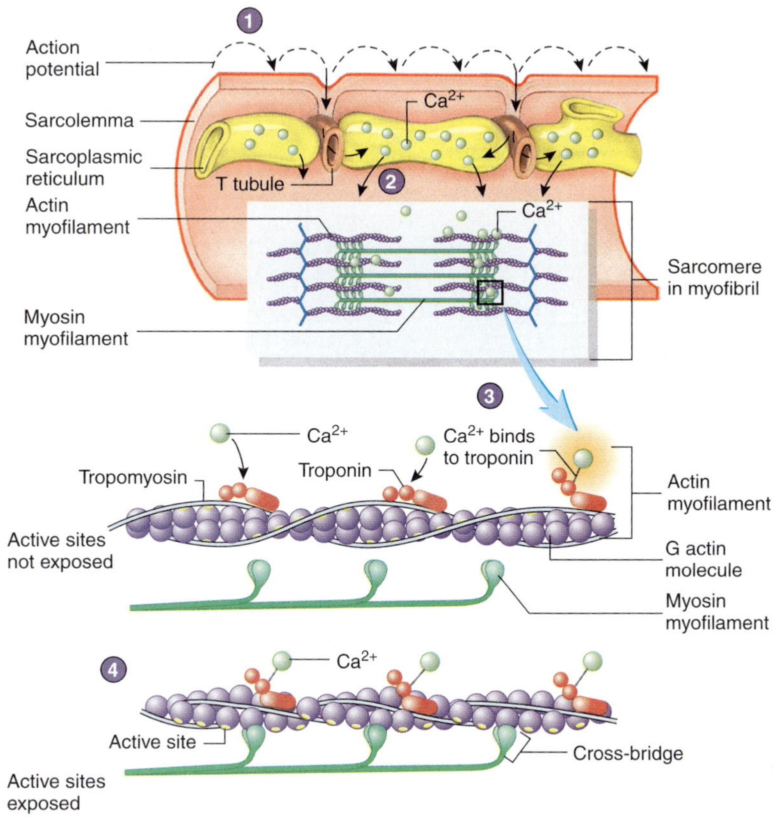

Action Potentials and Muscle Contraction
Process Figure 9.14

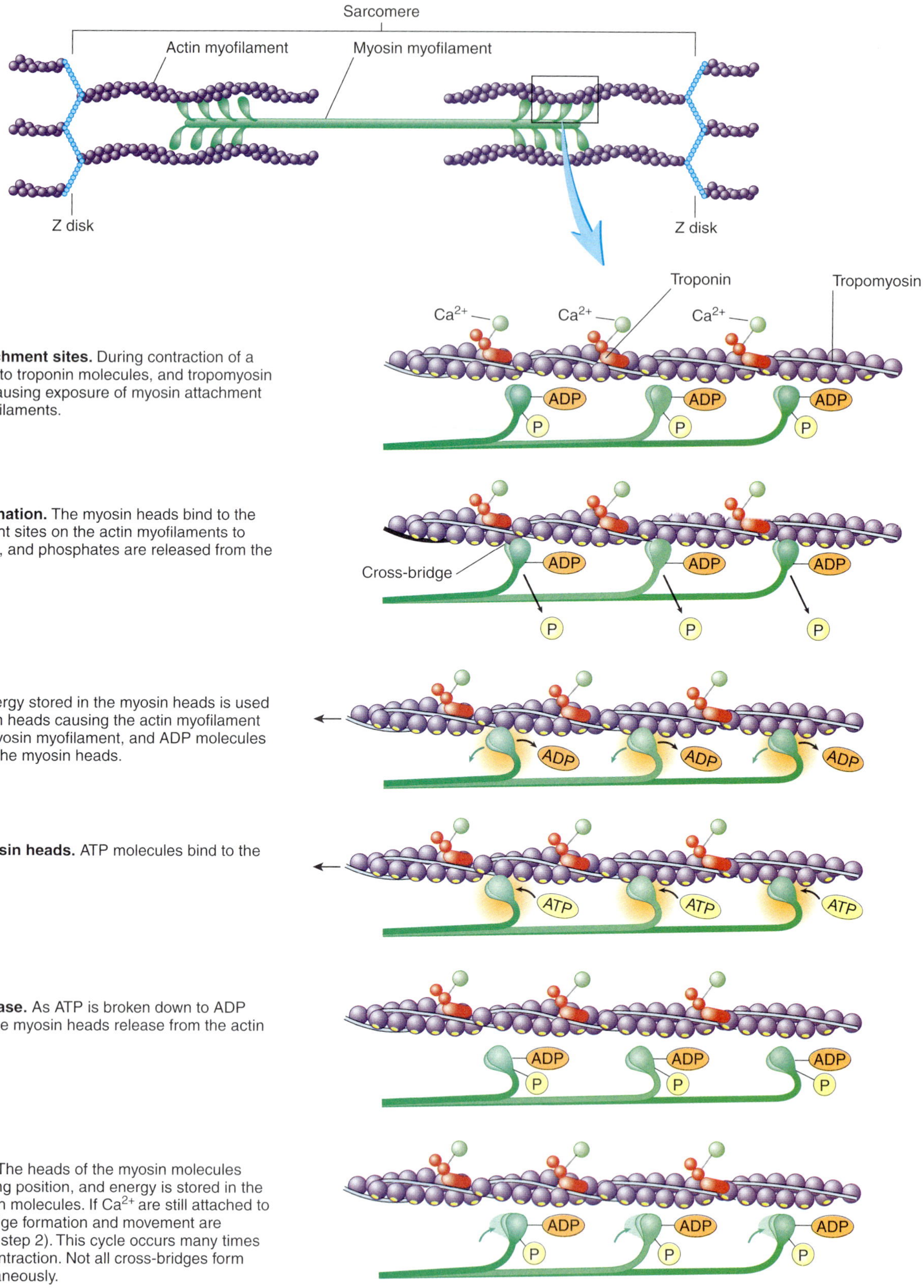

1. **Exposure of attachment sites.** During contraction of a muscle, Ca^{2+} bind to troponin molecules, and tropomyosin molecules move causing exposure of myosin attachment sites on actin myofilaments.

2. **Cross-bridge formation.** The myosin heads bind to the exposed attachment sites on the actin myofilaments to form cross-bridges, and phosphates are released from the myosin heads.

3. **Power stroke.** Energy stored in the myosin heads is used to move the myosin heads causing the actin myofilament to slide past the myosin myofilament, and ADP molecules are released from the myosin heads.

4. **ATP binds to myosin heads.** ATP molecules bind to the myosin heads.

5. **Cross-bridge release.** As ATP is broken down to ADP and phosphates, the myosin heads release from the actin attachment sites.

6. **Recovery stroke.** The heads of the myosin molecules return to their resting position, and energy is stored in the heads of the myosin molecules. If Ca^{2+} are still attached to troponin, cross-bridge formation and movement are repeated (return to step 2). This cycle occurs many times during a muscle contraction. Not all cross-bridges form and release simultaneously.

Breakdown of ATP and Cross-Bridge Movement During Muscle Contraction
Process Figure 9.15

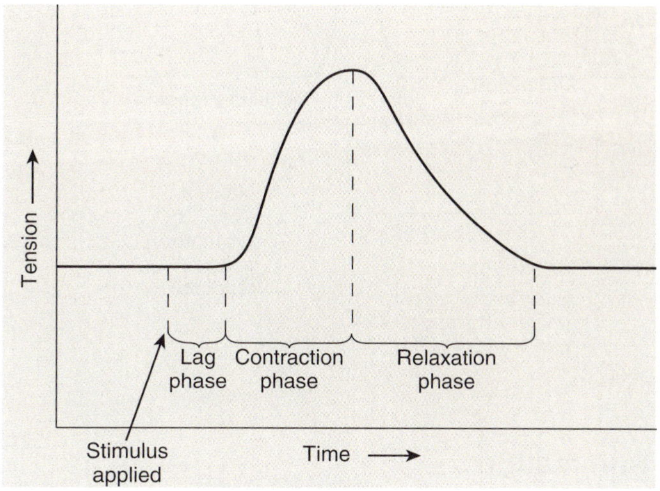

Phases of a Muscle Twitch
Figure 9.16

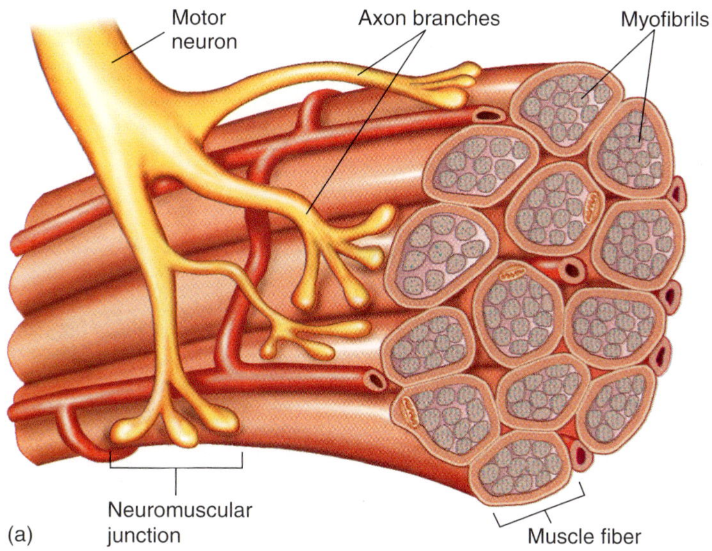

The Motor Unit
Figure 9.17

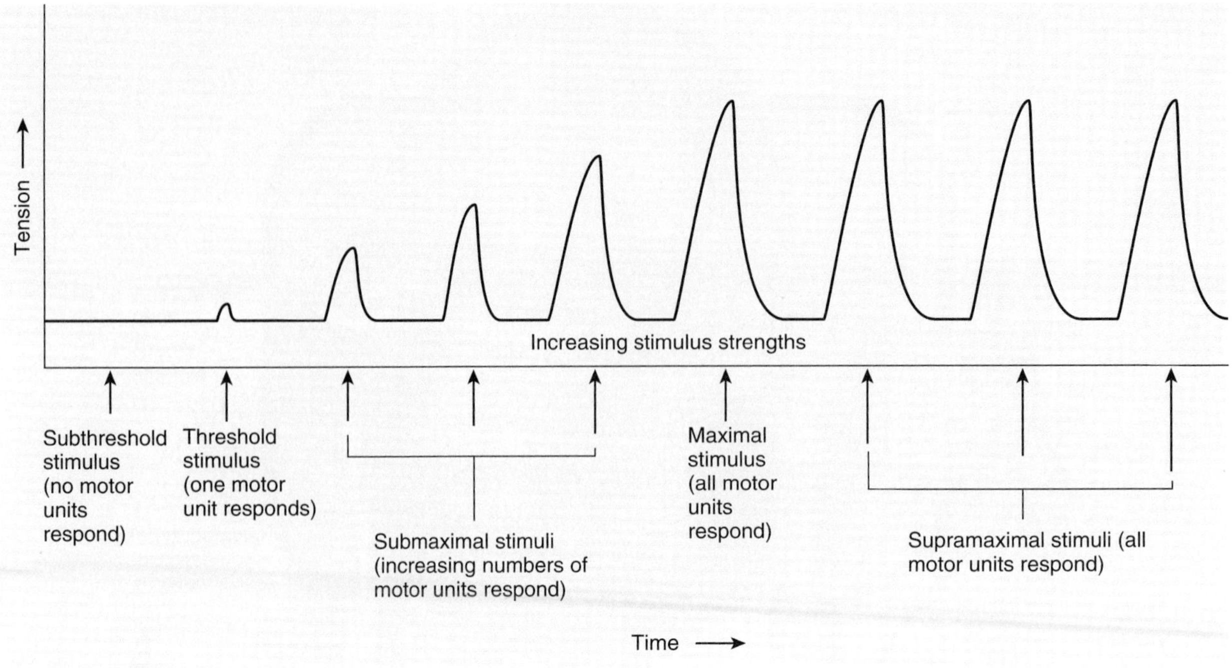

Multiple Motor Unit Summation
Figure 9.18

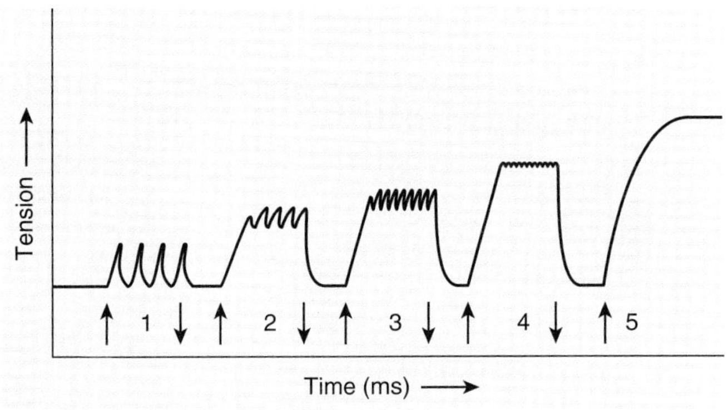

Multiple-Wave Summation
Figure 9.19

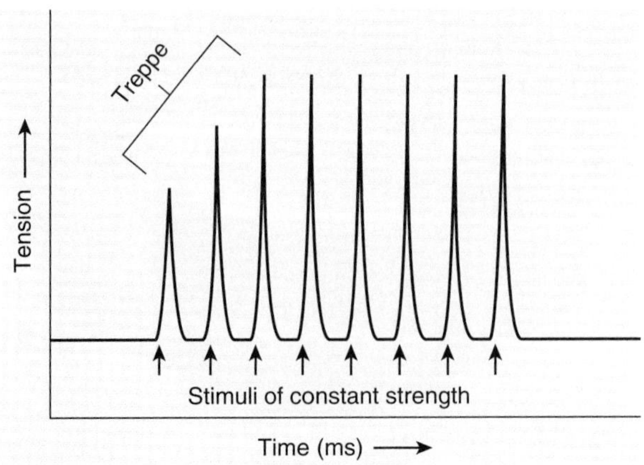

Treppe
Figure 9.20

There is an optimal muscle length at which the muscle produces a maximal tension in response to a maximal stimulus.

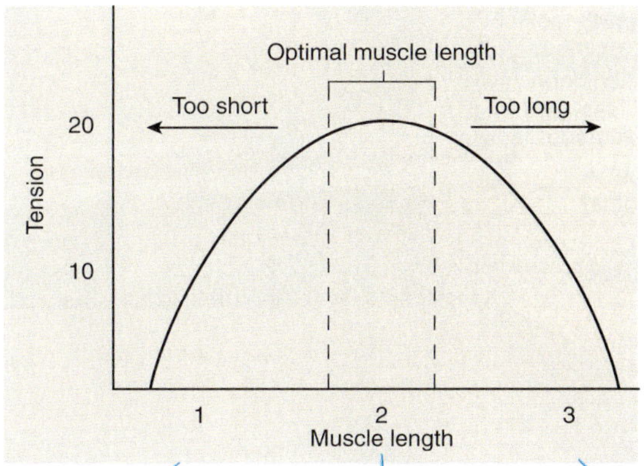

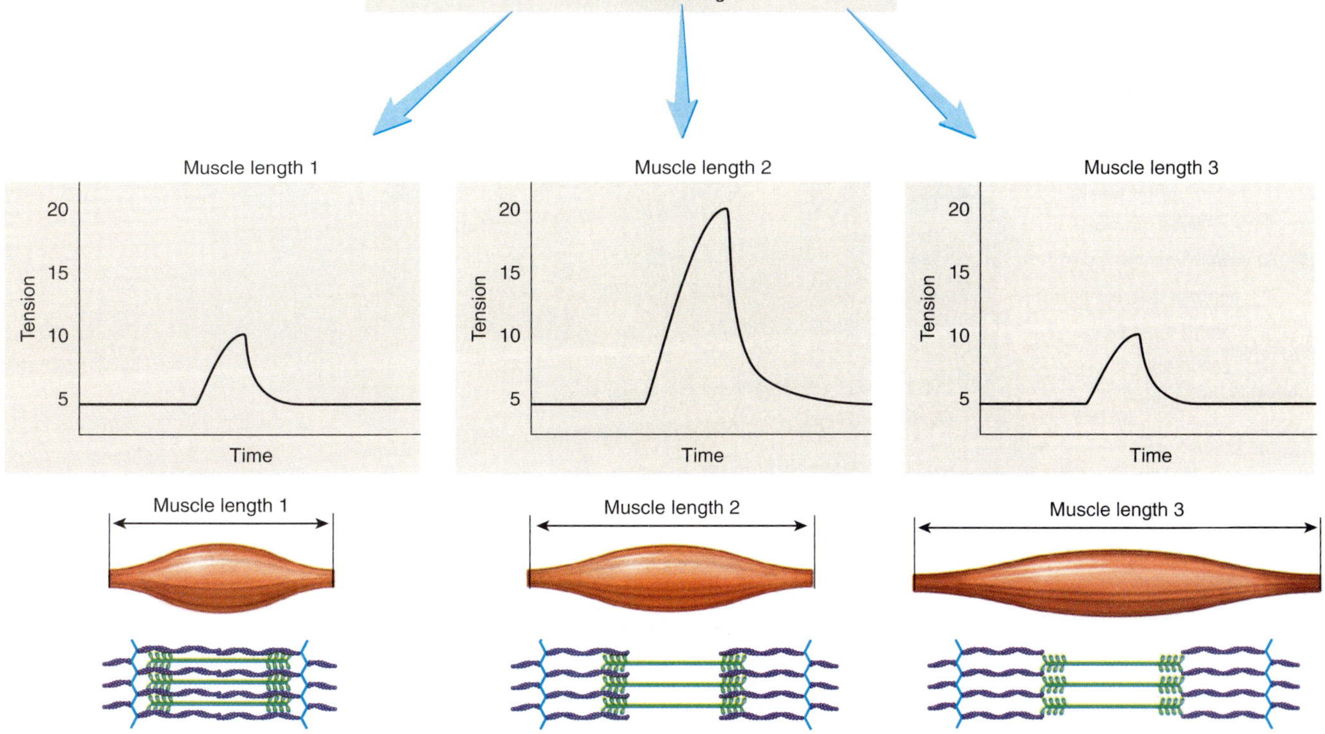

At muscle length 1, the muscle is not stretched, and the tension produced when the muscle contracts is small because the actin and myosin myofilaments are already overlapping nearly as much as they can and the sacromere cannot shorten much more.

At muscle length 2, the muscle is optimally stretched, and the tension produced when the muscle contracts is maximal because the number of cross-bridges that can form is maximal.

At muscle length 3, the muscle is stretched severely, and the tension produced is small because the actin and myosin myofilaments only slightly overlap and the number of cross-bridges that can form is small.

Muscle Length and Tension
Figure 9.21

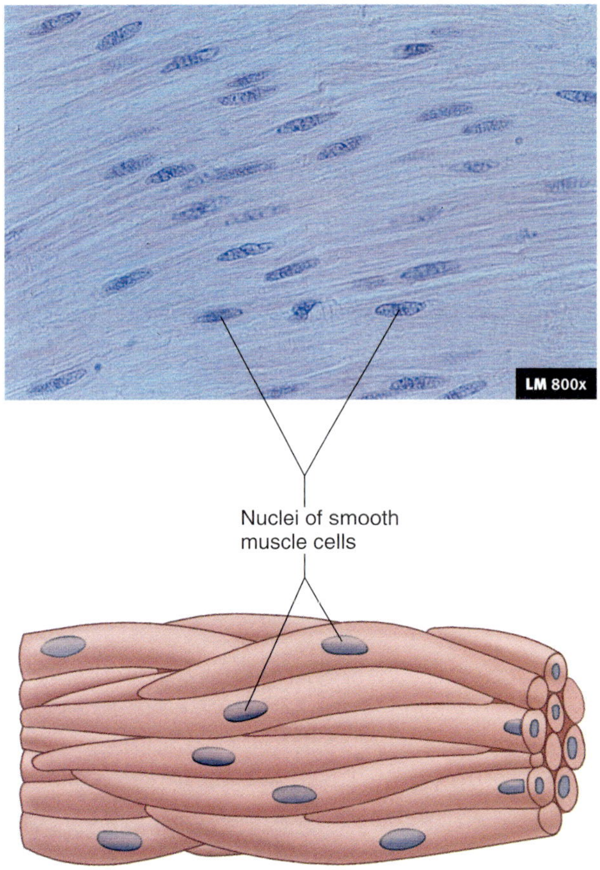

Smooth Muscle Histology
Figure 9.22
© Victor Eroschenko

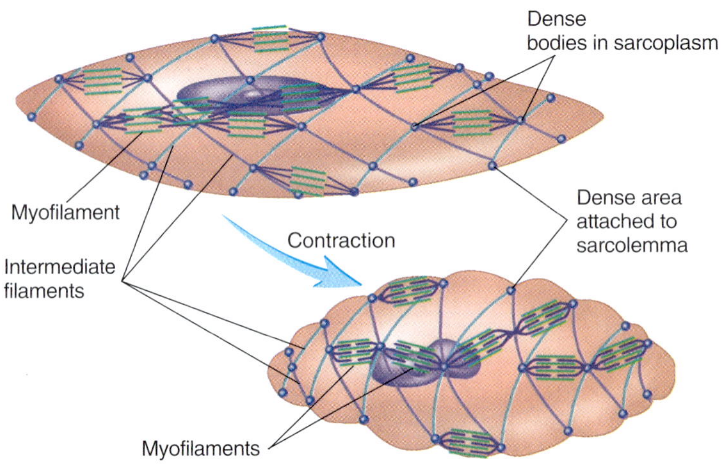

Contractile Proteins in a Smooth Muscle Cell
Figure 9.23

1. A hormone combines with a hormone receptor and activates a G protein mechanism, or depolarization of the plasma membrane occurs.

2. An α subunit opens the Ca^{2+} channel in the plasma membrane, or depolarization opens Ca^{2+} channels. Ca^{2+} diffuses through the Ca^{2+} channels and combine with calmodulin.

3. Calmodulin with a Ca^{2+} bound to it binds with myosin kinase and activates it.

4. Activated myosin kinase attaches phosphate from ATP to myosin heads to activate the contractile process.

5. A cycle of cross-bridge formation, movement, detachment, and cross-bridge formation occurs.

6. Relaxation occurs when myosin phosphatase removes phosphate from myosin.

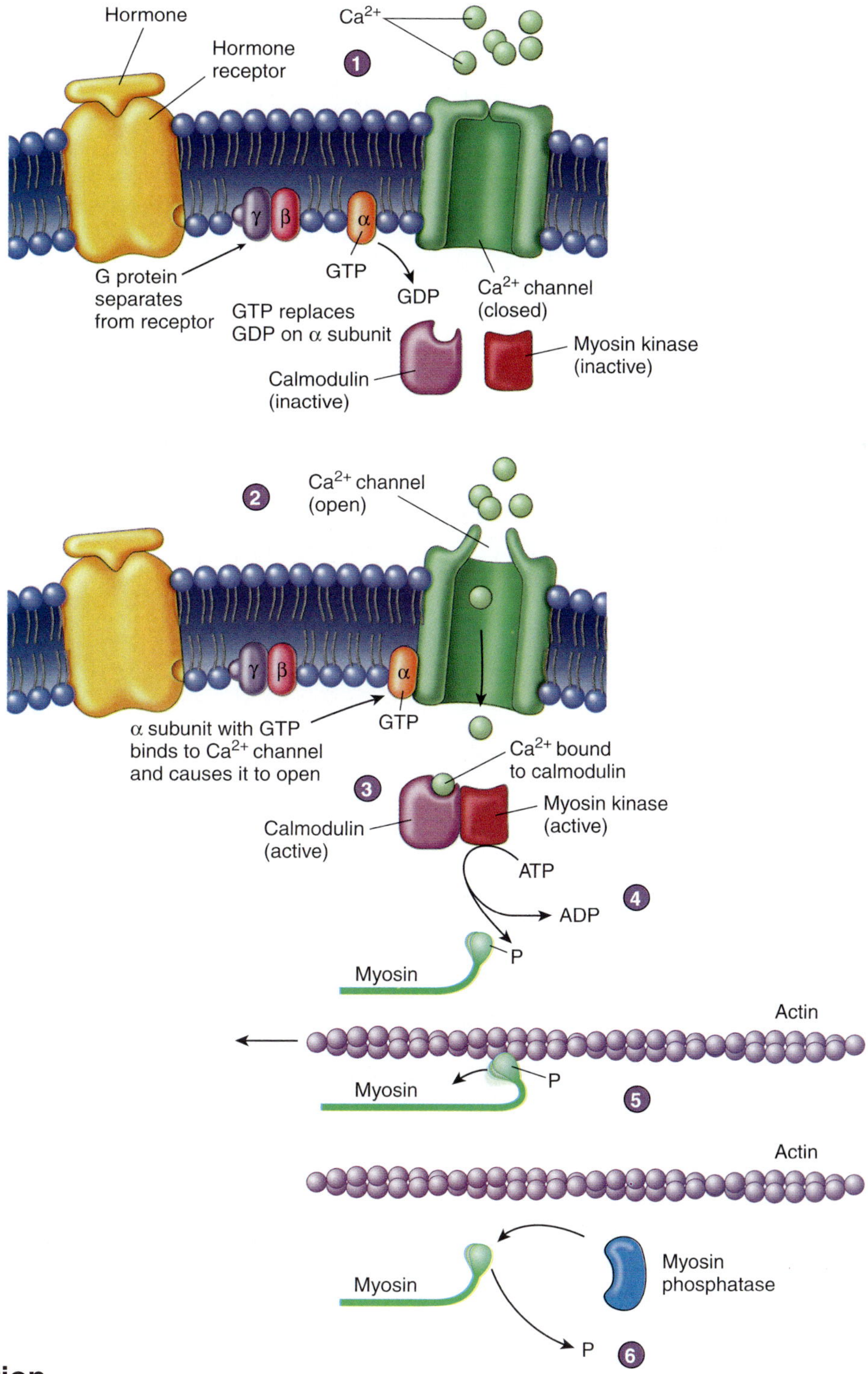

Smooth Muscle Contraction
Process Figure 9.24

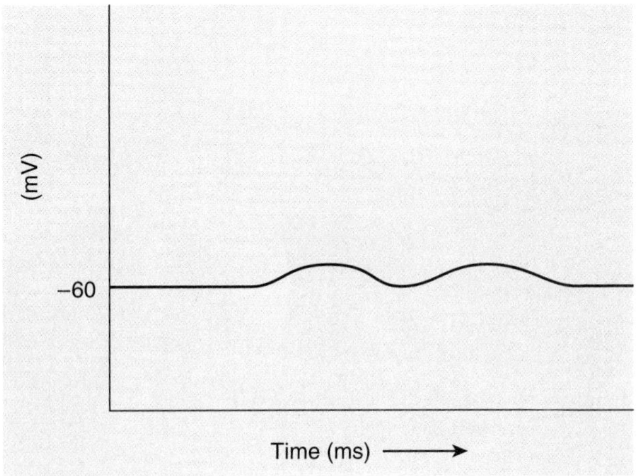

(a) **Slow waves of depolarization**

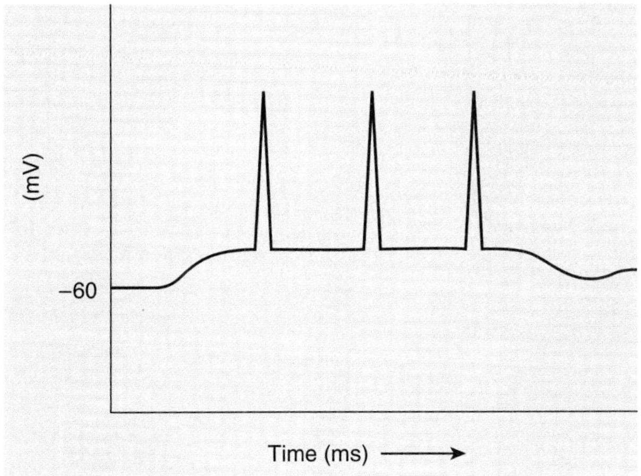

(b) **Action potentials superimposed on a slow wave of depolarization**

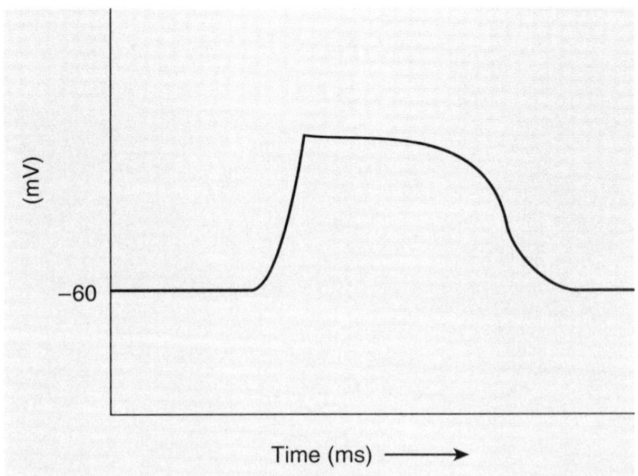

(c) **Action potential with prolonged depolarization (plateau)**

Membrane Potential from a Smooth Muscle Preparation

Figure 9.25

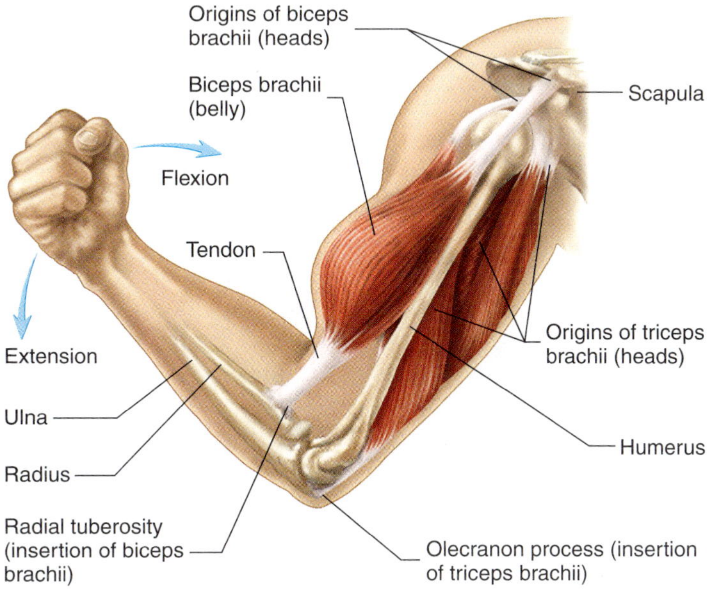

Muscle Attachment
Figure 10.1

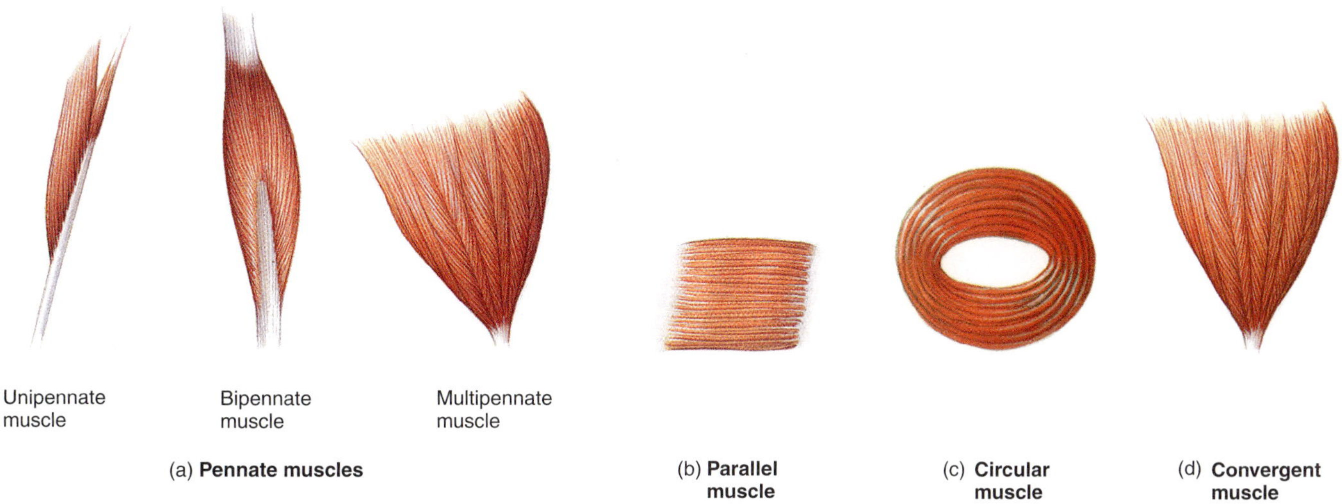

Examples of Muscle Types
Figure 10.2

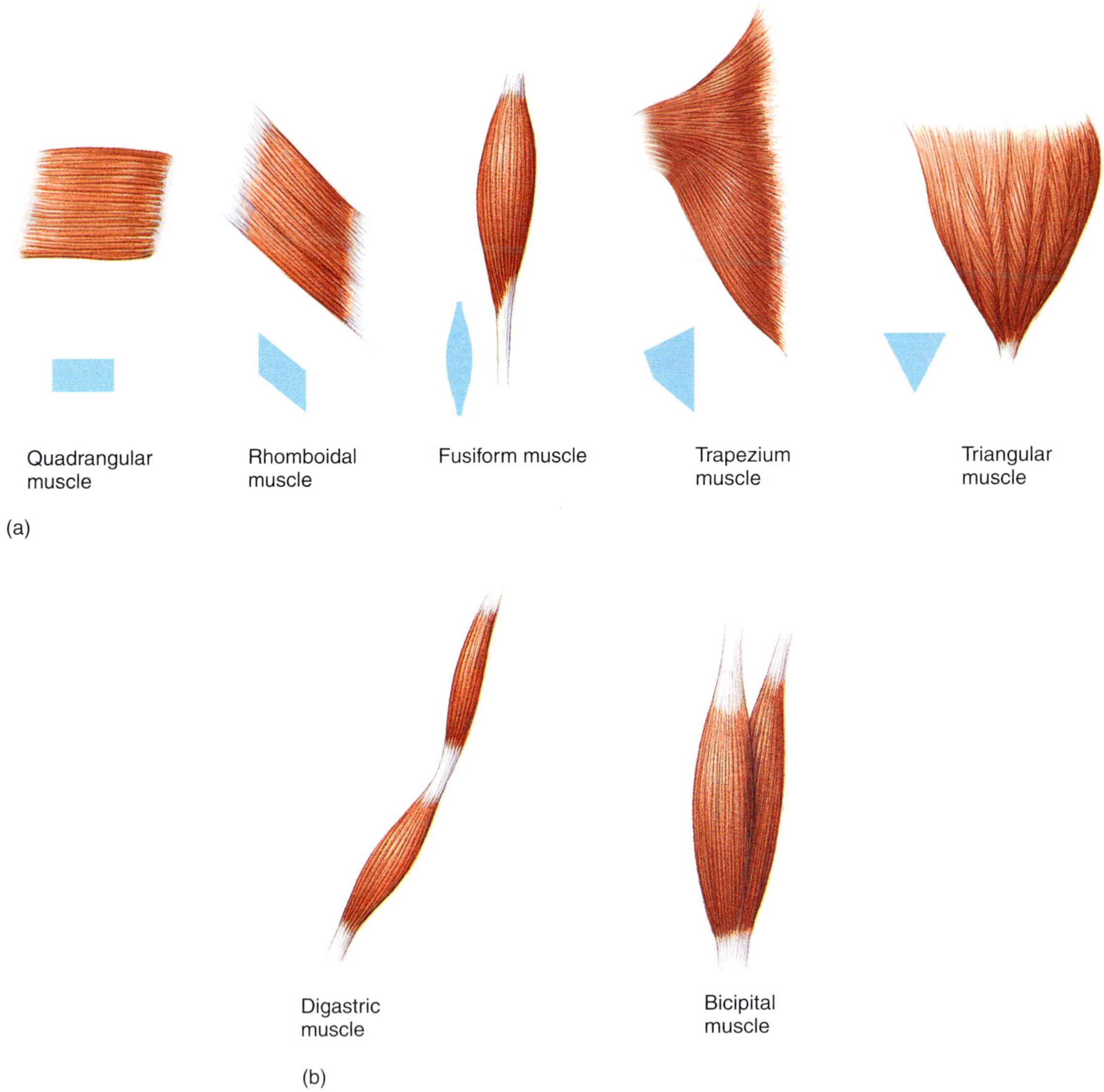
Quadrangular muscle
Rhomboidal muscle
Fusiform muscle
Trapezium muscle
Triangular muscle
(a)
Digastric muscle
Bicipital muscle
(b)

Examples of Muscle Shapes
Figure 10.3

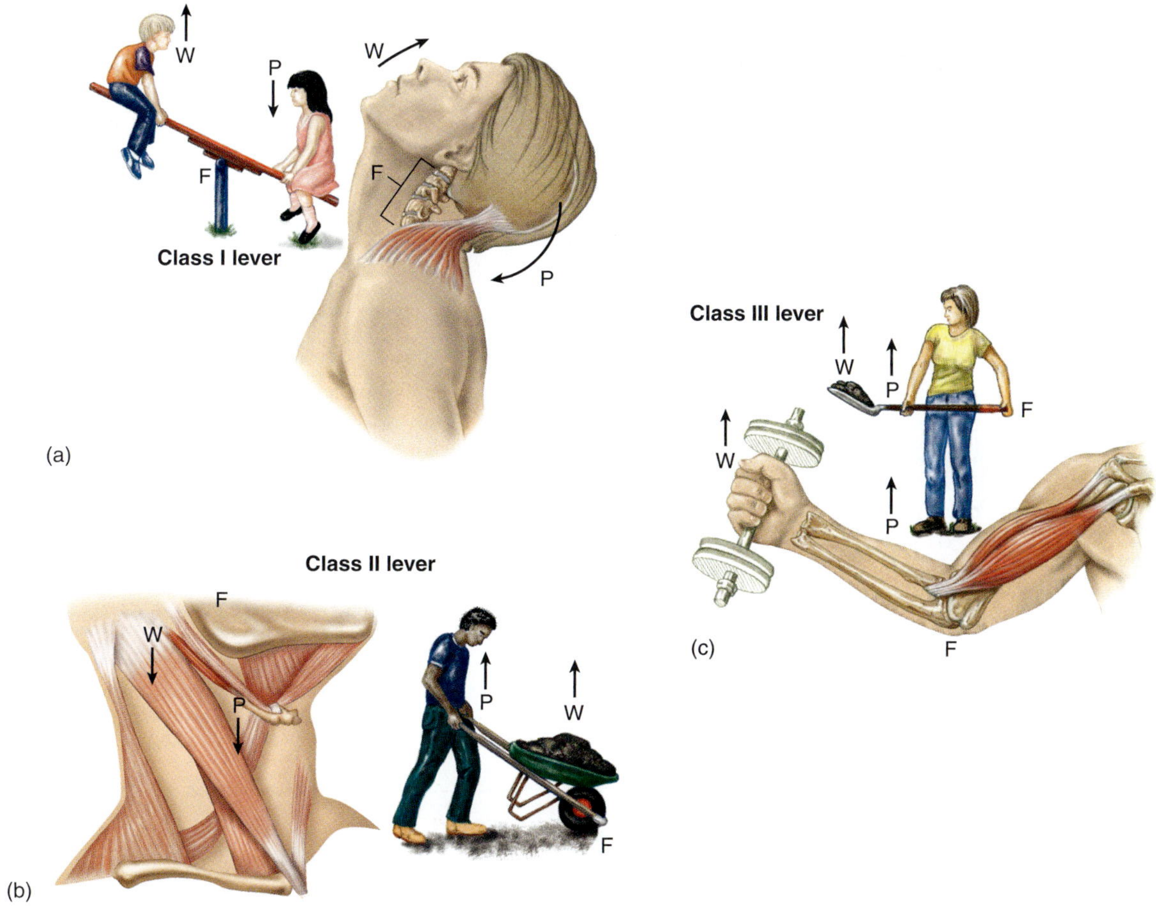
W
P
F
Class I lever
W
F
P
(a)
Class III lever
W
P
F
W
P
F
(c)
Class II lever
F
W
P
P
W
F
(b)

Lever Classes
Figure 10.4

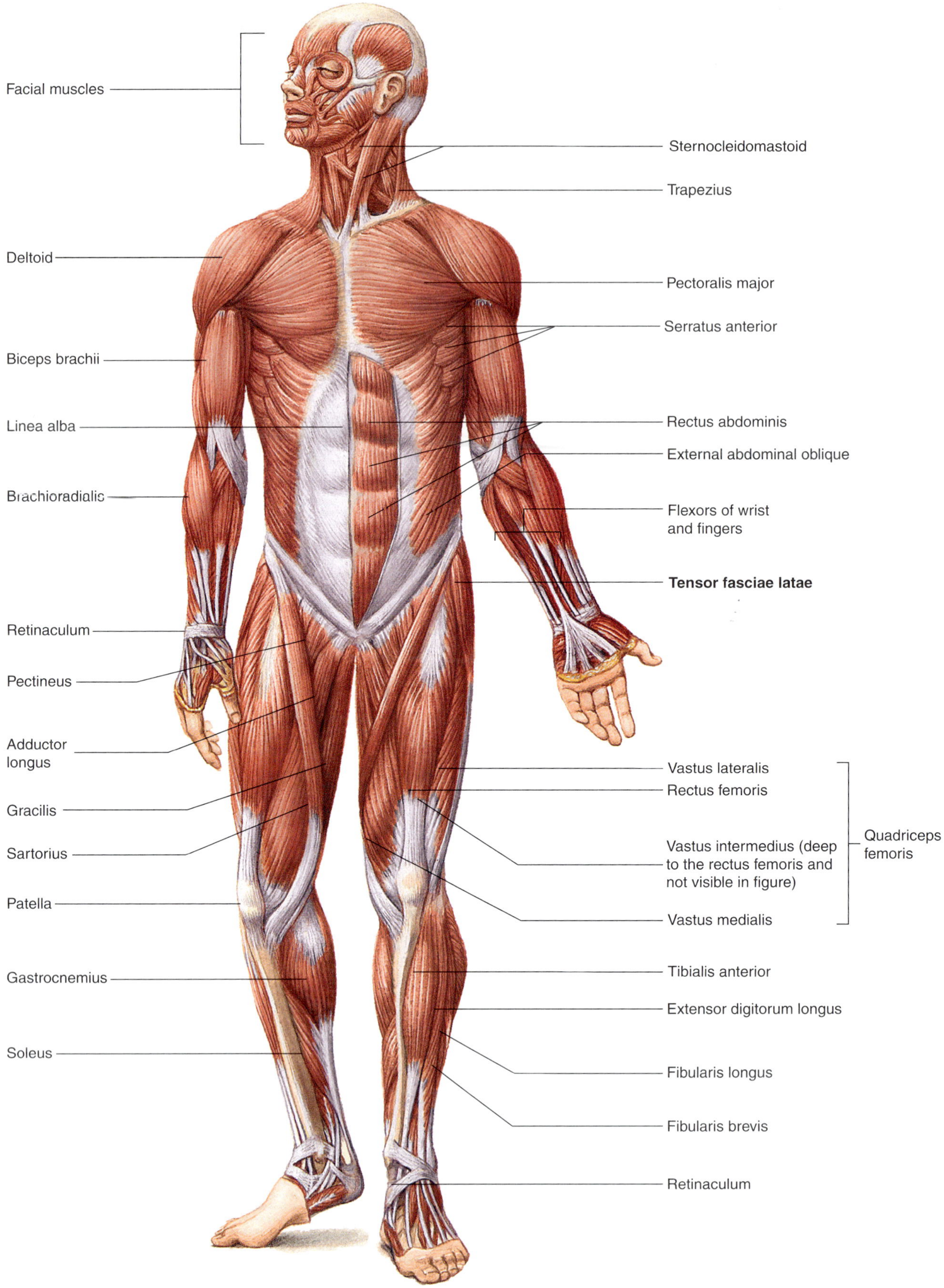

(a) **Anterior view**

General Overview of the Superficial Body Musculature
Figure 10.5

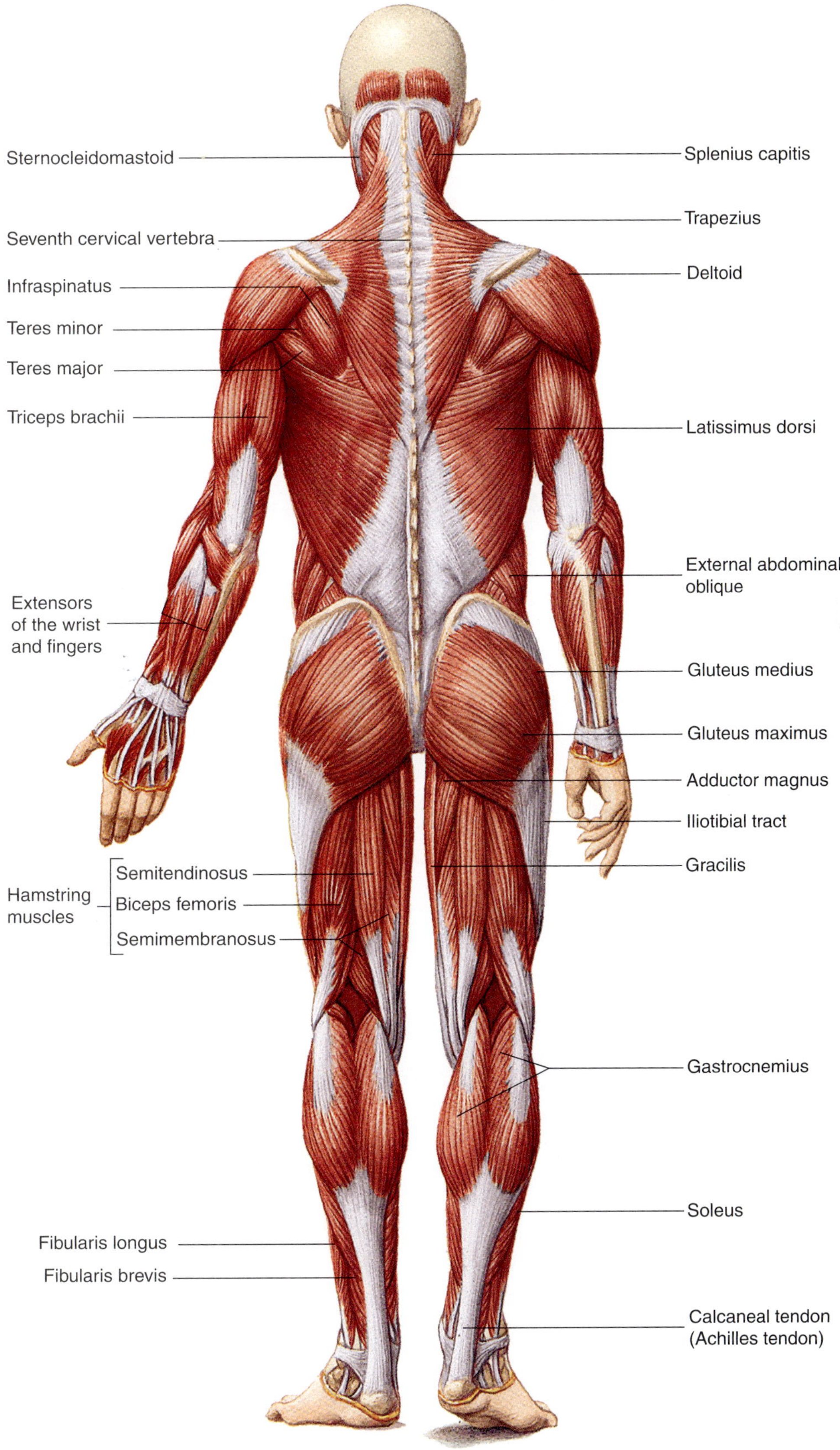

(b) **Posterior view**

General Overview of the Superficial Body Musculature
Figure 10.5 (continued)

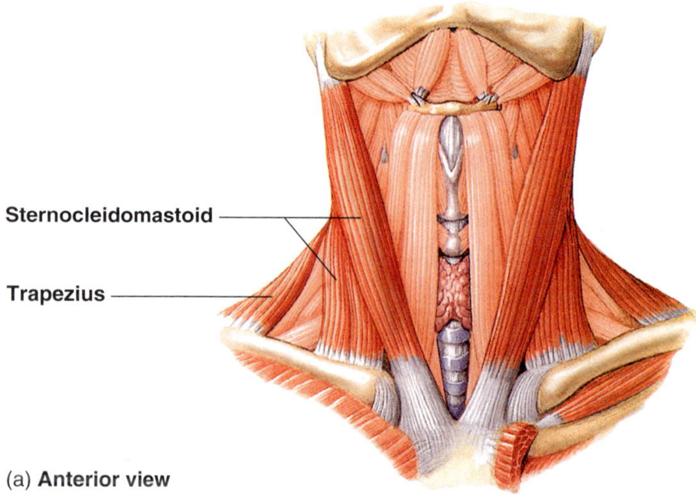

(a) **Anterior view**

Anterior Neck Muscles
Figure 10.6

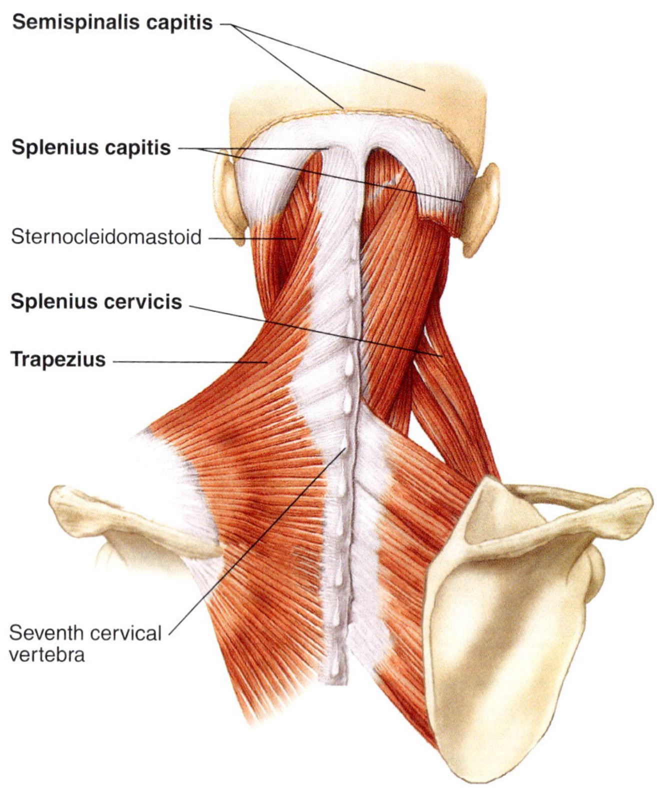

(a) **Posterior view**

Posterior Neck Muscles
Figure 10.7

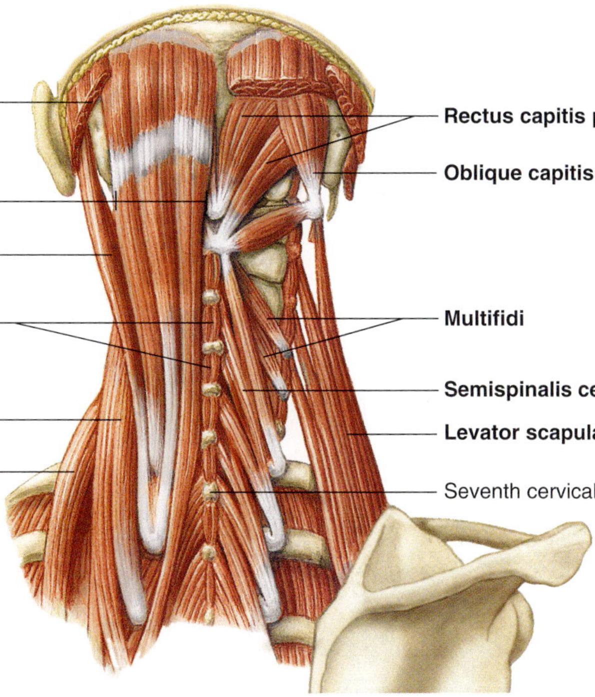

Posterior view

Posterior Deep Neck Muscles

Figure 10.8

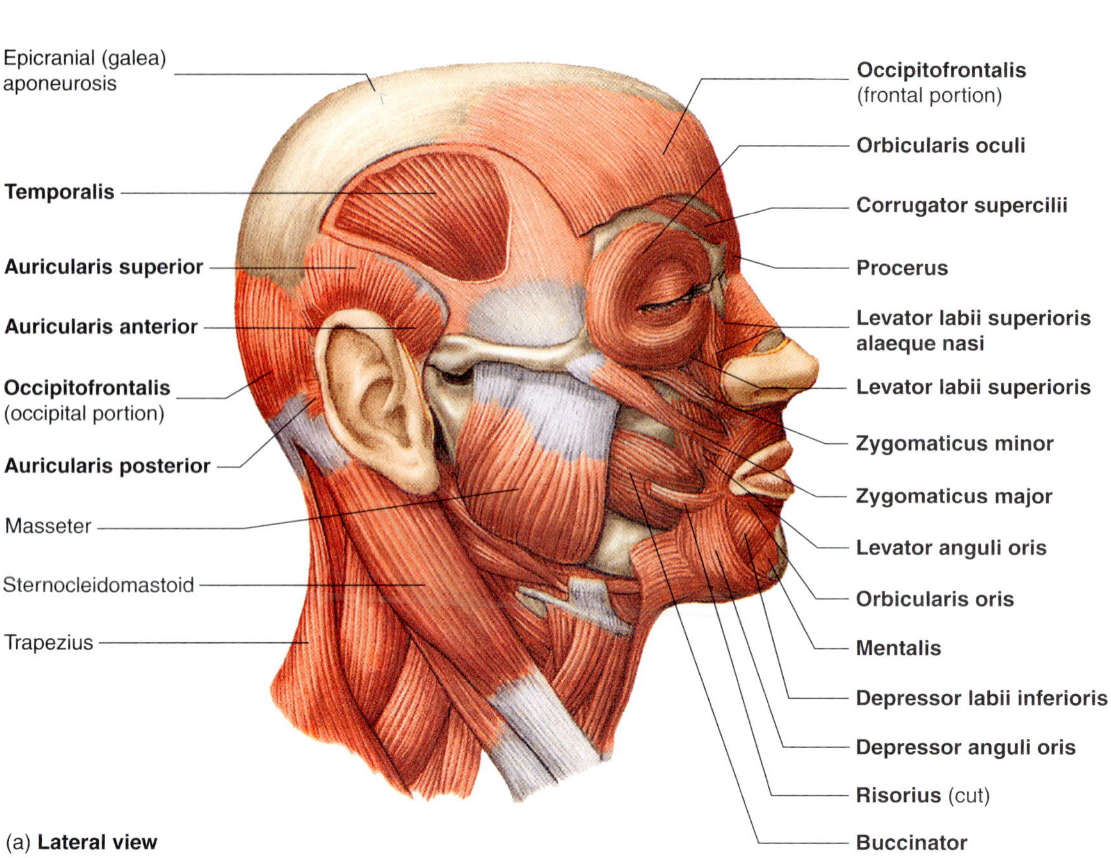

(a) **Lateral view**

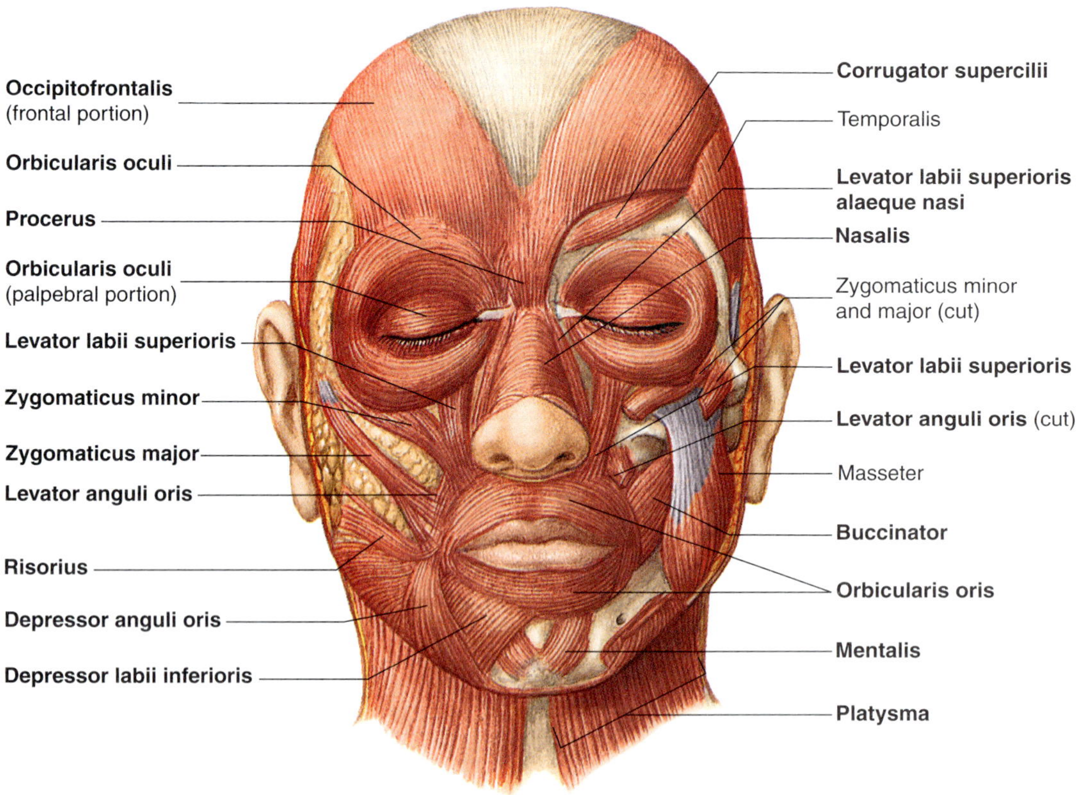

(b) **Anterior view**

Muscles of Facial Expression

Figure 10.9

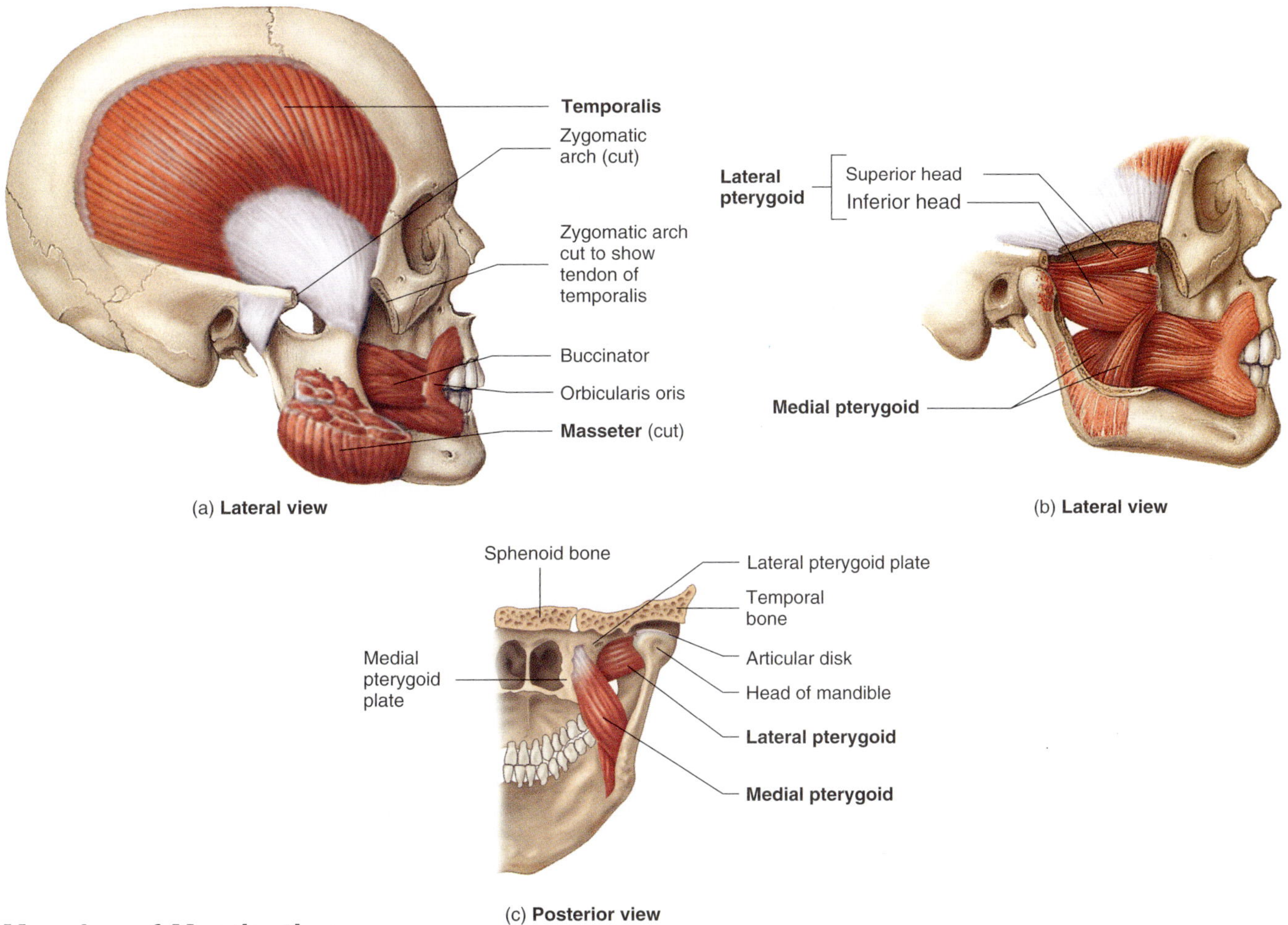

Temporalis
Zygomatic arch (cut)
Zygomatic arch cut to show tendon of temporalis
Buccinator
Orbicularis oris
Masseter (cut)
(a) Lateral view
Lateral pterygoid
Superior head
Inferior head
Medial pterygoid
(b) Lateral view
Sphenoid bone
Lateral pterygoid plate
Temporal bone
Medial pterygoid plate
Articular disk
Head of mandible
Lateral pterygoid
Medial pterygoid
(c) Posterior view

Muscles of Mastication
Figure 10.11

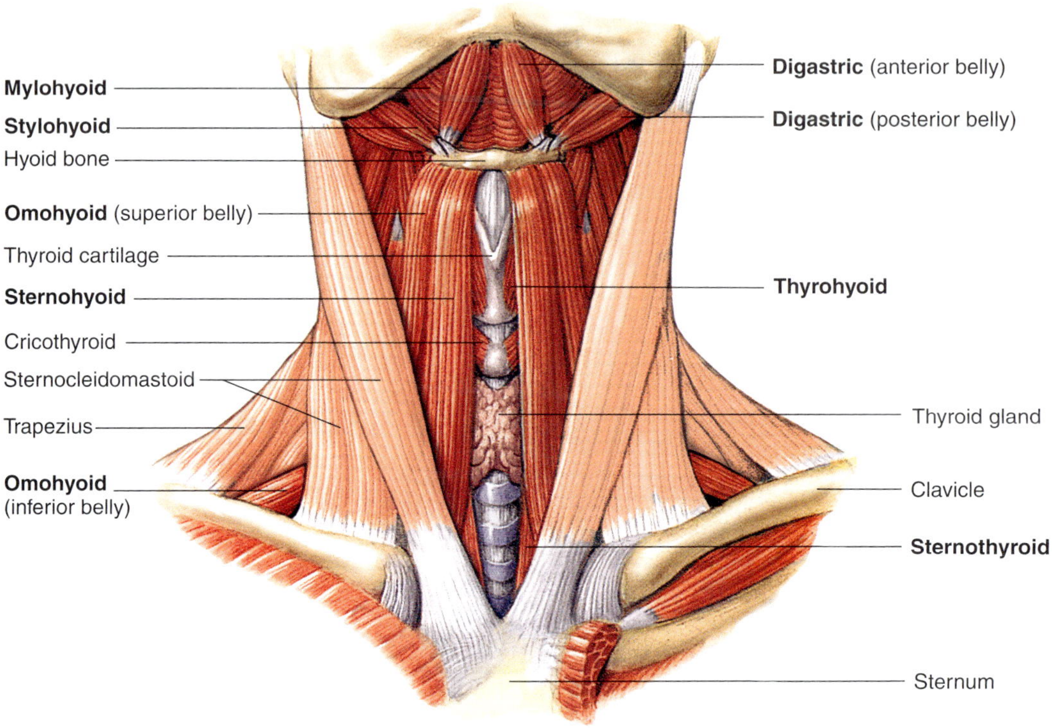
Mylohyoid
Stylohyoid
Hyoid bone
Omohyoid (superior belly)
Thyroid cartilage
Sternohyoid
Cricothyroid
Sternocleidomastoid
Trapezius
Omohyoid
(inferior belly)
Digastric (anterior belly)
Digastric (posterior belly)
Thyrohyoid
Thyroid gland
Clavicle
Sternothyroid
Sternum

(a) **Anterior view**

Hyoid Muscles
Figure 10.12

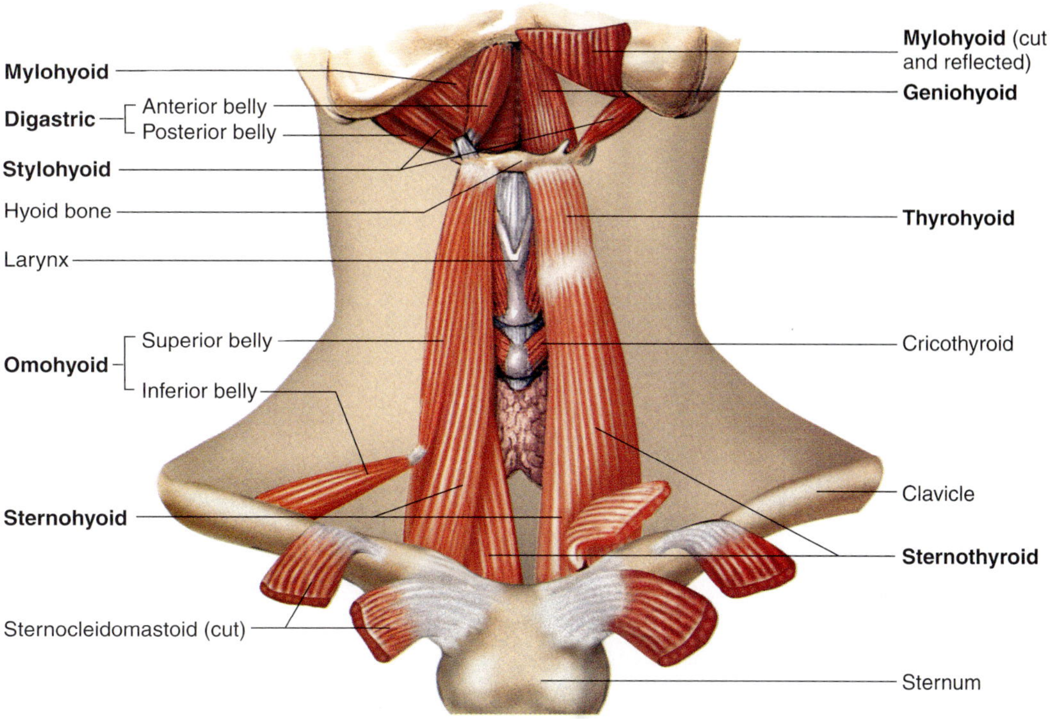

(b) **Anterior view**

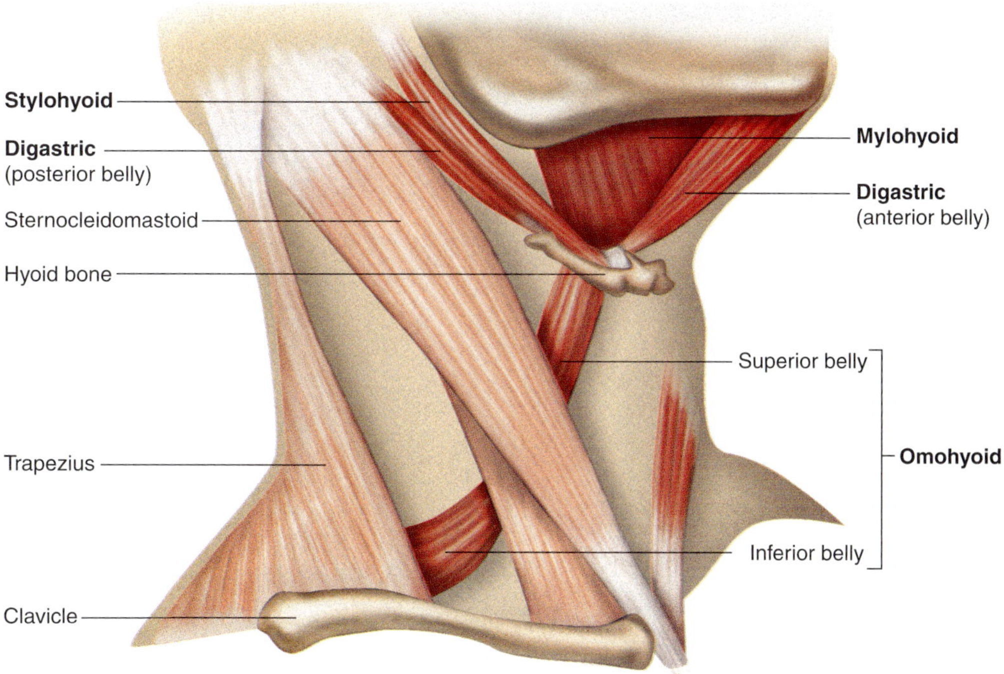

(c) **Anterolateral view**

Hyoid Muscles
Figure 10.12 (continued)

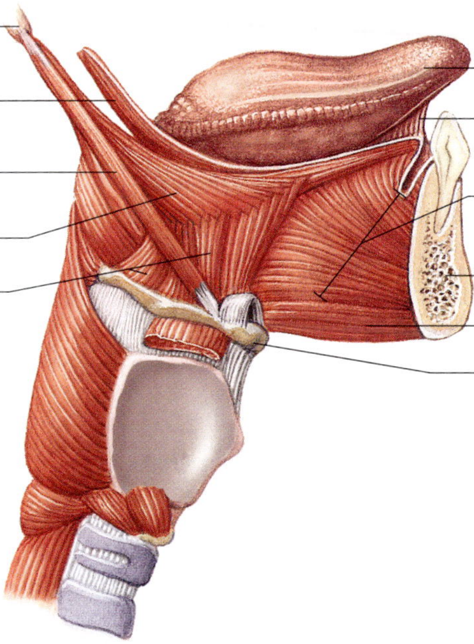
Styloid process
Palatoglossus
Stylohyoid
Styloglossus
Hyoglossus
Tongue
Frenulum
Genioglossus
Mandible
Geniohyoid
Hyoid bone
Lateral view

Tongue Muscles
Figure 10.13

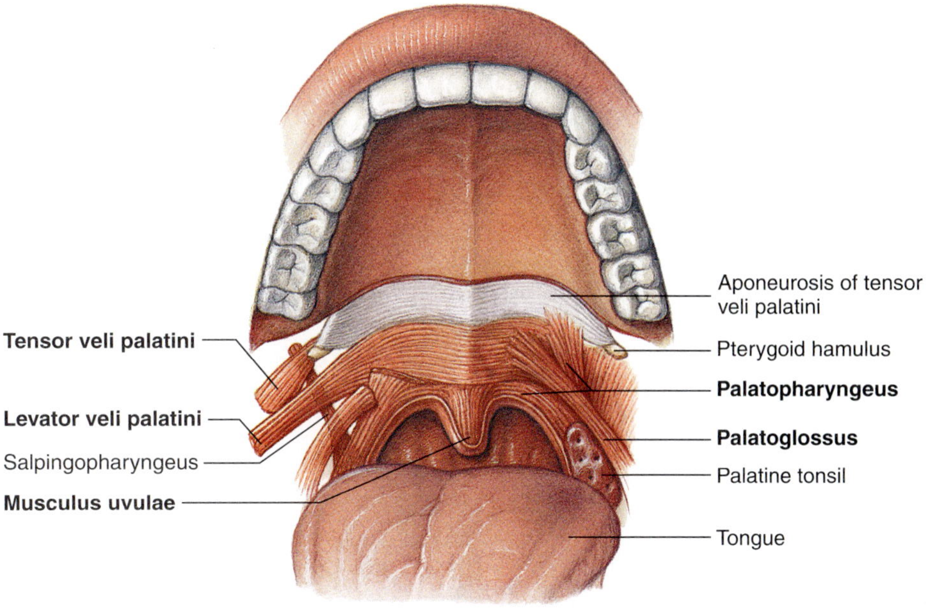

(a) **Anterior view**

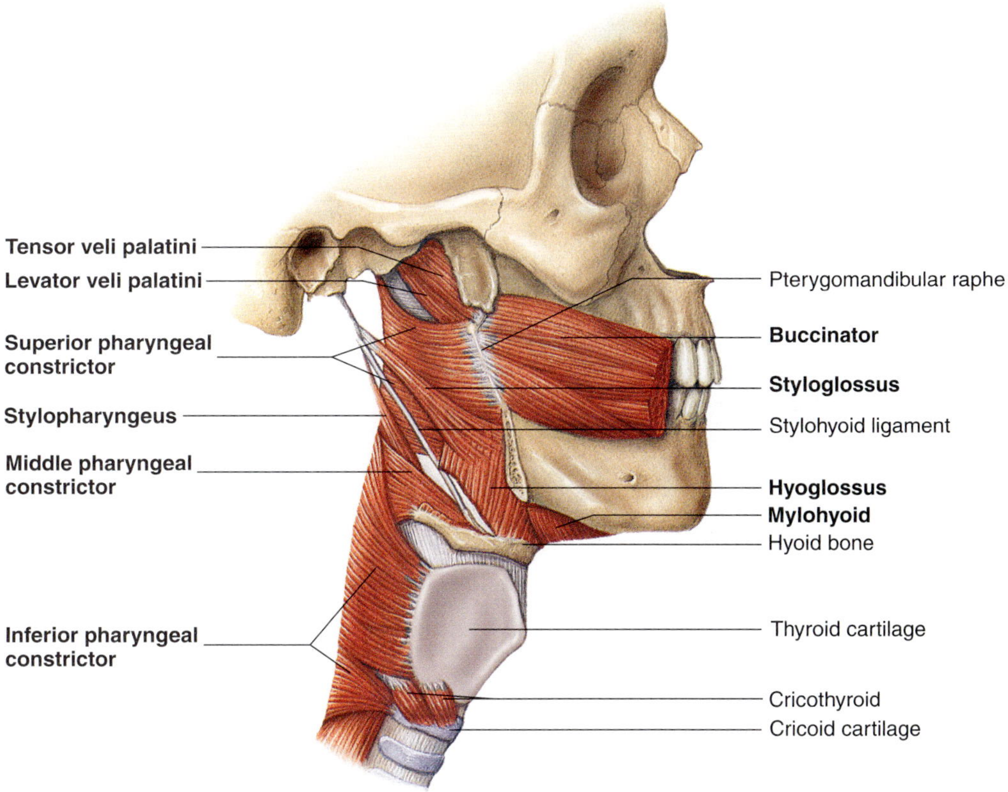

(b) **Lateral view**

Muscles of the Palate, Pharynx, and Larynx
Figure 10.14

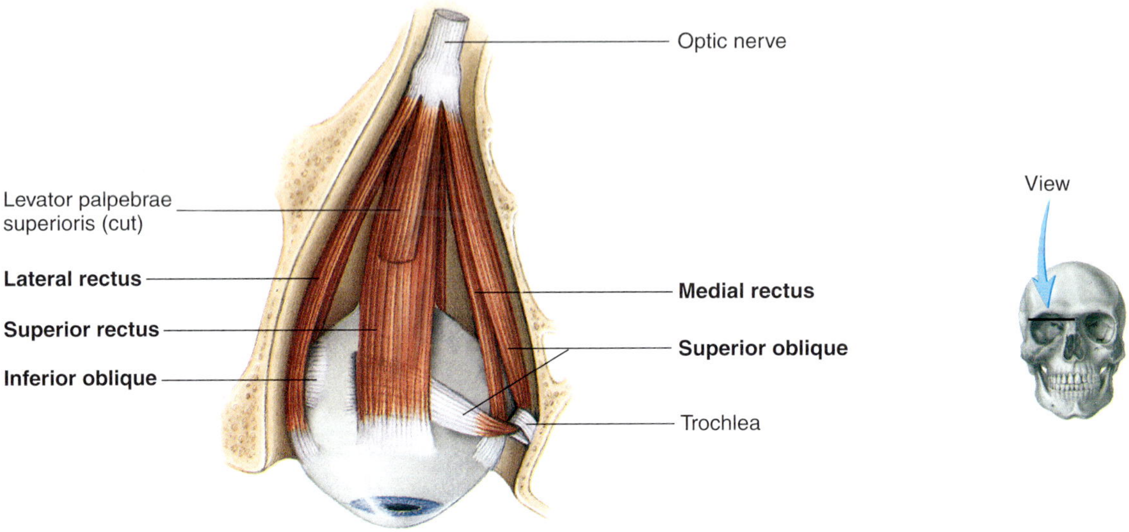

(a) **Superior view**

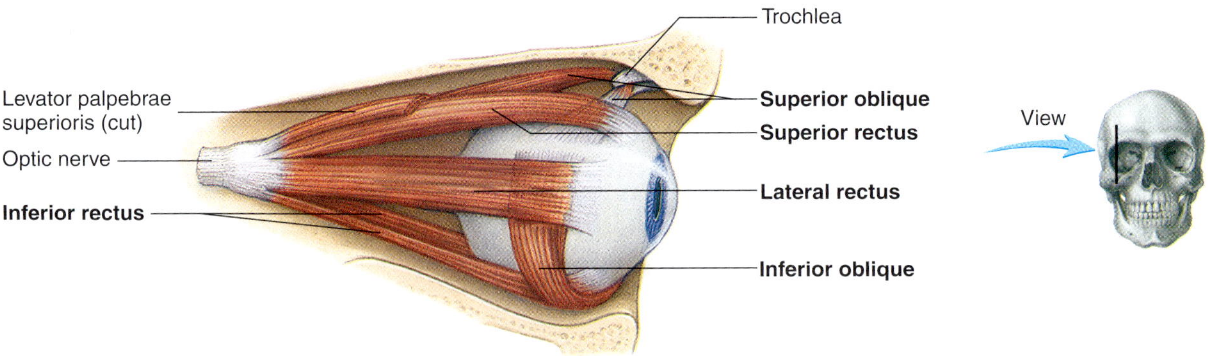

(b) **Lateral view**

Muscles Moving the Right Eyeball
Figure 10.15

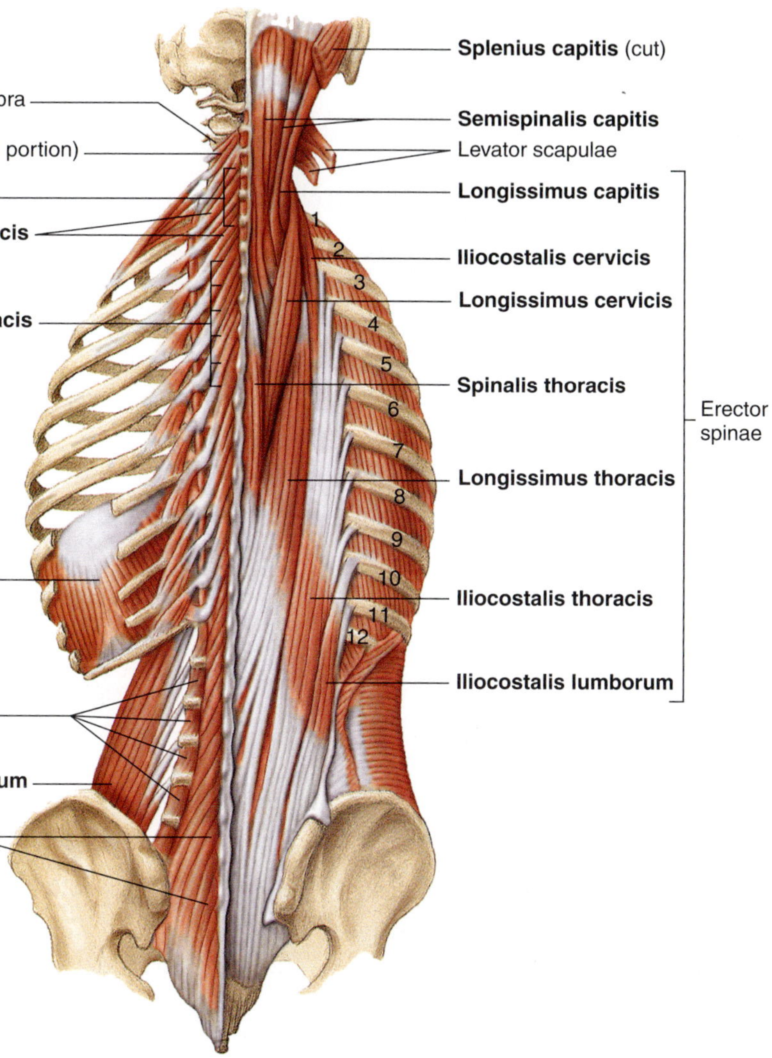

Splenius capitis (cut)
Third cervical vertebra
Semispinalis capitis
Multifidus (cervical portion)
Levator scapulae
Interspinalis
Longissimus capitis
Semispinalis cervicis
1
2
Iliocostalis cervicis
3
Longissimus cervicis
Semispinalis thoracis
4
5
Spinalis thoracis
6
Erector spinae
7
Longissimus thoracis
8
9
Diaphragm
10
Iliocostalis thoracis
11
12
Iliocostalis lumborum
Intertransversarii
Quadratus lumborum
Multifidus
(lumbar portion)
Posterior view

Deep Back Muscles
Figure 10.16

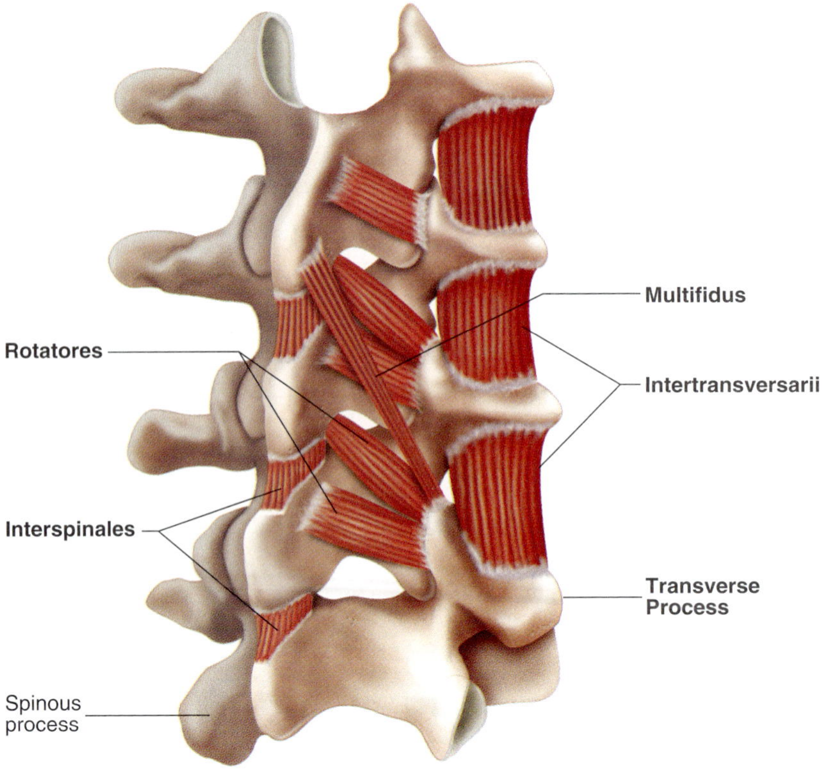

Posterolateral view

Deep Muscles Associated with the Vertebrae
Figure 10.17

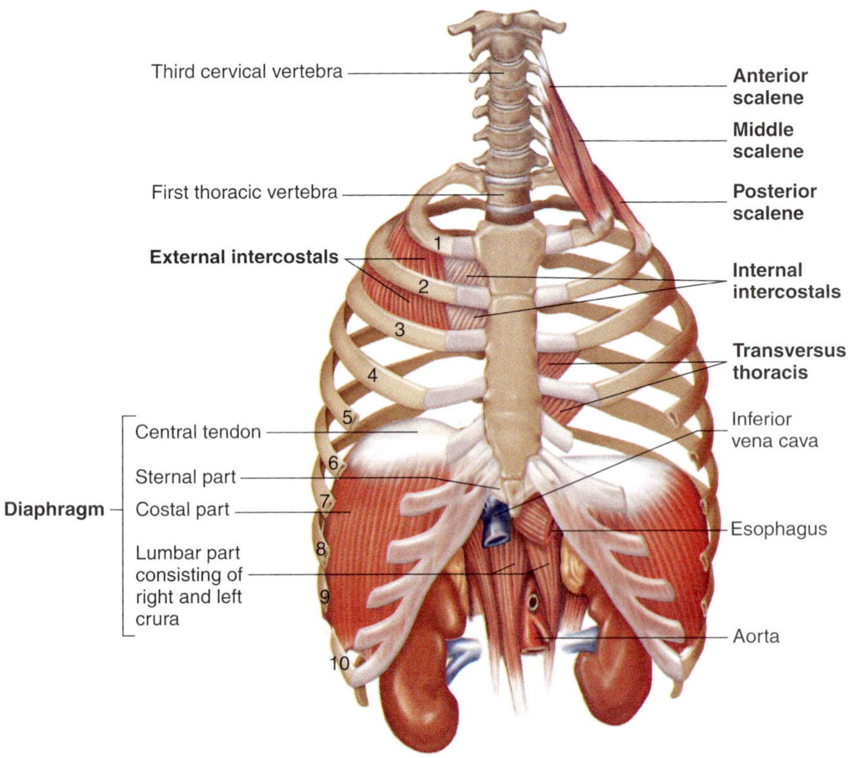

(a) **Anterior view**

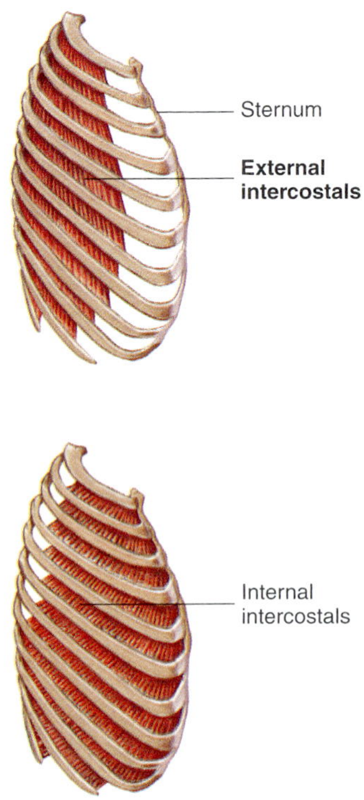

(b) **Lateral view**

Muscles of the Thorax
Figure 10.18

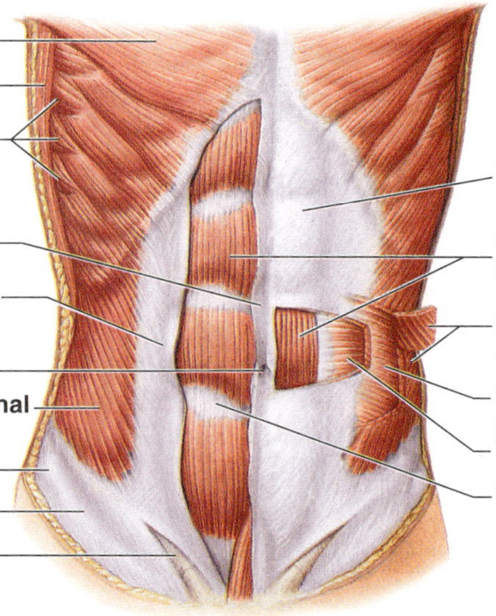

(a) **Anterior view**

Anterior Abdominal Wall Muscles
Figure 10.19

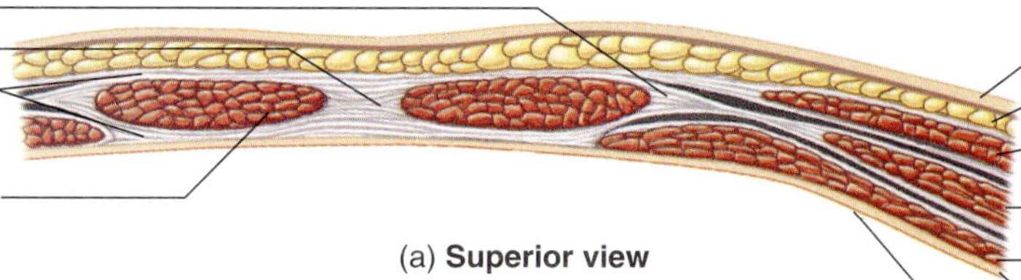

(a) **Superior view**

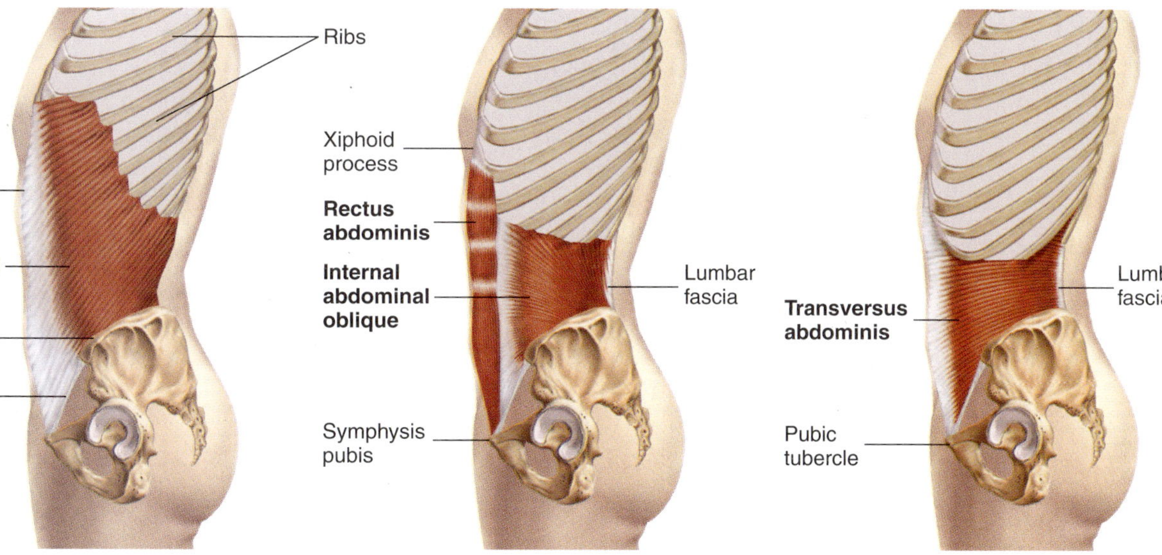

(b) **Lateral views**

Anterior Abdominal Wall Muscles
Figure 10.20

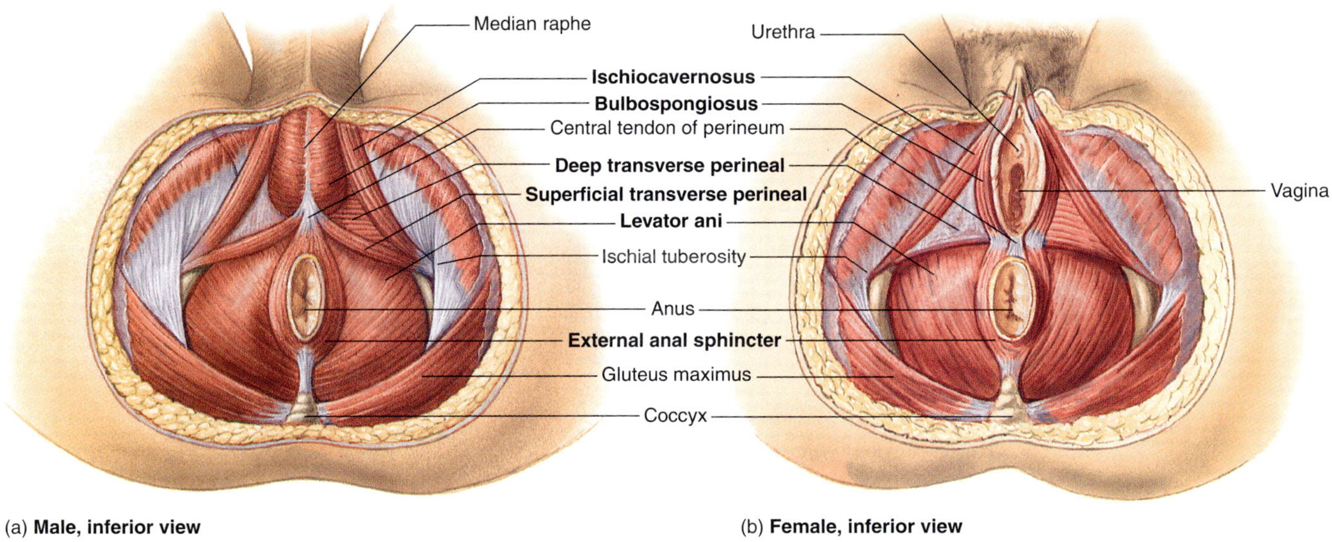

Muscles of the Pelvic Floor and Perineum
Figure 10.21

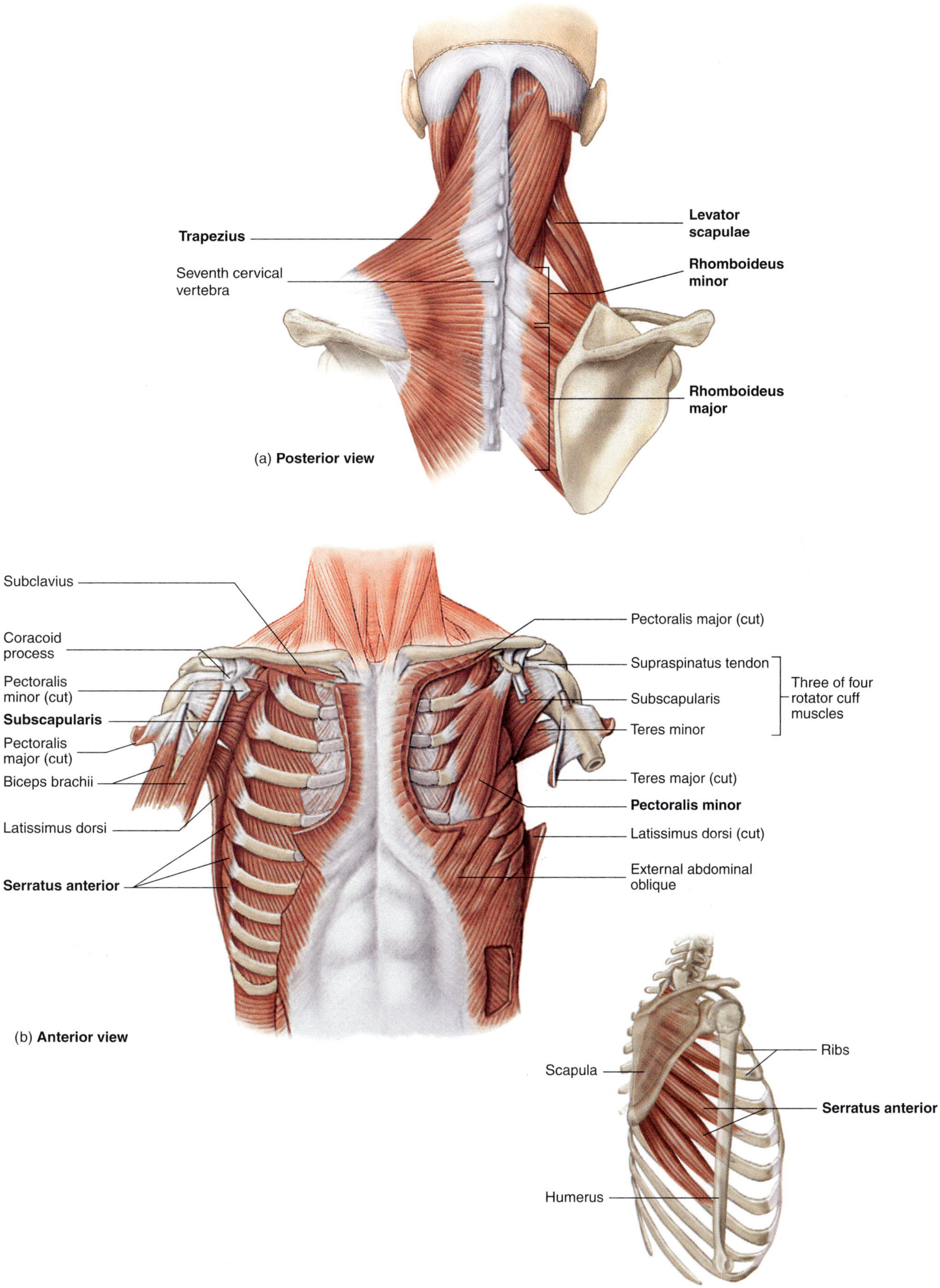

(a) **Posterior view**

(b) **Anterior view**

(c) **Lateral view**

Muscles Acting on the Scapula
Figure 10.22

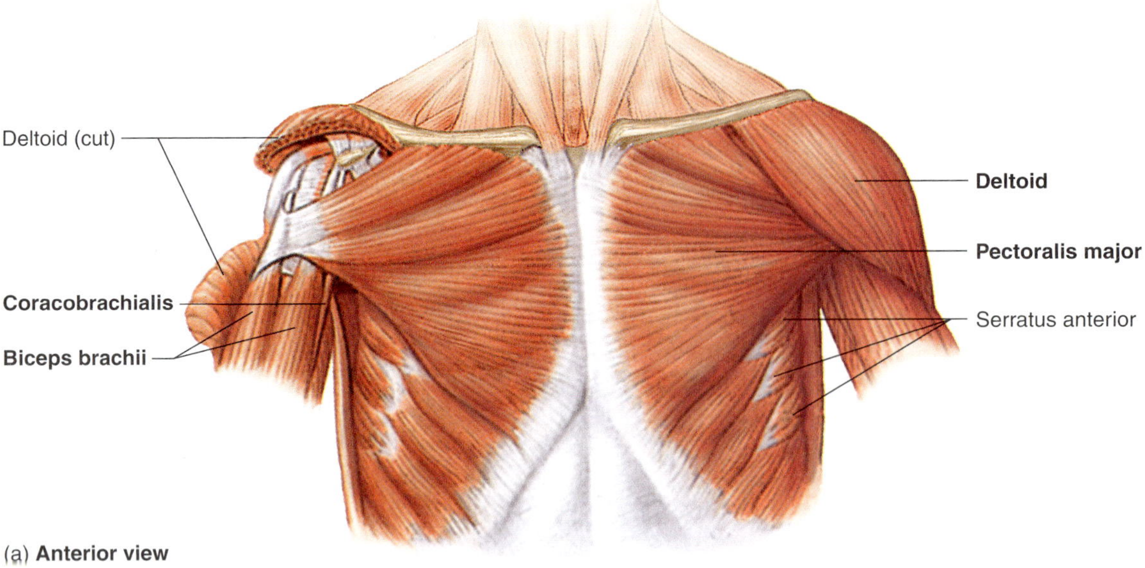

(a) **Anterior view**

Anterior Muscles Attaching the Upper Limb to the Body
Figure 10.23

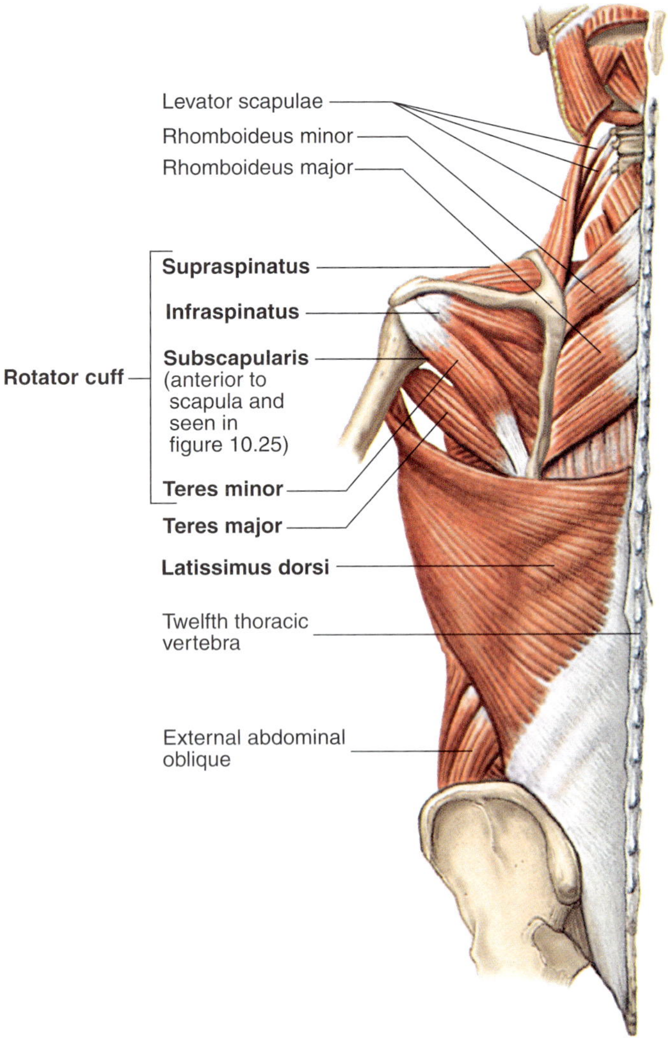

(a) **Posterior view**

Posterior Muscles Attaching the Upper Limb to the Body
Figure 10.24

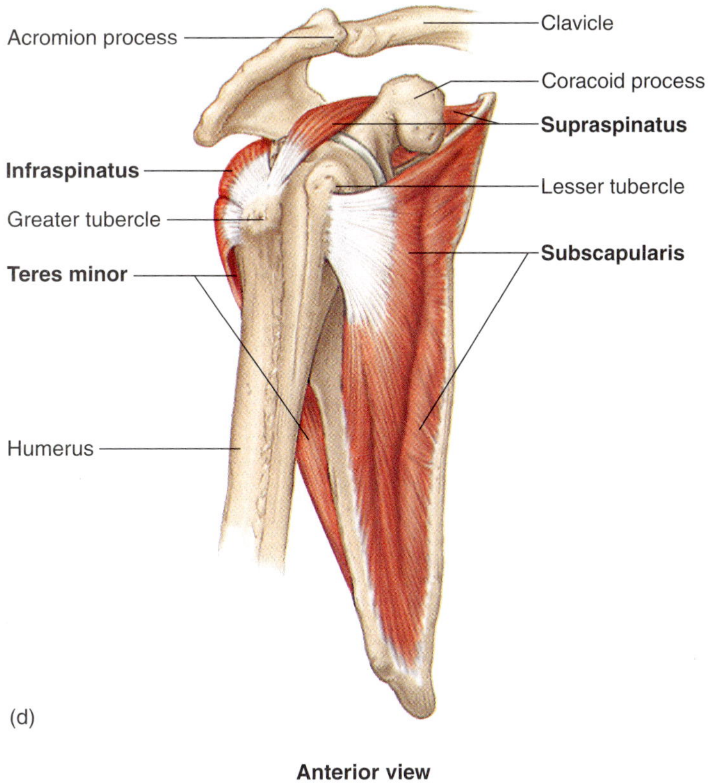

Right Rotator Cuff Muscles
Figure 10.25

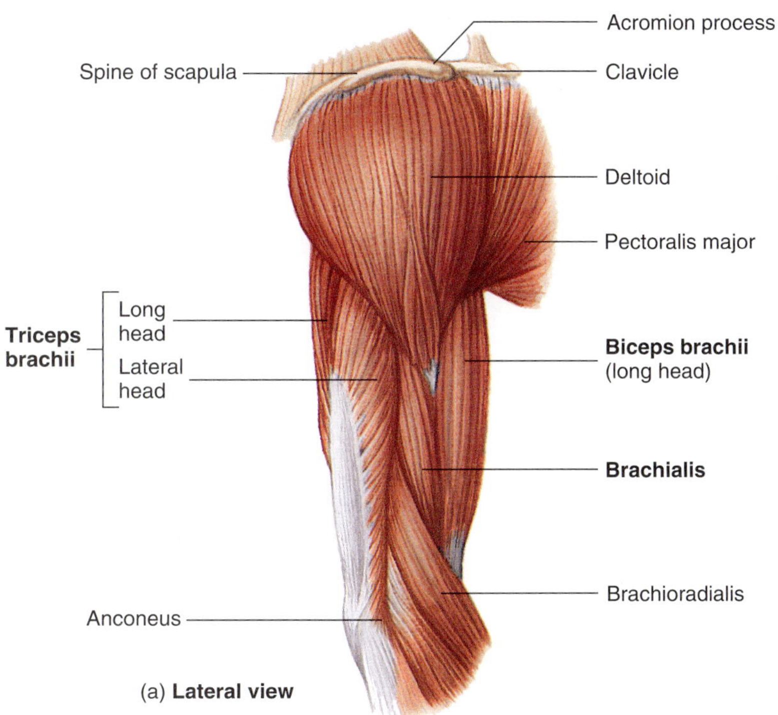

Lateral Right Arm Muscles
Figure 10.26

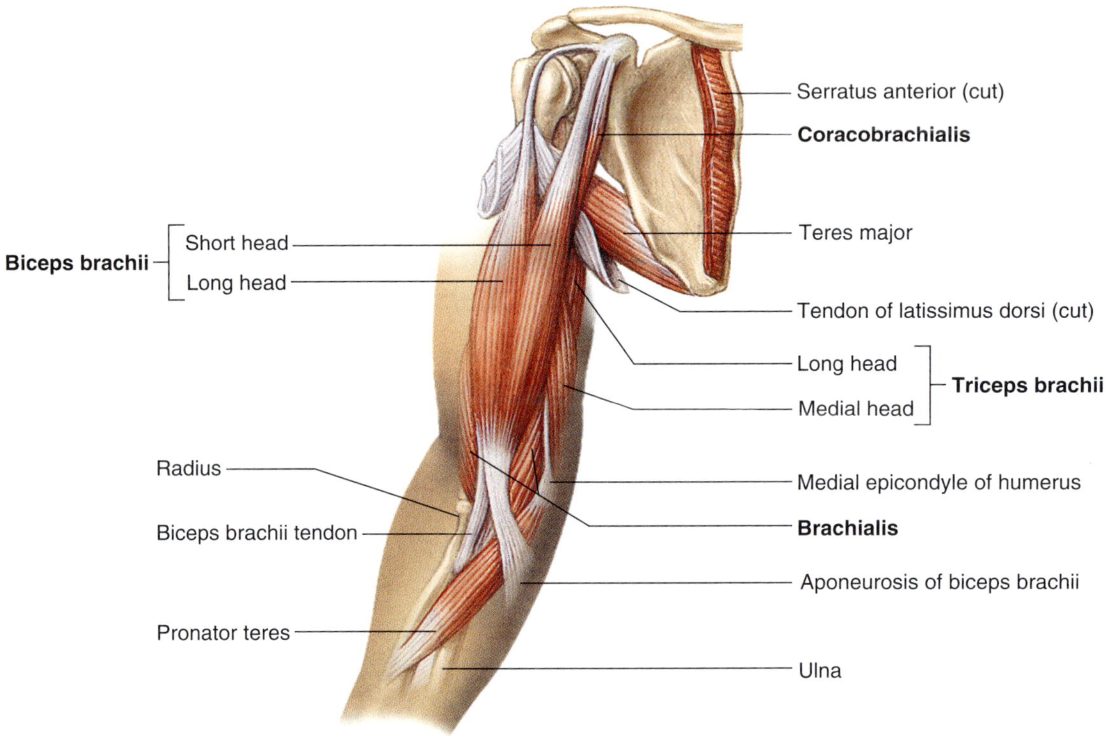
Serratus anterior (cut)
Coracobrachialis
Teres major
Biceps brachii
Short head
Long head
Tendon of latissimus dorsi (cut)
Long head
Medial head
Triceps brachii
Radius
Medial epicondyle of humerus
Biceps brachii tendon
Brachialis
Aponeurosis of biceps brachii
Pronator teres
Ulna
Anteromedial view

Anterior Right Arm Muscles
Figure 10.27

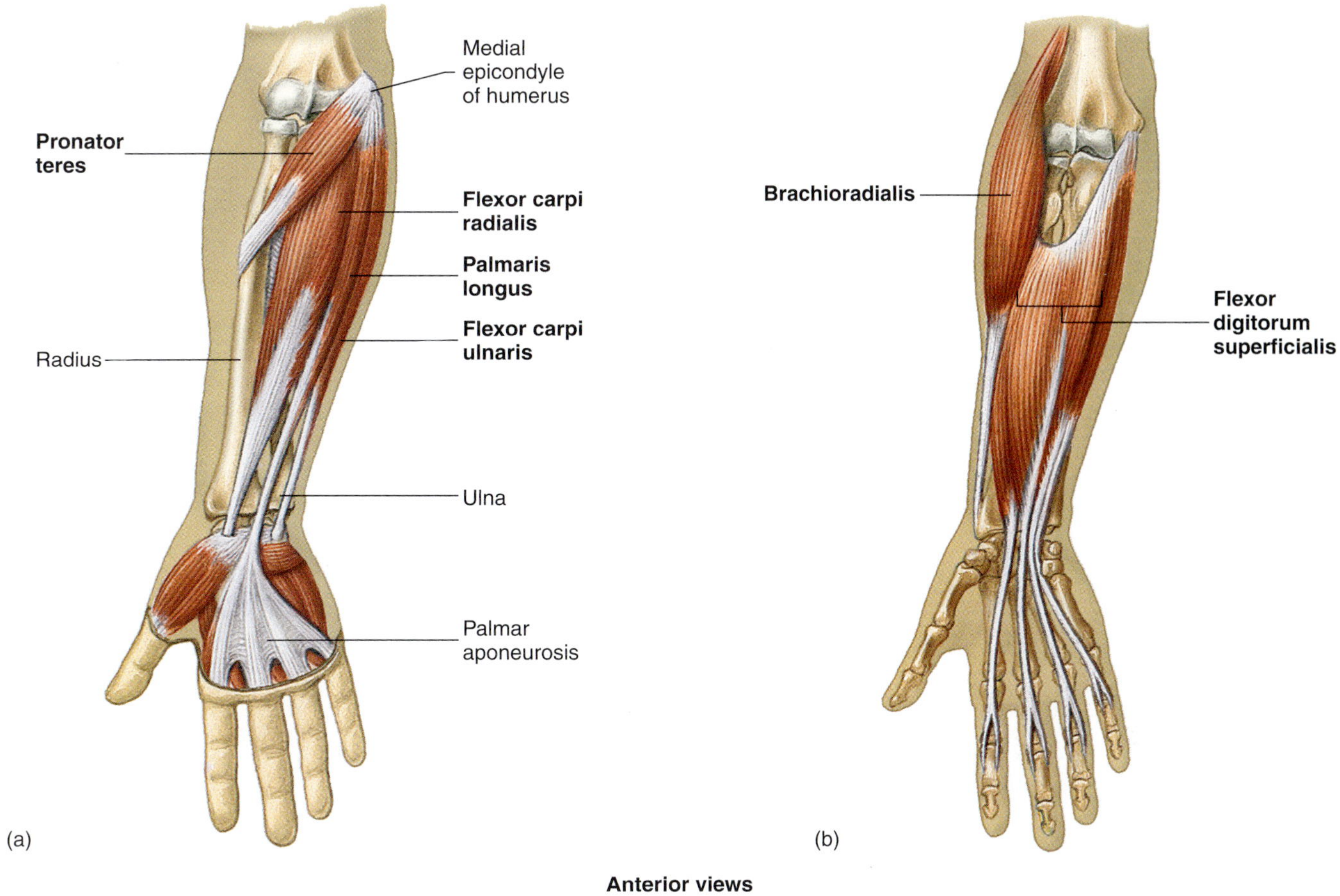
Medial
epicondyle
of humerus
Pronator
teres
Flexor carpi
radialis
Palmaris
longus
Flexor carpi
ulnaris
Radius
Ulna
Palmar
aponeurosis
(a)
Brachioradialis
Flexor
digitorum
superficialis
(b)
Anterior views

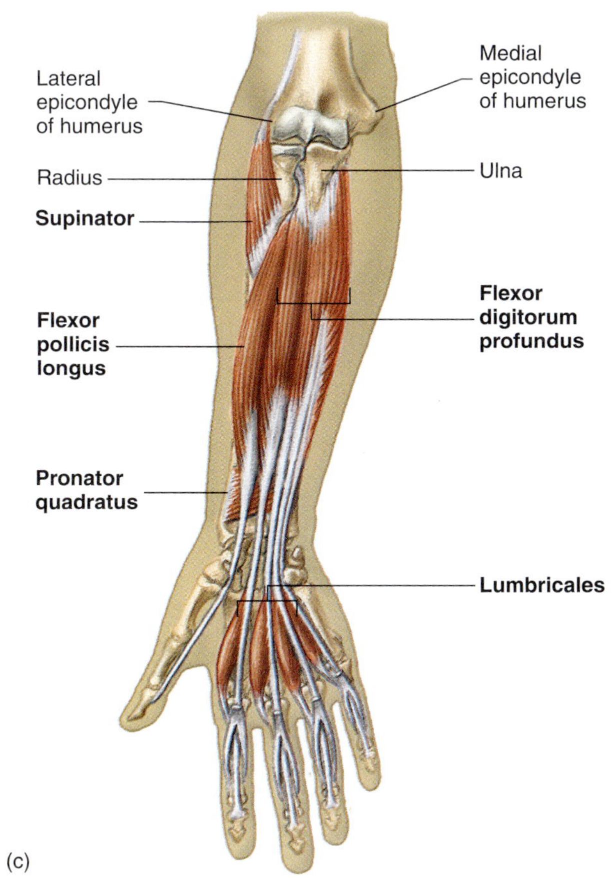
Lateral
epicondyle
of humerus
Medial
epicondyle
of humerus
Radius
Ulna
Supinator
Flexor
digitorum
profundus
Flexor
pollicis
longus
Pronator
quadratus
Lumbricales
(c)

Anterior Right Forearm Muscles
Figure 10.28

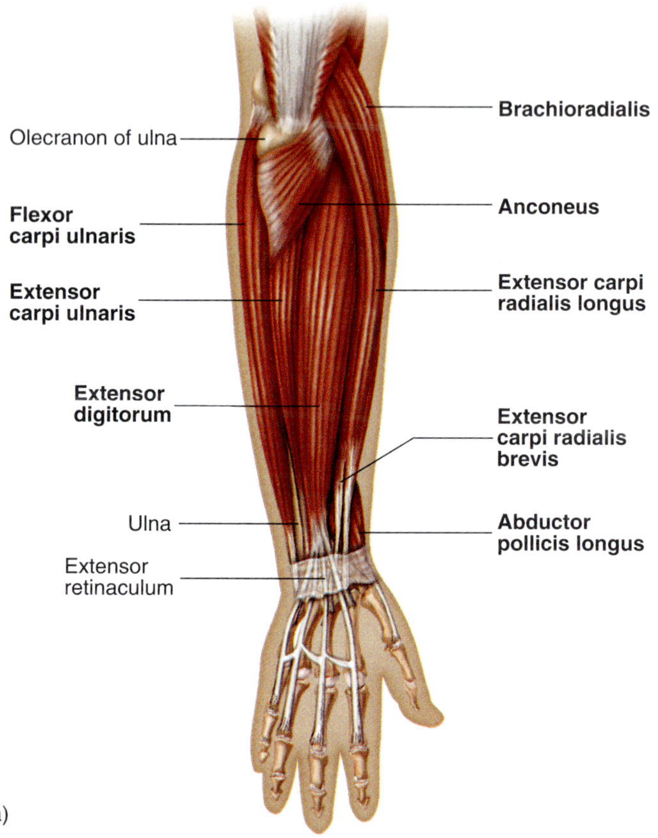
Olecranon of ulna
Flexor carpi ulnaris
Extensor carpi ulnaris
Extensor digitorum
Ulna
Extensor retinaculum
Brachioradialis
Anconeus
Extensor carpi radialis longus
Extensor carpi radialis brevis
Abductor pollicis longus
(a)

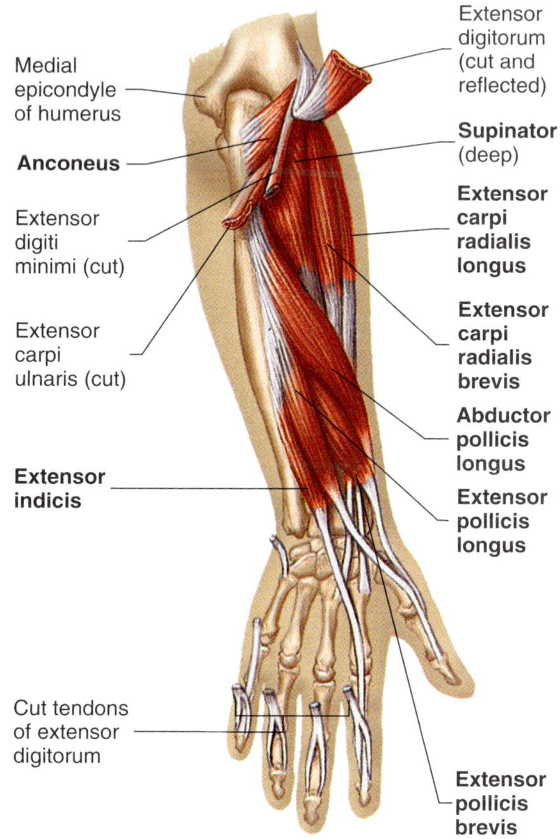
Medial epicondyle of humerus
Anconeus
Extensor digiti minimi (cut)
Extensor carpi ulnaris (cut)
Extensor indicis
Cut tendons of extensor digitorum
Extensor digitorum (cut and reflected)
Supinator (deep)
Extensor carpi radialis longus
Extensor carpi radialis brevis
Abductor pollicis longus
Extensor pollicis longus
Extensor pollicis brevis
(b)

Posterior Right Forearm Muscles
Figure 10.29

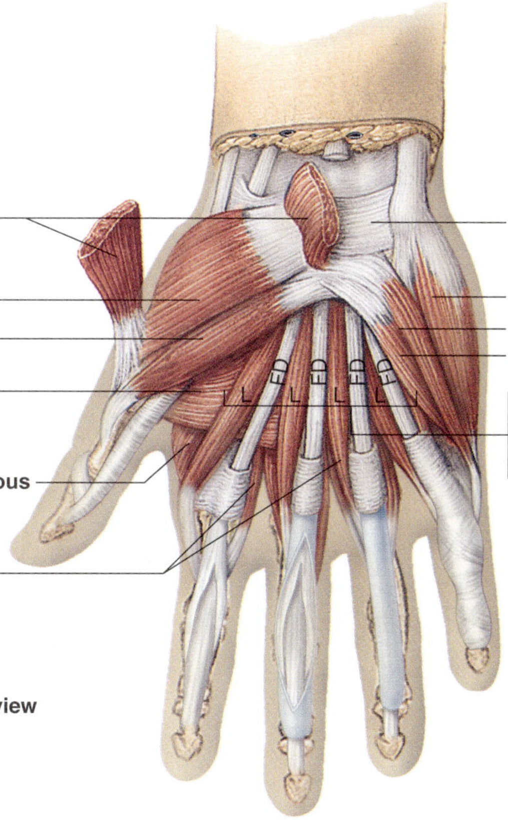

(a) **Anterior view**

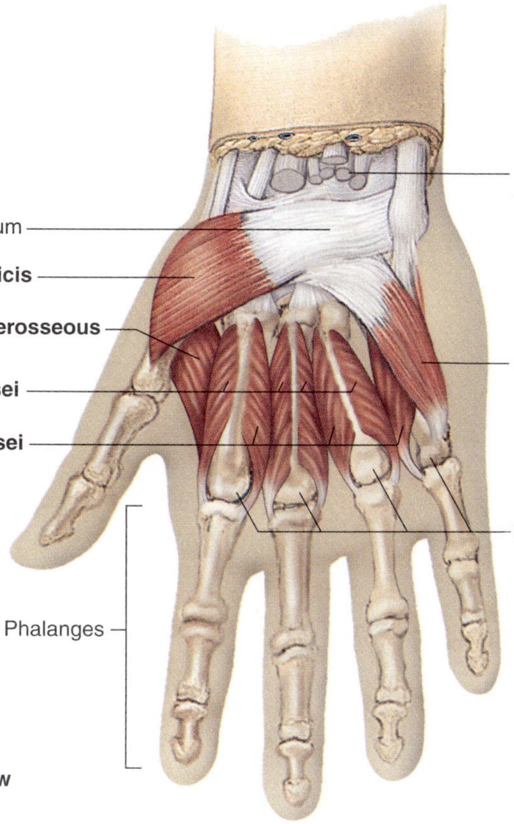

(b) **Anterior view**

Right Hand Muscles
Figure 10.30

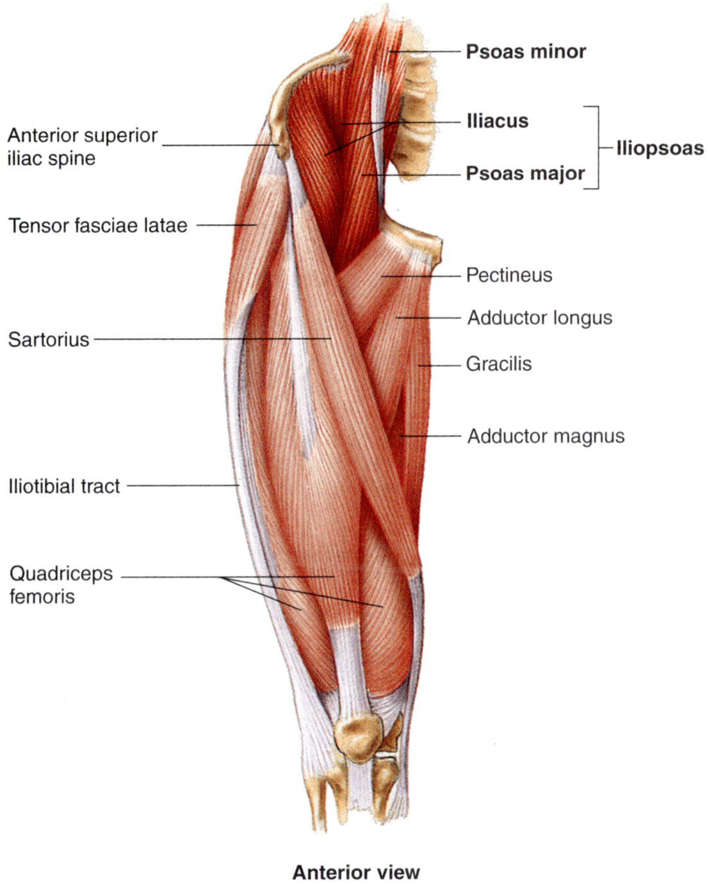

Right Proximal Anterior Thigh Muscles
Figure 10.31

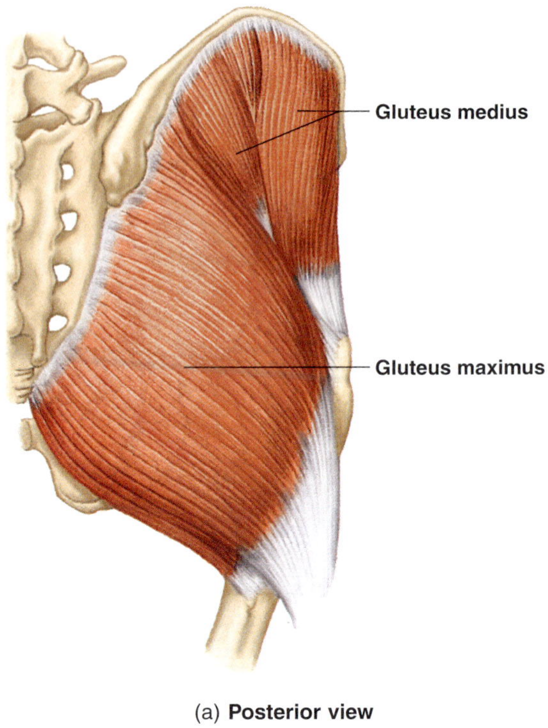

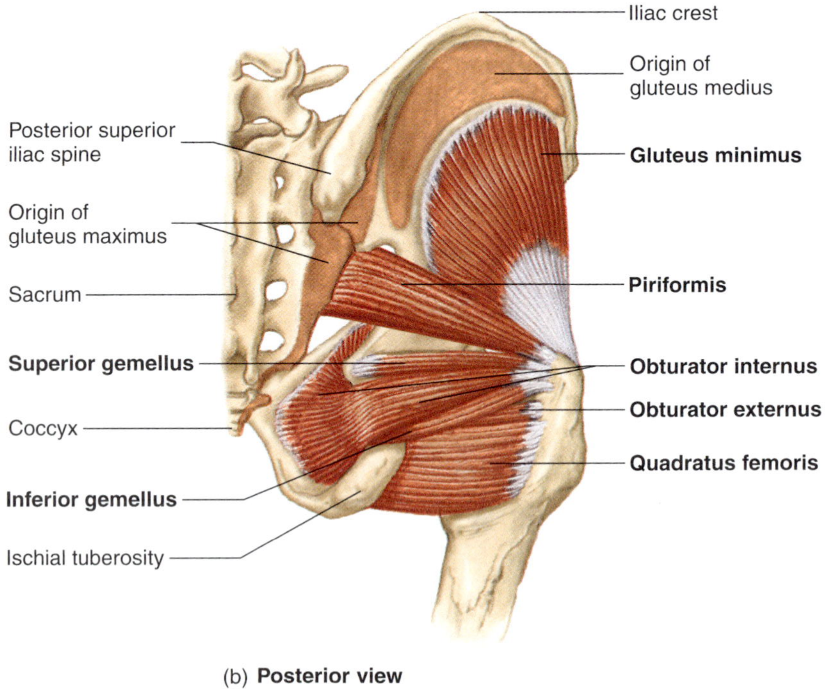

Right Posterior Hip Muscles
Figure 10.32

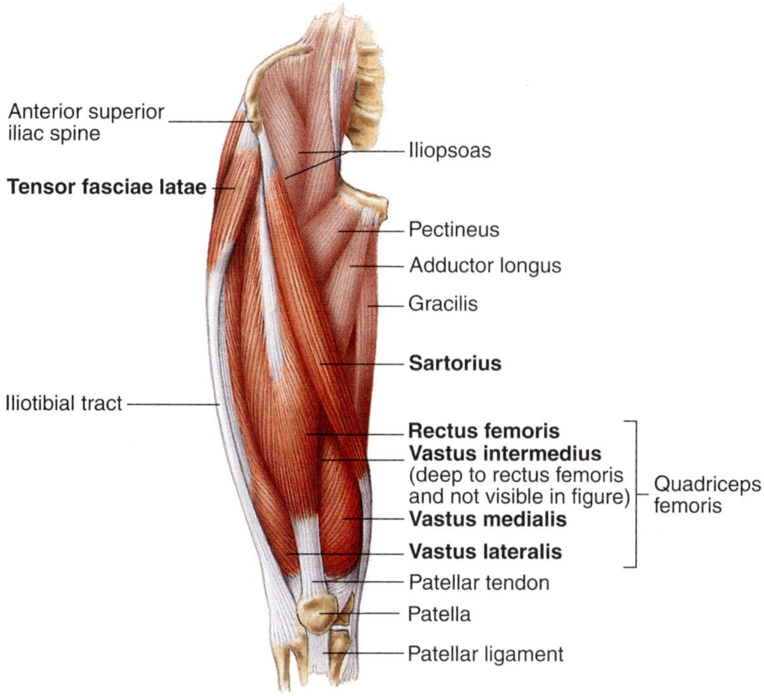

(a) **Anterior view**

Right Anterior Thigh Muscles
Figure 10.33

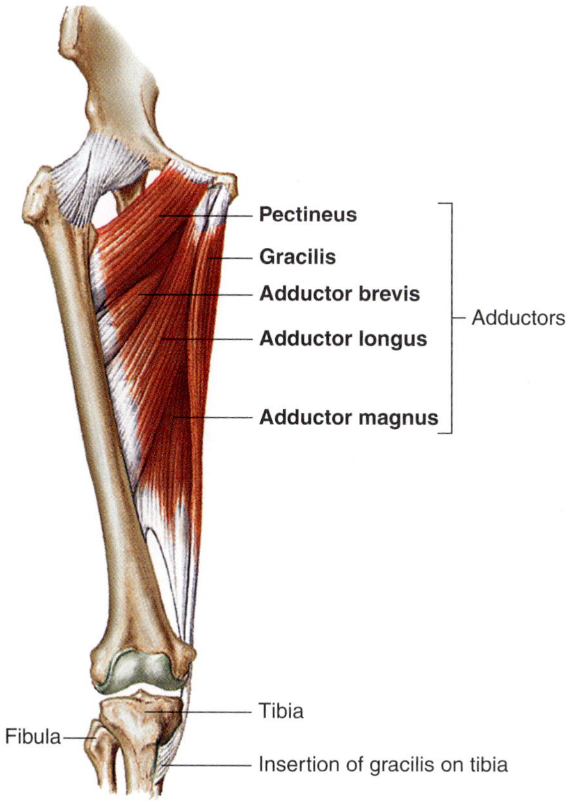

Anterior view

Right Medial Thigh Muscles
Figure 10.34

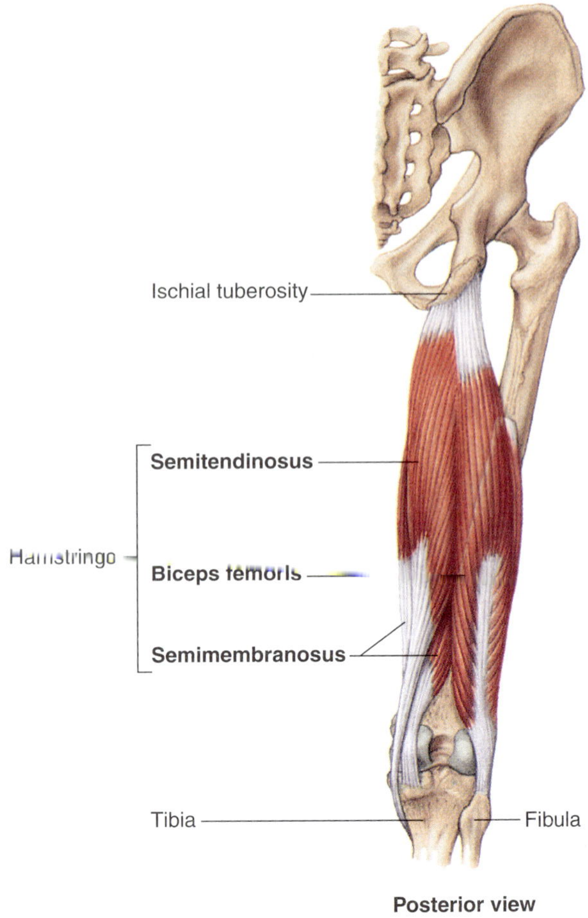

Right Posterior Thigh Muscles
Figure 10.35

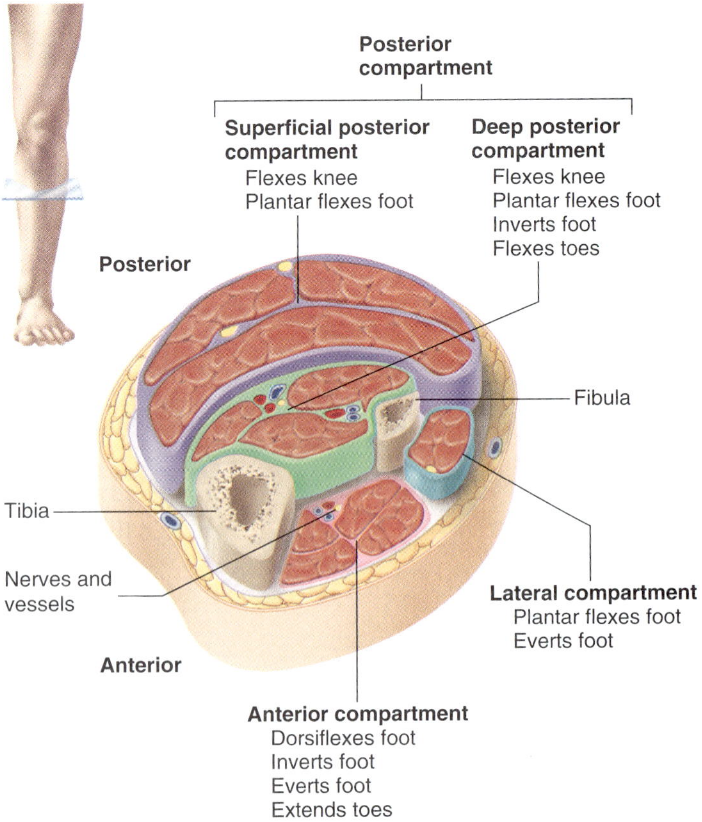

Cross Section Through the Left Leg
Figure 10.37

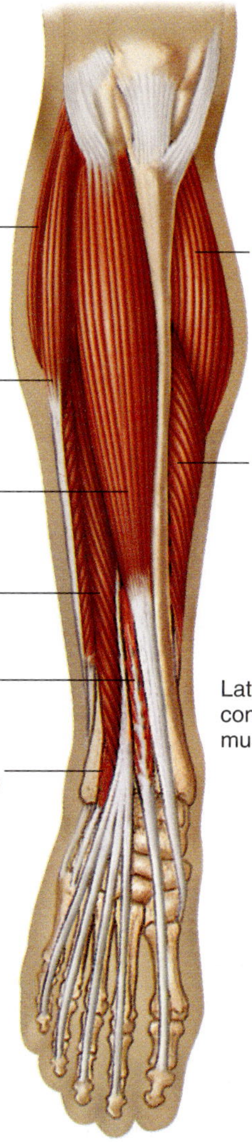

(a) **Anterior view**

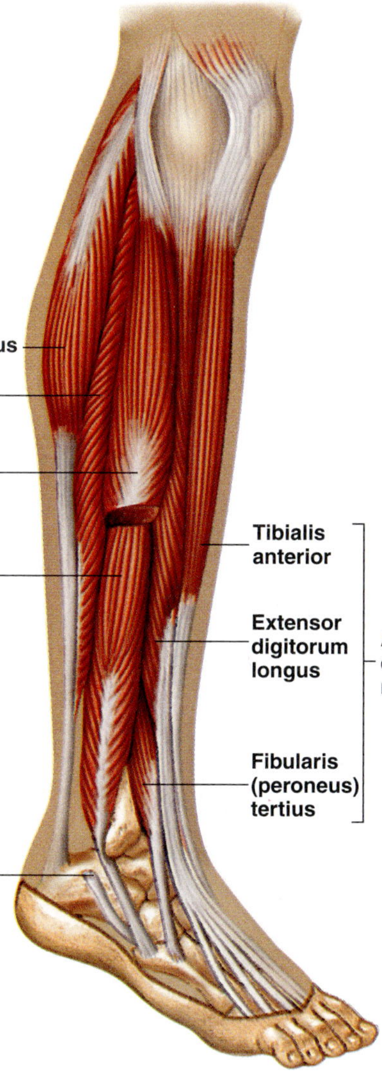

(b) **Lateral view**

Right Lateral Leg Muscles
Figure 10.38

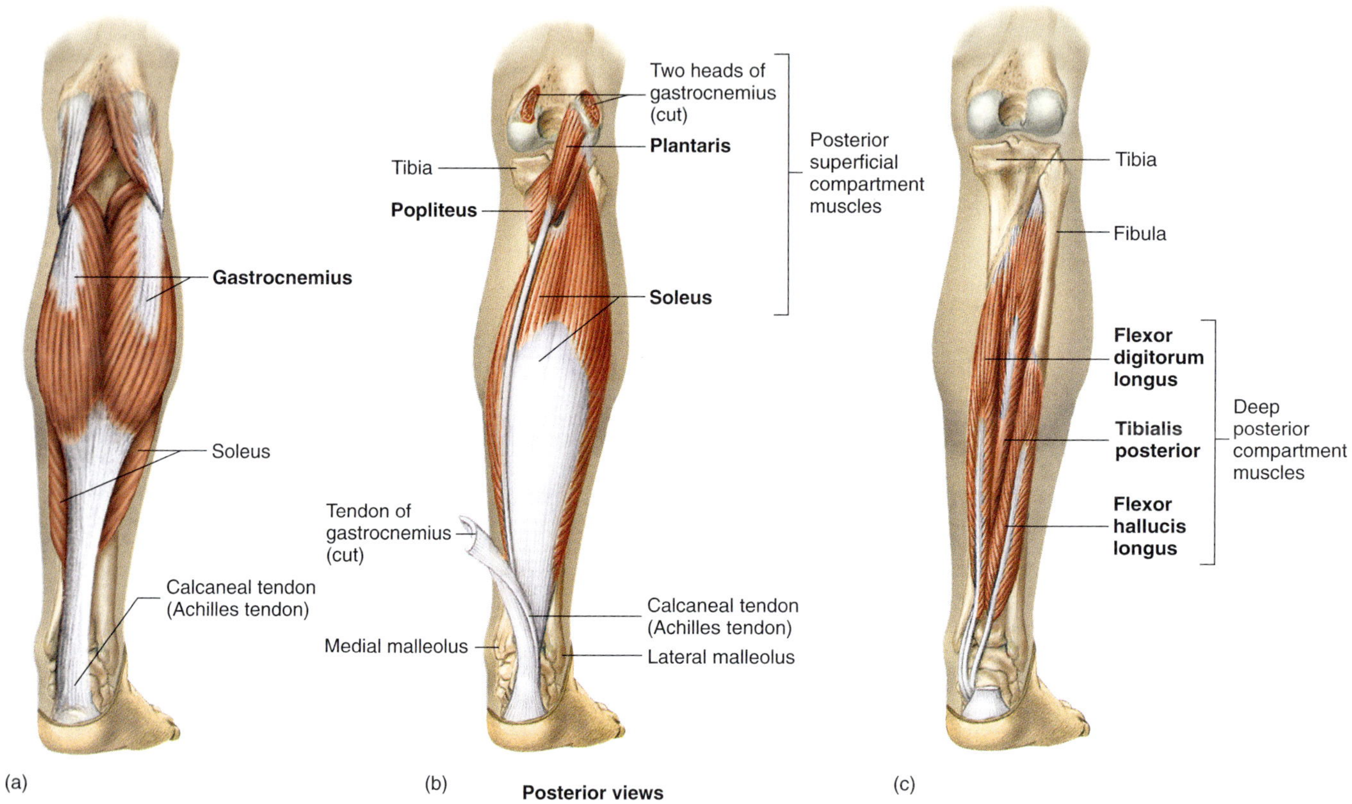
Gastrocnemius
Soleus
Calcaneal tendon
(Achilles tendon)
(a)
Two heads of
gastrocnemius
(cut)
Plantaris
Tibia
Popliteus
Posterior
superficial
compartment
muscles
Soleus
Tendon of
gastrocnemius
(cut)
Calcaneal tendon
(Achilles tendon)
Medial malleolus
Lateral malleolus
(b)
Posterior views
Tibia
Fibula
Flexor
digitorum
longus
Tibialis
posterior
Flexor
hallucis
longus
Deep
posterior
compartment
muscles
(c)

Right Posterior Leg Muscles
Figure 10.39

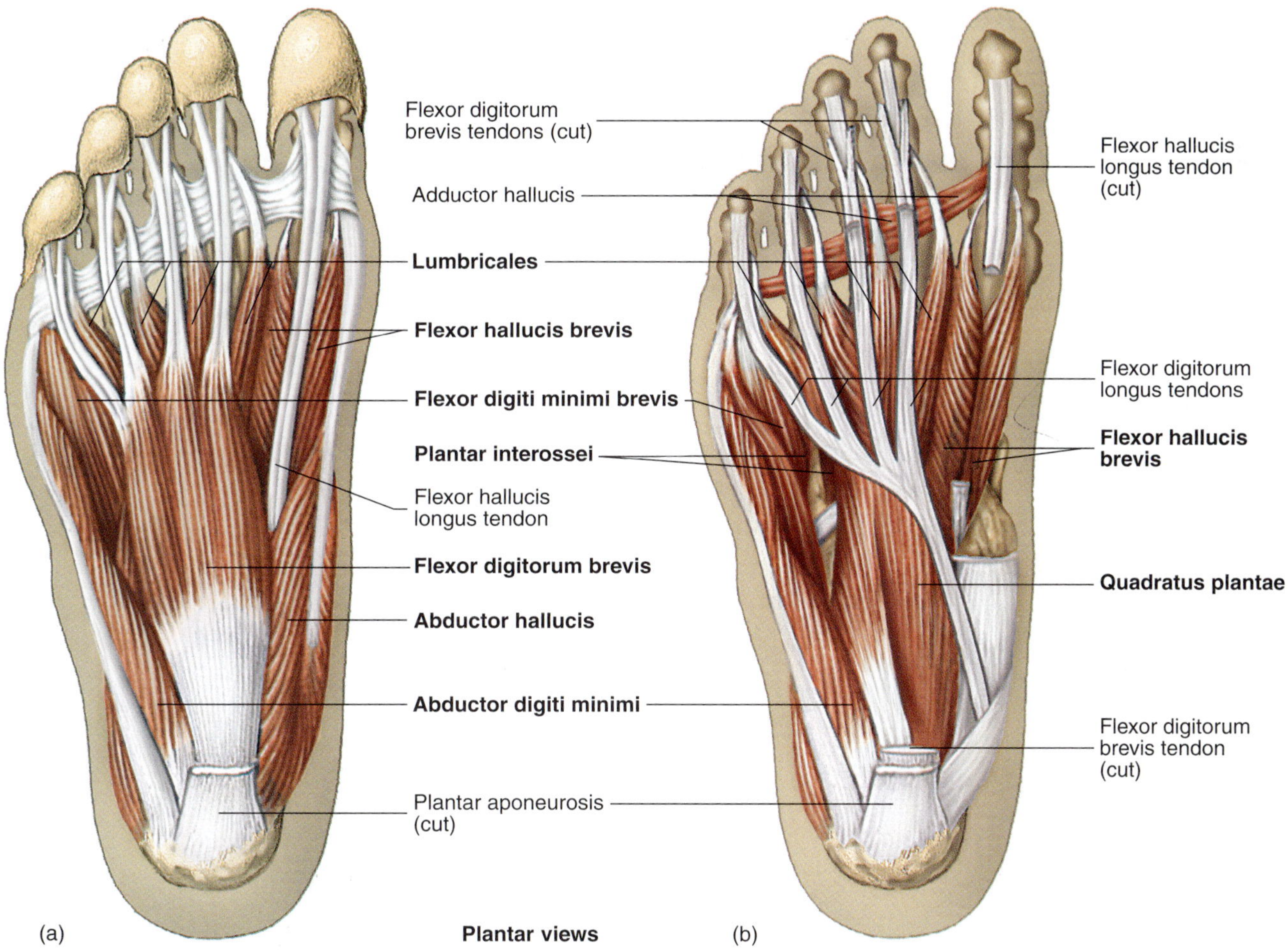

Flexor digitorum brevis tendons (cut)
Adductor hallucis
Lumbricales
Flexor hallucis brevis
Flexor digiti minimi brevis
Plantar interossei
Flexor hallucis longus tendon
Flexor digitorum brevis
Abductor hallucis
Abductor digiti minimi
Plantar aponeurosis (cut)
Flexor hallucis longus tendon (cut)
Flexor digitorum longus tendons
Flexor hallucis brevis
Quadratus plantae
Flexor digitorum brevis tendon (cut)
(a)
Plantar views
(b)

Right Foot Muscles
Figure 10.40

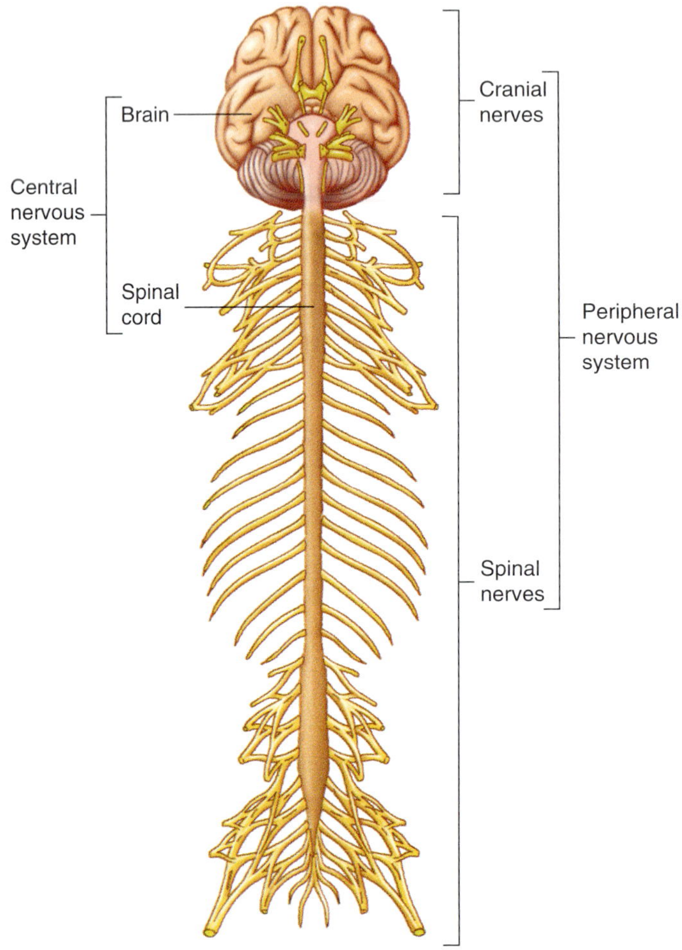

The Nervous System
Figure 11.1

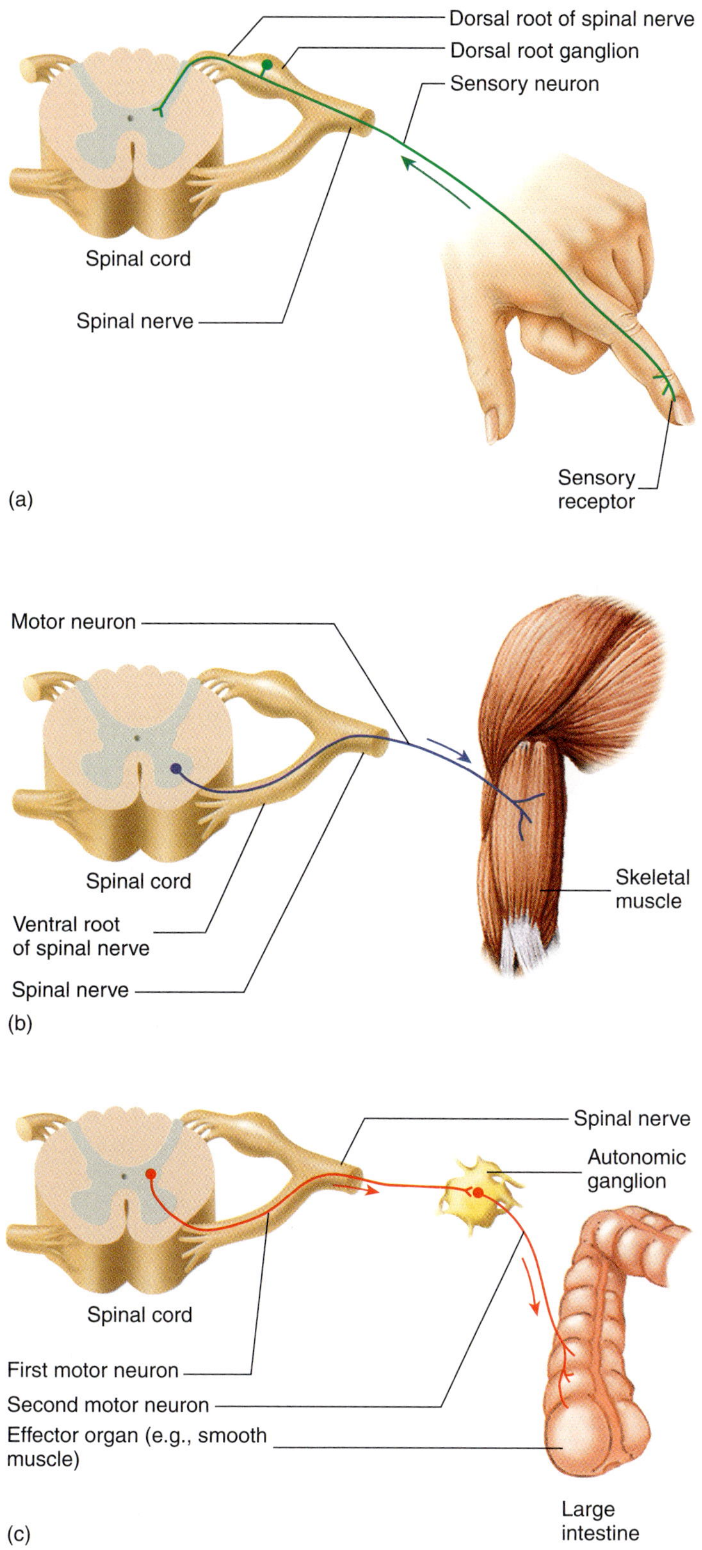

Divisions of the Peripheral Nervous System
Figure 11.2

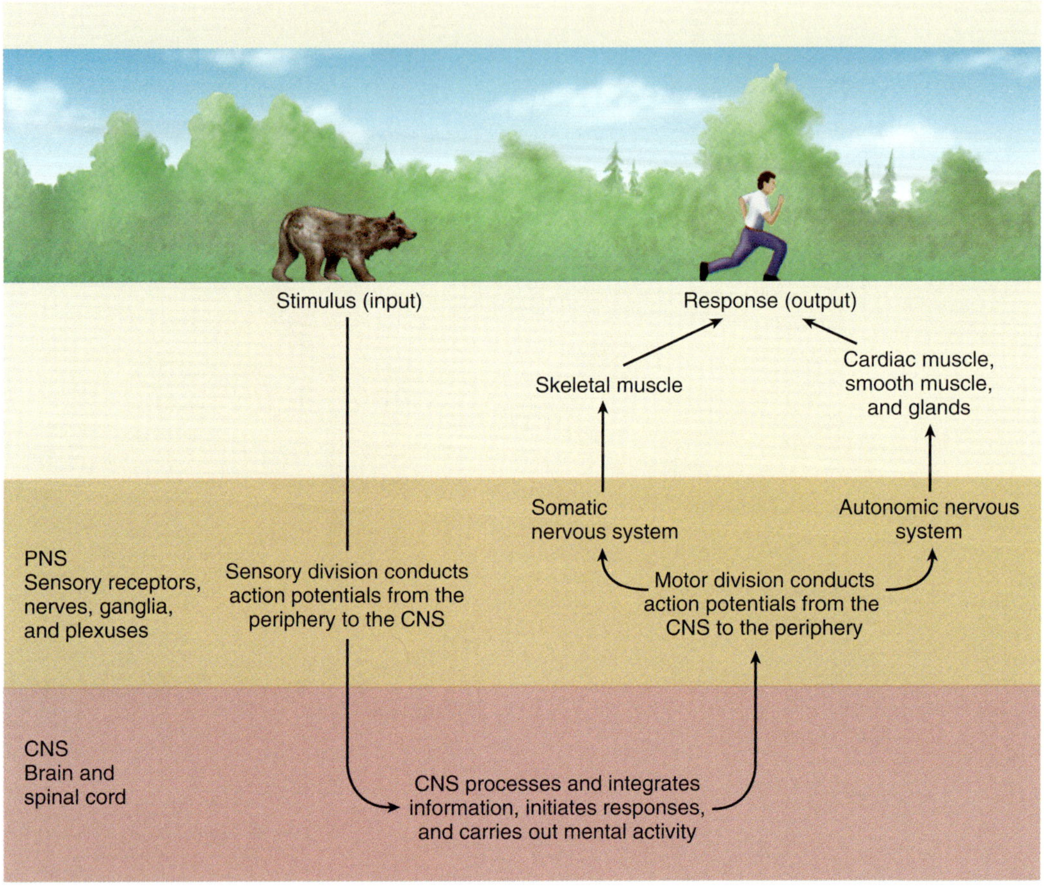

Stimulus (input)
Response (output)
Skeletal muscle
Cardiac muscle, smooth muscle, and glands
Somatic nervous system
Autonomic nervous system
PNS
Sensory receptors, nerves, ganglia, and plexuses
Sensory division conducts action potentials from the periphery to the CNS
Motor division conducts action potentials from the CNS to the periphery
CNS
Brain and spinal cord
CNS processes and integrates information, initiates responses, and carries out mental activity

Organization of the Nervous System
Figure 11.3

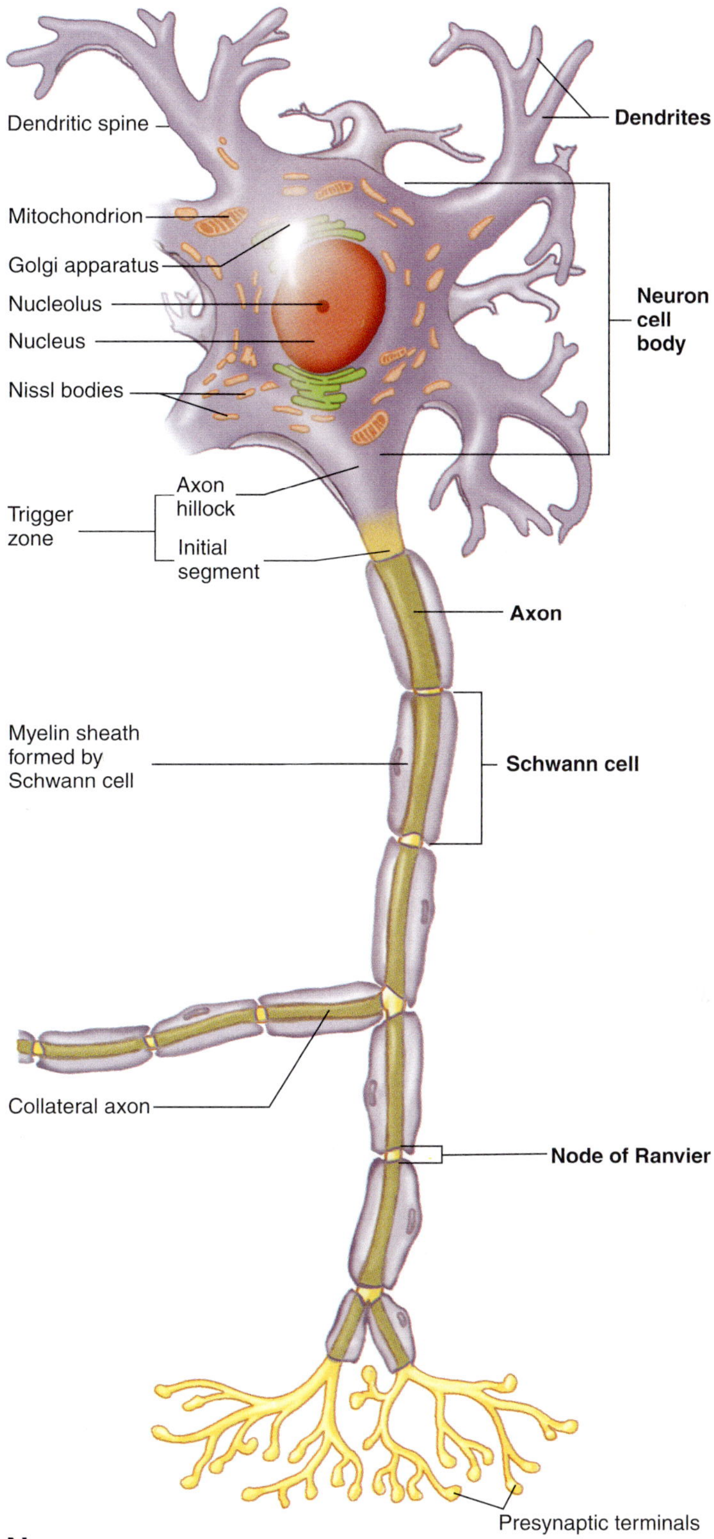

Dendritic spine
Dendrites
Mitochondrion
Golgi apparatus
Nucleolus
Nucleus
Nissl bodies
Neuron
cell
body
Axon
hillock
Trigger
zone
Initial
segment
Axon
Myelin sheath
formed by
Schwann cell
Schwann cell
Collateral axon
Node of Ranvier
Presynaptic terminals

Neuron
Figure 11.4

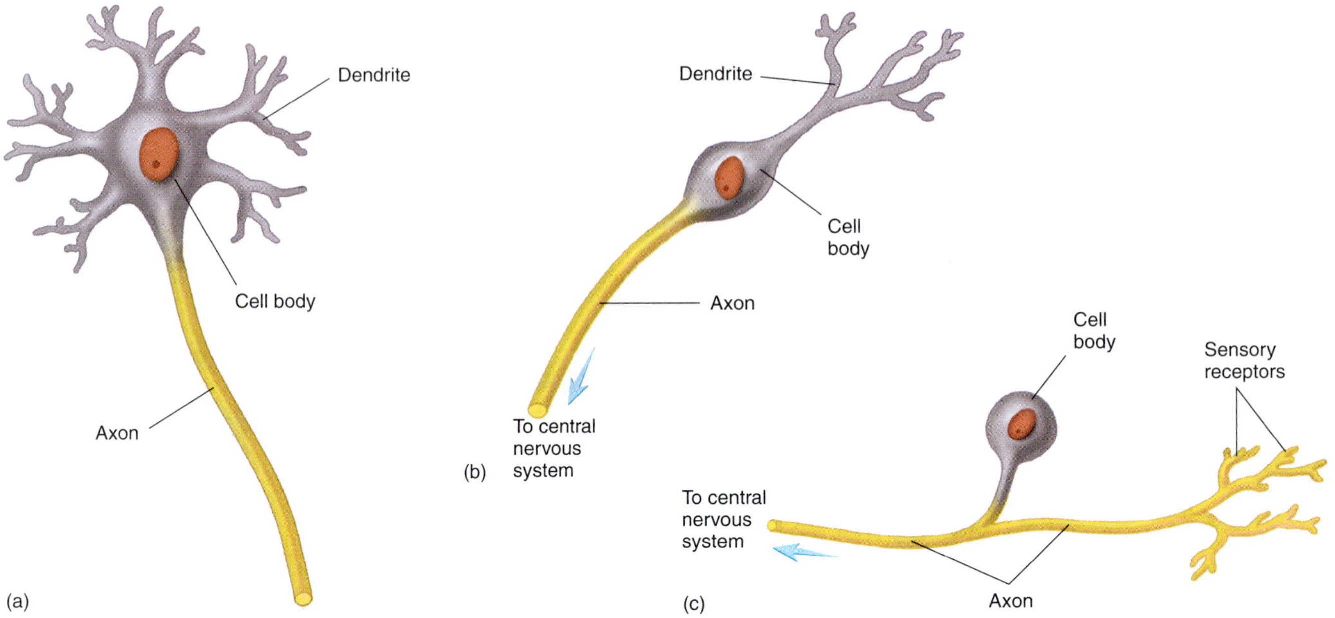

Types of Neurons
Figure 11.5

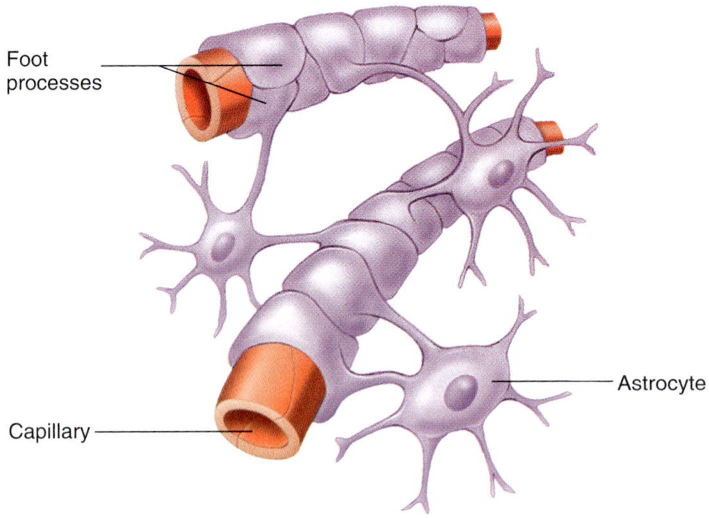

Astrocytes
Figure 11.6

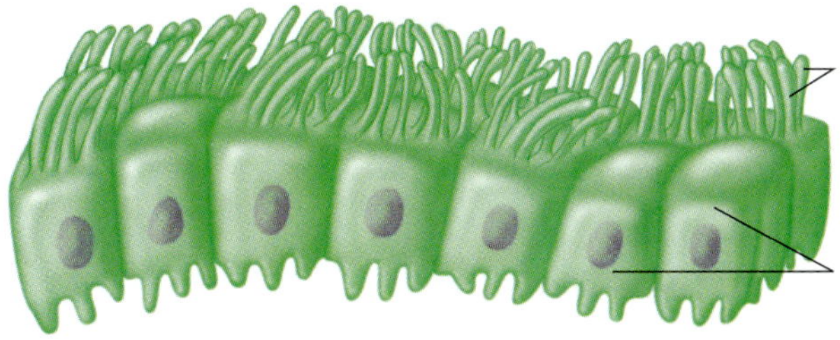

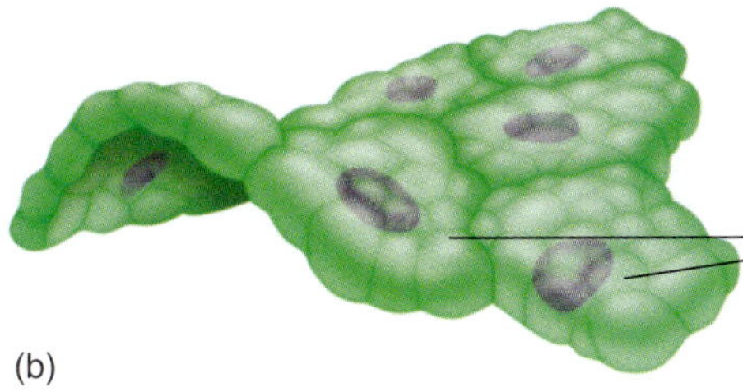

Ependymal Cells
Figure 11.7

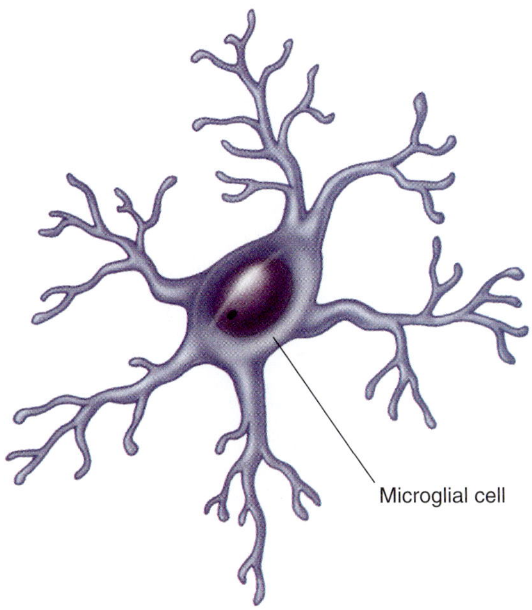

Microglia
Figure 11.8

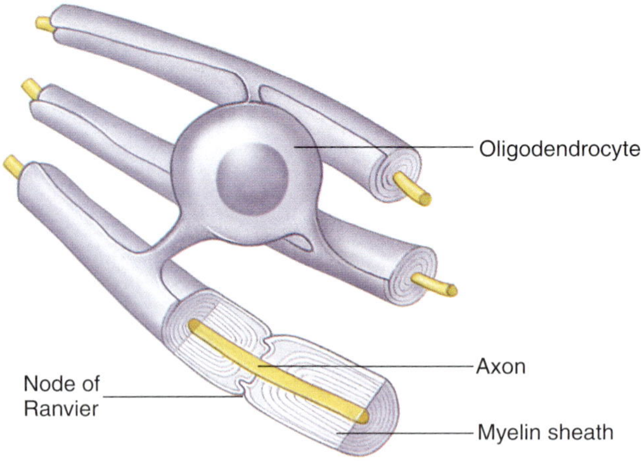

Oligodendrocyte
Figure 11.9

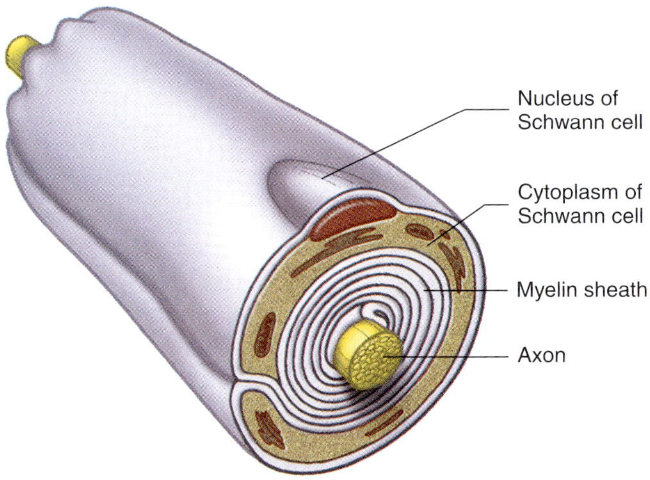

Schwann Cell
Figure 11.10

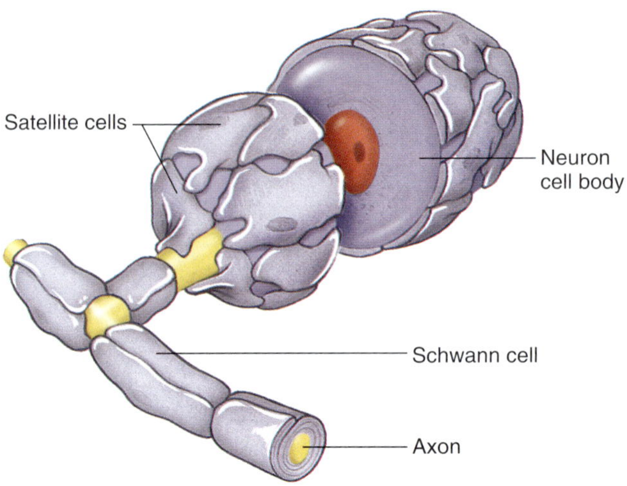

Satellite Cells
Figure 11.11

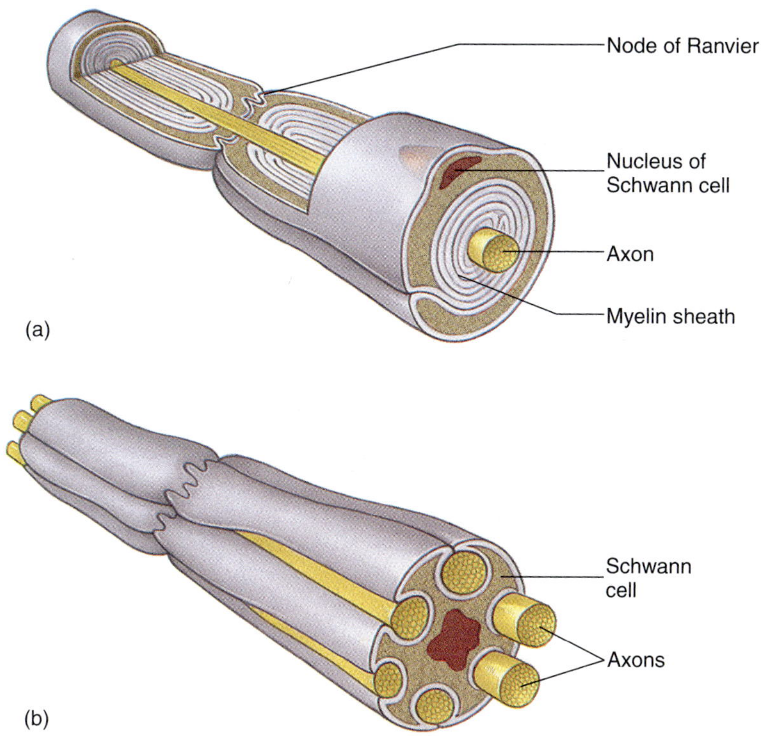

Comparison of Myelinated and Unmyelinated Axons
Figure 11.12

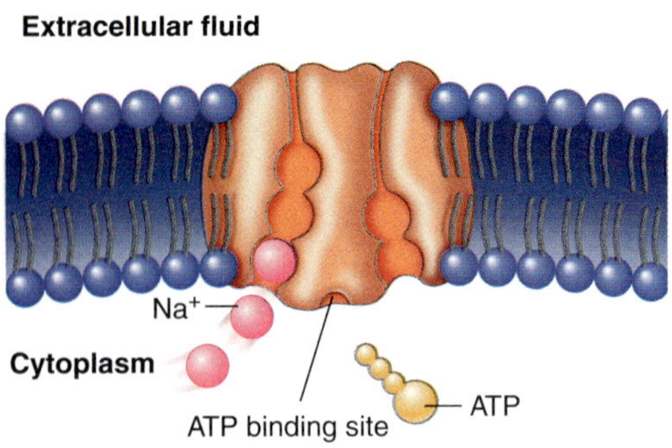

1. Three Na$^+$ and ATP bind to the carrier molecule.

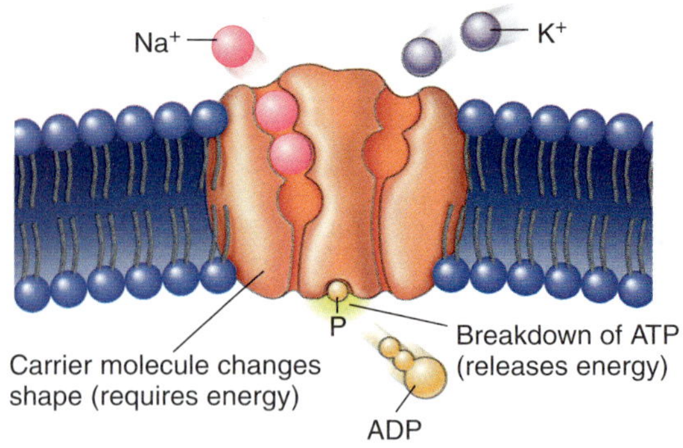

2. The ATP breaks down to ADP and phosphate and releases energy. The carrier molecule changes shape, and Na$^+$ are transported across the membrane.

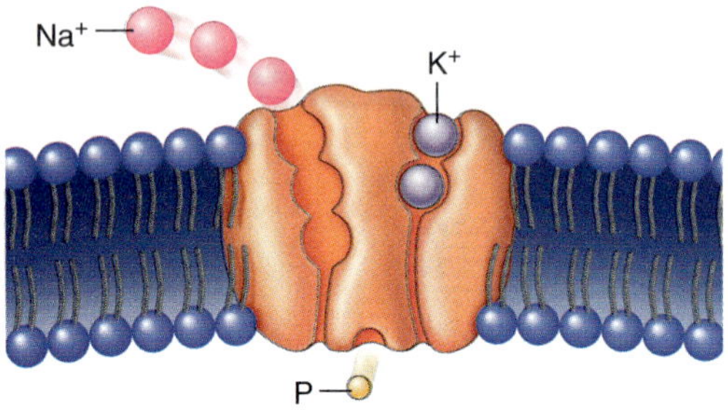

3. Na$^+$ diffuse away from the carrier molecule, two K$^+$ bind to the carrier molecule, and the phosphate is released.

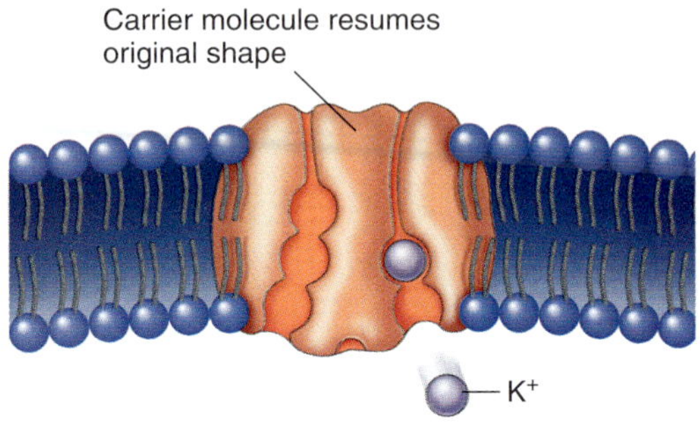

4. The carrier molecule resumes original shape, transporting K$^+$ across the membrane, and K$^+$ diffuse away from the carrier molecule. The carrier molcule can again bind to Na$^+$ and ATP.

The Na$^+$–K$^+$ Pump
Process Figure 11.13

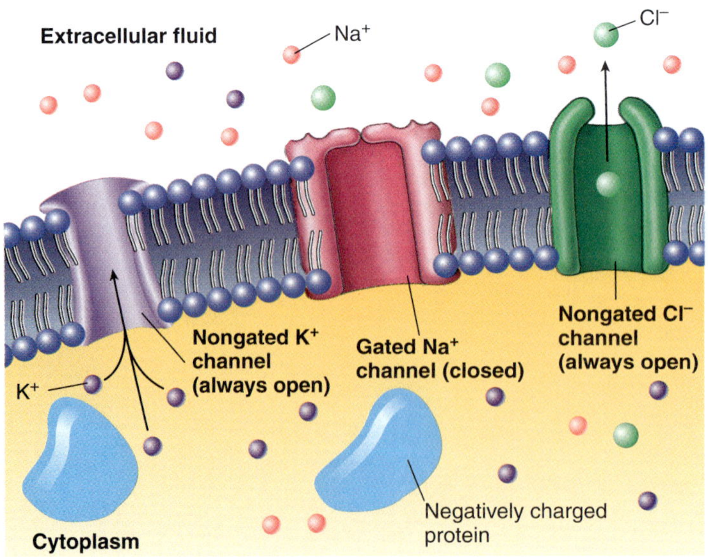

Membrane Permeability and Ion Channels
Figure 11.14

Ligand-Gated Ion Channel
Process Figure 11.15

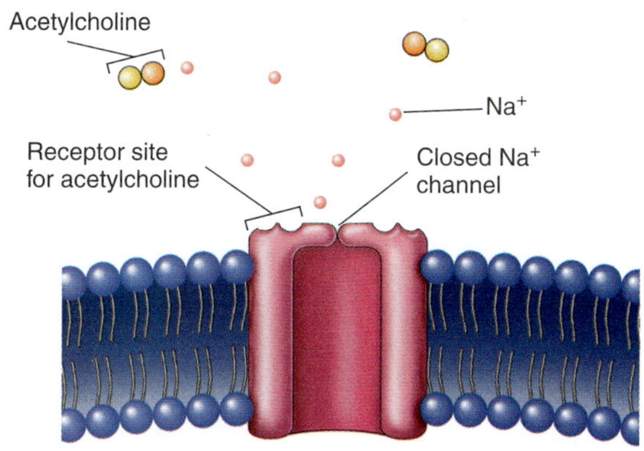

1. The Na^+ channel has receptor sites for the ligand, acetylcholine. When the receptor sites are not occupied by acetylcholine, the Na^+ channel remains closed.

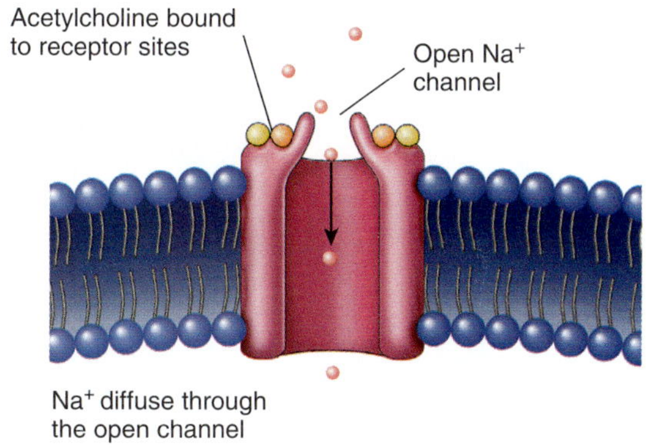

2. When two acetylcholine molecules bind to their receptor sites on the Na^+ channel, the channel opens to allow Na^+ to diffuse through the channel into the cell.

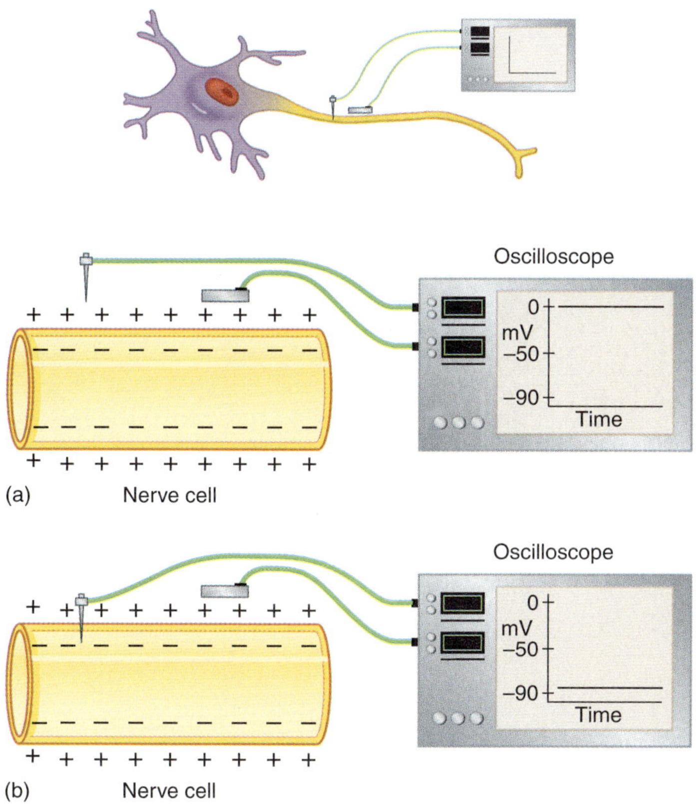

Measuring the Resting Membrane Potential
Figure 11.16

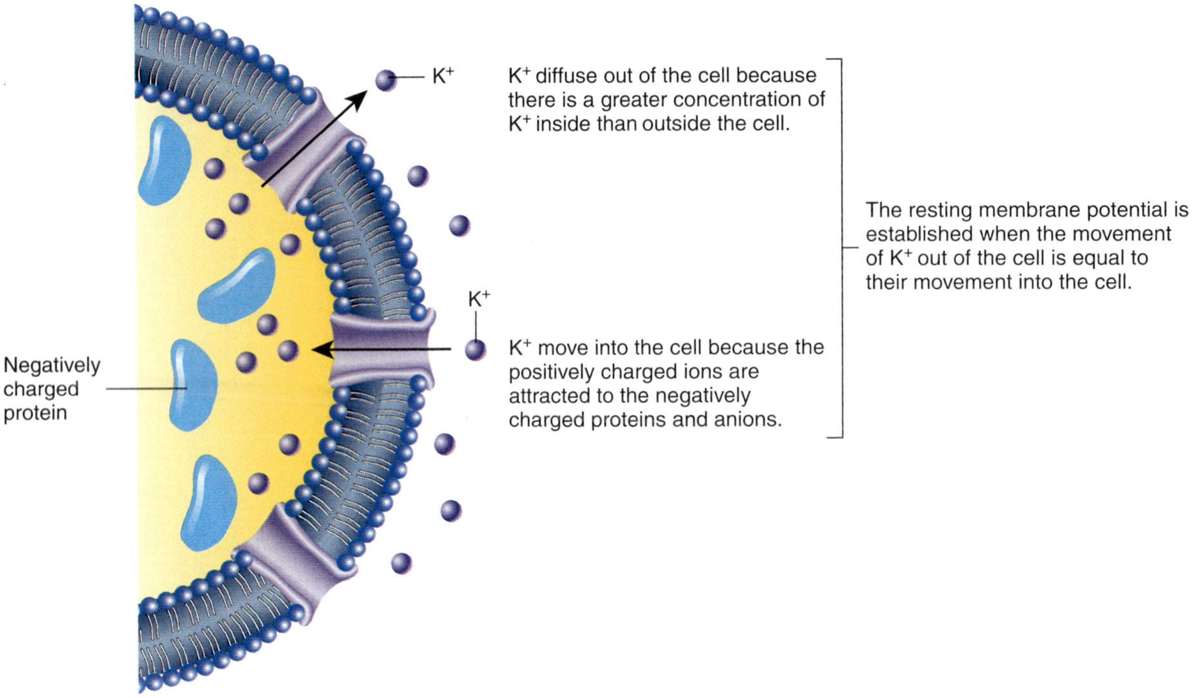

Potassium Ions and the Resting Membrane Potential
Figure 11.17

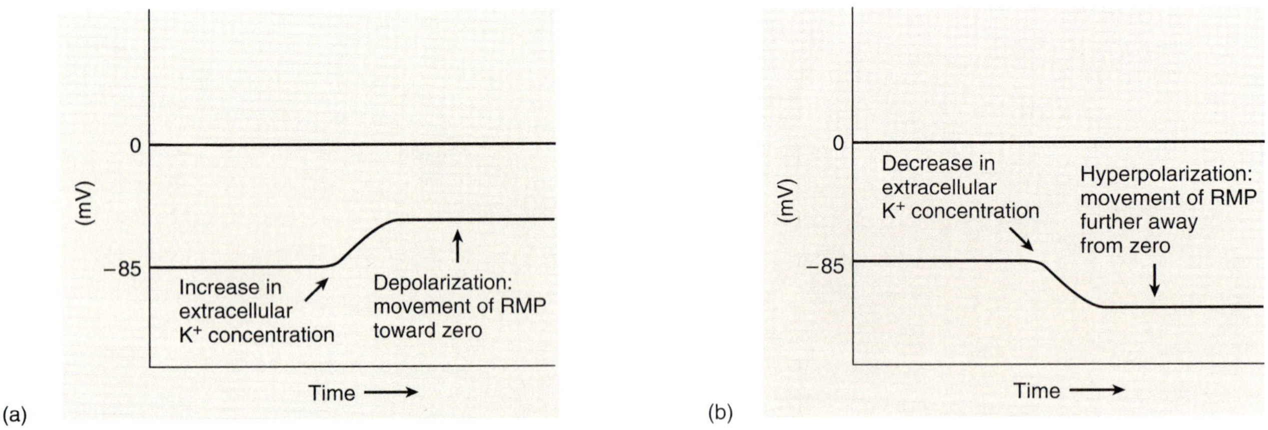

Changes in the Resting Membrane Potential Caused by Changes in Extracellular K^+ Concentration
Figure 11.18

(a)

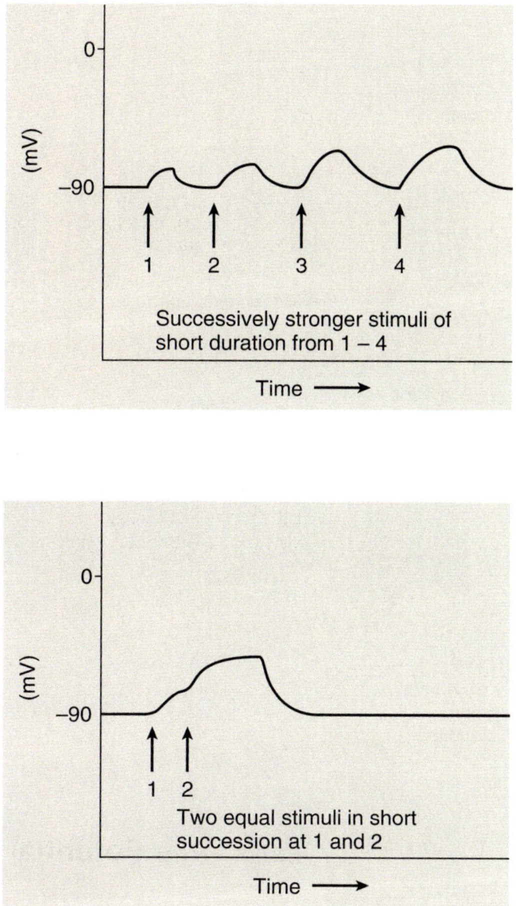

(b)

Local Potentials
Figure 11.19

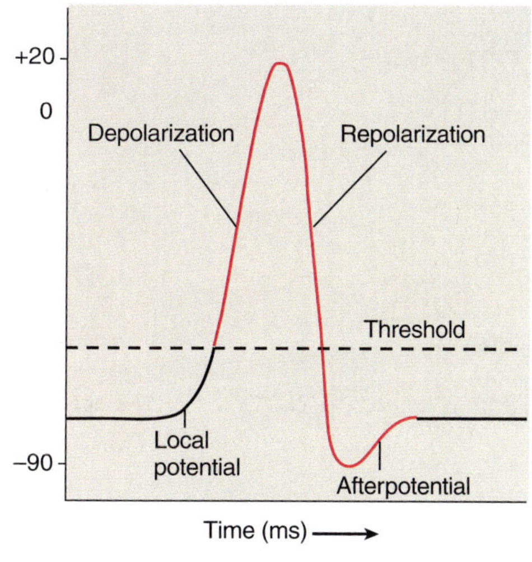

The Action Potential
Figure 11.20

1. **Resting membrane potential.** Voltage-gated Na^+ channels (*pink*) are closed (the activation gates are closed and the inactivation gates are open). Voltage-gated K^+ channels (*purple*) are closed.
2. **Depolarization.** Voltage-gated Na^+ channels open because the activation gates open. Voltage-gated K^+ channels start to open. Depolarization results because the inward diffusion of Na^+ is much greater than the outward diffusion of K^+.
3. **Repolarization.** Voltage-gated Na^+ channels are closed because the inactivation gates close. Voltage-gated K^+ channels are now open. Na^+ diffusion into the cell stops and K^+ diffuse out of the cell, causing repolarizaton.
4. **End of repolarization and afterpotential.** Voltage-gated Na^+ channels are closed. Closure of the activation gates and opening of the inactivation gates reestablish the resting condition for Na^+ channels (see step 1). Diffusion of K^+ through voltage-gated channels produces the afterpotential.
5. **Resting membrane potential.** The resting membrane potential is reestablished after the voltage-gated K^+ channels close.

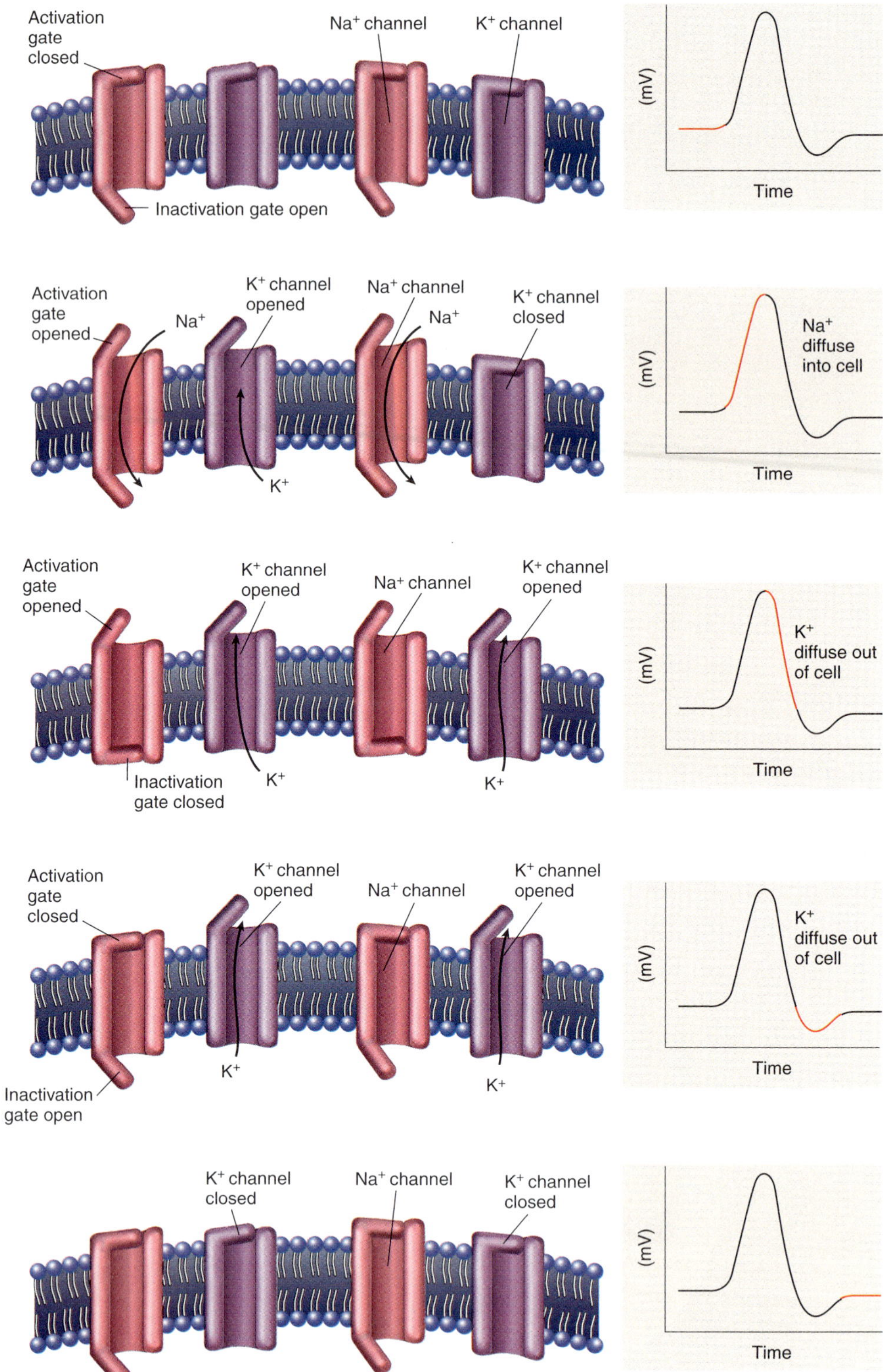

Voltage-Gated Ion Channels and the Action Potential
Process Figure 11.21

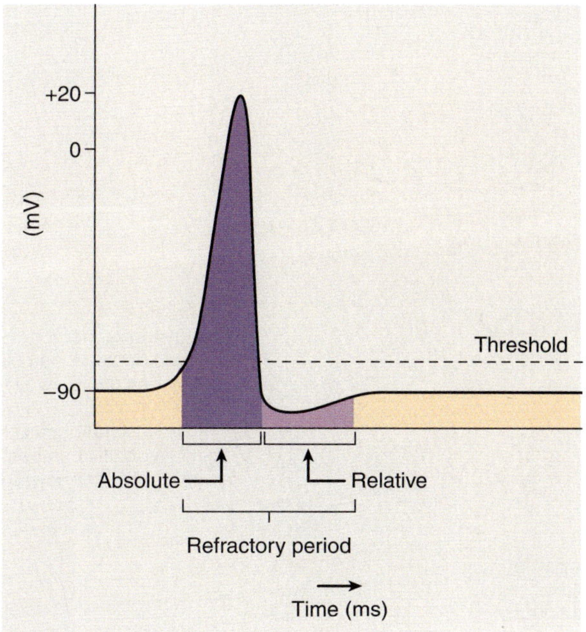

Refractory Period
Figure 11.22

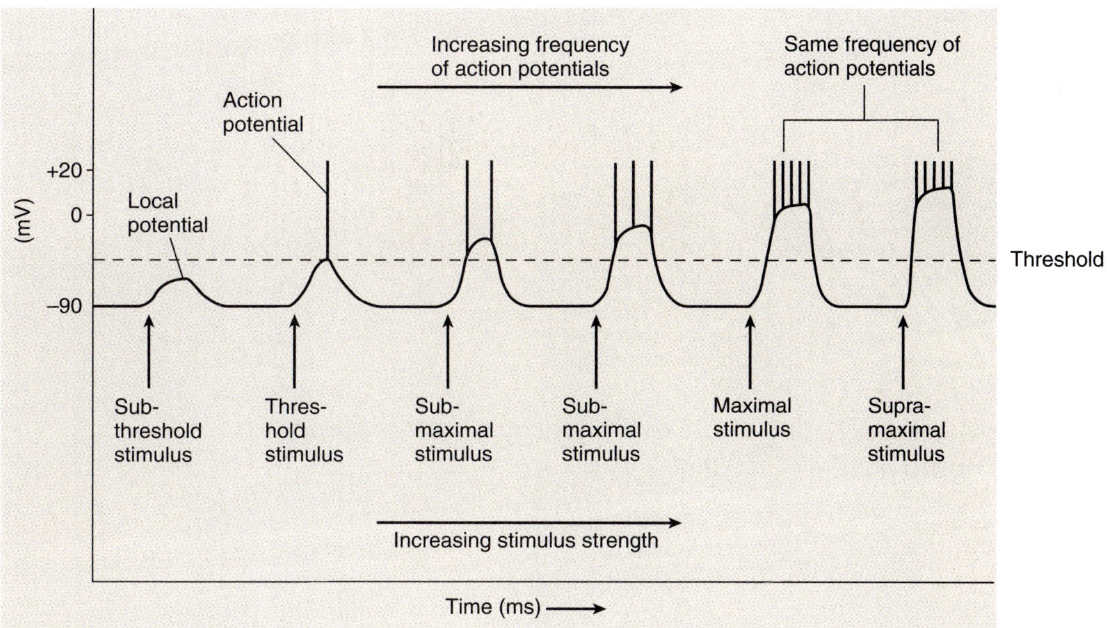

Stimulus Strength and Action Potential Frequency
Figure 11.23

1. Action potentials propagate in one direction along the axon.

2. An action potential (*orange* part of the membrane) generates local currents (*black arrows*) that tend to depolarize the membrane immediately adjacent to the action potential.

3. When depolarization caused by the local currents reaches threshold, a new action potential is produced adjacent to where the original action potential occurred.

4. Action potential propagation occurs in one direction because the absolute refractory period of the previous action potential prevents generation of an action potential in the reverse direction.

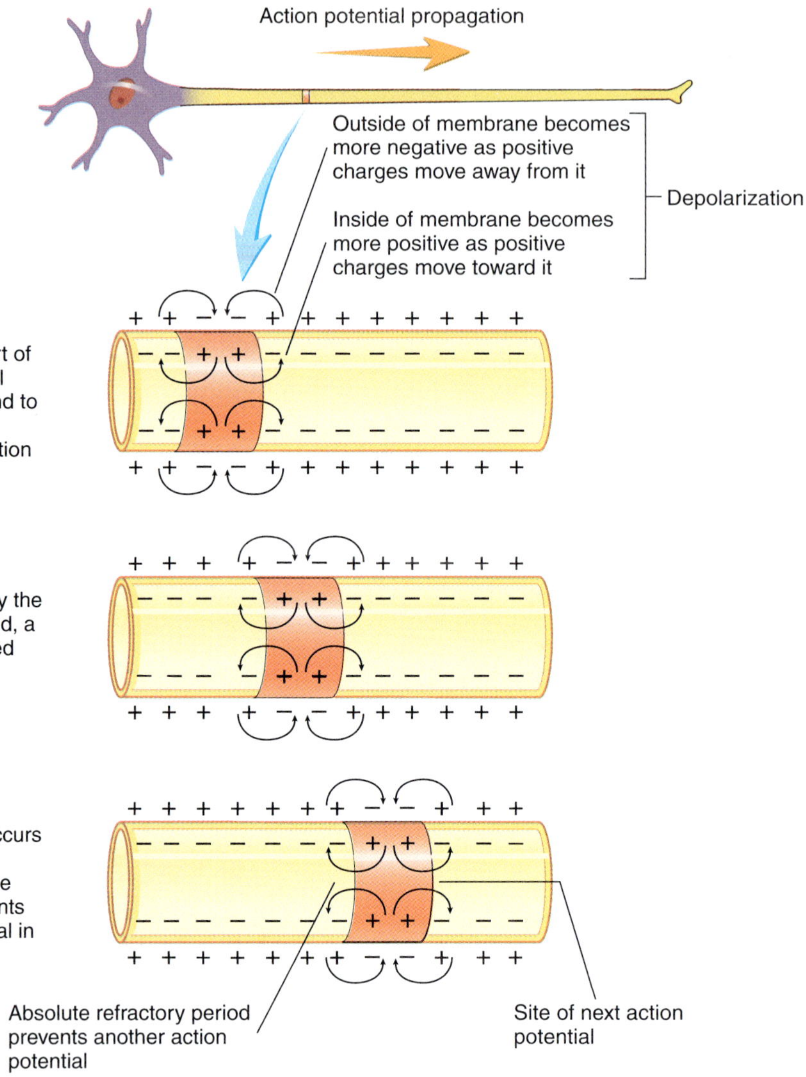

Action Potential Propagation in an Unmyelinated Axon
Process Figure 11.24

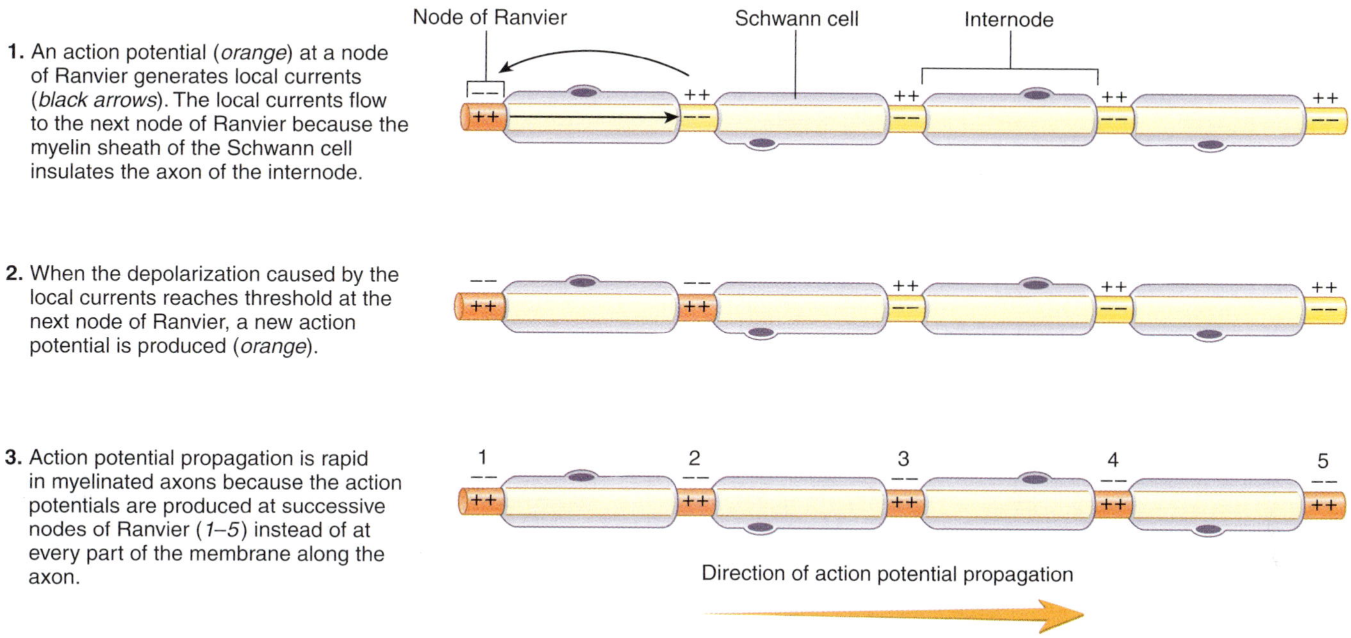

Saltatory Conduction: Action Potential Propagation in a Myelinated Axon
Process Figure 11.25

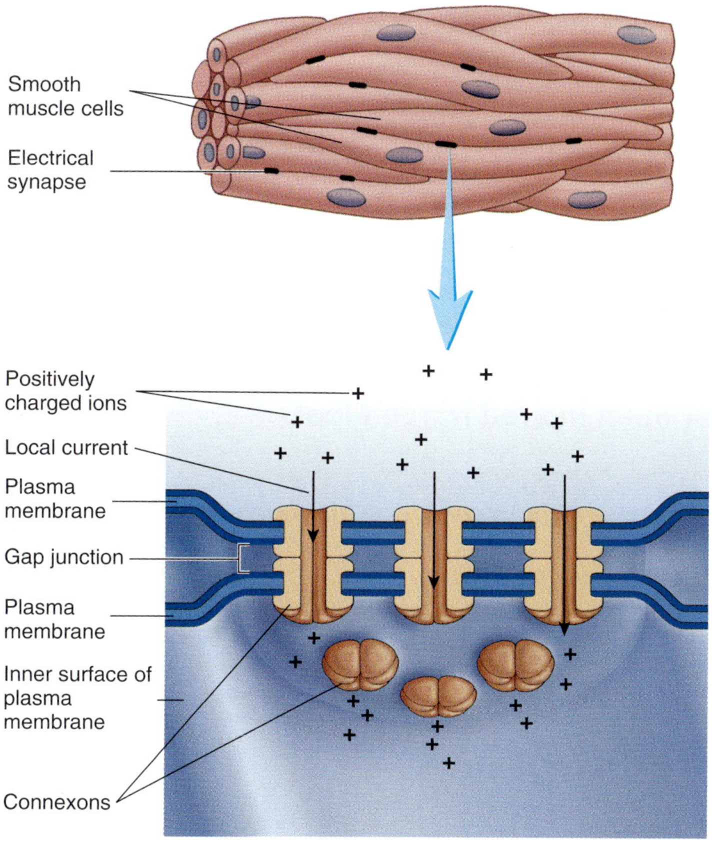

Electrical Synapse
Figure 11.26

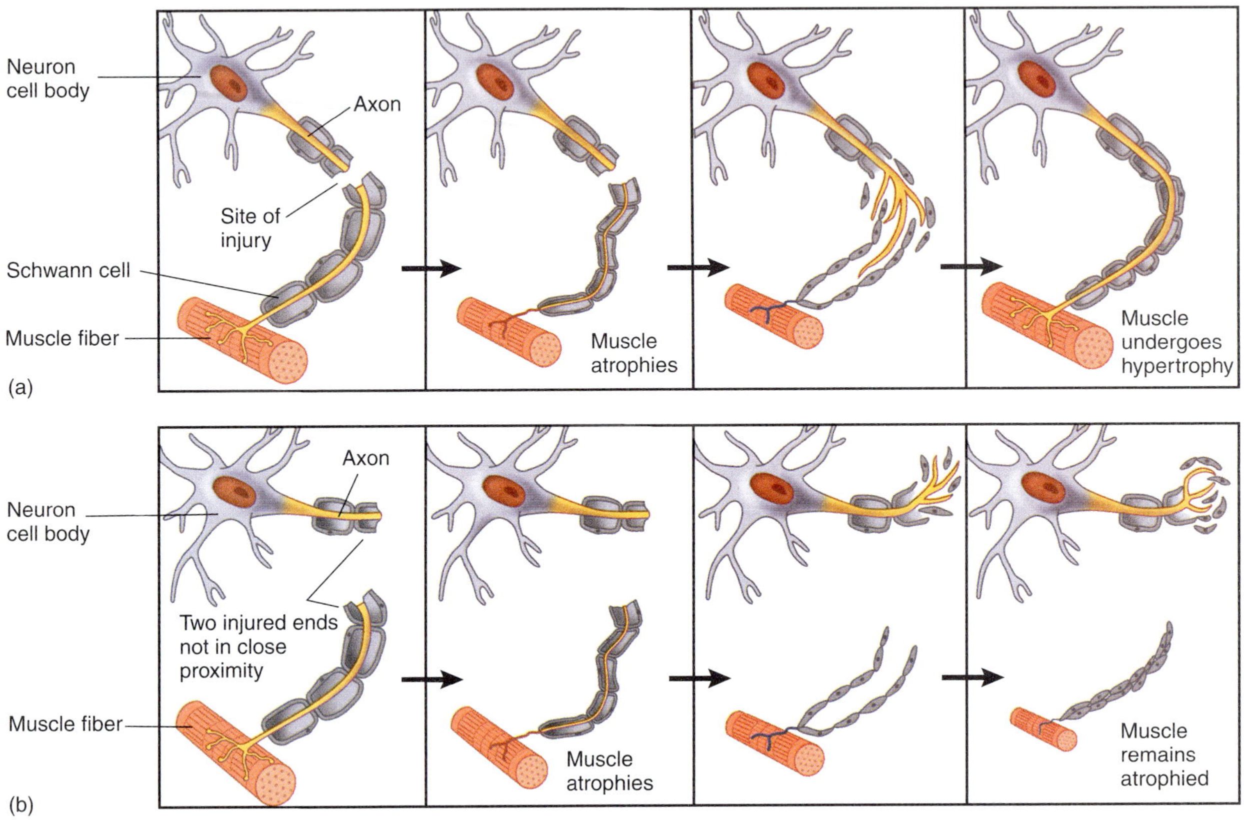

Changes That Occur in an Injured Nerve Fiber
Figure A

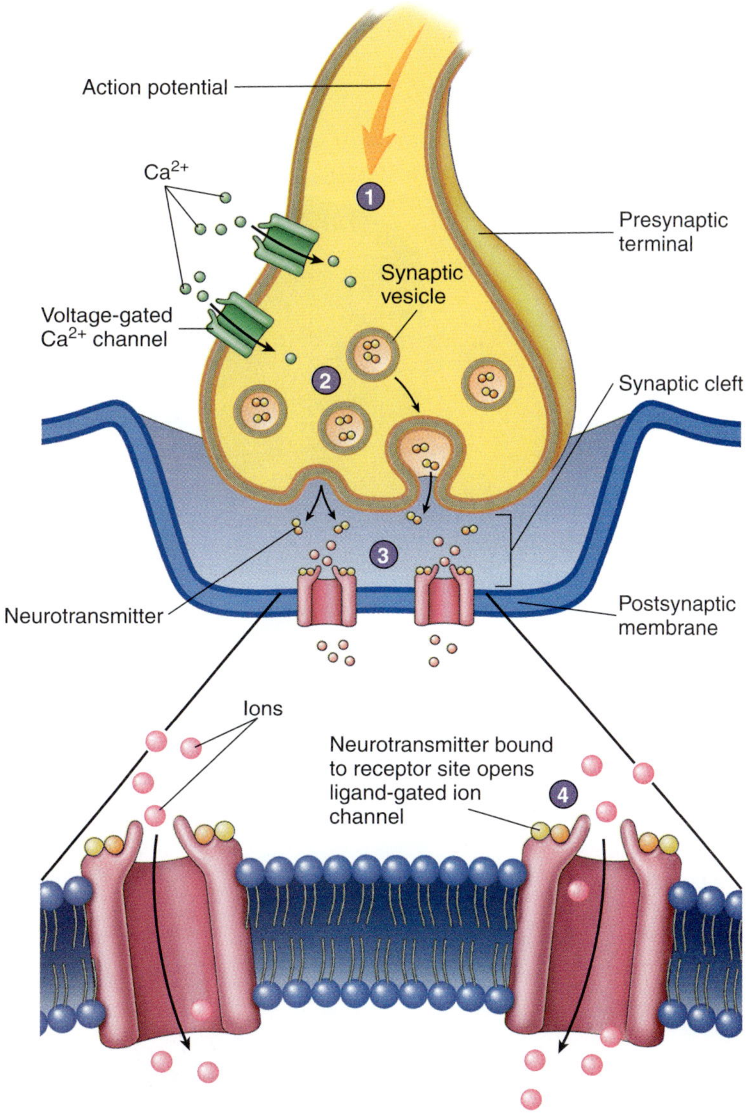

1. Action potentials arriving at the presynaptic terminal cause voltage-gated Ca^{2+} channels to open.
2. Ca^{2+} diffuse into the cell and cause synaptic vesicles to release neurotransmitter molecules.
3. Neurotransmitter molecules diffuse from the presynaptic terminal across the synaptic cleft.
4. Neurotransmitter molecules combine with their receptor sites and cause ligand-gated ion channels to open. Ions diffuse into the cell (*shown in illustration*) or out of the cell (*not shown*) and cause a change in membrane potential.

Chemical Synapse
Process Figure 11.27

1. Acetylcholine molecules bind to their receptors.
2. Acetylcholine molecules unbind from their receptors.
3. Acetylcholinesterase splits acetylcholine into choline and acetic acid, which prevents acetylcholine from again binding to its receptors. Choline is taken up by the presynaptic terminal.
4. Choline is used to make new acetylcholine molecules that are packaged into synaptic vesicles.

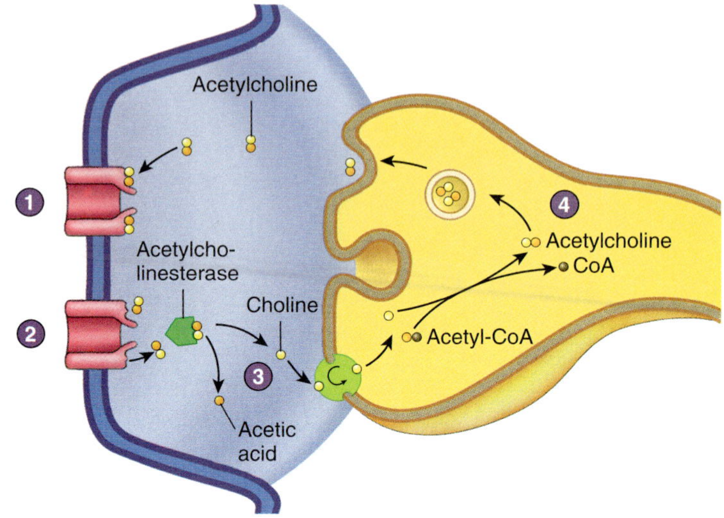

(a)

1. Norepinephrine binds to its receptor.
2. Norepinephrine unbinds from its receptor.
3. Norepinephrine is taken up by the presynaptic terminal, which prevents norepinephrine from again binding to its receptor.
4. Norepinephrine is repackaged into synaptic vesicles or is broken down by monoamine oxidase (MAO).

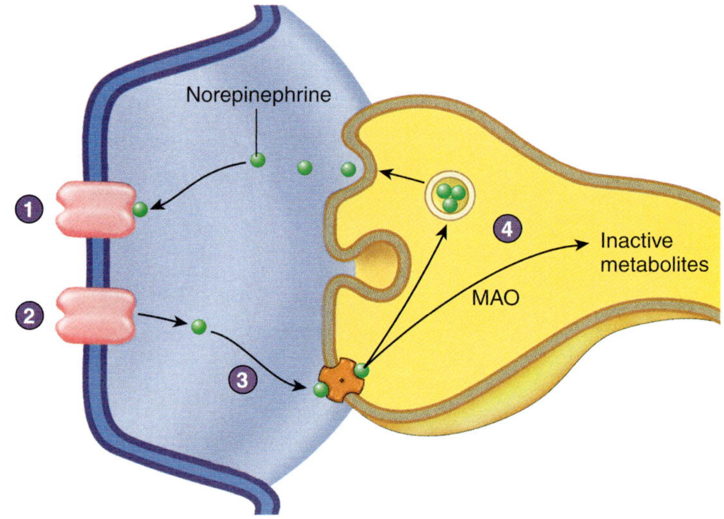

(b)

Removal of Neurotransmitter from the Synaptic Cleft
Process Figure 11.28

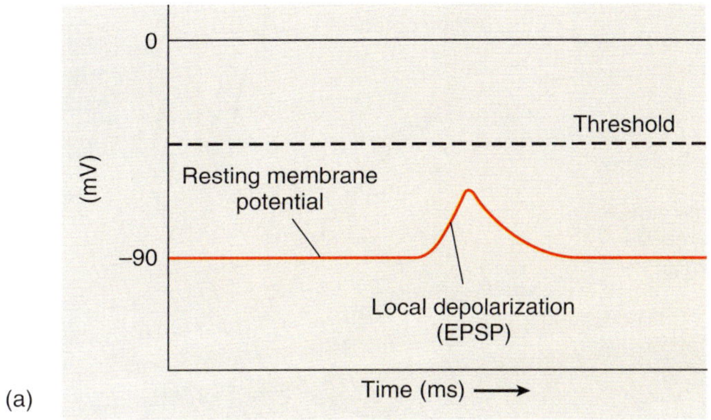

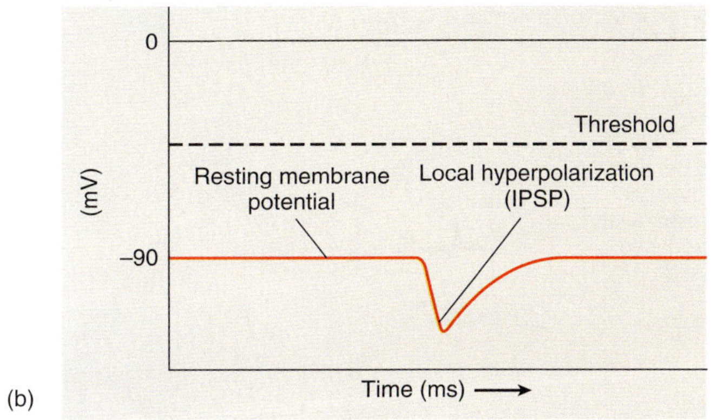

Postsynaptic Potentials
Figure 11.29

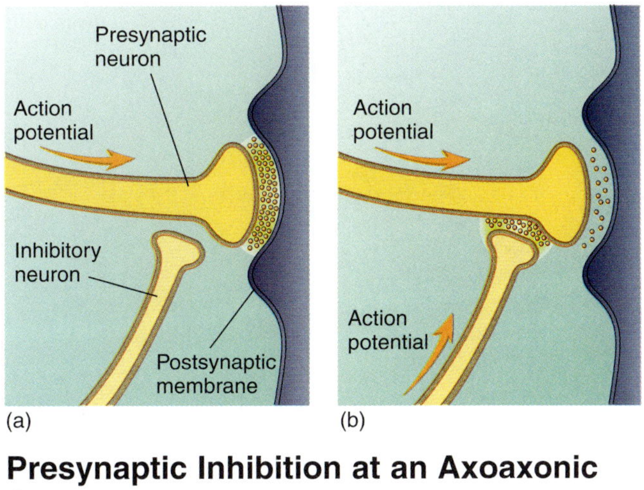

Presynaptic Inhibition at an Axoaxonic Synapse
Figure 11.30

(a) **Spatial summation.** Action potentials 1 and 2 cause the production of local depolarizations at two different dendrites. These local depolarizations summate at the trigger zone to produce a local depolarization that exceeds threshold, resulting in an action potential.

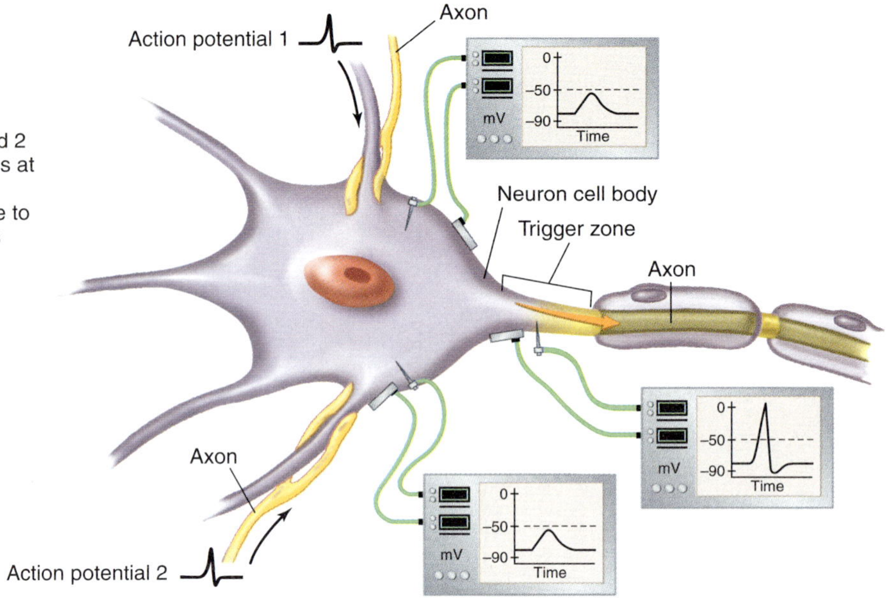

(b) **Temporal summation.** Two action potentials arrive in close succession at the presynaptic membrane. Before the first local depolarization returns to threshold, the second is produced. They summate to exceed threshold and produce an action potential.

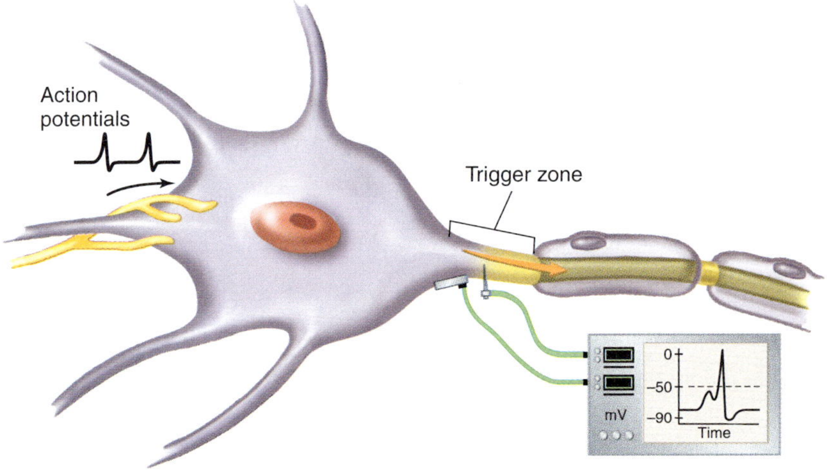

(c) Combined spatial and temporal summation with both excitatory postsynaptic potentials and inhibitory postsynaptic potentials. The outcome, which is the product of summation, is determined by which influence is greater.

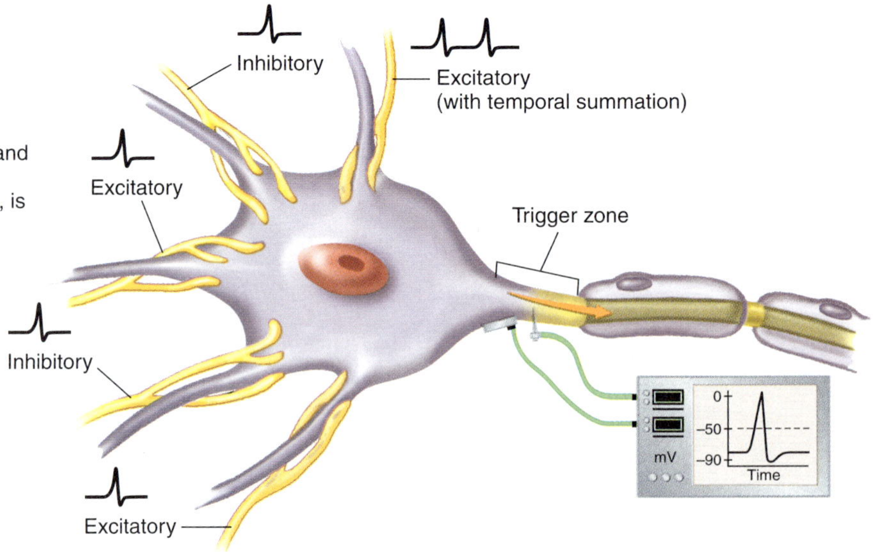

Summation
Figure 11.31

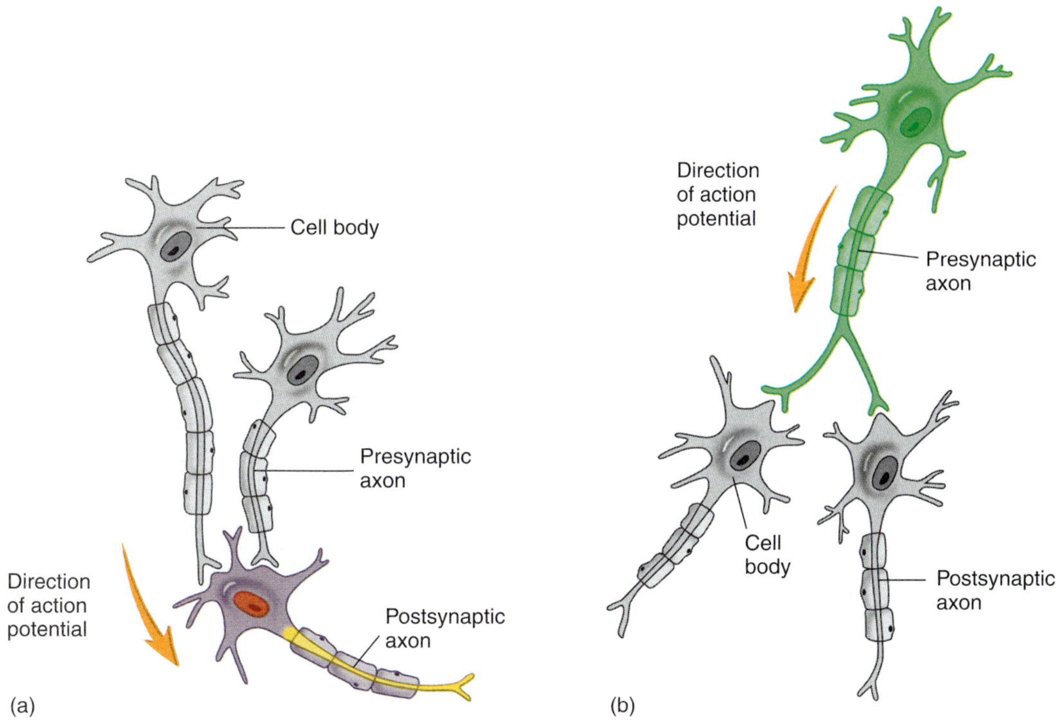

Convergent and Divergent Pathways
Figure 11.32

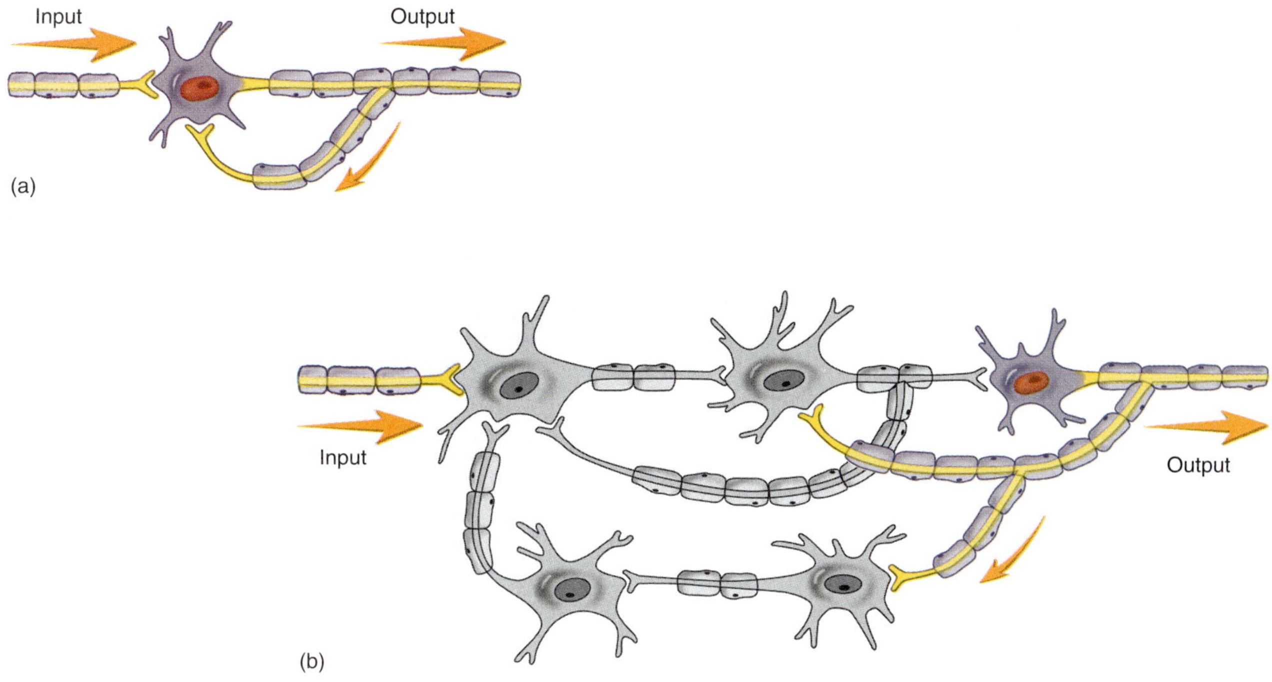

Oscillating Circuits
Figure 11.33

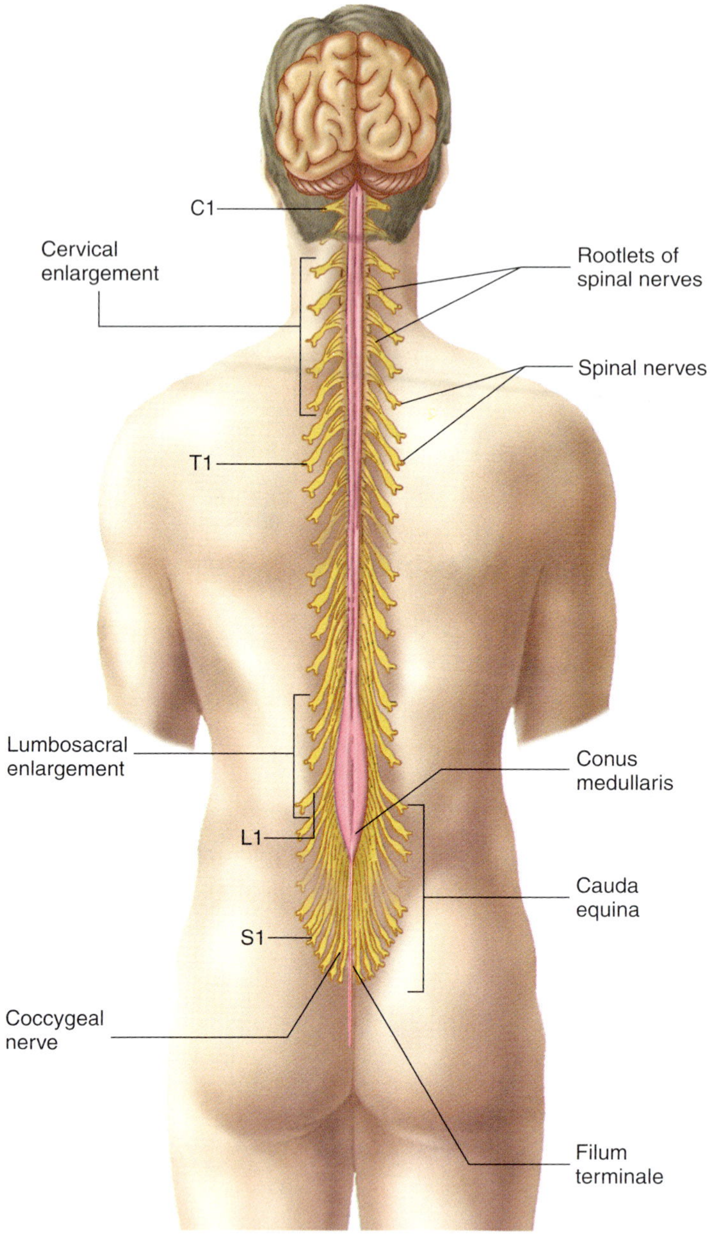

Posterior view

Spinal Cord and Spinal Nerve Roots
Figure 12.1

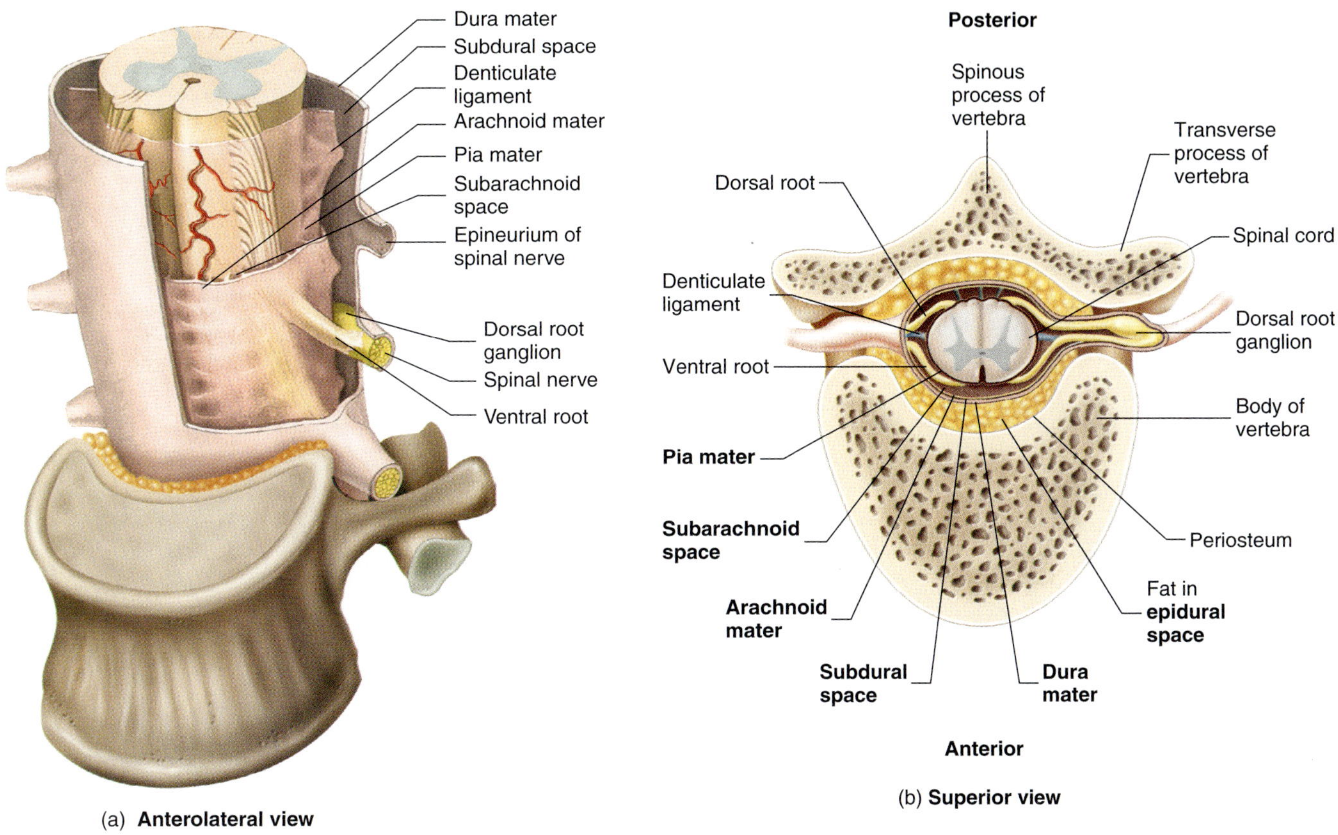

Meningeal Membranes Surrounding the Spinal Cord
Figure 12.2

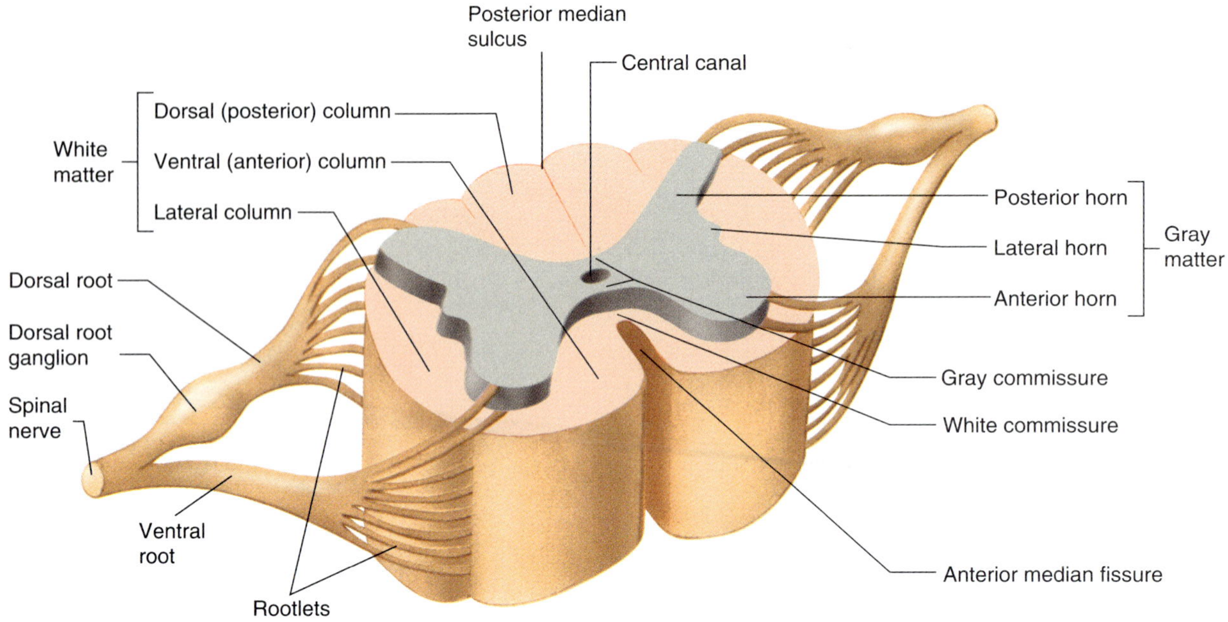

(a) **Anterolateral view**

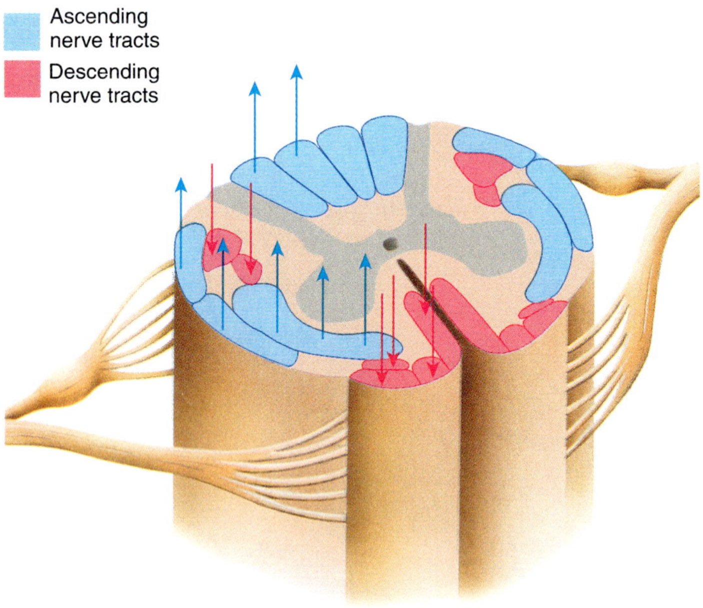

(c) **Anterolateral view**

Cross Section of the Spinal Cord
Figure 12.3

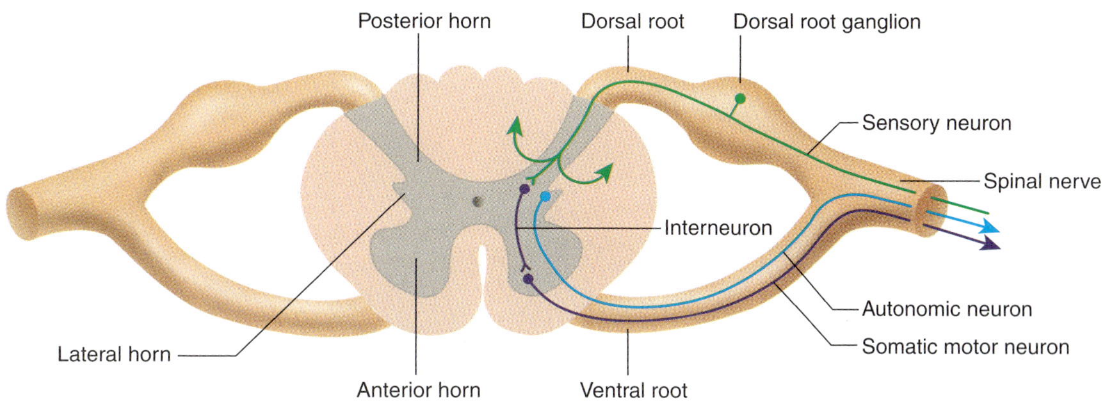

Superior view

Relationship of Sensory and Motor Neurons to the Spinal Cord
Figure 12.4

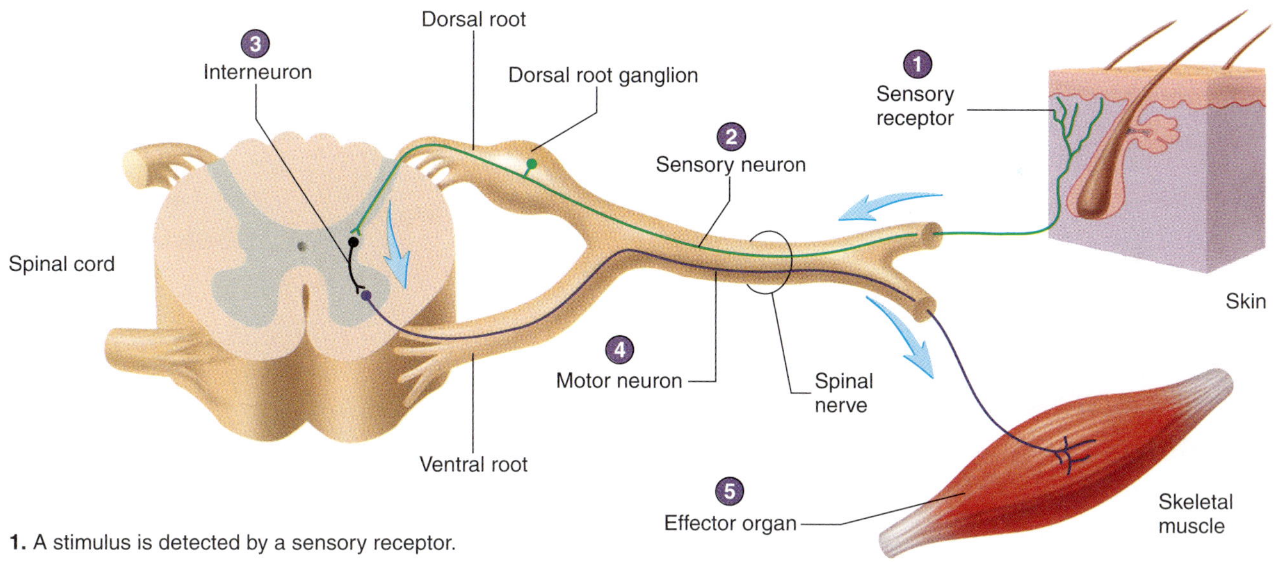

1. A stimulus is detected by a sensory receptor.
2. A sensory neuron conducts an action potential to the spinal cord.
3. In the spinal cord, the sensory neuron synapses with an interneuron.
4. The interneuron synapses with a motor neuron.
5. A motor neuron axon projects to an effector organ.

Reflex Arc
Process Figure 12.5

Sudden stretch of a muscle results in:

1. Muscle spindles detect stretch of the muscle.
2. Sensory neurons conduct action potentials to the spinal cord.
3. Sensory neurons synapse directly with alpha motor neurons.
4. Stimulation of the alpha motor neurons causes the muscle to contract and resist being stretched.

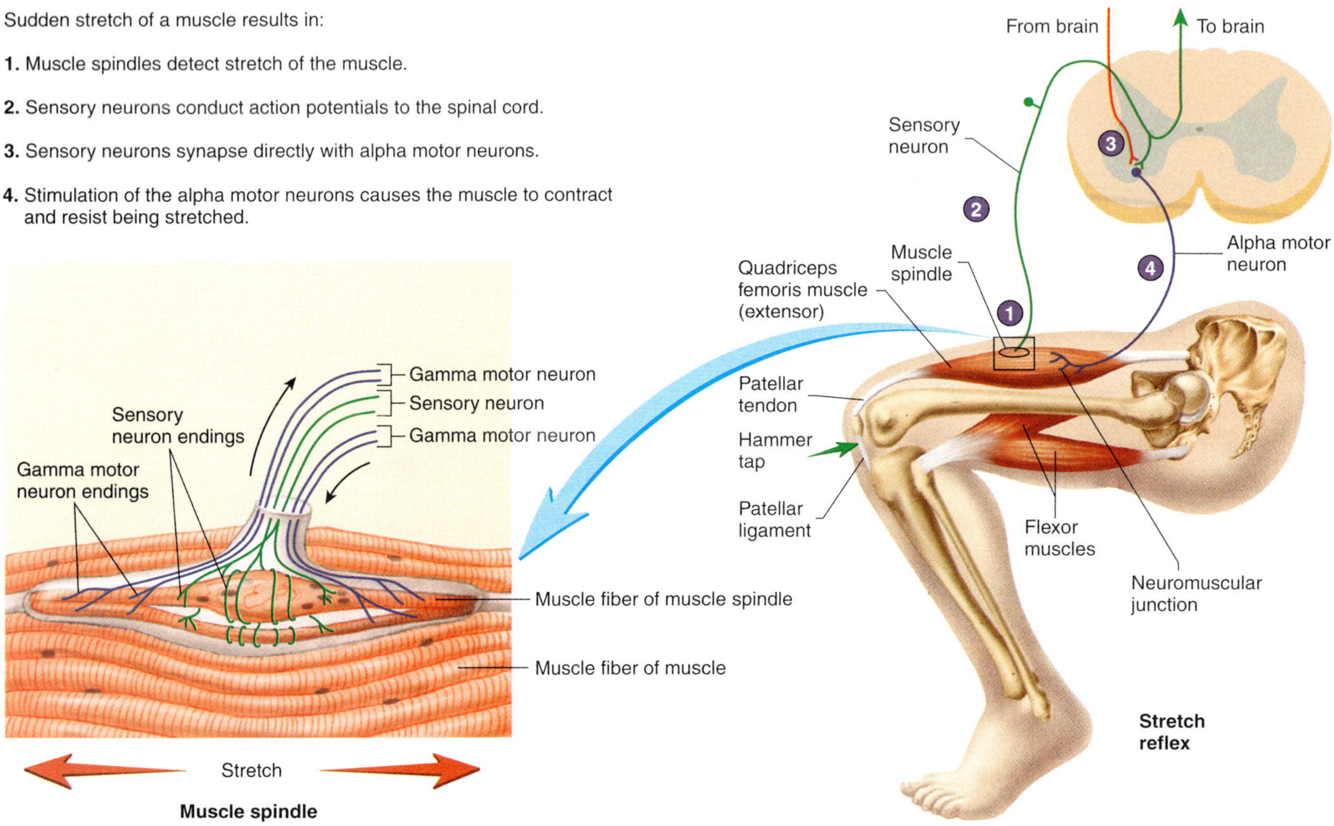

Stretch Reflex
Process Figure 12.6

Intense stretch of a skeletal muscle results in:

1. Golgi tendon organs detect tension applied to a tendon.
2. Sensory neurons conduct action potentials to the spinal cord.
3. Sensory neurons synapse with inhibitory interneurons that synapse with alpha motor neurons.
4. Inhibition of the alpha motor neurons causes muscle relaxation, relieving the tension applied to the tendon.

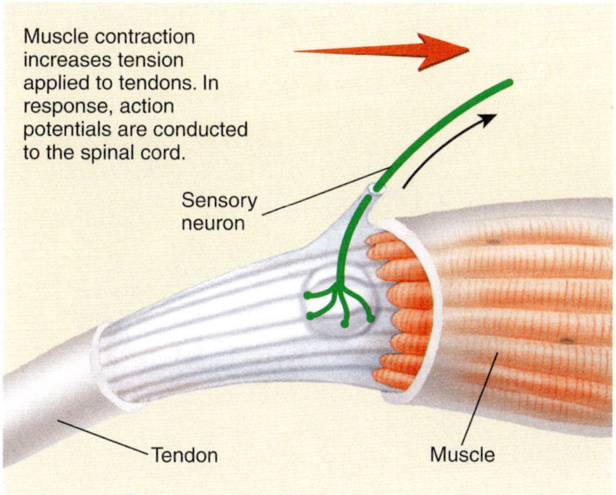

Golgi tendon organ

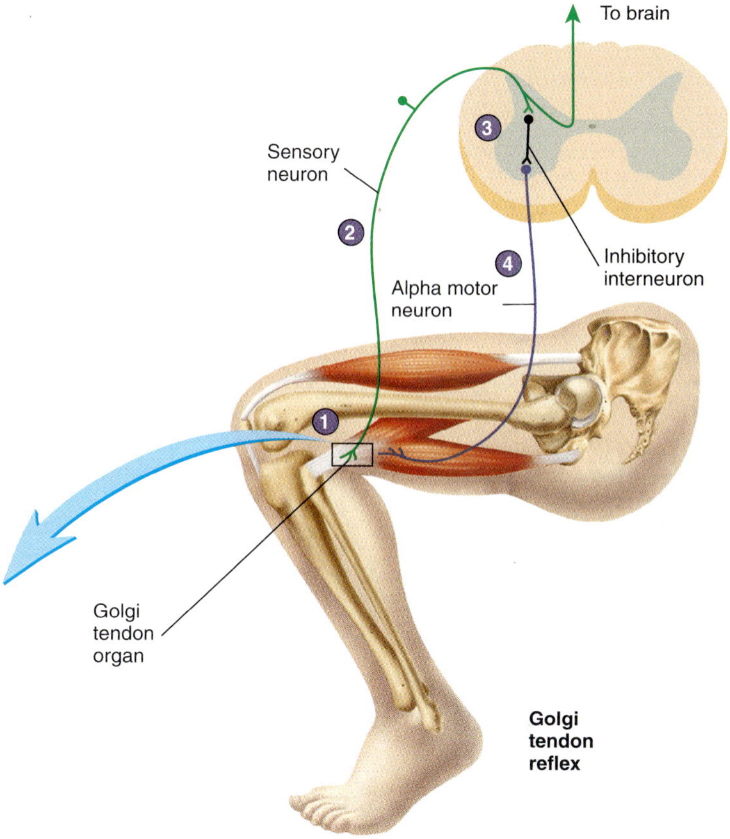

Golgi tendon reflex

Golgi Tendon Reflex
Process Figure 12.7

Stimulation of pain receptors results in:

1. Pain receptors detect a painful stimulus.
2. Sensory neurons conduct action potentials to the spinal cord.
3. Sensory neurons synapse with excitatory interneurons that synapse with alpha motor neurons.
4. Excitation of the alpha motor neurons results in contraction of the flexor muscles and withdrawal of the limb from the painful stimulus.

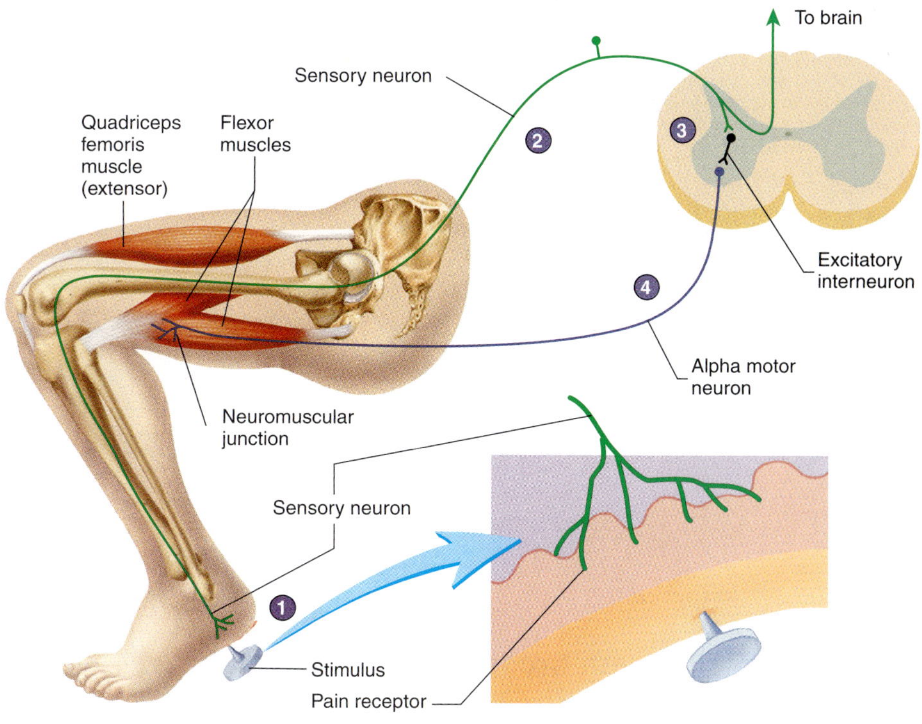

Withdrawal reflex

Withdrawal Reflex
Process Figure 12.8

Reciprocal innervation:

1. During the withdrawal reflex, sensory neurons conduct action potentials to the spinal cord.
2. Sensory neurons synapse with excitatory interneurons that are part of the withdrawal reflex.
3. Collateral branches also synapse with inhibitory interneurons that are part of reciprocal innervation.
4. Inhibition of the alpha motor neurons supplying the extensor muscles causes them to relax and not oppose the flexor muscles of the withdrawal reflex.

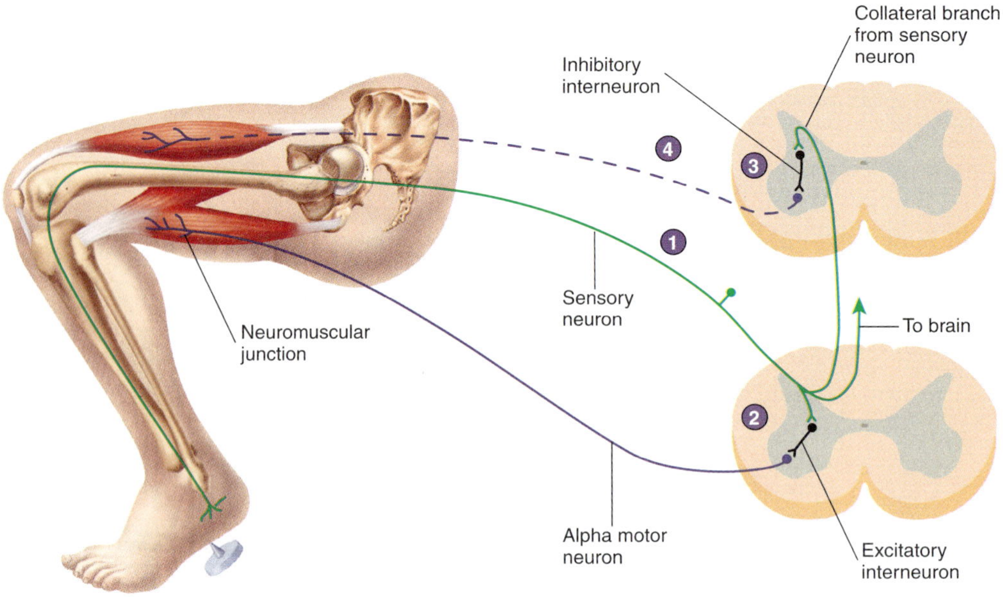

Withdrawal Reflex with Reciprocal Innervation
Process Figure 12.9

Crossed extensor reflex:

1. During the withdrawal reflex, sensory neurons conduct action potentials to the spinal cord.
2. Sensory neurons synapse with excitatory interneurons that are part of the withdrawal reflex. Collateral branches also synapse with excitatory interneurons that cross over to the opposite side of the spinal cord as part of the crossed extensor reflex.
3. Stimulation of the alpha motor neurons cause contraction of flexor muscles and stimulation of alpha motor neurons supplying extensor muscles in the opposite limb causes them to contract and support body weight during the withdrawal reflex.

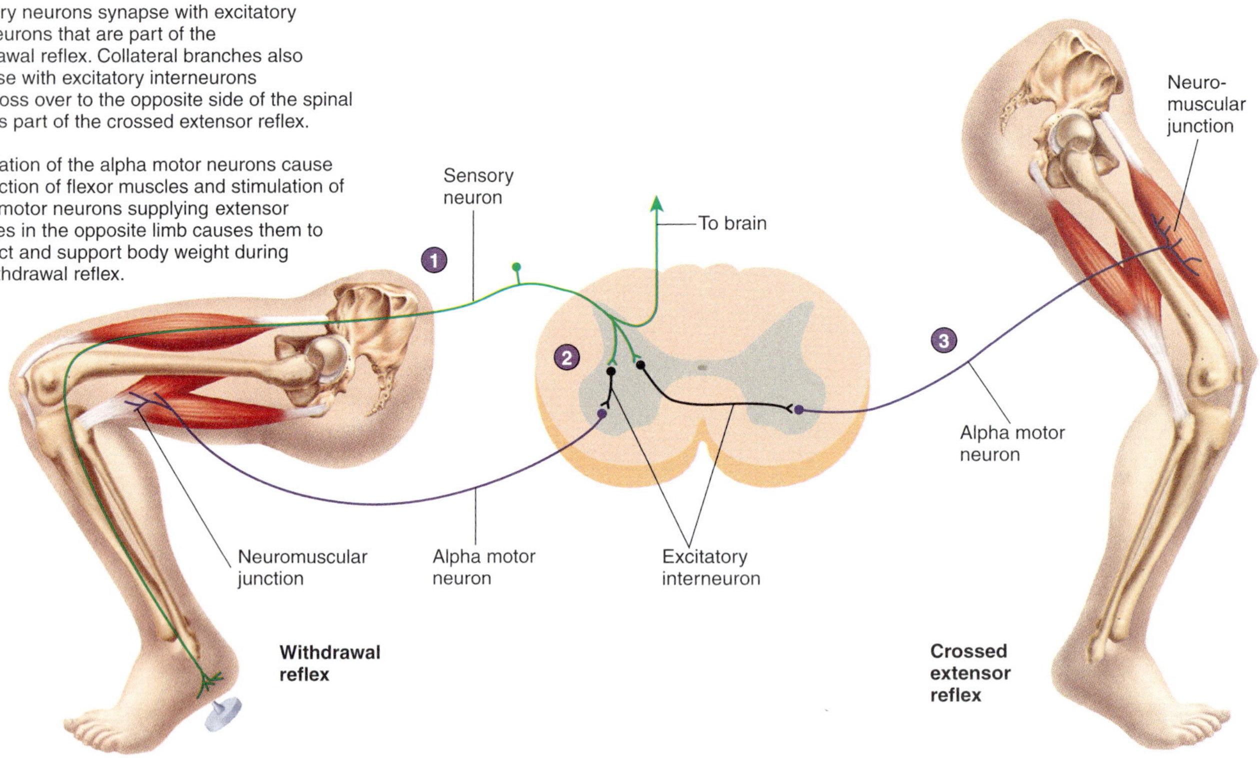

Withdrawal Reflex with Crossed Extensor Reflex
Process Figure 12.10

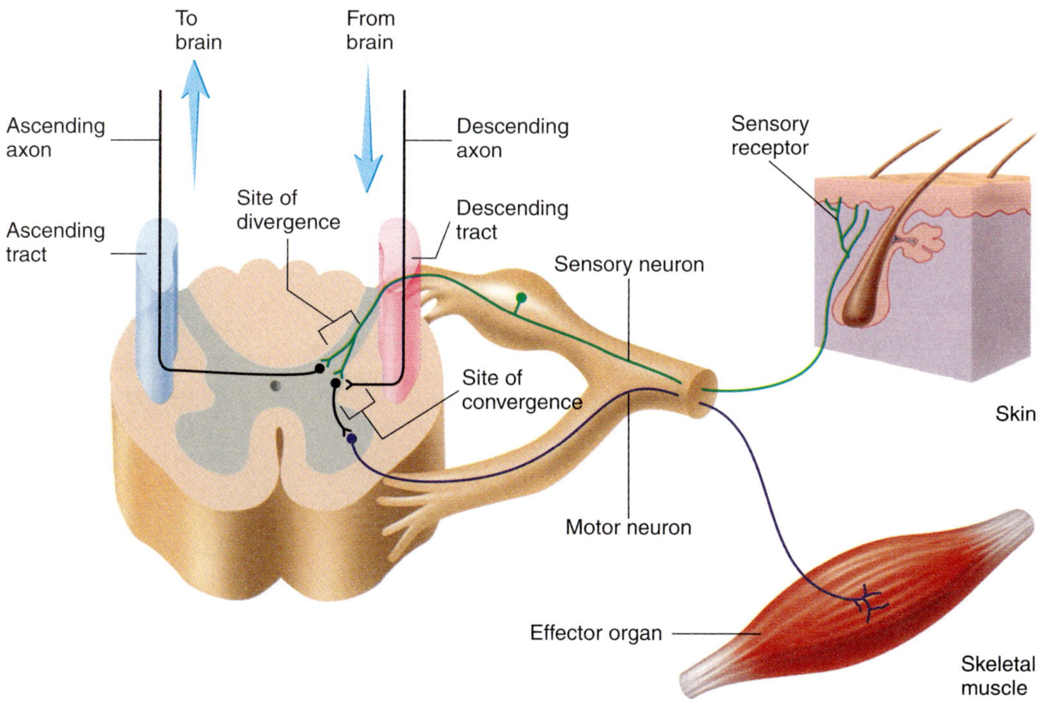

Spinal Reflex, with Ascending and Descending Axons
Figure 12.11

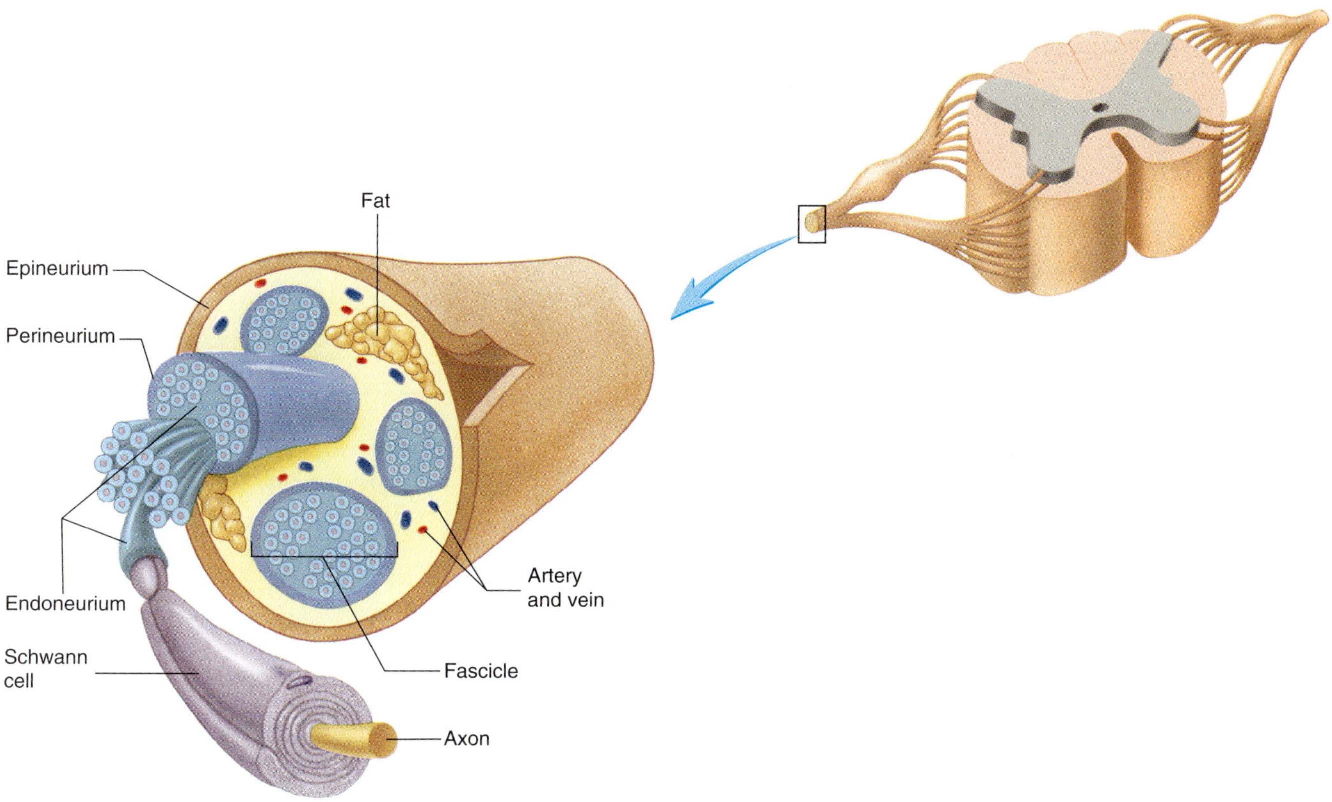

Structure of a Peripheral Nerve
Figure 12.12

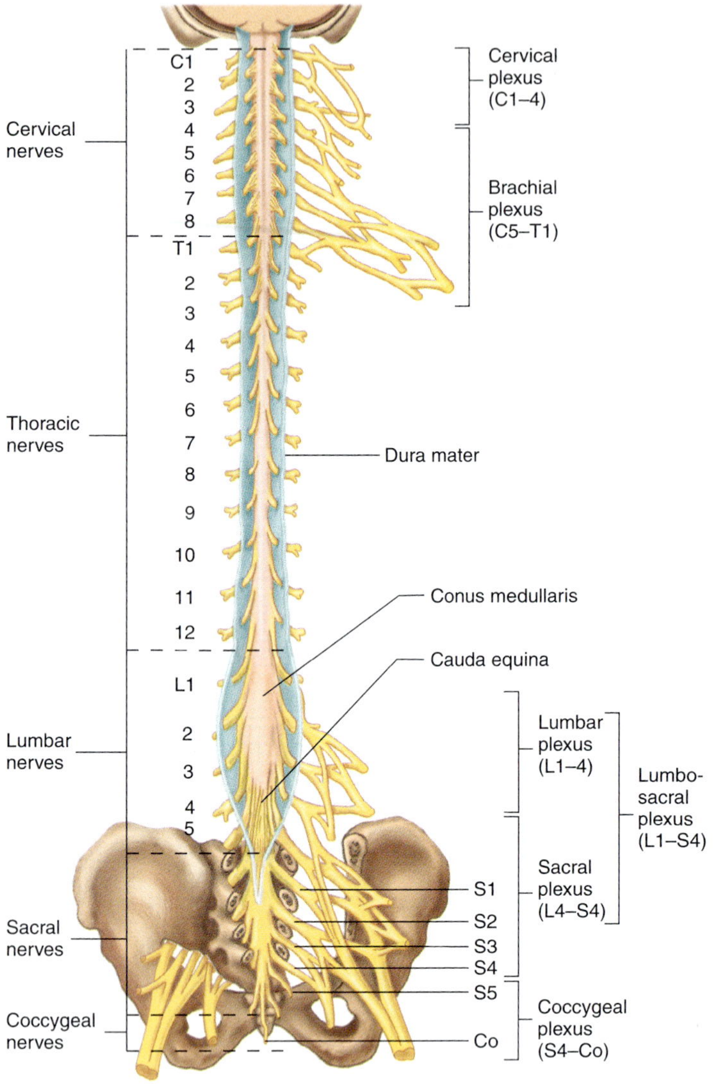

(a) **Posterior view**

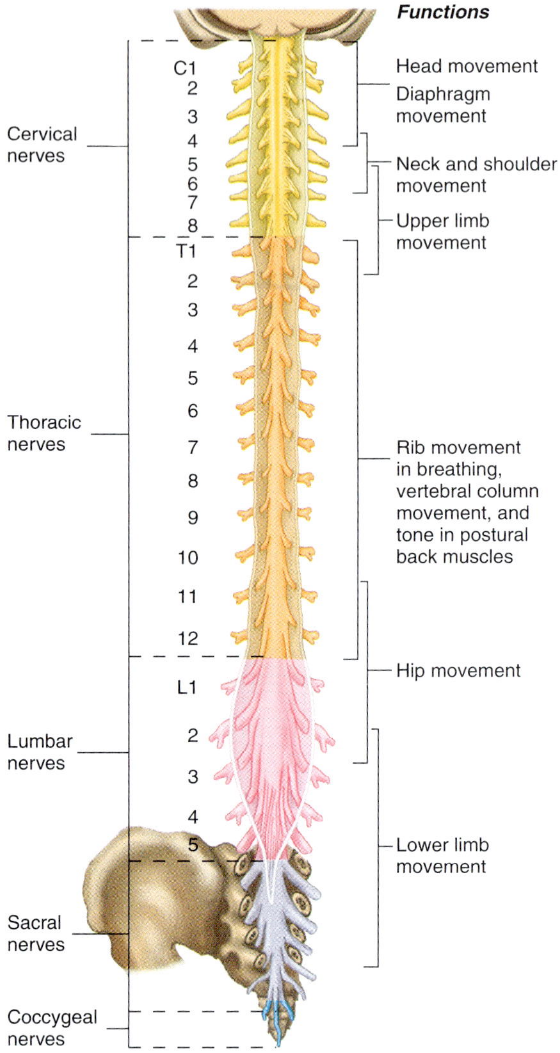

(b) **Posterior view**

Spinal Nerves

Figure 12.13

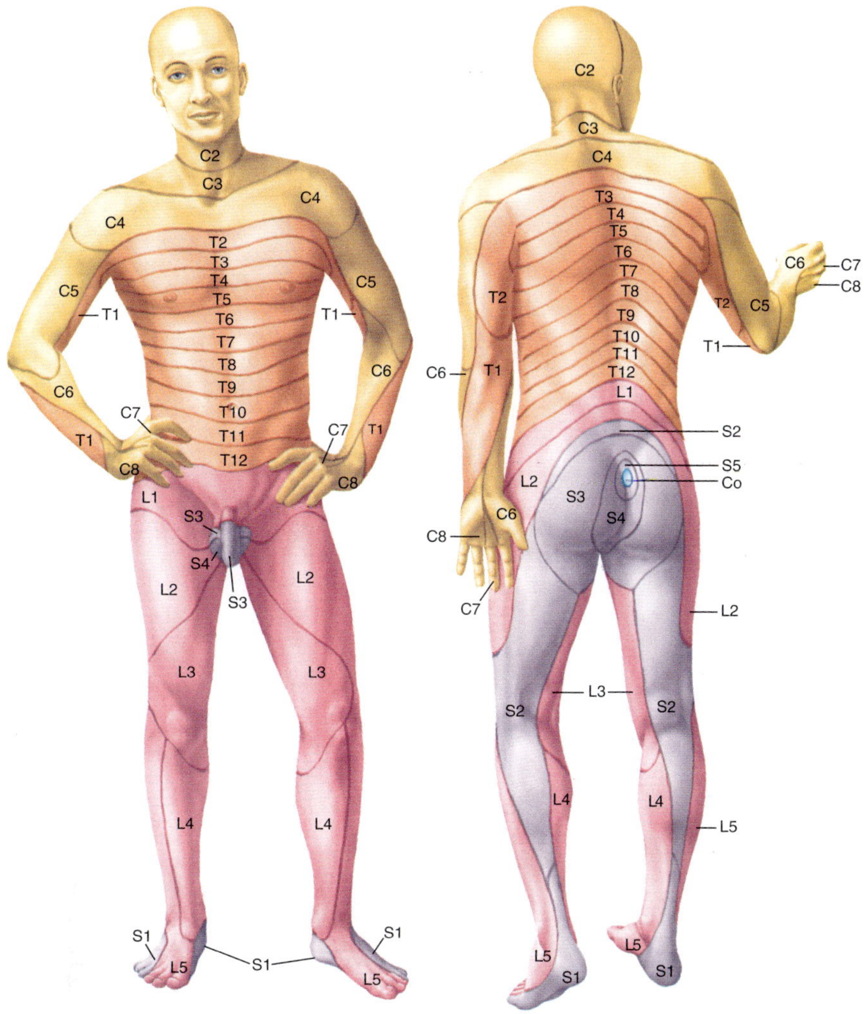
C2
C3
C4
C4
T2
T3
T4
T5
T6
T7
T8
T9
T10
T11
T12
C5
T1
T1
C5
C6
C6
C7
C7
T1
T1
C8
C8
L1
S3
S4
S3
L2
L2
L3
L3
L4
L4
S1
S1
L5
S1
L5
C2
C3
C4
T3
T4
T5
T6
T7
T8
T9
T10
T11
T12
L1
T2
T2
C5
C6
C7
C8
T1
C6
T1
S2
S5
Co
L2
S3
S4
C6
C8
C7
L2
L3
S2
S2
L4
L4
L5
L5
L5
S1
S1

Dermatomal Map
Figure 12.14

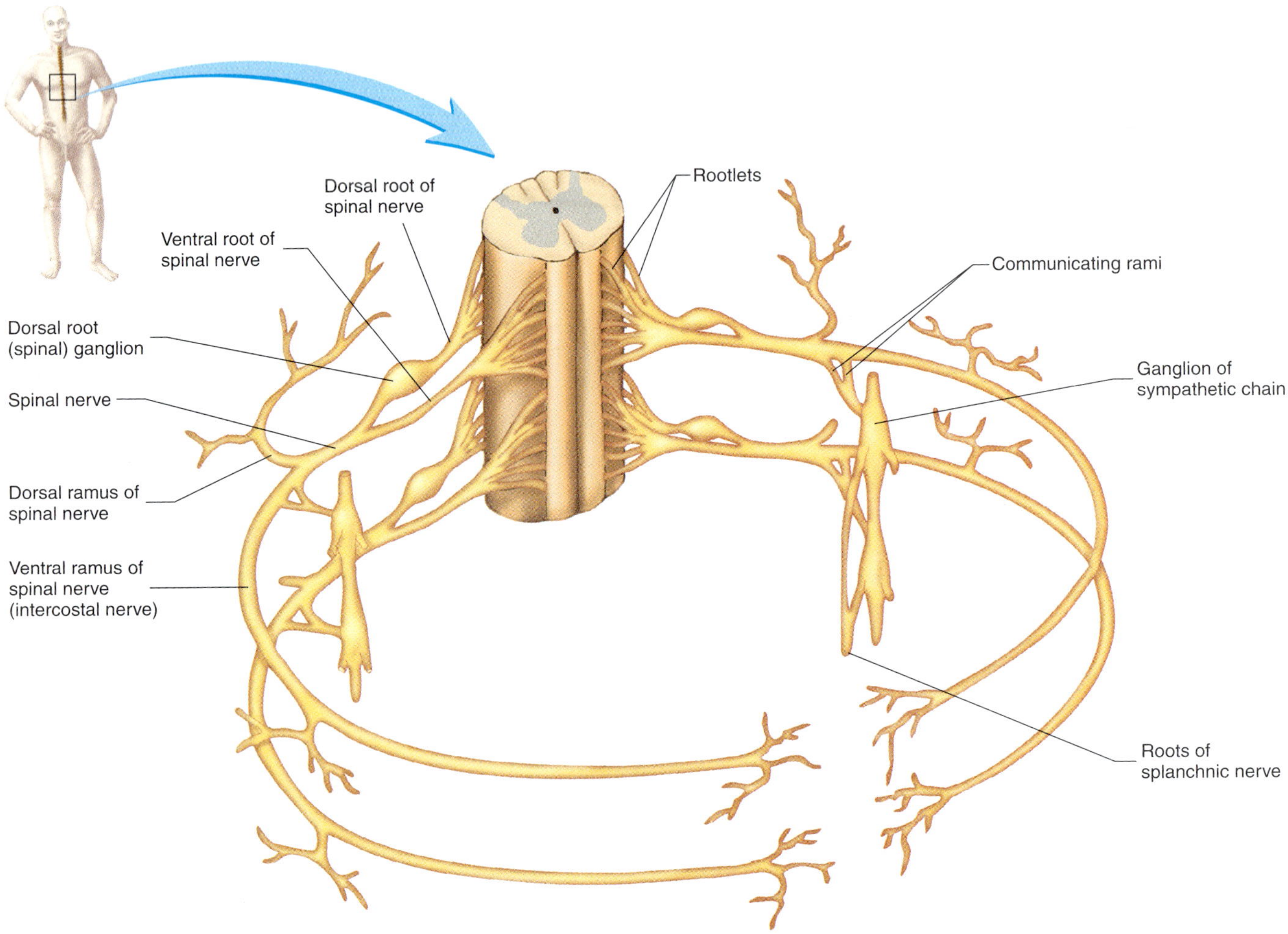
Dorsal root of spinal nerve
Ventral root of spinal nerve
Rootlets
Communicating rami
Dorsal root (spinal) ganglion
Spinal nerve
Ganglion of sympathetic chain
Dorsal ramus of spinal nerve
Ventral ramus of spinal nerve (intercostal nerve)
Roots of splanchnic nerve

(a) **Anterolateral view**

Spinal Nerves
Figure 12.15

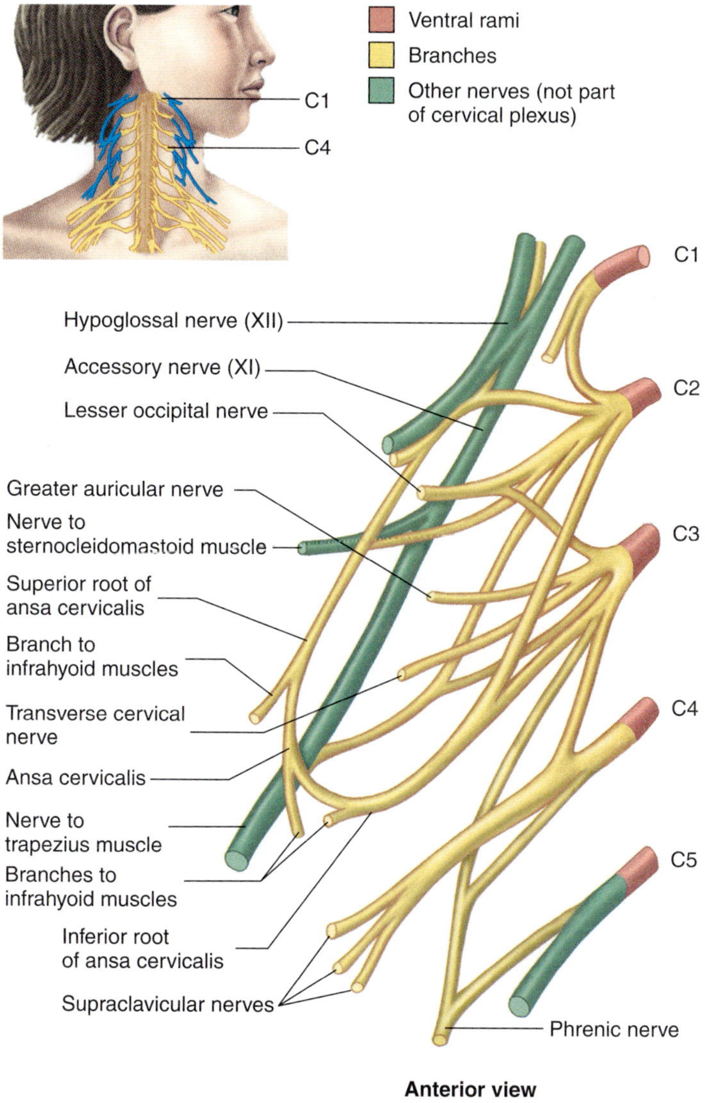

C1
C4
Ventral rami
Branches
Other nerves (not part of cervical plexus)
C1
C2
C3
C4
C5
Hypoglossal nerve (XII)
Accessory nerve (XI)
Lesser occipital nerve
Greater auricular nerve
Nerve to sternocleidomastoid muscle
Superior root of ansa cervicalis
Branch to infrahyoid muscles
Transverse cervical nerve
Ansa cervicalis
Nerve to trapezius muscle
Branches to infrahyoid muscles
Inferior root of ansa cervicalis
Supraclavicular nerves
Phrenic nerve
Anterior view

Cervical Plexus, Anterior View
Figure 12.16

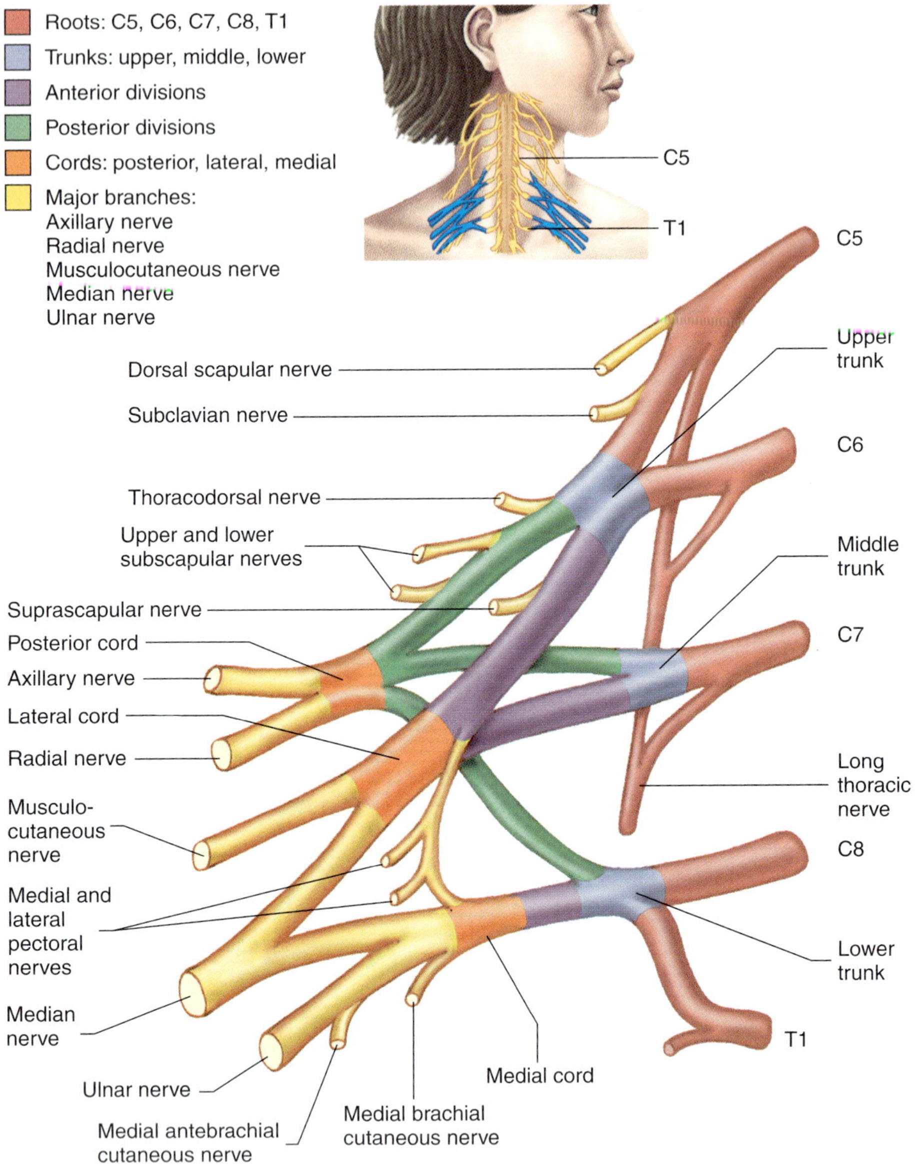

Anterior view

Brachial Plexus, Anterior View
Figure 12.17

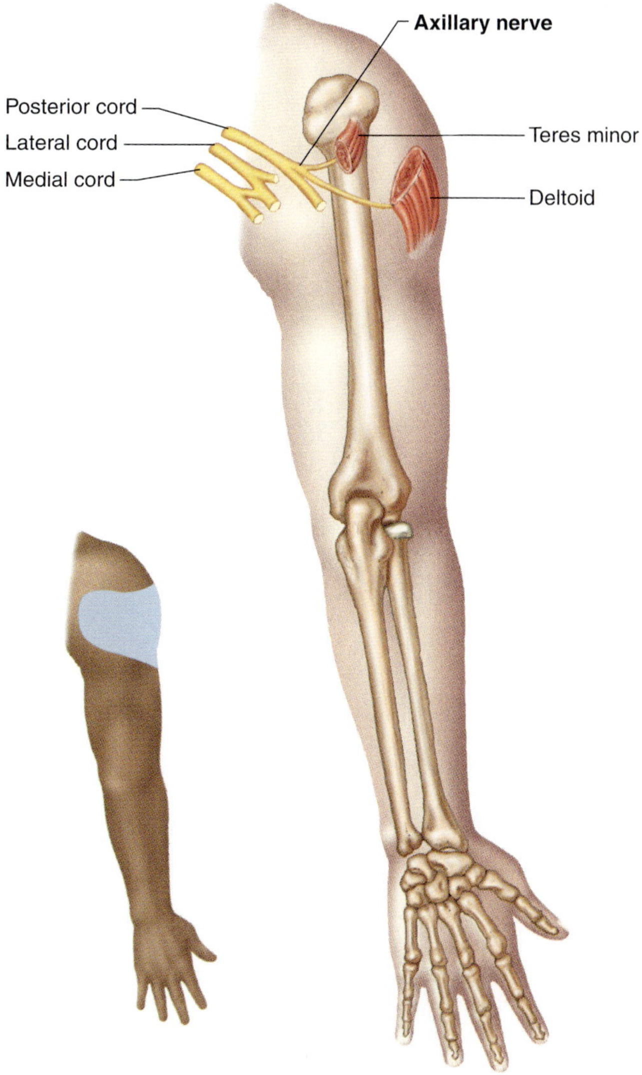
Axillary nerve
Posterior cord
Lateral cord
Medial cord
Teres minor
Deltoid

Posterior views

Axillary Nerve
Figure 12.18

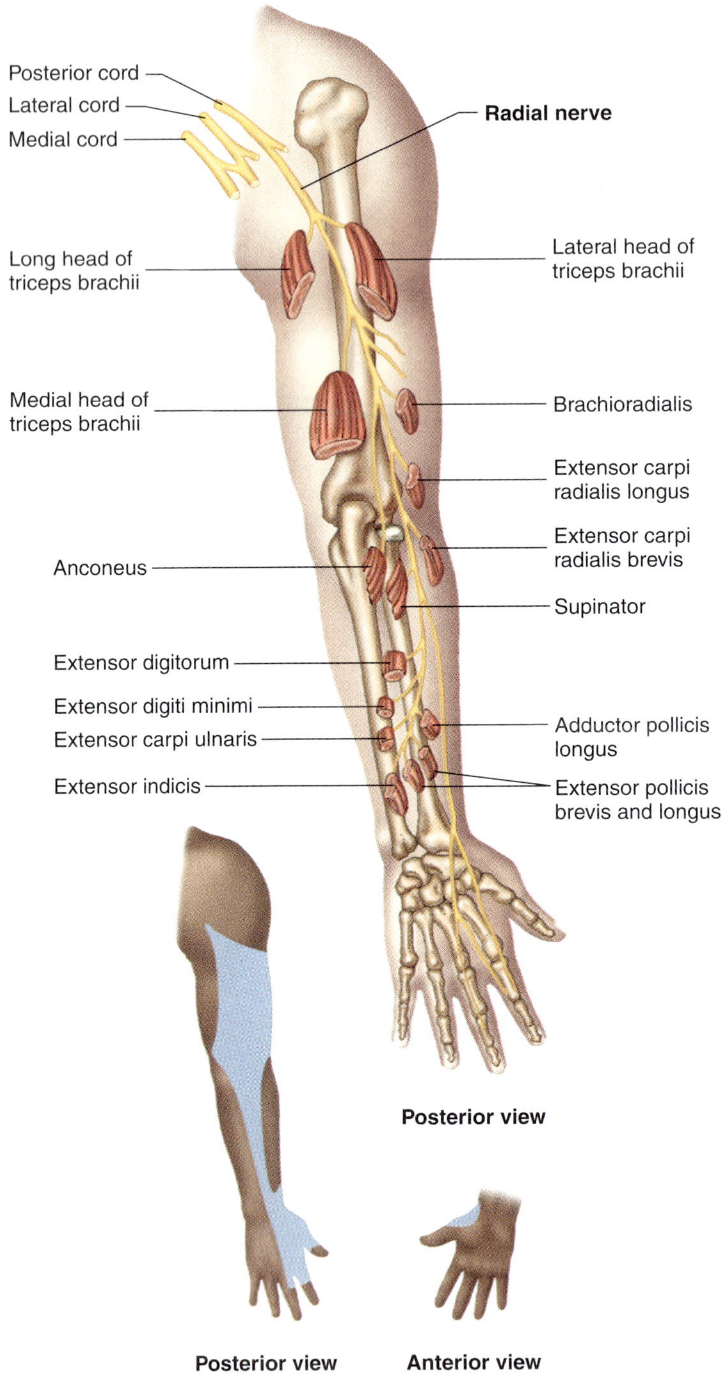
Posterior cord
Lateral cord
Medial cord
Radial nerve
Long head of triceps brachii
Lateral head of triceps brachii
Medial head of triceps brachii
Brachioradialis
Extensor carpi radialis longus
Extensor carpi radialis brevis
Anconeus
Supinator
Extensor digitorum
Extensor digiti minimi
Extensor carpi ulnaris
Extensor indicis
Adductor pollicis longus
Extensor pollicis brevis and longus
Posterior view
Posterior view
Anterior view

Radial Nerve
Figure 12.19

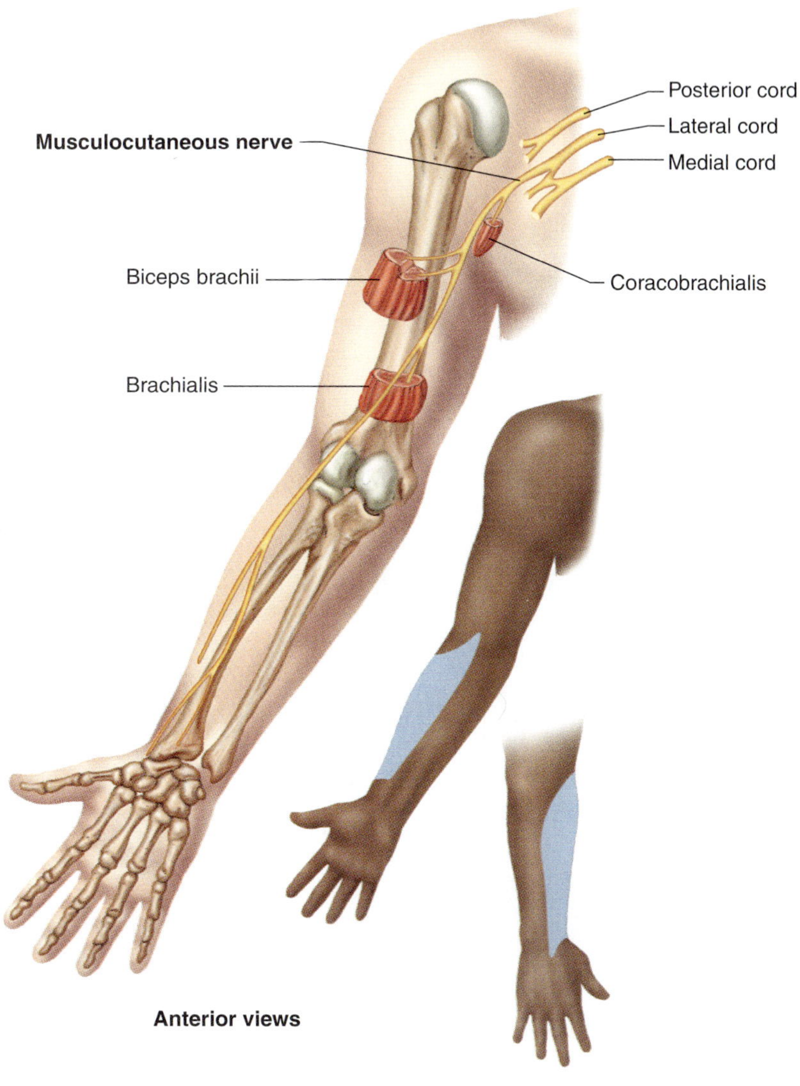
Posterior cord
Lateral cord
Medial cord
Musculocutaneous nerve
Biceps brachii
Coracobrachialis
Brachialis
Anterior views
Posterior view

Musculocutaneous Nerve
Figure 12.20

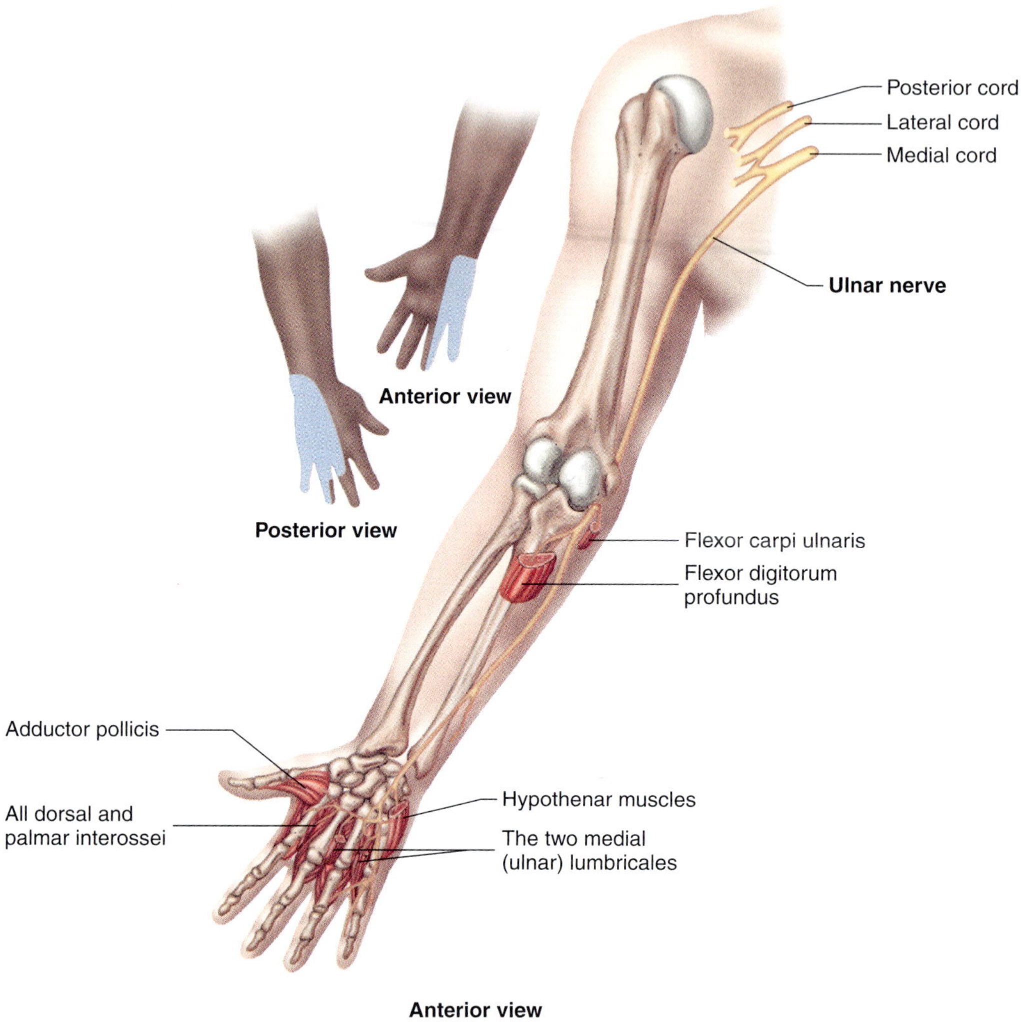
Posterior cord
Lateral cord
Medial cord
Ulnar nerve
Anterior view
Posterior view
Flexor carpi ulnaris
Flexor digitorum profundus
Adductor pollicis
All dorsal and palmar interossei
Hypothenar muscles
The two medial (ulnar) lumbricales
Anterior view

Ulnar Nerve
Figure 12.21

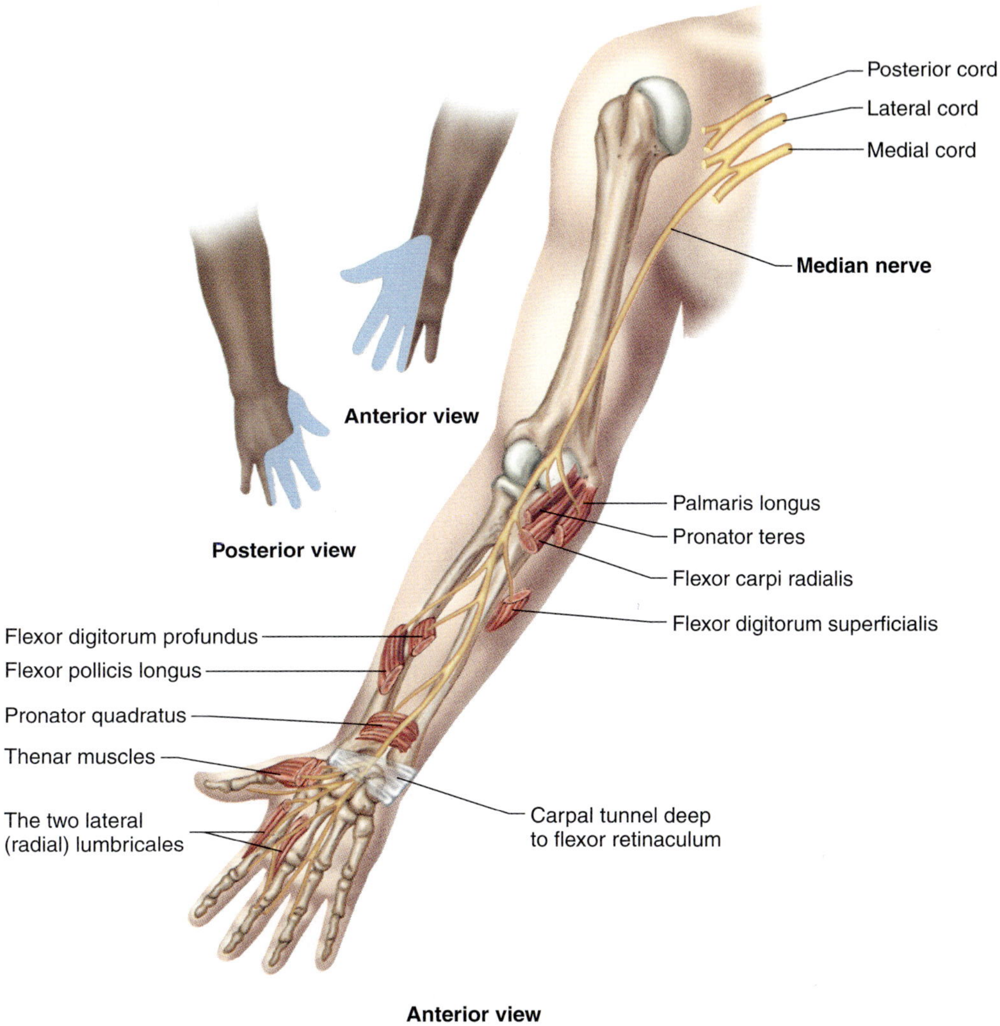
Posterior cord
Lateral cord
Medial cord
Median nerve
Anterior view
Posterior view
Palmaris longus
Pronator teres
Flexor carpi radialis
Flexor digitorum superficialis
Flexor digitorum profundus
Flexor pollicis longus
Pronator quadratus
Thenar muscles
The two lateral
(radial) lumbricales
Carpal tunnel deep
to flexor retinaculum
Anterior view

Median Nerve
Figure 12.22

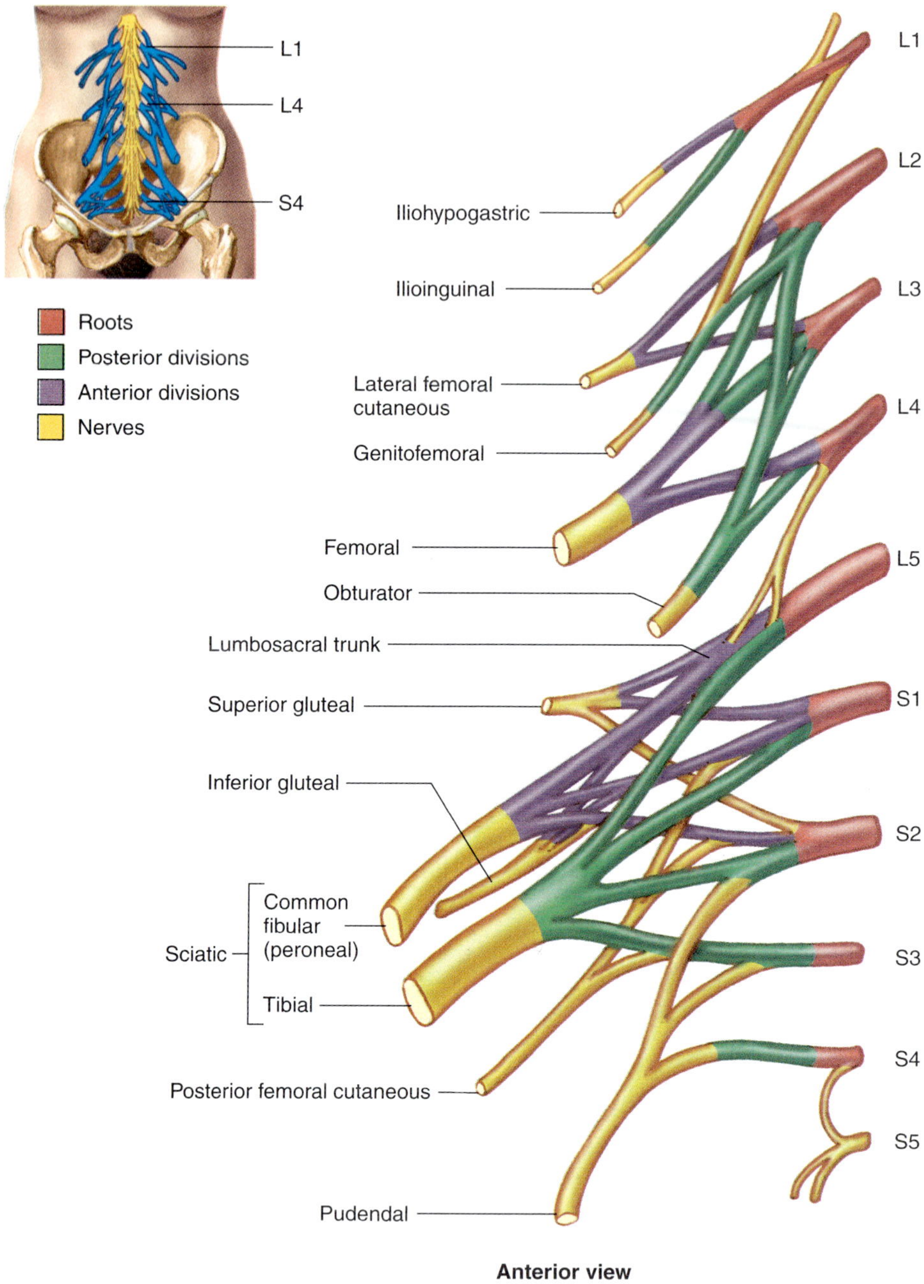
L1
L4
S4
Roots
Posterior divisions
Anterior divisions
Nerves
L1
L2
L3
L4
L5
S1
S2
S3
S4
S5
Iliohypogastric
Ilioinguinal
Lateral femoral cutaneous
Genitofemoral
Femoral
Obturator
Lumbosacral trunk
Superior gluteal
Inferior gluteal
Sciatic
Common fibular (peroneal)
Tibial
Posterior femoral cutaneous
Pudendal
Anterior view

Lumbosacral Plexus, Anterior View
Figure 12.23

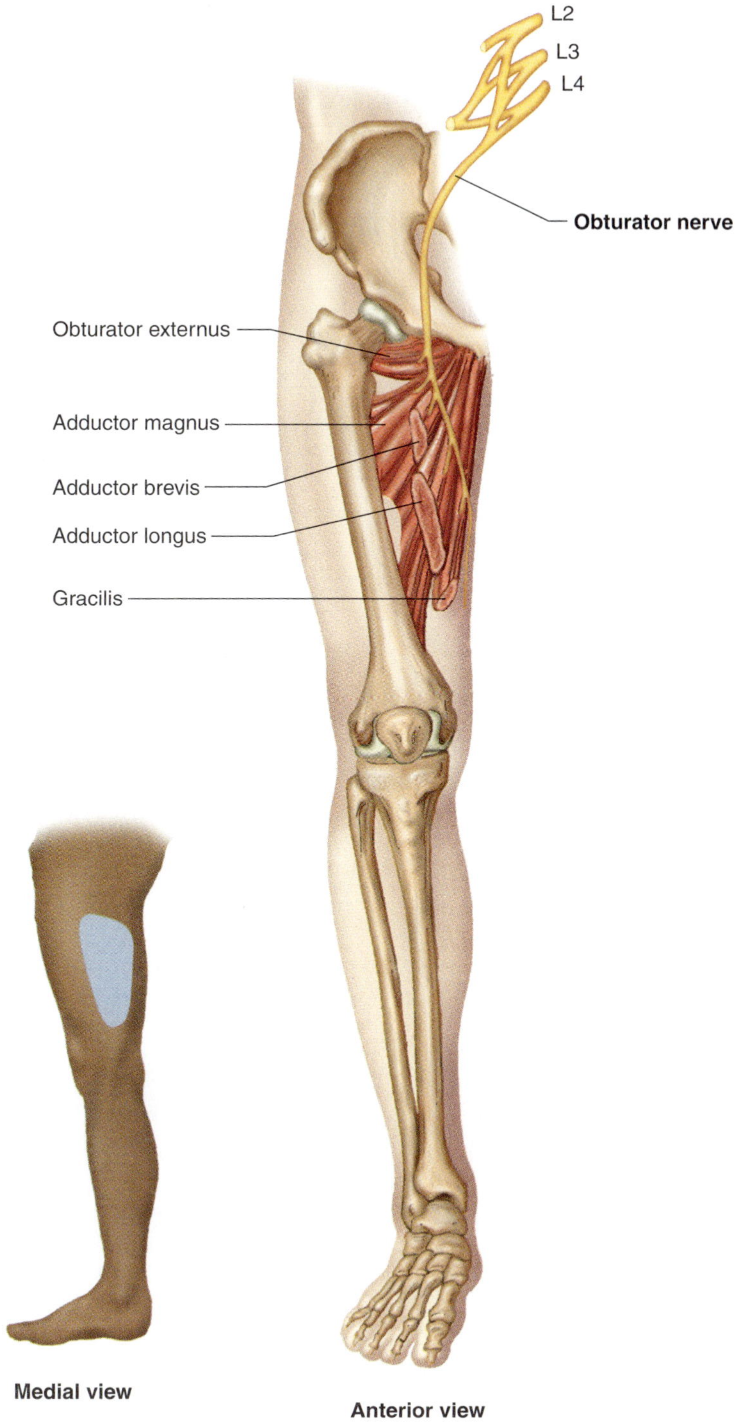
L2
L3
L4
Obturator nerve
Obturator externus
Adductor magnus
Adductor brevis
Adductor longus
Gracilis
Medial view
Anterior view

Obturator Nerve
Figure 12.24

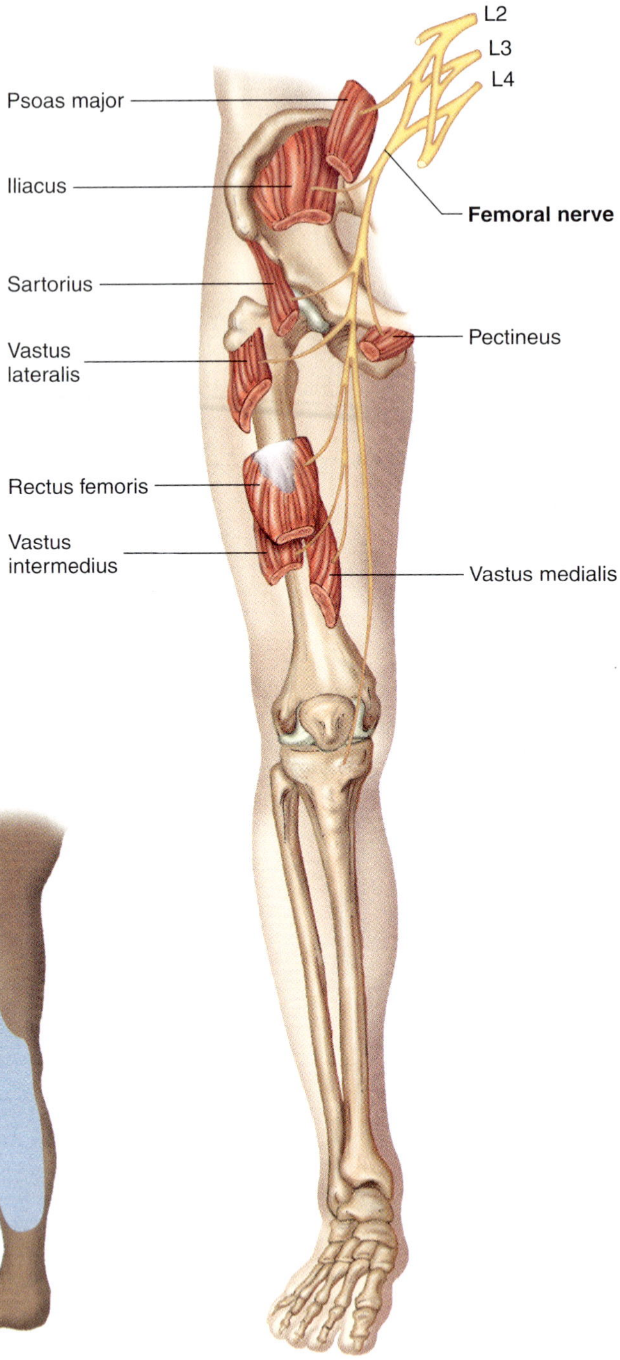

Anterior view

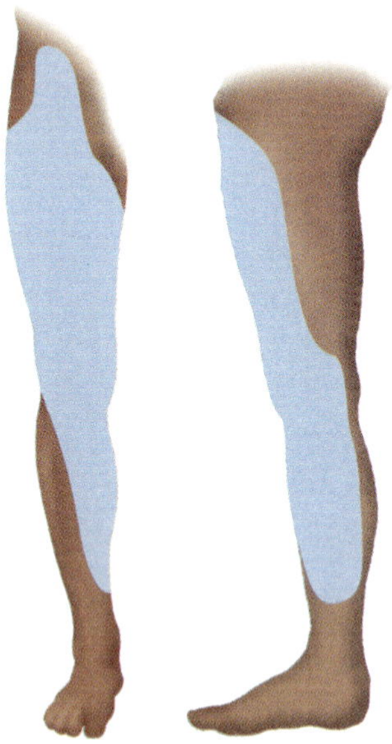

Anterior view **Medial view**

Femoral Nerve
Figure 12.25

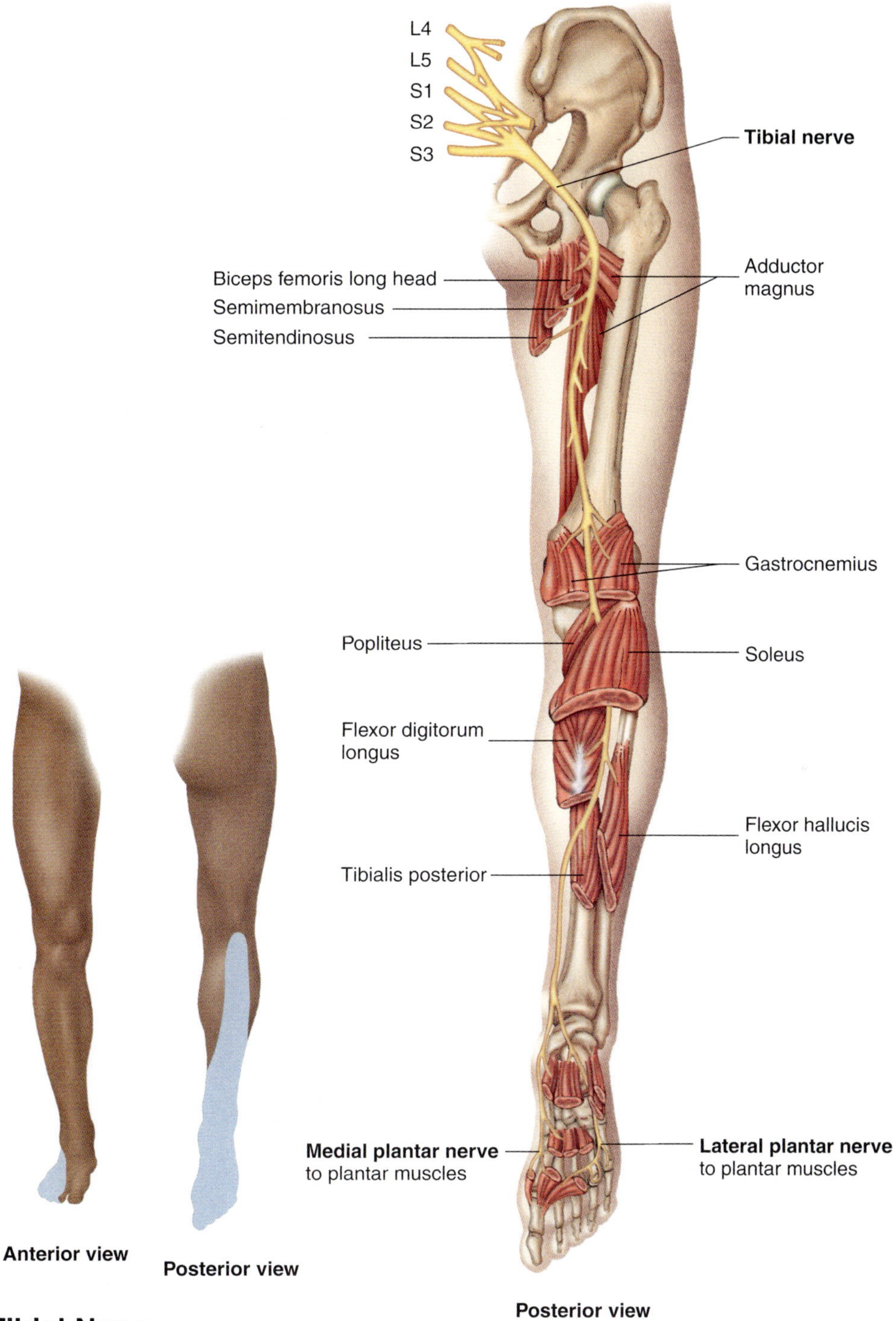
L4
L5
S1
S2
S3
Tibial nerve
Biceps femoris long head
Semimembranosus
Semitendinosus
Adductor magnus
Gastrocnemius
Popliteus
Soleus
Flexor digitorum longus
Flexor hallucis longus
Tibialis posterior
Medial plantar nerve to plantar muscles
Lateral plantar nerve to plantar muscles
Anterior view
Posterior view
Posterior view

Tibial Nerve
Figure 12.26

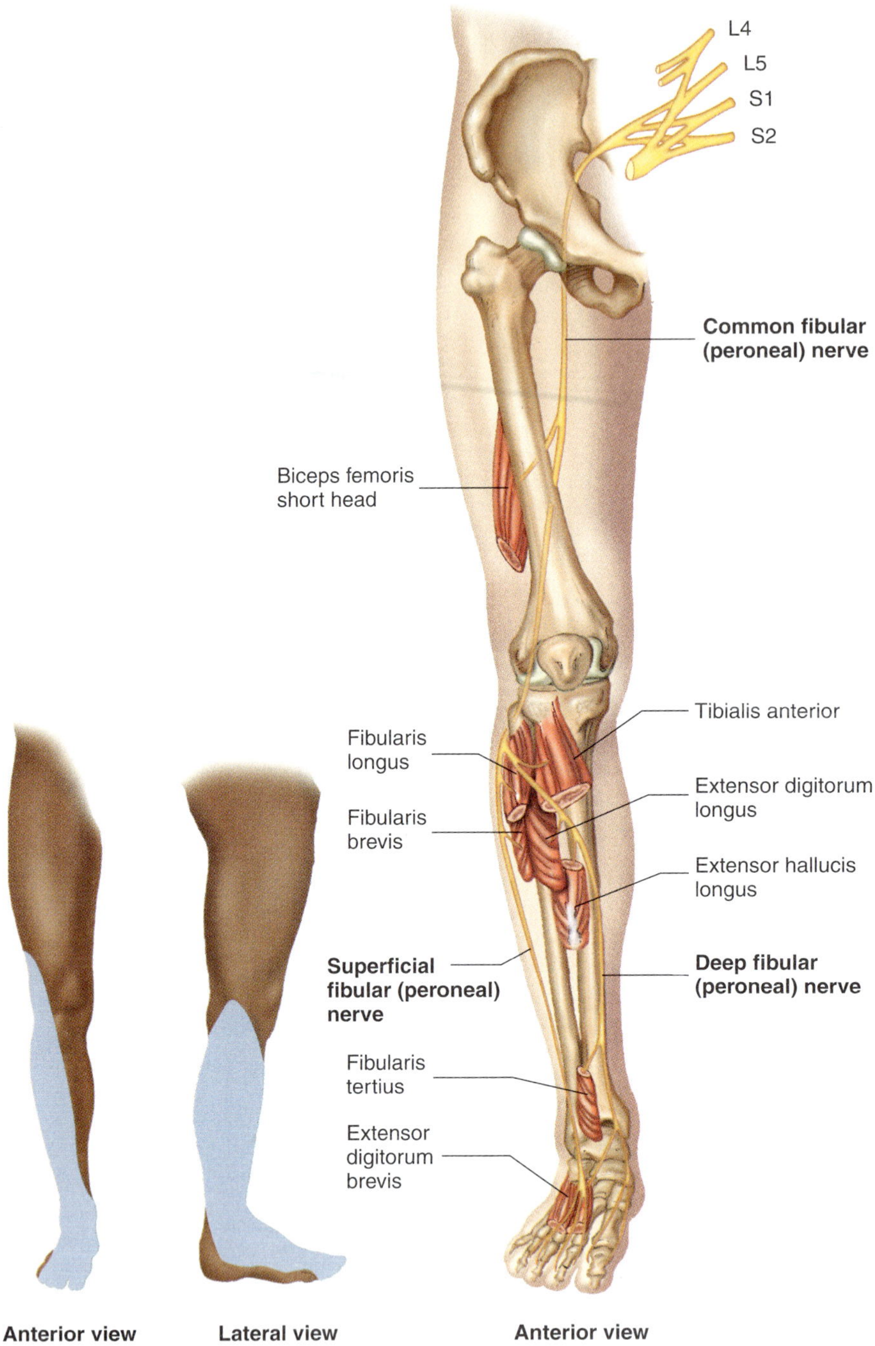

L4
L5
S1
S2
Common fibular (peroneal) nerve
Biceps femoris short head
Tibialis anterior
Fibularis longus
Extensor digitorum longus
Fibularis brevis
Extensor hallucis longus
Superficial fibular (peroneal) nerve
Deep fibular (peroneal) nerve
Fibularis tertius
Extensor digitorum brevis
Anterior view
Lateral view
Anterior view

Fibular Nerve

Figure 12.27

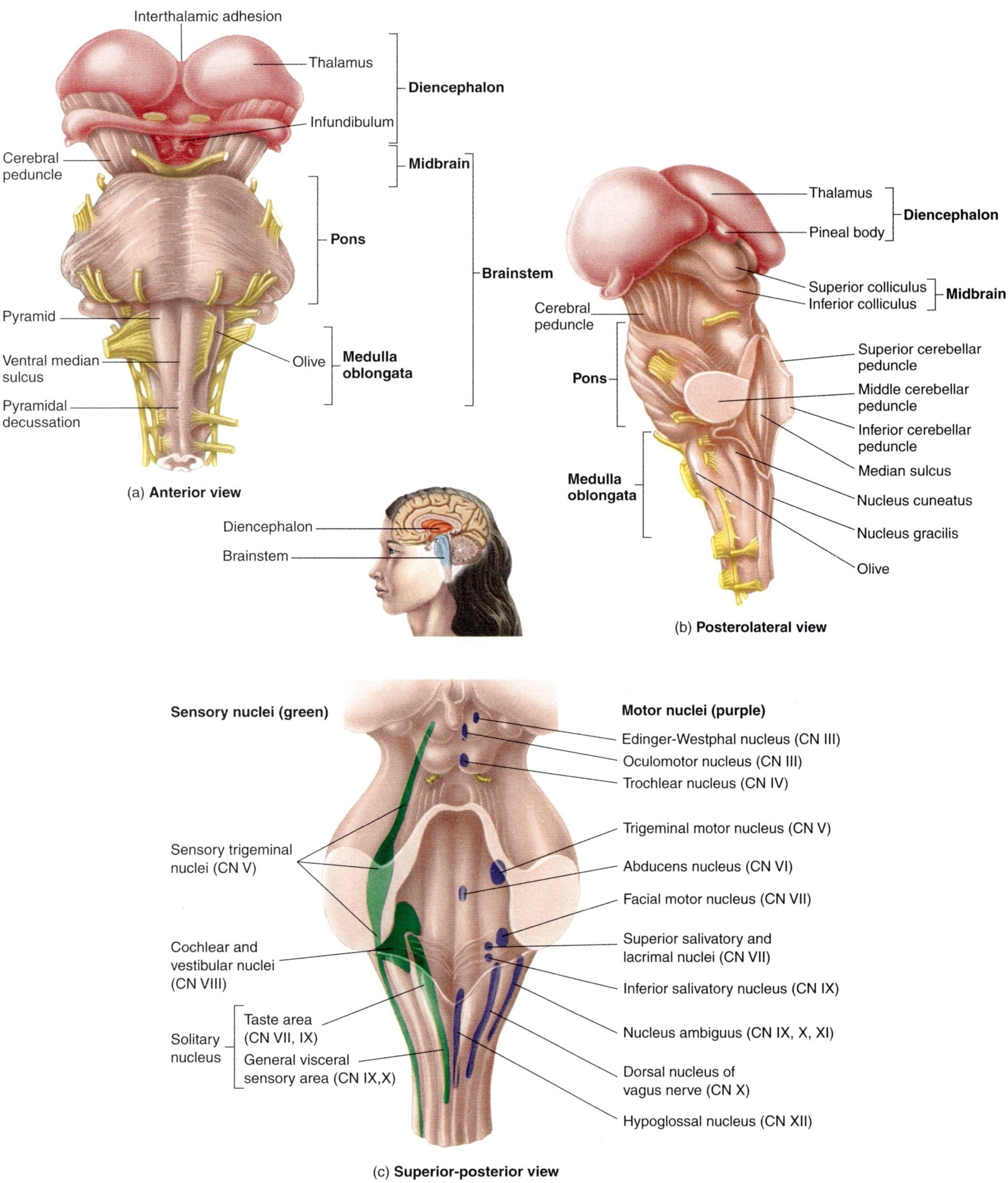
Interthalamic adhesion
Thalamus
Diencephalon
Infundibulum
Cerebral peduncle
Midbrain
Pons
Brainstem
Pyramid
Ventral median sulcus
Olive
Medulla oblongata
Pyramidal decussation
(a) Anterior view
Diencephalon
Brainstem
Thalamus
Pineal body
Diencephalon
Superior colliculus
Inferior colliculus
Midbrain
Cerebral peduncle
Superior cerebellar peduncle
Pons
Middle cerebellar peduncle
Inferior cerebellar peduncle
Median sulcus
Medulla oblongata
Nucleus cuneatus
Nucleus gracilis
Olive
(b) Posterolateral view
Sensory nuclei (green)
Motor nuclei (purple)
Edinger-Westphal nucleus (CN III)
Oculomotor nucleus (CN III)
Trochlear nucleus (CN IV)
Trigeminal motor nucleus (CN V)
Sensory trigeminal nuclei (CN V)
Abducens nucleus (CN VI)
Facial motor nucleus (CN VII)
Cochlear and vestibular nuclei (CN VIII)
Superior salivatory and lacrimal nuclei (CN VII)
Inferior salivatory nucleus (CN IX)
Solitary nucleus
Taste area (CN VII, IX)
General visceral sensory area (CN IX,X)
Nucleus ambiguus (CN IX, X, XI)
Dorsal nucleus of vagus nerve (CN X)
Hypoglossal nucleus (CN XII)
(c) Superior-posterior view

Brainstem and Diencephalon
Figure 13.2

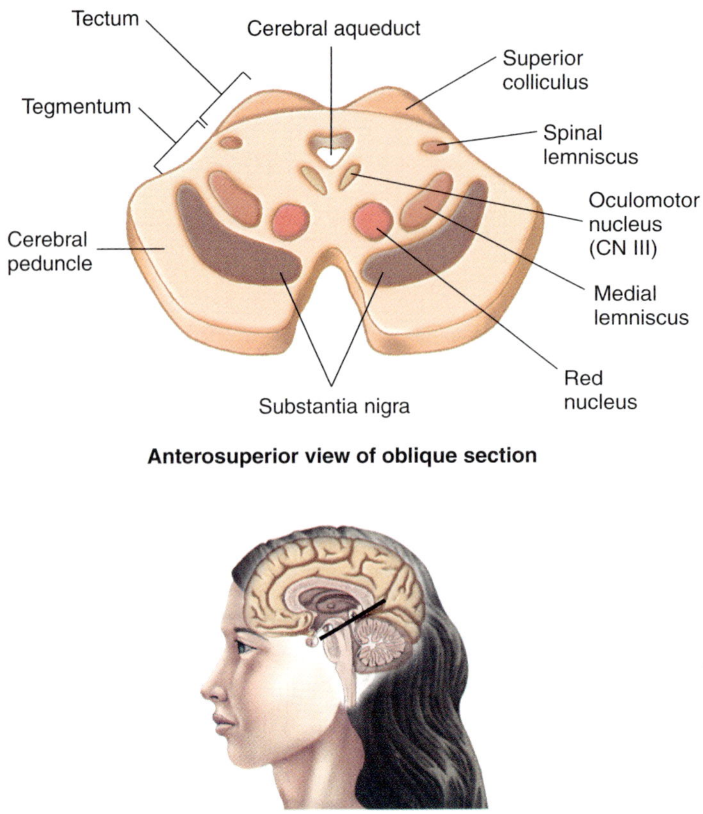

Oblique Section Through the Midbrain
Figure 13.3

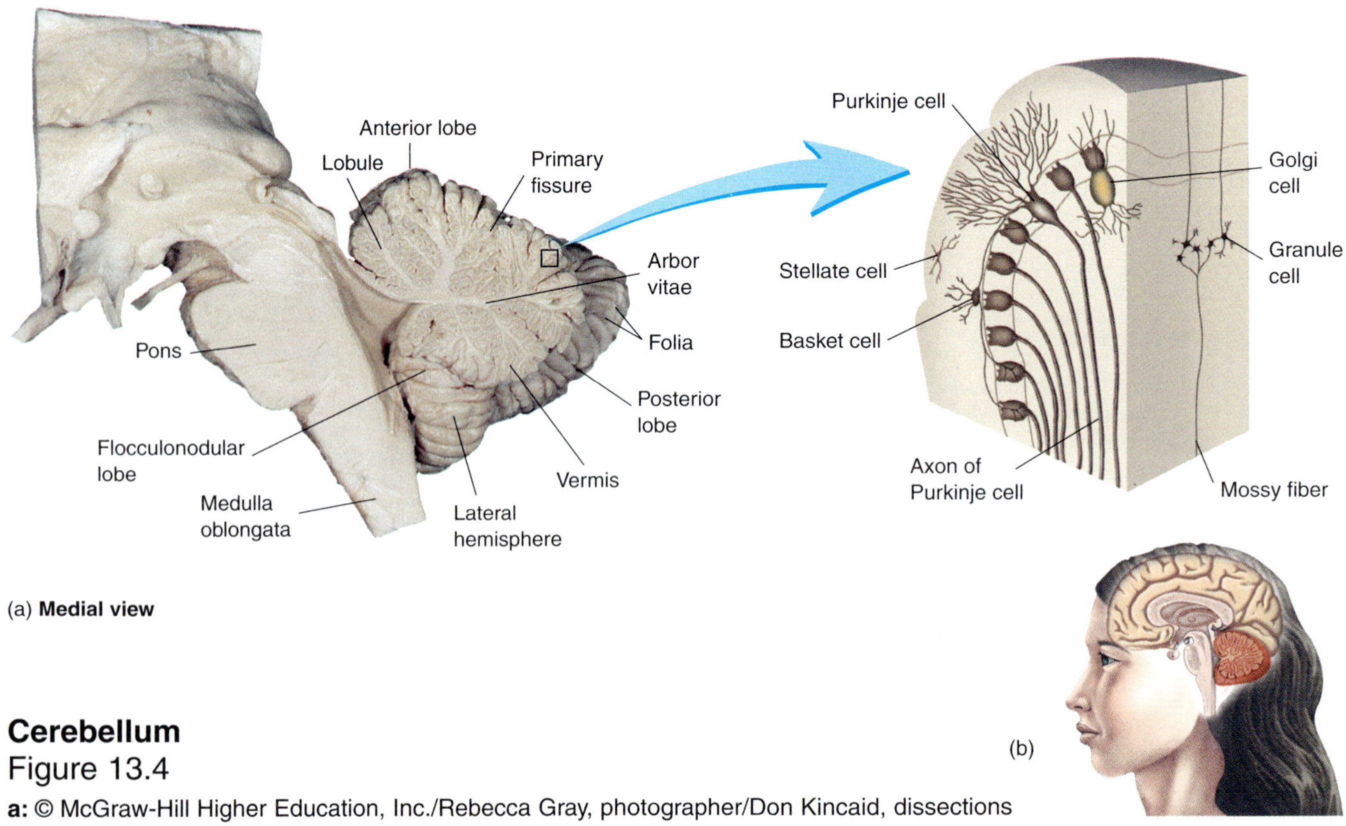

Cerebellum
Figure 13.4

a: © McGraw-Hill Higher Education, Inc./Rebecca Gray, photographer/Don Kincaid, dissections

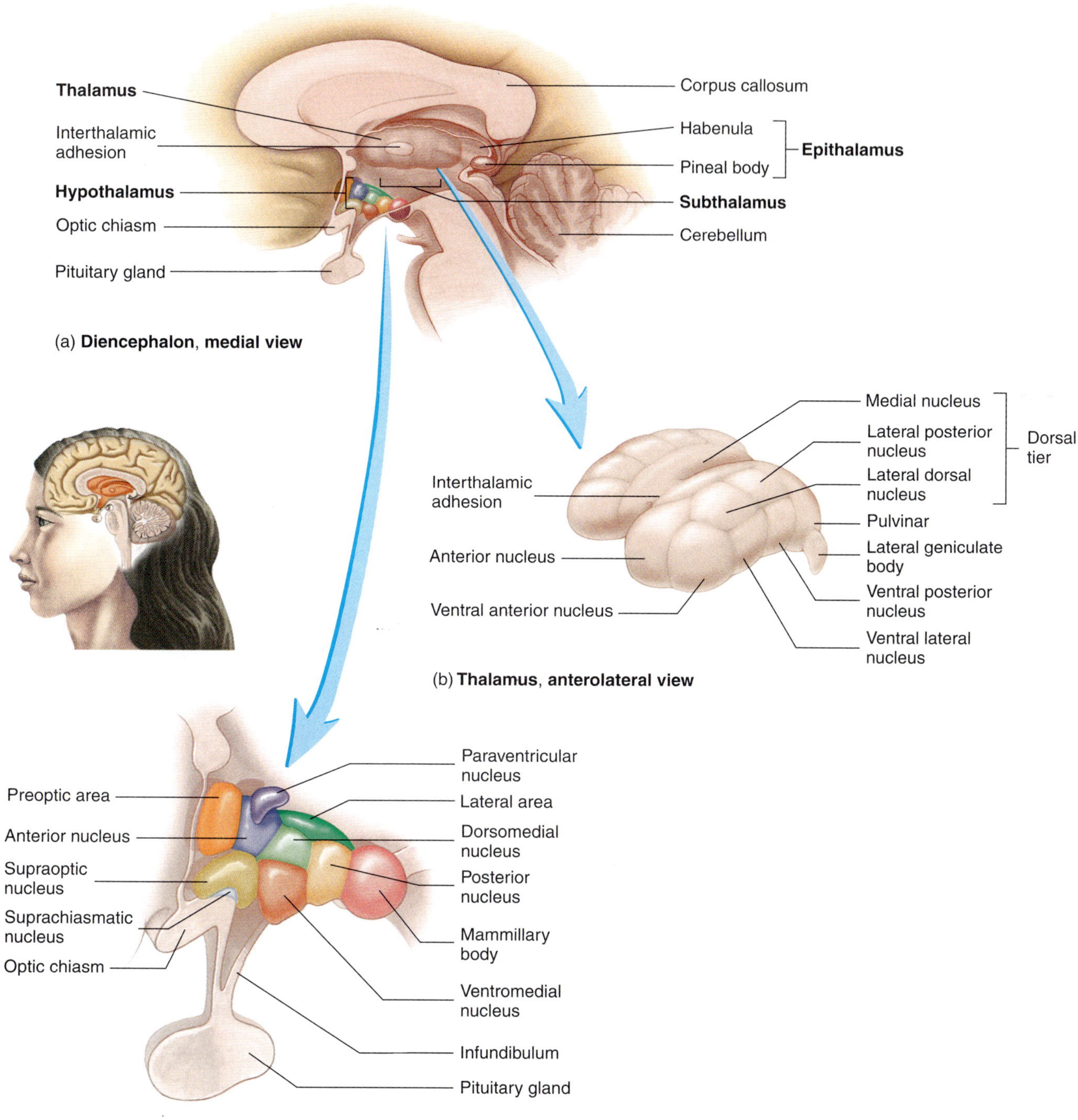

(c) **Hypothalamus**, **medial view**

Diencephalon
Figure 13.5

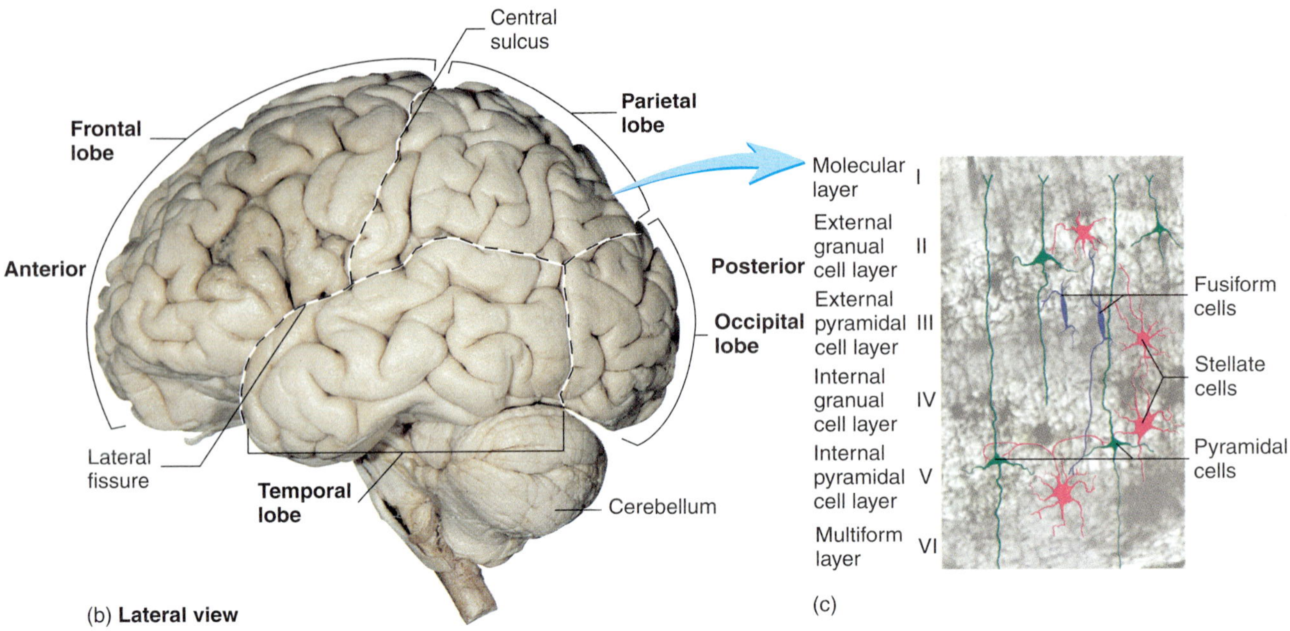

The Brain
Figure 13.6

b: © McGraw-Hill Higher Education, Inc./Rebecca Gray, photographer/Don Kincaid, dissections

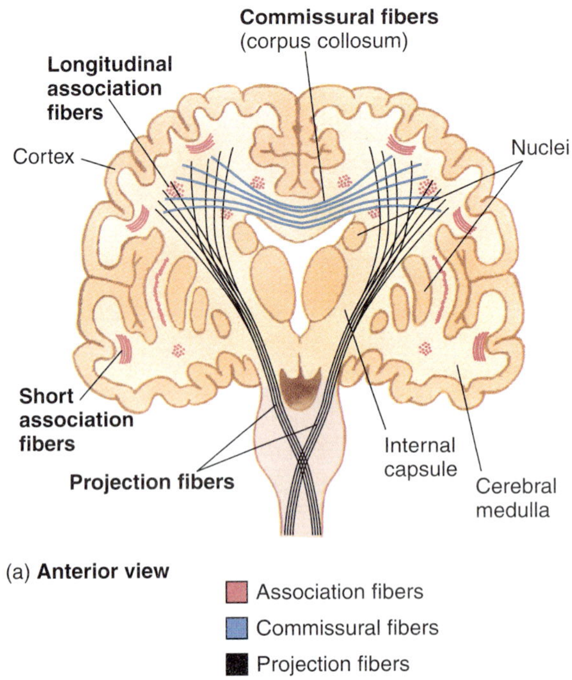

Cerebral Medullary Tracts
Figure 13.7

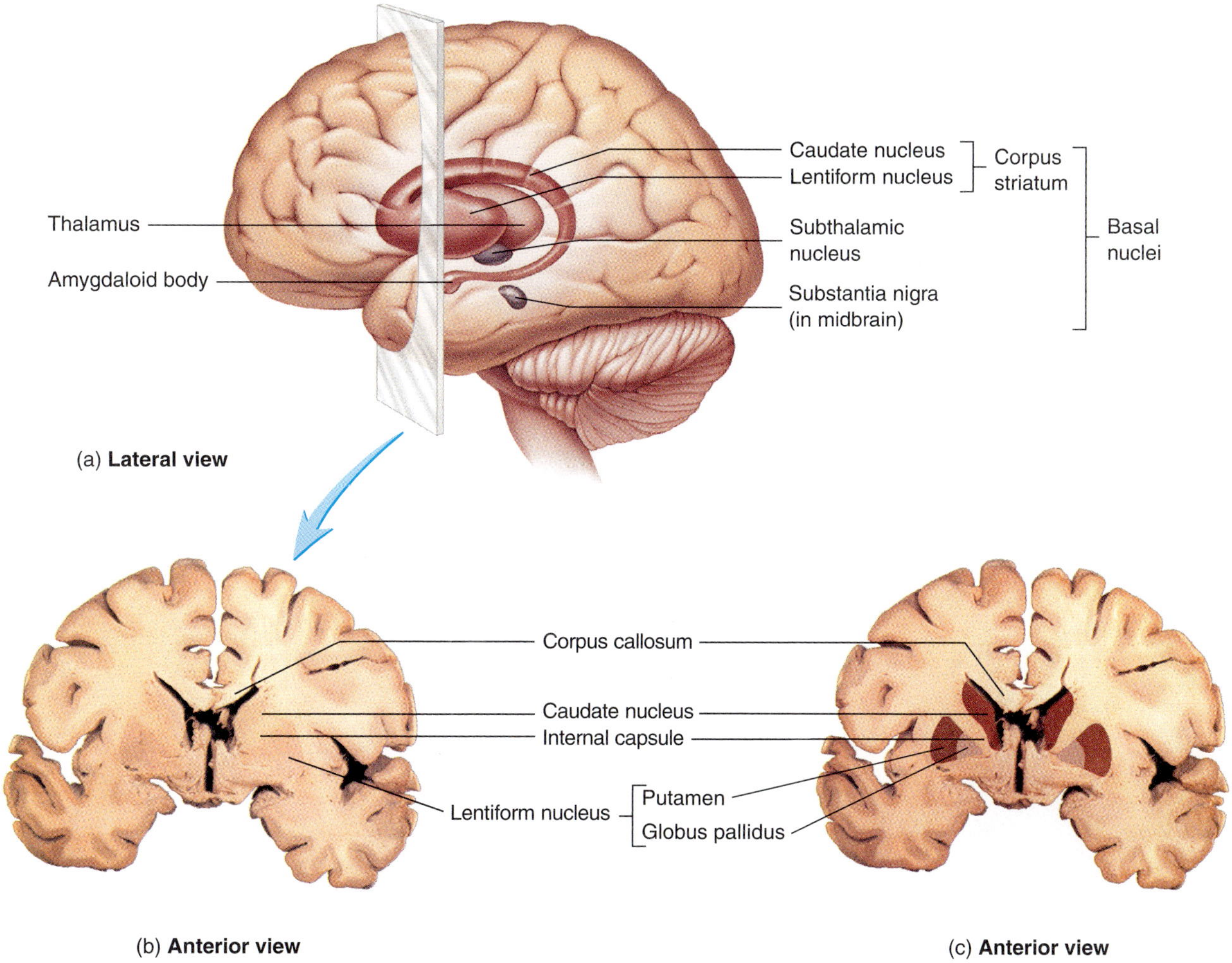

Basal Nuclei (Ganglia) of the Left Hemisphere
Figure 13.8
b–c: © Trent Stephens

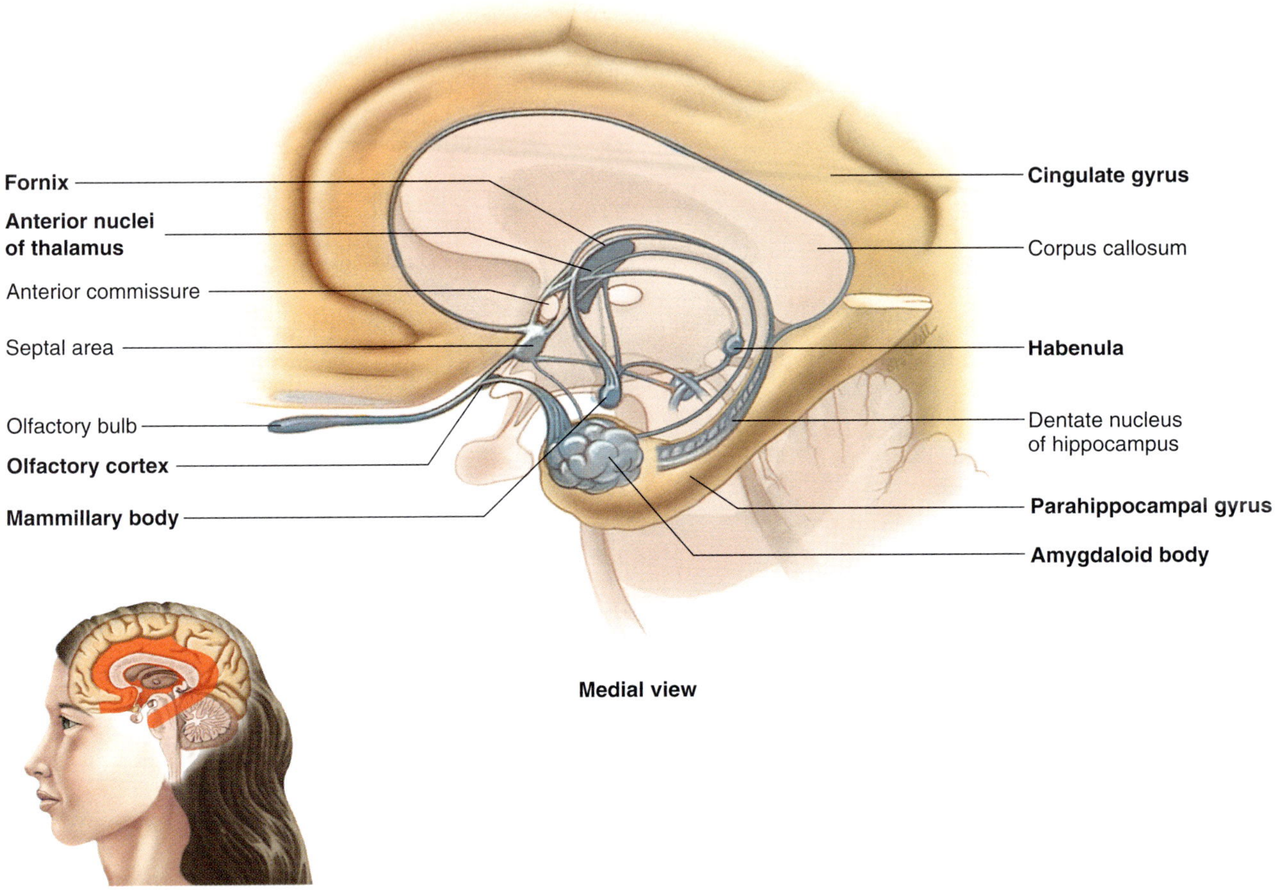

Limbic System and Associated Structures of the Right Hemisphere as Seen in a Midsagittal Section
Figure 13.9

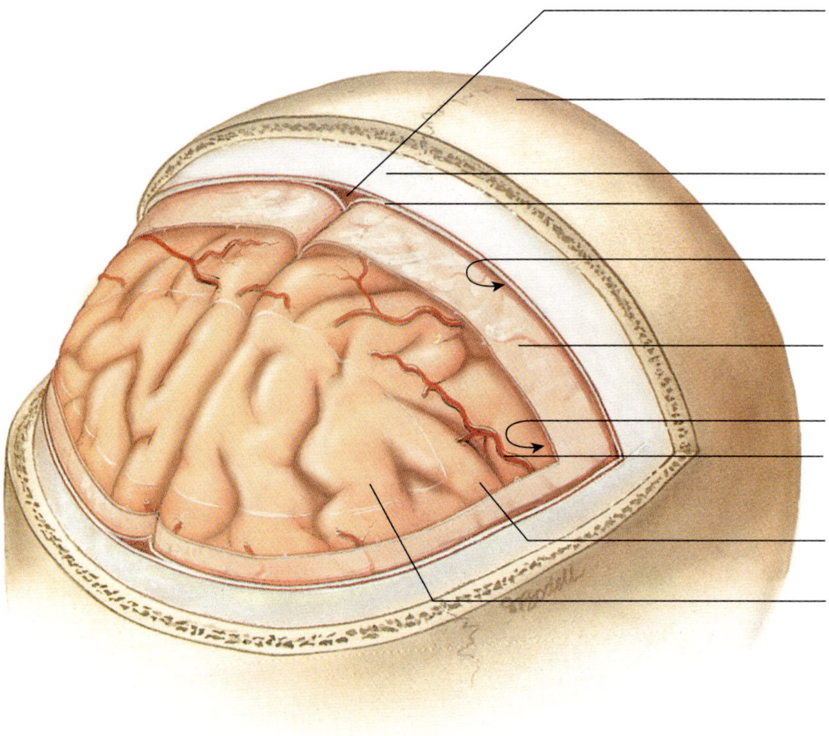

(a) **Anterosuperior view**

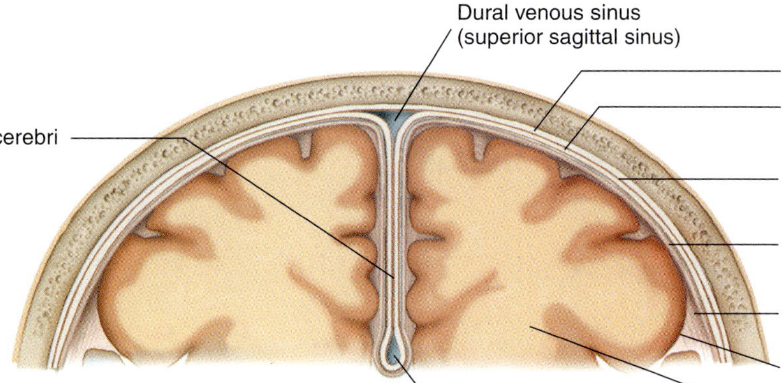

(b) **Anterior view**

Meninges
Figure 13.10

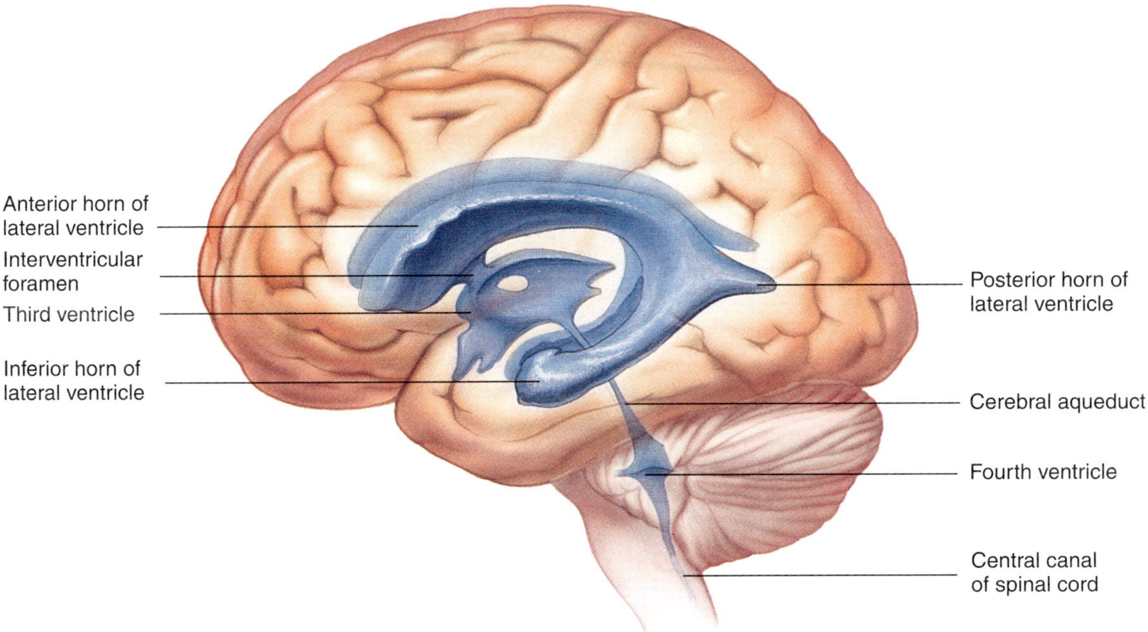

Lateral view

Ventricles of the Brain Viewed from the Left
Figure 13.11

1. Cerebrospinal fluid (CSF) is produced by the choroid plexuses of each of the four ventricles (*inset*).
2. CSF from the lateral ventricles flows through the interventricular foramina to the third ventricle.
3. CSF flows from the third ventricle through the cerebral aqueduct to the fourth ventricle.
4. CSF exits the fourth ventricle through the lateral and median apertures and enters the subarachnoid space. Some CSF enters the central canal of the spinal cord.
5. CSF flows through the subarachnoid space to the arachnoid granulations in the superior sagittal sinus, where it enters the venous circulation (*inset*).

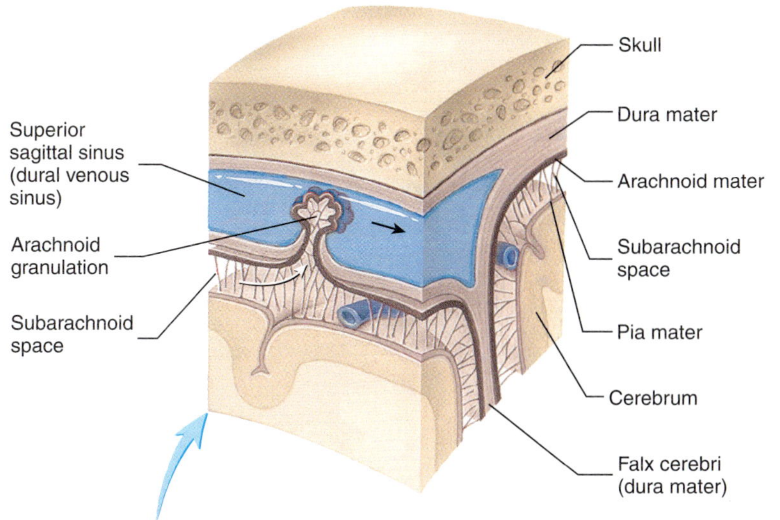

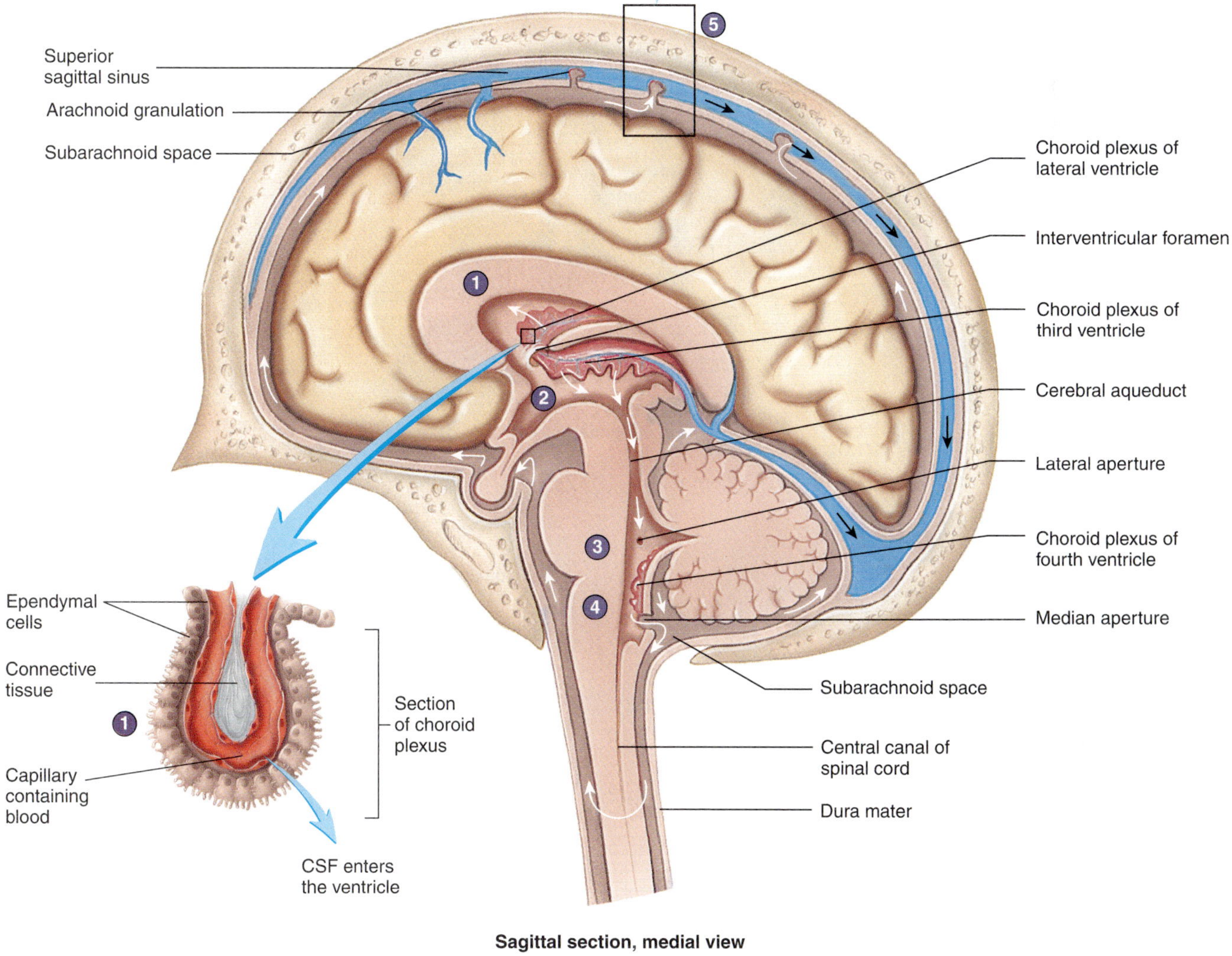

Flow of CSF
Process Figure 13.12

1. The neural plate is formed from ectoderm.

2. Neural folds form as parallel ridges along the embryo.

3. Neural crest cells break away from the crest of the neural folds. Neural crest cells give rise to a number of structures: sensory and autonomic neurons in the PNS, facial pigment cells, facial bones, and dentin of the teeth.

4. The neural folds meet at the midline to form the neural tube. The neural tube becomes the brain and spinal cord.

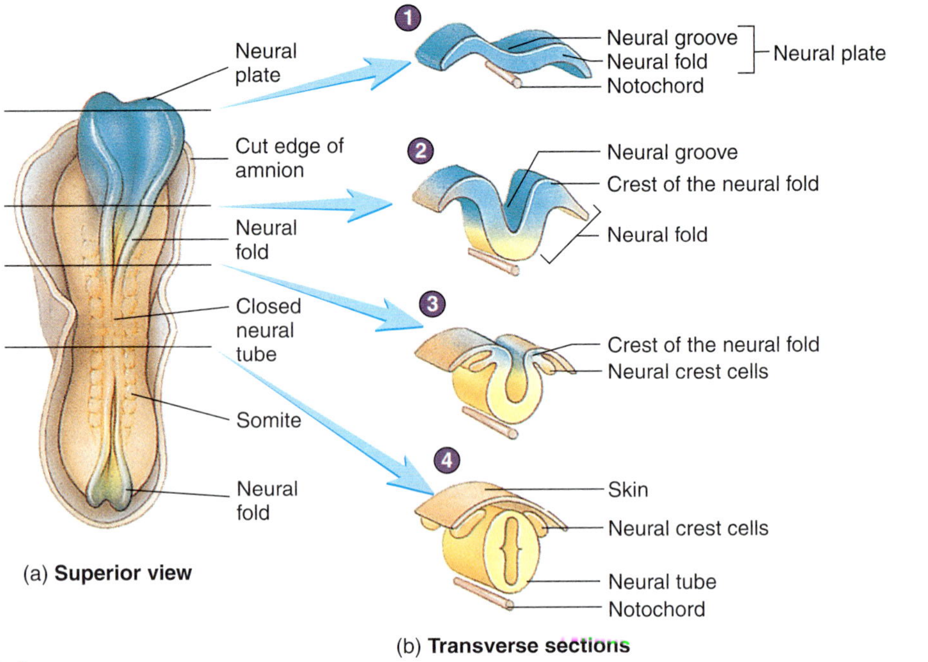

Formation of the Neural Tube
Process Figure 13.13

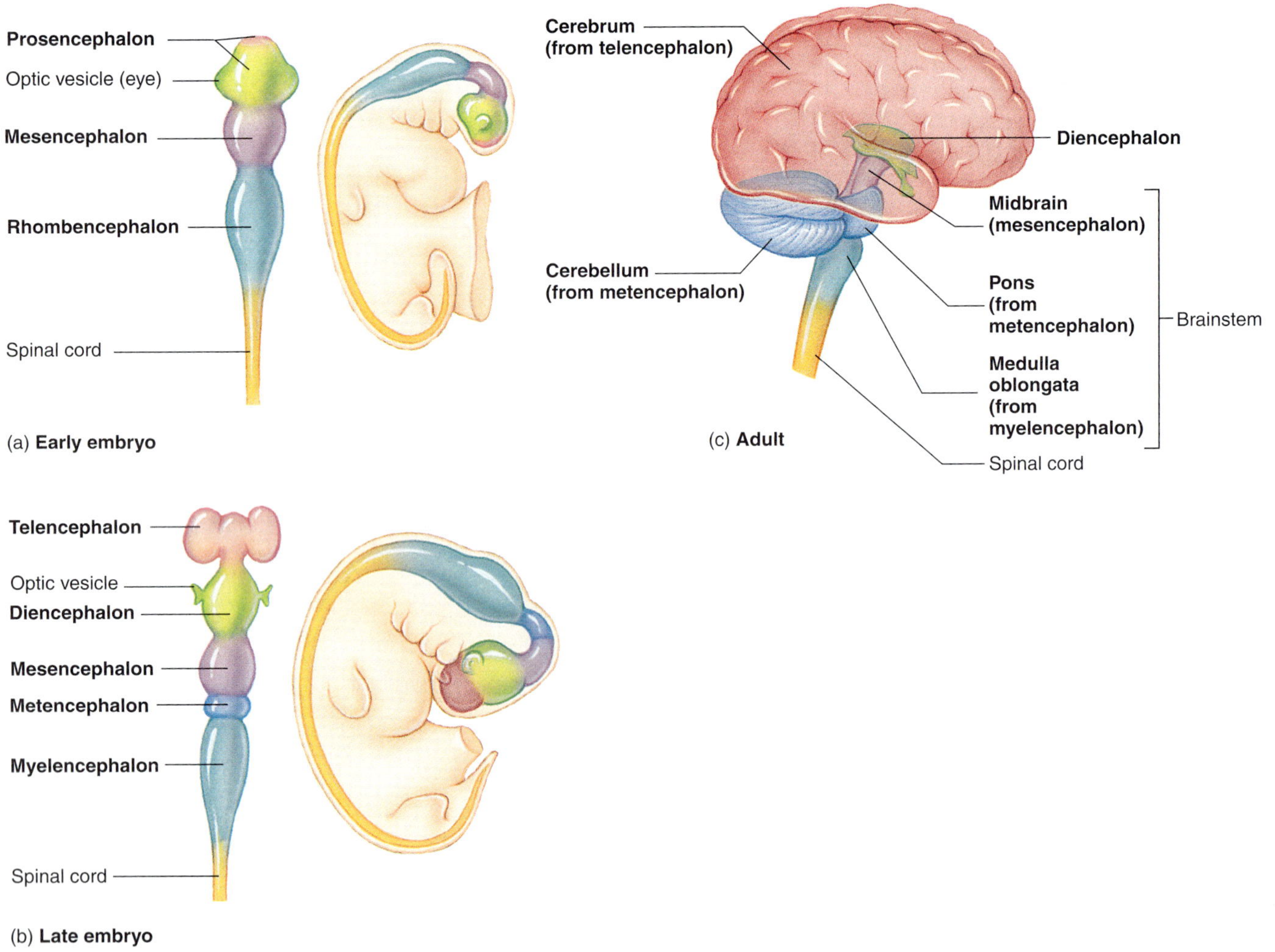

Development of the Brain Segments and Ventricles
Figure 13.14

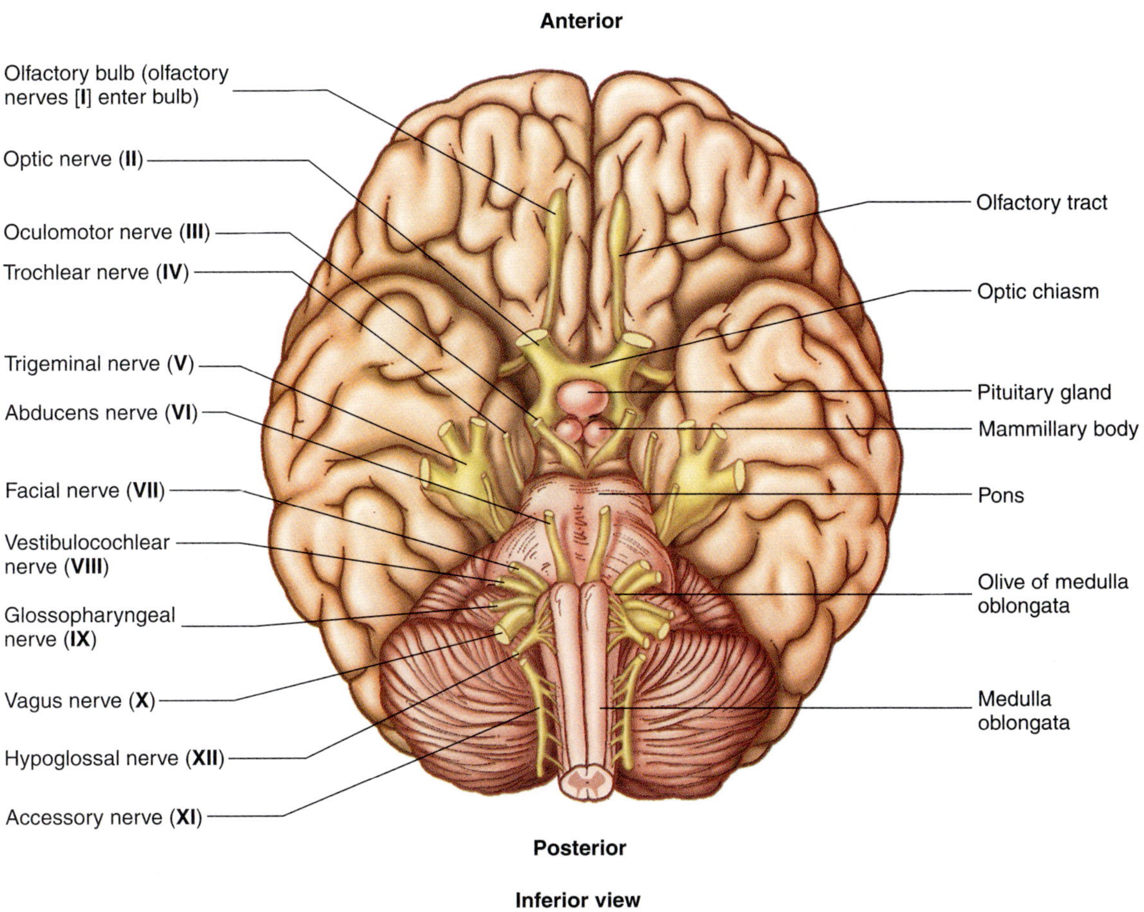

Inferior Surface of the Brain Showing the Origin of the Cranial Nerves
Figure 13.15

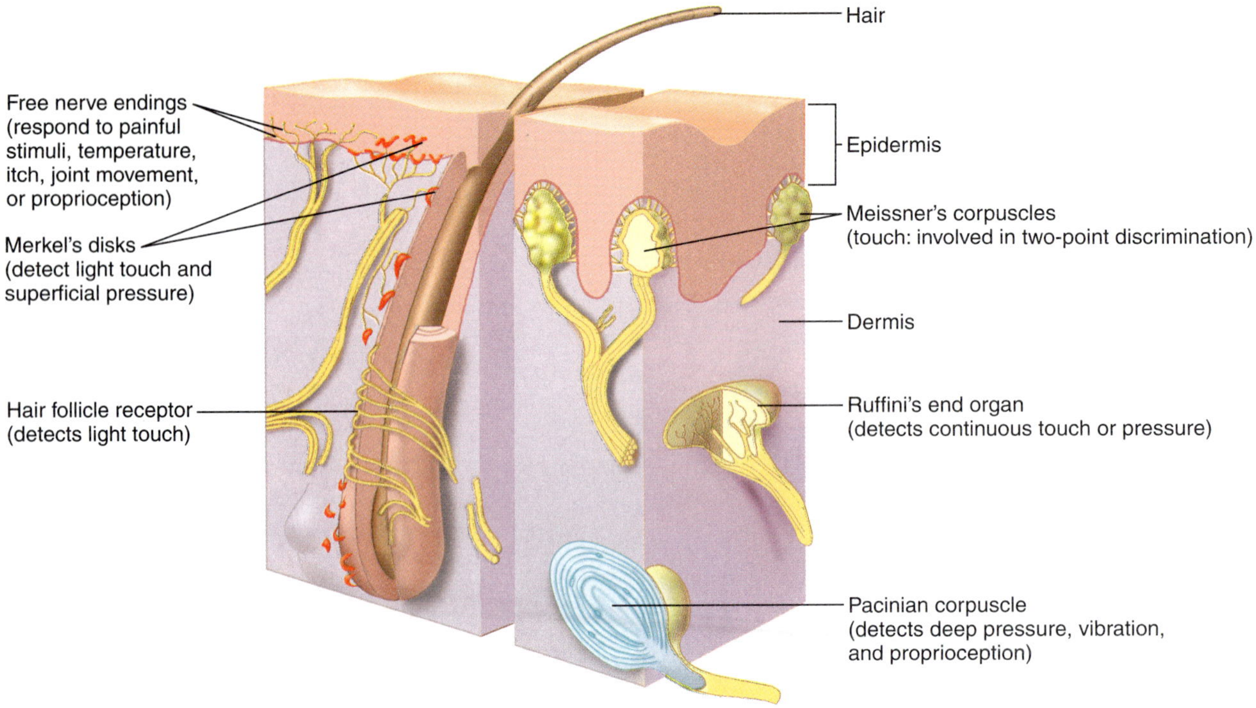

Sensory Nerve Endings in the Skin
Figure 14.1

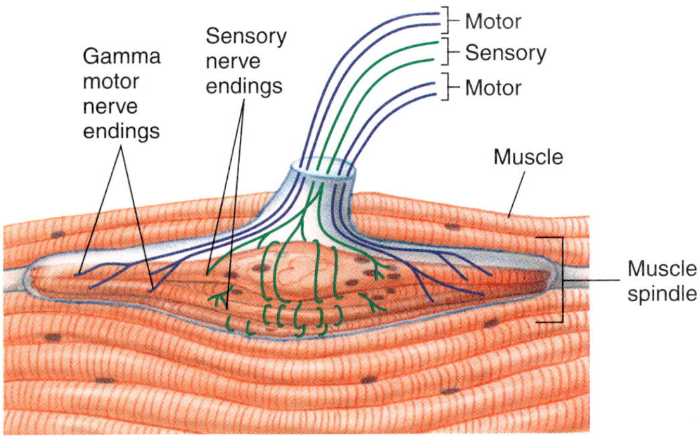

Muscle Spindle
Figure 14.3

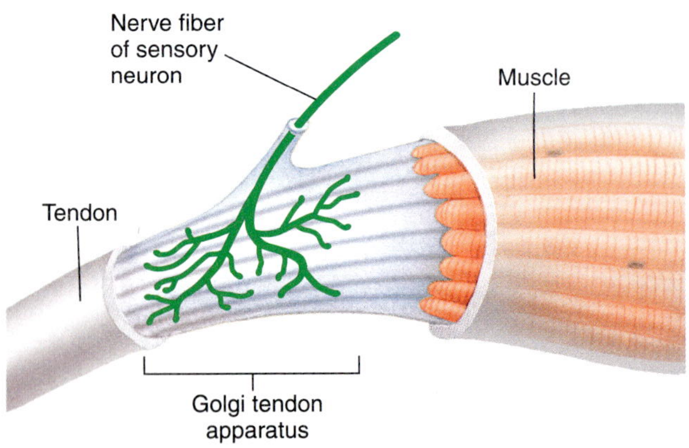

Golgi Tendon Organ
Figure 14.4

Cross Section of the Spinal Cord at the Cervical Level Depicting the Ascending Pathways

Figure 14.5

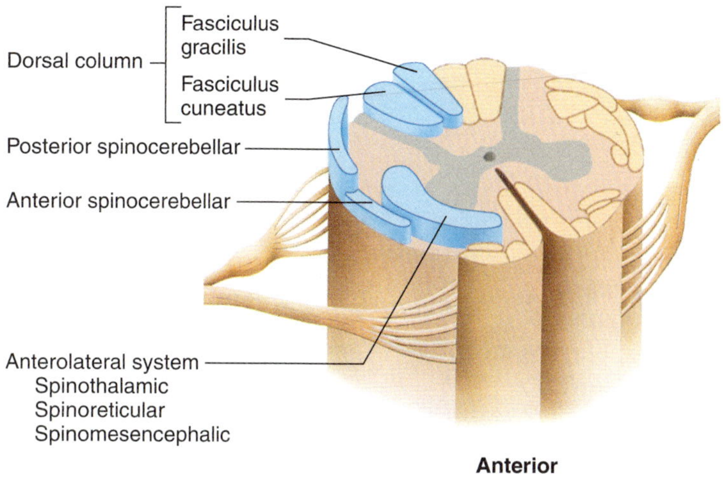

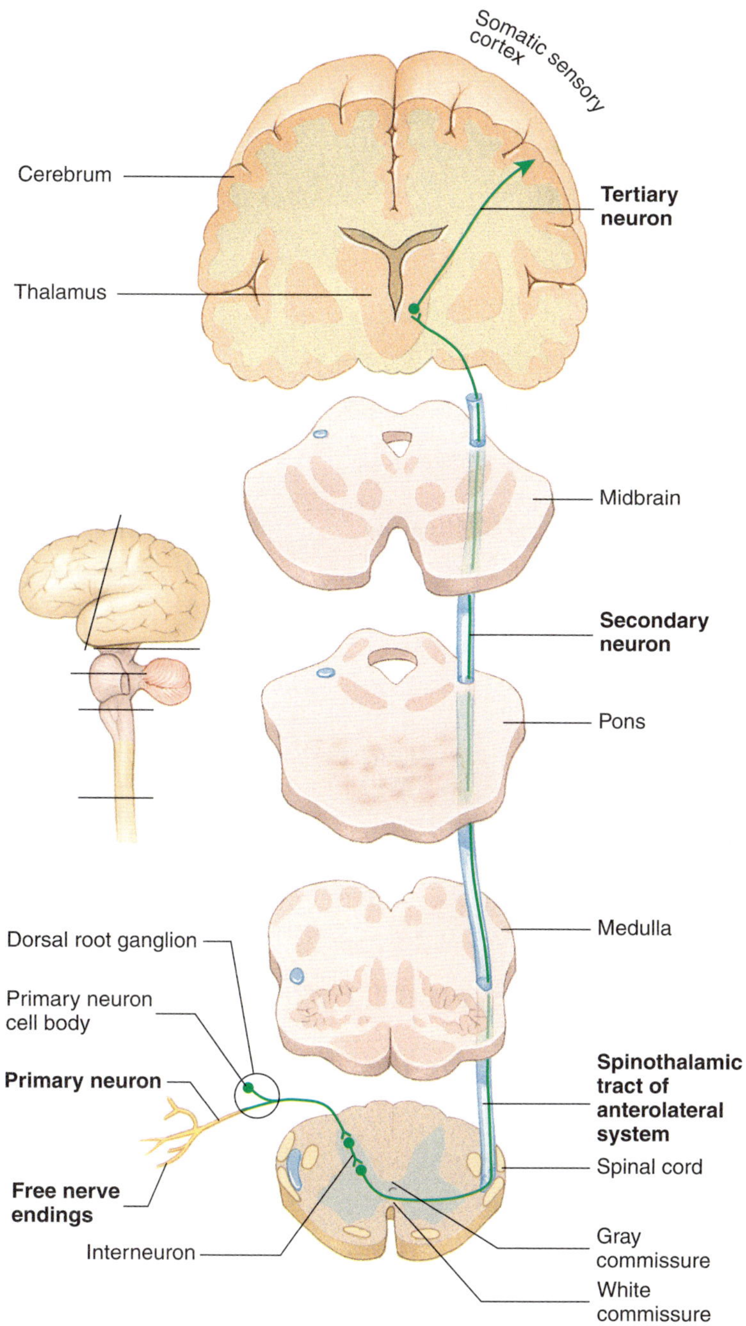

The Spinothalamic Portion of the Anterolateral System

Figure 14.6

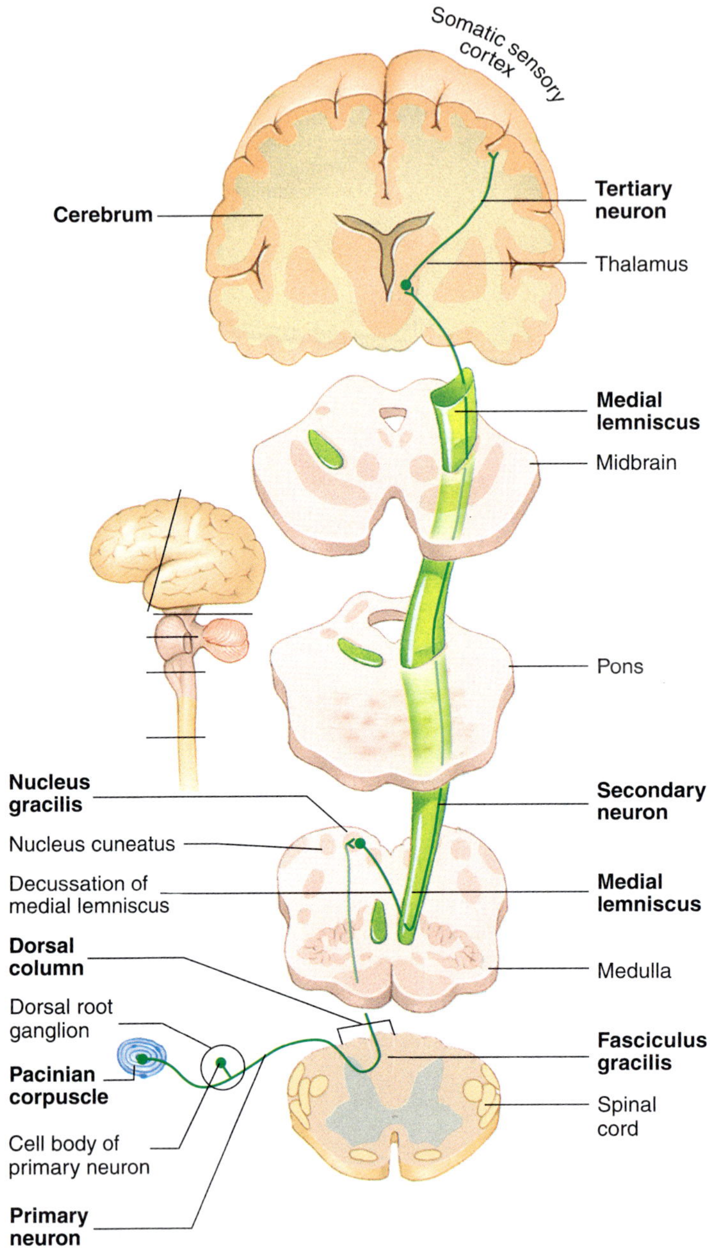

Dorsal-Column/Medial-Lemniscal System

Figure 14.7

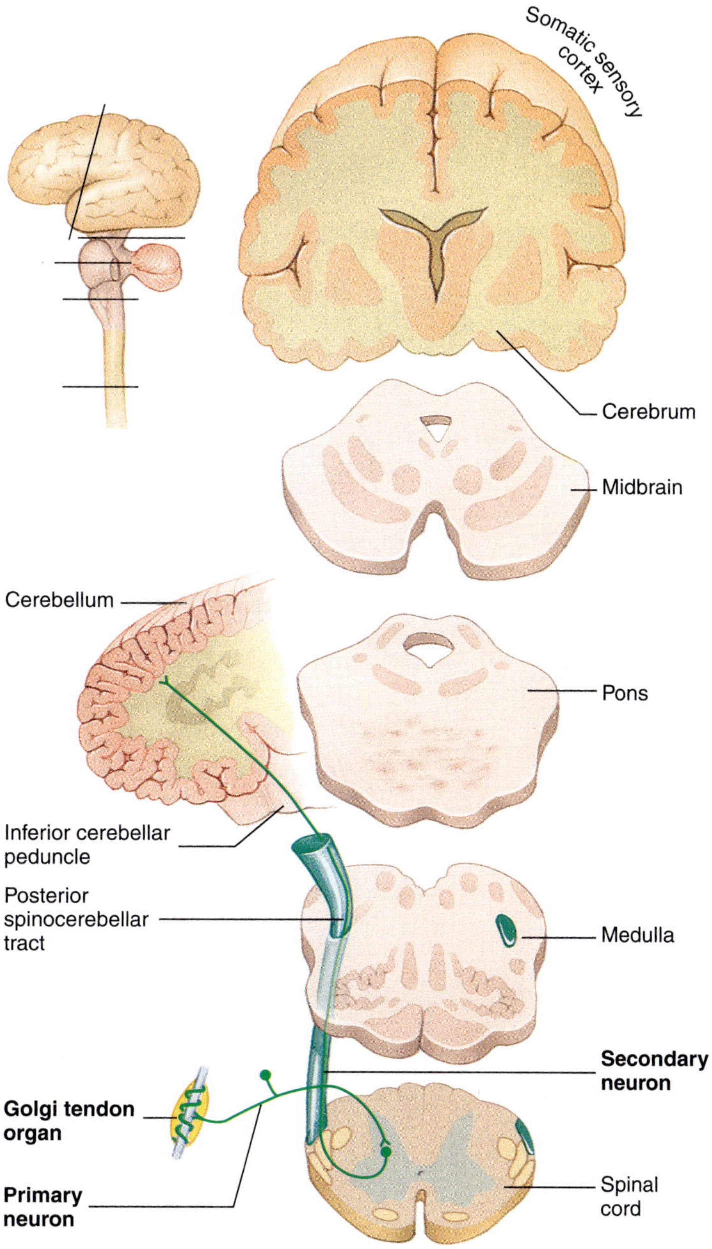
Somatic sensory cortex
Cerebrum
Midbrain
Cerebellum
Pons
Inferior cerebellar peduncle
Posterior spinocerebellar tract
Medulla
Secondary neuron
Golgi tendon organ
Primary neuron
Spinal cord

Posterior Spinocerebellar Tract
Figure 14.8

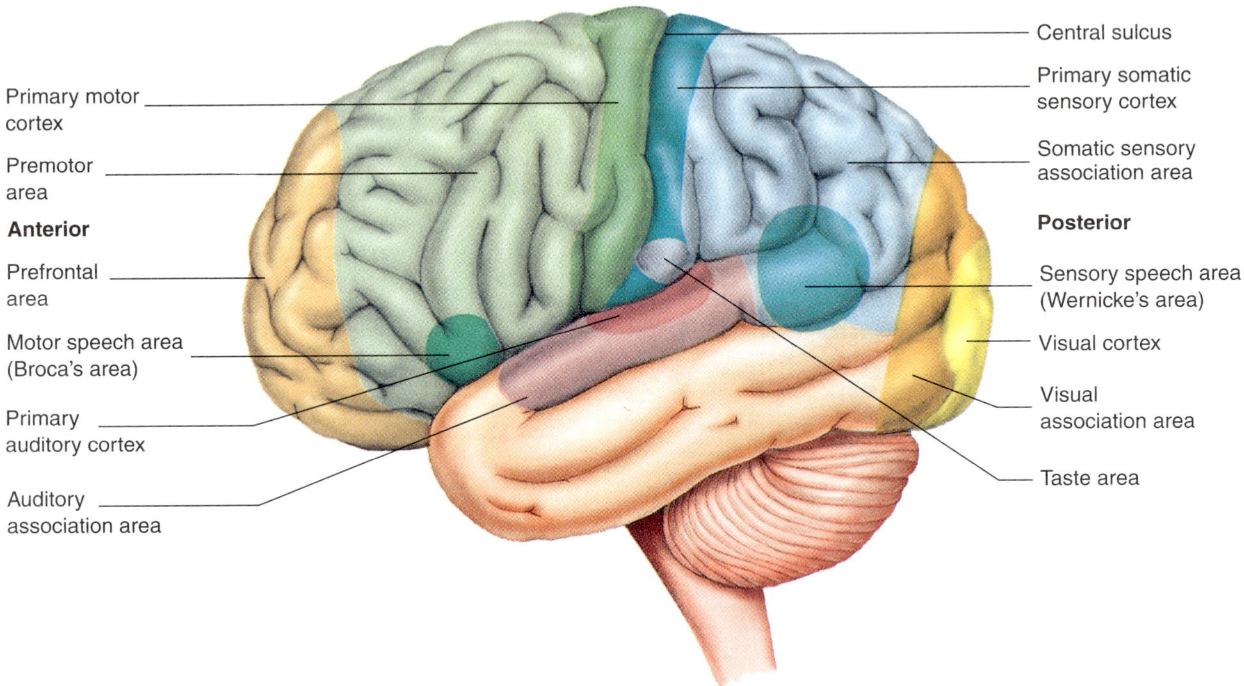

Functional Regions of the Lateral Side of the Left Cerebral Cortex
Figure 14.9

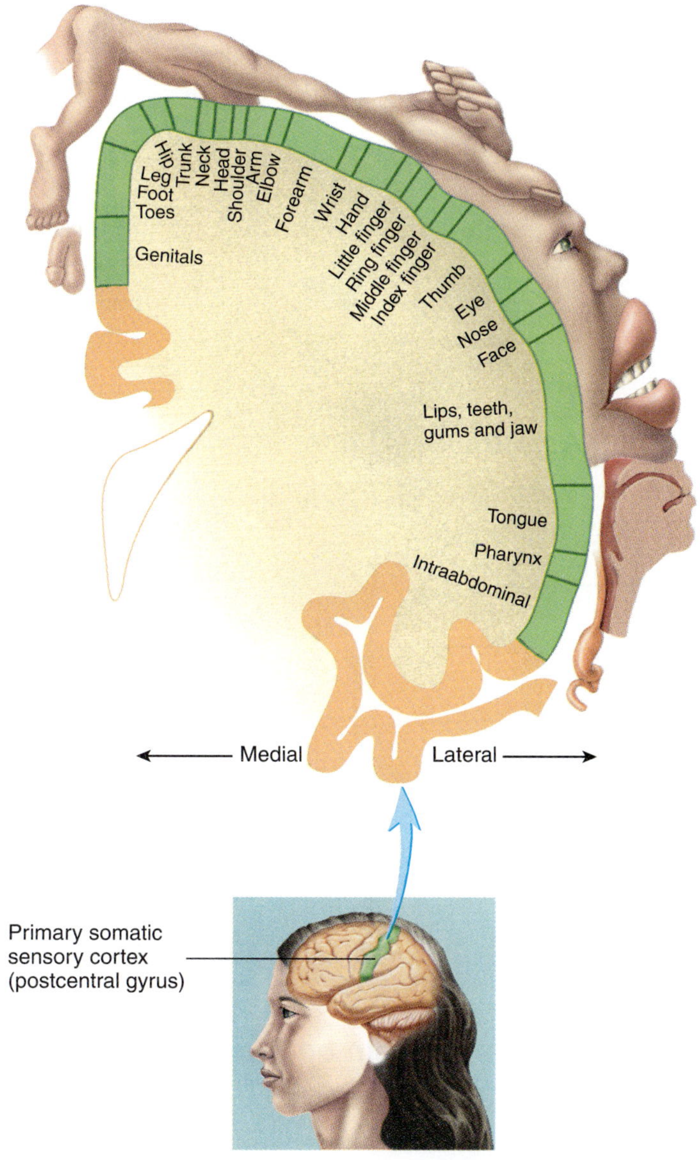

Topography of the Somatic Sensory Cortex
Figure 14.10

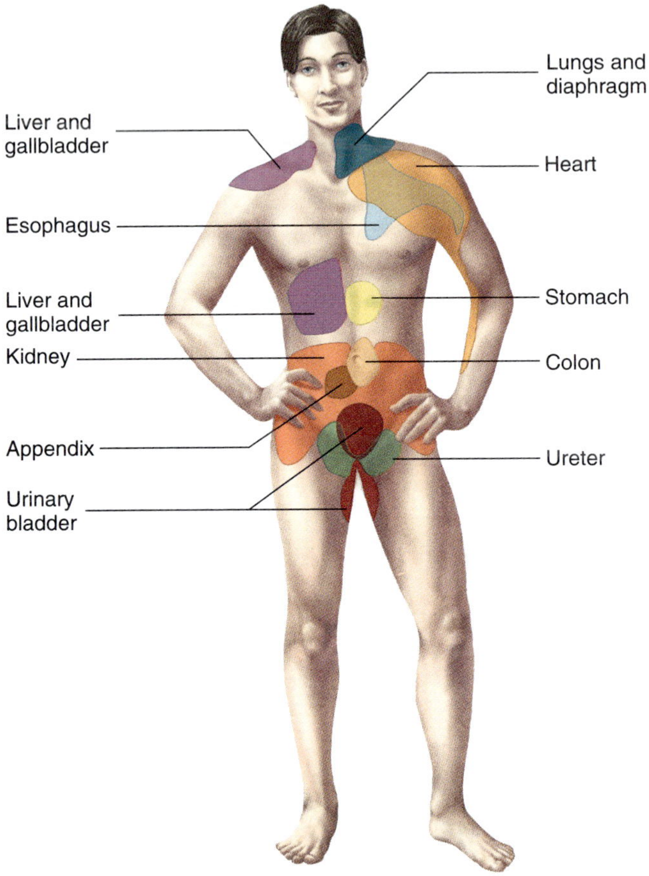

Areas of Referred Pain on the Body Surface
Figure A

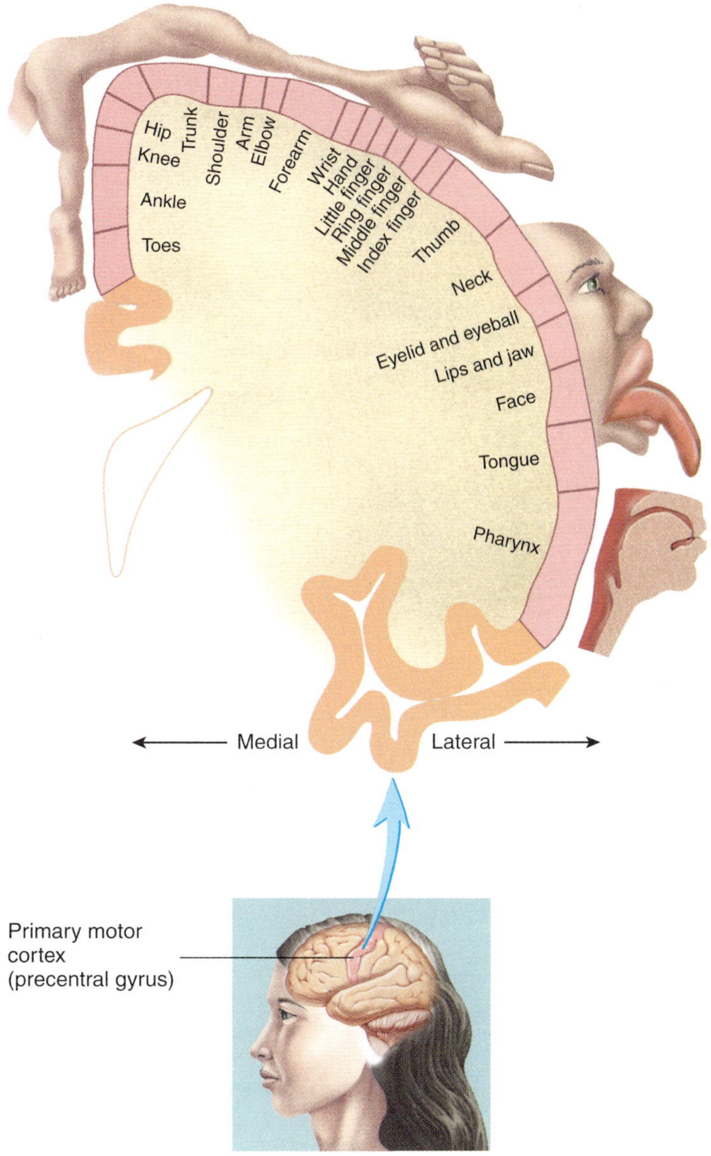

Topography of the Primary Motor Cortex
Figure 14.11

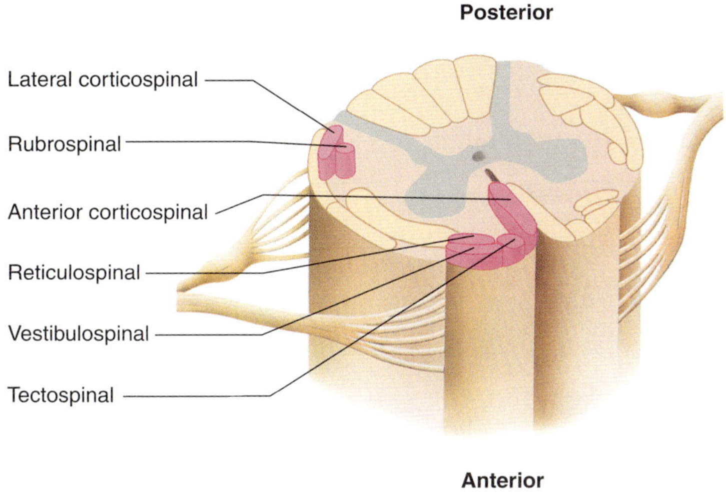

Cross Section of the Spinal Cord at the Cervical Level Depicting the Descending Pathways
Figure 14.12

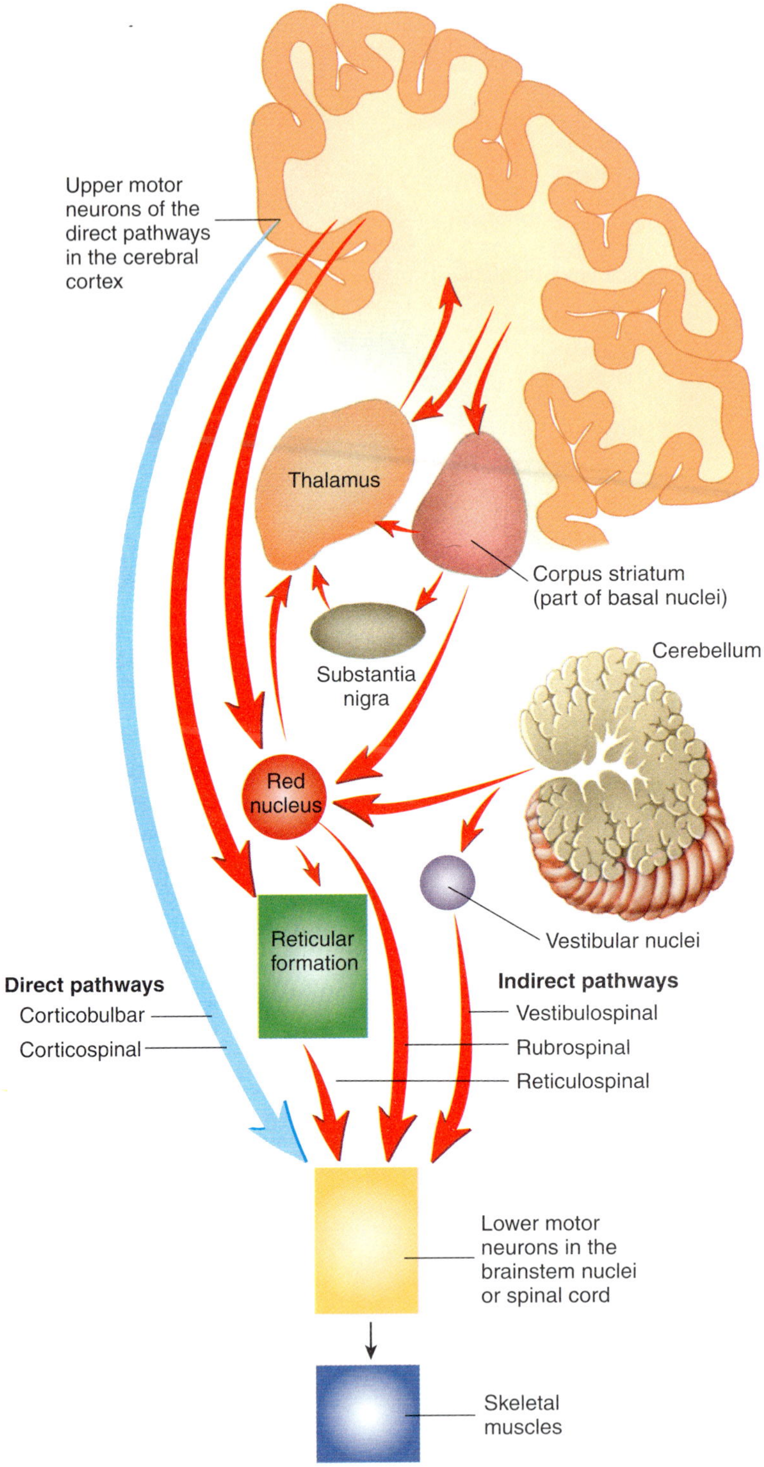
Upper motor neurons of the direct pathways in the cerebral cortex
Thalamus
Corpus striatum (part of basal nuclei)
Substantia nigra
Cerebellum
Red nucleus
Vestibular nuclei
Reticular formation
Direct pathways
Corticobulbar
Corticospinal
Indirect pathways
Vestibulospinal
Rubrospinal
Reticulospinal
Lower motor neurons in the brainstem nuclei or spinal cord
Skeletal muscles

Descending Pathways
Figure 14.13

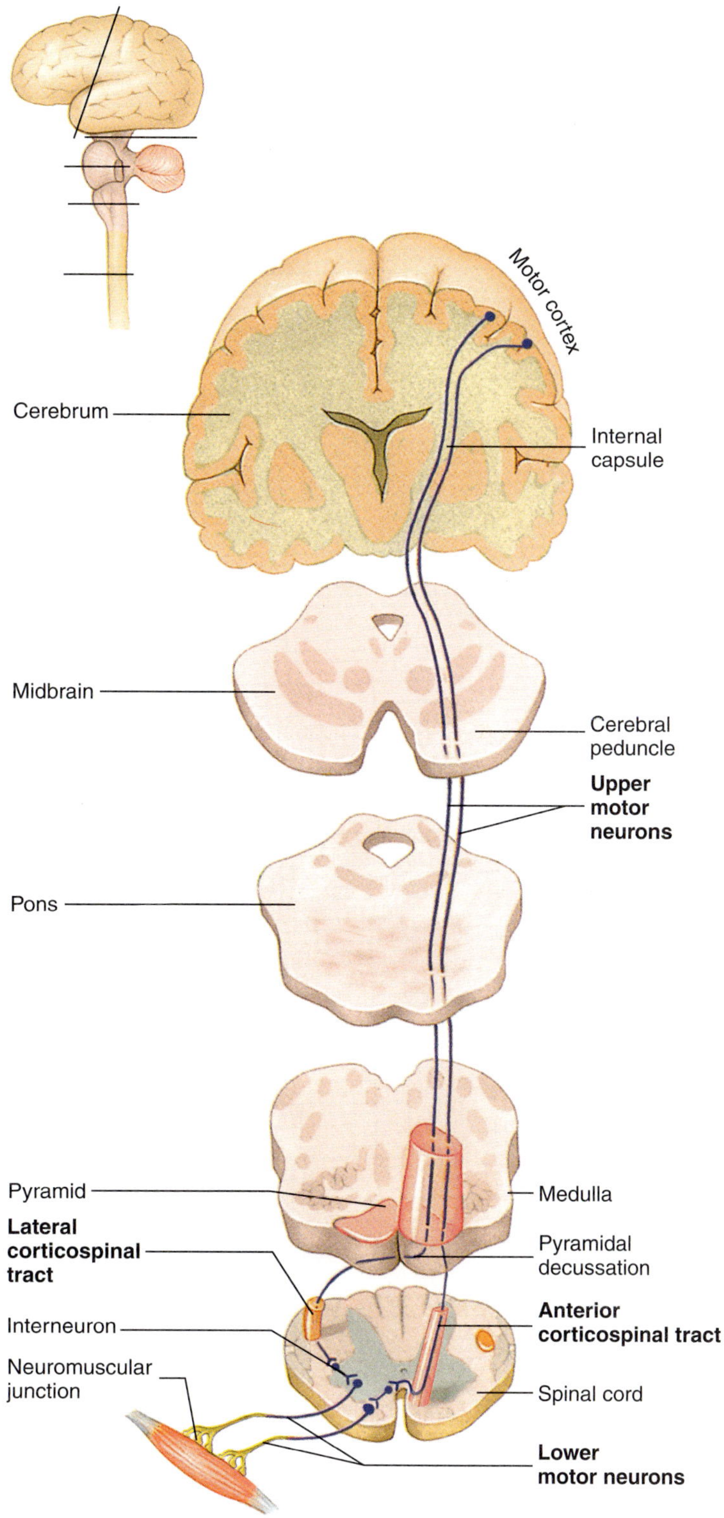
Motor cortex
Cerebrum
Internal capsule
Midbrain
Cerebral peduncle
Upper motor neurons
Pons
Pyramid
Medulla
Lateral corticospinal tract
Pyramidal decussation
Interneuron
Anterior corticospinal tract
Neuromuscular junction
Spinal cord
Lower motor neurons

Direct Pathways
Figure 14.14

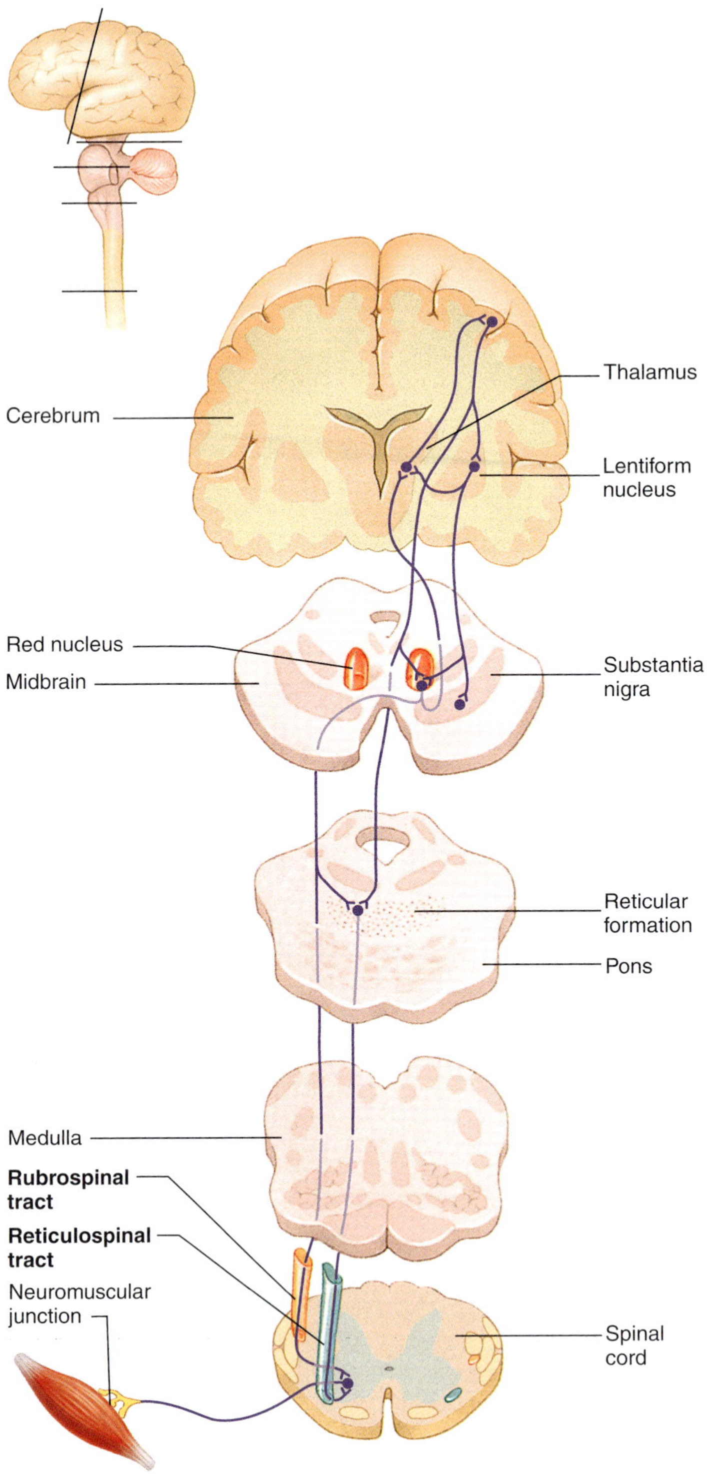
Thalamus
Cerebrum
Lentiform nucleus
Red nucleus
Midbrain
Substantia nigra
Reticular formation
Pons
Medulla
Rubrospinal tract
Reticulospinal tract
Neuromuscular junction
Spinal cord

Indirect Pathways
Figure 14.15

1. The motor cortex sends action potentials to lower motor neurons in the spinal cord.
2. Action potentials from the motor cortex inform the cerebellum of the intended movement.
3. Lower motor neurons in the spinal cord send action potentials to skeletal muscles, causing them to contract.
4. Proprioceptive signals from the skeletal muscles and joints to the cerebellum convey information concerning the status of the muscles and the structure being moved during contraction.
5. The cerebellum compares the information from the motor cortex to the proprioceptive information from the skeletal muscles and joints.
6. Action potentials from the cerebellum to the spinal cord modify the stimulation from the motor cortex to the lower motor neurons.
7. Action potentials from the cerebellum are sent to the motor cortex, which modify its motor activity.

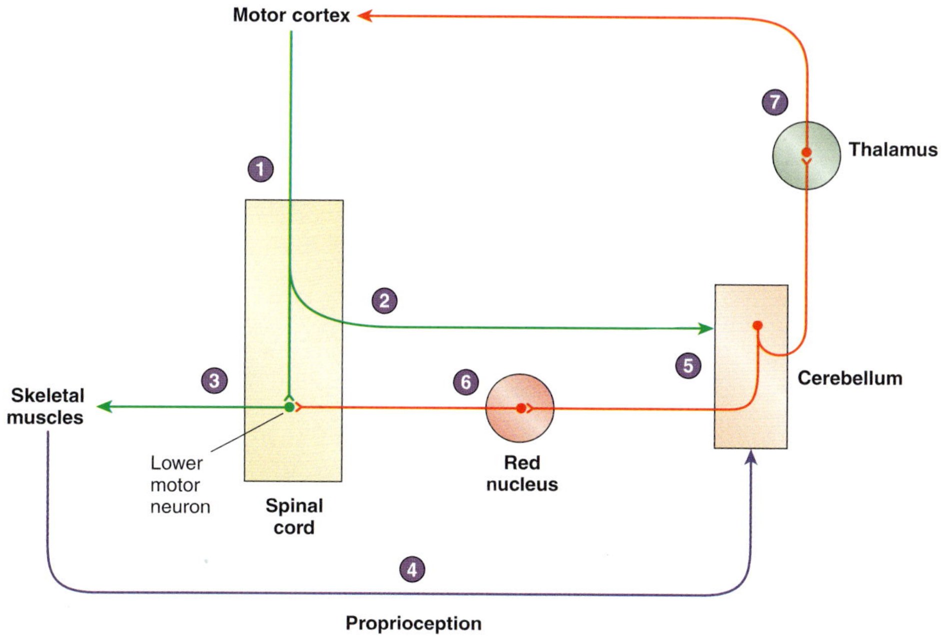

Cerebellar Comparator Function
Process Figure 14.16

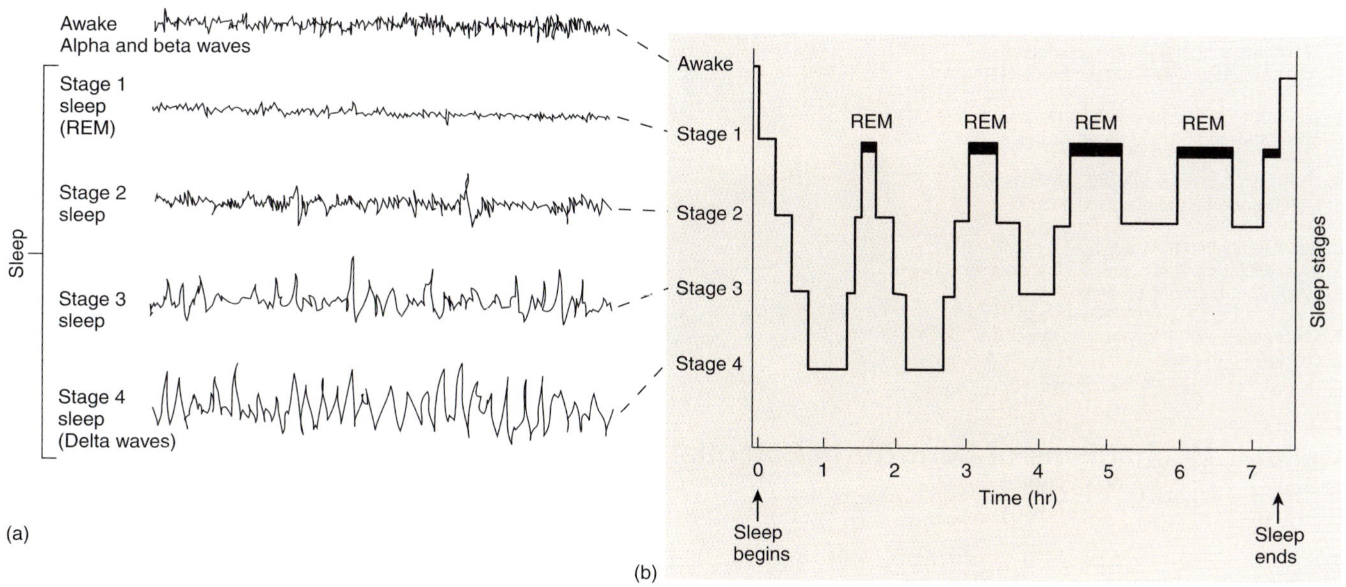

Electroencephalograms (EEGs) Showing Brain Waves
Figure 14.18

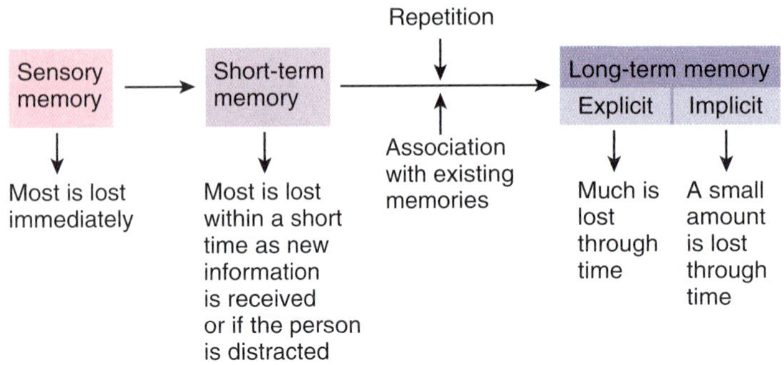

Memory Processing
Figure 14.19

1. The amount of the neurotransmitter glutamate produced by the presynaptic neuron increases.
2. The amount of glutamate released by the presynaptic neuron also increases.
3. The number of glutamate receptors on the postsynaptic neuron membrane increases.
4. Ca^{2+} channels in the postsynaptic membrane open, allowing Ca^{2+} to enter the cell.
5. The Ca^{2+} that enter the cell associate with the intracellular molecule calmodulin.
6. Activated calmodulin activates a cAMP intracellular mediator, which stimulates synthesis of specific proteins.
7. The cellular effect may involve structural changes in the cell.

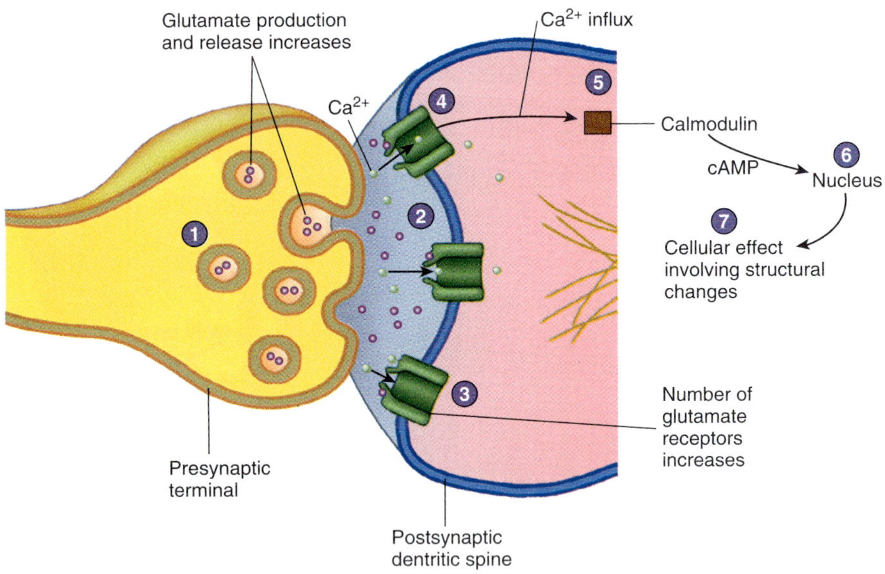

Cellular Mechanisms of Long-Term Potentiation
Process Figure 14.20

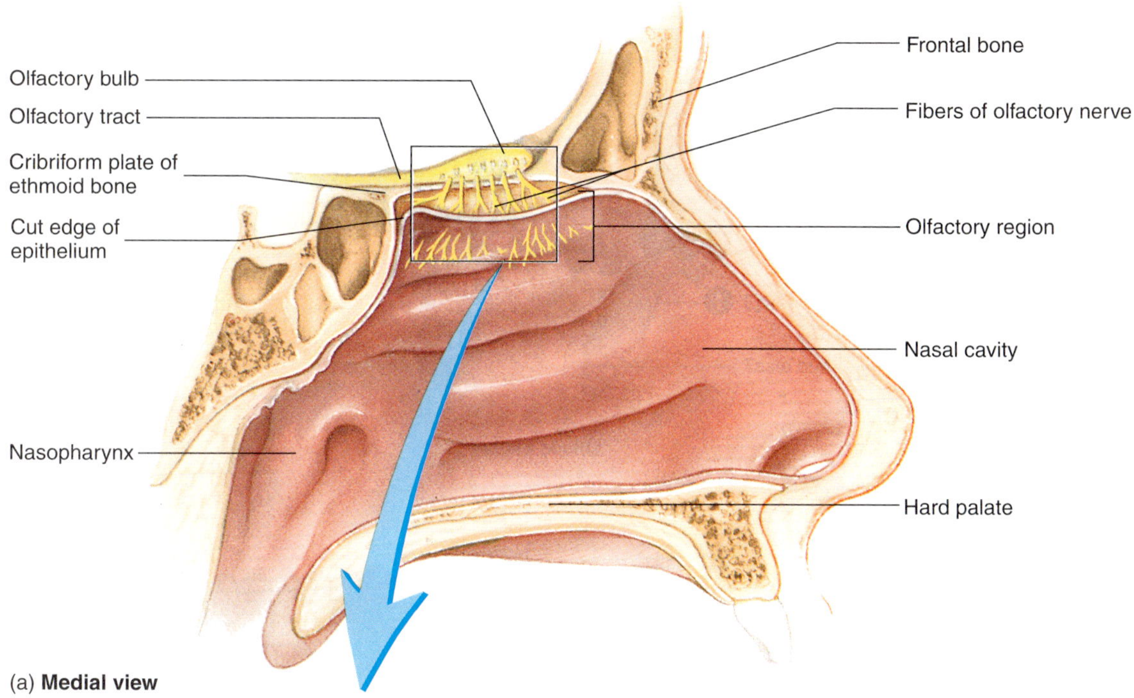

(a) **Medial view**

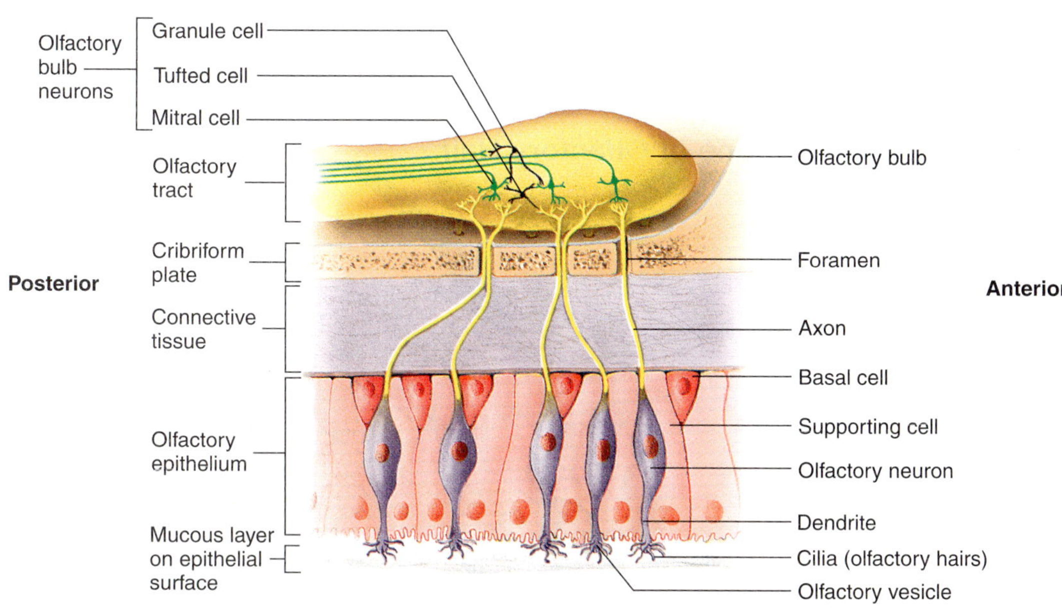

(b) **Medial view**

Olfactory Region, Epithelium, and Bulb

Figure 15.1

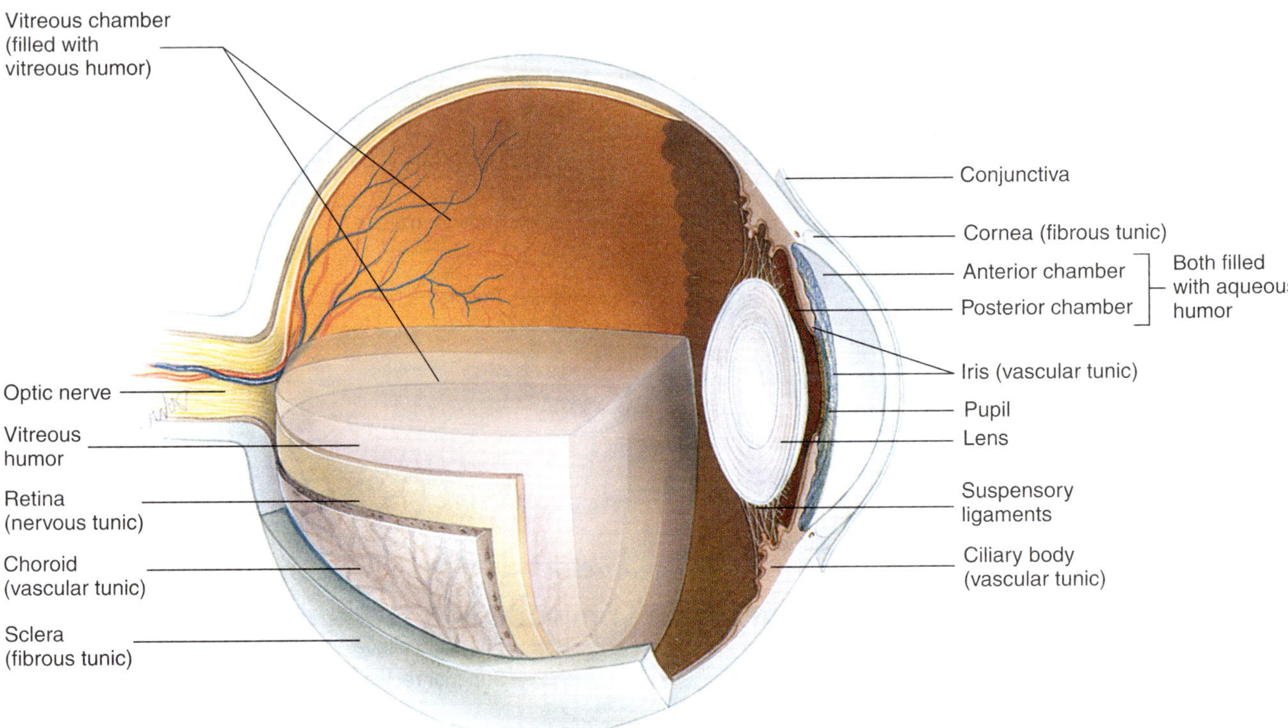

Sagittal Section of the Eye, Demonstrating Its Layers
Figure 15.12

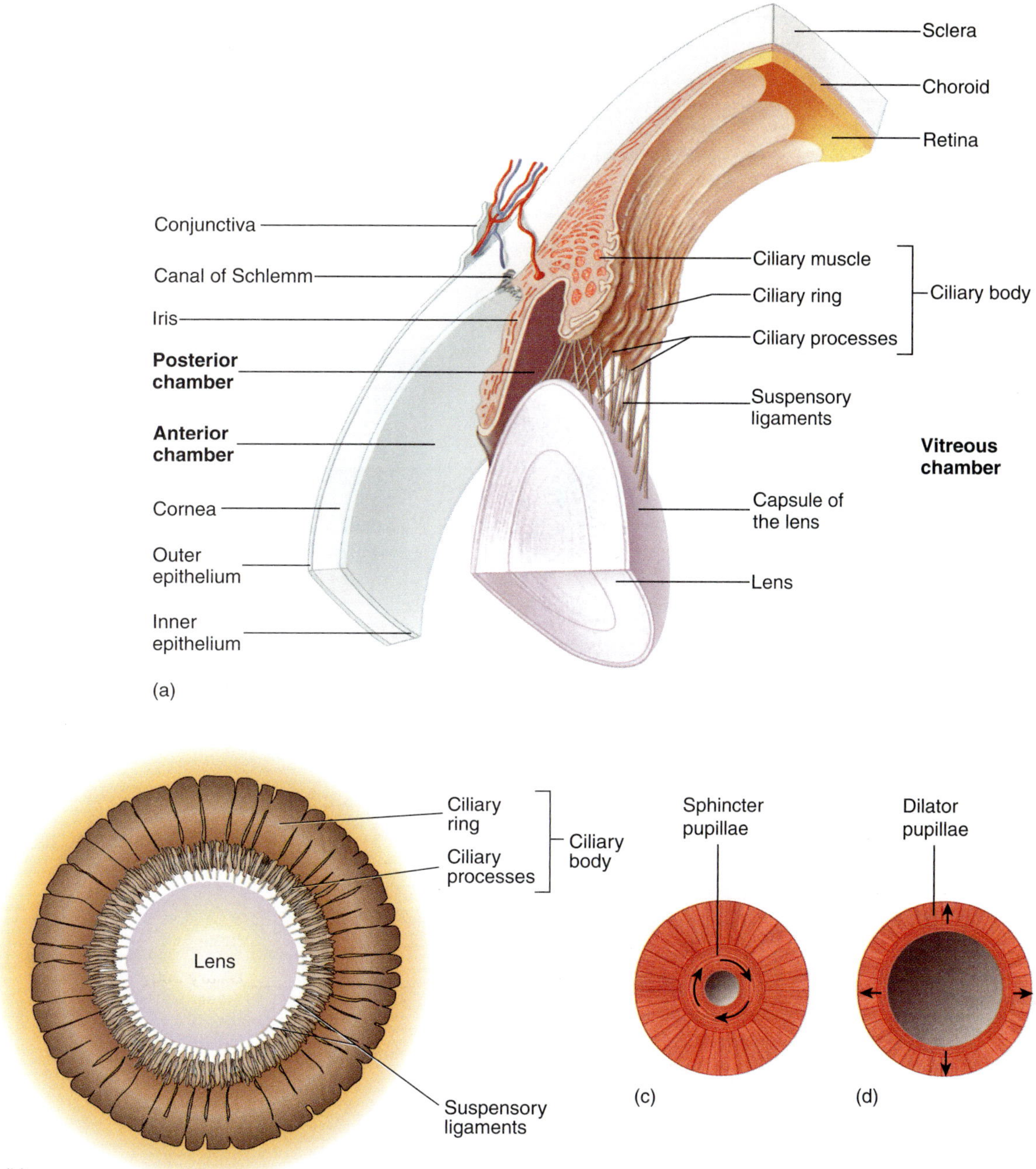

Lens, Cornea, Iris, and Ciliary Body
Figure 15.13

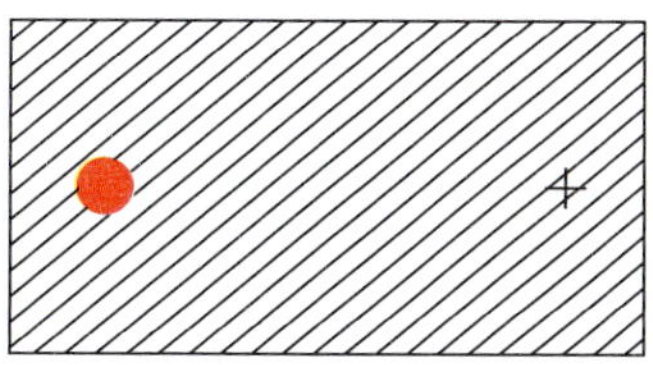

(b)

Ophthalmoscopic View of the Left Retina
Figure 15.14

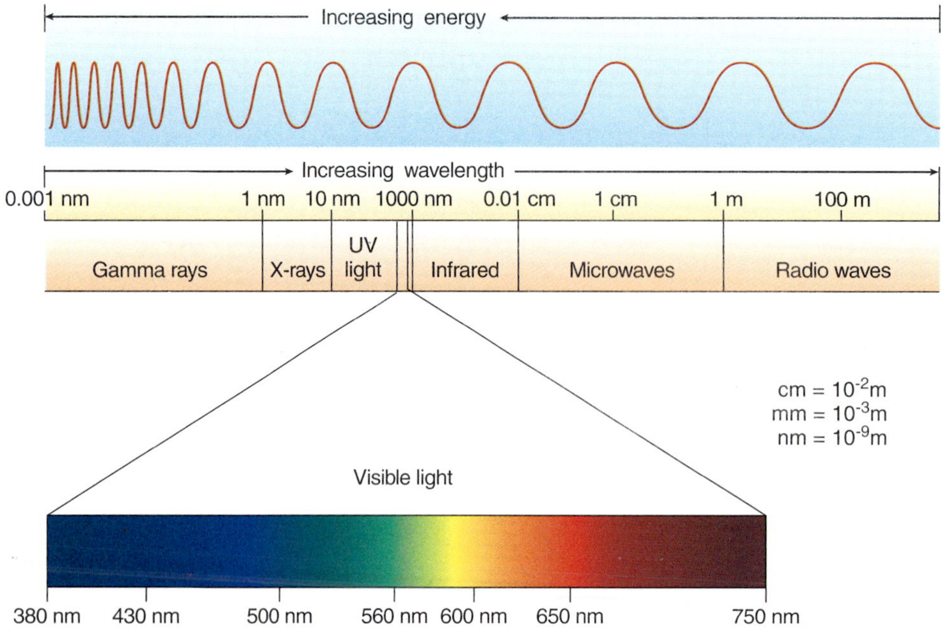

The Electromagnetic Spectrum
Figure 15.15

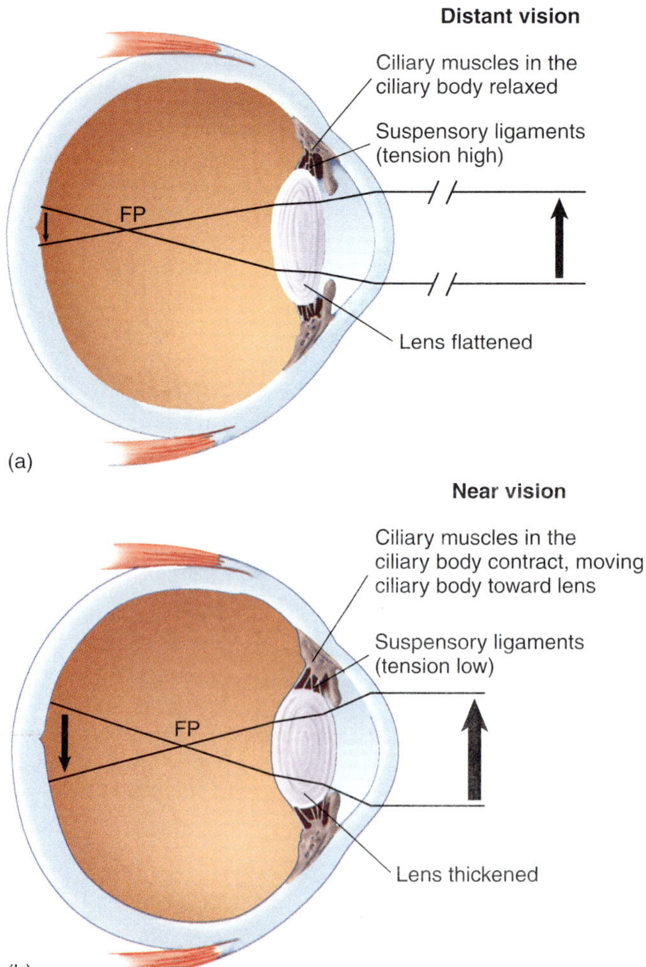

Focus and Accommodation by the Eye
Figure 15.16

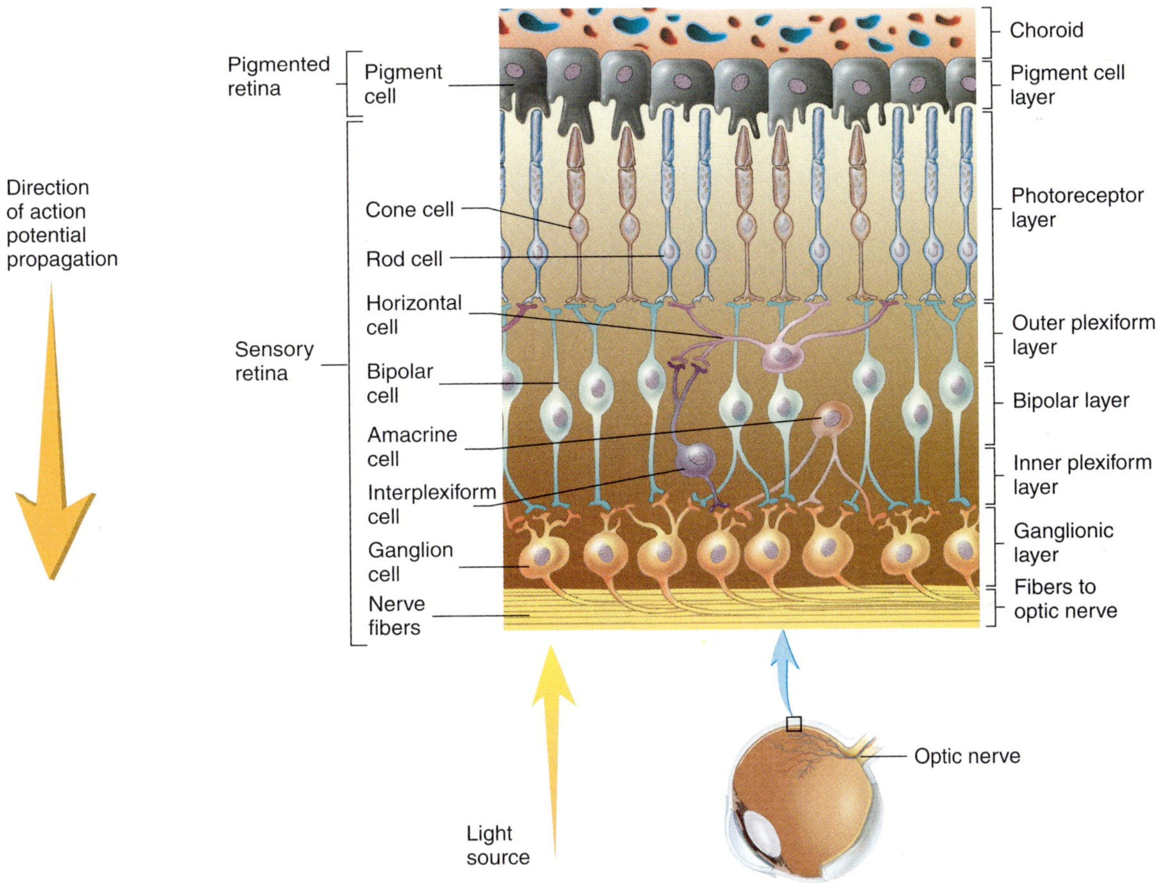
Direction
of action
potential
propagation
Pigmented
retina
Pigment
cell
Sensory
retina
Cone cell
Rod cell
Horizontal
cell
Bipolar
cell
Amacrine
cell
Interplexiform
cell
Ganglion
cell
Nerve
fibers
Choroid
Pigment cell
layer
Photoreceptor
layer
Outer plexiform
layer
Bipolar layer
Inner plexiform
layer
Ganglionic
layer
Fibers to
optic nerve
Optic nerve
Light
source

Retina
Figure 15.17

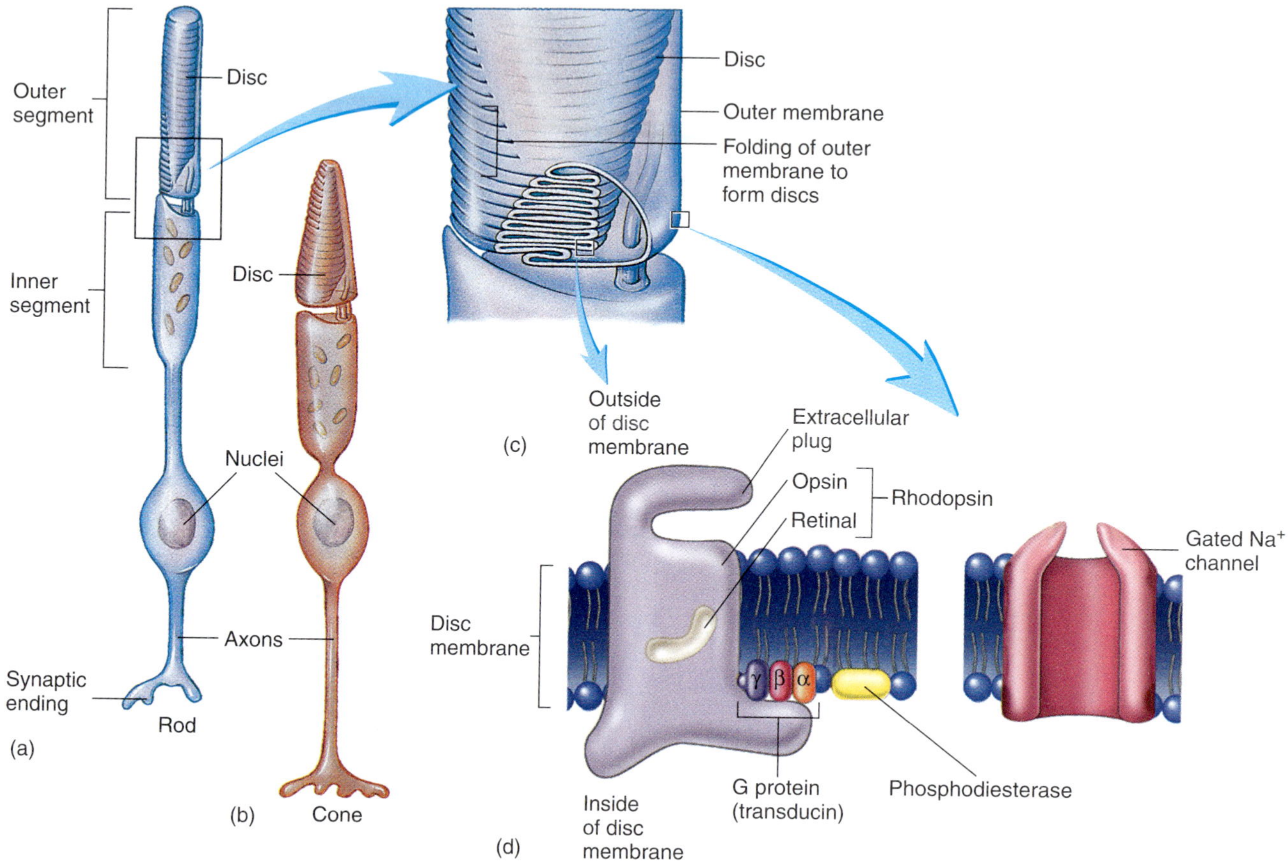

Sensory Receptor Cells of the Retina
Figure 15.18

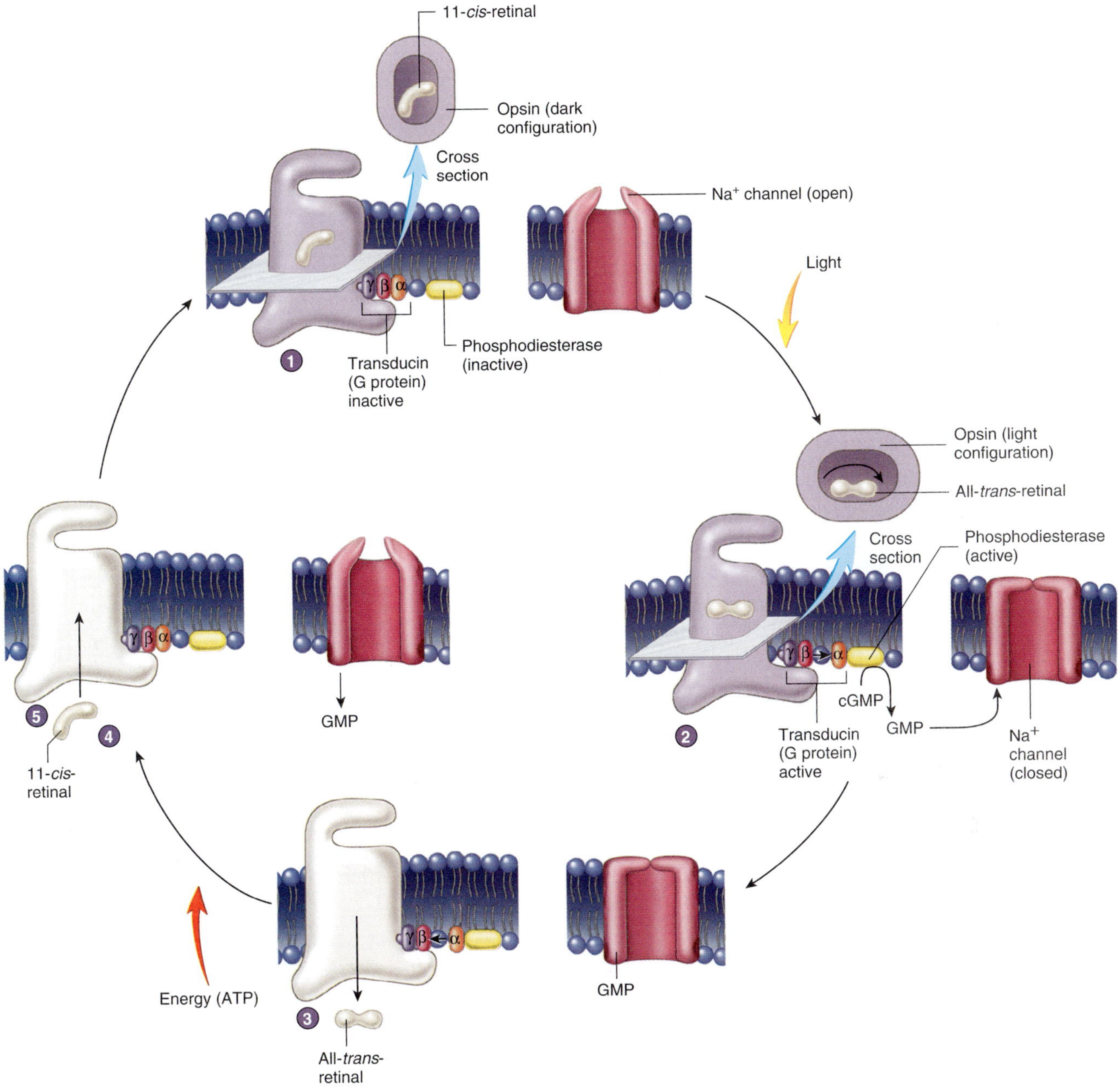

1. Retinal (in an inactive configuration called 11-*cis*-) is attached inside opsin to make rhodopsin.
2. Light causes retinal to change shape from 11-*cis*-retinal to all-*trans*-retinal. This change causes opsin also to change shape (from its dark to light configuration). This activated rhodopsin activates a G protein (called transducin), which activates a phosphodiesterase. The phosphodiesterase catalyzes the conversion of cGMP to GMP, which closes Na^+ channels.
3. All-*trans*-retinal detaches from opsin. Transducin is reassociated.
4. All-*trans*-retinal is converted to 11-*cis*-retinal, a process that requires energy.
5. 11-*cis*-retinal attaches to opsin, which returns to its original (dark) configuration. GMP is displaced from the Na^+ channel and the channel opens.

Rhodopsin Cycle
Process Figure 15.19

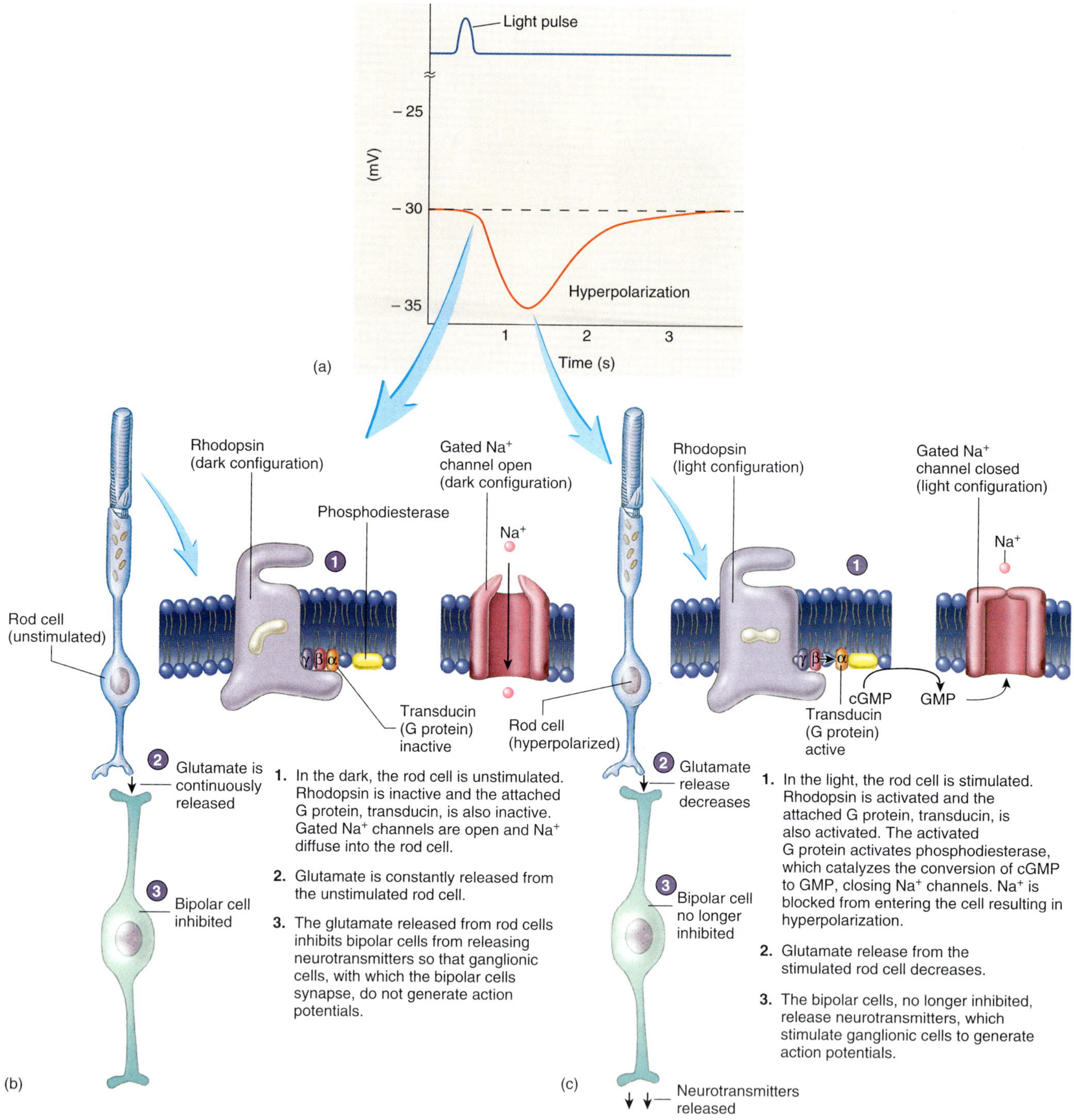

Rod Cell Hyperpolarization
Process Figure 15.20

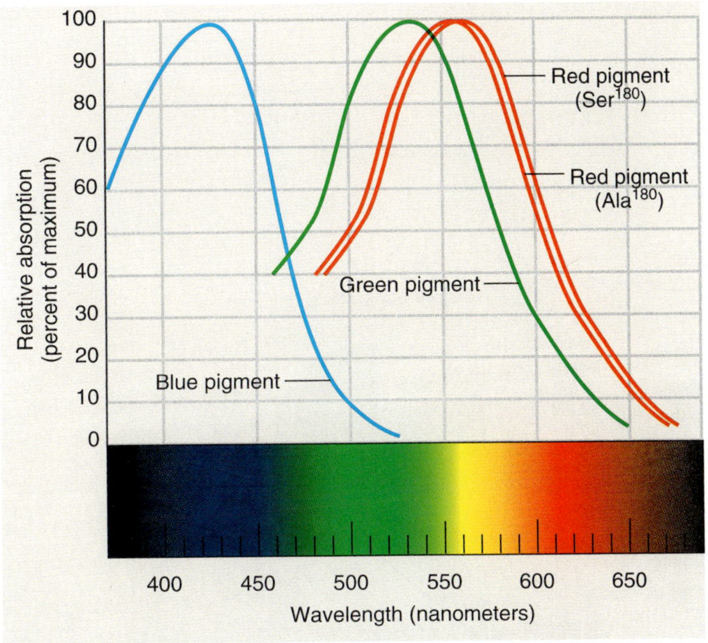

Wavelengths to Which Each of the Three Visual Pigments are Sensitive: Blue, Green, Red
Figure 15.21

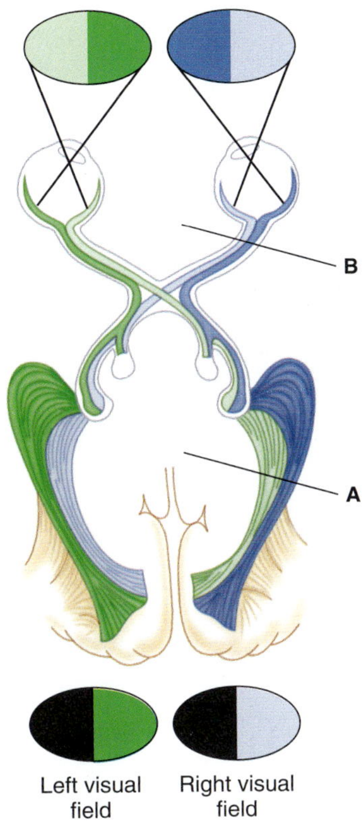

Lesions of Visual Pathways
Figure A

1. Each visual field is divided into a temporal and nasal half.
2. After passing through the lens, light from each half of a visual field projects to the opposite side of the retina.
3. An optic nerve consists of axons extending from the retina to the optic chiasm.
4. In the optic chiasm, axons from the nasal part of the retina cross and project to the opposite side of the brain. Axons from the temporal part of the retina do not cross.
5. An optic tract consists of axons that have passed through the optic chiasm (with or without crossing) to the thalamus.
6. The axons synapse in the lateral geniculate nuclei of the thalamus. Collateral branches of the axons in the optic tracts synapse in the superior colliculi.
7. An optic radiation consists of axons from thalamic neurons that project to the visual cortex.
8. (*b*) The right part of each visual field (*dark green* and *light blue*) projects to the left side of the brain, and the left part of each visual field projects to the right side of the brain (*light green* and *dark blue*).

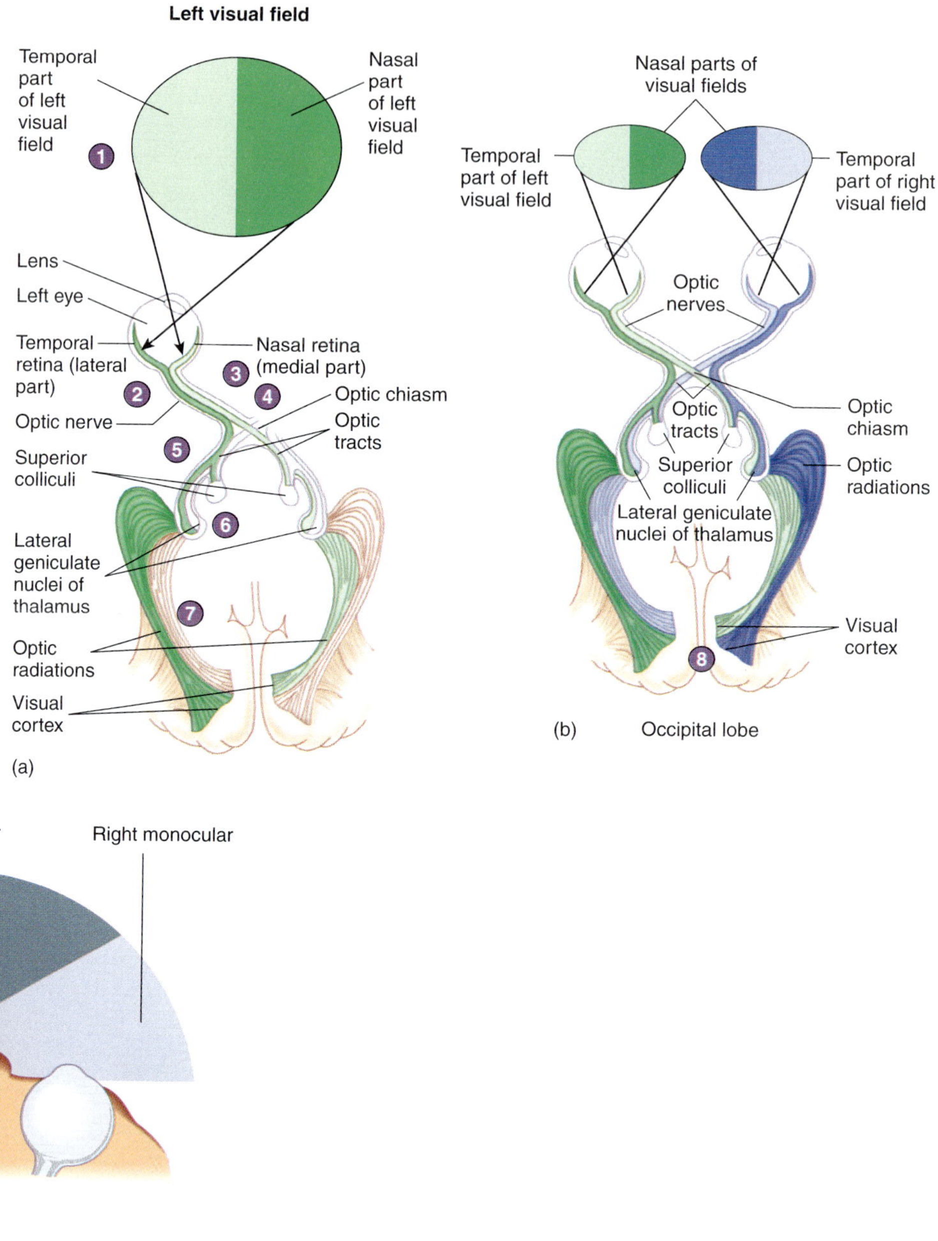

Visual Pathways
Process Figure 15.22

Visual Disorders and Their Correction by Various Lenses
Figure B

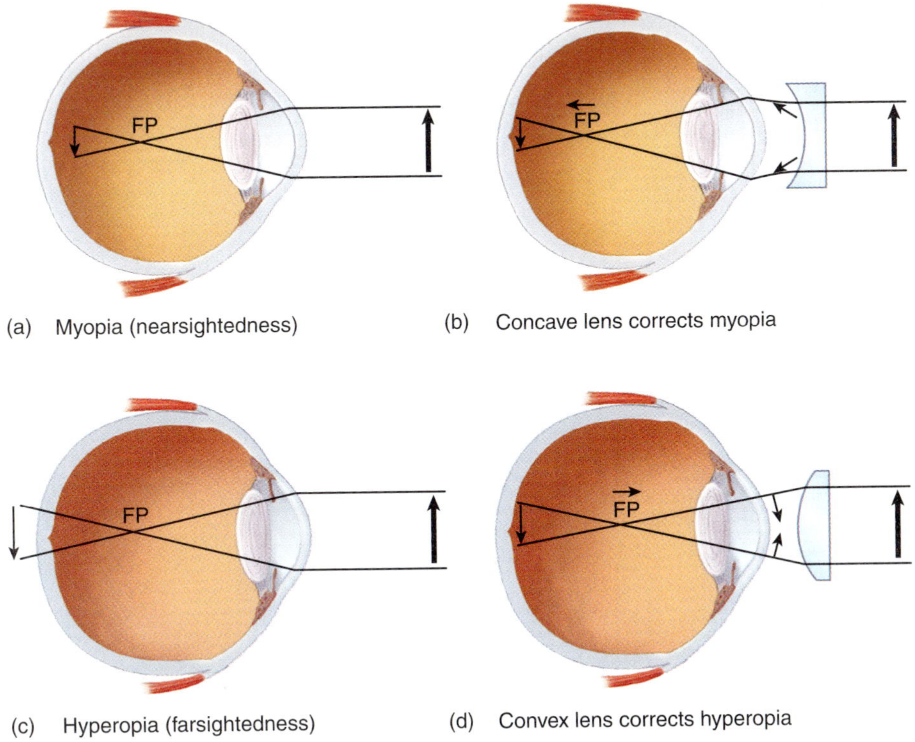

(a) Myopia (nearsightedness)

(b) Concave lens corrects myopia

(c) Hyperopia (farsightedness)

(d) Convex lens corrects hyperopia

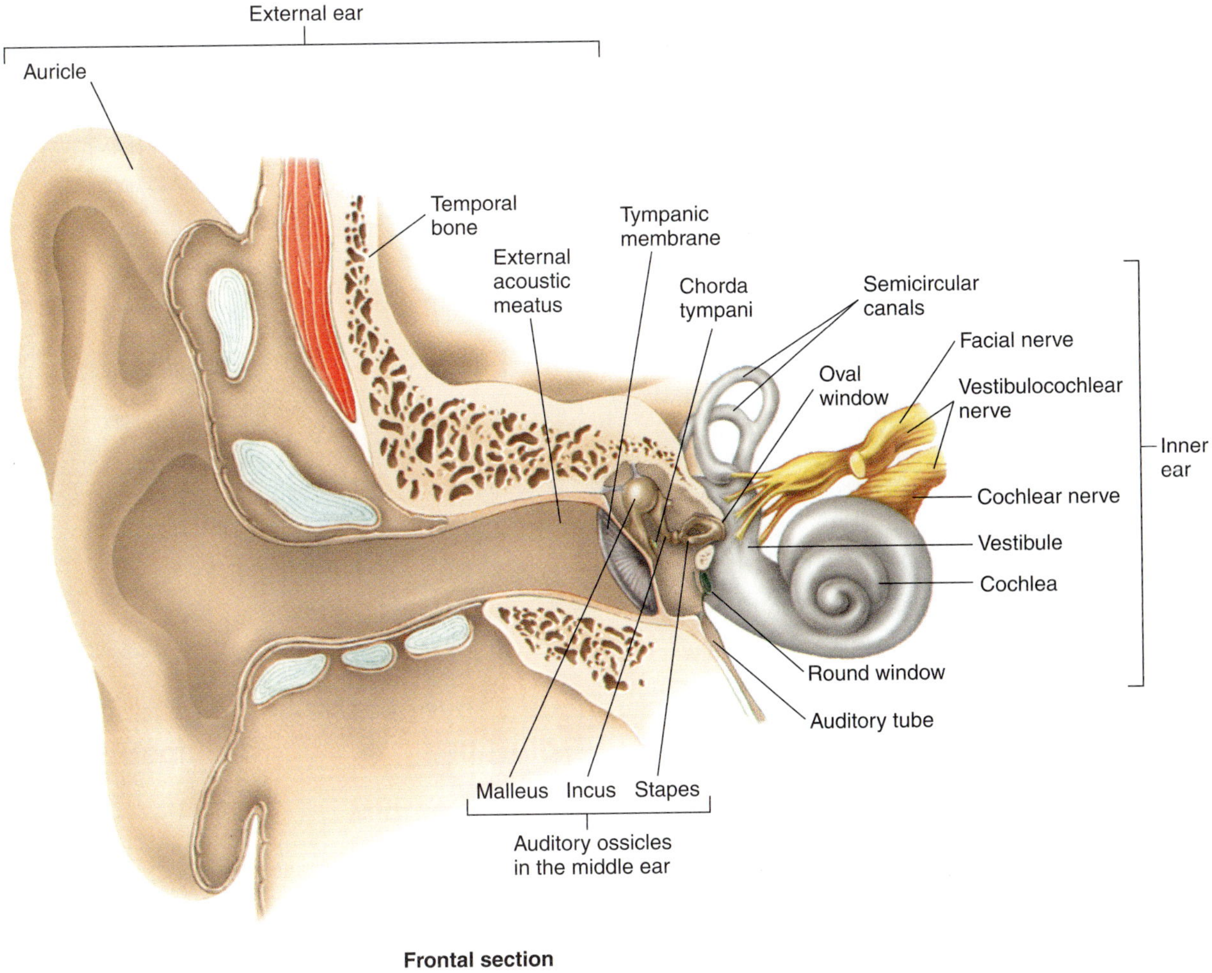

External, Middle, and Inner Ear
Figure 15.23

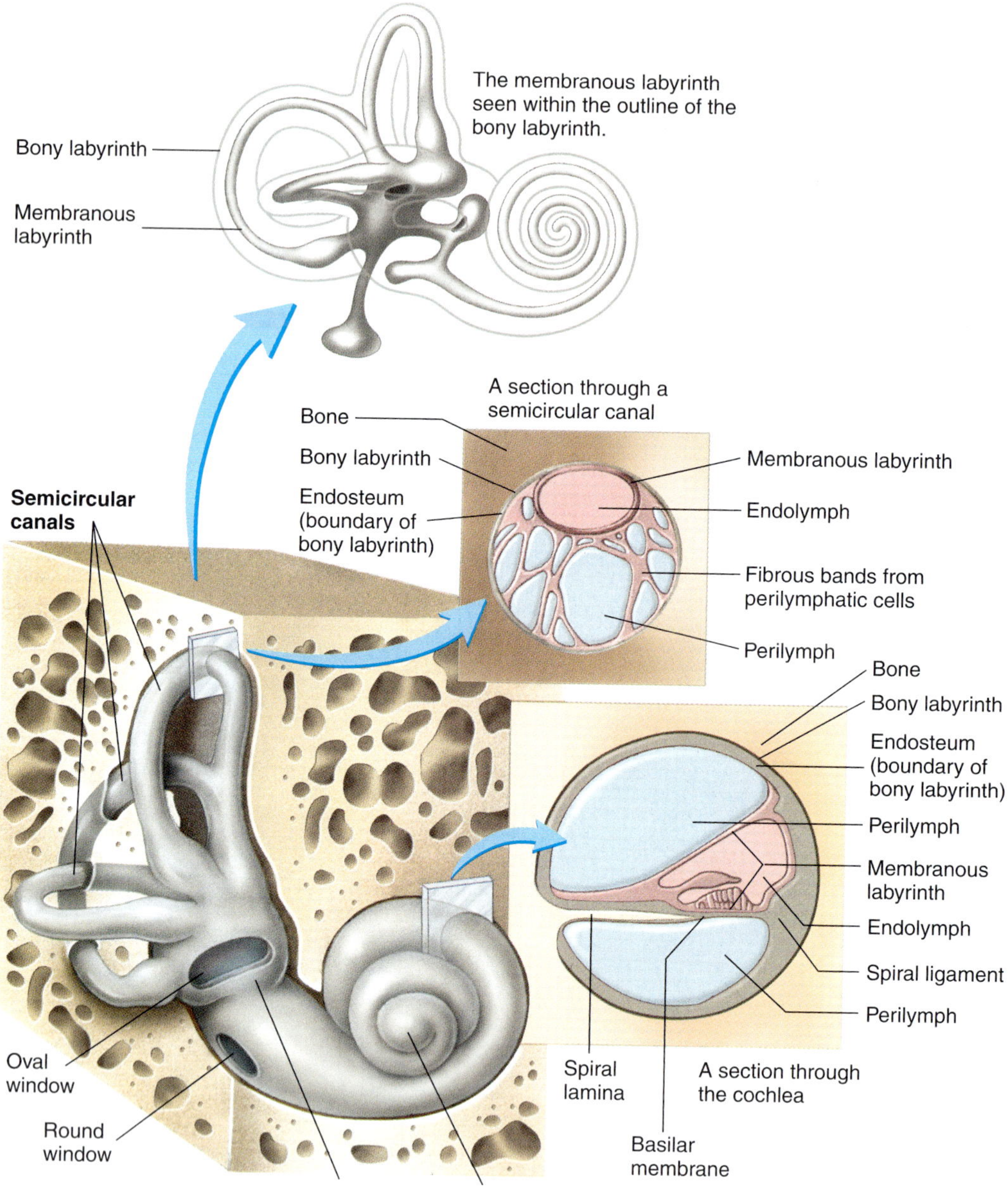

The Inner Ear: Bony and Membranous Labyrinths
Figure 15.25

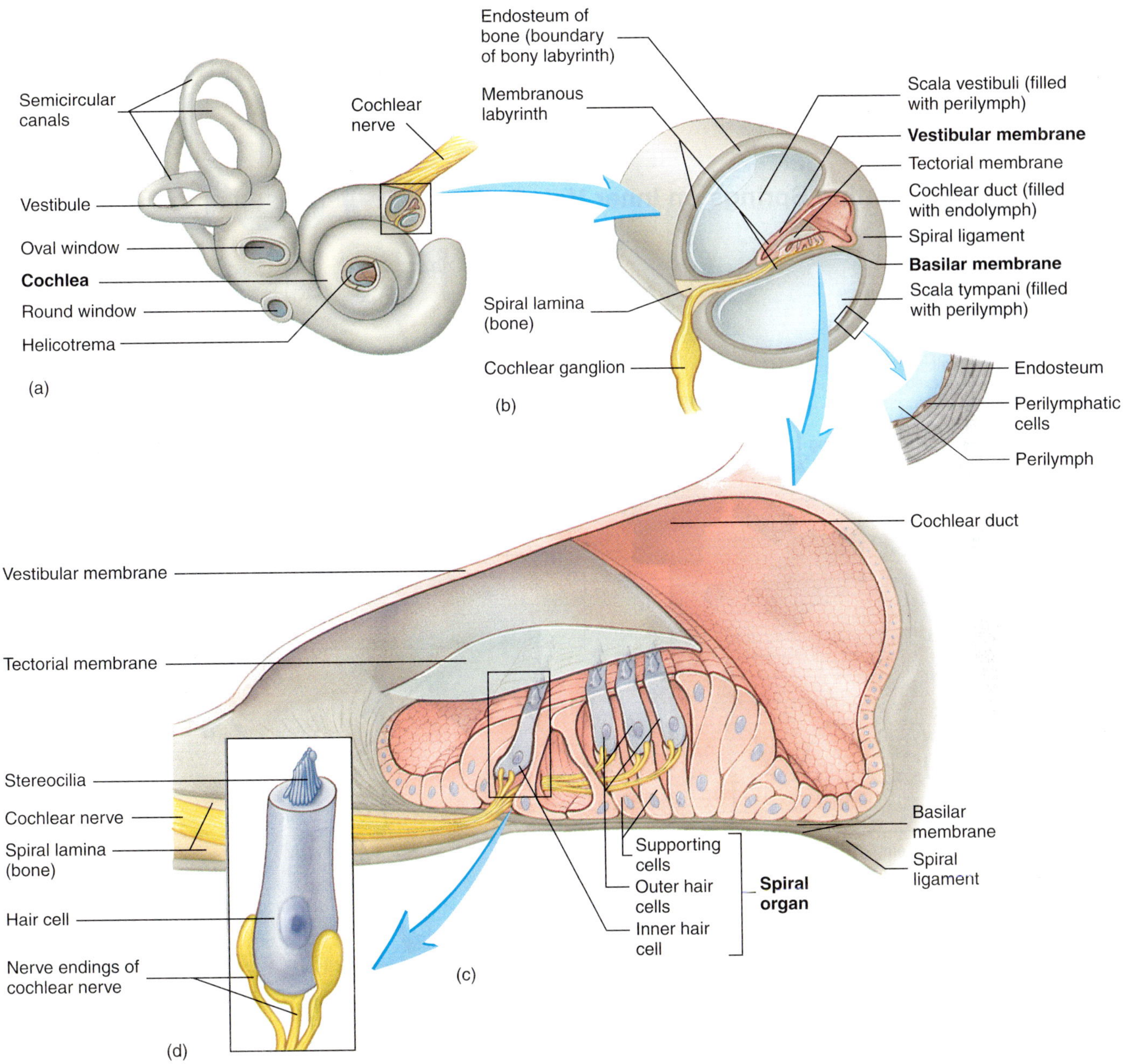

Semicircular canals
Cochlear nerve
Vestibule
Oval window
Cochlea
Round window
Helicotrema
(a)
Endosteum of bone (boundary of bony labyrinth)
Membranous labyrinth
Scala vestibuli (filled with perilymph)
Vestibular membrane
Tectorial membrane
Cochlear duct (filled with endolymph)
Spiral ligament
Basilar membrane
Scala tympani (filled with perilymph)
Spiral lamina (bone)
Cochlear ganglion
(b)
Endosteum
Perilymphatic cells
Perilymph
Cochlear duct
Vestibular membrane
Tectorial membrane
Stereocilia
Cochlear nerve
Spiral lamina (bone)
Hair cell
Nerve endings of cochlear nerve
(d)
Supporting cells
Outer hair cells
Inner hair cell
Spiral organ
Basilar membrane
Spiral ligament
(c)

Structure of the Cochlea
Figure 15.26

1. Sensory axons from the cochlear ganglion terminate in the cochlear nucleus in the brainstem.
2. Axons from the neurons in the cochlear nucleus project to the superior olivary nucleus or to the inferior colliculus.
3. Axons from the inferior colliculus project to the medial geniculate nucleus of the thalamus.
4. Thalamic neurons project to the auditory cortex.
5. Neurons in the superior olivary nucleus send axons to the inferior colliculus, back to the inner ear, or to motor nuclei in the brainstem that send efferent fibers to the middle ear muscles.

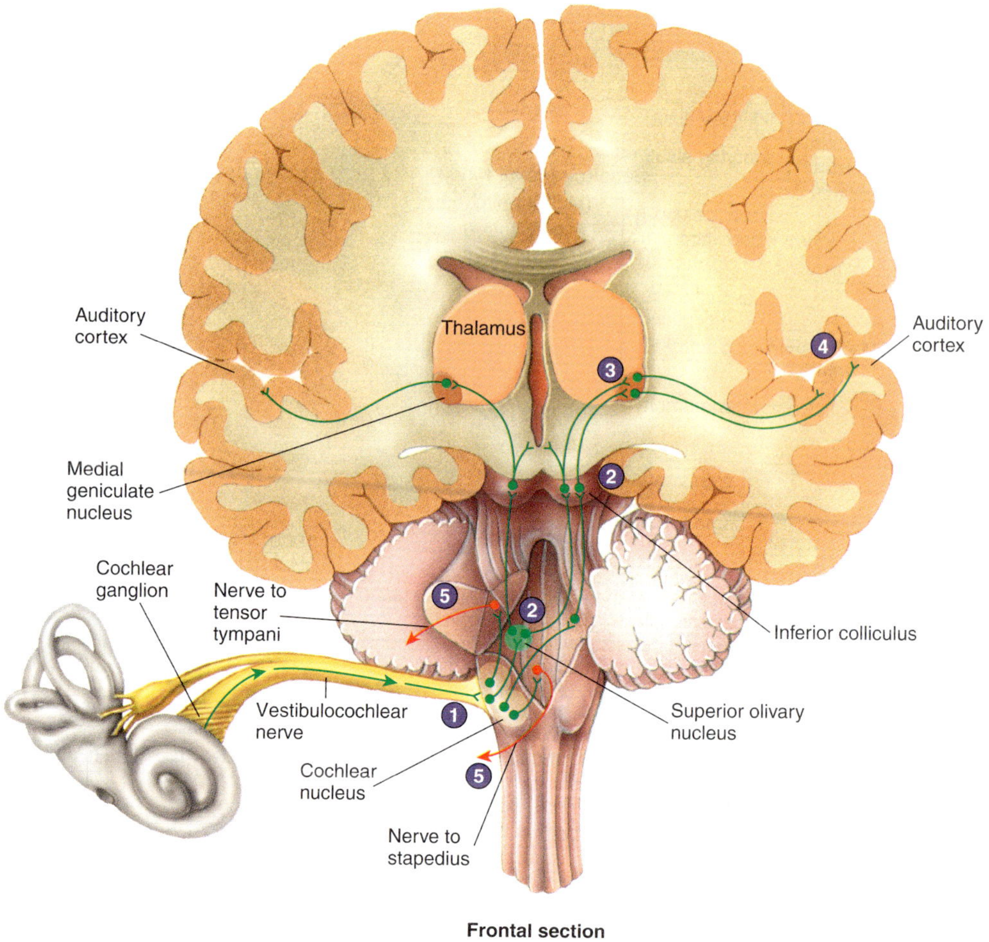

Central Nervous System Pathways for Hearing
Process Figure 15.34

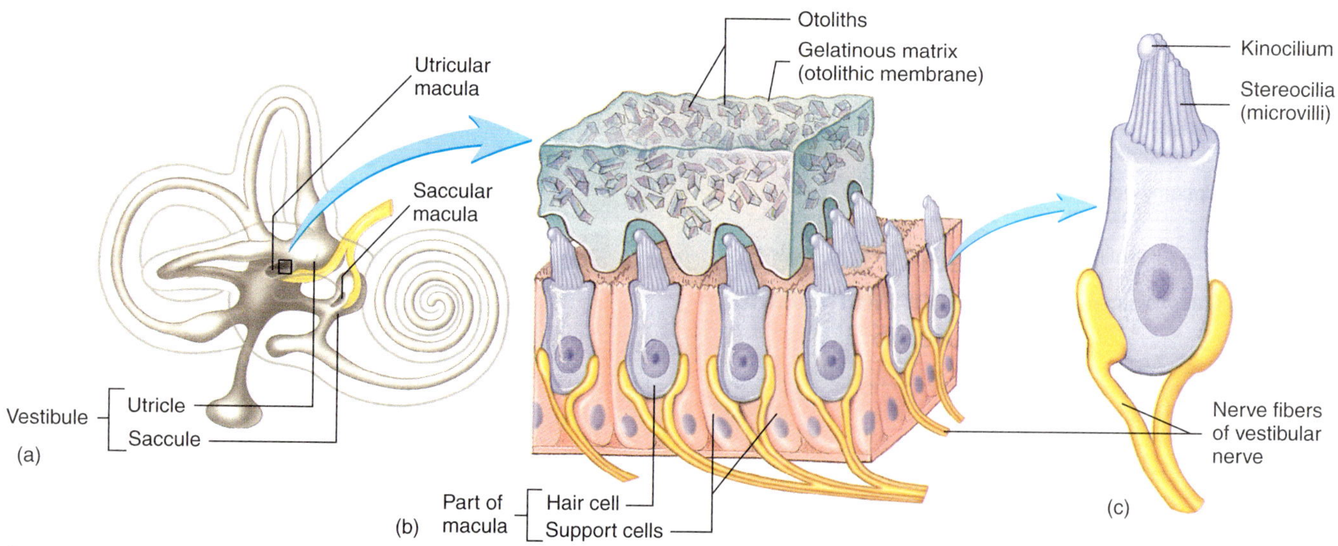

Structure of the Macula
Figure 15.35

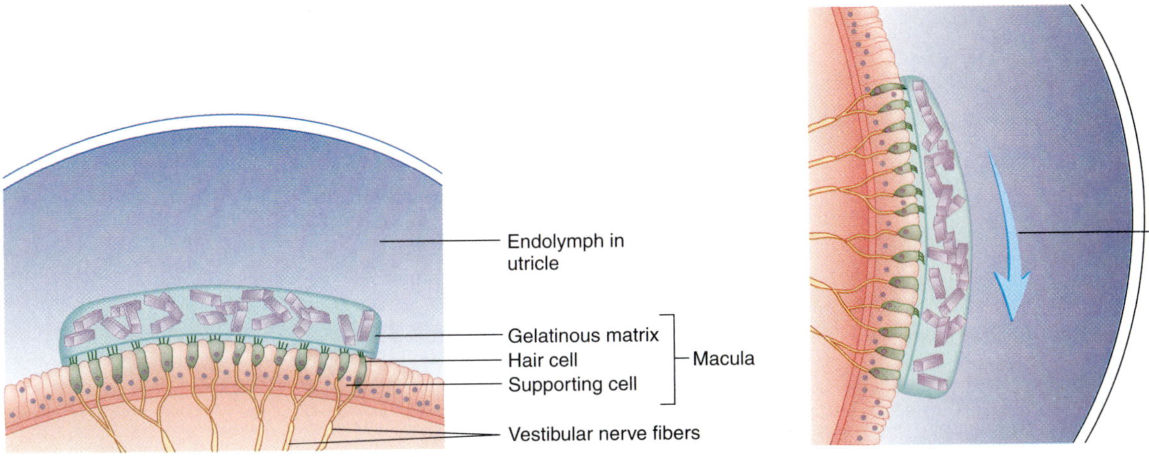

Function of the Vestibule in Maintaining Balance
Figure 15.36

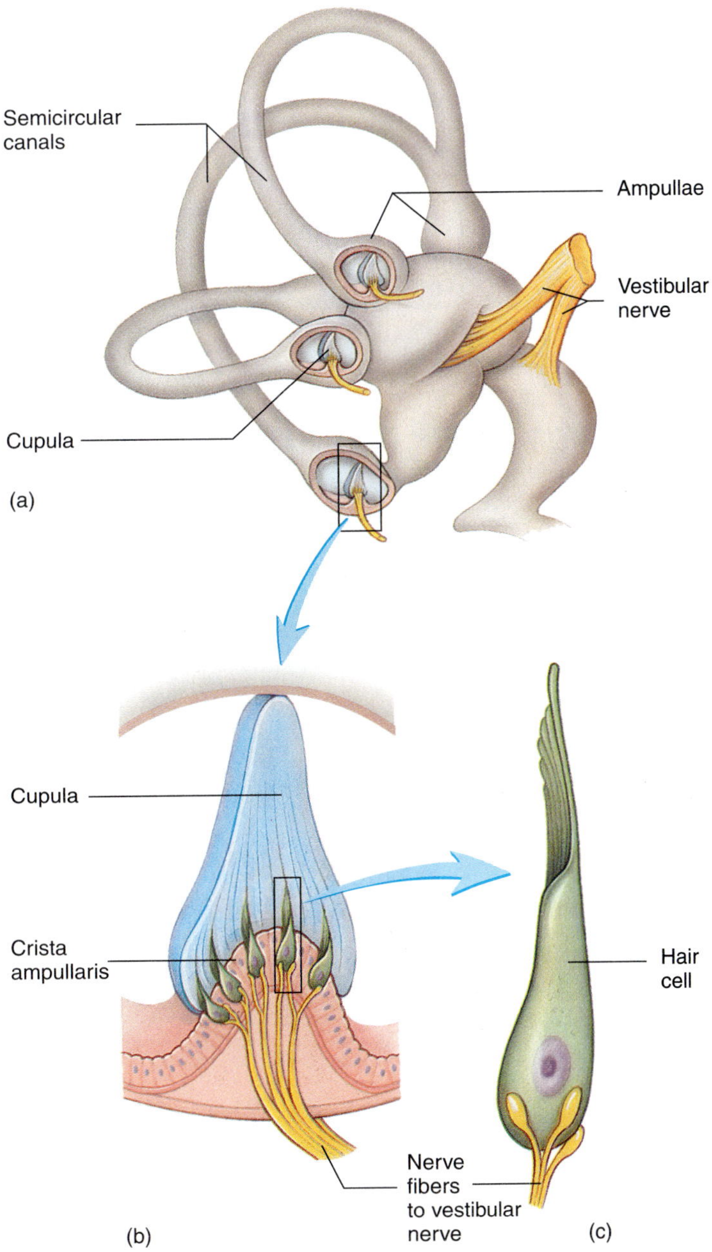

Semicircular Canals
Figure 15.37

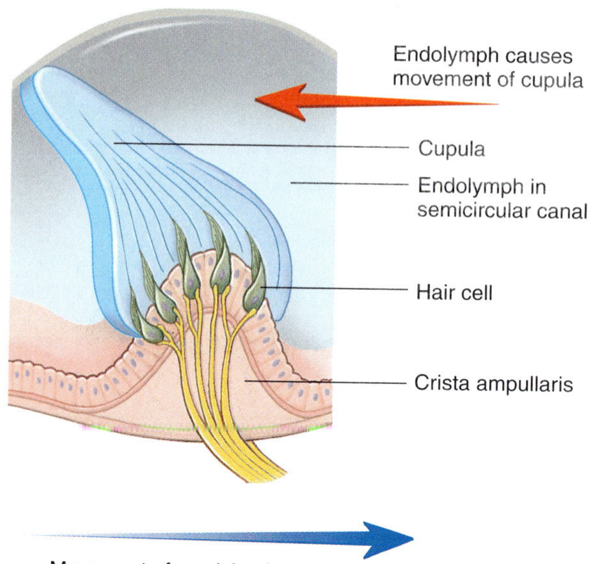

(a)

Function of the Semicircular Canals
Figure 15.38

1. A receiver, transmitter, and antenna are implanted under the skin near the auricle.

2. A small lead from the transmitter is fed through the external acoustic meatus, tympanic membrane, and middle ear into the cochlea.

3. In the cochlea, the cochlear nerve can be directly stimulated by electric impulses from the receiver.

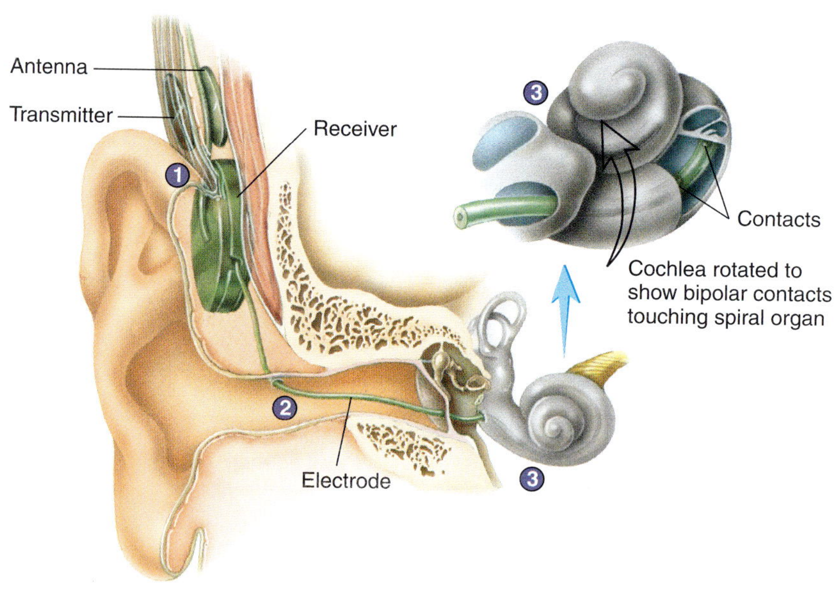

Cochlear Implant
Figure E

1. Sensory axons from the vestibular ganglion pass through the vestibular nerve to the vestibular nucleus, which also receives input from several other sources, such as proprioception from the legs.
2. Vestibular neurons send axons to the cerebellum, which influences postural muscles, and to the motor nuclei (oculomotor, trochlear, and abducens), which control extrinsic eye muscles.
3. Vestibular neurons also send axons to the posterior ventral nucleus of the thalamus.
4. Thalamic neurons project to the vestibular area of the cortex.

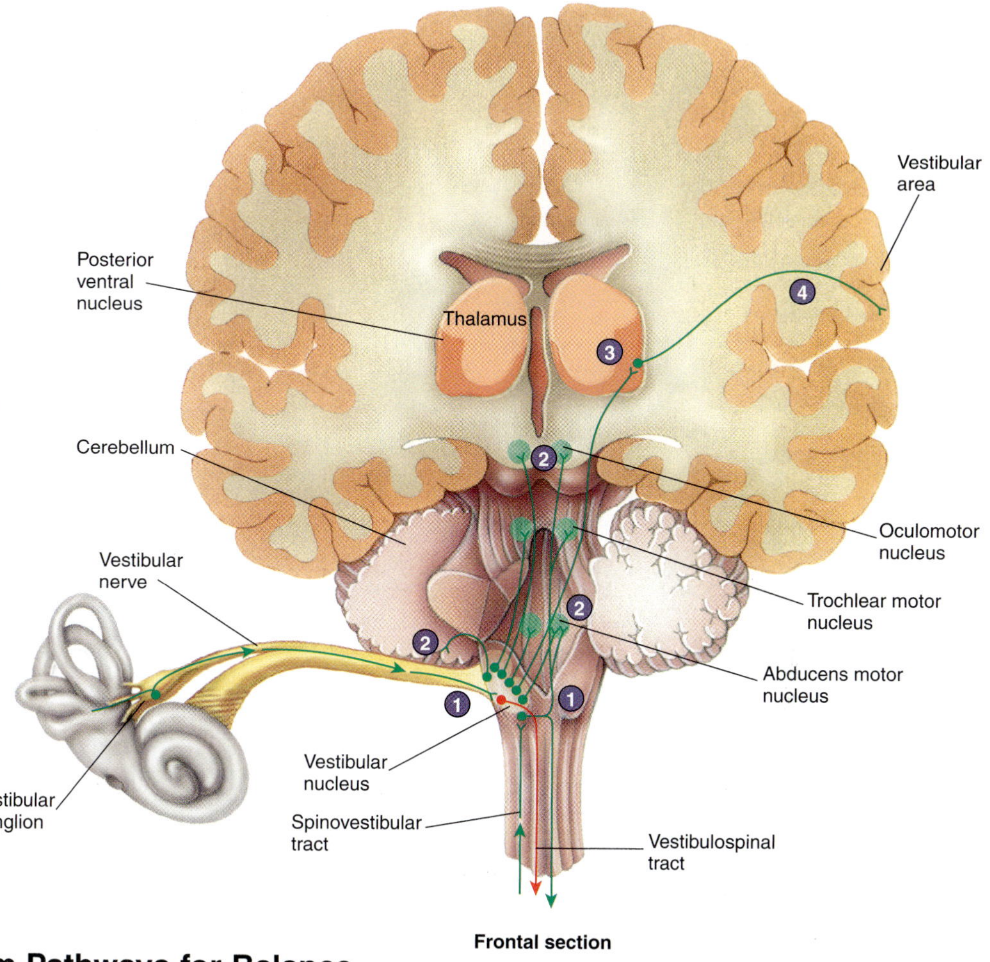

Central Nervous System Pathways for Balance
Process Figure 15.39

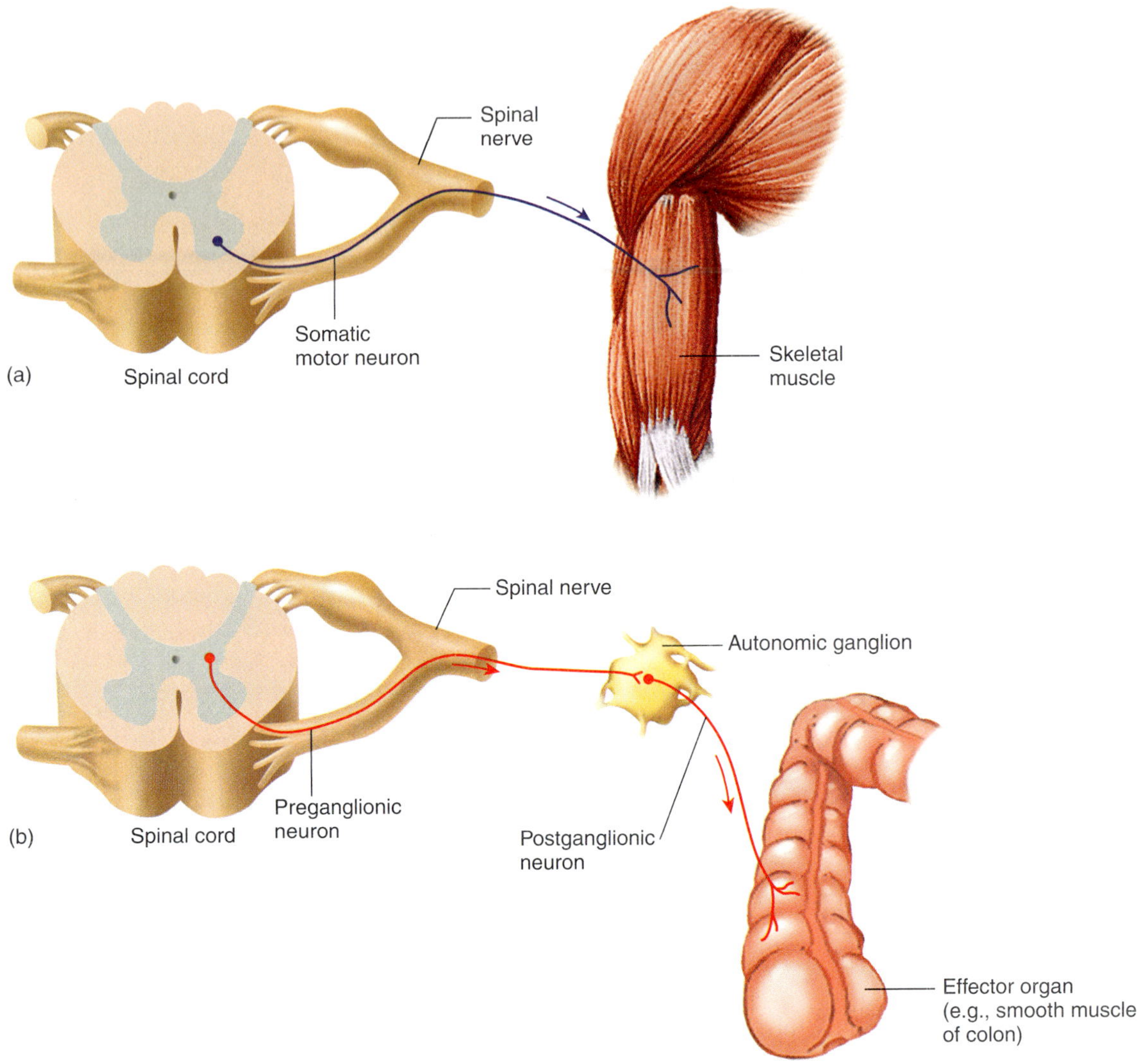

Organization of Somatic and Autonomic Nervous System Neurons
Figure 16.1

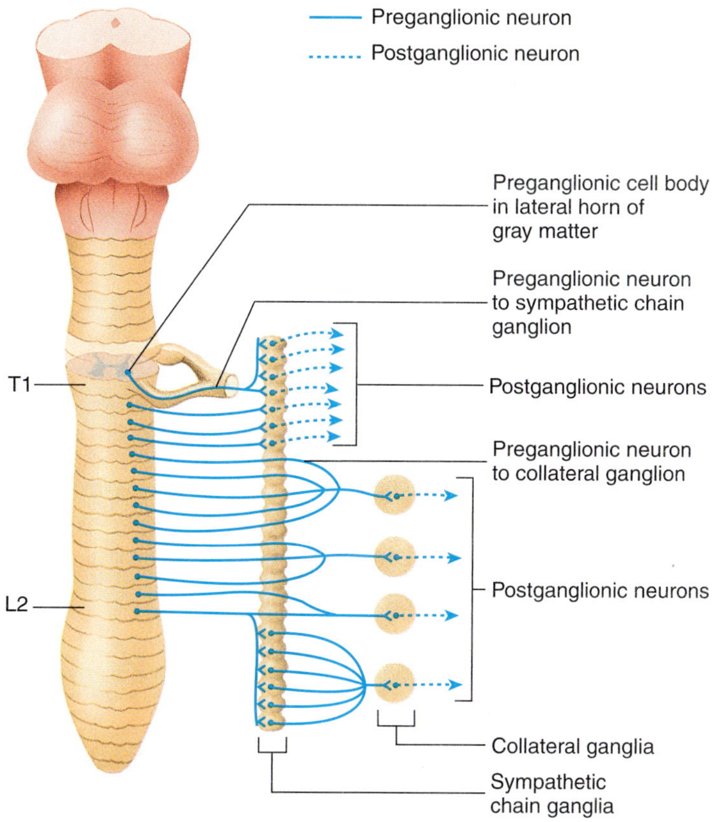
Preganglionic neuron
Postganglionic neuron
Preganglionic cell body in lateral horn of gray matter
Preganglionic neuron to sympathetic chain ganglion
T1
Postganglionic neurons
Preganglionic neuron to collateral ganglion
Postganglionic neurons
L2
Collateral ganglia
Sympathetic chain ganglia

Sympathetic Division
Figure 16.2

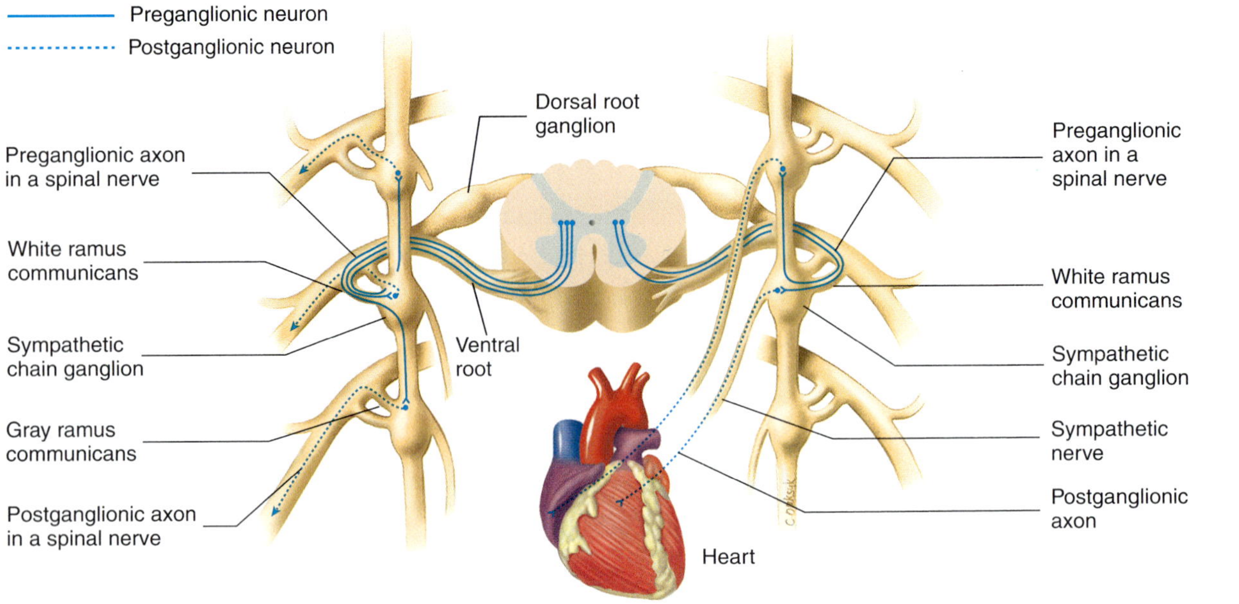

(a) Preganglionic axons from a spinal nerve pass through a white ramus communicans into a sympathetic chain ganglion. Some axons synapse with a postganglionic neuron at the level of entry; others ascend or descend to other levels before synapsing. Each postganglionic axon exits the sympathetic chain through a gray ramus communicans and enters a spinal nerve.

(b) Like part (*a*), except that each postganglionic neuron exits a sympathetic chain ganglion through a sympathetic nerve.

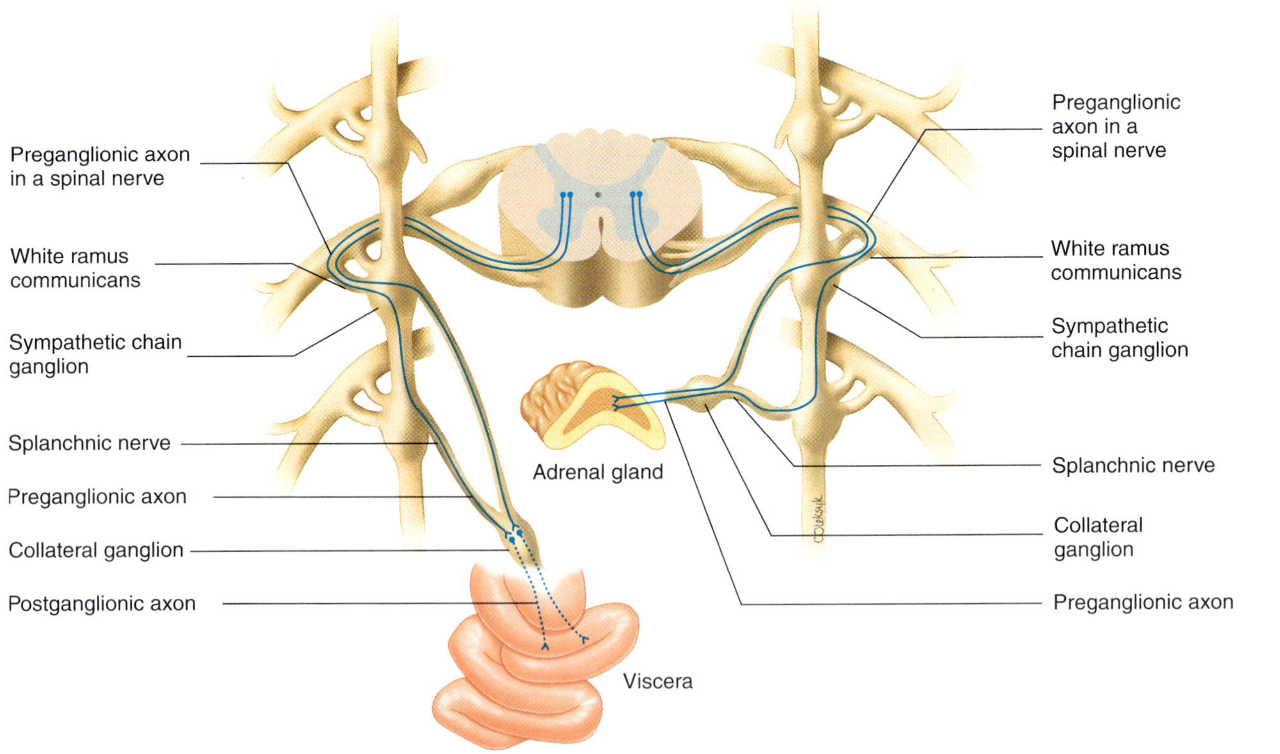

(c) Preganglionic neurons do not synapse in the sympathetic chain ganglia, but exit in splanchnic nerves and extend to a collateral ganglion, where they synapse with postganglionic neurons.

(d) Like part (*c*), except that the preganglionic axons extend to the adrenal medulla, where they synapse with specialized adrenal medullary cells.

Routes Taken by Sympathetic Axons

Figure 16.3

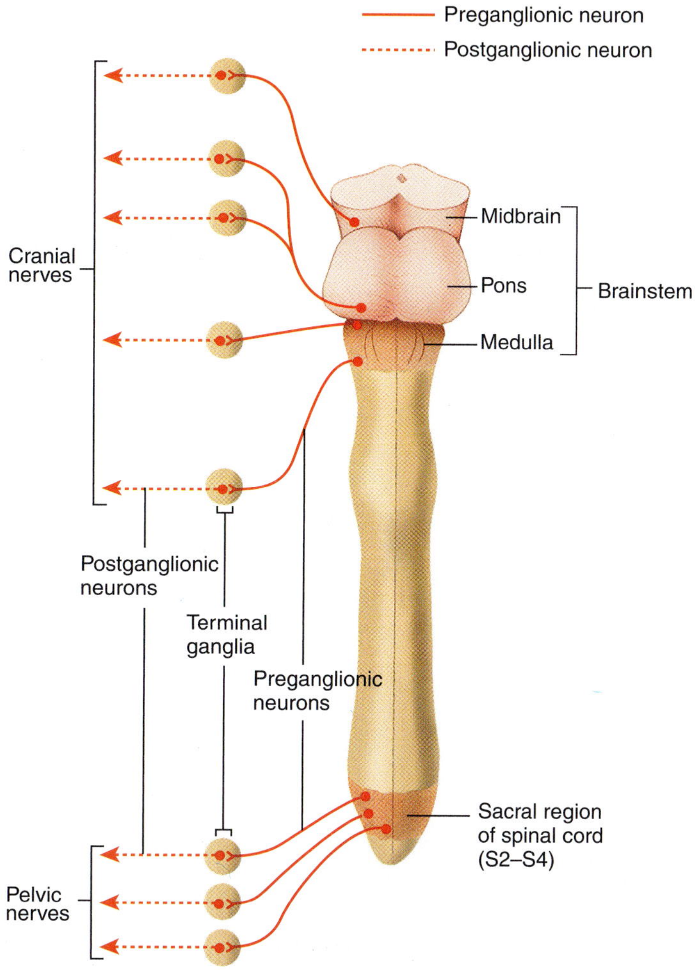
Preganglionic neuron
Postganglionic neuron
Cranial nerves
Midbrain
Pons
Medulla
Brainstem
Postganglionic neurons
Terminal ganglia
Preganglionic neurons
Sacral region of spinal cord (S2–S4)
Pelvic nerves

Parasympathetic Division
Figure 16.4

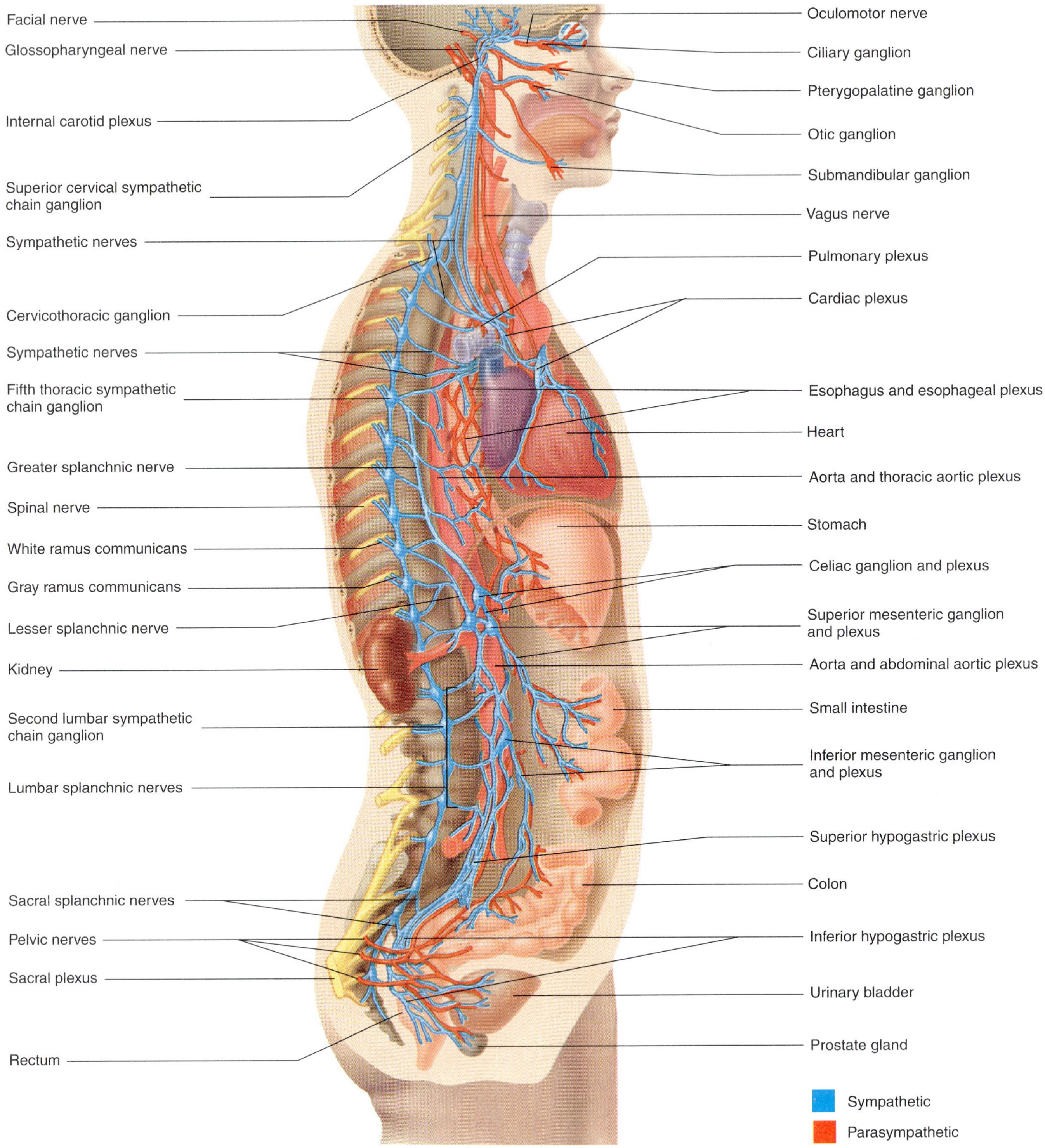

Distribution of Autonomic Nerve Fibers
Figure 16.5

Sympathetic division

Most target tissues innervated by the sympathetic division have adrenergic receptors. When norepinephrine (NE) binds to adrenergic receptors, some target tissues are stimulated, and others are inhibited. For example, smooth muscle cells in blood vessels are stimulated to constrict, and stomach glands are inhibited.

Sympathetic division

Some sympathetic target tissues, such as sweat glands, have muscarinic receptors, which respond to acetylcholine (ACh). Stimulation of sweat glands results in increased sweat production.

Parasympathetic division

All parasympathetic target tissues have muscarinic receptors. The general response to ACh is excitatory, but some target tissues, such as the heart, are inhibited.

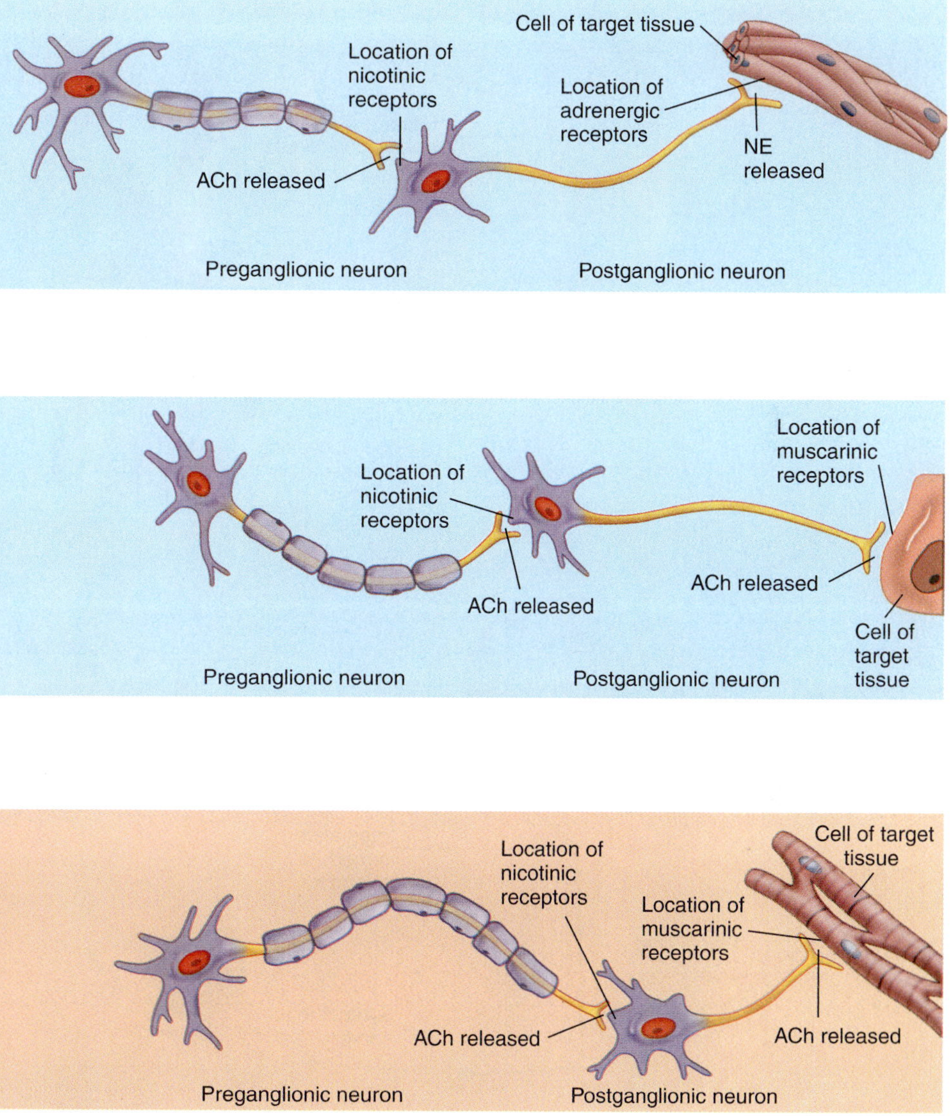

Location of ANS Receptors
Figure 16.6

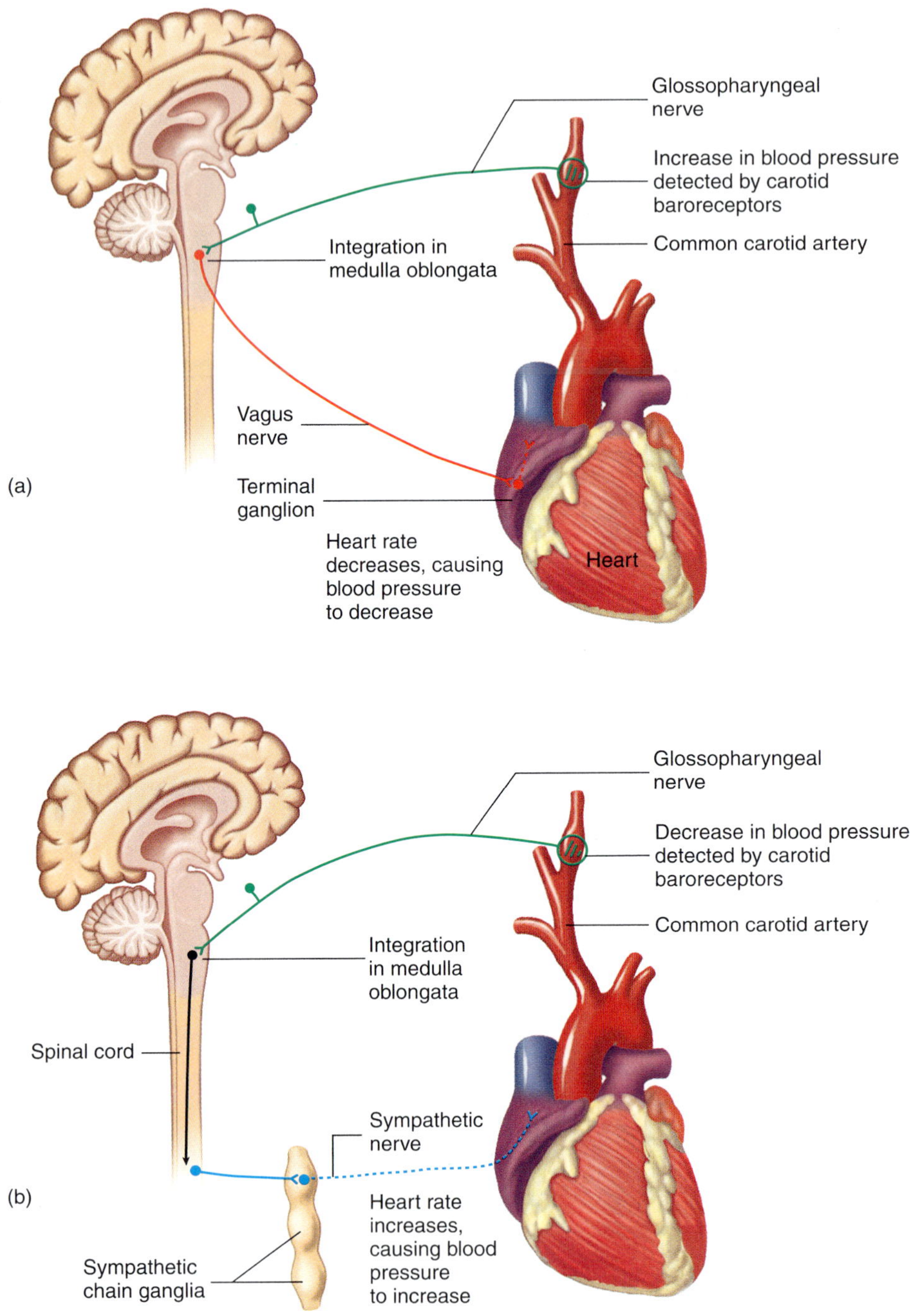
Glossopharyngeal nerve
Increase in blood pressure detected by carotid baroreceptors
Common carotid artery
Integration in medulla oblongata
Vagus nerve
(a)
Terminal ganglion
Heart rate decreases, causing blood pressure to decrease
Heart
Glossopharyngeal nerve
Decrease in blood pressure detected by carotid baroreceptors
Common carotid artery
Integration in medulla oblongata
Spinal cord
Sympathetic nerve
(b)
Heart rate increases, causing blood pressure to increase
Sympathetic chain ganglia

Autonomic Reflexes
Figure 16.7

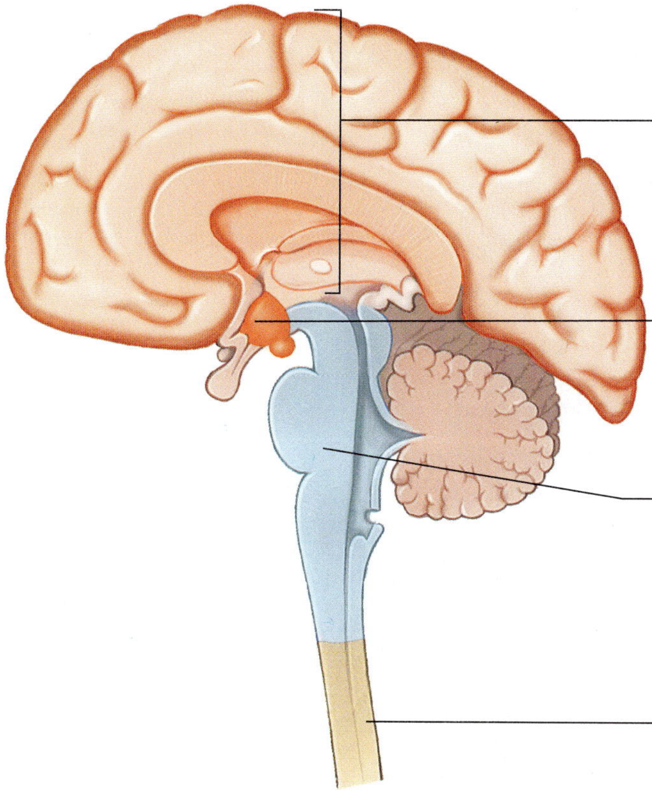

Influence of Higher Parts of the Brain on Autonomic Functions
Figure 16.8

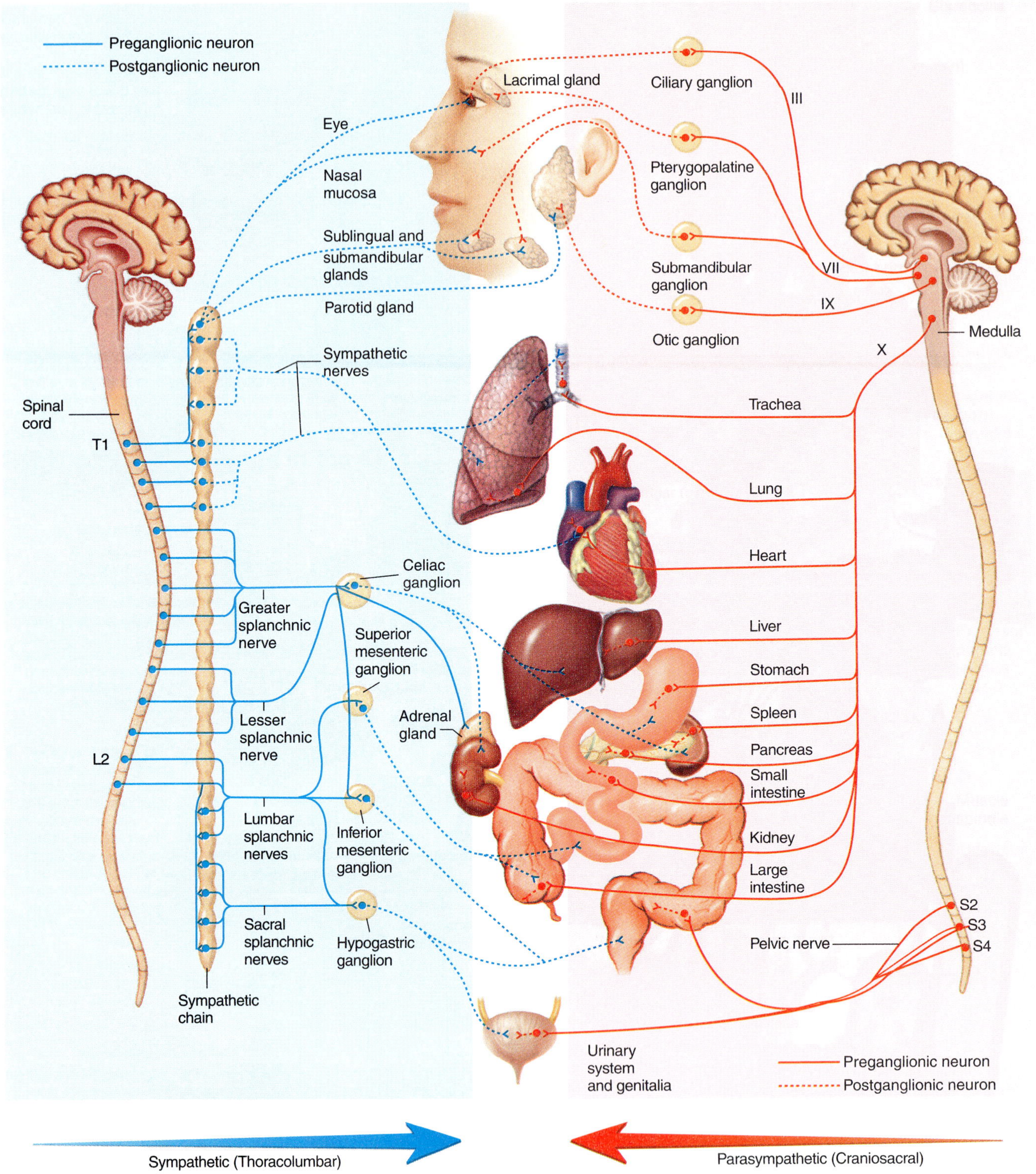
Preganglionic neuron
Postganglionic neuron
Lacrimal gland
Ciliary ganglion
III
Eye
Nasal mucosa
Pterygopalatine ganglion
Sublingual and submandibular glands
Submandibular ganglion
VII
Parotid gland
IX
Otic ganglion
Medulla
X
Sympathetic nerves
Spinal cord
T1
Trachea
Lung
Heart
Celiac ganglion
Liver
Greater splanchnic nerve
Superior mesenteric ganglion
Stomach
Spleen
Adrenal gland
Lesser splanchnic nerve
Pancreas
Small intestine
L2
Lumbar splanchnic nerves
Inferior mesenteric ganglion
Kidney
Large intestine
S2
S3
S4
Sacral splanchnic nerves
Hypogastric ganglion
Pelvic nerve
Sympathetic chain
Urinary system and genitalia
Preganglionic neuron
Postganglionic neuron
Sympathetic (Thoracolumbar)
Parasympathetic (Craniosacral)

Innervation of Organs by the ANS
Figure 16.9

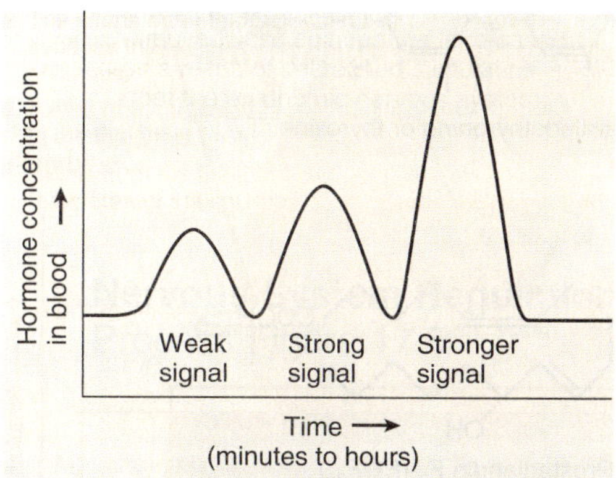

Amplitude-modulated system. The concentration of the hormone determines the strength of the signal and the magnitude of the response. For most hormones, a small concentration of a hormone is a weak signal and produces a small response, whereas a larger concentration is a stronger signal and results in a greater response.

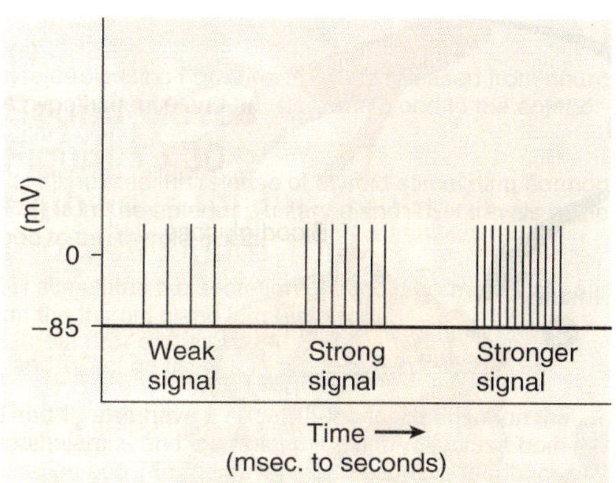

Frequency-modulated system. The strength of the signal depends on the frequency, not the size, of the action potentials. All action potentials are the same size in a given tissue. A low frequency of action potentials is a weak stimulus, and a higher frequency is a stronger stimulus.

Regulatory Systems
Figure 17.2

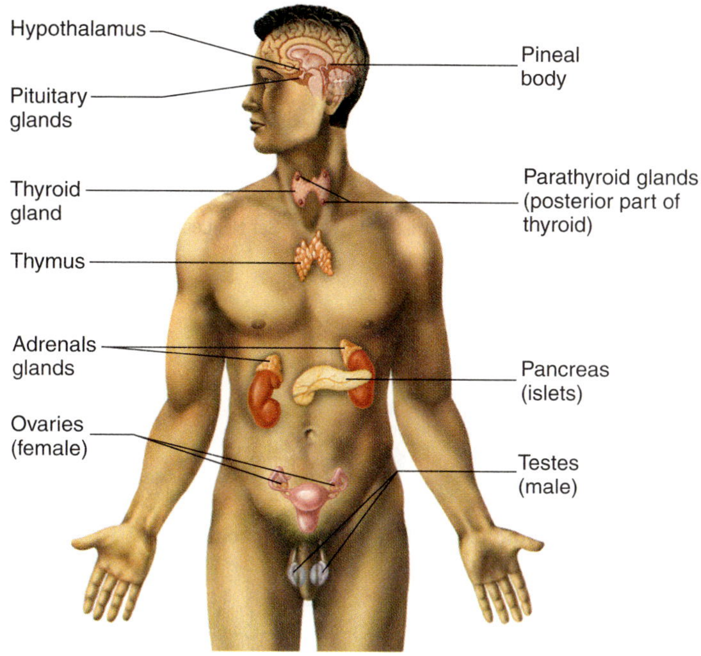

Endocrine Glands
Figure 17.1

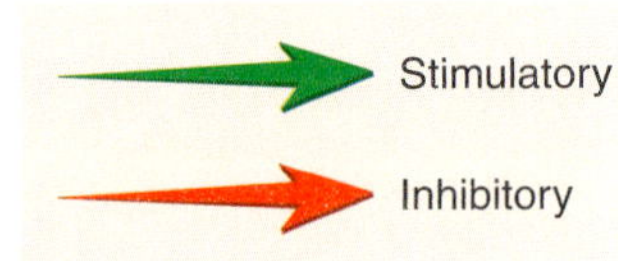

1. During the menstrual cycle, before ovulation, small amounts of estrogen are secreted from the ovary.
2. Estrogen stimulates the release of gonadotropin-releasing hormone (GnRH) from the hypothalamus and luteinizing hormone (LH) from the anterior pituitary.
3. GnRH also stimulates the release of LH from the anterior pituitary.
4. LH causes the release of additional estrogen from the ovary. The GnRH and LH levels in the blood increase because of this positive-feedback effect.

(a)

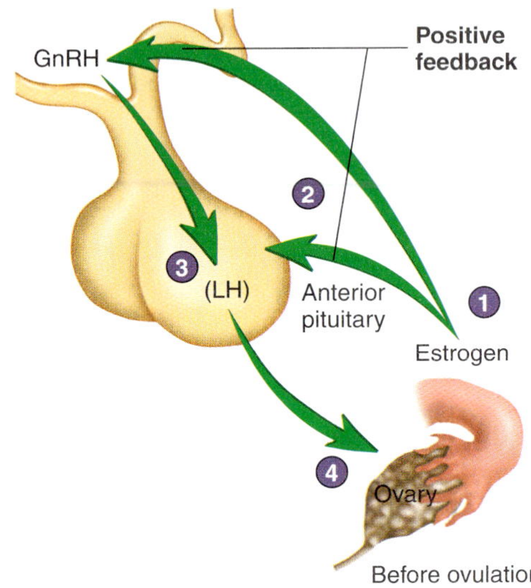

1. During the menstrual cycle, after ovulation, the ovary begins to secrete progesterone in response to LH.
2. Progesterone inhibits the release of GnRH from the hypothalamus and LH from the anterior pituitary.
3. Decreased GnRH release from the hypothalamus reduces LH secretion from the anterior pituitary. GnRH and LH levels in the blood decrease because of this negative-feedback effect.

(b)

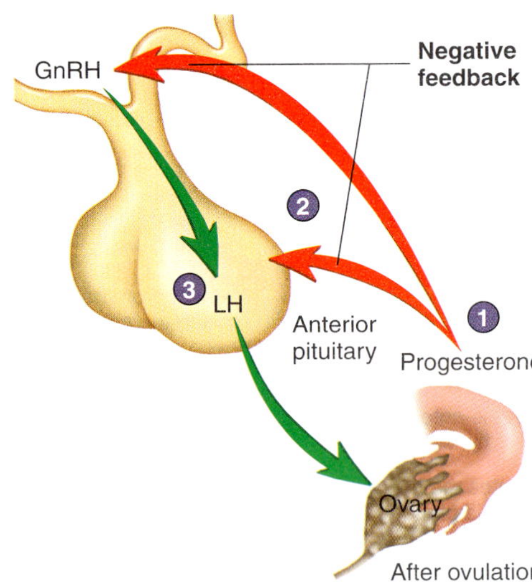

Positive and Negative Feedback
Process Figure 17.7

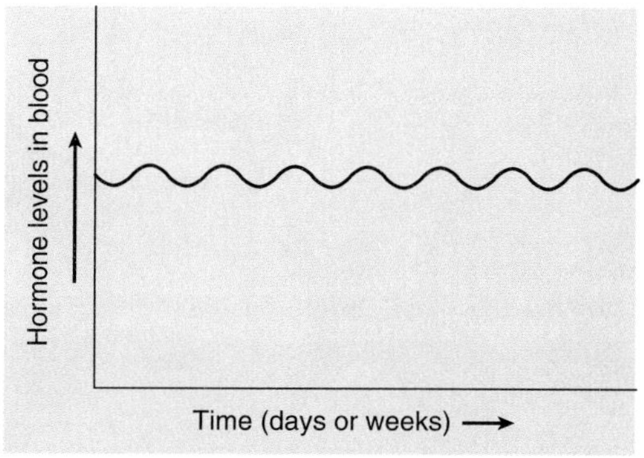

(a) **Chronic hormone regulation.** The maintenance of a relatively constant concentration of hormone in the circulating blood over a relatively long period.

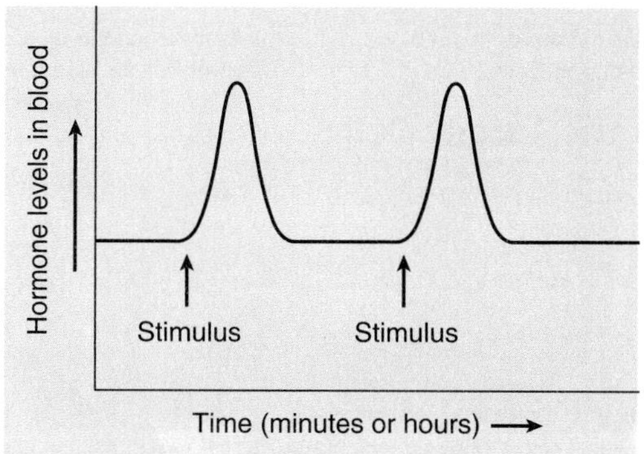

(b) **Acute hormone regulation.** A hormone rapidly increases in the blood for a short time in response to a stimulus.

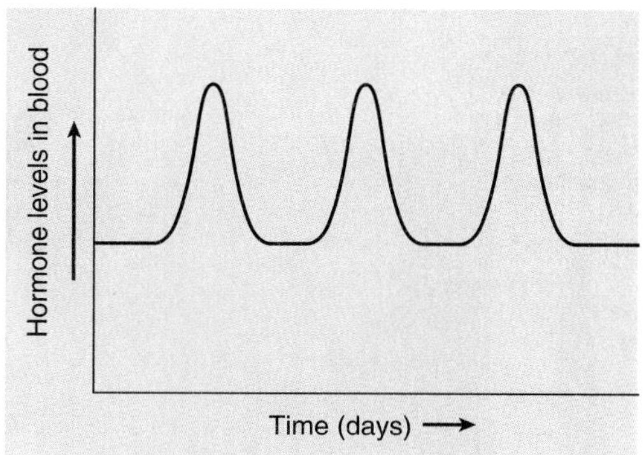

(c) **Cyclic hormone regulation.** A hormone is regulated so that it increases and decreases in the blood at a relatively constant time and to roughly the same amount.

Changes in Hormone Secretion Through Time
Figure 17.8

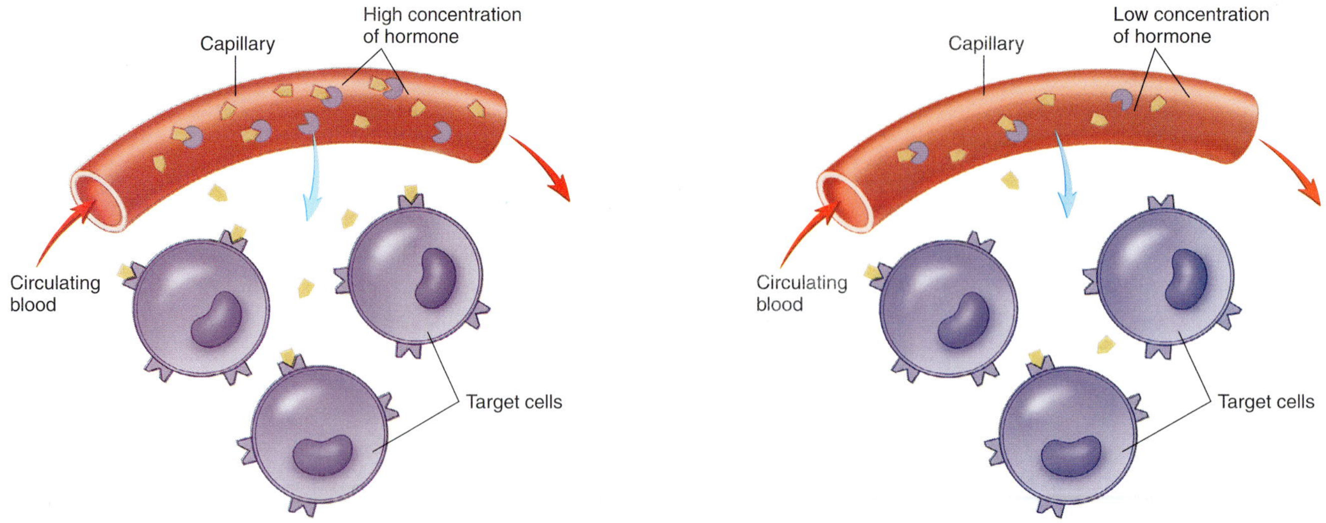

(a) As the concentration of free hormone molecules increases in the blood, more molecules diffuse from the capillary to the target cells.

(b) As the concentration of hormone molecules decreases in the blood, fewer diffuse from the capillary to the target cells.

Hormone Concentrations at the Target Cell
Figure 17.9

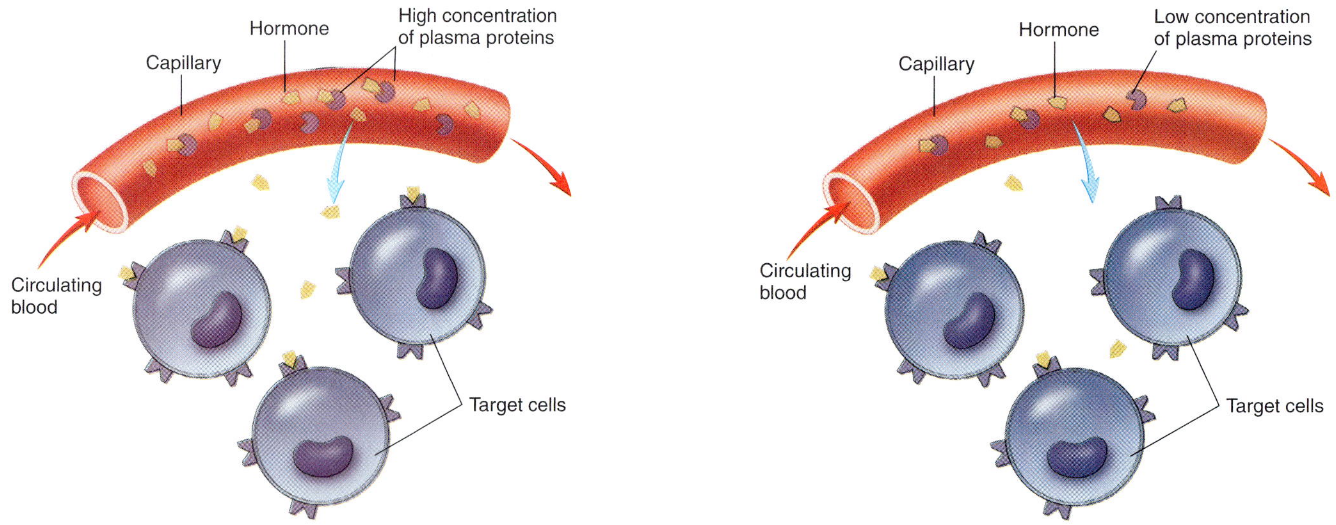

(a) An equilibrium exists between free hormone molecules and hormone molecules bound to plasma proteins. The free hormone molecules can diffuse from the capillaries to the interstitial spaces. Free hormones are also more readily eliminated from the body by the kidneys and liver. Hormones bound to plasma proteins remain in the blood for a longer period of time.

(b) A decrease in plasma protein concentration reduces the number of hormone molecules bound to plasma proteins. This increases the rate at which free hormone molecules diffuse from the capillaries, and increases the rate at which hormones are eliminated from the blood by the kidneys and liver. The rapid loss of hormone from the circulatory system reduces the hormone concentration in the blood.

Effect of Changes in Plasma Protein Concentration on the Concentration of Free Hormone
Figure 17.10

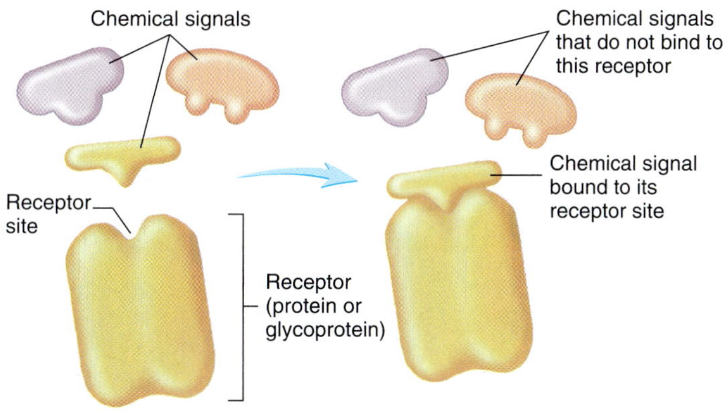

Receptors and Specificity of Receptor Sites
Figure 17.11

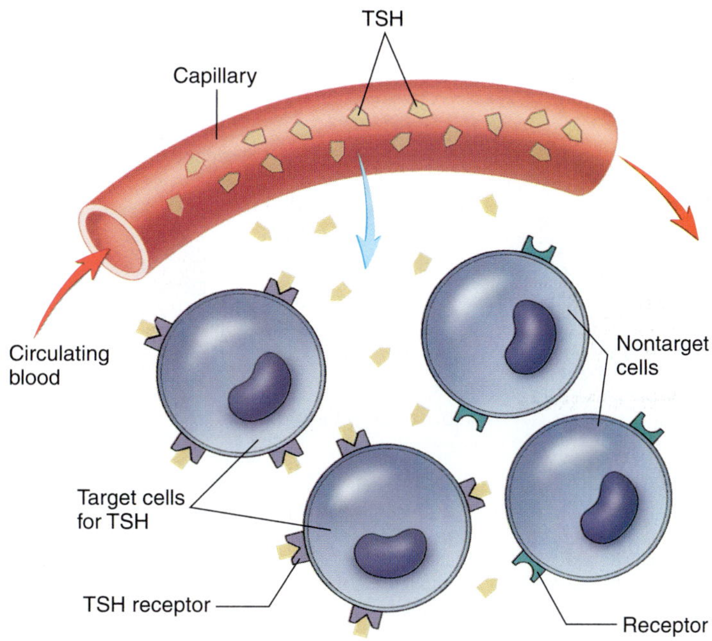

Response of Target Cells to Hormones
Figure 17.12

Down-regulation

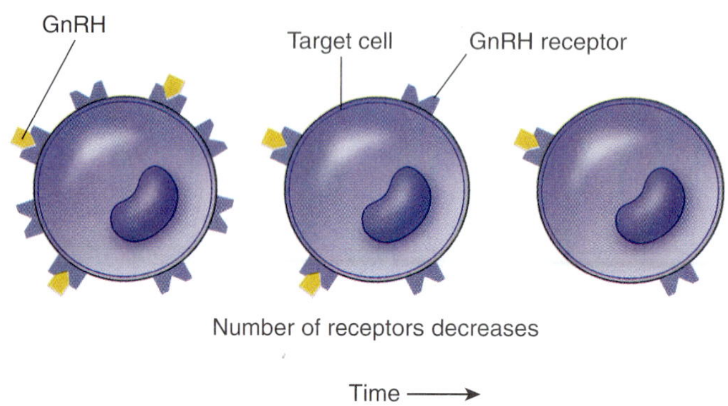

(a) Down-regulation occurs when the number of receptors for a hormone decreases within target cells. For example, gonadotropin-releasing hormone (GnRH) released from the hypothalamus binds to GnRH receptors in the anterior pituitary. GnRH bound to its receptors causes down-regulation of the GnRH receptors so that eventually the target cells become less sensitive to the GnRH.

Up-regulation

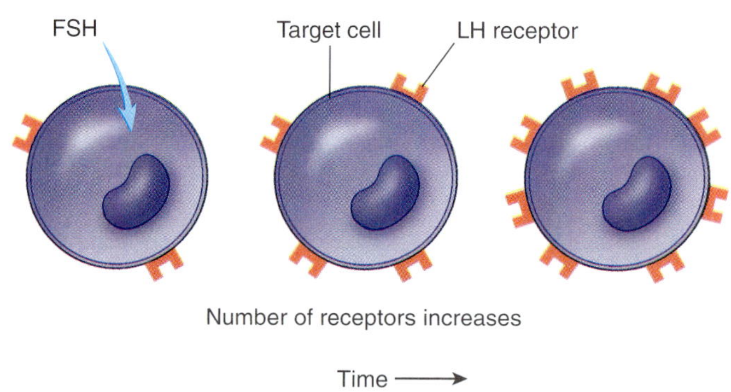

(b) Up-regulation occurs when some stimulus causes the number of receptors for a hormone to increase within a target cell. For example, FSH acts on cells of the ovary to up-regulate the number of receptors for LH. Thus the ovary becomes more sensitive to the effect of LH.

Down-Regulation and Up-Regulation
Figure 17.13

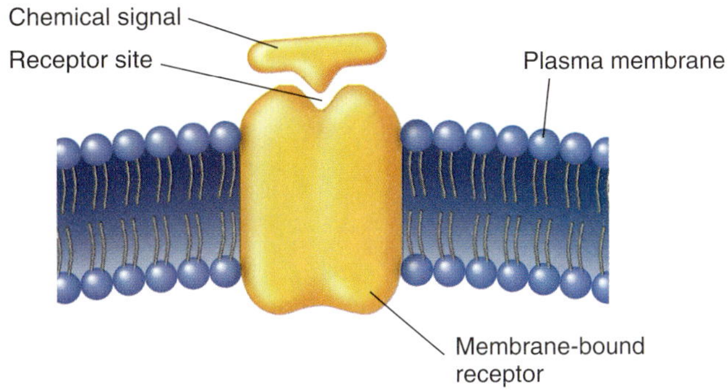

(a) Chemical signals that are large and water-soluble interact with membrane-bound receptors that extend across the plasma membrane and have receptor sites exposed to the outer surface of the plasma membrane. The portion of the receptor inside the cell produces the response.

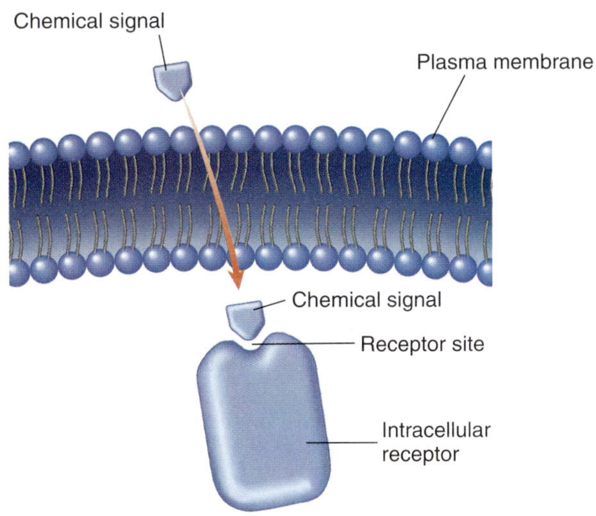

(b) Small lipid-soluble chemical signals diffuse through the plasma membrane and combine with the receptor sites of intracellular receptors. The combination of chemical signals and intracellular receptors produces a response.

Membrane-Bound and Intracellular Receptors
Figure 17.14

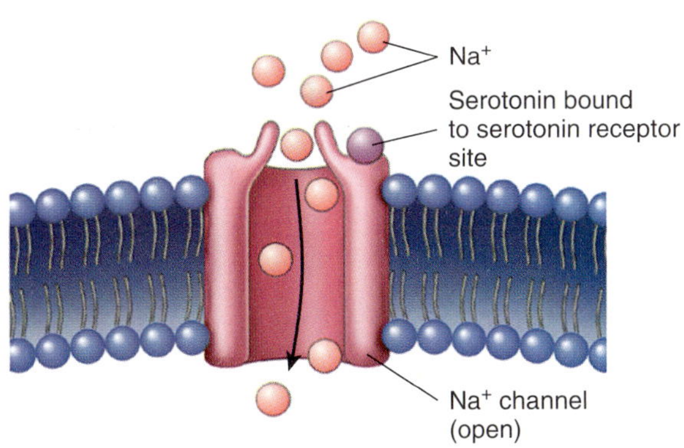

Membrane-Bound Receptors That Directly Control Membrane Channels
Figure 17.15

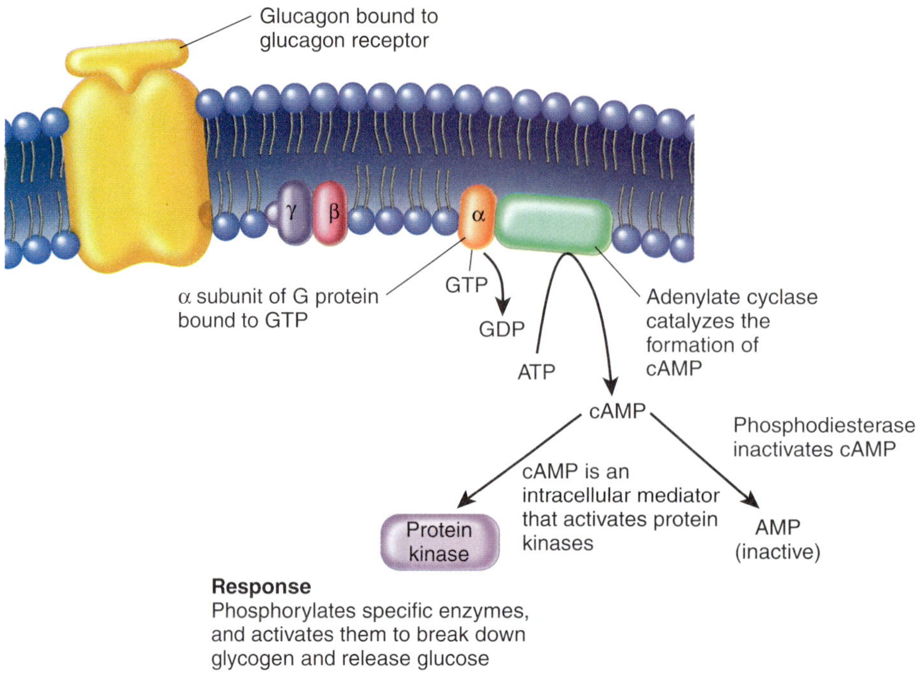

Membrane-Bound Receptors That Activate G Proteins and Increase the Synthesis of cAMP
Figure 17.18

1. Epinephrine binds to its receptor in the smooth muscle plasma membrane.
2. G proteins dissociate and GTP replaces GDP on the α subunit.
3. The active α subunit then binds with phospholipase C, which acts on phosphoinositol (PIP_2) and produces inositol triphosphate (IP_3) and diacylglycerol (DAG).
4. IP_3 releases Ca^{2+} from the endoplasmic reticulum or opens Ca^{2+} channels in the plasma membrane. Ca^{2+} then regulates enzyme activity.
5. DAG regulates enzymes such as those that synthesize prostaglandin.

These responses increase smooth muscle contraction.

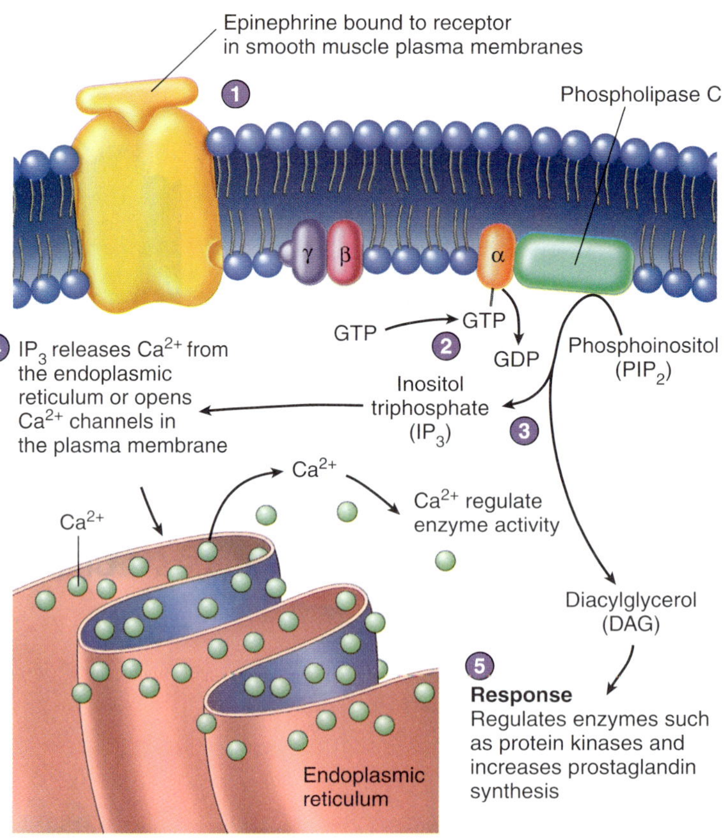

Membrane-Bound Receptors That Activate G Proteins and Increase the Synthesis of IP_3 and DAG
Process Figure 17.19

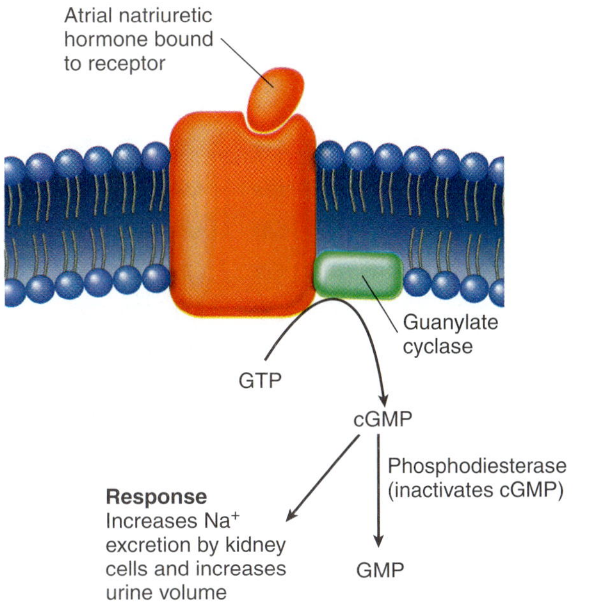

Membrane-Bound Receptor That Directly Activates an Intracellular Mediator
Figure 17.20

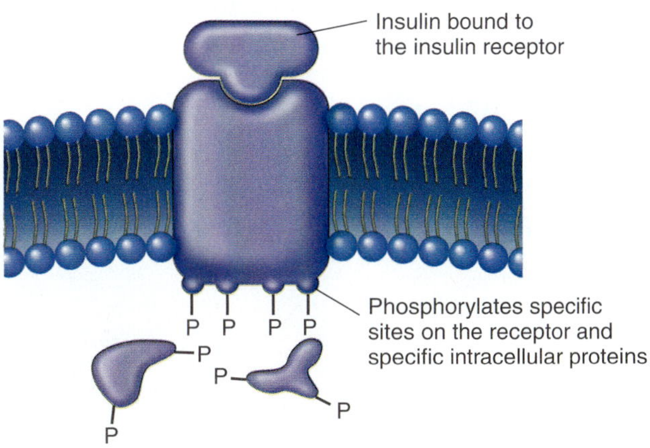

Membrane-Bound Receptors That Phosphorylate Intracellular Proteins
Figure 17.21

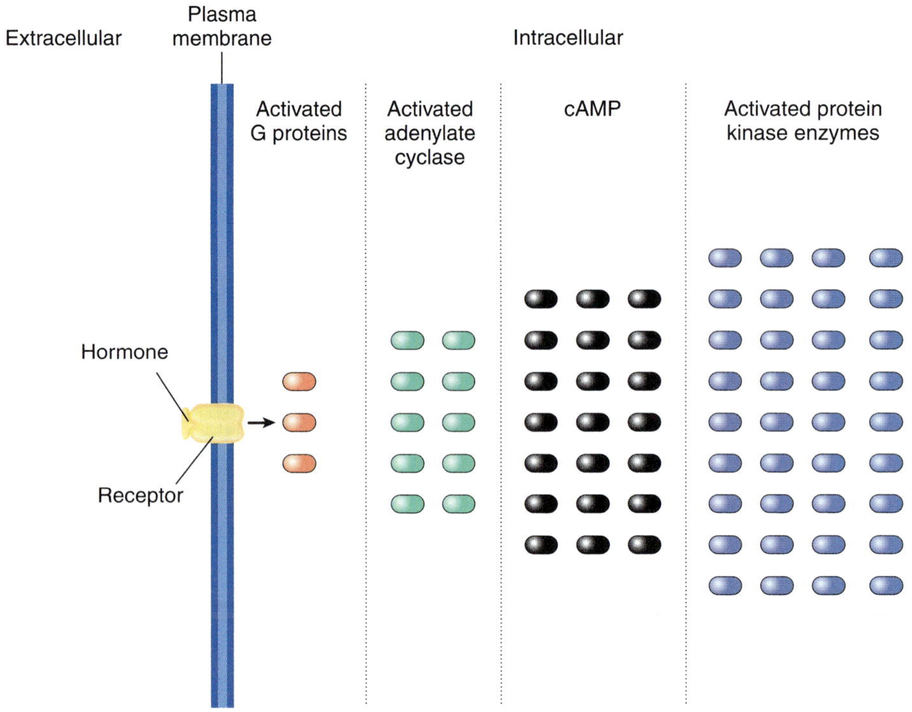

The Cascade Effect
Figure 17.22

1. Aldosterone is a lipid-soluble hormone and can easily diffuse through the plasma membrane.
2. Once aldosterone is inside of the cell, it binds with an aldosterone receptor molecule in the cytoplasm.
3. The aldosterone–receptor complex moves into the nucleus and binds to DNA.
4. The binding of the aldosterone–receptor complex to DNA stimulates the synthesis of messenger RNA (mRNA), which codes for specific proteins.
5. The mRNA leaves the nucleus, passes into the cytoplasm of the cell, and binds to ribosomes, where it directs the synthesis of the specific proteins.
6. The proteins synthesized on the ribosomes produce the response of the cell to aldosterone. For example, aldosterone increases the synthesis of proteins that transport Na^+ and K^+ across epithelial plasma membranes in the kidney.

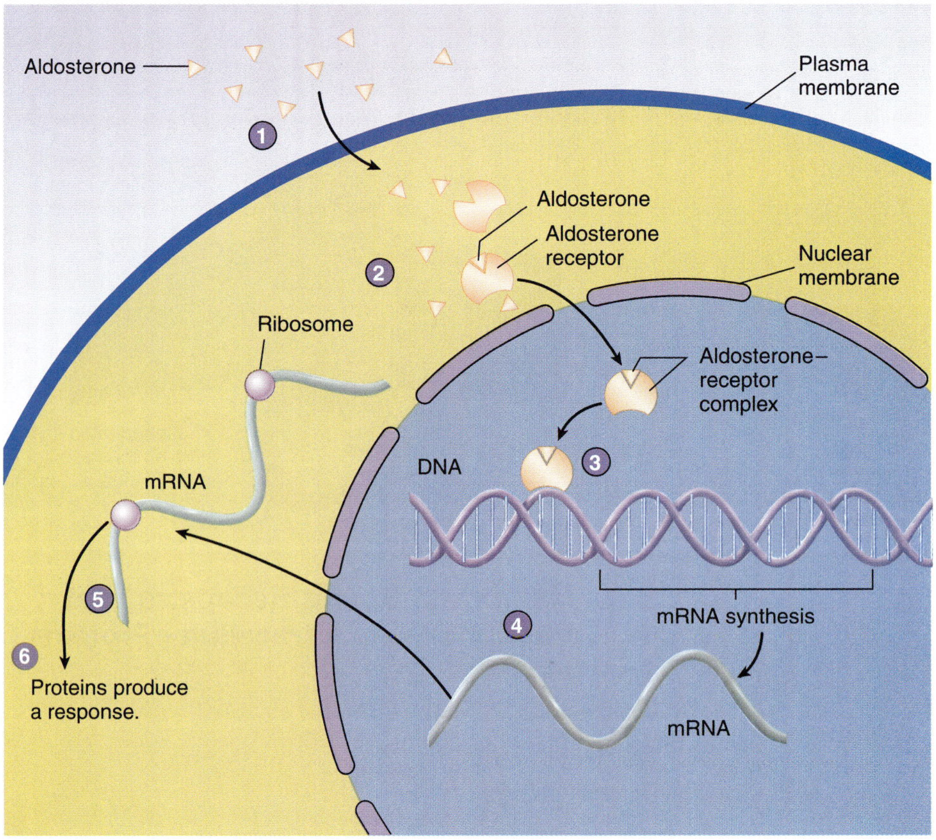

Intracellular Receptor Model
Process Figure 17.23

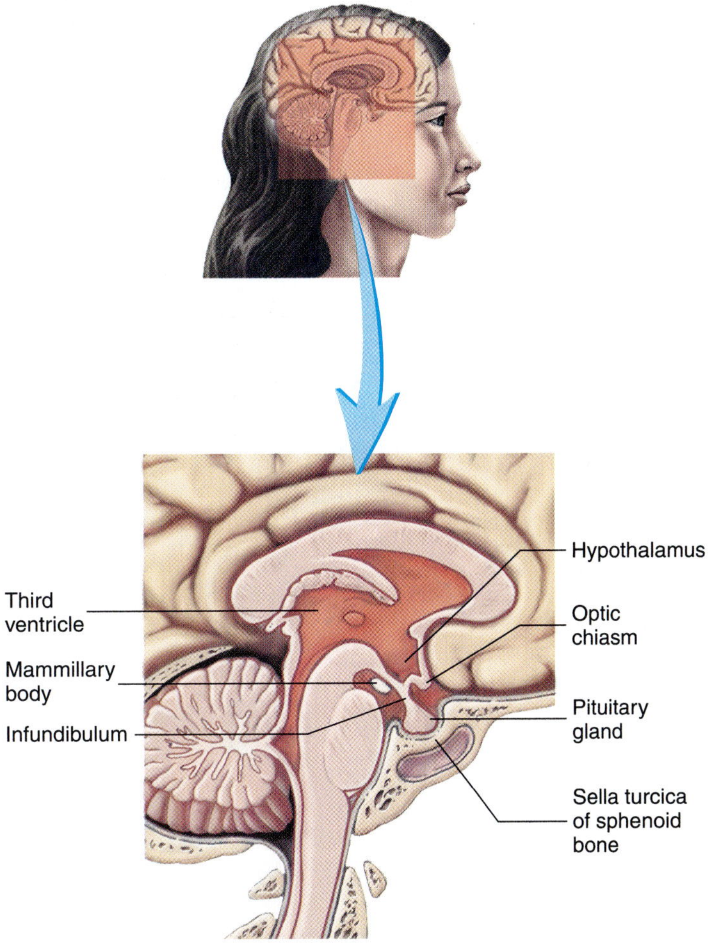

The Hypothalamus and Pituitary Gland
Figure 18.1

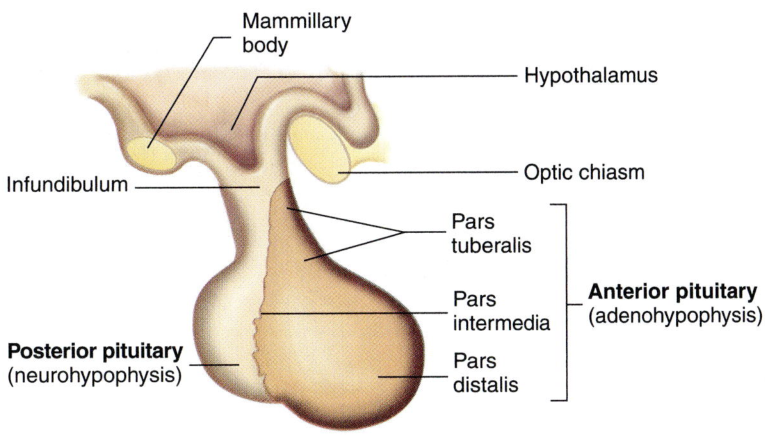

Subdivisions of the Pituitary Gland
Figure 18.2

1. Stimuli within the nervous system cause releasing hormones (*blue balls*) to be secreted from neurons of the hypothalamus.

2. Releasing hormones pass through the hypothalamohypophysial portal system to the anterior pituitary.

3. Releasing hormones leave capillaries, bind to membrane-bound receptors, and stimulate the release of hormones from anterior pituitary cells.

4. Anterior pituitary hormones (*yellow squares*) are carried in the blood to their target tissues (*green arrow*), which, in some cases, are other endocrine glands.

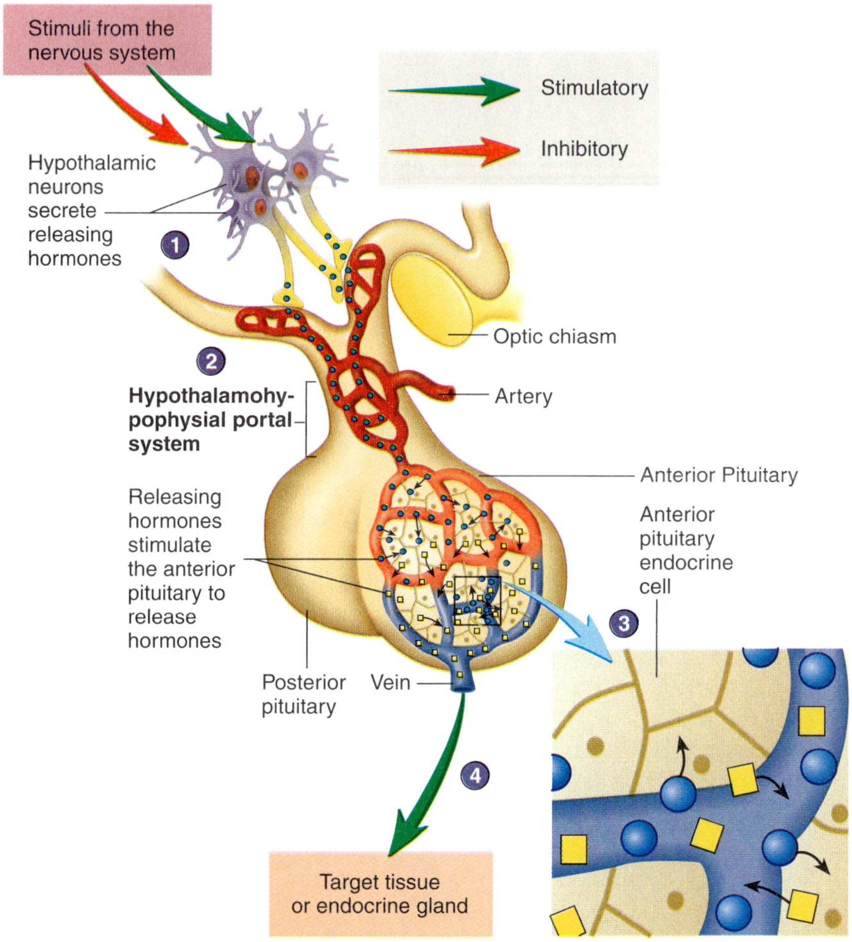

Relationship Among the Hypothalamus, Anterior Pituitary, and Target Tissues
Process Figure 18.3

1. Stimuli within the nervous system stimulate hypothalamic neurons to either increase or decrease their action potential frequency.

2. Action potentials are carried by axons through the hypothalamohypophysial tract to the posterior pituitary. The axons of neurons store hormones in the posterior pituitary.

3. In the posterior pituitary gland, action potentials cause the release of neurohormones (*red balls*) from axon terminals into the circulatory system.

4. The neurohormones pass through the circulatory system and influence the activity of their target tissues (*green arrow*).

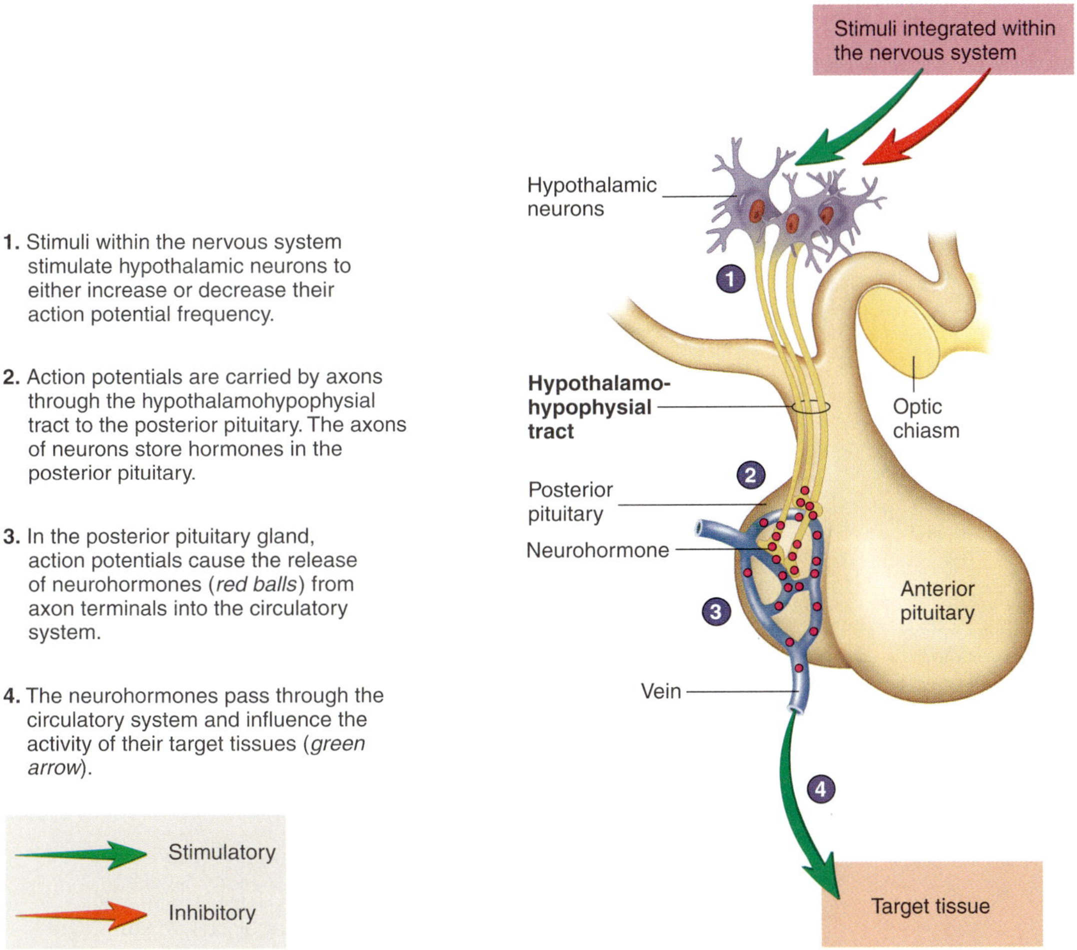

Relationship Among the Hypothalamus, Posterior Pituitary, and Target Tissues
Process Figure 18.4

1. Osmoreceptors in the hypothalamus detect changes in blood osmolality, and sensory neurons that send action potentials through the vagus nerves to the hypothalamus detect changes in blood presure.

2. An increase in osmolality and a decrease in blood pressure increase action potentials in ADH-secreting neurons.

3. Action potentials are carried by axons of ADH-secreting neurons through the hypothalamohypophysial tract to the posterior pituitary.

4. In the posterior pituitary, increased action potentials increase the release of ADH from the axon terminals into the circulatory system.

5. Increasing ADH acts on the kidney tubules to increase water reabsorption, resulting in a reduced urine volume, increased urine osmolality, and decreased blood osmolality. This helps to maintain blood volume.

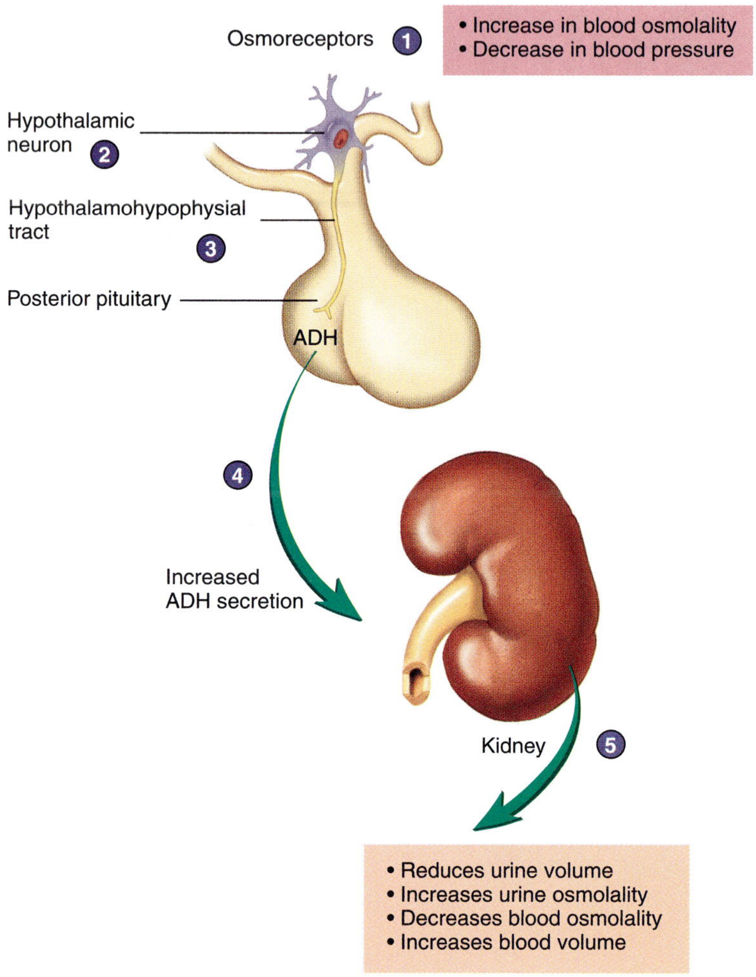

Control of Antidiuretic Hormone (ADH) Secretion
Process Figure 18.5

1. Action potentials are carried by sensory neurons from the uterus and breast to the spinal cord, and up spinal nerve tracts to the hypothalamus.

2. Stretch of the uterus and the uterine cervix or nipples of the breasts increase action potentials in oxytocin-secreting neurons.

3. Action potentials are carried by oxytocin-secreting neurons in the hypothalamohypophysial tract to the posterior pituitary gland.

4. Oxytocin enters the circulation. It increases contractions of the uterus and milk ejection from the lactating breast.

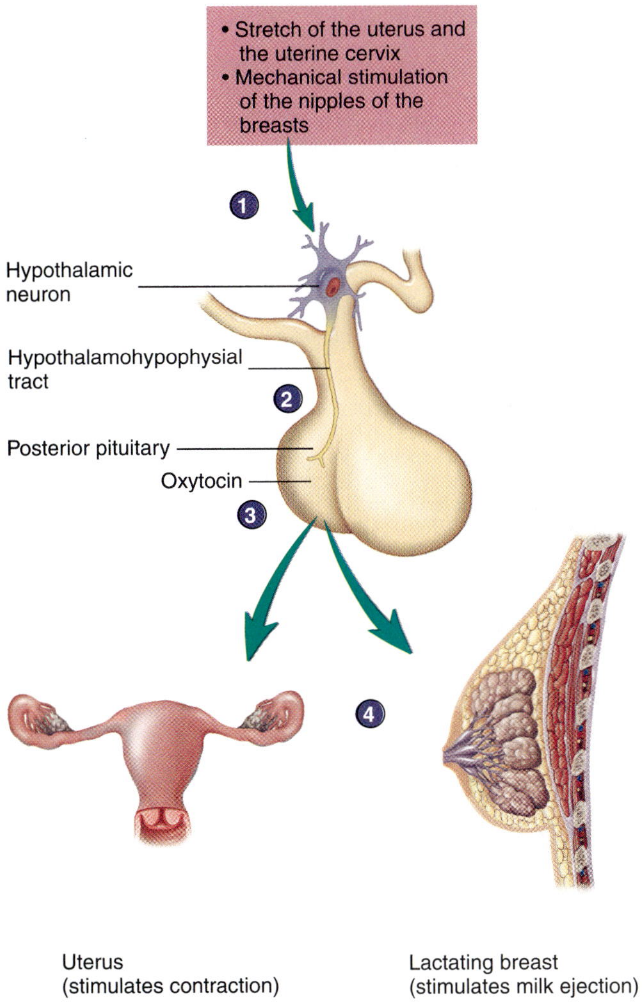

Control of Oxytocin Secretion
Process Figure 18.6

1. Stress and decreased blood glucose levels increase the release of growth hormone-releasing hormone (GHRH) and decrease the release of growth hormone-inhibiting hormone (GHIH).

2. GHRH and GHIH travel through the hypothalamohypophysial portal system to the anterior pituitary.

3. Increased GHRH and reduced GHIH act on the anterior pituitary and result in increased GH secretion.

4. GH acts on target tissues.

5. Increasing GH levels have a negative-feedback effect on the hypothalamus, resulting in decreased GHRH and decreased GHIH.

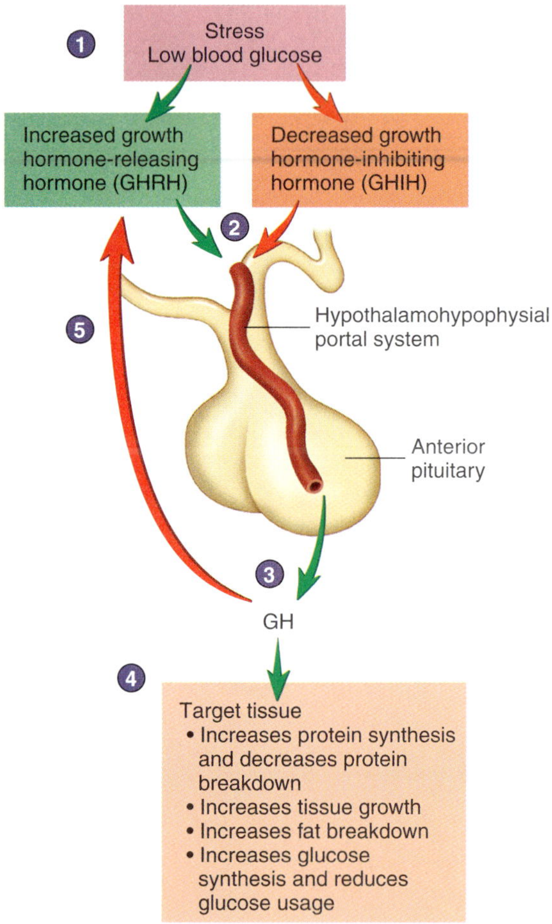

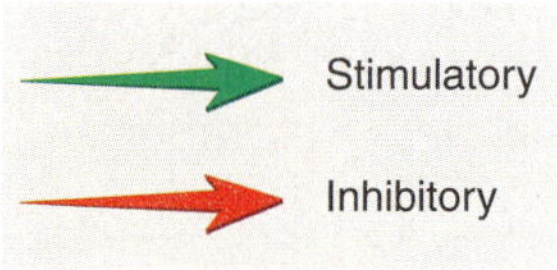

Control of Growth Hormone (GH) Secretion
Process Figure 18.7

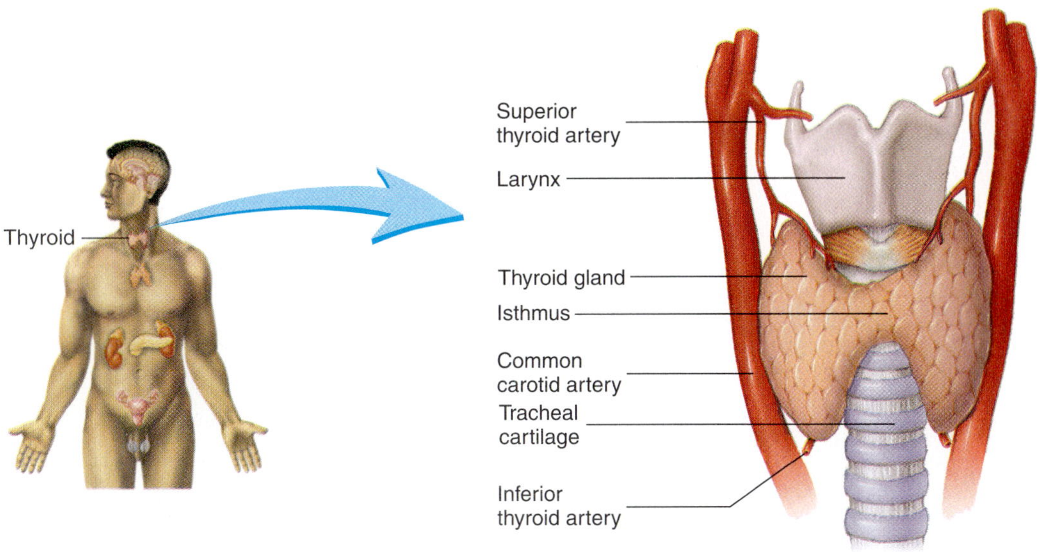

(a) **Anterior view**

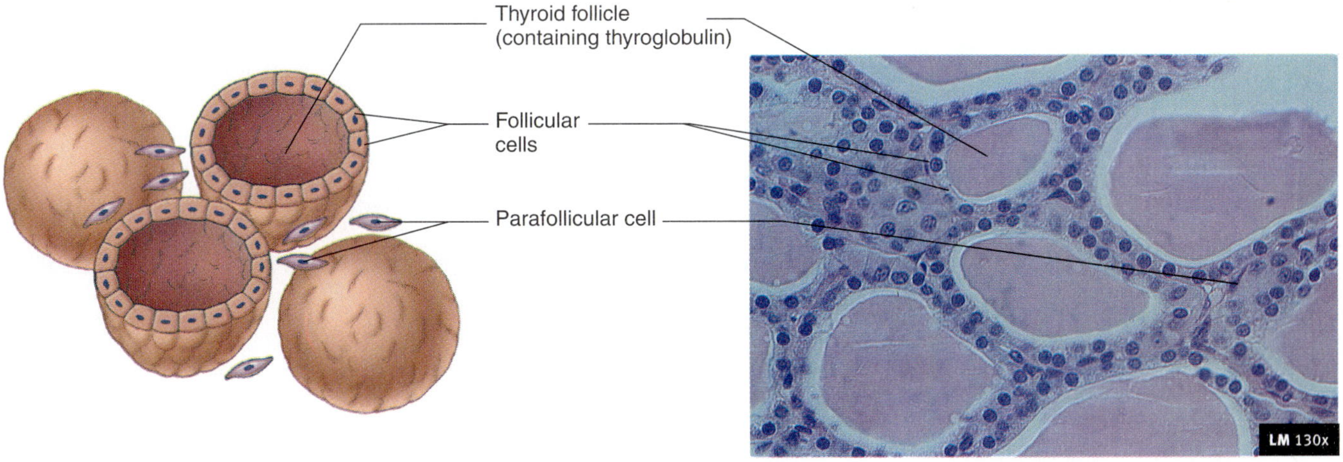

(b)

(c)

Anatomy and Histology of the Thyroid Gland

Figure 18.8

c: © Victor Eroschenko

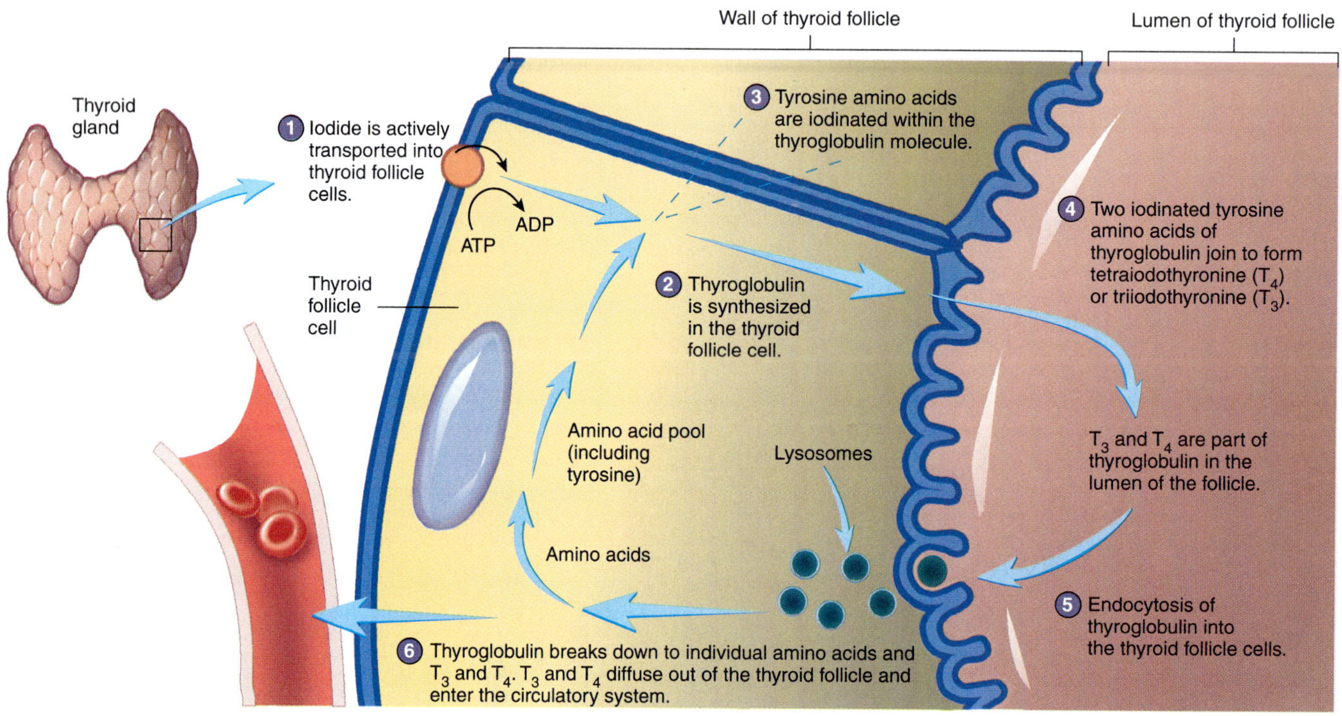

Biosynthesis of Thyroid Hormones
Process Figure 18.9

1. TRH is released from neurons within the hypothalamus. It passes through the hypothalamohypophysial portal system to the anterior pituitary.
2. TRH causes cells of the anterior pituitary to secrete TSH, which passes through the general circulation to the thyroid gland.
3. TSH causes increased release of T_3 and T_4 into the general circulation.
4. T_3 and T_4 act on target tissues to produce a response.
5. T_3 and T_4 also have an inhibitory effect on the secretion of TRH from the hypothalamus and TSH from the anterior pituitary.

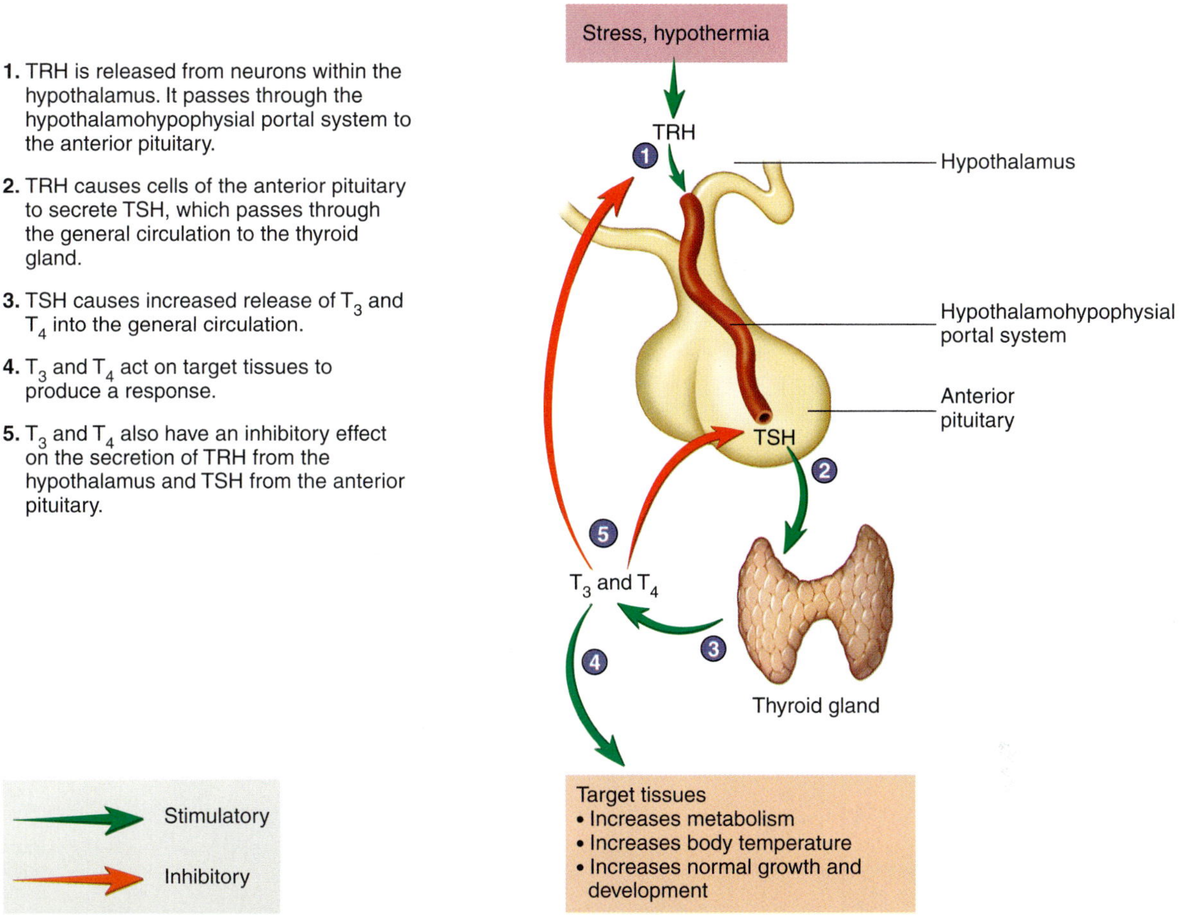

Regulation of Thyroid Hormone (T_3 and T_4) Secretion
Process Figure 18.10

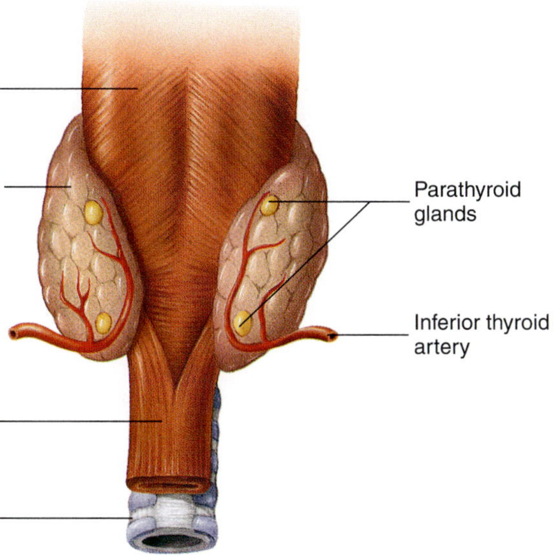

Anatomy and Histology of the Parathyroid Glands
Figure 18.11

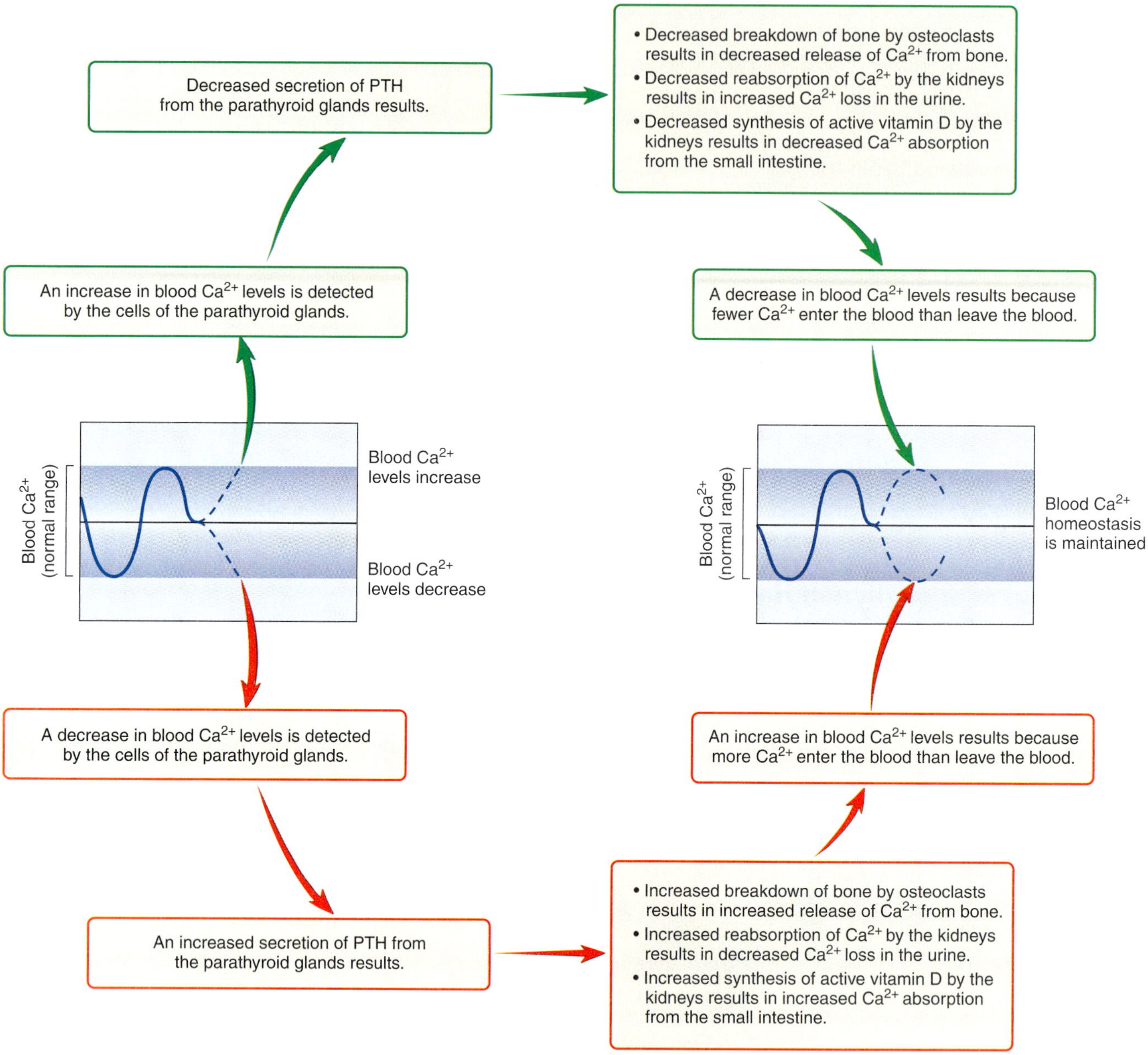

Parathyroid Hormone (PTH) Regulation of Blood Ca^{2+} Levels
Homeostasis Figure 18.12

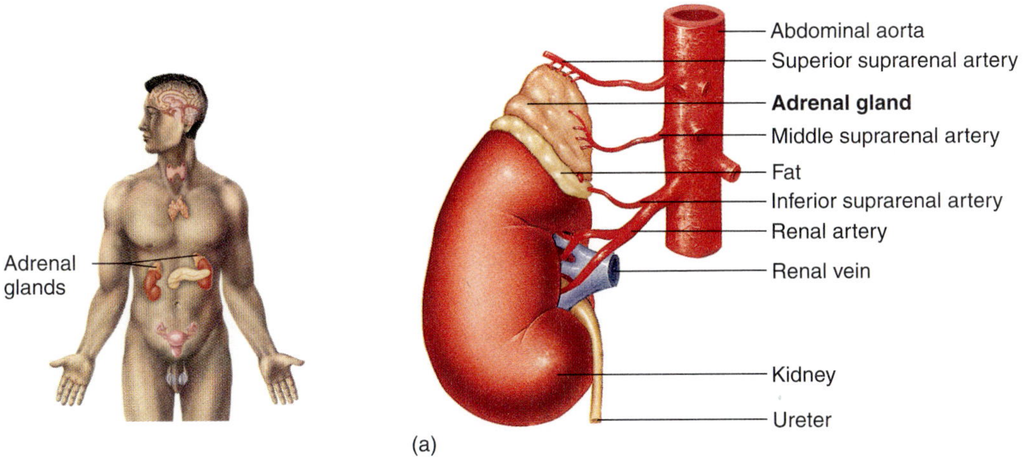

Anatomy and Histology of the Adrenal Gland
Figure 18.13

1. Stress, physical activity, and low blood glucose levels act as stimuli to the hypothalamus, resulting in increased sympathetic nervous system activity.
2. An increased frequency of action potentials conducted through the sympathetic division of the autonomic nervous system stimulates the adrenal medulla to secrete epinephrine and some norepinephrine into the circulatory system.
3. Epinephrine and norepinephrine act on their target tissues to produce responses.

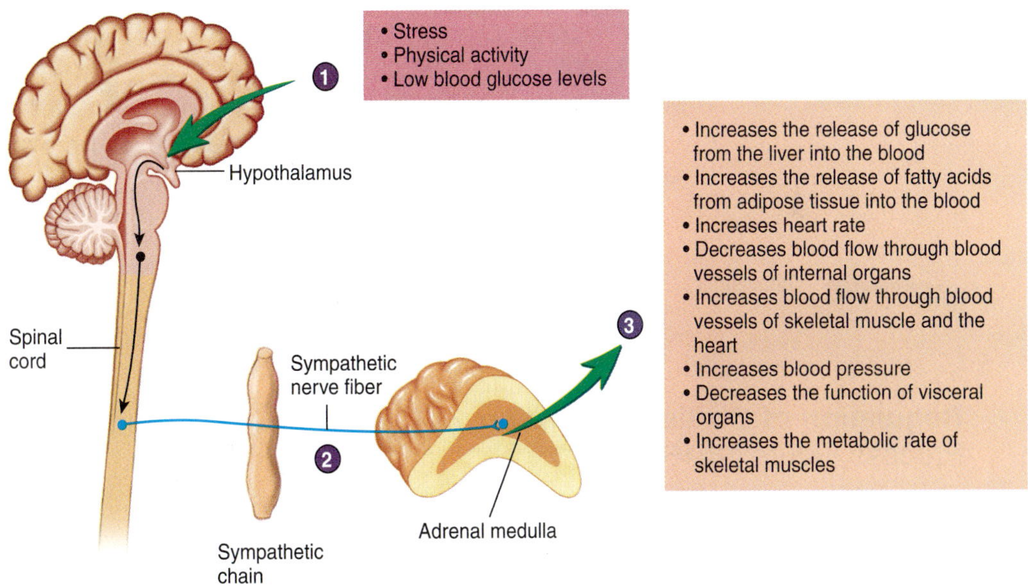

Regulation of Adrenal Medullary Secretions
Process Figure 18.14

1. Corticotropin-releasing hormone (CRH) is released from hypothalamic neurons in response to stress or low blood glucose and passes, by way to the hypothalamohypophysial portal system, to the anterior pituitary.
2. In the anterior pituitary CRH binds to and stimulates cells that secrete adrenocorticotropic hormone (ACTH).
3. ACTH binds to membrane-bound receptors on cells of the adrenal cortex and stimulates the secretion of glucocorticoids, primarily cortisol.
4. Cortisol acts on target tissues, resulting in increased fat and protein breakdown, increased glucose levels, and anti-inflammatory effects.
5. Cortisol has a negative-feedback effect because it inhibits CRH from the hypothalamus and ACTH secretion from the anterior pituitary.

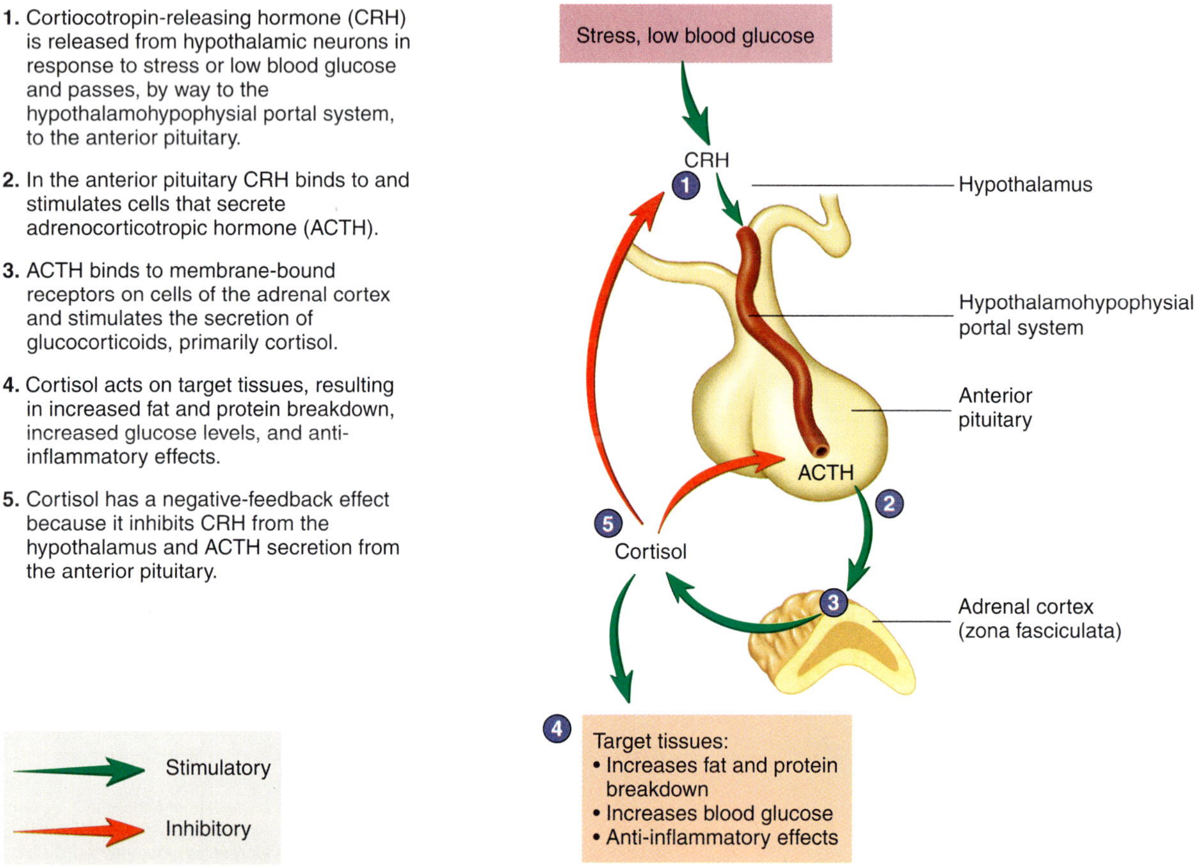

Regulation of Cortisol Secretion
Process Figure 18.15

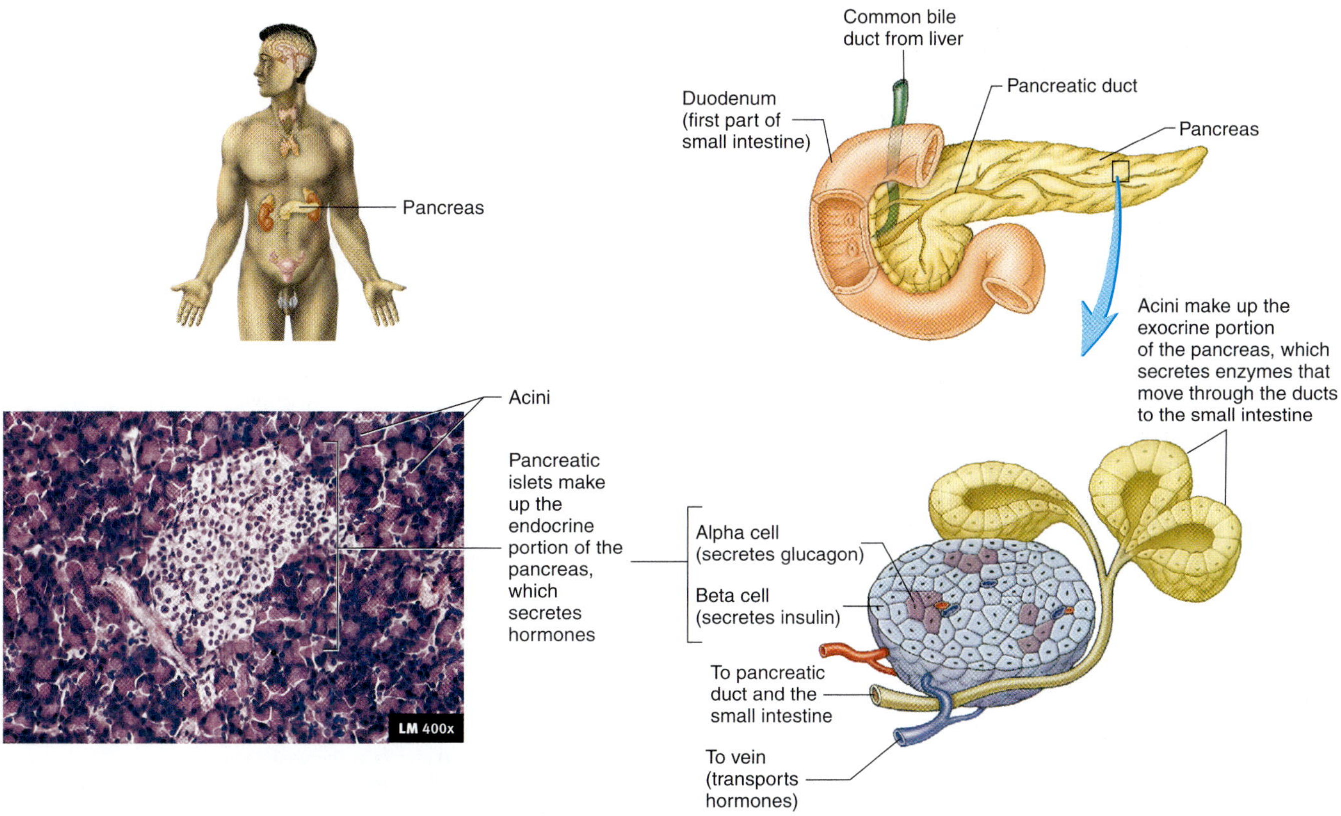

Histology of the Pancreatic Islets
Figure 18.16

© Bio-Photo Assocs/Photo Researchers, Inc.

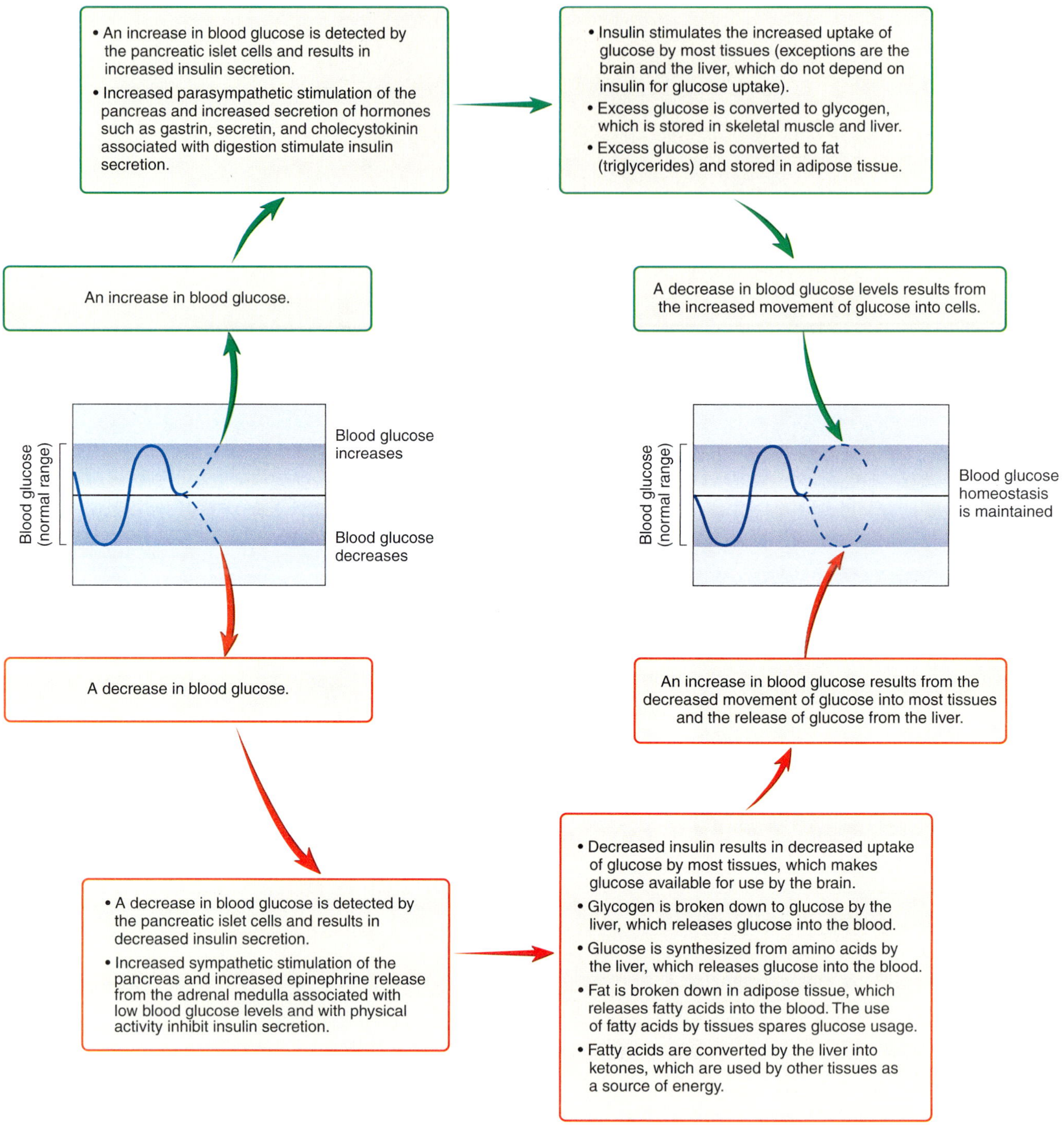

Regulation of Insulin Secretion
Homeostasis Figure 18.17

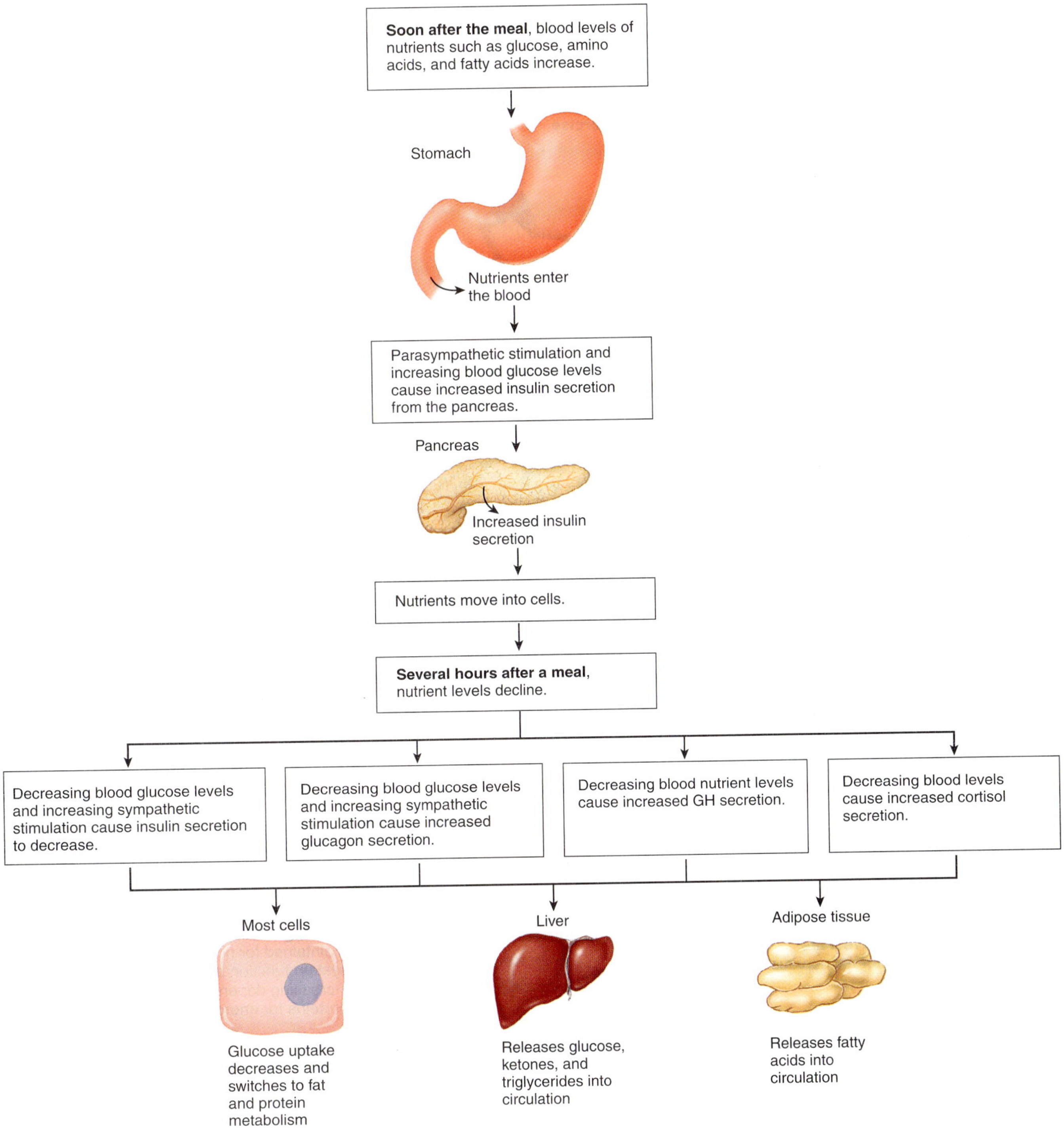

Regulation of Blood Nutrient Levels
Figure 18.18

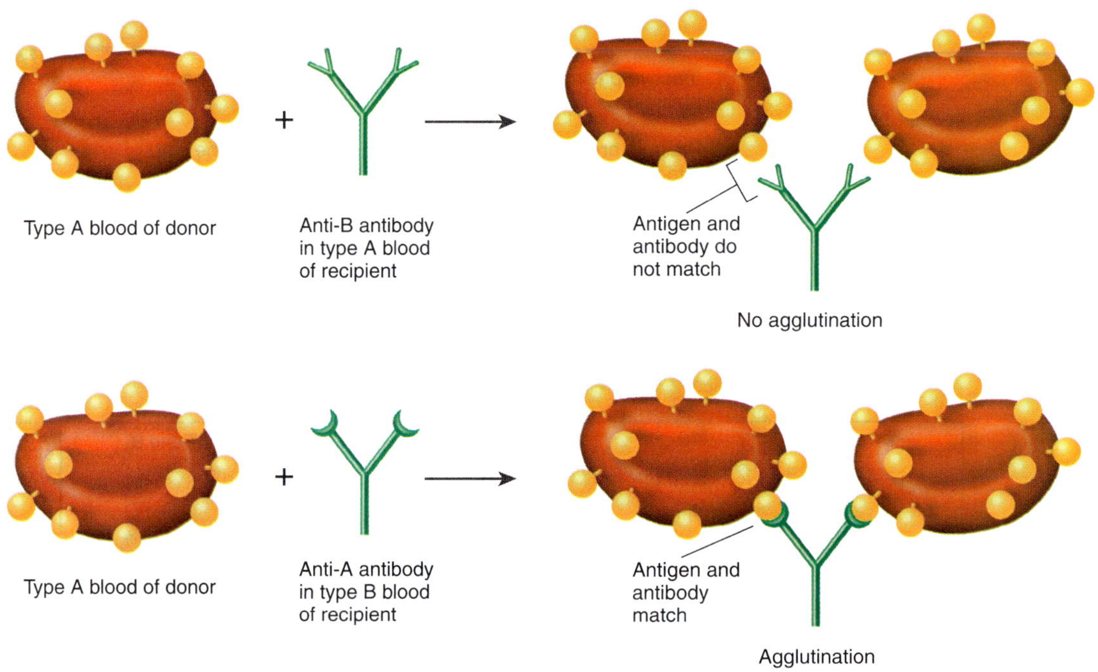

(a) **No agglutination reaction.** Type A blood donated to a type A recipient does not cause an agglutination reaction because the anti-B antibodies in the recipient do not combine with the type A antigens on the red blood cells in the donated blood.

(b) **Agglutination reaction.** Type A blood donated to a type B recipient causes an agglutination reaction because the anti-A antibodies in the recipient combine with the type A antigens on the red blood cells in the donated blood.

Agglutination Reaction
Figure 19.14

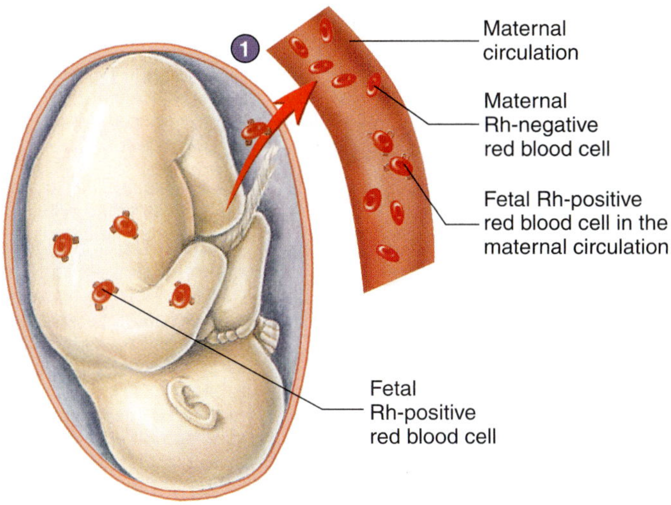

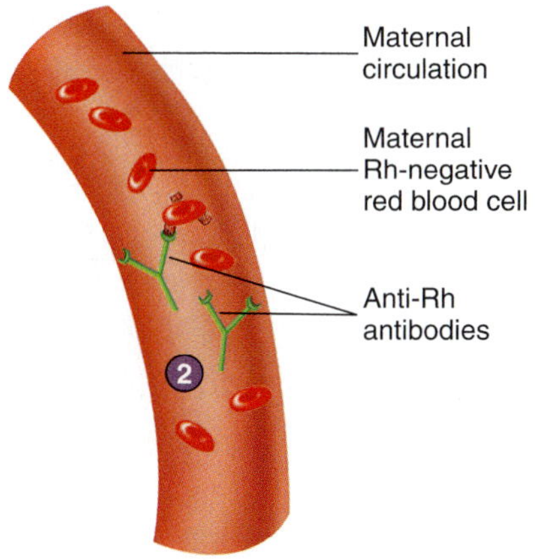

1. Before or during delivery, Rh-positive red blood cells from the fetus enter the blood of an Rh-negative woman through a tear in the placenta.

2. The mother is sensitized to the Rh antigen and produces anti-Rh antibodies. Because this usually happens after delivery, there is no effect on the fetus in the first pregnancy.

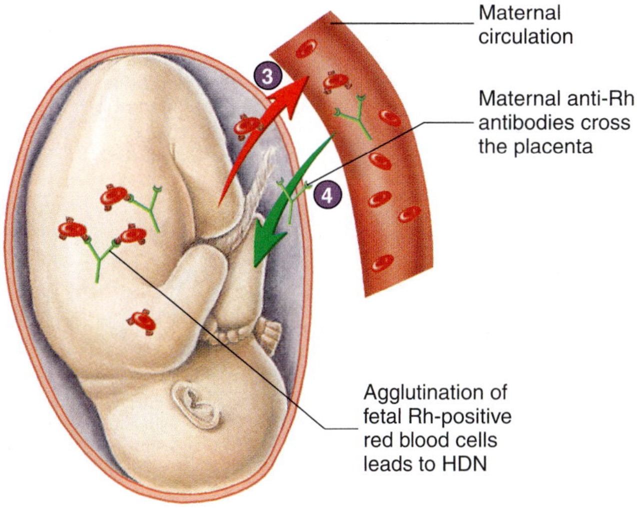

3. During a subsequent pregnancy with an Rh-positive fetus, Rh-positive red blood cells cross the placenta, enter the maternal circulation, and stimulate the mother to produce antibodies against the Rh antigen. Antibody production is rapid because the mother has been sensitized to the Rh antigen.

4. The anti-Rh antibodies from the mother cross the placenta, causing agglutination and hemolysis of fetal red blood cells, and hemolytic disease of the newborn (HDN) develops.

Hemolytic Disease of the Newborn (HDN)
Process Figure 19.15

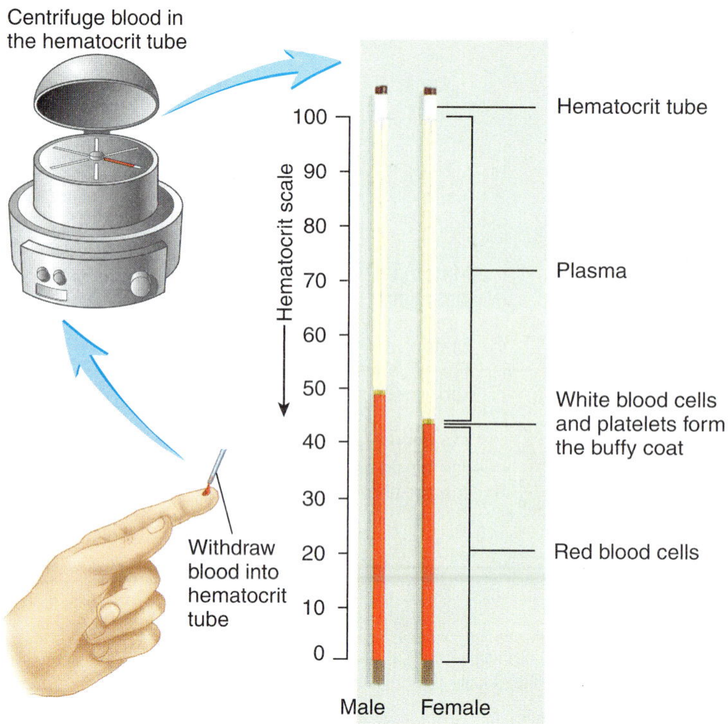

Hematocrit
Figure 19.16

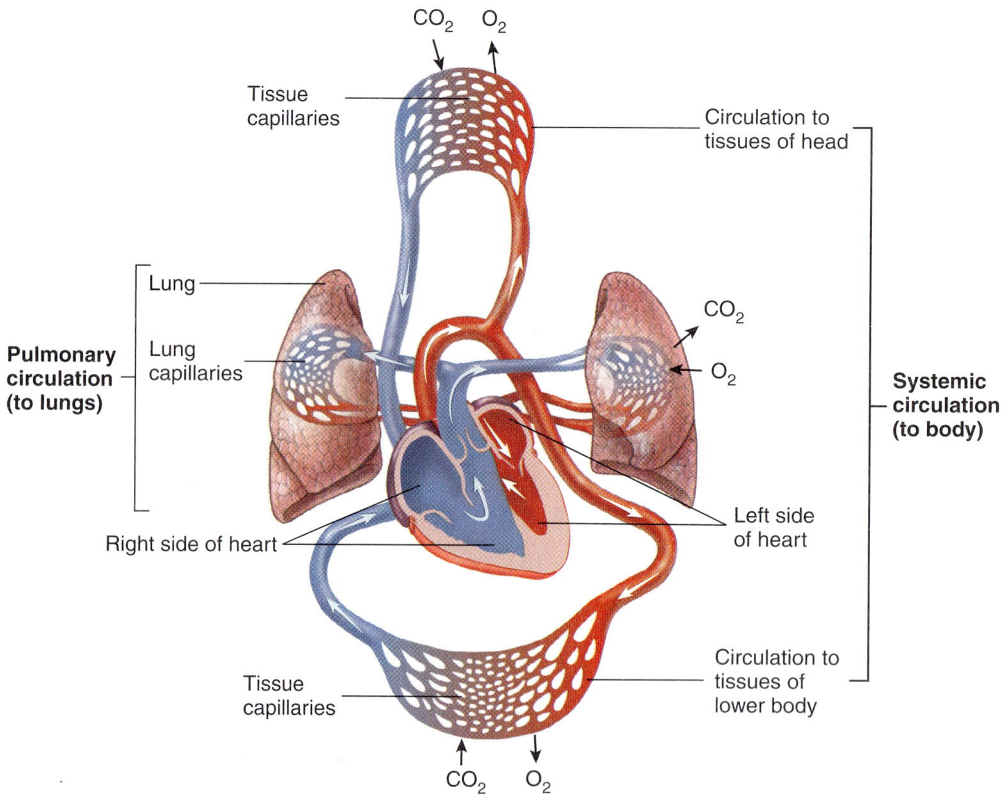

Systemic and Pulmonary Circulation
Figure 20.1

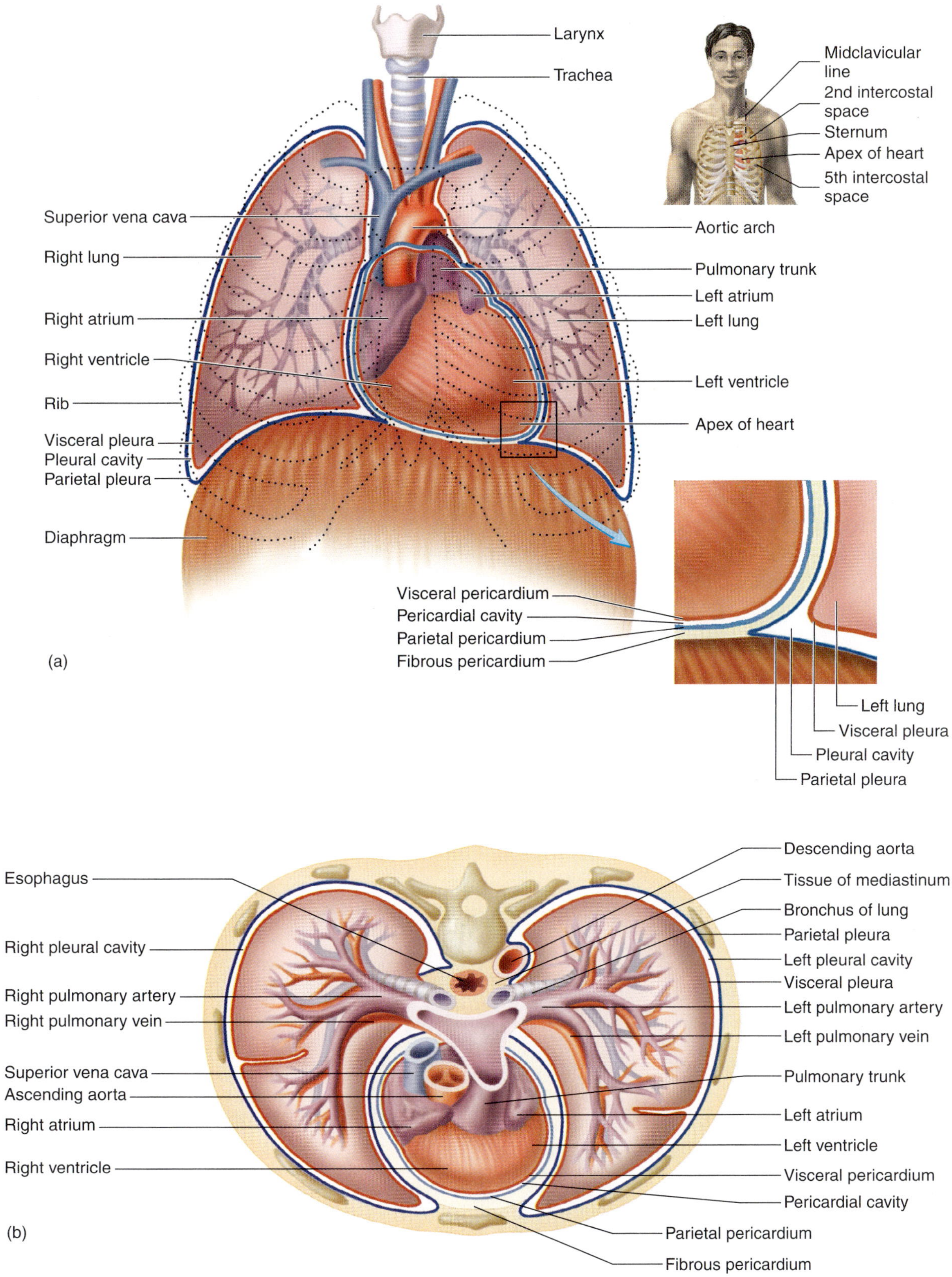
Larynx
Trachea
Midclavicular line
2nd intercostal space
Sternum
Apex of heart
5th intercostal space
Superior vena cava
Right lung
Right atrium
Right ventricle
Rib
Visceral pleura
Pleural cavity
Parietal pleura
Diaphragm
Aortic arch
Pulmonary trunk
Left atrium
Left lung
Left ventricle
Apex of heart
Visceral pericardium
Pericardial cavity
Parietal pericardium
Fibrous pericardium
(a)
Left lung
Visceral pleura
Pleural cavity
Parietal pleura
Esophagus
Right pleural cavity
Right pulmonary artery
Right pulmonary vein
Superior vena cava
Ascending aorta
Right atrium
Right ventricle
(b)
Descending aorta
Tissue of mediastinum
Bronchus of lung
Parietal pleura
Left pleural cavity
Visceral pleura
Left pulmonary artery
Left pulmonary vein
Pulmonary trunk
Left atrium
Left ventricle
Visceral pericardium
Pericardial cavity
Parietal pericardium
Fibrous pericardium

Location of the Heart in the Thorax
Figure 20.2

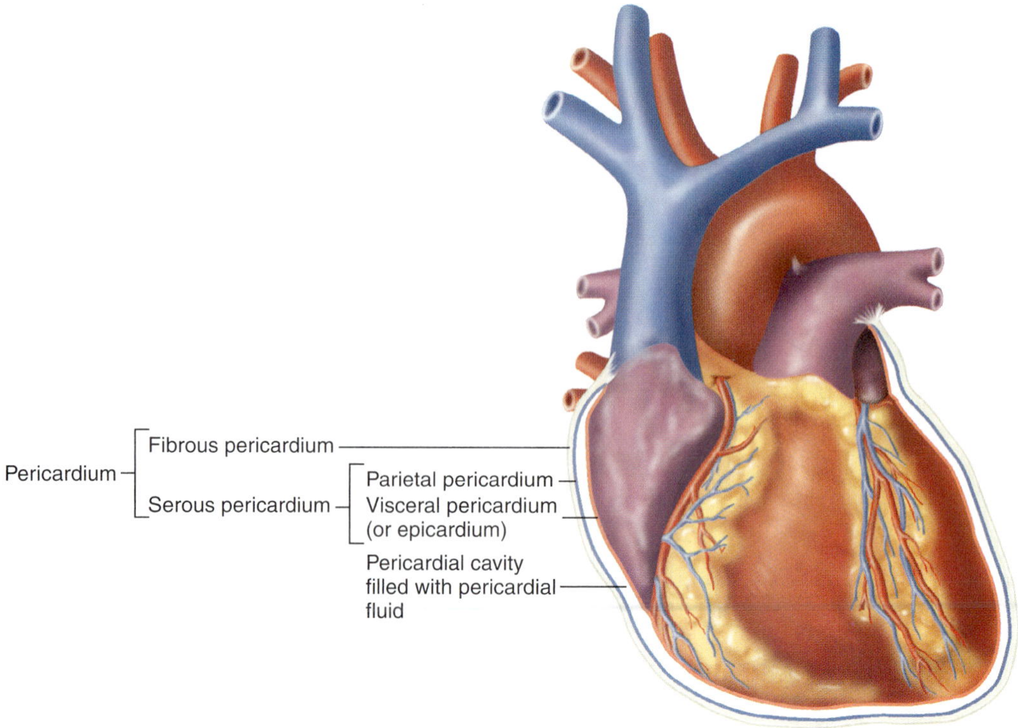

Heart in the Pericardium
Figure 20.3

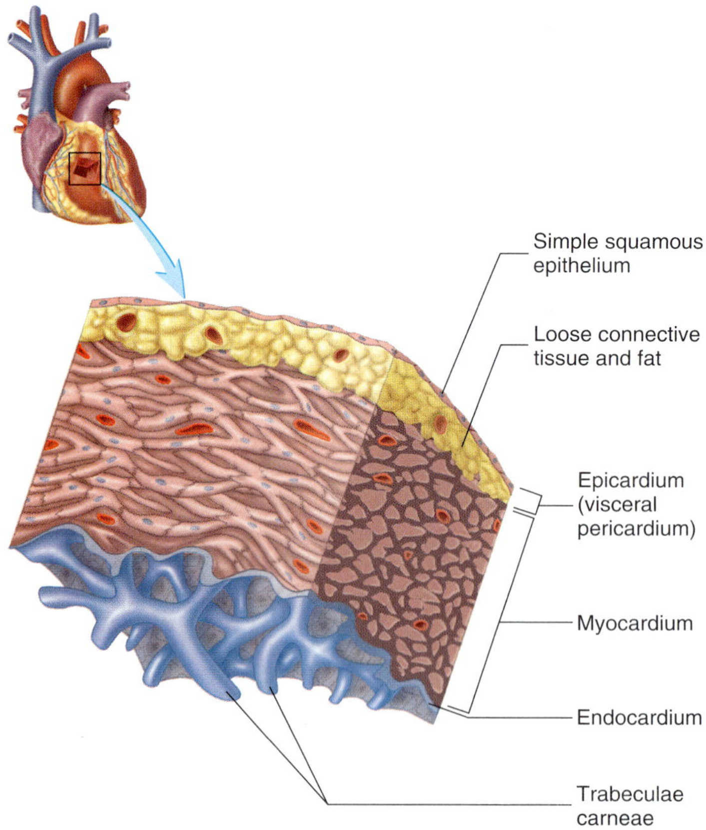

Heart Wall
Figure 20.4

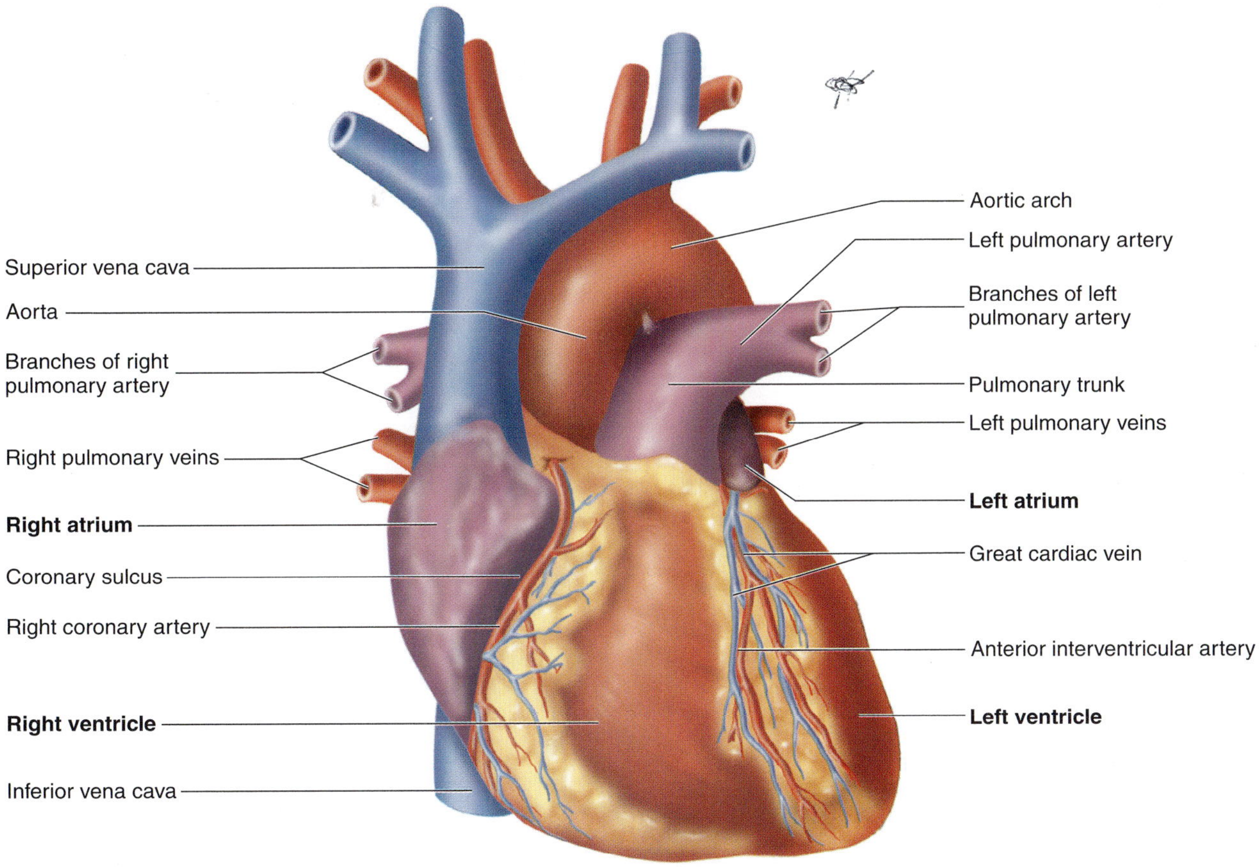

(a) **Anterior view**

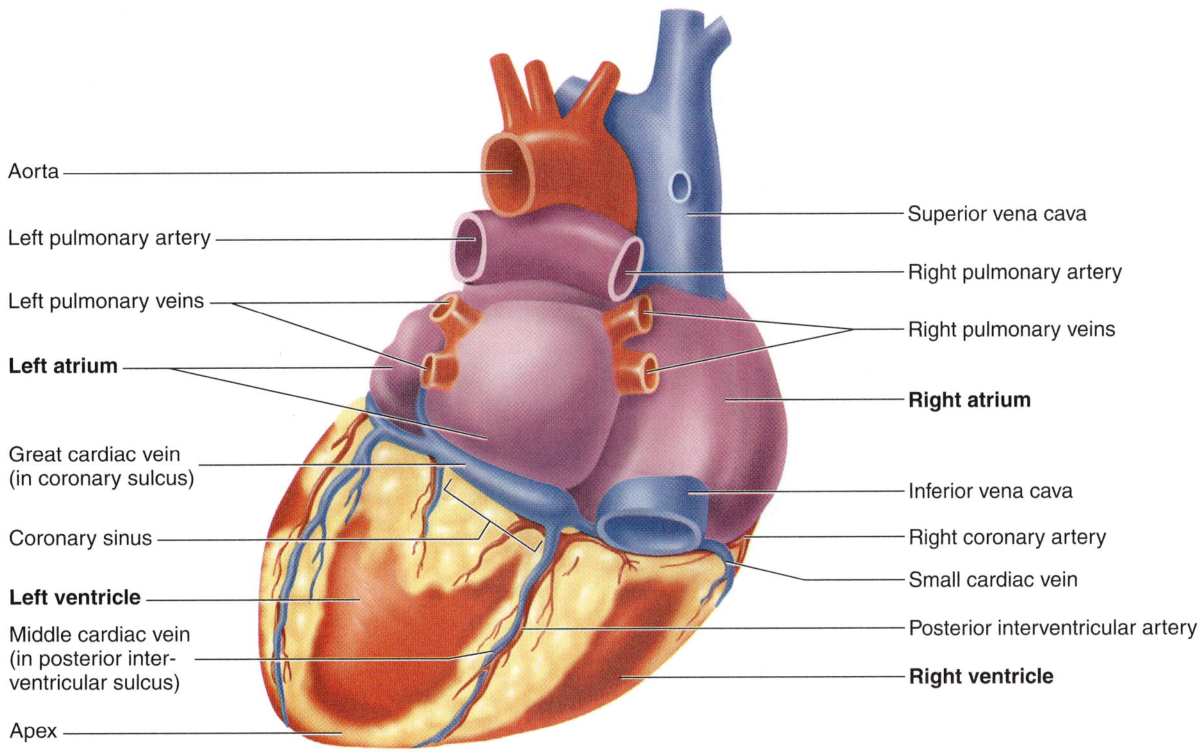

(c) **Posterior view**

Anterior Surface View of the Heart
Figure 20.5

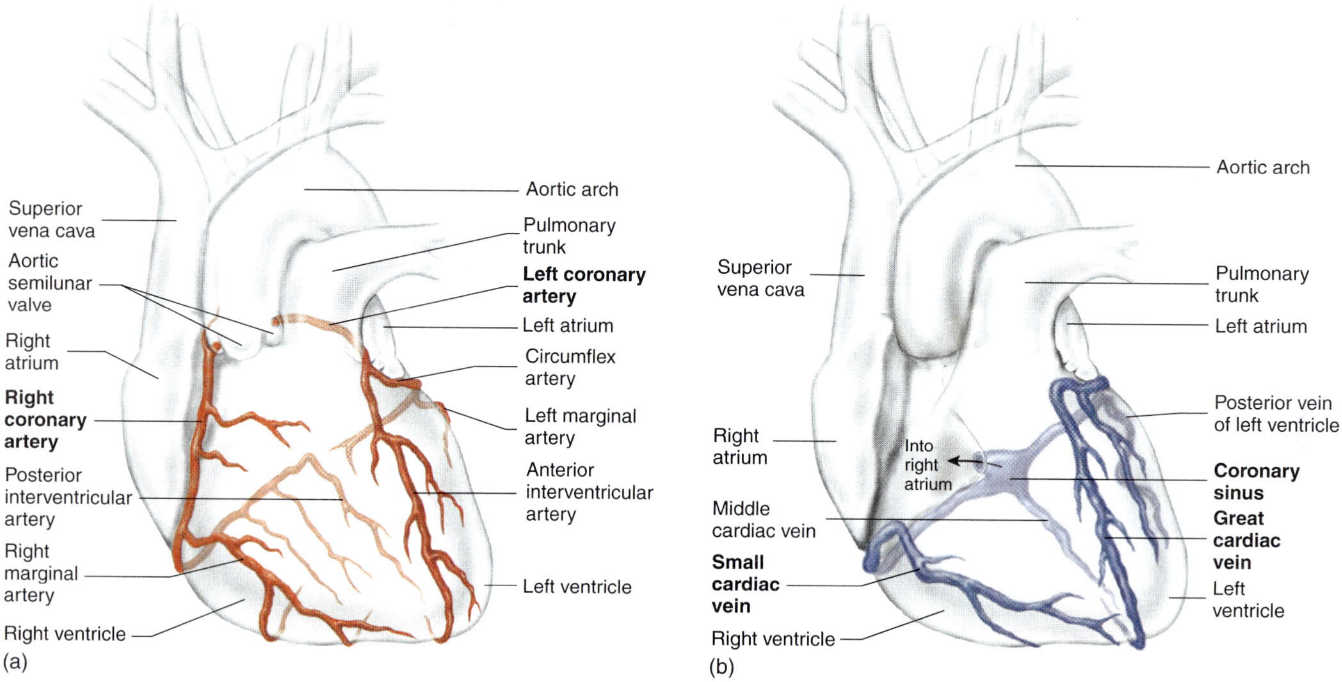
Superior vena cava
Aortic semilunar valve
Right atrium
Right coronary artery
Posterior interventricular artery
Right marginal artery
Right ventricle
(a)
Aortic arch
Pulmonary trunk
Left coronary artery
Left atrium
Circumflex artery
Left marginal artery
Anterior interventricular artery
Left ventricle
Superior vena cava
Right atrium
Into right atrium
Middle cardiac vein
Small cardiac vein
Right ventricle
(b)
Aortic arch
Pulmonary trunk
Left atrium
Posterior vein of left ventricle
Coronary sinus
Great cardiac vein
Left ventricle

Coronary Circulation

Figure 20.6

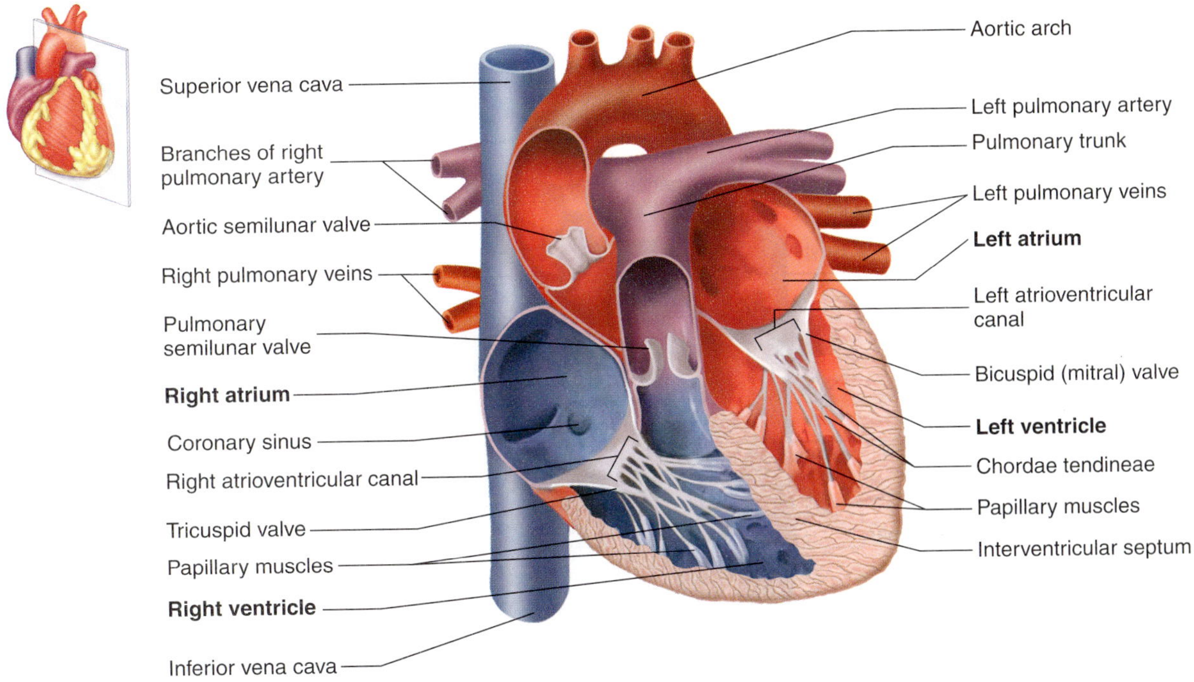

Internal Anatomy of the Heart
Figure 20.7

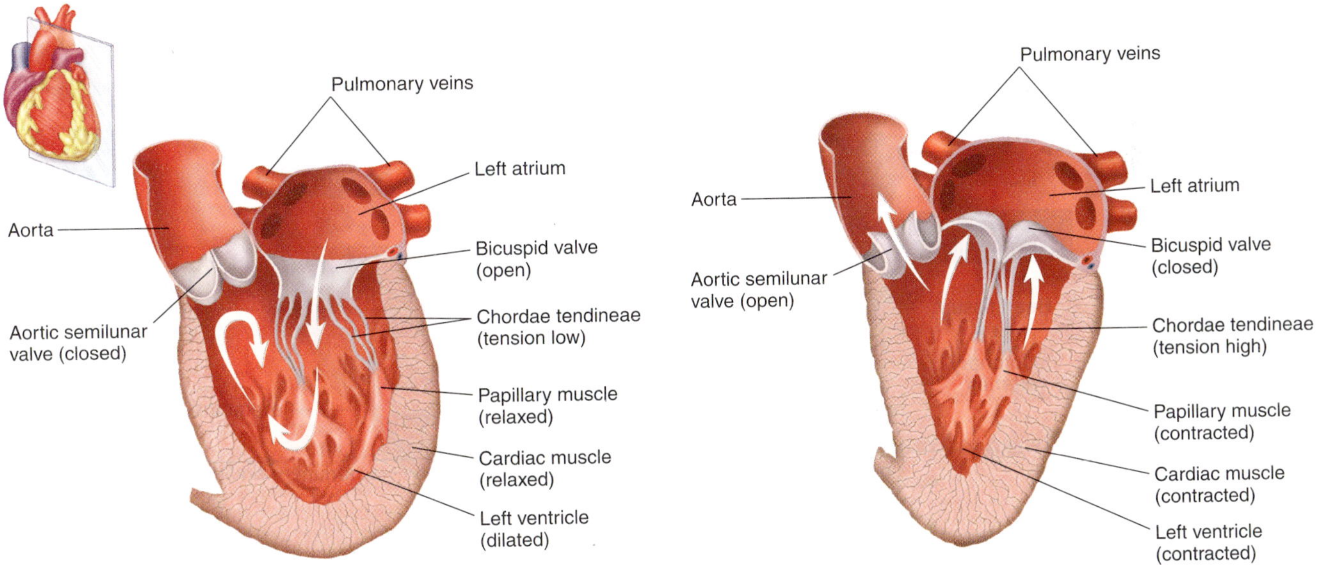

(a) When the bicuspid valve is open, the cusps of the valve are pushed by blood into the ventricle. Papillary muscles are relaxed and tension on the chordae tendineae is low. Blood flows from the left atrium into the left ventricle. When the aortic semilunar valve is closed, the cusps of the valve overlap as they are pushed by the blood in the aorta toward the ventricle. There is no blood flow from the aorta into the ventricle.

(b) When the bicuspid valve is closed, the cusps of the valves overlap as they are pushed by the blood toward the left atrium. There is no blood flow from the ventricle into the atrium. Papillary muscles are contracted and tension on the chordae tendineae is increased. When the aortic semilunar valve is open, the cusps of the valve are pushed by the blood toward the aorta. Blood then flows from the left ventricle into the aorta.

Function of the Heart Valves
Figure 20.9

1. Action potentials originate in the sinoatrial (SA) node (the pacemaker) and travel across the wall of the atrium (*arrows*) from the SA node to the atrioventricular (AV) node.

2. Action potentials pass through the AV node and along the atrioventricular (AV) bundle, which extends from the AV node, through the fibrous skeleton, into the interventricular septum.

3. The AV bundle divides into right and left bundle branches, and action potentials descend to the apex of each ventricle along the bundle branches.

4. Action potentials are carried by the Purkinje fibers from the bundle branches to the ventricular walls.

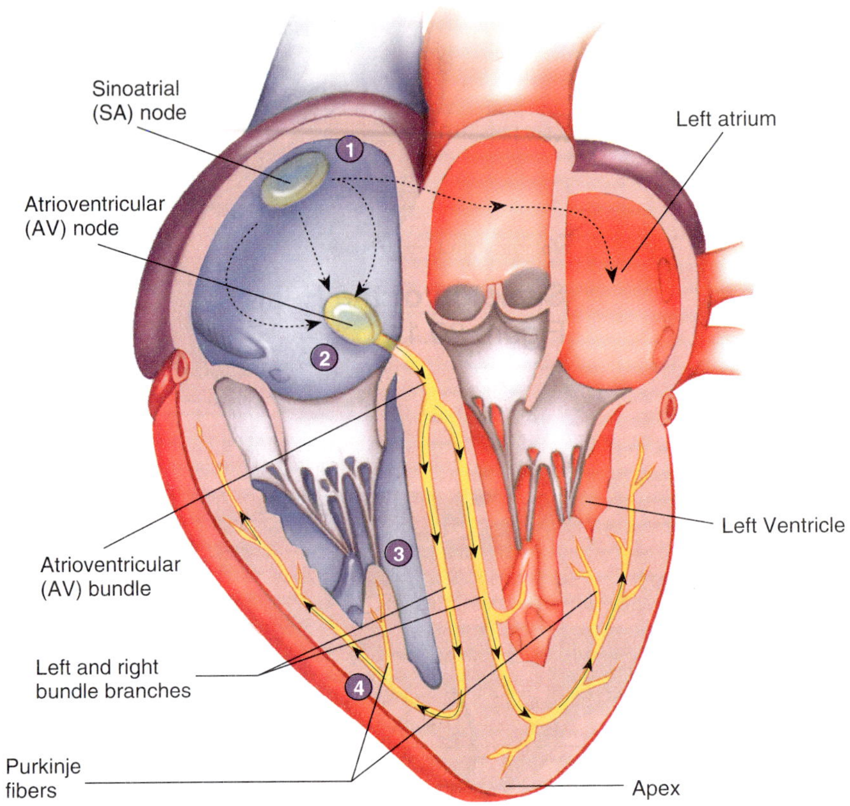

Conducting System of the Heart
Process Figure 20.13

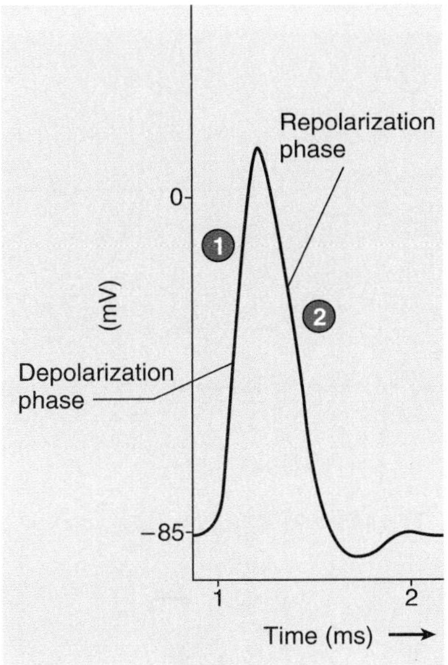

(a)

Permeability changes during an action potential in skeletal muscle:

1. **Depolarization phase**
   - Voltage-gated Na^+ channels open.
   - Voltage-gated K^+ channels begin to open.

2. **Repolarization phase**
   - Voltage-gated Na^+ channels close.
   - Voltage-gated K^+ channels continue to open.
   - Voltage-gated K^+ channels close at the end of repolarization and return the membrane potential to its resting value.

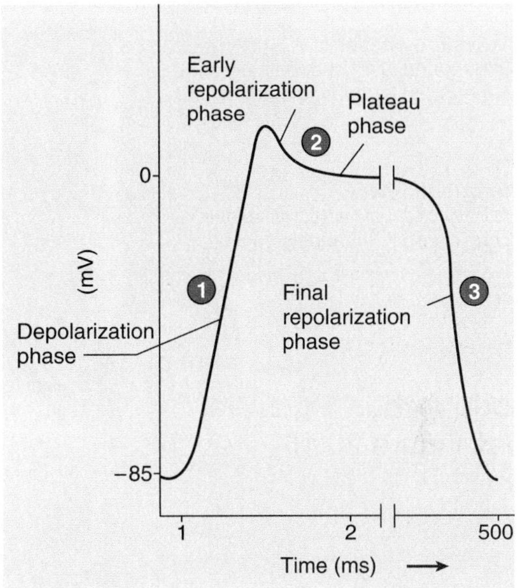

(b)

Permeability changes during an action potential in cardiac muscle:

1. **Depolarization phase**
   - Voltage-gated Na^+ channels open.
   - Voltage-gated K^+ channels close.
   - Voltage-gated Ca^{2+} channels begin to open.

2. **Early repolarization and plateau phases**
   - Voltage-gated Na^+ channels close.
   - Some voltage-gated K^+ channels open, causing early repolarization.
   - Voltage-gated Ca^{2+} channels are open, producing the plateau by slowing further repolarization.

3. **Final repolarization phase**
   - Voltage-gated Ca^{2+} channels close.
   - Many voltage-gated K^+ channels open.

Comparison of Action Potentials in Skeletal and Cardiac Muscle
Process Figure 20.14

Permeability changes in pacemaker cells:

1. **Prepotential**
   - A small number of Na^+ channels are open.
   - Voltage-gated K^+ channels that opened in the repolarization phase of the previous action potential are closing.
   - Voltage-gated Ca^{2+} channels begin to open.

2. **Depolarization phase**
   - Voltage-gated Ca^{2+} channels are open.
   - Voltage-gated K^+ channels are closed.

3. **Repolarization phase**
   - Voltage-gated Ca^{2+} channels close.
   - Voltage-gated K^+ channels open.

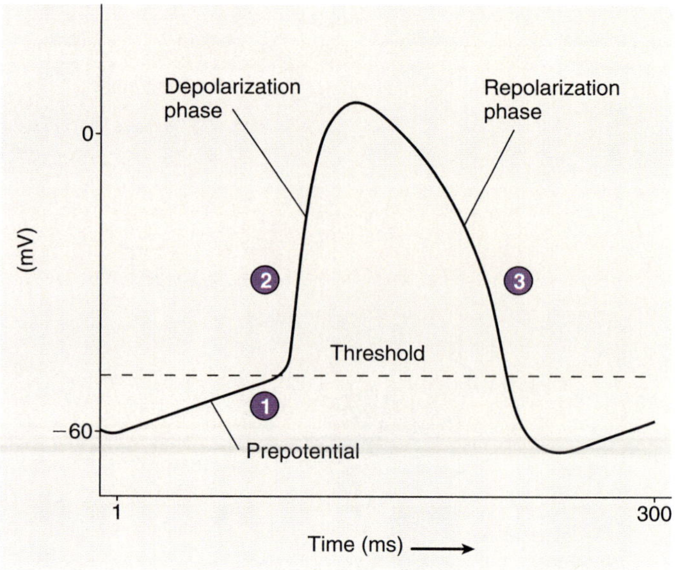

SA Node Action Potential
Process Figure 20.15

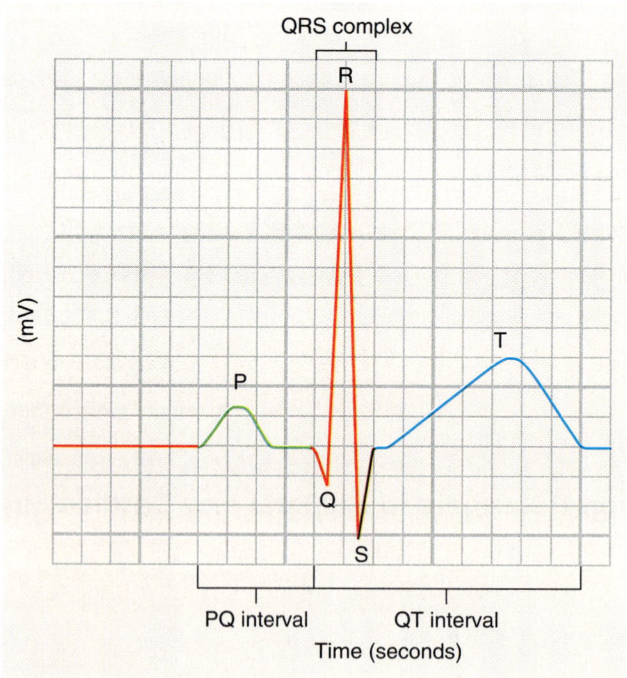

Electrocardiogram
Figure 20.16

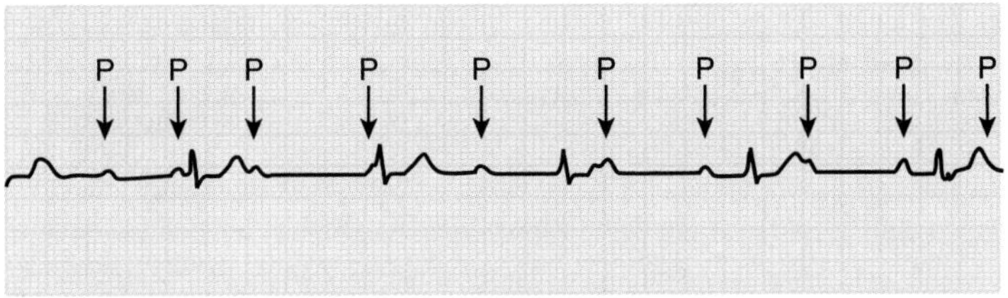

Complete heart block (P waves and QRS complexes are not coordinated)

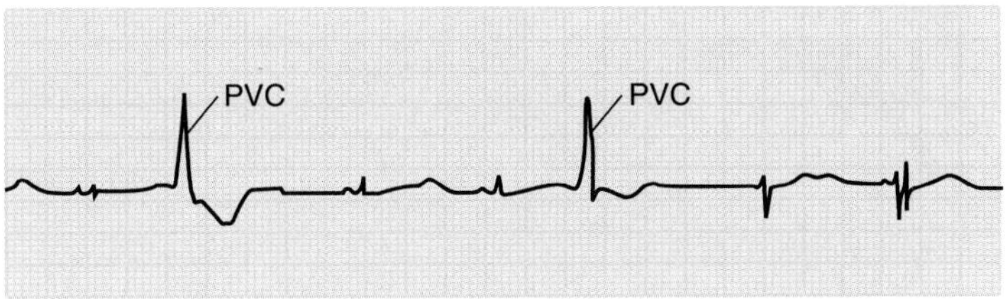

Premature ventricular contraction (PVC) (no P waves precede PVCs)

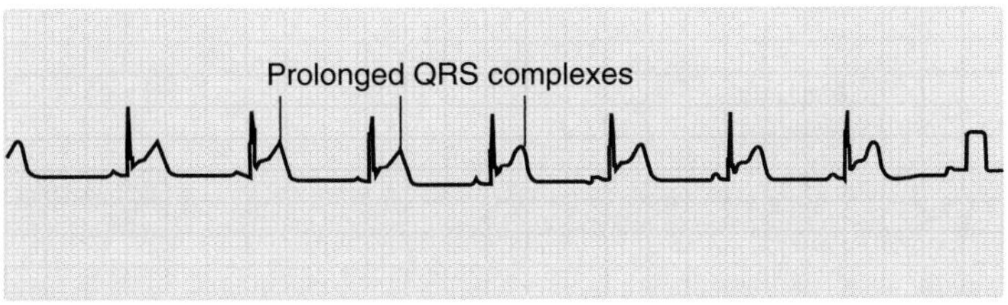

Bundle branch block

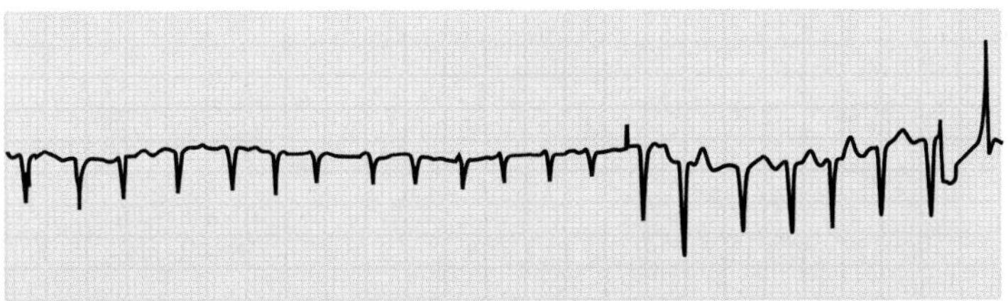

Atrial fibrillation (no clear P waves and rapid QRS complexes)

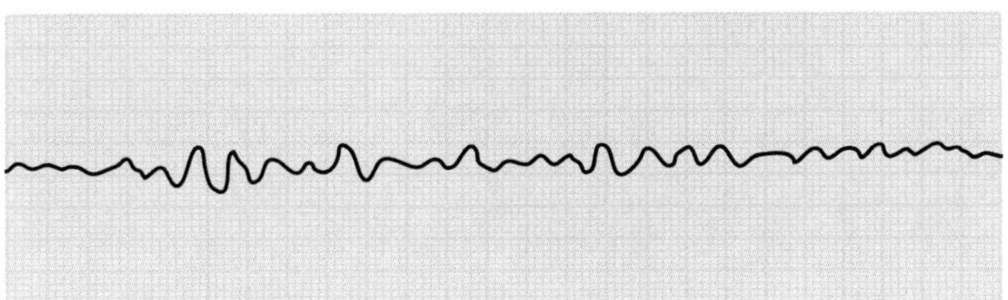

Ventricular fibrillation (no P, QRS, or T waves)

Examples of Alterations in the Electrocardiogram
Figure 20.17

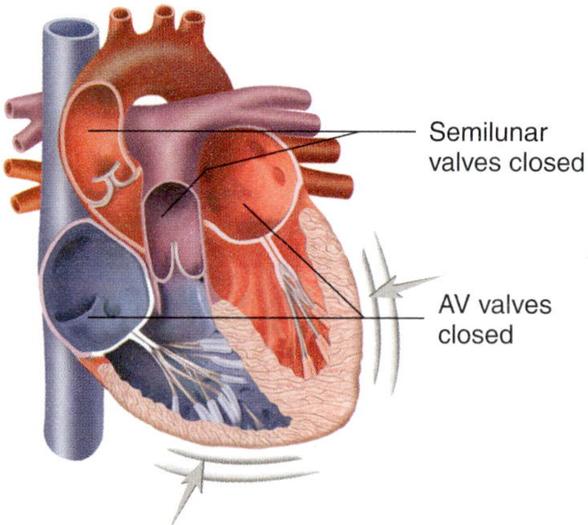

1. *Systole: Period of isovolumic contraction.* Ventricular contraction causes the AV valves to close, which is the beginning of ventricular systole. The semilunar valves were closed in the previous diastole and remain closed during this period.

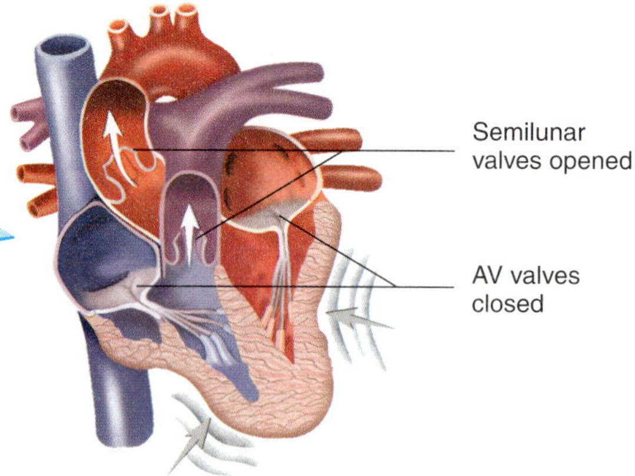

2. *Systole: Period of ejection.* Continued ventricular contraction pushes blood out of the ventricles, causing the semilunar valves to open.

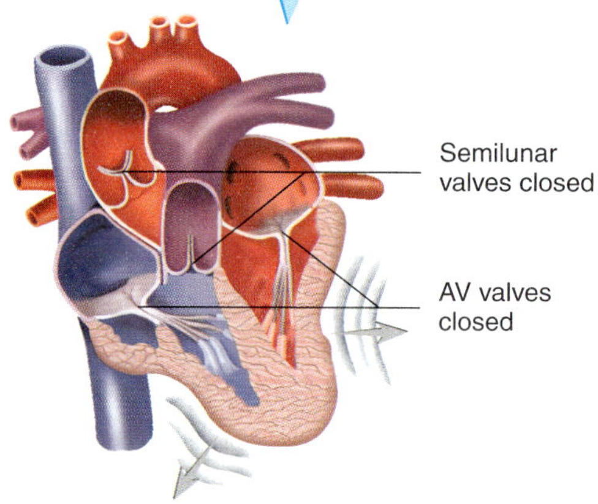

3. *Diastole: Period of isovolumic relaxation.* Blood flowing back toward the relaxed ventricles causes the semilunar valves to close, which is the beginning of ventricular diastole. Note that the AV valves are closed, also.

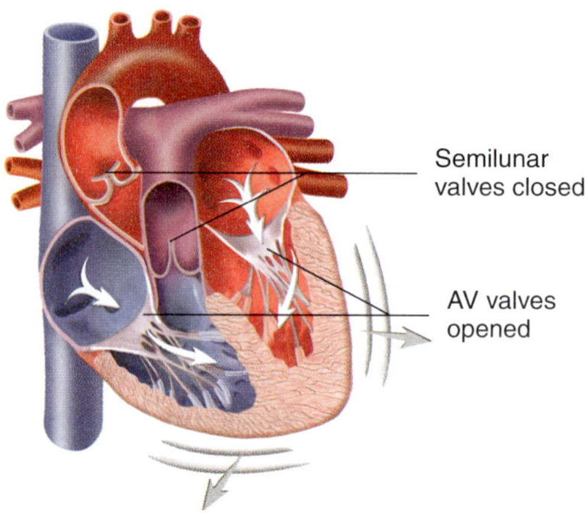

4. *Diastole: Passive ventricular filling.* The AV valves open and blood flows into the relaxed ventricles, accounting for most of the ventricular filling.

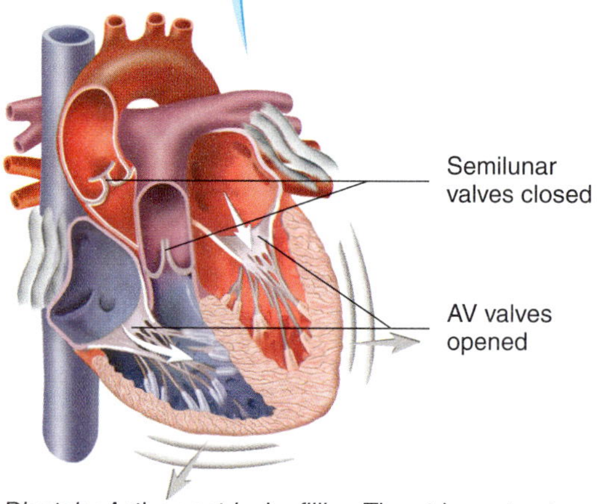

5. *Diastole: Active ventricular filling.* The atria contract and complete ventricular filling.

The Cardiac Cycle
Process Figure 20.18

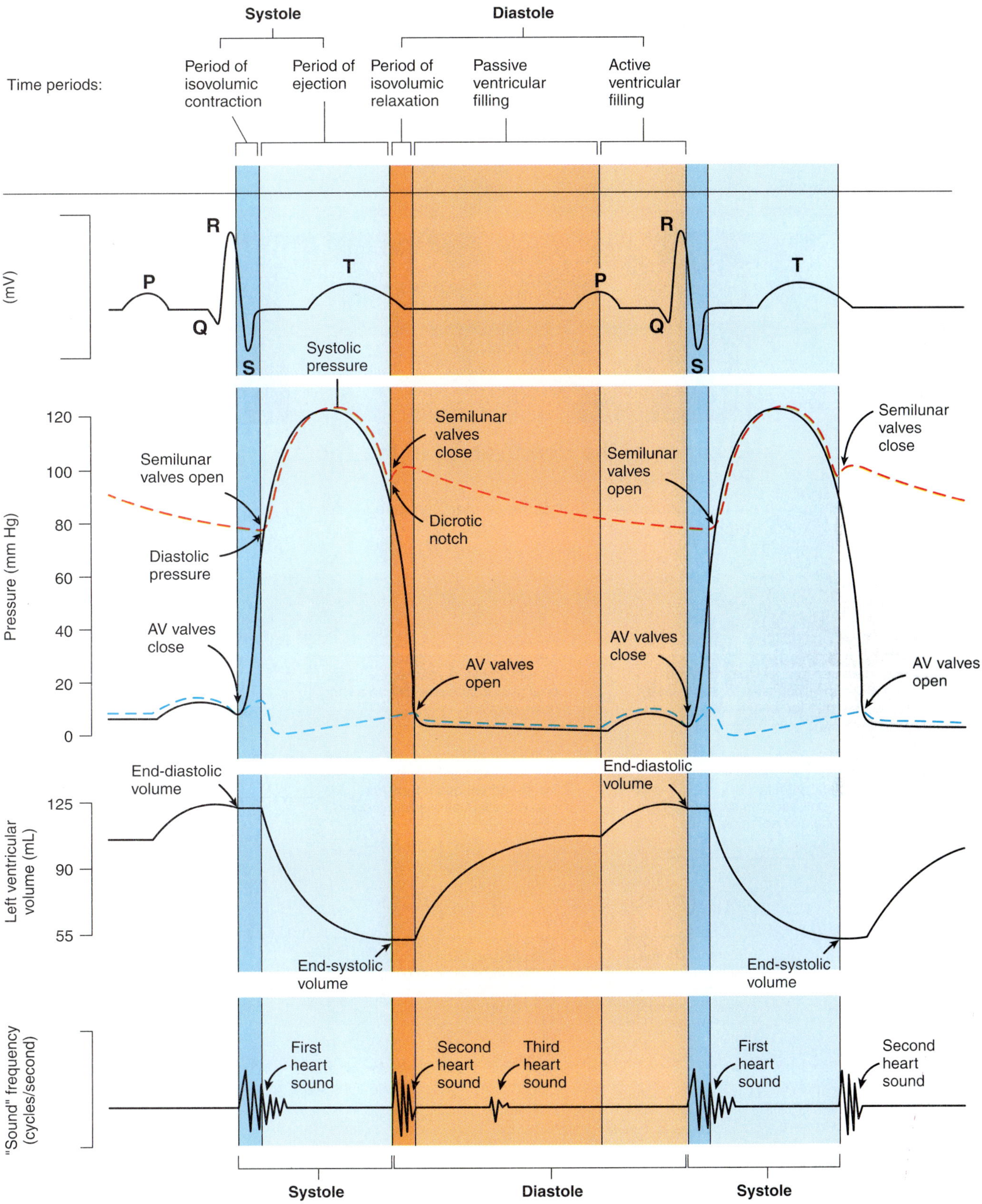

Events Occurring During the Cardiac Cycle
Figure 20.19

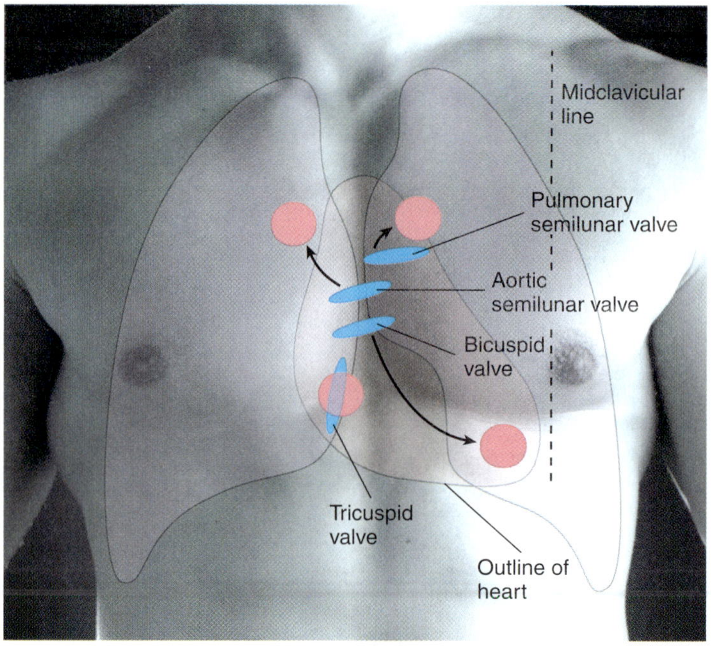

Location of the Heart Valves in the Thorax
Figure 20.20
© Terry Cockerham/Cynthia Alexander/Synapse Media Productions

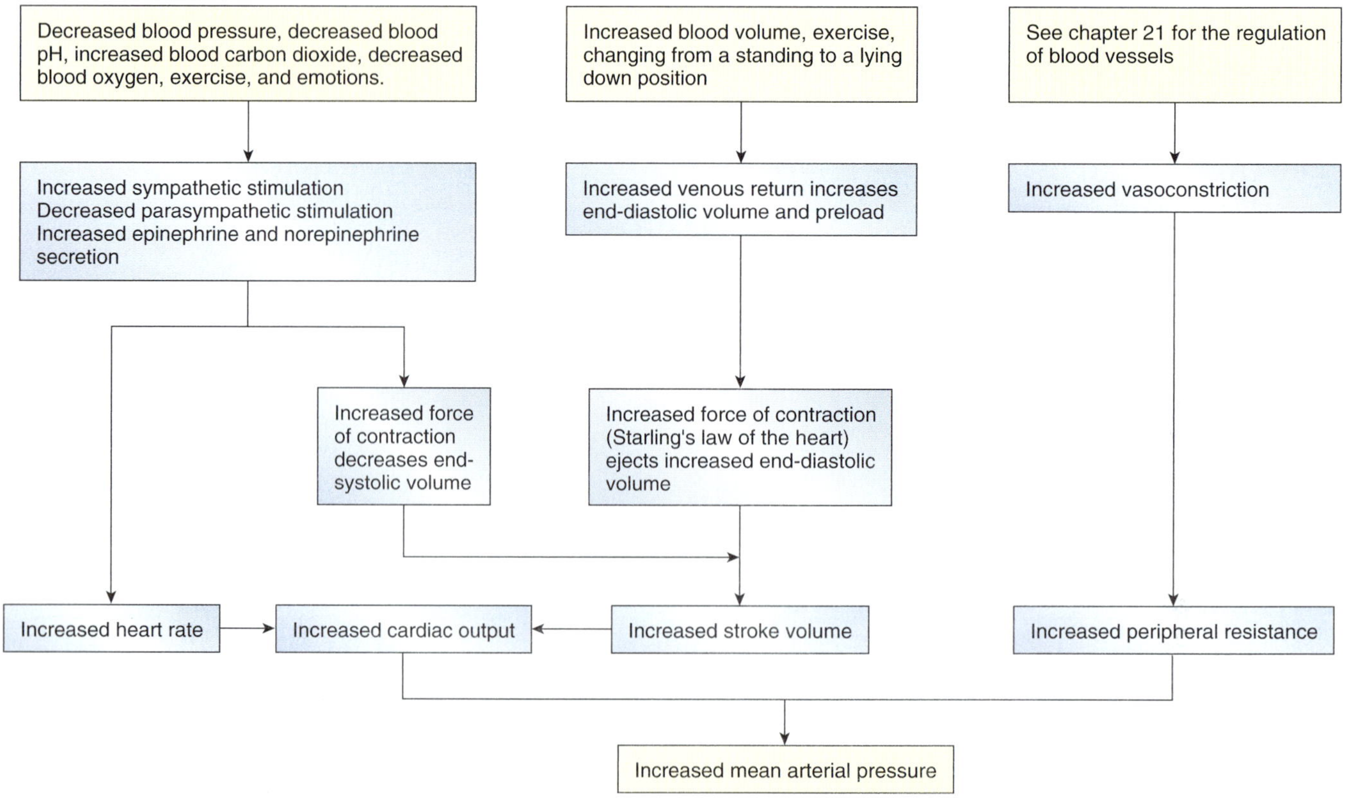

Factors Affecting Mean Arterial Pressure
Figure 20.21

1. Sensory (*green*) neurons carry action potentials from baroreceptors to the cardioregulatory center. Chemoreceptors in the medulla oblongata influence the cardioregulatory center.

2. The cardioregulatory center controls the frequency of action potentials in the parasympathetic (*red*) neurons extending to the heart through the vagus nerves. The parasympathetic neurons decrease the heart rate.

3. The cardioregulatory center controls the frequency of action potential in the sympathetic (*blue*) neurons. The sympathetic neurons extend through the cardiac nerves and increase the heart rate and the stroke volume.

4. The cardioregulatory center influences the frequency of action potentials in the sympathetic (*blue*) neurons extending to the adrenal medulla. The sympathetic neurons increase the secretion of epinephrine and some norepinephrine into the general circulation. Epinephrine and norepinephrine increase the heart rate and stroke volume.

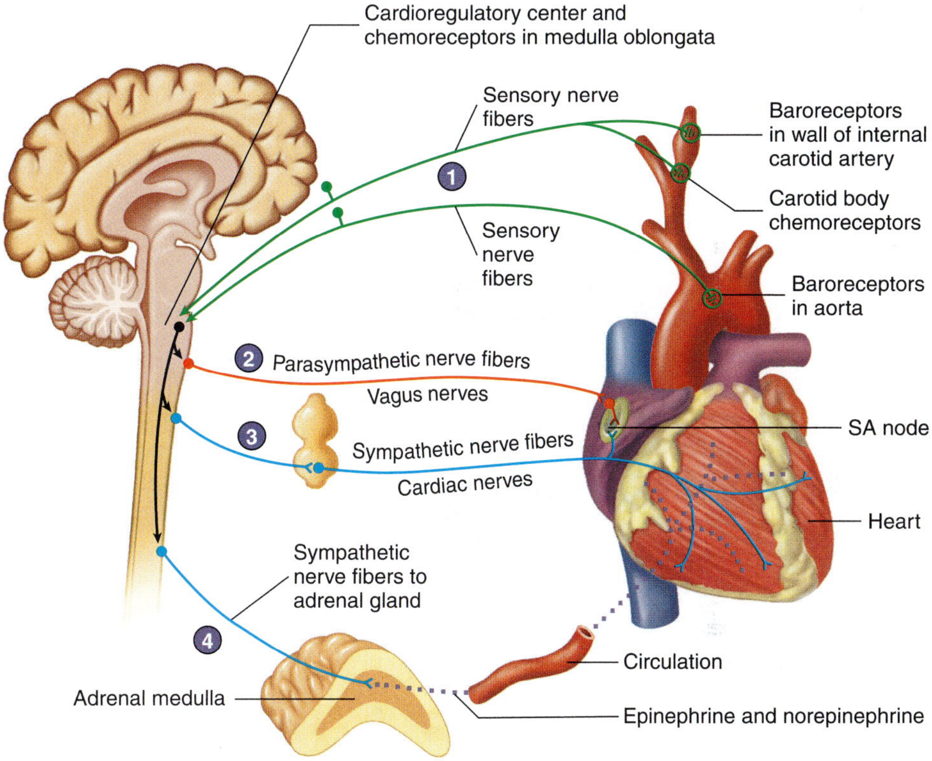

Baroreceptor and Chemoreceptor Reflexes

Process Figure 20.22

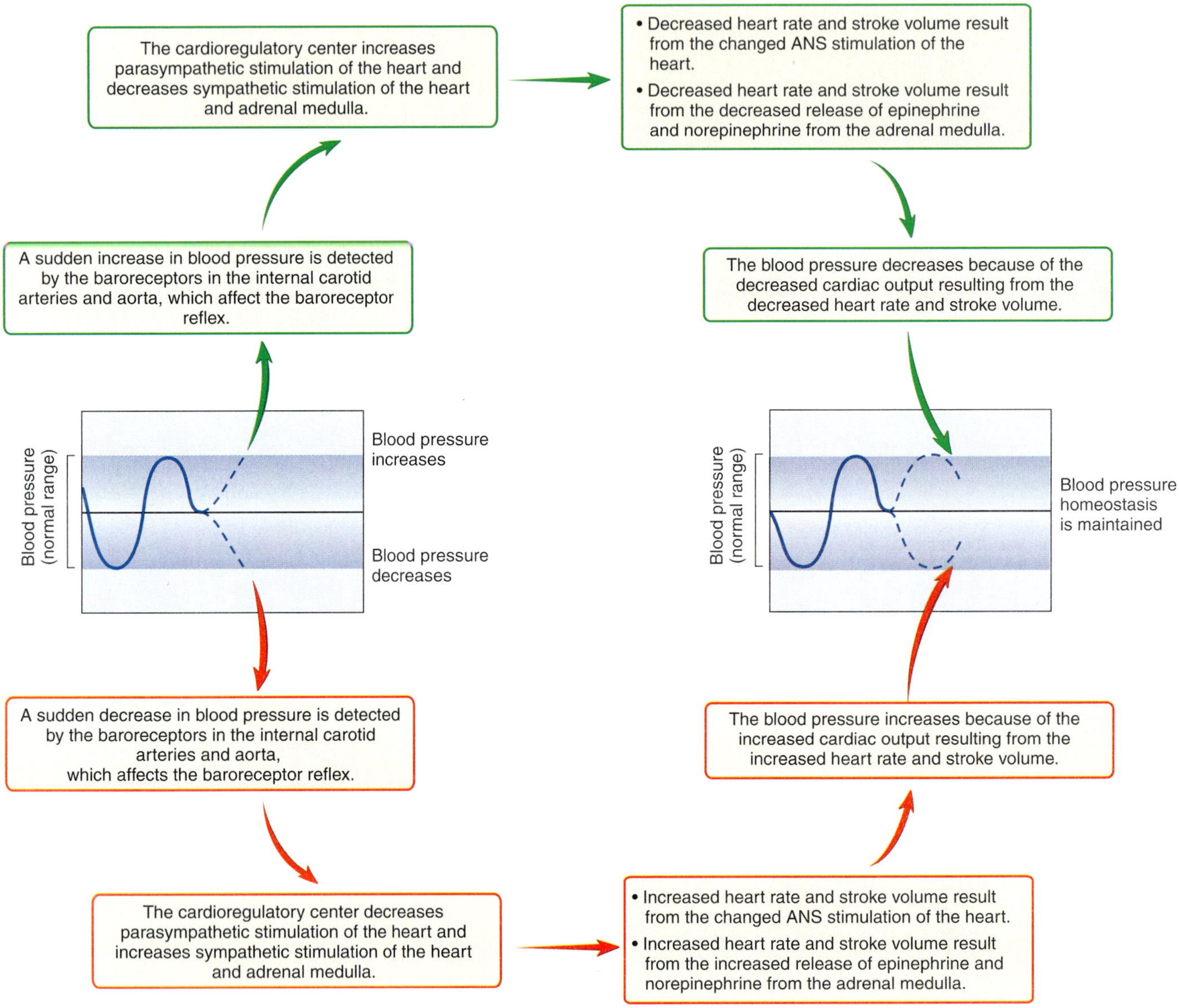

The cardioregulatory center increases parasympathetic stimulation of the heart and decreases sympathetic stimulation of the heart and adrenal medulla.
• Decreased heart rate and stroke volume result from the changed ANS stimulation of the heart.
• Decreased heart rate and stroke volume result from the decreased release of epinephrine and norepinephrine from the adrenal medulla.
A sudden increase in blood pressure is detected by the baroreceptors in the internal carotid arteries and aorta, which affect the baroreceptor reflex.
The blood pressure decreases because of the decreased cardiac output resulting from the decreased heart rate and stroke volume.
Blood pressure (normal range)
Blood pressure increases
Blood pressure decreases
Blood pressure (normal range)
Blood pressure homeostasis is maintained
A sudden decrease in blood pressure is detected by the baroreceptors in the internal carotid arteries and aorta, which affects the baroreceptor reflex.
The blood pressure increases because of the increased cardiac output resulting from the increased heart rate and stroke volume.
The cardioregulatory center decreases parasympathetic stimulation of the heart and increases sympathetic stimulation of the heart and adrenal medulla.
• Increased heart rate and stroke volume result from the changed ANS stimulation of the heart.
• Increased heart rate and stroke volume result from the increased release of epinephrine and norepinephrine from the adrenal medulla.

Baroreceptor Reflex
Homeostasis Figure 20.23

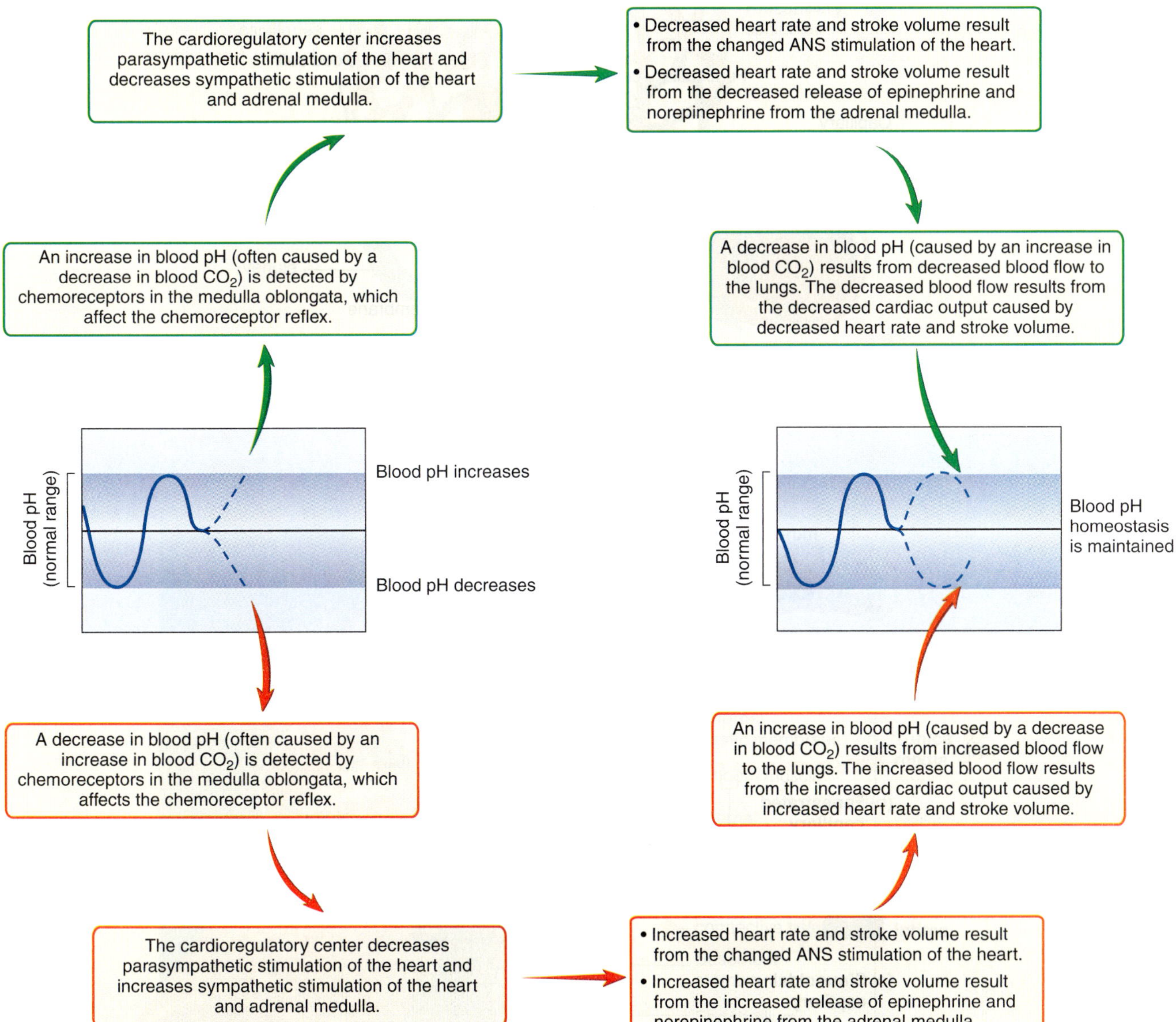

Chemoreceptor Reflex-pH
Homeostasis Figure 20.24

(a) **Elastic arteries.** The tunica media is mostly elastic connective tissue. Elastic arteries recoil when stretched, which prevents blood pressure from falling rapidly.

(b) **Muscular arteries.** The tunica media is a thick layer of smooth muscle. Muscular arteries regulate blood flow to different regions of the body.

(c) **Medium and large veins.** All three tunics are present. The tunica media is thin, but can regulate vessel diameter because blood pressure in the venous system is low. The predominant layer is the tunica adventitia.

(d) Folds in the endothelium form the valves of veins, which allow blood to flow toward the heart but not in the opposite direction.

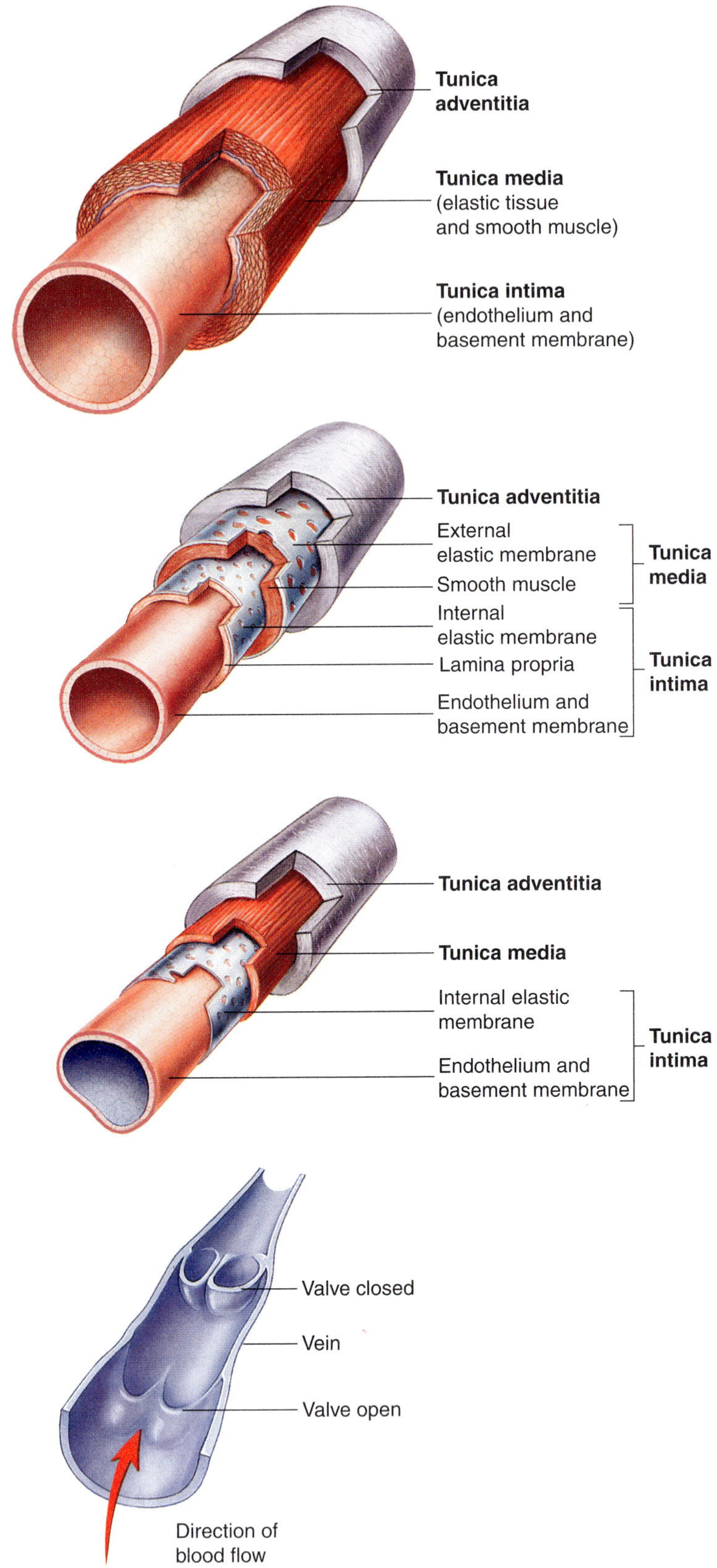

Structural Comparison of Blood Vessel Types
Figure 21.6

Mic

Par
ren
bra
mic

Pitu
(att

Pos
art

Ba

Ver

Par
rem
bra
pos

Th
Fi

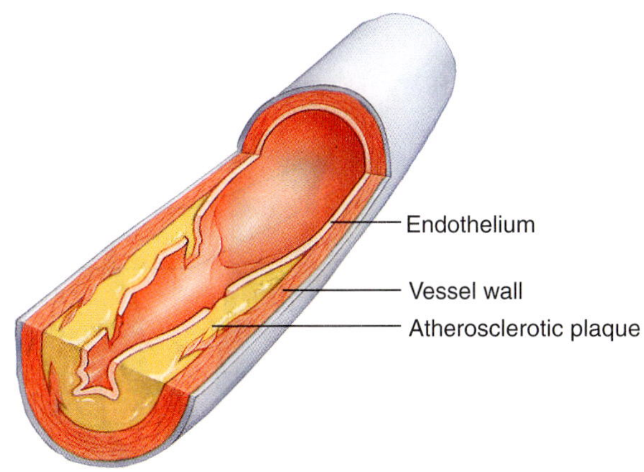

Atherosclerotic Plaque in an Artery
Figure 21.7

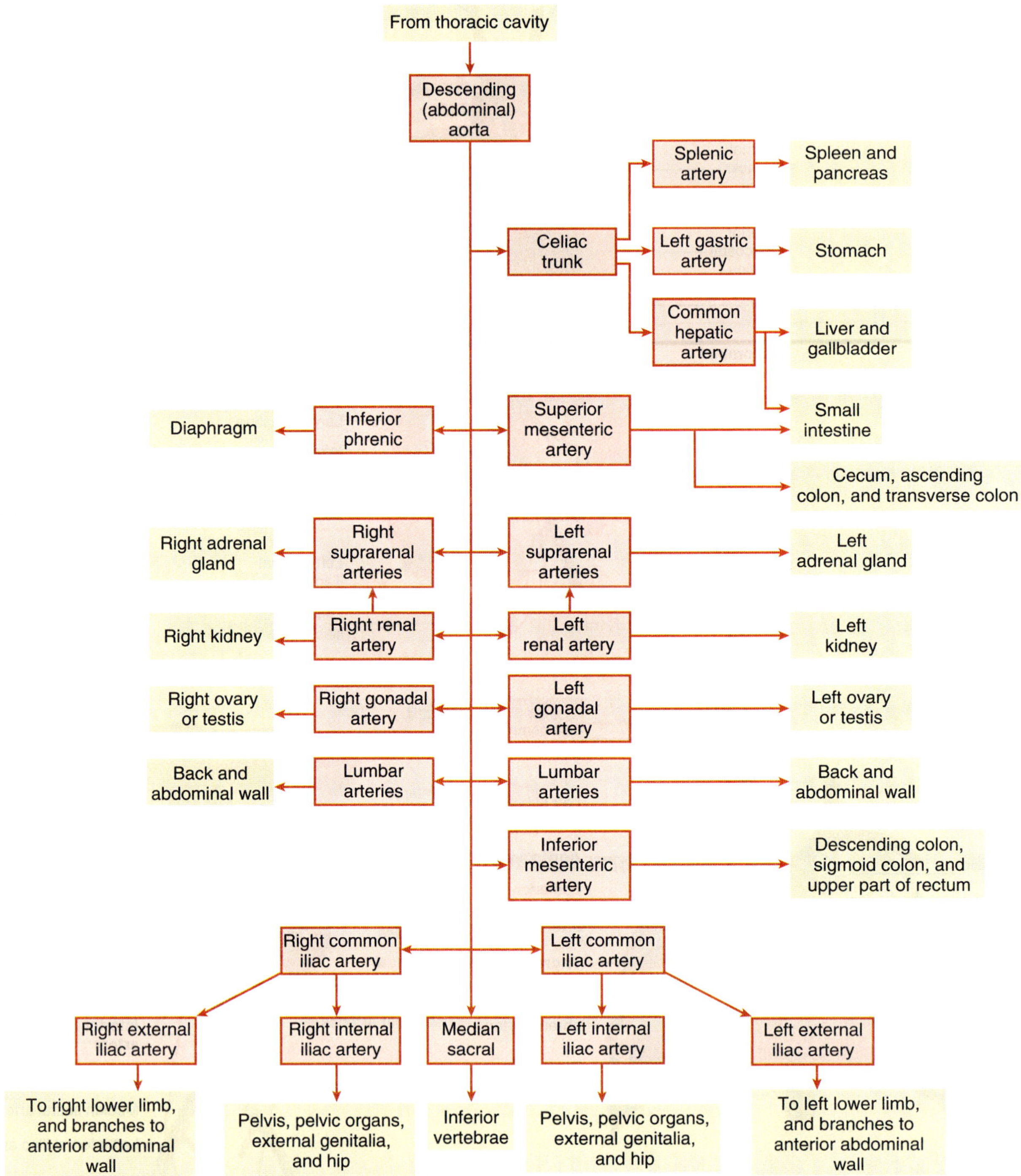

Major Arteries of the Abdomen and Pelvis
Figure 21.15

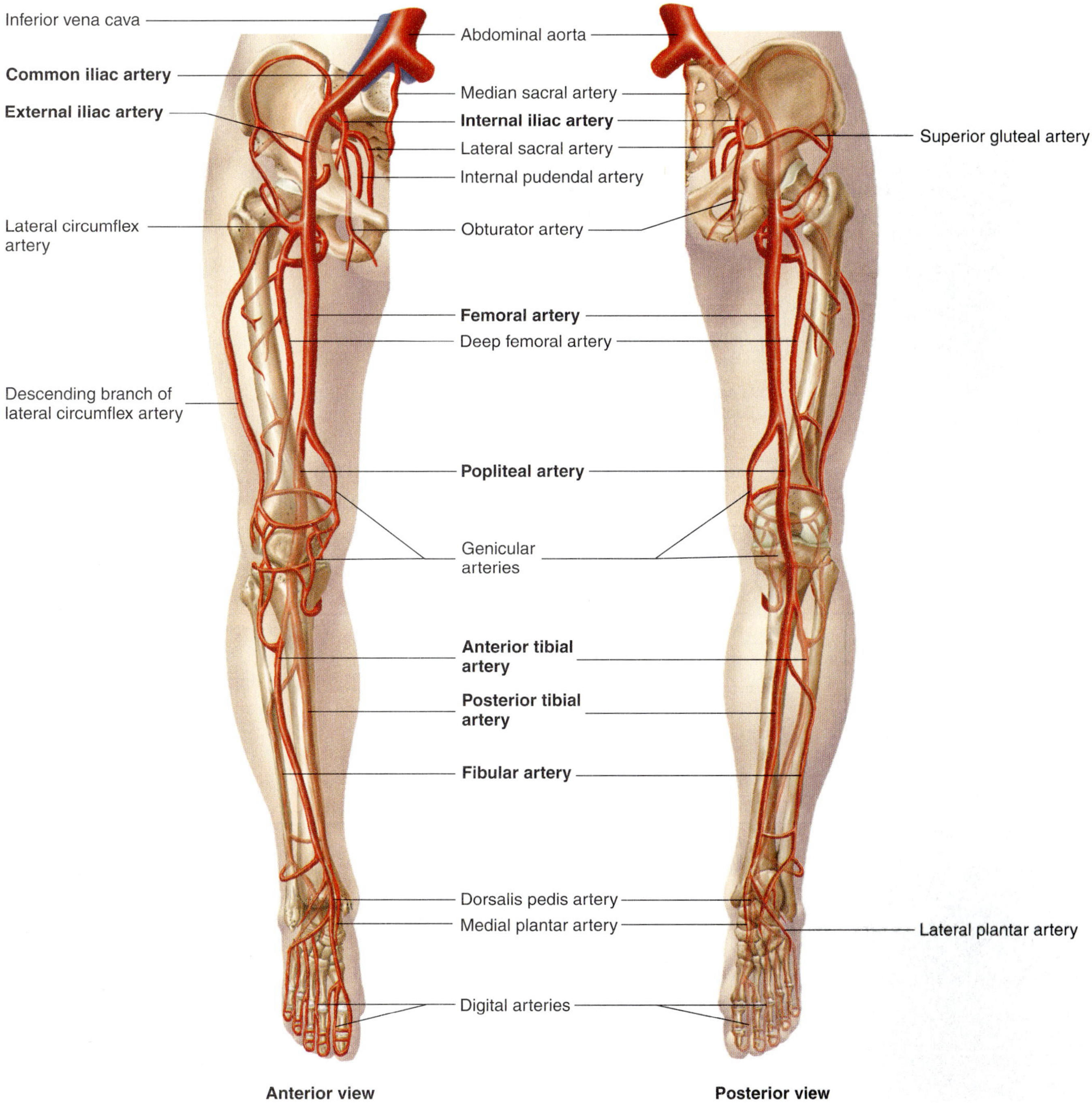

Arteries of the Pelvis and Lower Limb
Figure 21.16

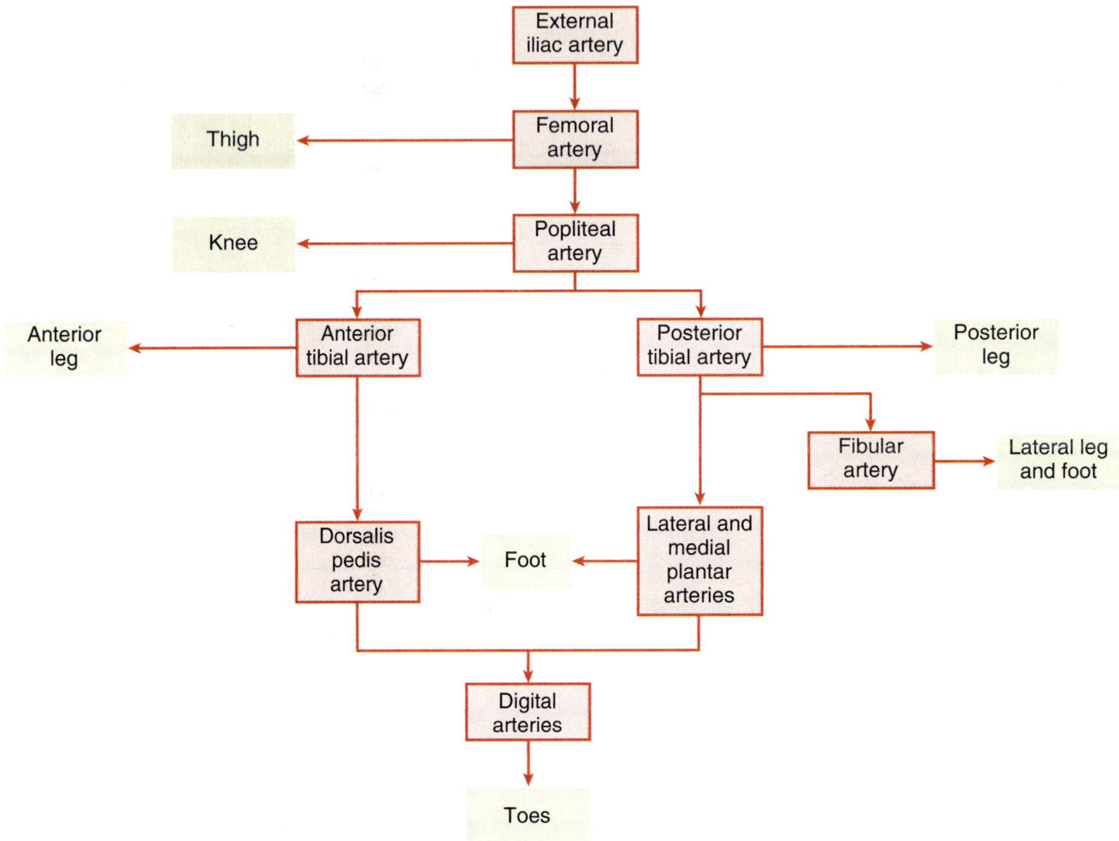
External
iliac artery
Thigh
Femoral
artery
Knee
Popliteal
artery
Anterior
leg
Anterior
tibial artery
Posterior
tibial artery
Posterior
leg
Fibular
artery
Lateral leg
and foot
Dorsalis
pedis
artery
Foot
Lateral and
medial
plantar
arteries
Digital
arteries
Toes

Major Arteries of the Lower Limb
Figure 21.17

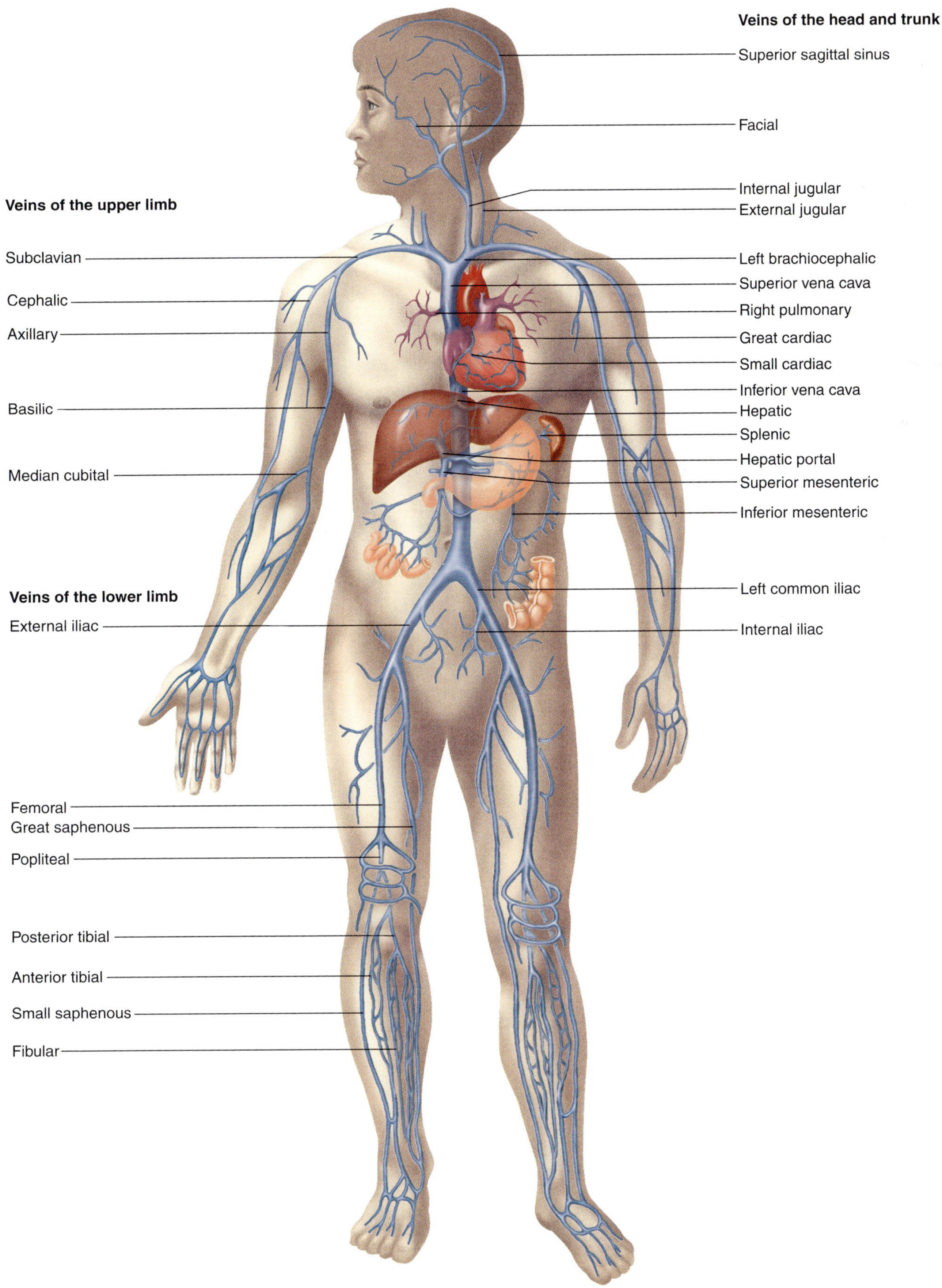
Veins of the head and trunk
Superior sagittal sinus
Facial
Internal jugular
External jugular
Left brachiocephalic
Superior vena cava
Right pulmonary
Great cardiac
Small cardiac
Inferior vena cava
Hepatic
Splenic
Hepatic portal
Superior mesenteric
Inferior mesenteric
Left common iliac
Internal iliac
Veins of the upper limb
Subclavian
Cephalic
Axillary
Basilic
Median cubital
Veins of the lower limb
External iliac
Femoral
Great saphenous
Popliteal
Posterior tibial
Anterior tibial
Small saphenous
Fibular

The Major Veins
Figure 21.18

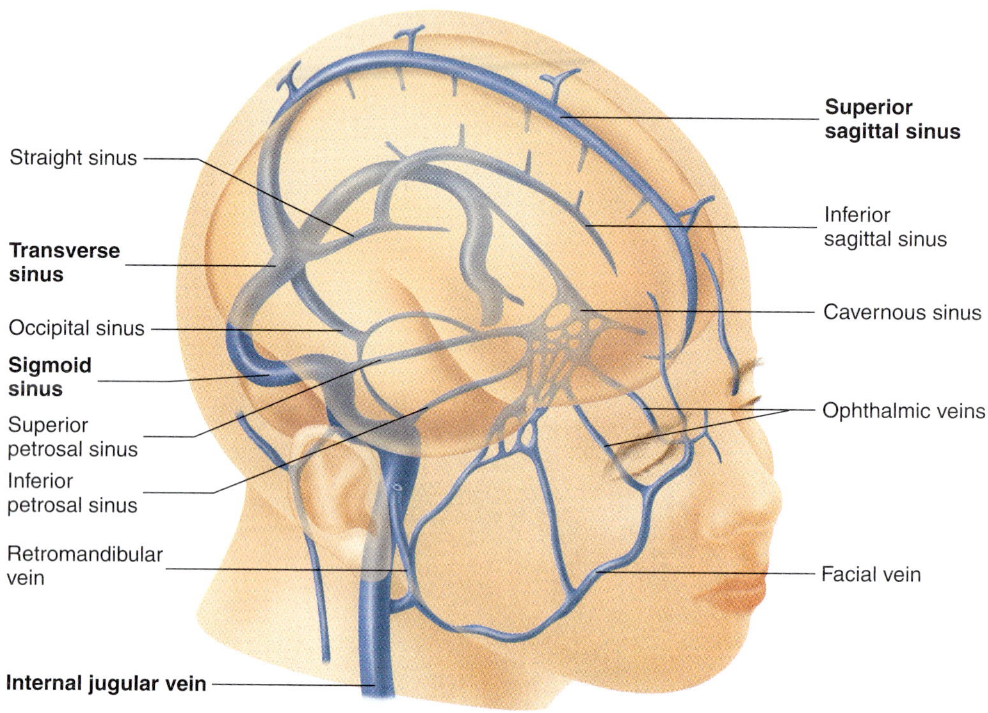

Venous Sinuses Associated with the Brain
Figure 21.19

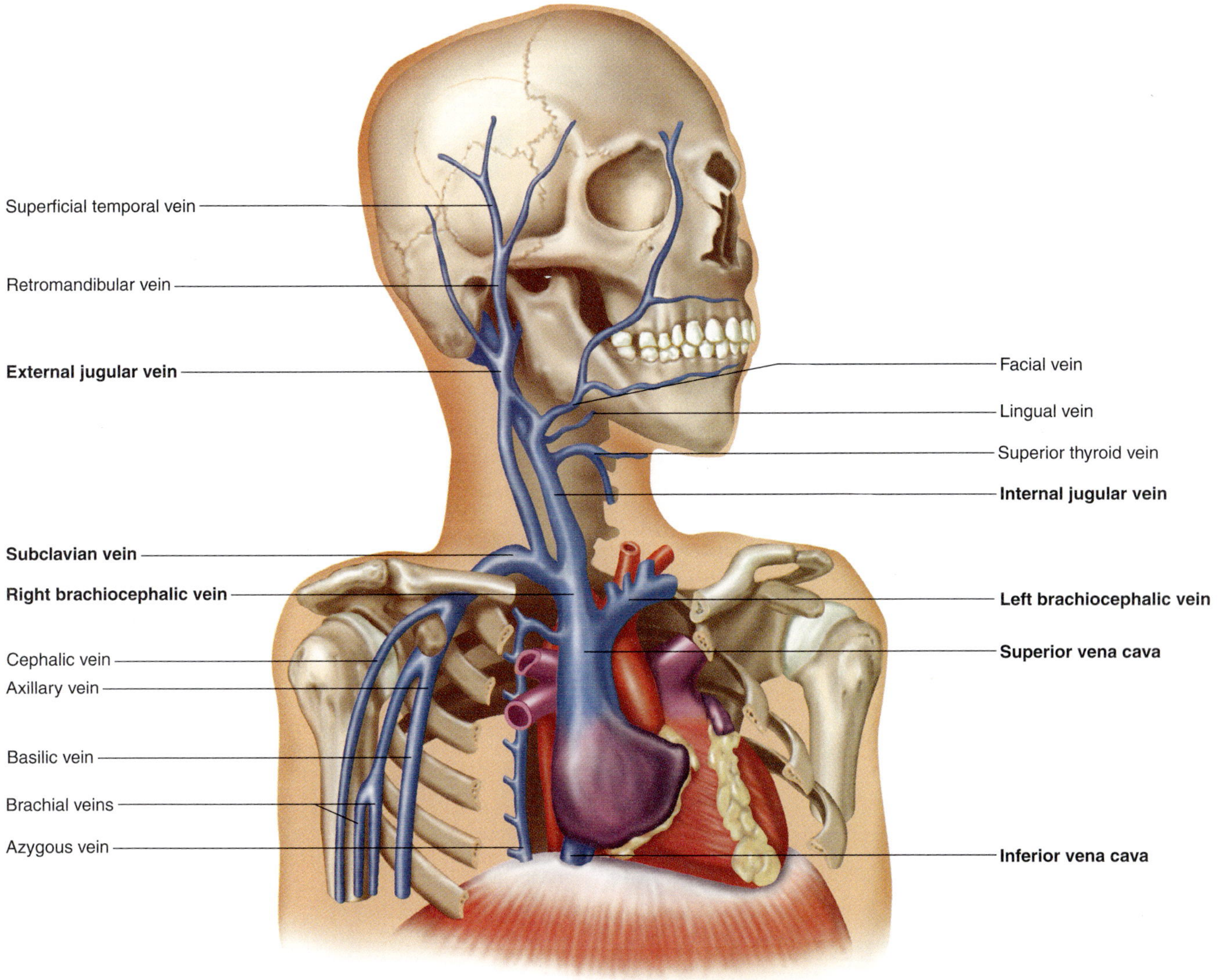
Superficial temporal vein
Retromandibular vein
External jugular vein
Subclavian vein
Right brachiocephalic vein
Cephalic vein
Axillary vein
Basilic vein
Brachial veins
Azygous vein
Facial vein
Lingual vein
Superior thyroid vein
Internal jugular vein
Left brachiocephalic vein
Superior vena cava
Inferior vena cava

Veins of the Head and Neck
Figure 21.20

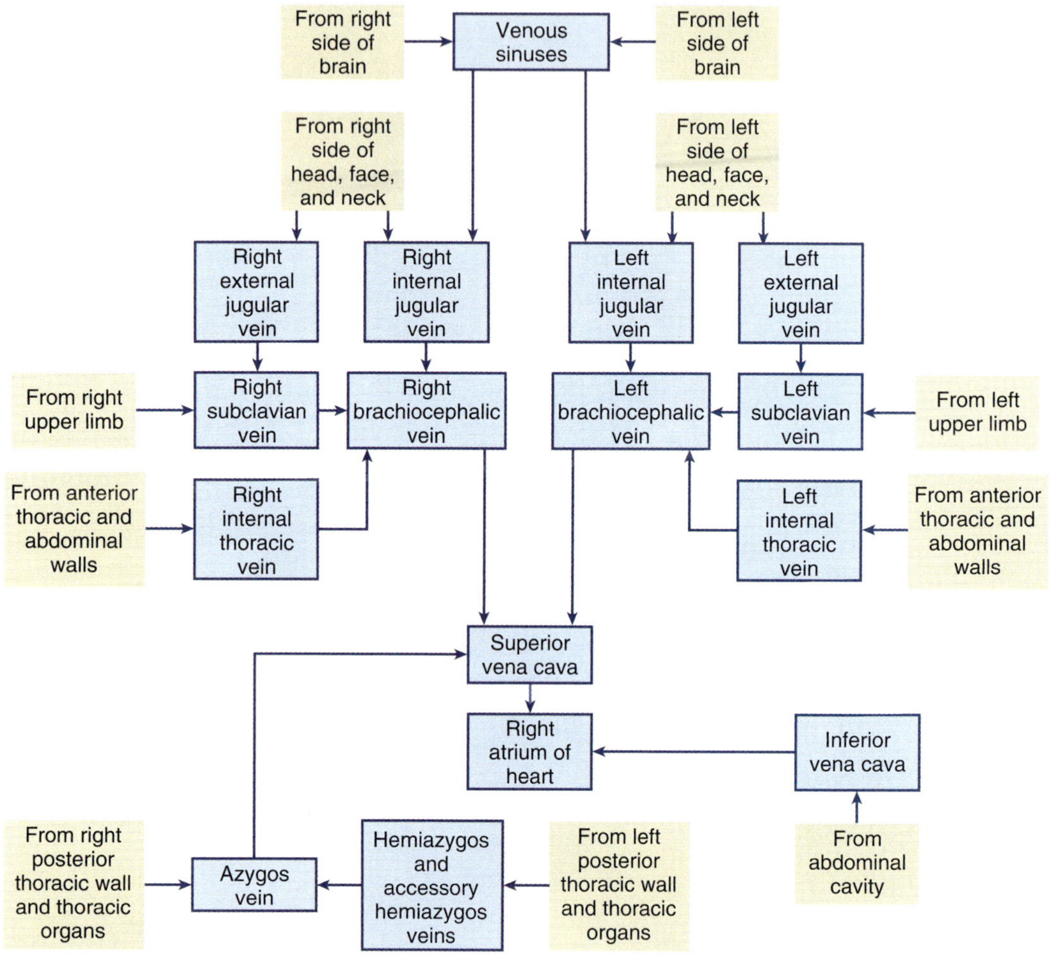
From right side of brain
Venous sinuses
From left side of brain
From right side of head, face, and neck
From left side of head, face, and neck
Right external jugular vein
Right internal jugular vein
Left internal jugular vein
Left external jugular vein
From right upper limb
Right subclavian vein
Right brachiocephalic vein
Left brachiocephalic vein
Left subclavian vein
From left upper limb
From anterior thoracic and abdominal walls
Right internal thoracic vein
Left internal thoracic vein
From anterior thoracic and abdominal walls
Superior vena cava
Right atrium of heart
Inferior vena cava
From abdominal cavity
From right posterior thoracic wall and thoracic organs
Azygos vein
Hemiazygos and accessory hemiazygos veins
From left posterior thoracic wall and thoracic organs

Major Veins of the Head and Thorax
Figure 21.21

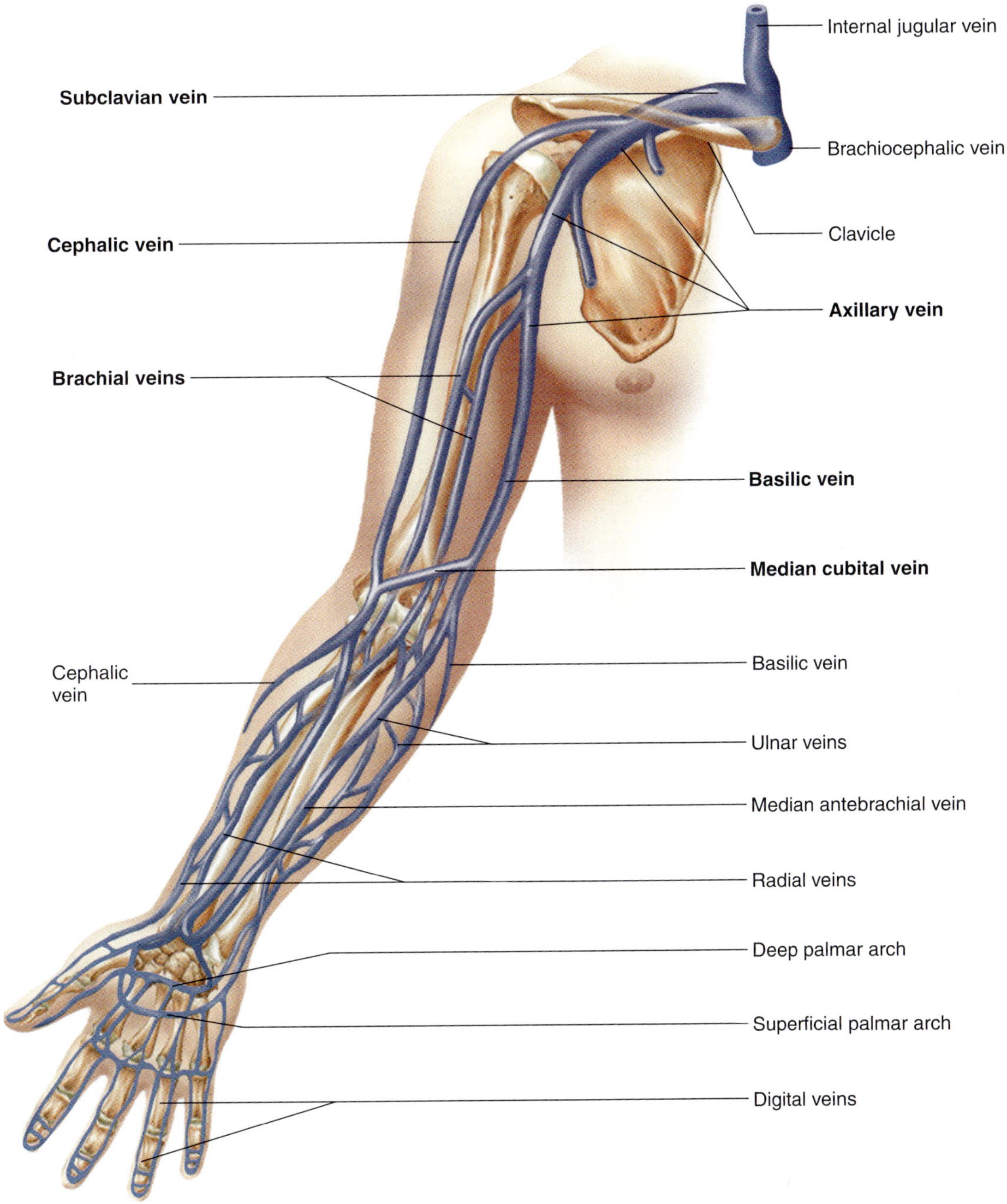
Internal jugular vein
Subclavian vein
Brachiocephalic vein
Clavicle
Cephalic vein
Axillary vein
Brachial veins
Basilic vein
Median cubital vein
Cephalic vein
Basilic vein
Ulnar veins
Median antebrachial vein
Radial veins
Deep palmar arch
Superficial palmar arch
Digital veins
Anterior view

Veins of the Upper Limb
Figure 21.22

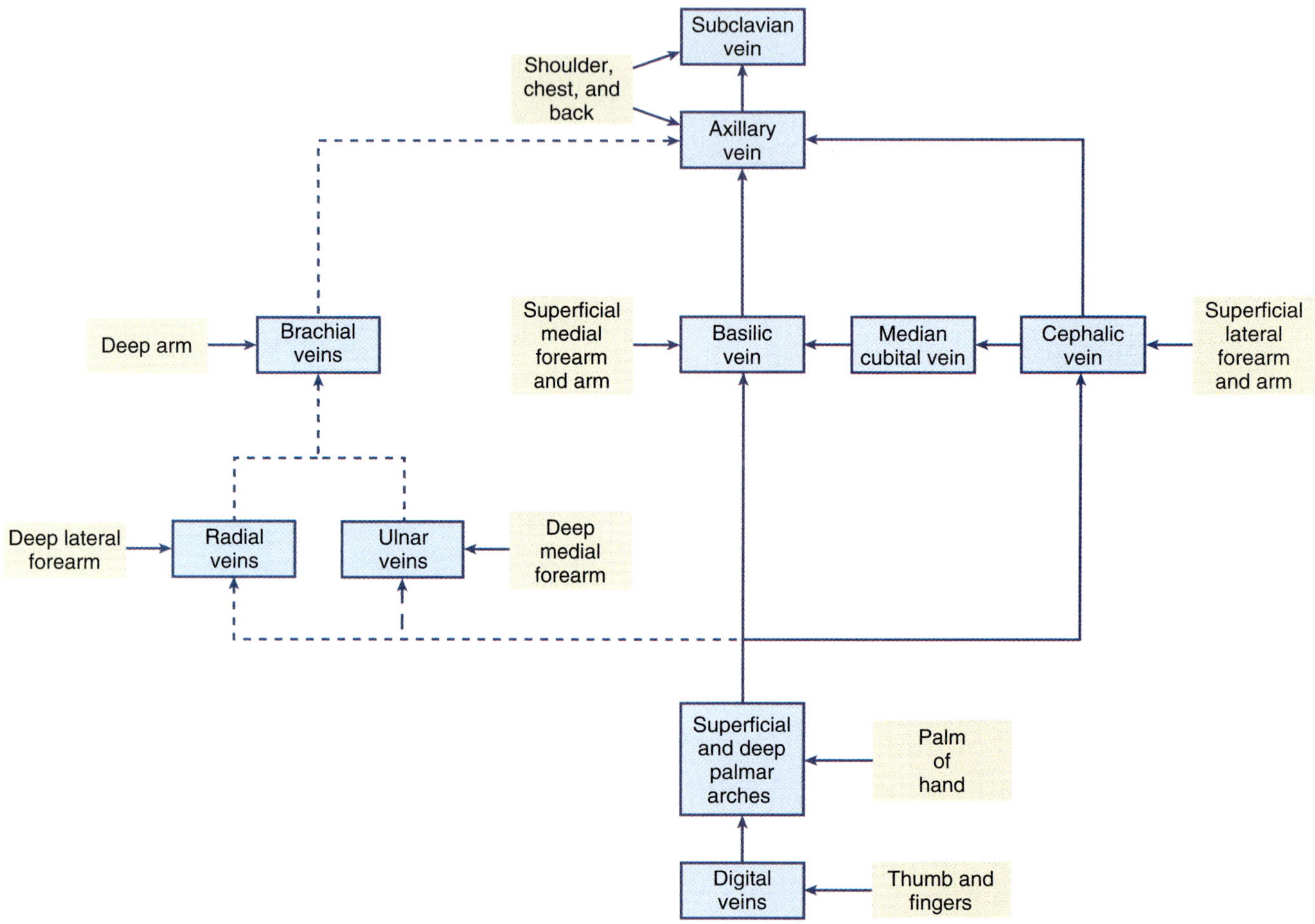

Major Veins of the Shoulder and Upper Limb
Figure 21.23

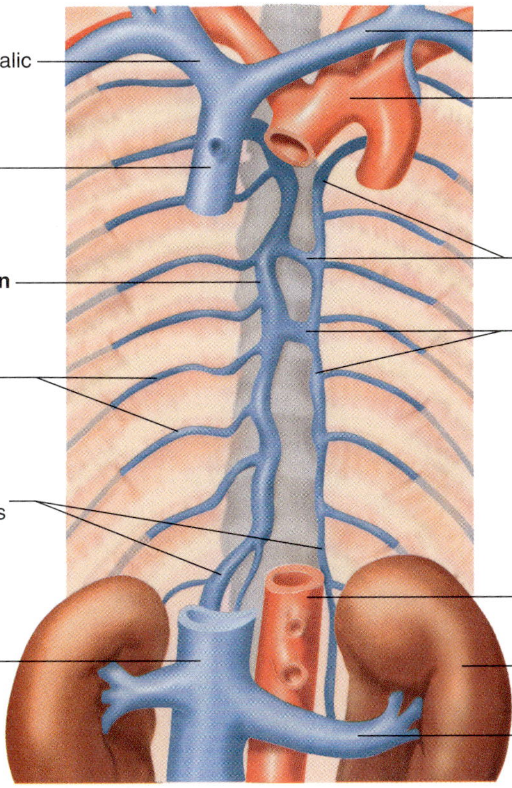

Veins of the Thorax
Figure 21.24

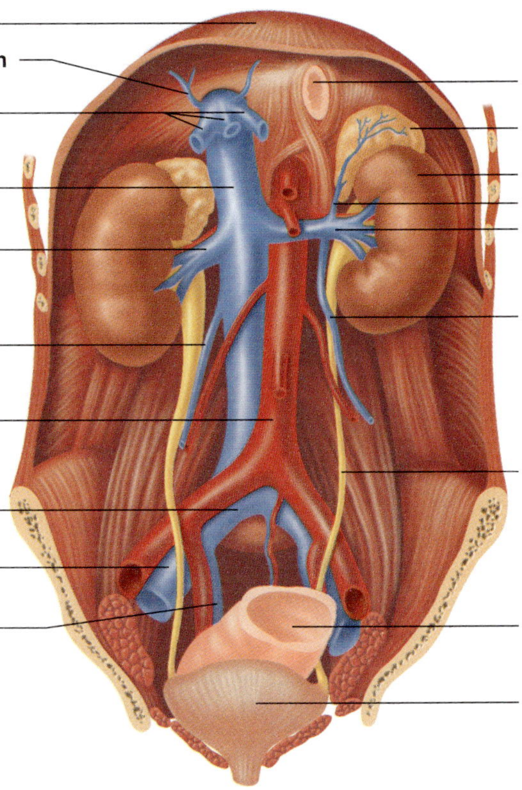

Inferior Vena Cava and Its Tributaries
Figure 21.25

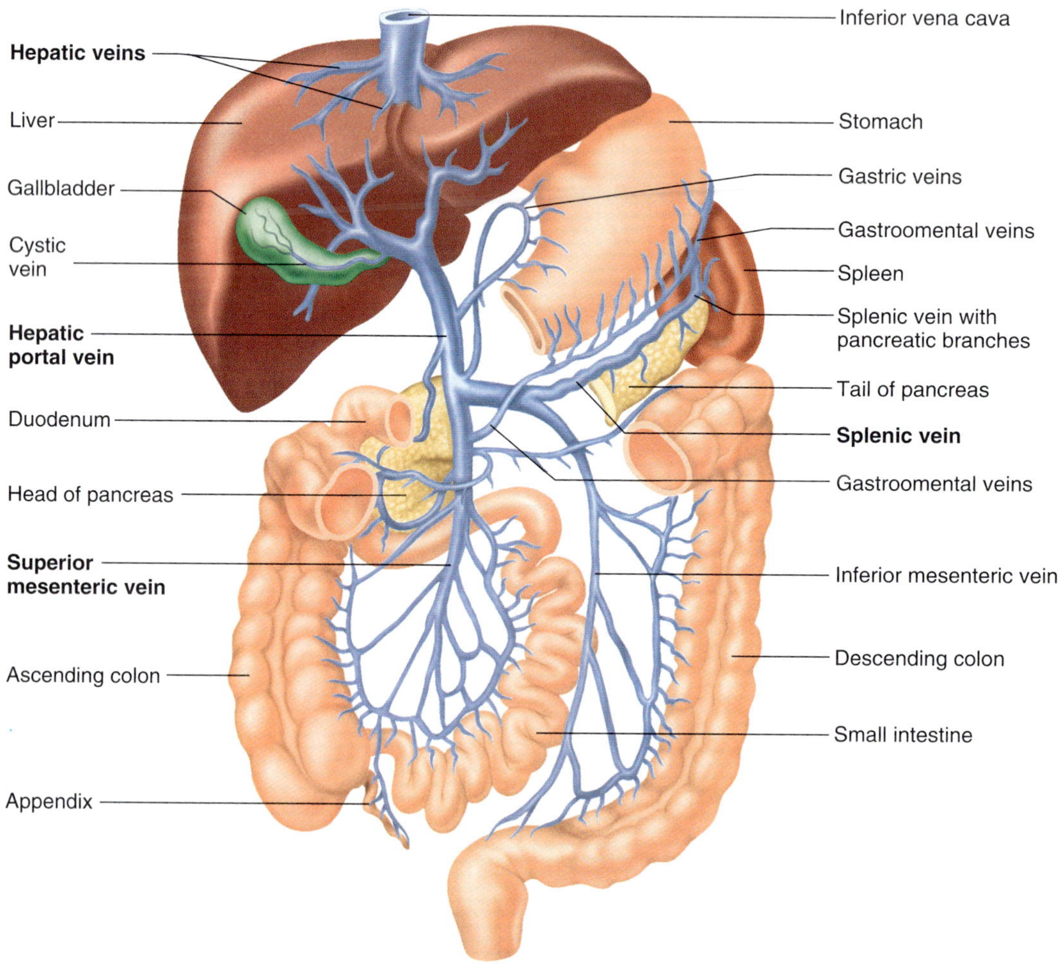

Inferior vena cava
Hepatic veins
Liver
Stomach
Gastric veins
Gallbladder
Gastroomental veins
Cystic vein
Spleen
Splenic vein with pancreatic branches
Hepatic portal vein
Tail of pancreas
Duodenum
Splenic vein
Gastroomental veins
Head of pancreas
Superior mesenteric vein
Inferior mesenteric vein
Ascending colon
Descending colon
Small intestine
Appendix

Veins of the Hepatic Portal System
Figure 21.26

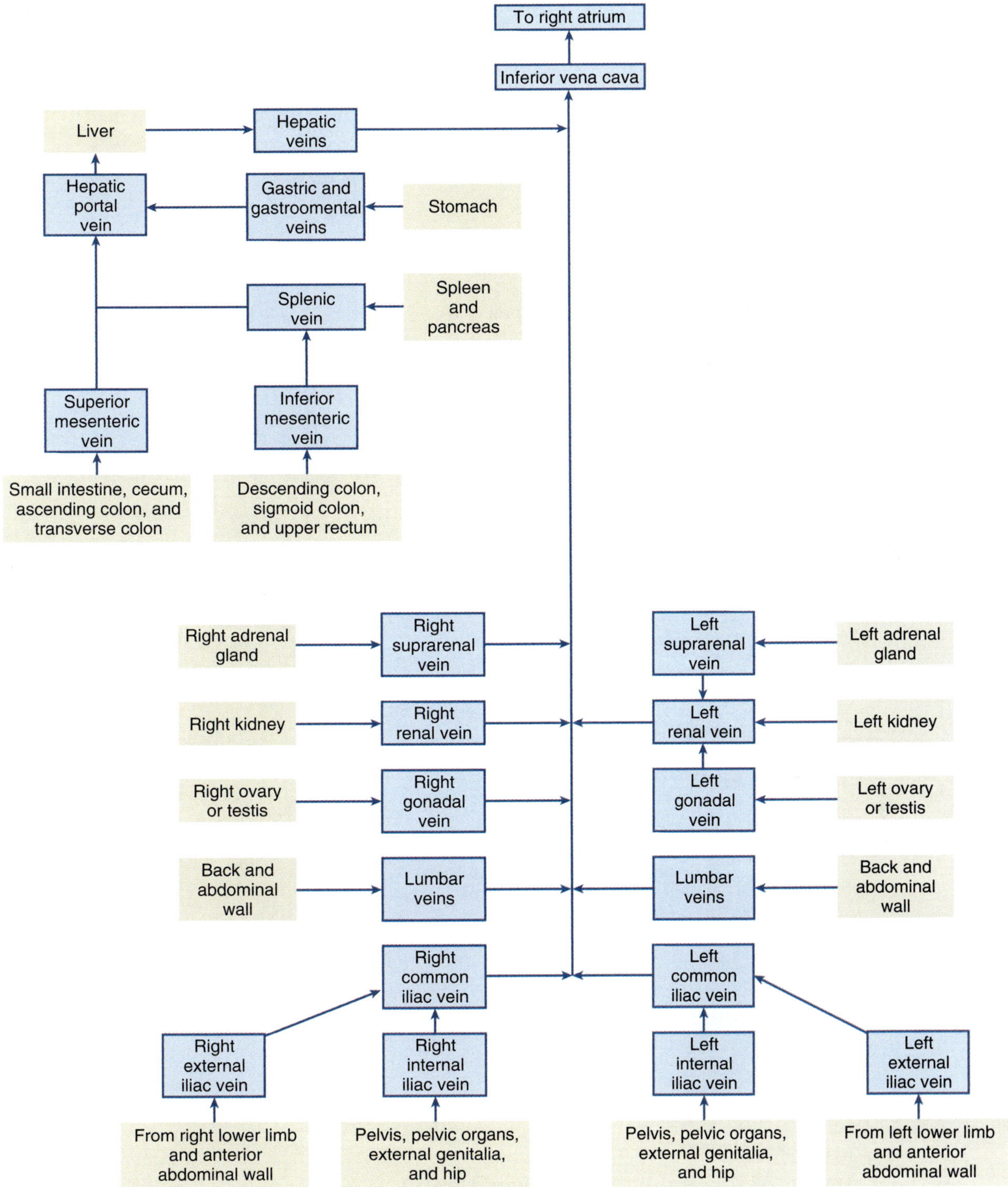

Major Veins of the Abdomen and Pelvis
Figure 21.27

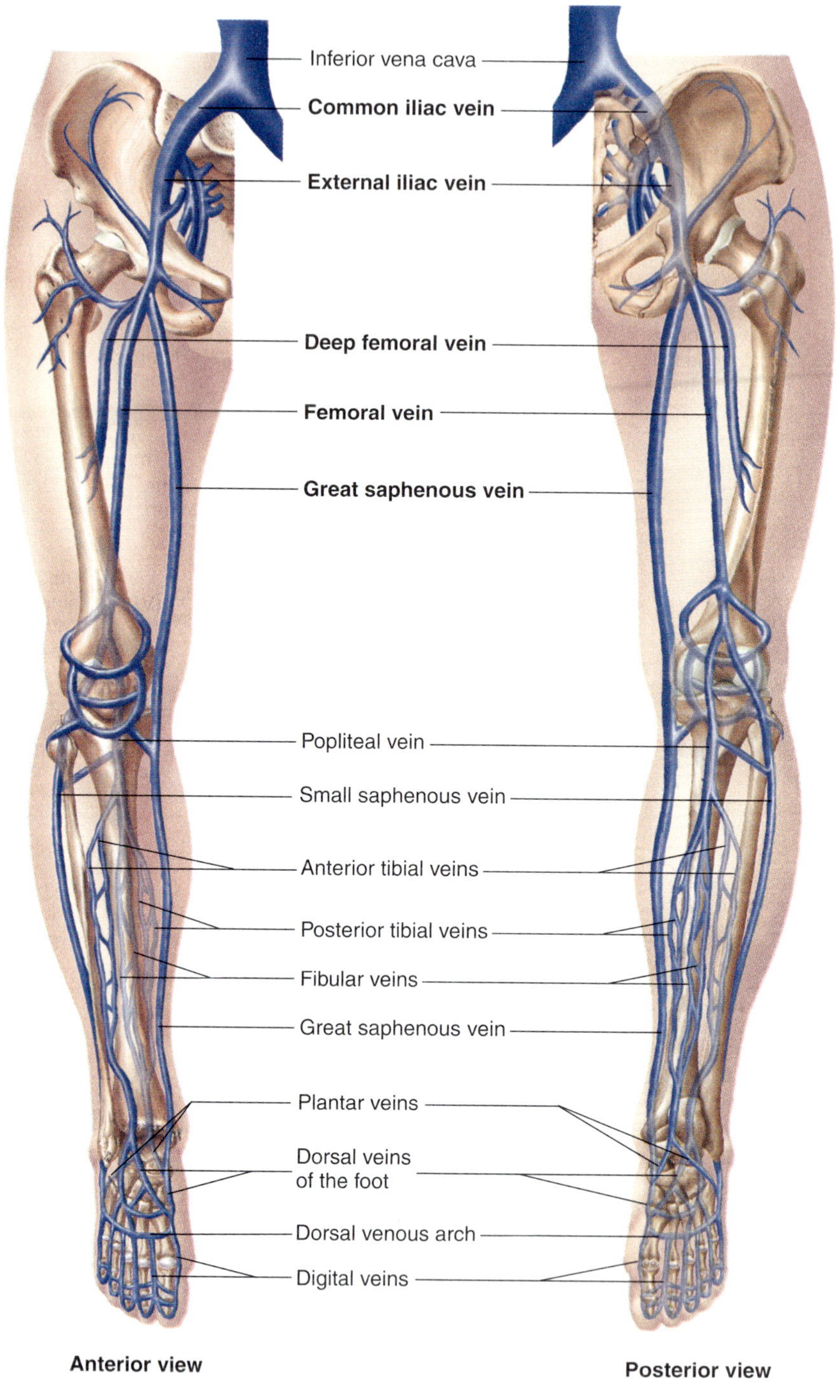
Inferior vena cava
Common iliac vein
External iliac vein
Deep femoral vein
Femoral vein
Great saphenous vein
Popliteal vein
Small saphenous vein
Anterior tibial veins
Posterior tibial veins
Fibular veins
Great saphenous vein
Plantar veins
Dorsal veins of the foot
Dorsal venous arch
Digital veins
Anterior view
Posterior view

Veins of the Pelvis and Lower Limb
Figure 21.28

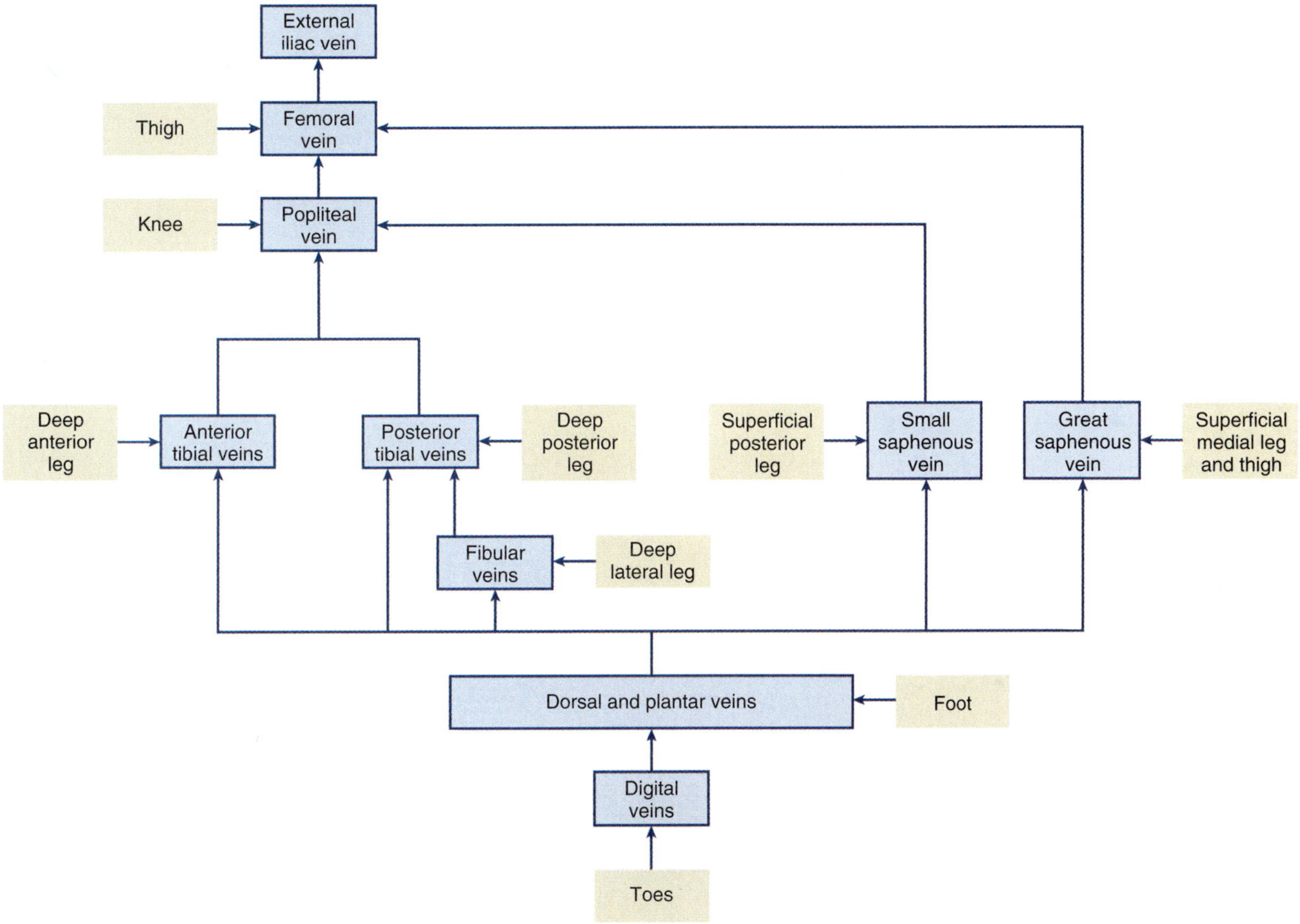
External iliac vein
Femoral vein
Thigh
Popliteal vein
Knee
Deep anterior leg
Anterior tibial veins
Posterior tibial veins
Deep posterior leg
Superficial posterior leg
Small saphenous vein
Great saphenous vein
Superficial medial leg and thigh
Fibular veins
Deep lateral leg
Dorsal and plantar veins
Foot
Digital veins
Toes

Major Veins of the Lower Limb
Figure 21.29

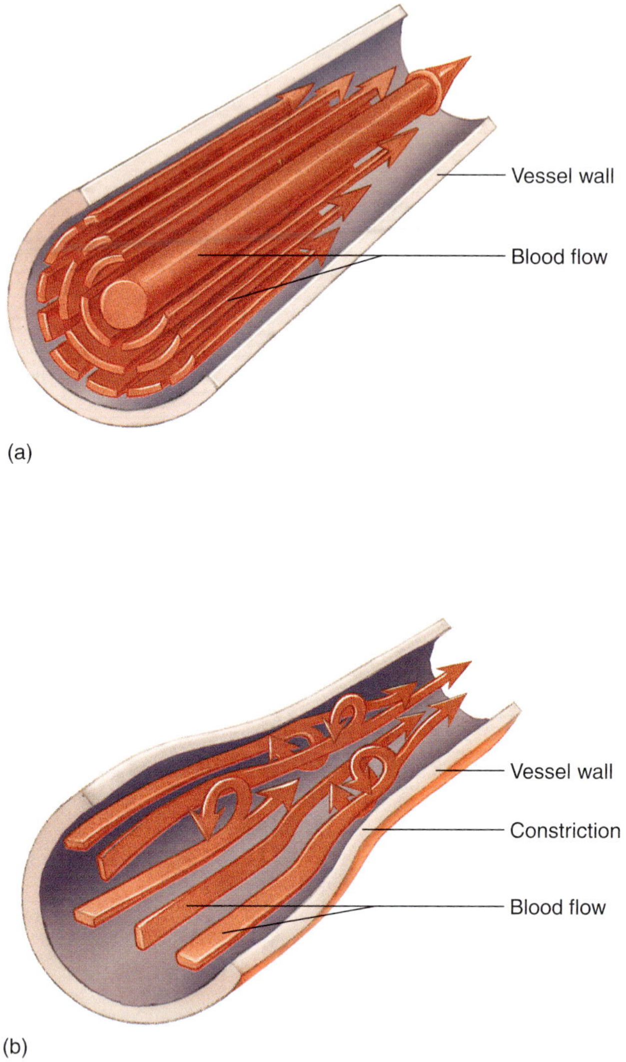
Vessel wall
Blood flow
(a)
Vessel wall
Constriction
Blood flow
(b)

Laminar and Turbulent Flow
Figure 21.30

1. No sound is heard because there is no blood flow when the cuff pressure is high enough to keep the brachial artery closed.
2. **Systolic pressure** is the pressure at which a Korotkoff sound is first heard. When cuff pressure decreases and is no longer able to keep the brachial artery closed during systole, blood is pushed through the partially opened brachial artery to produce turbulent blood flow and a sound. The brachial artery remains closed during diastole.
3. As cuff pressure continues to decrease, the brachial artery opens even more during systole. At first, the artery is closed during diastole, but as cuff pressure continues to decrease, the brachial artery partially opens during diastole. Turbulent blood flow during systole produces Korotkoff sounds, although the pitch of the sounds change as the artery becomes more open.
4. **Diastolic pressure** is the pressure at which the sound disappears. Eventually cuff pressure decreases below the pressure in the brachial artery and it remains open during systole and diastole. Nonturbulent flow is reestablished and no sounds are heard.

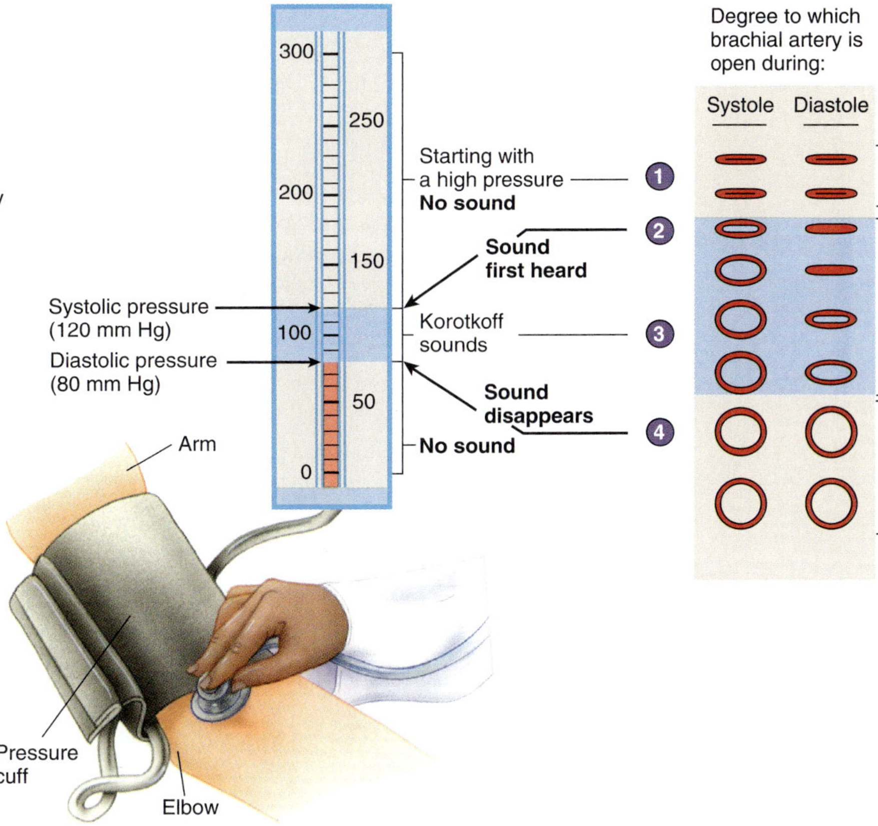

Blood Pressure Measurement
Process Figure 21.31

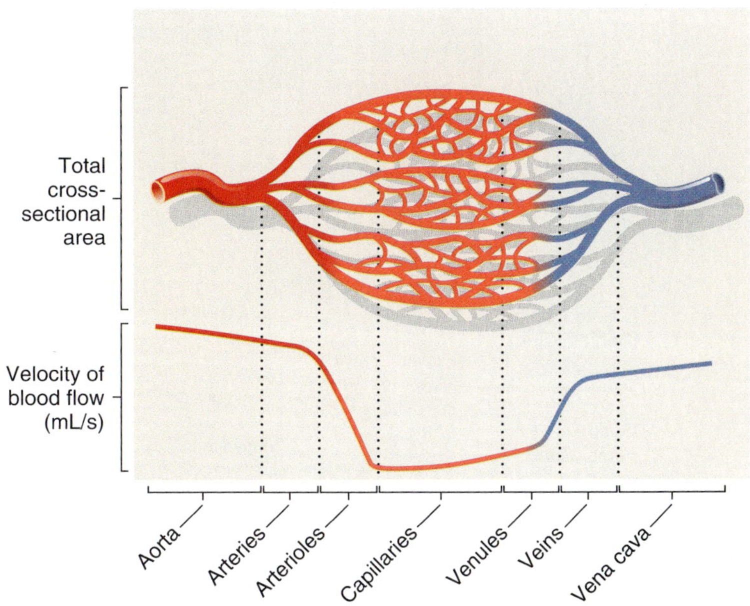

Blood Vessel Types and Velocity of Blood Flow
Figure 21.32

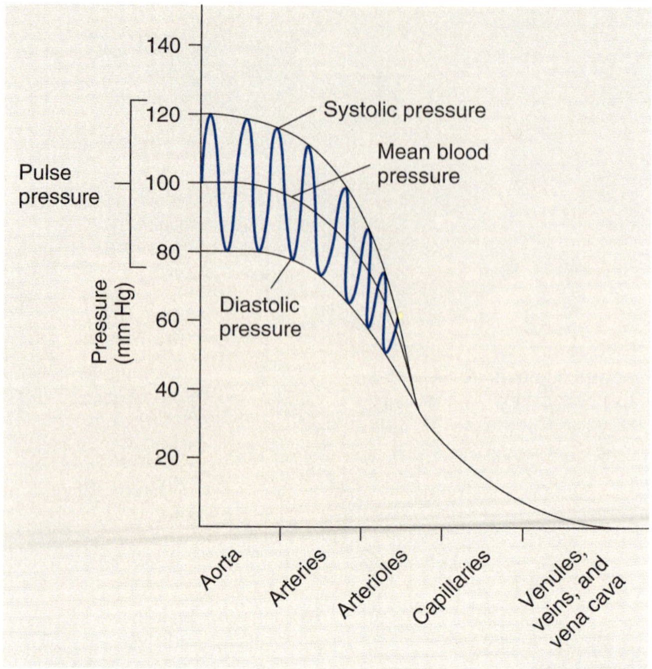

Blood Pressure in Major Blood Vessel Types
Figure 21.33

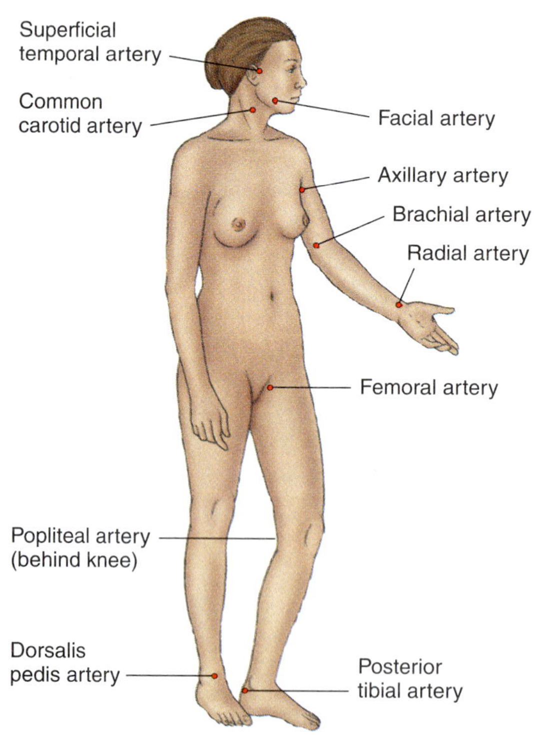

Location of Major Points at Which the Pulse Can Be Monitored
Figure A

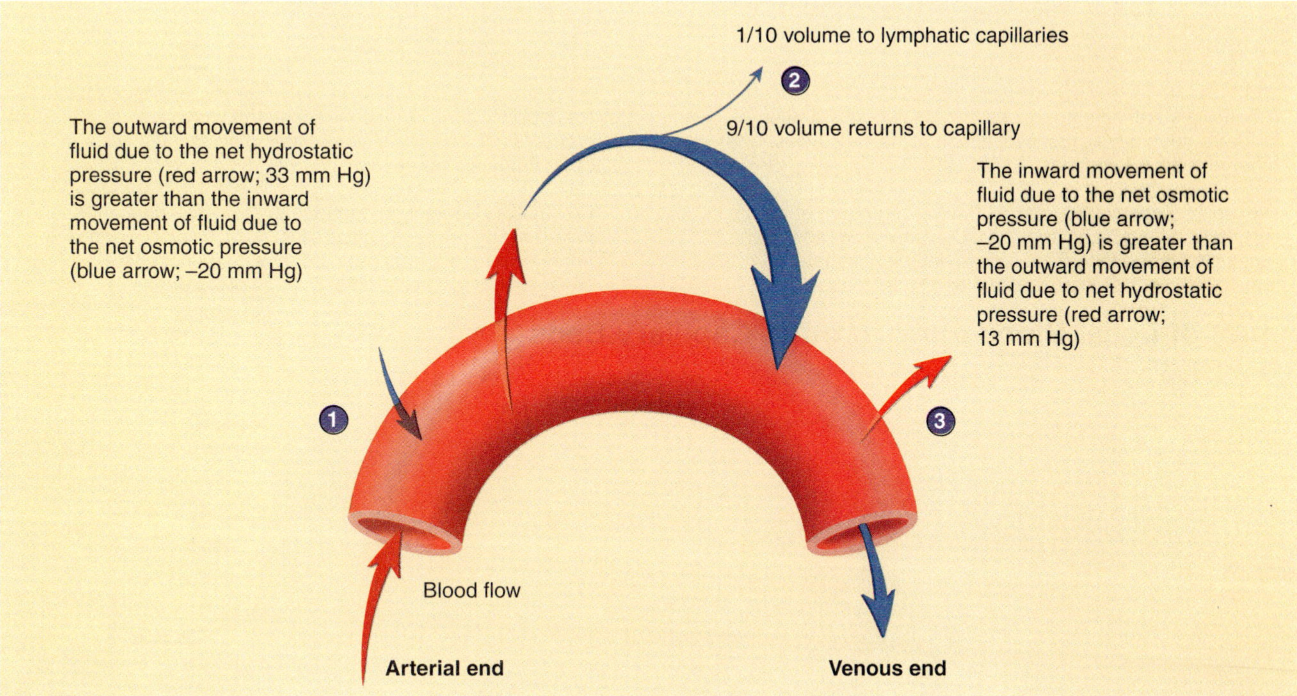

1. At the arterial end of the capillary, the **net hydrostatic pressure** is greater than the **net osmotic pressure,** resulting in a positive **net filtration pressure** that causes fluid to move out of the capillary.

Net hydrostatic pressure	= 33 mm Hg
Net osmotic pressure	= –20 mm Hg
Net filtration pressure	= 13 mm Hg

2. Approximately nine-tenths of the fluid that leaves the capillary at its arterial end reenters the capillary at its venous end. About one-tenth of the fluid passes into the lymphatic capillaries.

3. At the venous end of the capillary, the **net hydrostatic pressure** is less than the **net osmotic pressure,** resulting in a negative **net filtration pressure** that causes fluid to move into the capillary.

Net hydrostatic pressure	= 13 mm Hg
Net osmotic pressure	= –20 mm Hg
Net filtration pressure	= –7 mm Hg

Fluid Exchange Across the Walls of Capillaries
Process Figure 21.34

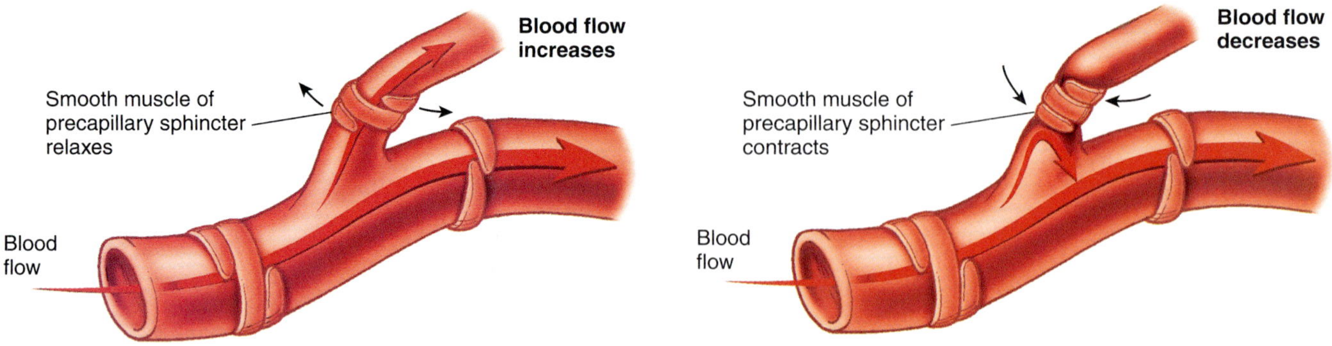

(a) Precapillary sphincters relax as the tissue concentration of nutrients such as O_2, glucose, amino acids, fatty acids, and other nutrients decreases.

Precapillary sphincters relax as the concentration of tissue metabolic by-products, such as CO_2, lactic acid, adenosine, adenosine monophosphate, adenosine diphosphate, nitric oxide, and K^+ increase, and as pH decreases.

(b) Precapillary sphincters contract as the tissue concentration of nutrients such as O_2, glucose, amino acids, fatty acids, and other nutrients increases.

Precapillary sphincters contract as the tissue concentration of metabolic by-products, such as CO_2, lactic acid, adenosine, adenosine monophosphate, adenosine diphosphate, nitric oxide decrease, and as K^+ and pH increase.

Control of Local Blood Flow Through Capillary Beds
Figure 21.35

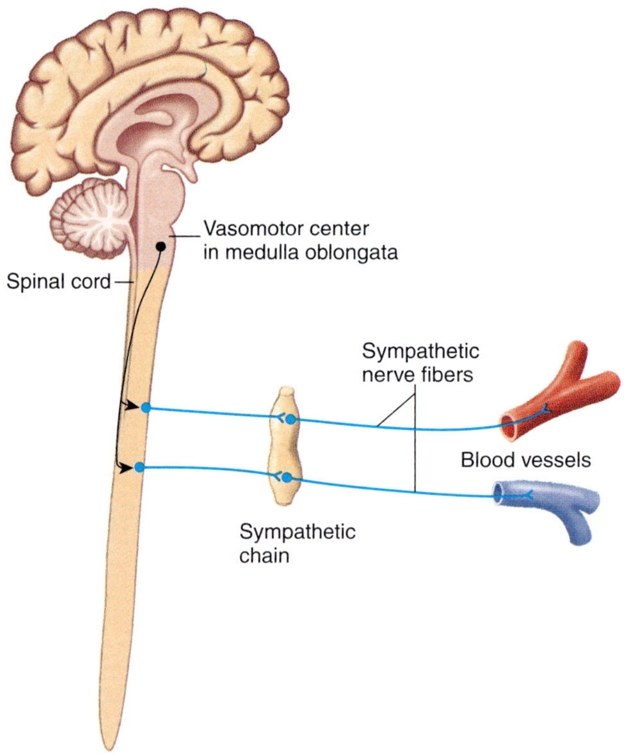

Nervous Regulation of Blood Vessels
Figure 21.36

1. Baroreceptors in the carotid sinus and aortic arch monitor blood pressure.
2. Action potentials are conducted by the glossopharyngeal and vagus nerves to the cardioregulatory and vasomotor centers in the medulla oblongata.
3. Increased parasympathetic stimulation of the heart decreases the heart rate.
4. Increased sympathetic stimulation of the heart increases the heart rate and stroke volume.
5. Increased sympathetic stimulation of blood vessels increases vasoconstriction.

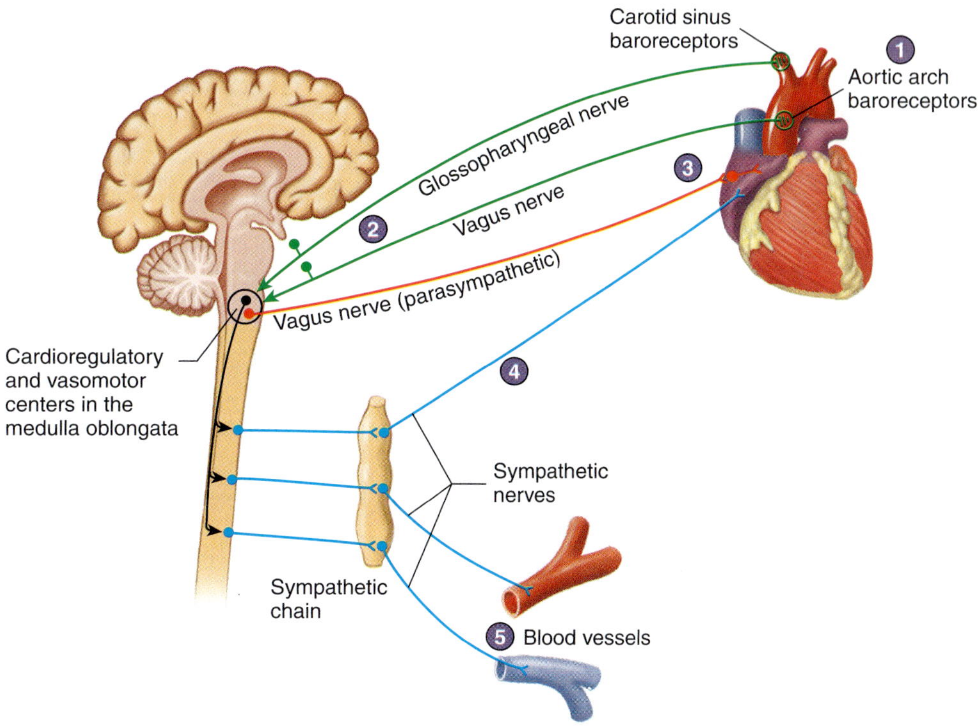

Baroreceptor Reflex Control of Blood Pressure
Process Figure 21.37

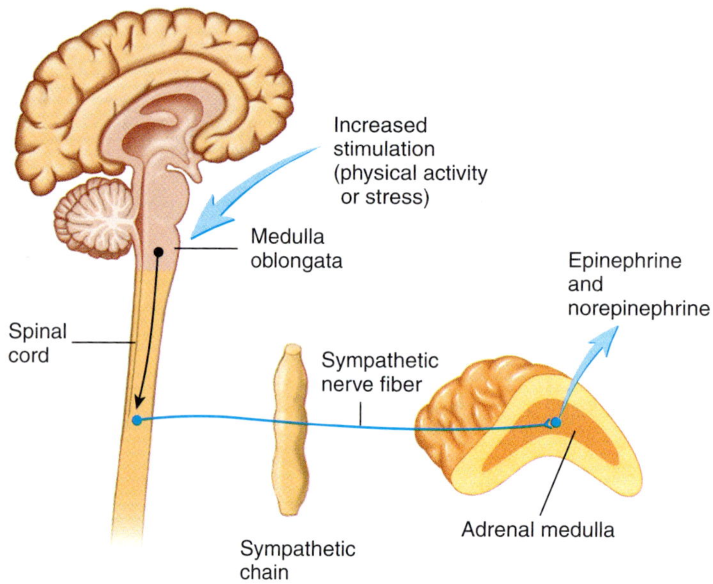

The Adrenal Medullary Mechanism
Figure 21.38

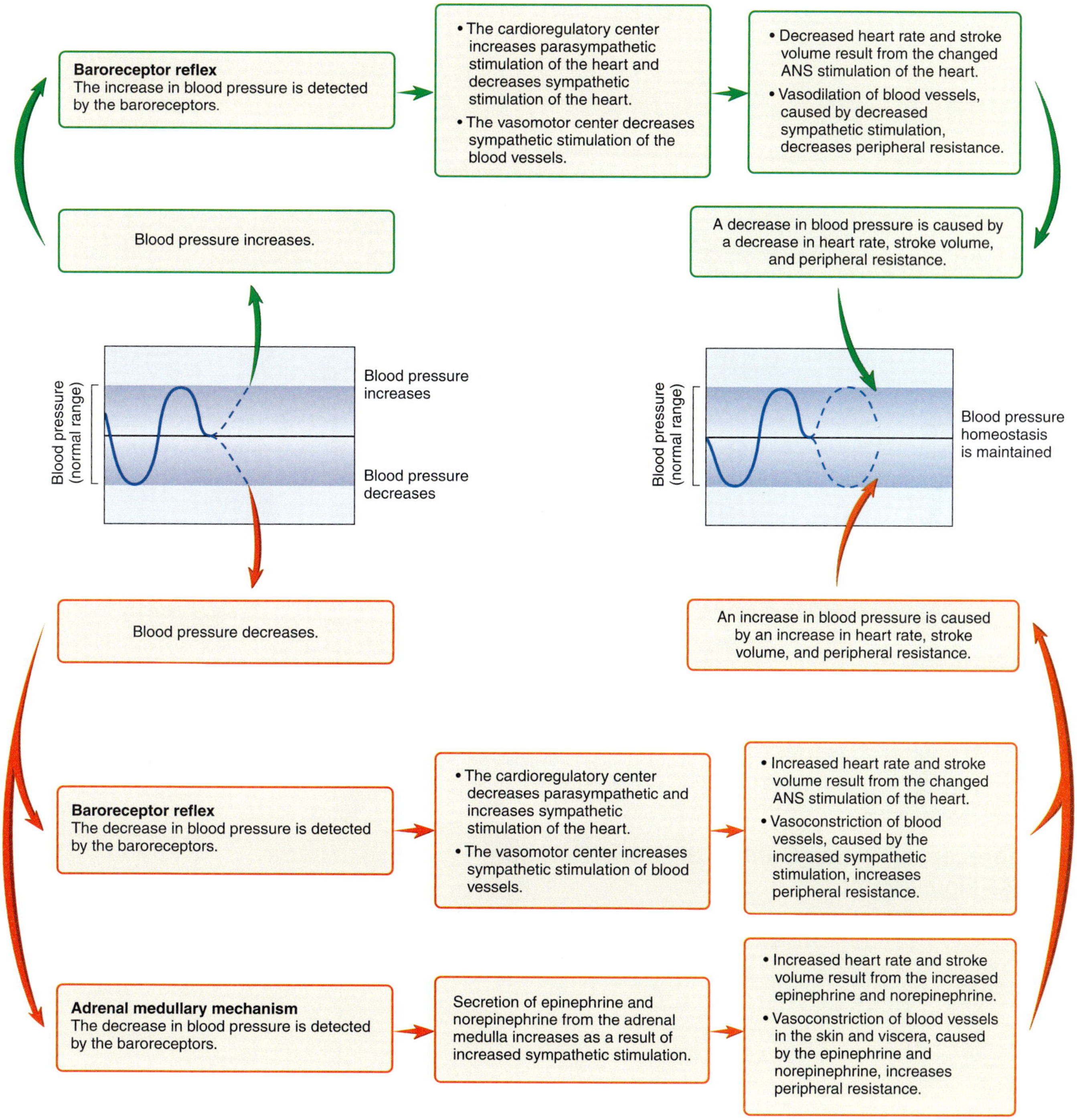

Baroreceptor Effects on Blood Pressure
Homeostasis Figure 21.39

1. Chemoreceptors in the carotid and aortic bodies monitor blood O_2, CO_2, and pH.
2. Chemoreceptors in the medulla oblongata monitor blood CO_2 and pH.
3. Decreased blood O_2, increased CO_2, and decreased pH decrease parasympathetic stimulation of the heart, which increases the heart rate.
4. Decreased blood O_2, increased CO_2, and decreased pH increase sympathetic stimulation of the heart, which increases the heart rate and stroke volume.
5. Increased sympathetic stimulation of blood vessels increases vasoconstriction.

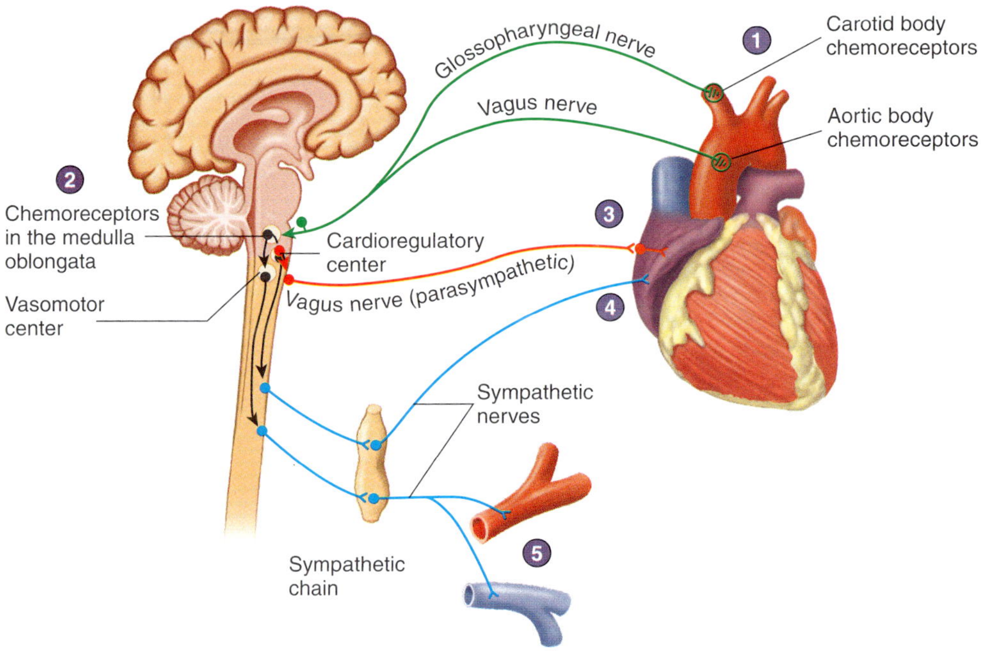

Chemoreceptor Reflex Control of Blood Pressure
Process Figure 21.40

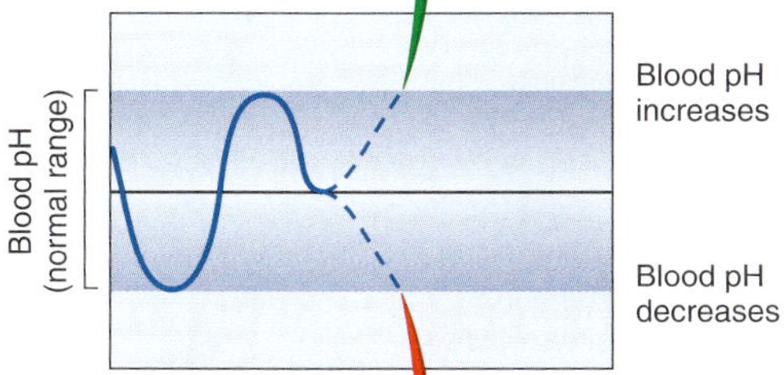

Blood pH increases (often caused by a decrease in blood CO_2).

Chemoreceptor reflex: Medulla oblongata
The increase in pH (decrease in CO_2) is detected by chemoreceptors in the medulla oblongata.

- The vasomotor center decreases sympathetic stimulation of blood vessels.
- The cardioregulatory center increases parasympathetic and decreases sympathetic stimulation of the heart.

- Vasodilation of blood vessels decreases peripheral resistance.
- Heart rate and stroke volume decrease, resulting in decreased cardiac output.

The decrease in blood pH (caused by an increase in blood CO_2) results from decreased blood flow to the lungs. The decreased blood flow results from the decreased blood pressure and cardiac output caused by decreased peripheral resistance, heart rate, and stroke volume.

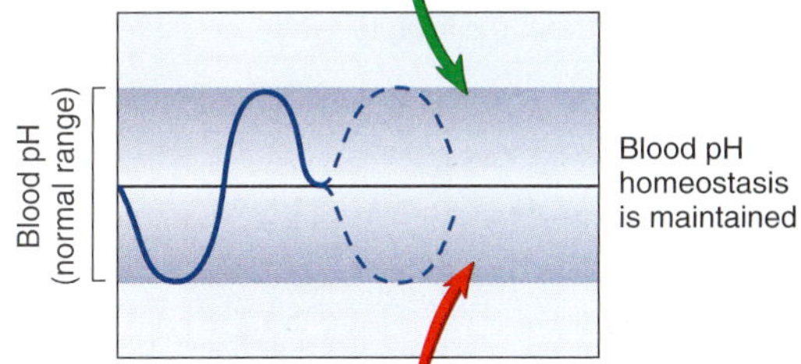

Blood pH decreases (often caused by an increase in blood CO_2) or there is a large decrease in blood O_2.

Chemoreceptor reflex: Carotid and aortic bodies
A large decrease in O_2 is detected by chemoreceptors in the carotid and aortic bodies.

- The vasomotor center increases sympathetic stimulation of blood vessels.
- Respiration rate increases, which results in decreased parasympathetic and increased sympathetic stimulation of the heart.

- Vasoconstriction of blood vessels increases peripheral resistance.
- Heart rate and stroke volume increase, resulting in increased cardiac output.

Chemoreceptor reflex: Medulla oblongata
A decrease in pH (increase in CO_2) is detected by chemoreceptors in the medulla oblongata.

The cardioregulatory center decreases parasympathetic and increases sympathetic stimulation of the heart.

Heart rate and stroke volume increase, resulting in increased cardiac output.

Central nervous system ischemic response
A large decrease in pH (increase in CO_2) is detected by chemoreceptors.

The vasomotor center increases sympathetic stimulation of blood vessels.

Vasoconstriction of blood vessels increases peripheral resistance.

The increase in blood pH (caused by a decrease in blood CO_2) or increase in blood O_2 results from increased blood flow to the lungs. The increased blood flow results from the increased blood pressure and cardiac output caused by increased peripheral resistance, heart rate, and stroke volume.

Effects of pH and Gases on Blood Pressure
Homeostasis Figure 21.41

1. Decreased blood pressure is detected by the kidneys, resulting in increased renin secretion.
2. Renin converts angiotensinogen, a protein secreted from the liver, to angiotensin I.
3. Angiotensin-converting enzyme in the lungs converts angiotensin I to angiotensin II.
4. Angiotensin II is a potent vasoconstrictor, resulting in increased blood pressure.
5. Angiotensin II stimulates the adrenal cortex to secrete aldosterone.
6. Aldosterone acts on the kidneys to increase Na^+ reabsorption. As a result, urine volume decreases and blood volume increases, resulting in increased blood pressure.

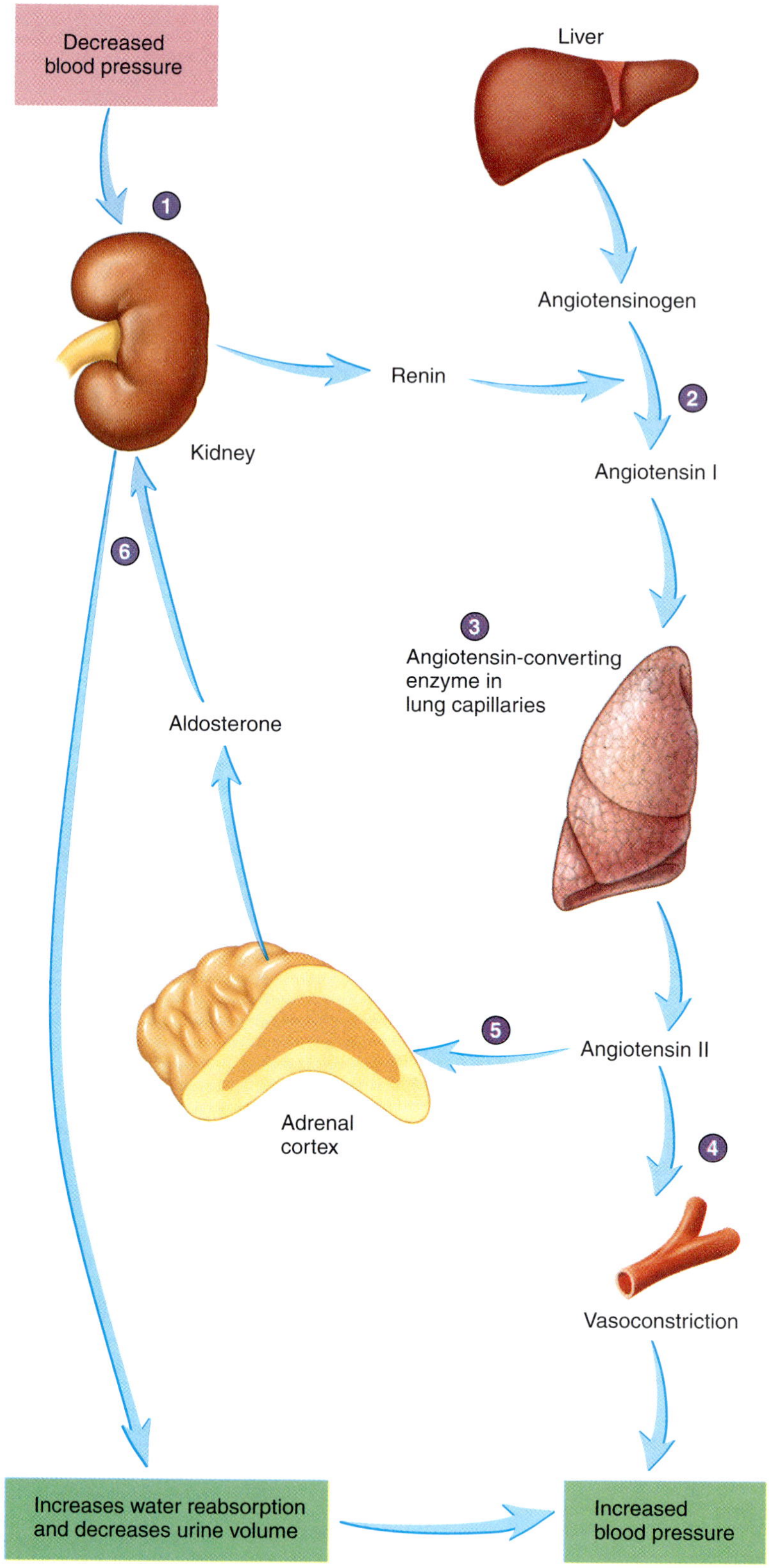

The Renin-Angiotensin-Aldosterone Mechanism
Process Figure 21.42

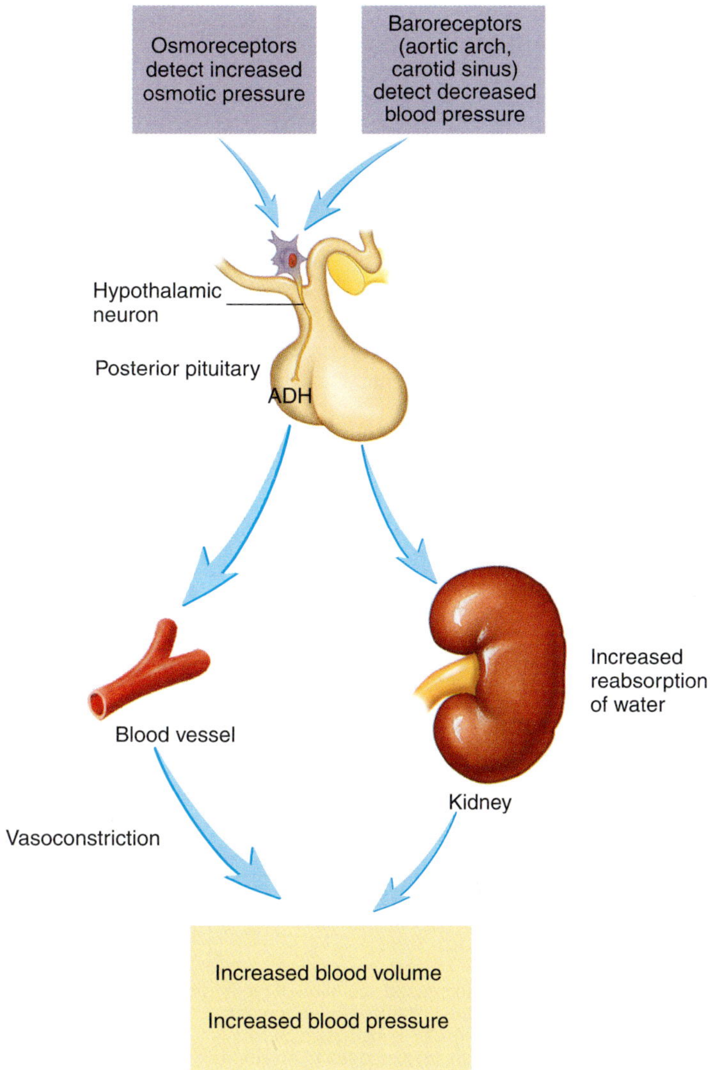
Osmoreceptors detect increased osmotic pressure
Baroreceptors (aortic arch, carotid sinus) detect decreased blood pressure
Hypothalamic neuron
Posterior pituitary
ADH
Blood vessel
Increased reabsorption of water
Kidney
Vasoconstriction
Increased blood volume
Increased blood pressure

The Vasopressin (ADH) Mechanism
Figure 21.43

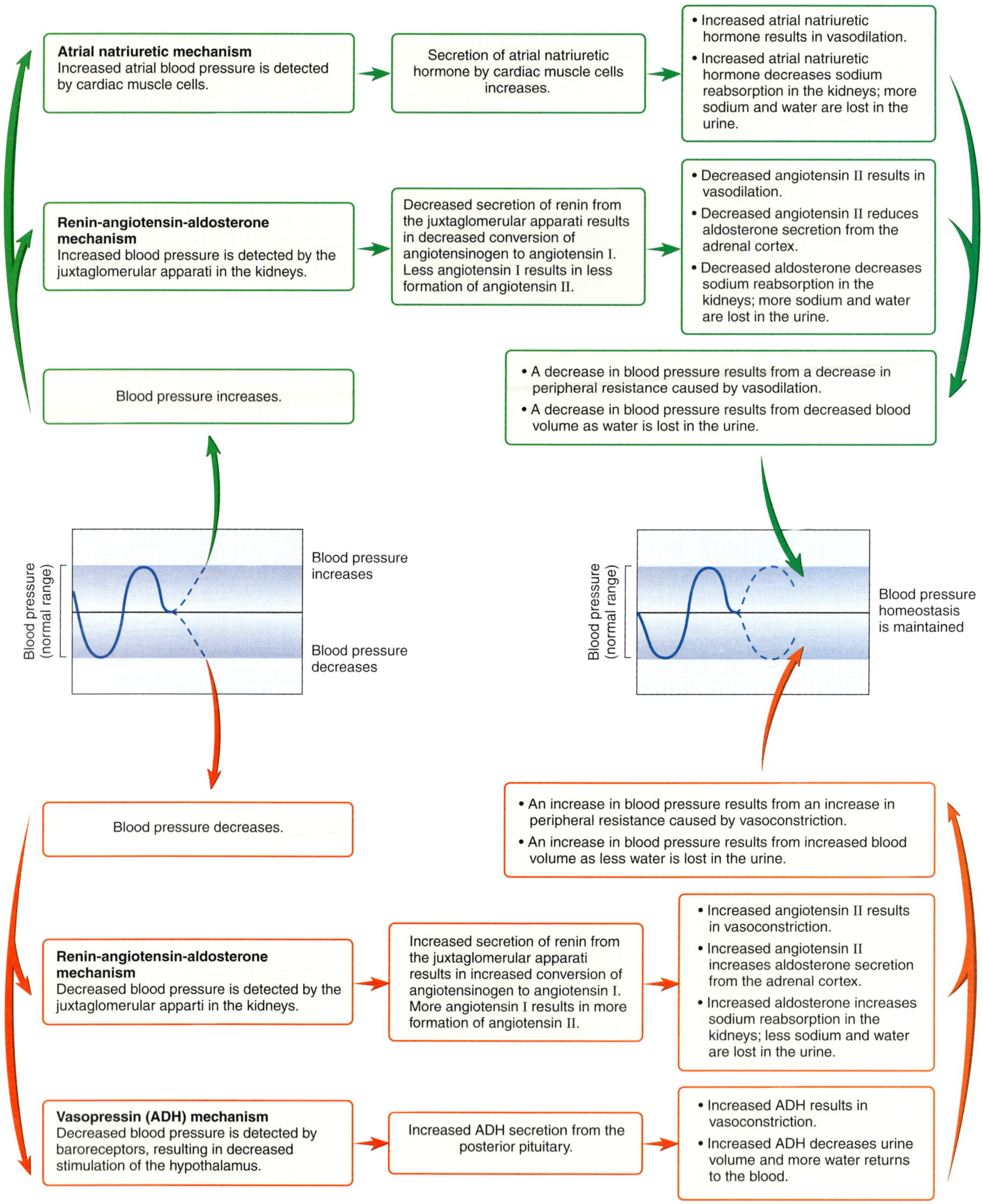

Control of Blood Pressure Long-Term (Slow-Acting) Mechanisms
Homeostasis Figure 21.44

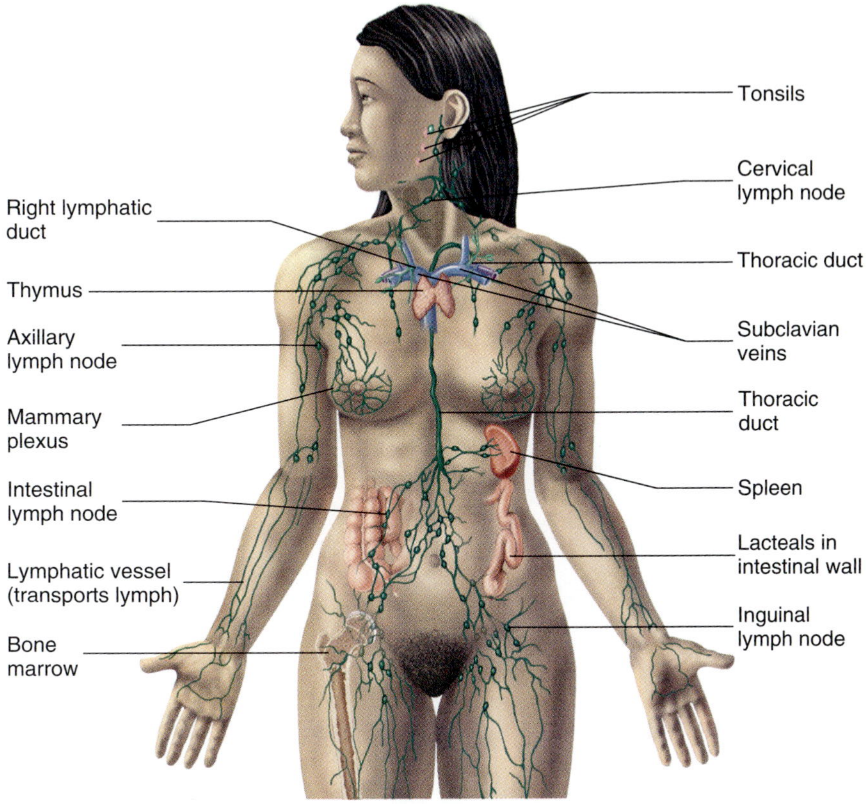

Lymphatic System
Figure 22.1

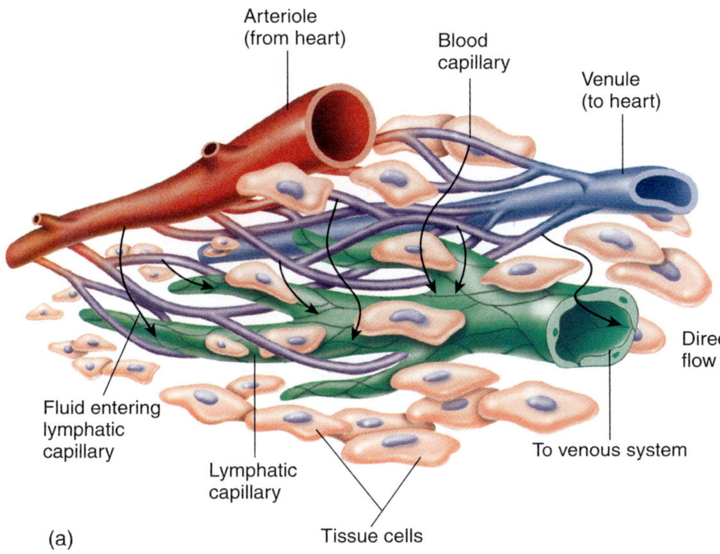

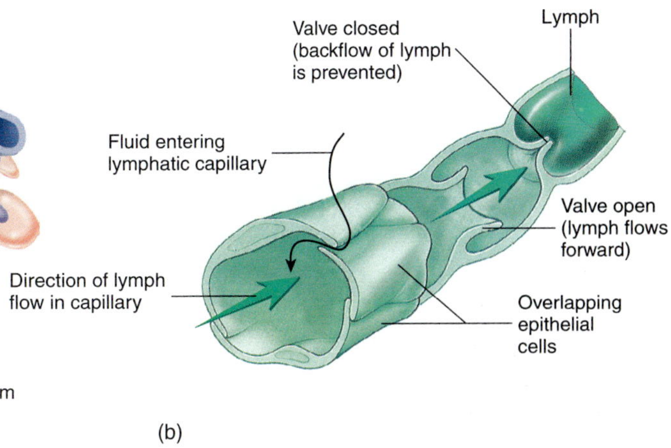

Lymph Formation and Movement
Figure 22.2

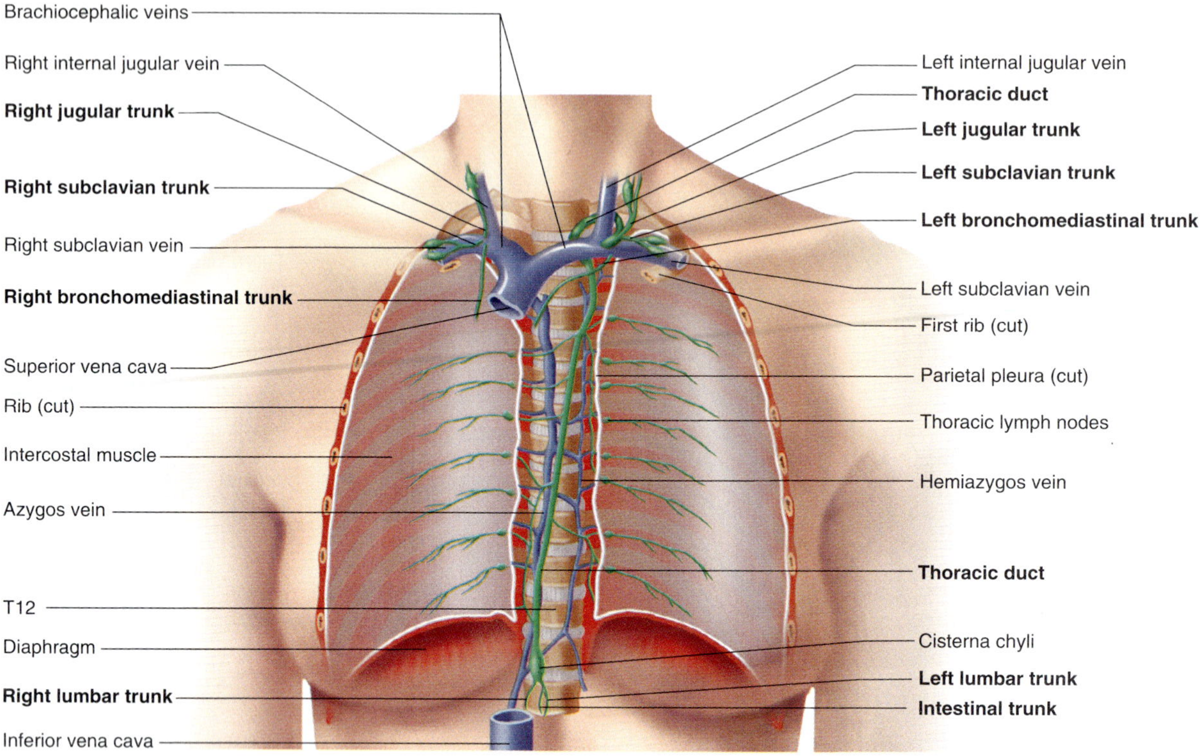
Brachiocephalic veins
Right internal jugular vein
Right jugular trunk
Right subclavian trunk
Right subclavian vein
Right bronchomediastinal trunk
Superior vena cava
Rib (cut)
Intercostal muscle
Azygos vein
T12
Diaphragm
Right lumbar trunk
Inferior vena cava
Left internal jugular vein
Thoracic duct
Left jugular trunk
Left subclavian trunk
Left bronchomediastinal trunk
Left subclavian vein
First rib (cut)
Parietal pleura (cut)
Thoracic lymph nodes
Hemiazygos vein
Thoracic duct
Cisterna chyli
Left lumbar trunk
Intestinal trunk

(a)

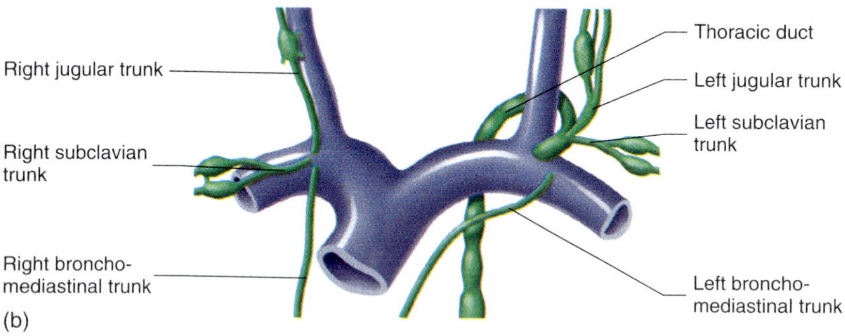
Right jugular trunk
Right subclavian trunk
Right broncho-mediastinal trunk
Thoracic duct
Left jugular trunk
Left subclavian trunk
Left broncho-mediastinal trunk

(b)

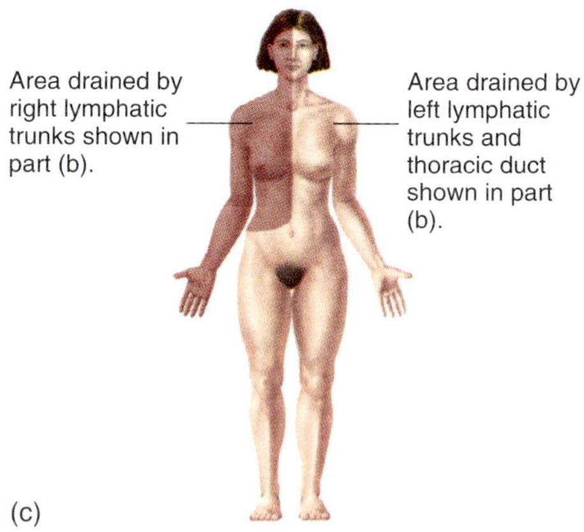
Area drained by right lymphatic trunks shown in part (b).
Area drained by left lymphatic trunks and thoracic duct shown in part (b).

(c)

Lymph Drainage Into Veins

Figure 22.3

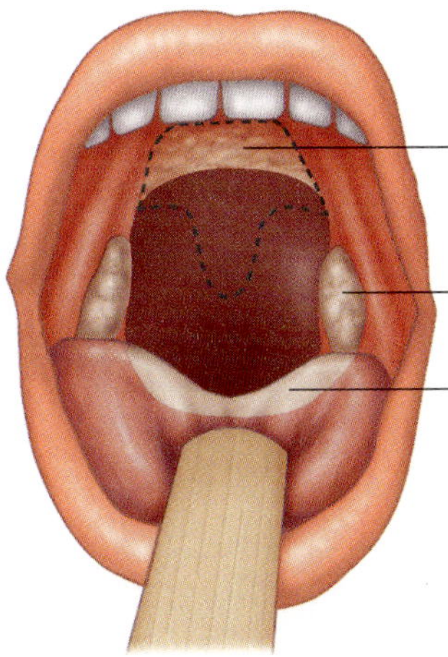

Location of the Tonsils
Figure 22.5

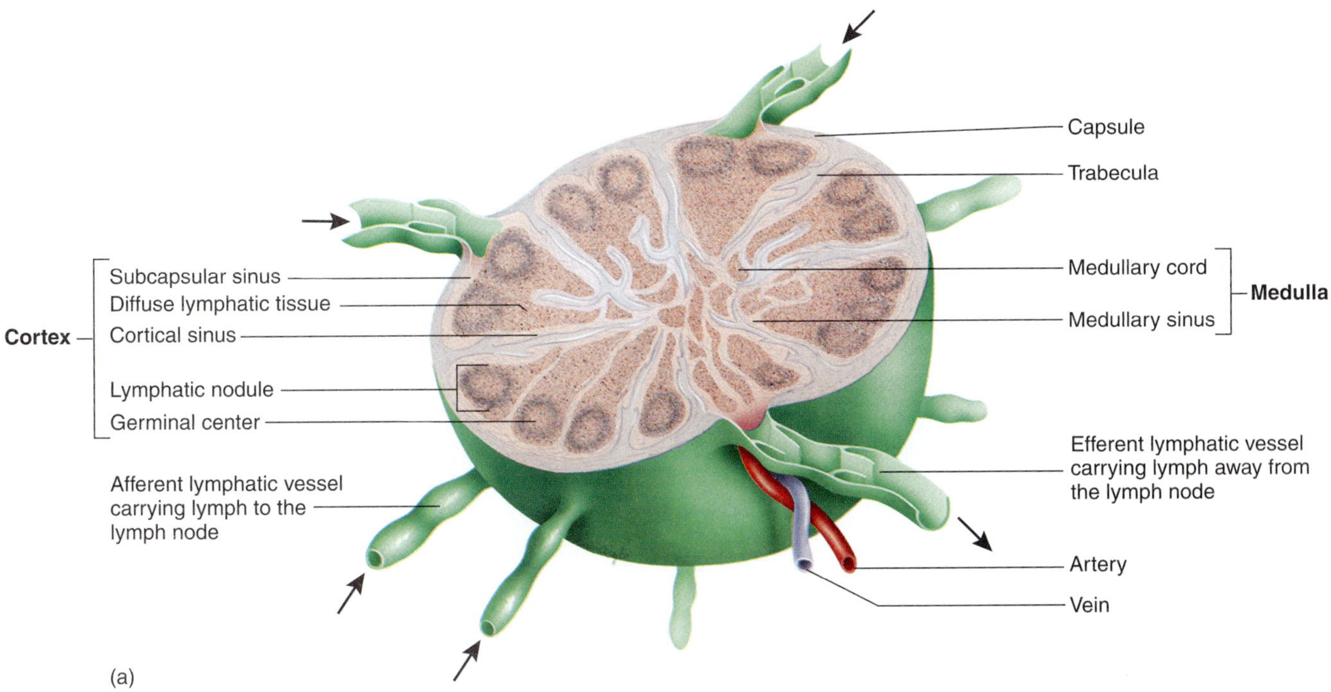

Lymph Node
Figure 22.6

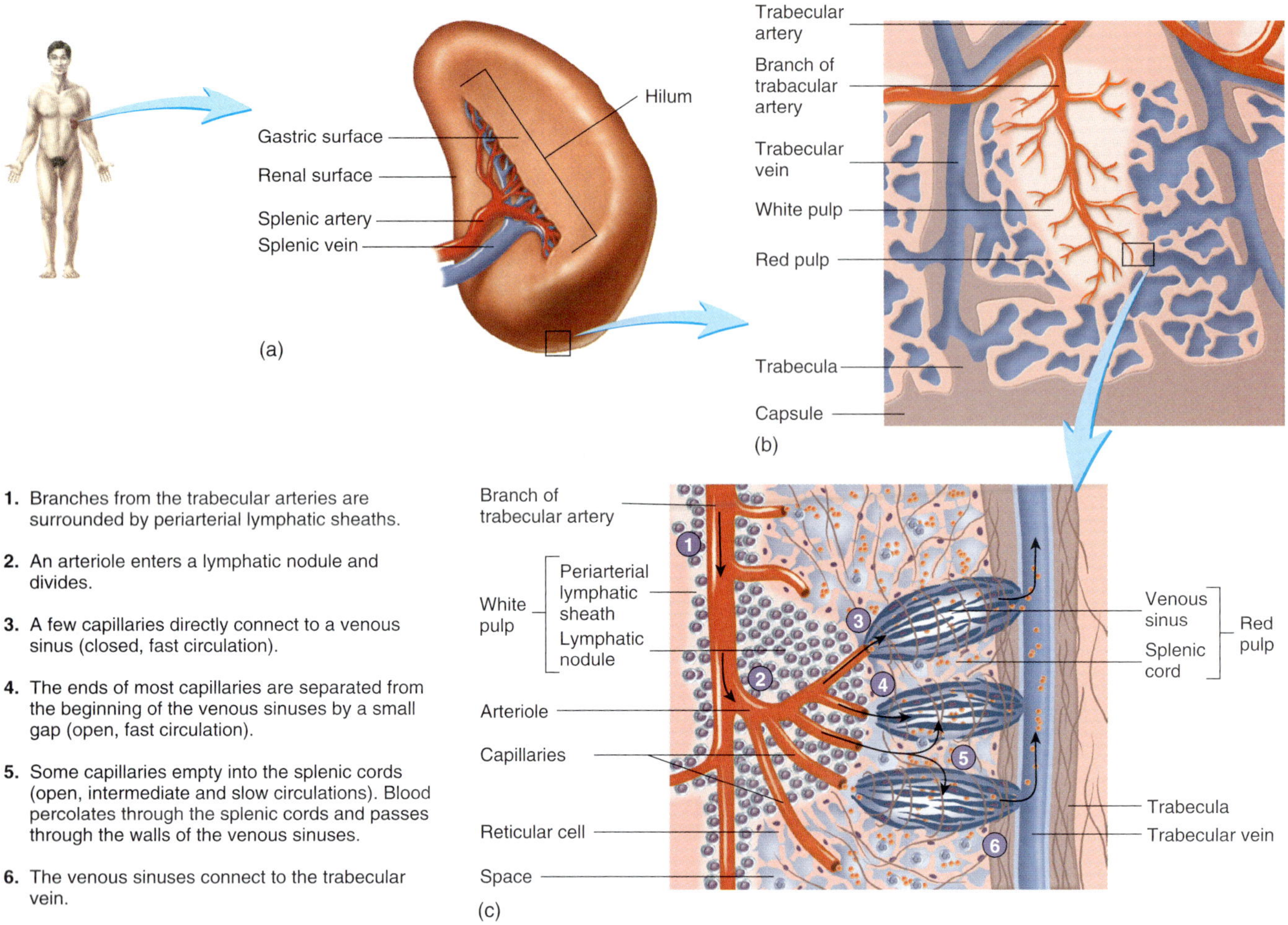

1. Branches from the trabecular arteries are surrounded by periarterial lymphatic sheaths.
2. An arteriole enters a lymphatic nodule and divides.
3. A few capillaries directly connect to a venous sinus (closed, fast circulation).
4. The ends of most capillaries are separated from the beginning of the venous sinuses by a small gap (open, fast circulation).
5. Some capillaries empty into the splenic cords (open, intermediate and slow circulations). Blood percolates through the splenic cords and passes through the walls of the venous sinuses.
6. The venous sinuses connect to the trabecular vein.

Spleen
Process Figure 22.7

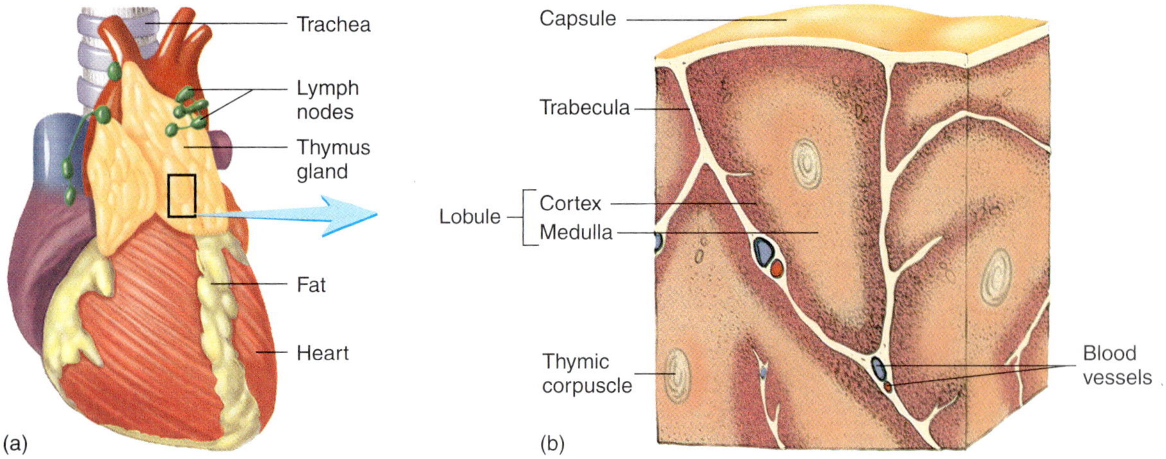

Thymus
Figure 22.8

(a) Lymphatic capillaries remove fluid from tissues. The fluid becomes lymph (see figure 22.2*a*).

(b) Lymph flows through lymphatic vessels, which have valves that prevent the backflow of lymph (see figure 22.2*b*).

(c) Lymph nodes filter lymph (see figure 22.6) and are sites where lymphocytes respond to infections, etc.

(d) Lymph enters lymphatic trunks and ducts (see figure 22.3*b*).

(e) Lymph enters the blood.

(f) Lacteals in the small intestine (see figure 24.16*c*) absorb fats, which enter the thoracic duct (see figure 22.3*a*).

(g) Chyle, which is lymph containing fats, enters the blood.

(h) The spleen (see figure 22.7) filters blood and is a site where lymphocytes respond to infections, etc.

(i) Lymphocytes (pre-B and pre-T cells) originate from stem cells in the red bone marrow (see figure 22.12). The pre-B cells become mature B cells in the red bone marrow and are released into the blood. The pre-T cells enter the blood and migrate to the thymus.

(j) The thymus (see figure 22.8) is where pre-T cells derived from red bone marrow increase in number and become mature T cells that are released into the blood (see figure 22.12).

(k) B and T cells from the blood enter and populate all lymphatic tissues. These lymphocytes can remain in the lymphatic tissues or pass through them and return to the blood. B and T cells can also respond to infections, etc., by dividing and increasing in number (see figures 22.18 and 22.22).

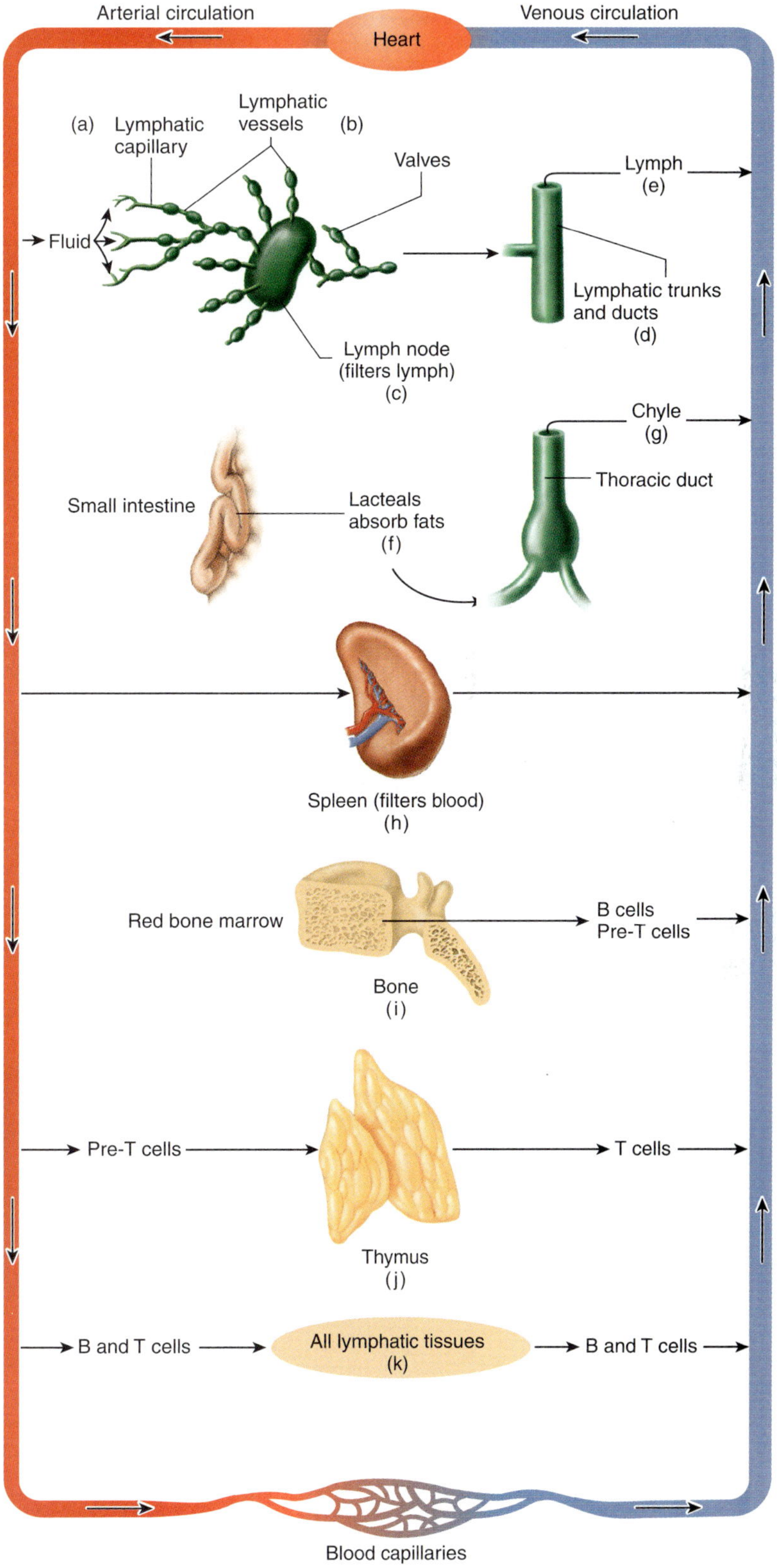

Overview of the Lymphatic System
Figure 22.9

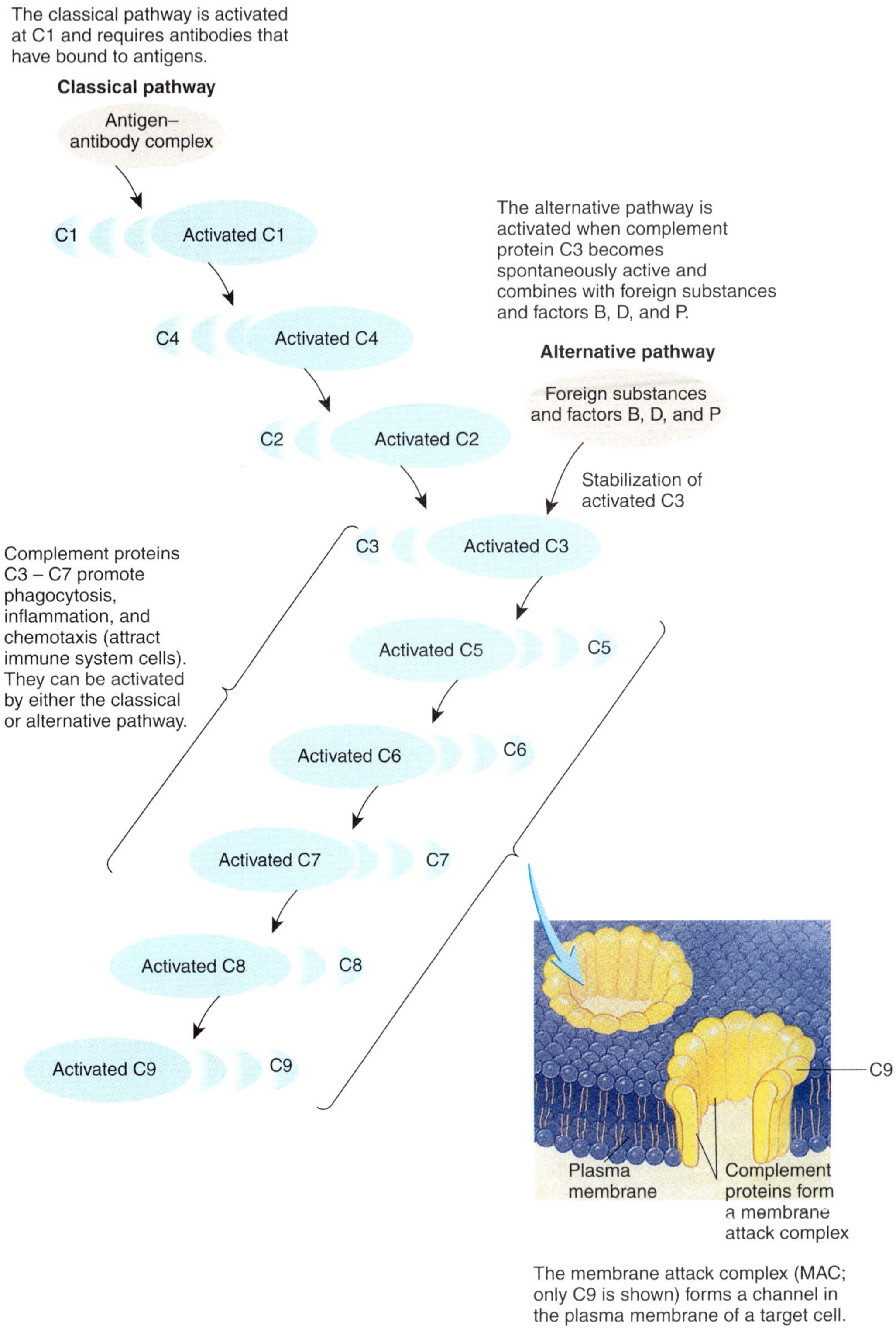
The classical pathway is activated at C1 and requires antibodies that have bound to antigens.
Classical pathway
Antigen–antibody complex
C1
Activated C1
C4
Activated C4
C2
Activated C2
The alternative pathway is activated when complement protein C3 becomes spontaneously active and combines with foreign substances and factors B, D, and P.
Alternative pathway
Foreign substances and factors B, D, and P
Stabilization of activated C3
C3
Activated C3
Complement proteins C3 – C7 promote phagocytosis, inflammation, and chemotaxis (attract immune system cells). They can be activated by either the classical or alternative pathway.
Activated C5
C5
Activated C6
C6
Activated C7
C7
Activated C8
C8
Activated C9
C9
C9
Plasma membrane
Complement proteins form a membrane attack complex
The membrane attack complex (MAC; only C9 is shown) forms a channel in the plasma membrane of a target cell.

Complement Cascade
Figure 22.10

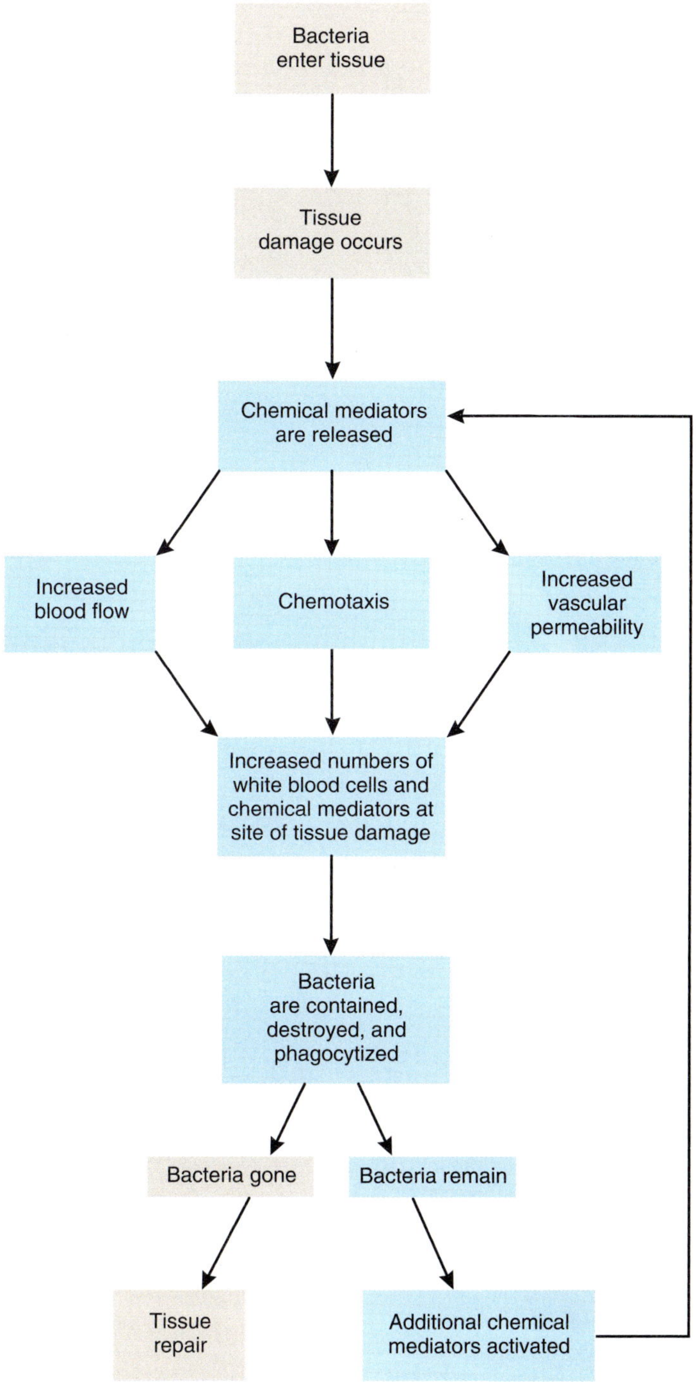
Bacteria enter tissue
Tissue damage occurs
Chemical mediators are released
Increased blood flow
Chemotaxis
Increased vascular permeability
Increased numbers of white blood cells and chemical mediators at site of tissue damage
Bacteria are contained, destroyed, and phagocytized
Bacteria gone
Bacteria remain
Tissue repair
Additional chemical mediators activated

Inflammatory Response
Figure 22.11

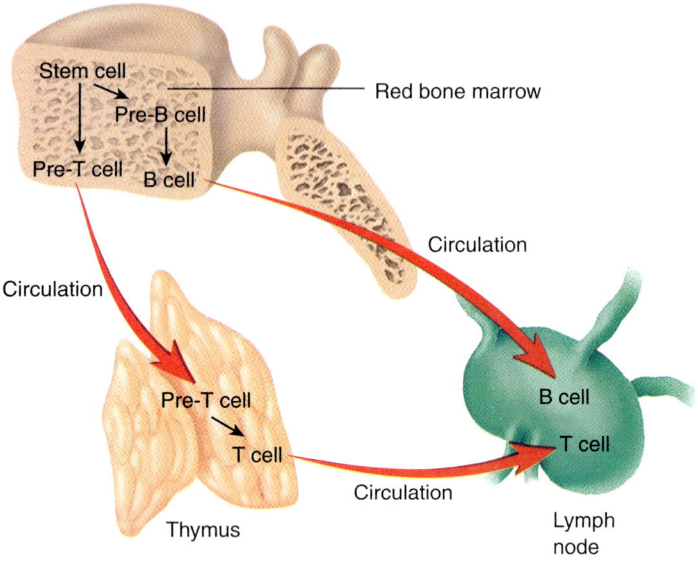

Origin and Processing of B and T Cells
Figure 22.12

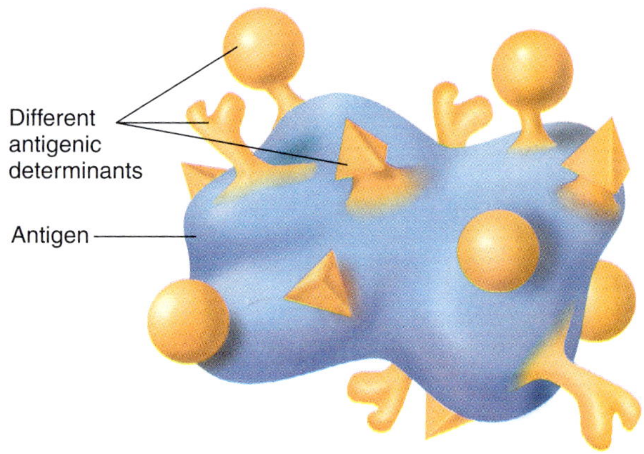

Antigenic Determinants
Figure 22.13

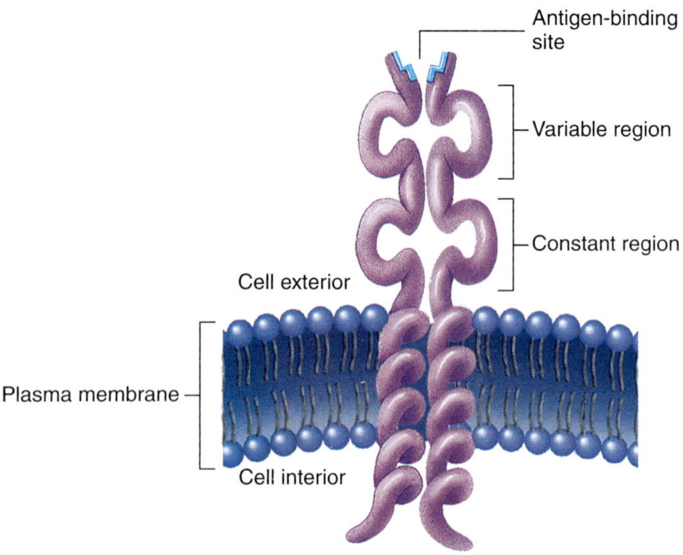

The T-Cell Receptor
Figure 22.14

1. Foreign proteins or self-proteins within the cytosol are broken down into fragments that are antigens.
2. Antigens are transported into the rough endoplasmic reticulum.
3. Antigens combine with MHC class I molecules.
4. The MHC class I/antigen complex is transported to the Golgi apparatus, packaged into a vesicle, and transported to the plasma membrane.
5. Foreign antigens combined with MHC class I molecules stimulate cell destruction.
6. Self-antigens combined with MHC class I molecules do not stimulate cell destruction.

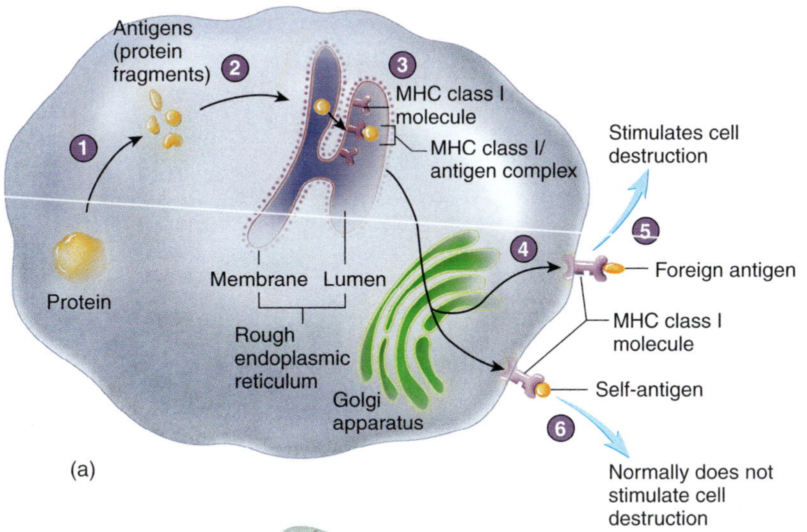

(a)

1. A foreign antigen is ingested by endocytosis and is within a vesicle.
2. The antigen is broken down into fragments to form processed foreign antigens.
3. The vesicle containing the processed foreign antigens fuses with vesicles produced by the Golgi apparatus that contain MHC class II molecules. Processed foreign antigens and MHC class II molecules combine.
4. The MHC class II/antigen complex is transported to the plasma membrane.
5. The displayed MHC class II/antigen complex can stimulate immune cells.

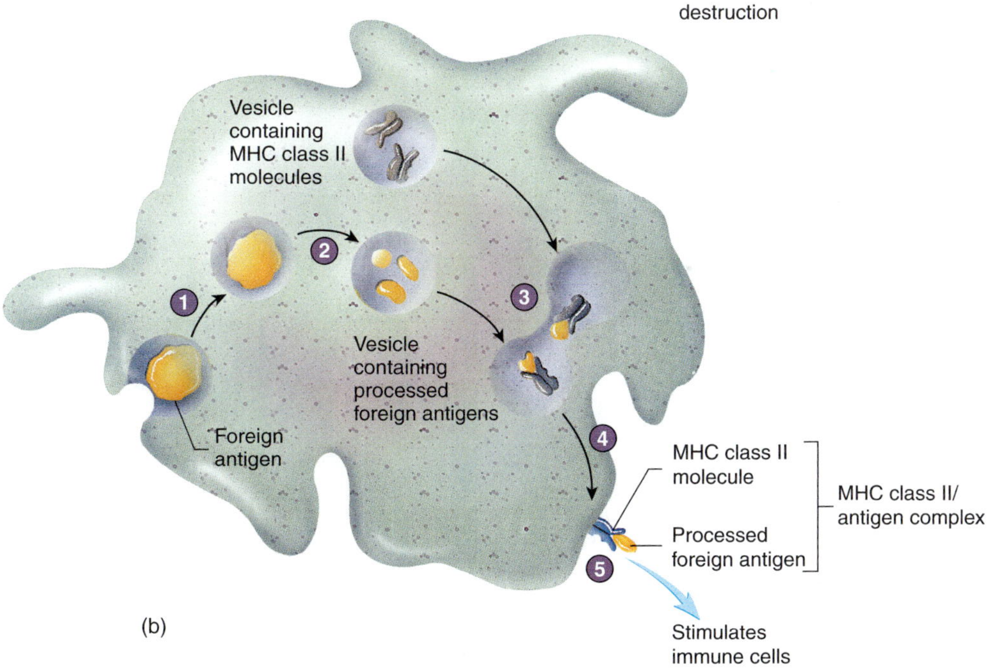

(b)

Antigen Processing
Process Figure 22.15

(a) Helper T cells are activated by a first signal and by costimulation. The first signal is the binding of the MHC class II/antigen complex to the T-cell receptor. Costimulation is an additional signal, such as molecules released from another cell. For example, the release by the macrophage of cytokines that bind to receptors on the helper T cell.

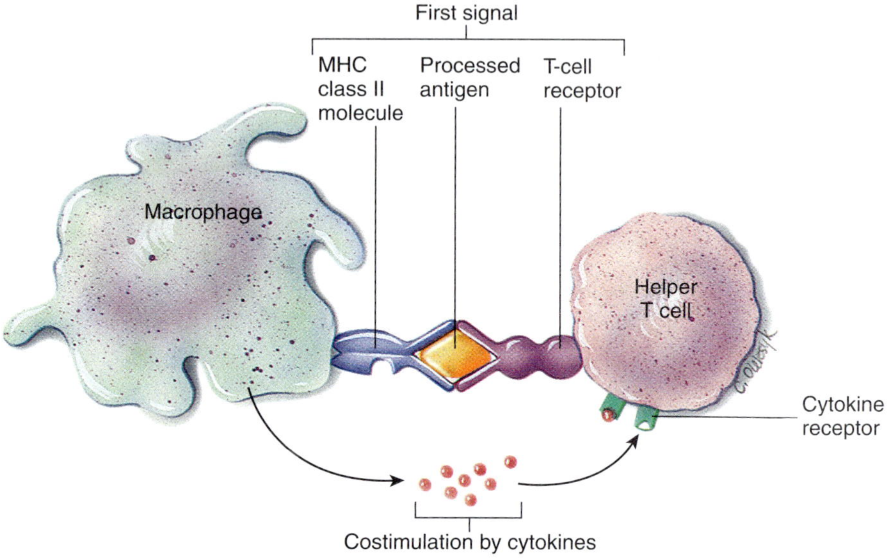

(b) Other costimulatory signals are the combining of surface molecules between cells, such as the binding of a B7 molecule of the macrophage with a CD28 molecule of the helper T cell. The CD4 molecule of the helper T cell binds to the macrophage's MHC class II molecule and helps to hold the cells together.

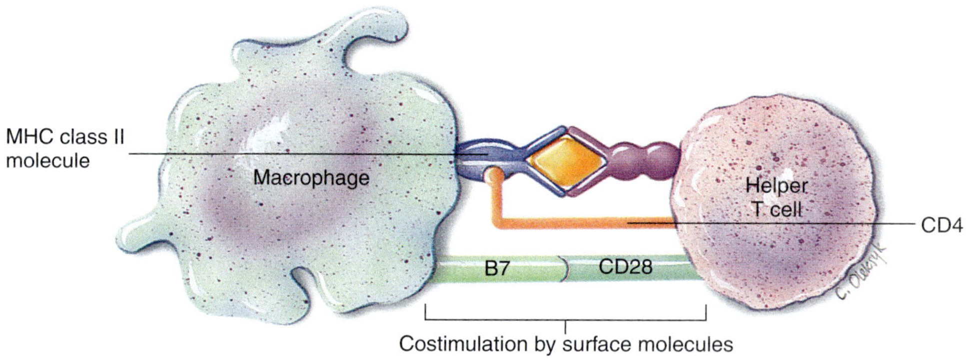

Costimulation
Figure 22.16

1. Antigen-presenting cells such as macrophages take in, process, and display antigens on the cell's surface.
2. The antigen is bound to a MHC class II molecule, which functions to present the processed antigen to a T-cell receptor of a helper T cell for recognition.
3. Costimulation occurs by a CD4 glycoprotein of the helper T cell or by cytokines. The macrophage secretes a cytokine called interleukin-1.
4. Interleukin-1 attaches to interleukin receptors and stimulates the helper T cell to secrete the cytokine interleukin-2 and to produce interleukin-2 receptors.
5. The helper T cell stimulates itself to divide when interleukin-2 binds to interleukin-2 receptors.
6. The "daughter" helper T cells resulting from this division can be stimulated to divide again if they are exposed to the same antigen that stimulated the "parent" helper T cell. This greatly increases the number of helper T cells.
7. The increased number of helper T cells can facilitate the activation of B cells or effector T cells.

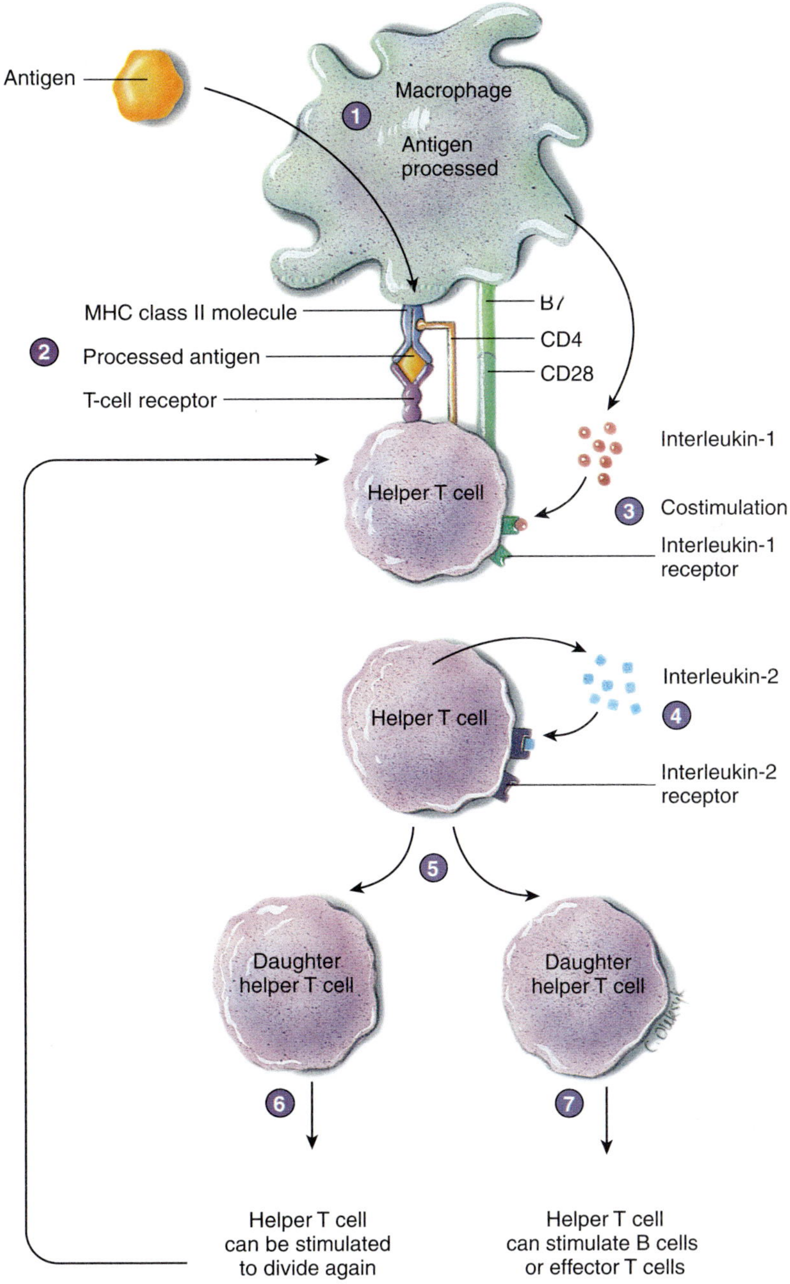

Proliferation of Helper T Cells
Process Figure 22.17

1. Before a B cell can be activated by a helper T cell, the B cell must phagocytize and process the same antigen that activated the helper T cell. The antigen binds to a B-cell receptor, and both the receptor and antigen are taken into the cell by endocytosis.

2. The B cell uses a MHC class II molecule to present the processed antigen to the helper T cell.

3. The T-cell receptor binds to the MHC class II/antigen complex.

4. There is costimulation of the B cell by CD4 and other surface molecules.

5. There is costimulation by interleukins (cytokines) released from the helper T cell.

6. The B cell divides, and the resulting daughter cells divide, and so on, eventually producing many cells (only two are shown here).

7. The daughter cells produce antibodies, which are part of the antibody-mediated immune system response that eliminates the antigen.

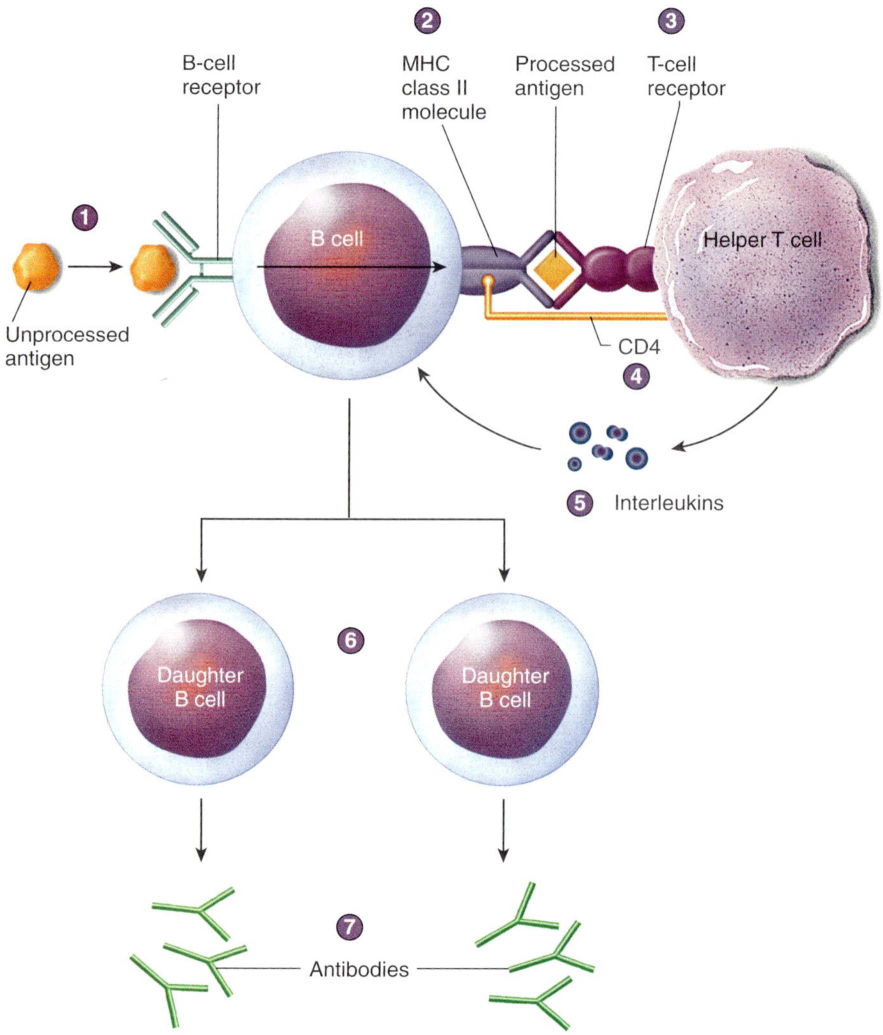

Proliferation of B Cells
Process Figure 22.18

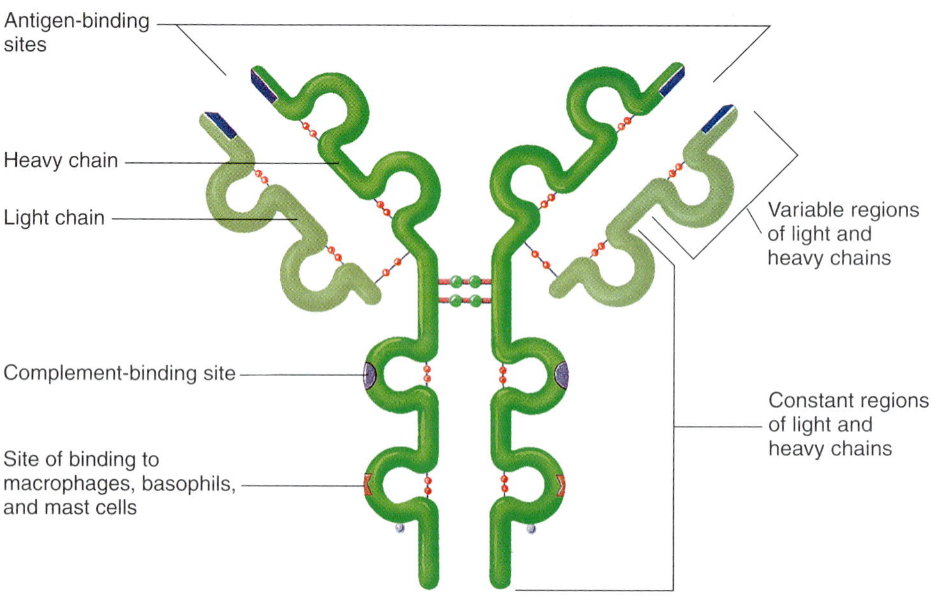

Structure of an Antibody
Figure 22.19

(a) **Inactivates the antigen.** An antibody binds to an antigen and inactivates it.

(b) **Binds antigens together.** Antibodies bind several antigens together.

(c) **Activates the complement cascade.** An antigen binds to an antibody. As a result, the antibody can activate complement proteins, which can produce inflammation, chemotaxis, and lysis.

(d) **Initiates the release of inflammatory chemicals.** An antibody binds to a mast cell or basophil. When an antigen binds to the antibody, it triggers a release of chemicals that cause inflammation.

(e) **Facilitates phagocytosis.** An antibody binds to an antigen and then to a macrophage, which phagocytizes the antibody and antigen.

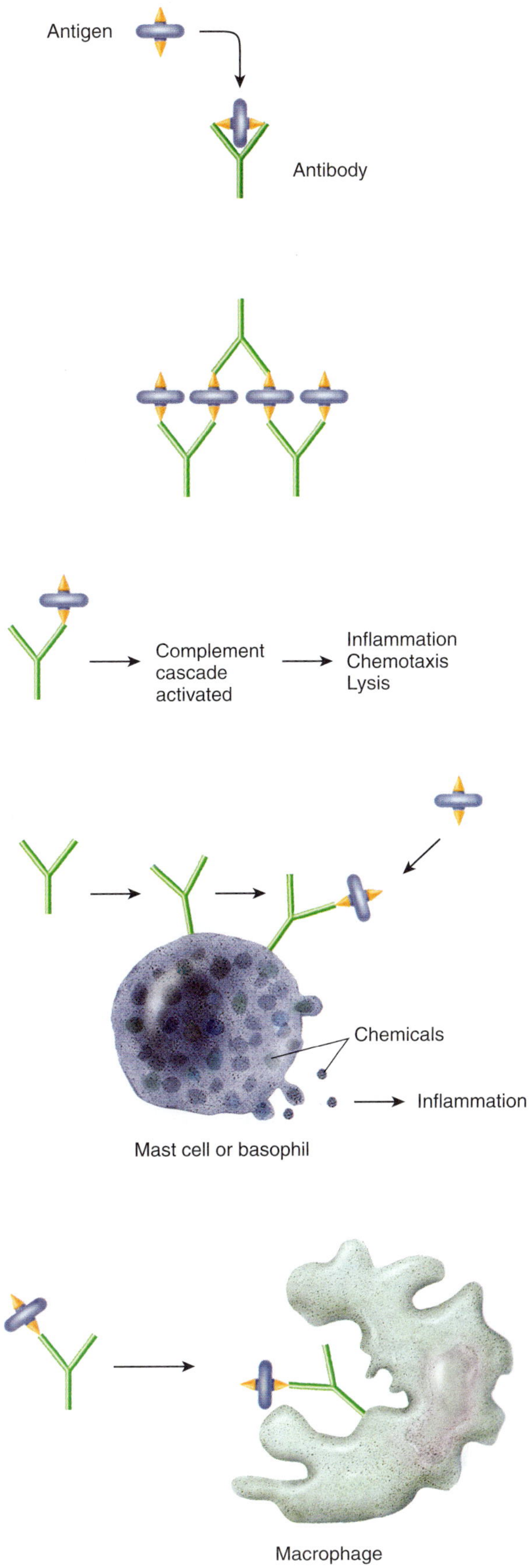

Effects of Antibodies
Figure 22.20

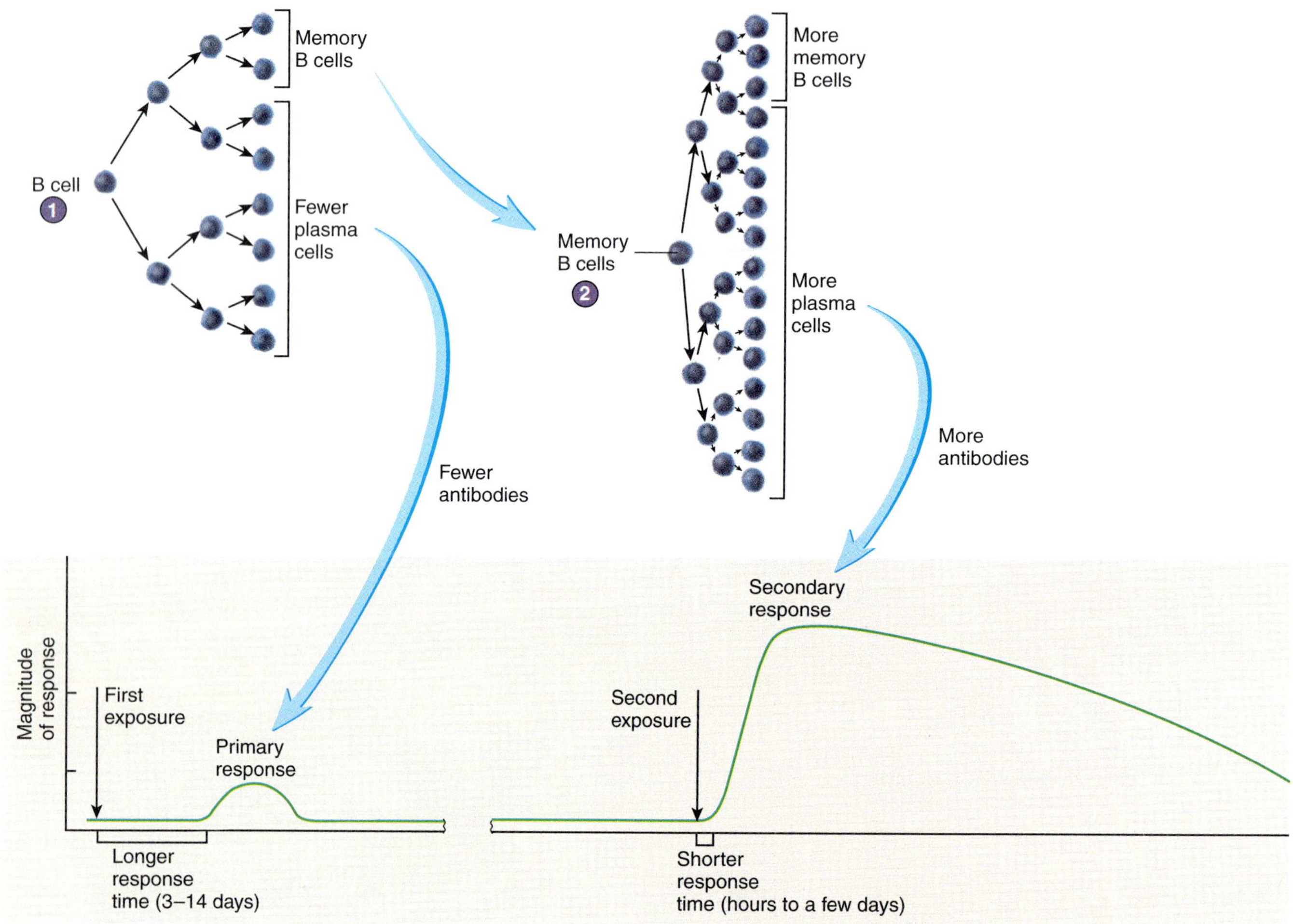

1. **Primary response.** The primary response occurs when a B cell is first activated by an antigen. The B cell proliferates to form plasma cells and memory cells. The plasma cells produce antibodies.

2. **Secondary response.** The secondary response occurs when another exposure to the same antigen causes the memory cells to rapidly form plasma cells and additional memory cells. The secondary response is faster and produces more antibodies than the primary response.

Antibody Production
Process Figure 22.21

1. A MHC class I molecule displays an antigen, such as a viral protein, on the surface of a target cell.
2. Activation of a cytotoxic T cell begins when a T-cell receptor binds to the MHC class I/antigen complex.
3. There is costimulation of the cytotoxic T cell by CD8 and other surface molecules.
4. There is costimulation by cytokines, such as interleukin-2, released from helper T cells.
5. The activated cytotoxic T cell divides, and the resulting daughter cells divide, and so on, eventually producing many cytotoxic T cells (only two are shown here).

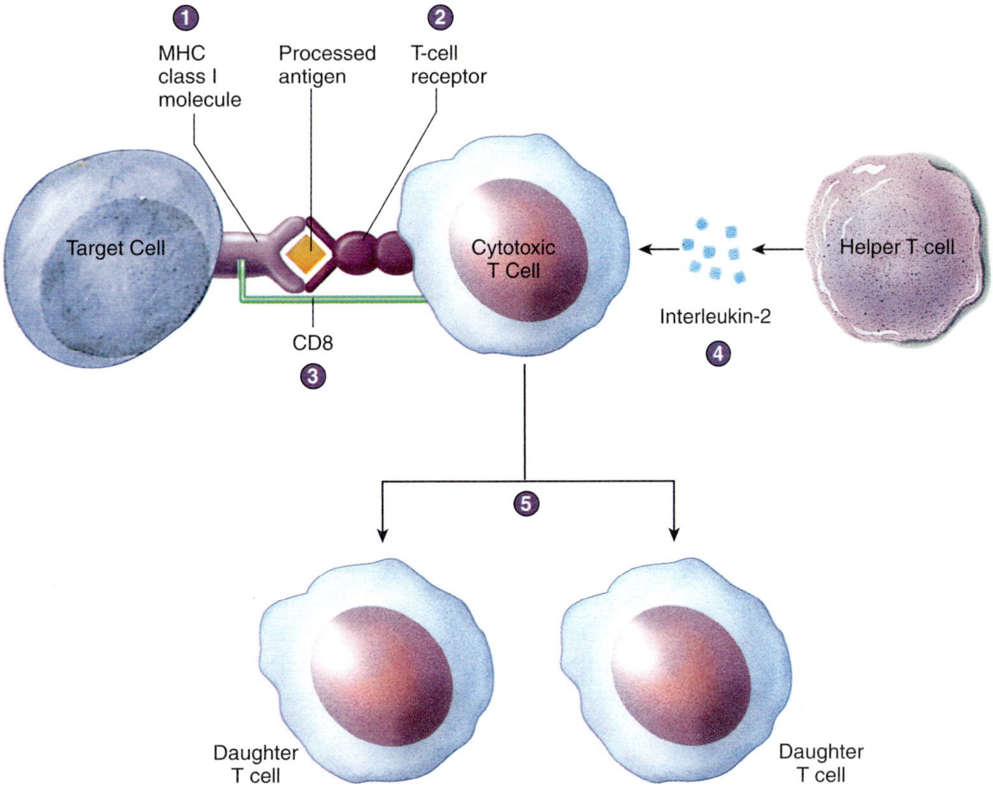

Proliferation of Cytotoxic T Cells
Process Figure 22.22

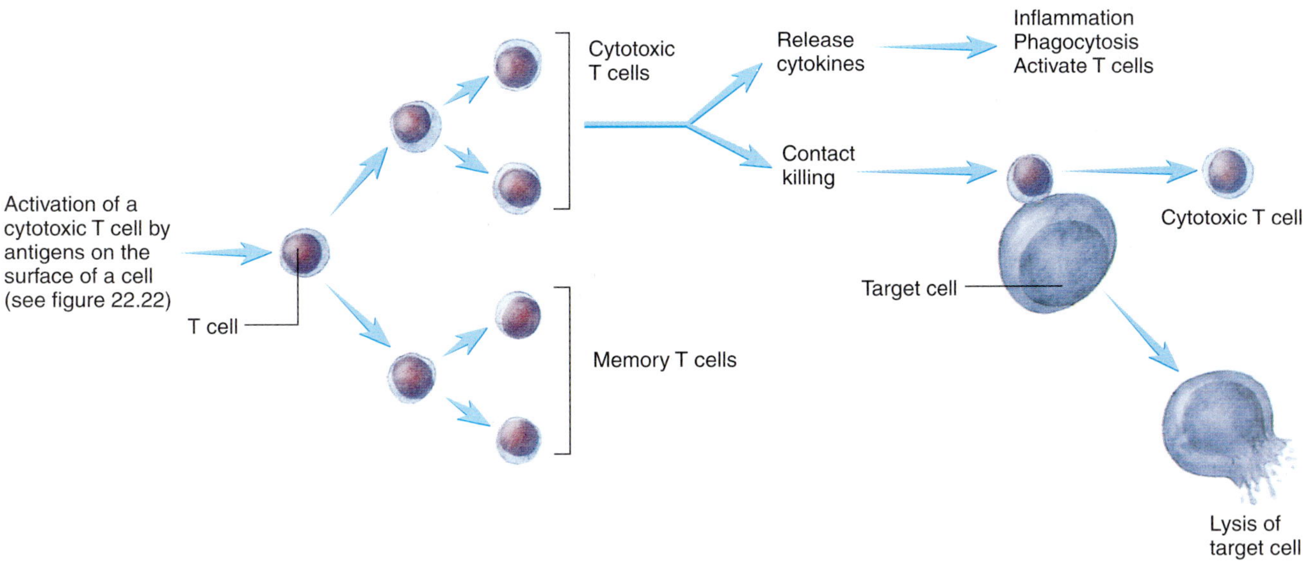

Stimulation and Effects of T Cells
Figure 22.23

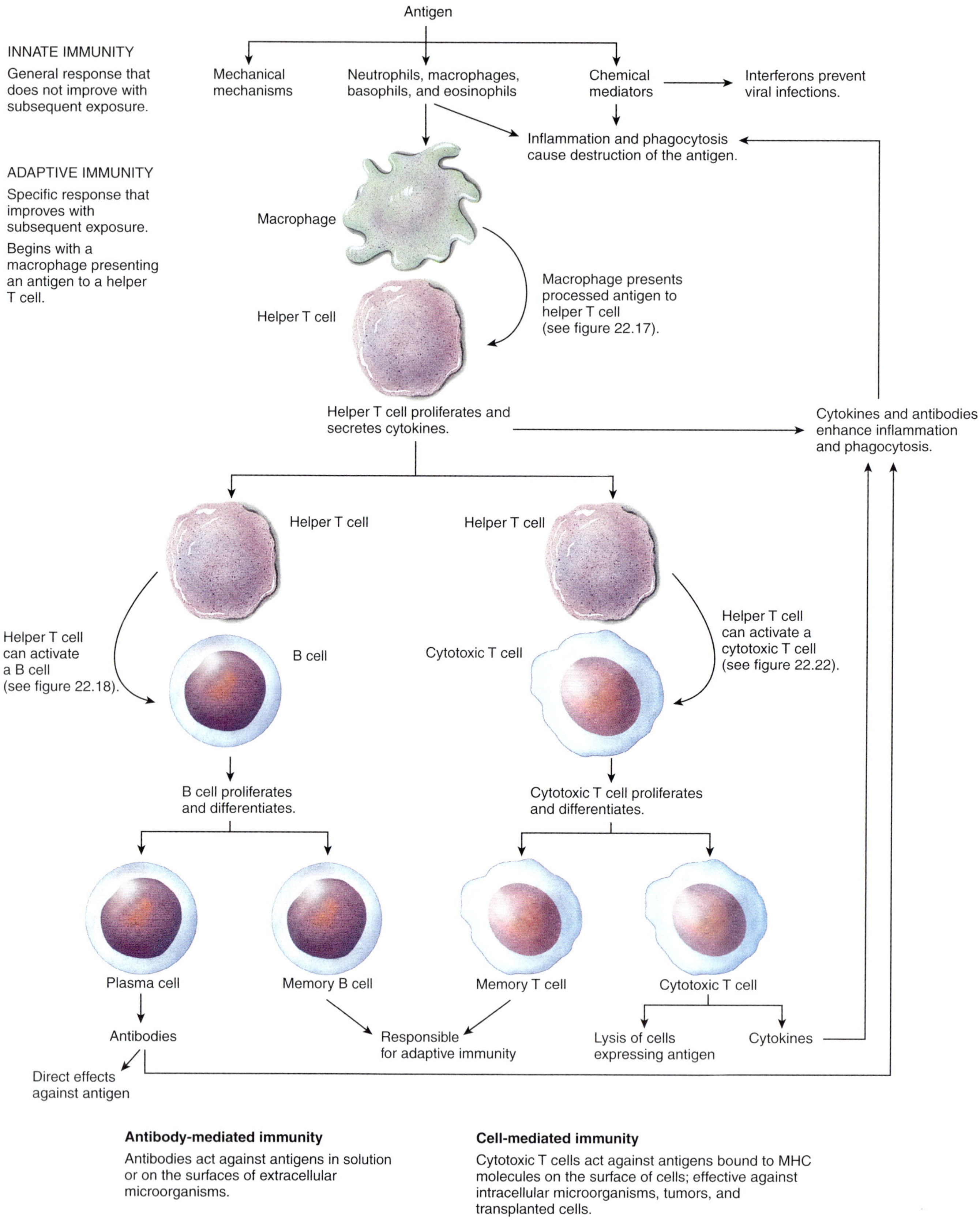
Antigen
INNATE IMMUNITY
General response that does not improve with subsequent exposure.
Mechanical mechanisms
Neutrophils, macrophages, basophils, and eosinophils
Chemical mediators
Interferons prevent viral infections.
Inflammation and phagocytosis cause destruction of the antigen.
ADAPTIVE IMMUNITY
Specific response that improves with subsequent exposure.
Begins with a macrophage presenting an antigen to a helper T cell.
Macrophage
Helper T cell
Macrophage presents processed antigen to helper T cell (see figure 22.17).
Helper T cell proliferates and secretes cytokines.
Cytokines and antibodies enhance inflammation and phagocytosis.
Helper T cell
Helper T cell
Helper T cell can activate a B cell (see figure 22.18).
B cell
Cytotoxic T cell
Helper T cell can activate a cytotoxic T cell (see figure 22.22).
B cell proliferates and differentiates.
Cytotoxic T cell proliferates and differentiates.
Plasma cell
Memory B cell
Memory T cell
Cytotoxic T cell
Antibodies
Responsible for adaptive immunity
Lysis of cells expressing antigen
Cytokines
Direct effects against antigen
Antibody-mediated immunity
Antibodies act against antigens in solution or on the surfaces of extracellular microorganisms.
Cell-mediated immunity
Cytotoxic T cells act against antigens bound to MHC molecules on the surface of cells; effective against intracellular microorganisms, tumors, and transplanted cells.

Immune Interactions
Figure 22.24

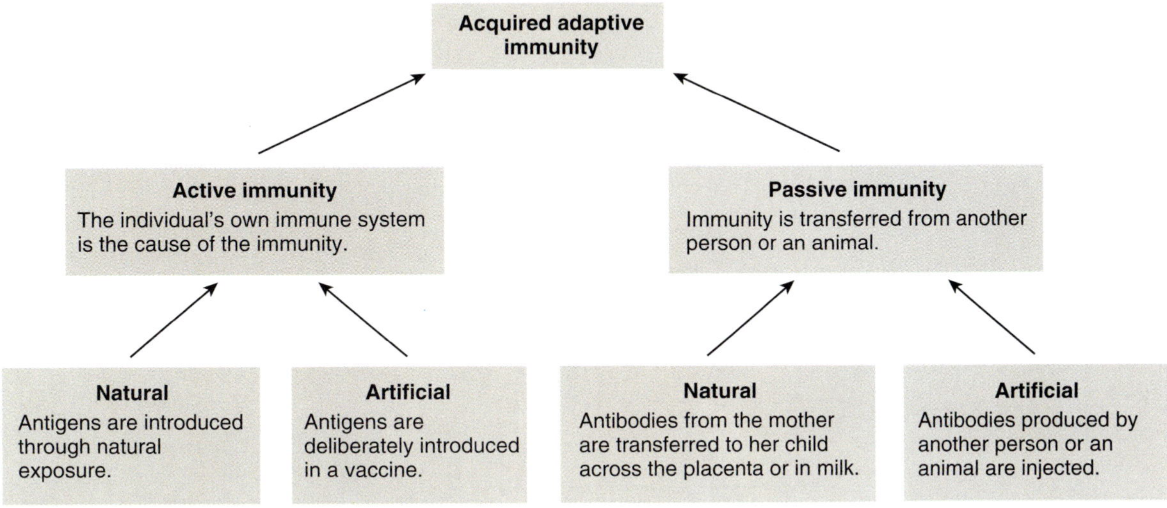

Ways to Acquire Adaptive Immunity
Figure 22.25

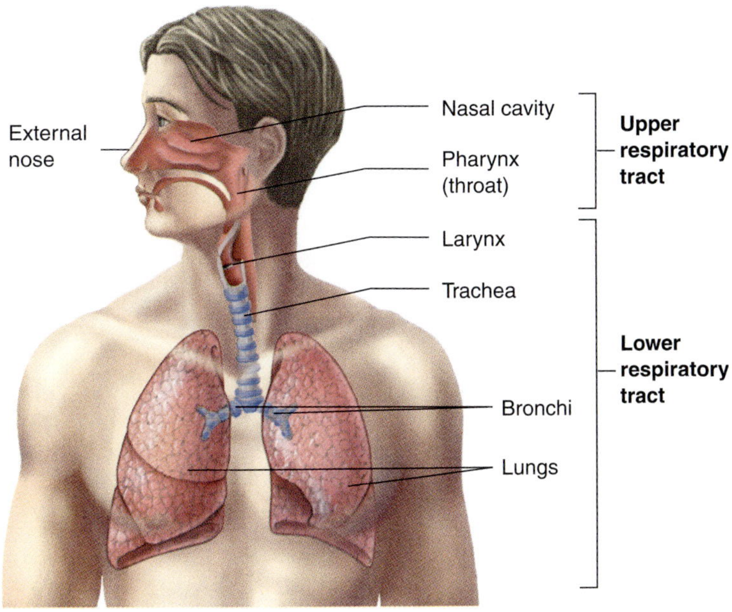

The Respiratory System
Figure 23.1

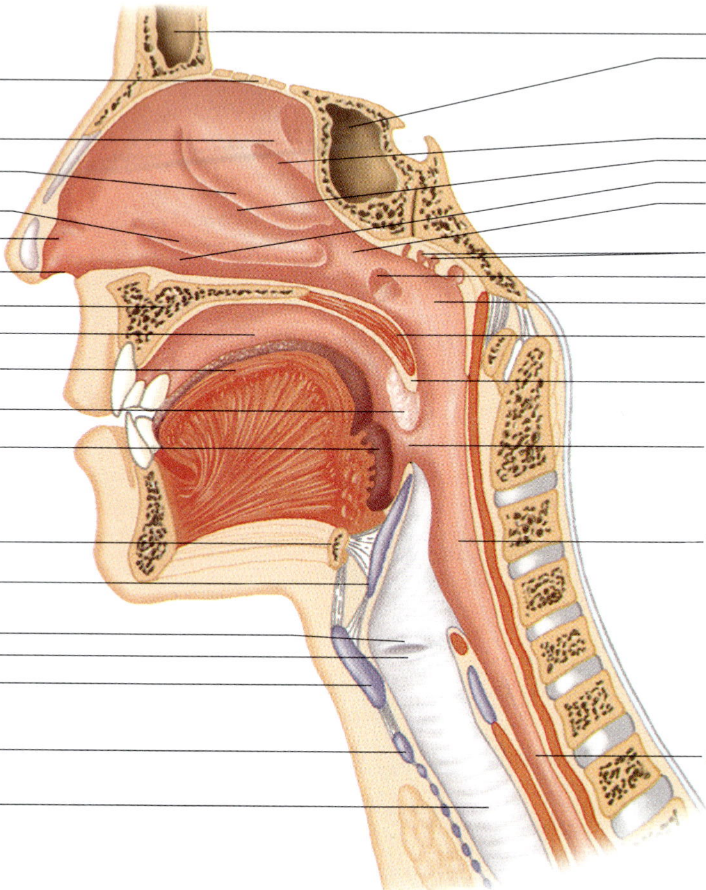

Frontal sinus
Sphenoidal sinus
Paranasal sinuses
Cribriform plate
Superior concha
Middle concha
Nasal cavity
Inferior concha
Vestibule
Naris
Superior meatus
Middle meatus
Inferior meatus
Choana
Nasal cavity
Pharyngeal tonsil
Opening of auditory tube
Hard palate
Oral cavity
Tongue
Palatine tonsil
Lingual tonsil
Nasopharynx
Soft palate
Uvula
Pharynx
Oropharynx
Hyoid bone
Laryngopharynx
Epiglottis
Vestibular fold
Vocal fold
Larynx
Thyroid cartilage
Cricoid cartilage
Trachea
Esophagus
(a)
Medial view

Nasal Cavity and Pharynx
Figure 23.2

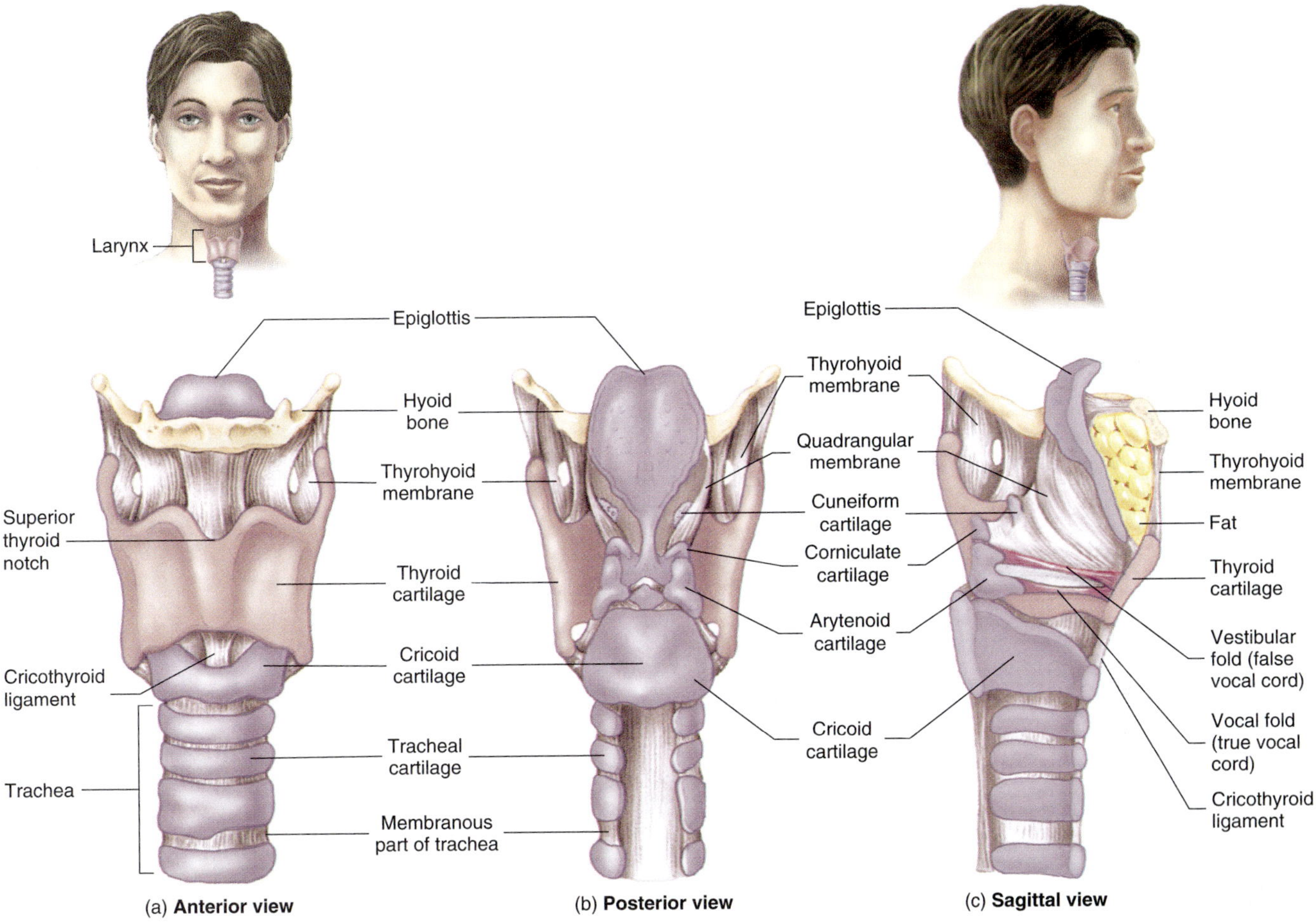
Larynx
Epiglottis
Hyoid bone
Thyrohyoid membrane
Superior thyroid notch
Thyroid cartilage
Cricothyroid ligament
Cricoid cartilage
Trachea
Tracheal cartilage
Membranous part of trachea
(a) Anterior view
(b) Posterior view
Epiglottis
Thyrohyoid membrane
Quadrangular membrane
Cuneiform cartilage
Corniculate cartilage
Arytenoid cartilage
Cricoid cartilage
Hyoid bone
Thyrohyoid membrane
Fat
Thyroid cartilage
Vestibular fold (false vocal cord)
Vocal fold (true vocal cord)
Cricothyroid ligament
(c) Sagittal view

Anatomy of the Larynx
Figure 23.3

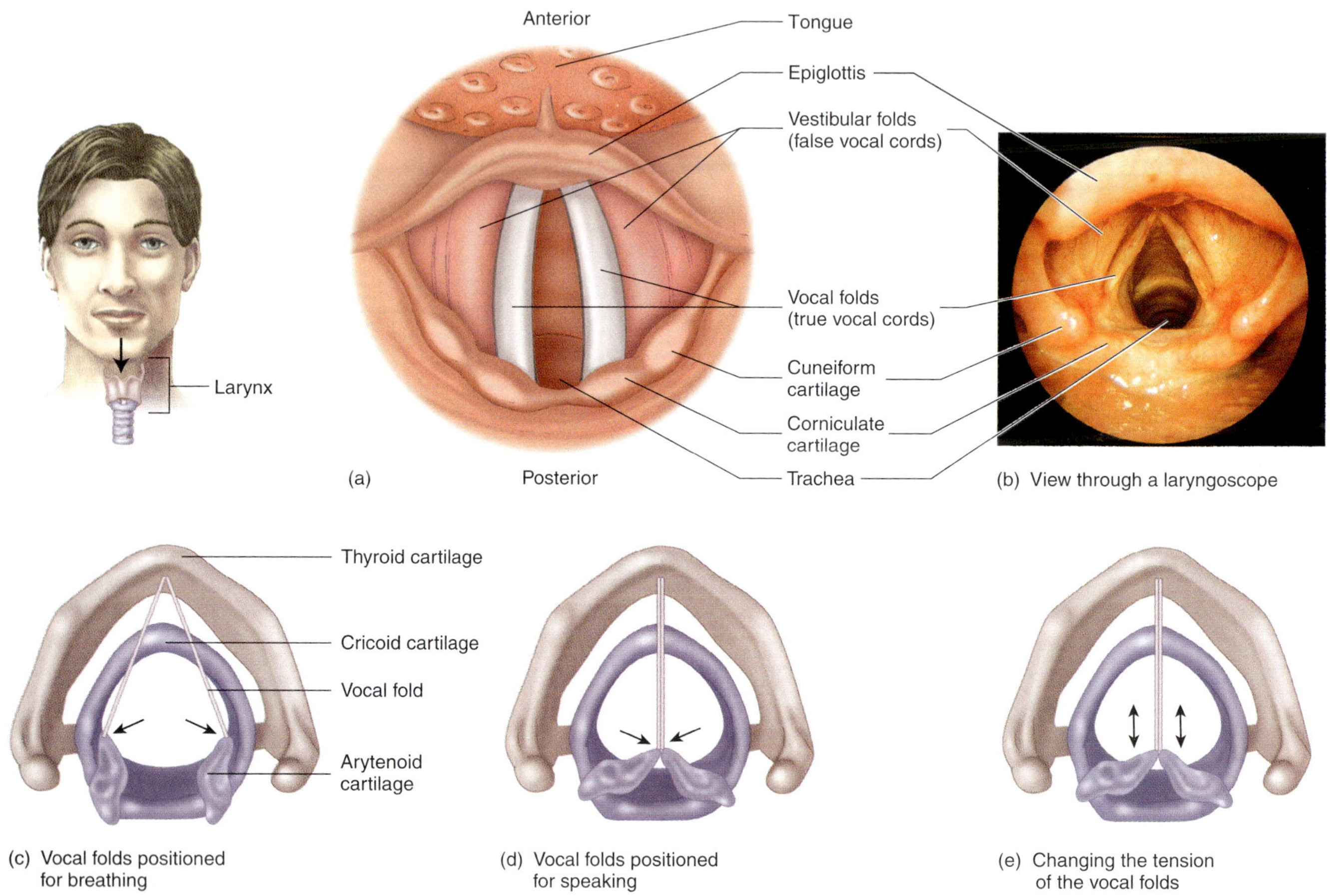

Vestibular Vocal Folds

Figure 23.4

b: © CNRI/Phototake

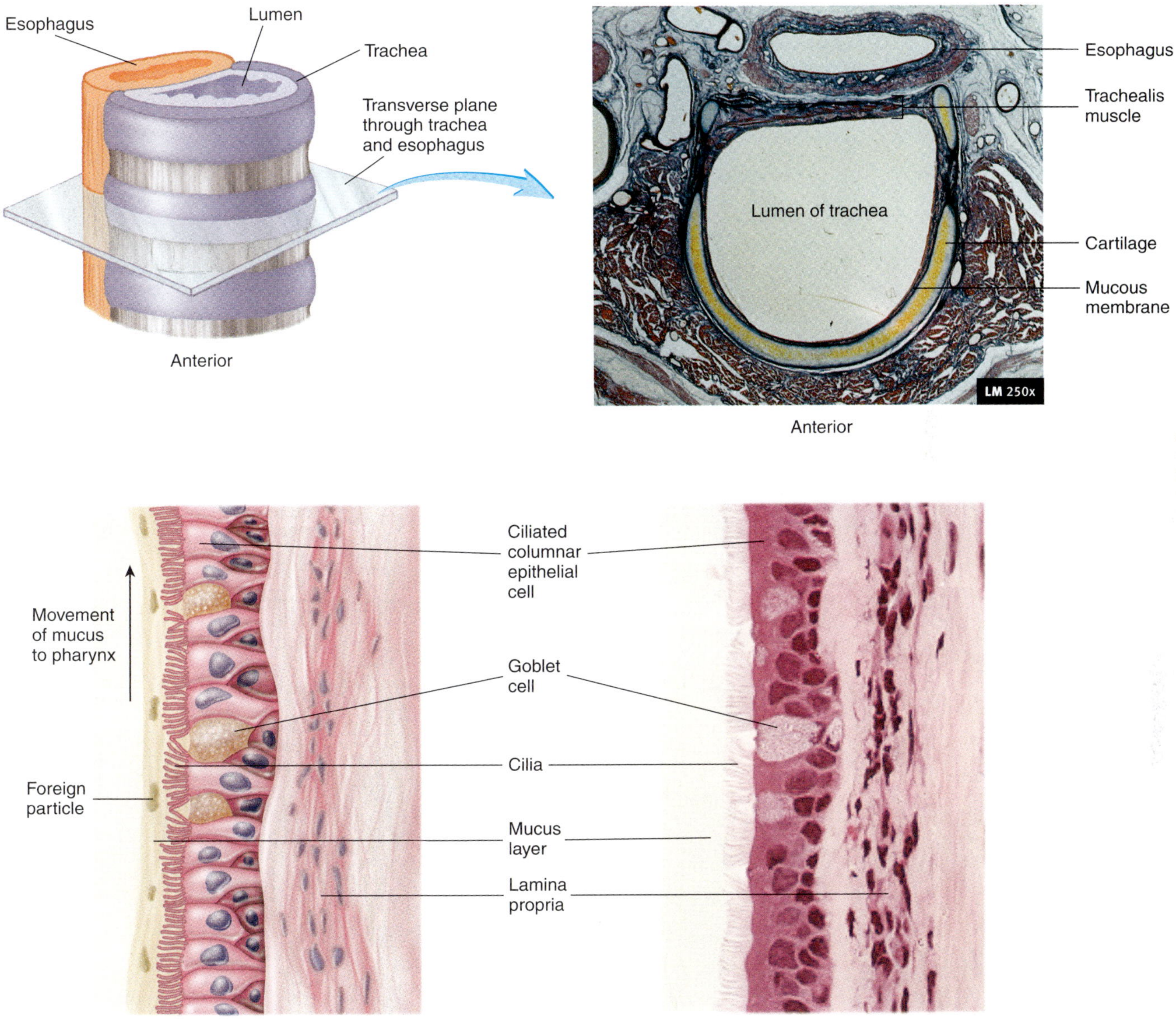

Trachea
Figure 23.5
top: © John Cunningham/Visuals Unlimited; **bottom:** © Ed Reschke/Peter Arnold, Inc.

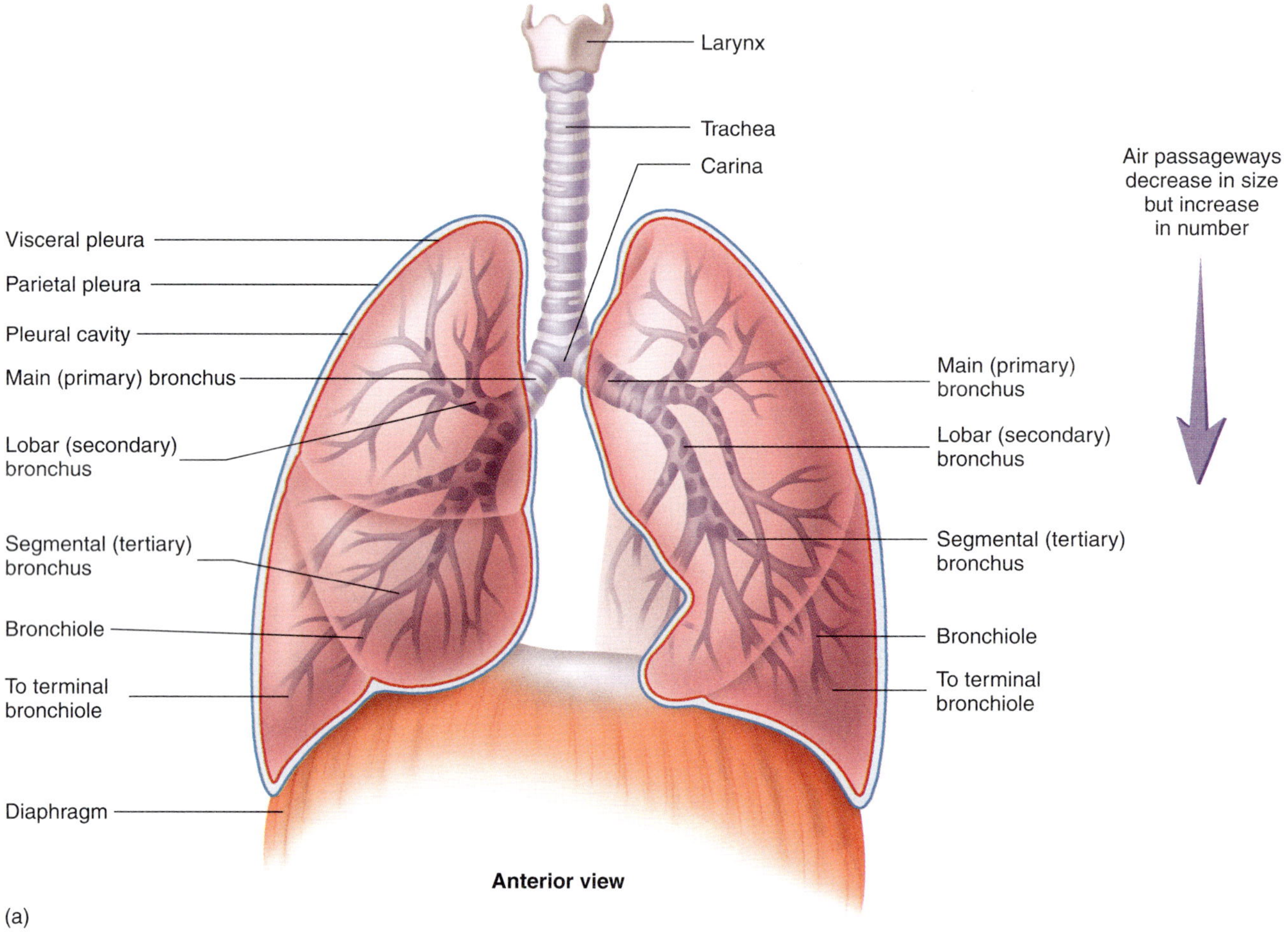
Larynx
Trachea
Carina
Air passageways decrease in size but increase in number
Visceral pleura
Parietal pleura
Pleural cavity
Main (primary) bronchus
Lobar (secondary) bronchus
Segmental (tertiary) bronchus
Bronchiole
To terminal bronchiole
Diaphragm
Main (primary) bronchus
Lobar (secondary) bronchus
Segmental (tertiary) bronchus
Bronchiole
To terminal bronchiole
Anterior view

(a)

Tracheobronchial Tree
Figure 23.6

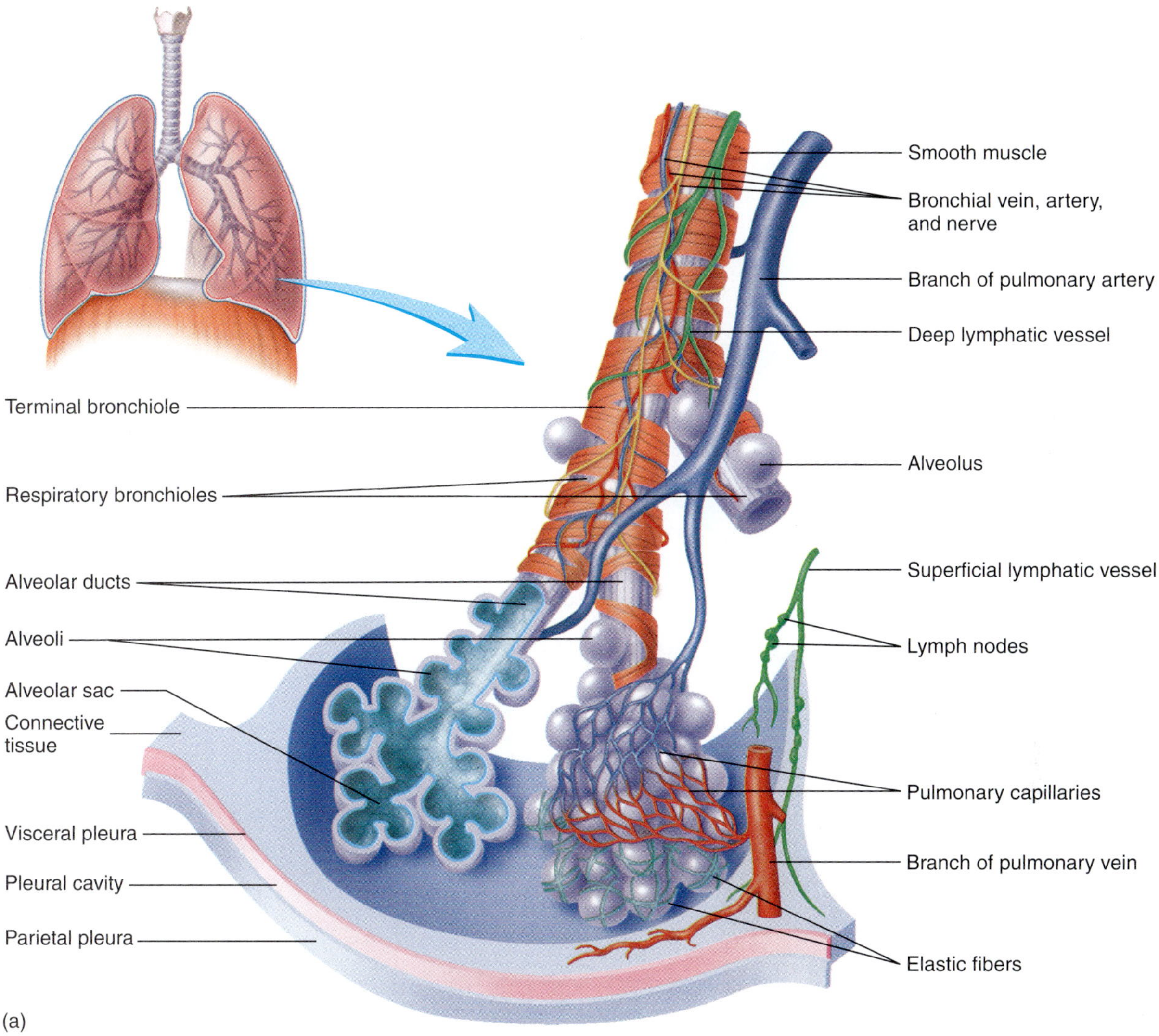
Smooth muscle
Bronchial vein, artery, and nerve
Branch of pulmonary artery
Deep lymphatic vessel
Terminal bronchiole
Alveolus
Respiratory bronchioles
Superficial lymphatic vessel
Alveolar ducts
Alveoli
Lymph nodes
Alveolar sac
Connective tissue
Pulmonary capillaries
Visceral pleura
Branch of pulmonary vein
Pleural cavity
Parietal pleura
Elastic fibers
(a)

Bronchioles and Alveoli
Figure 23.7

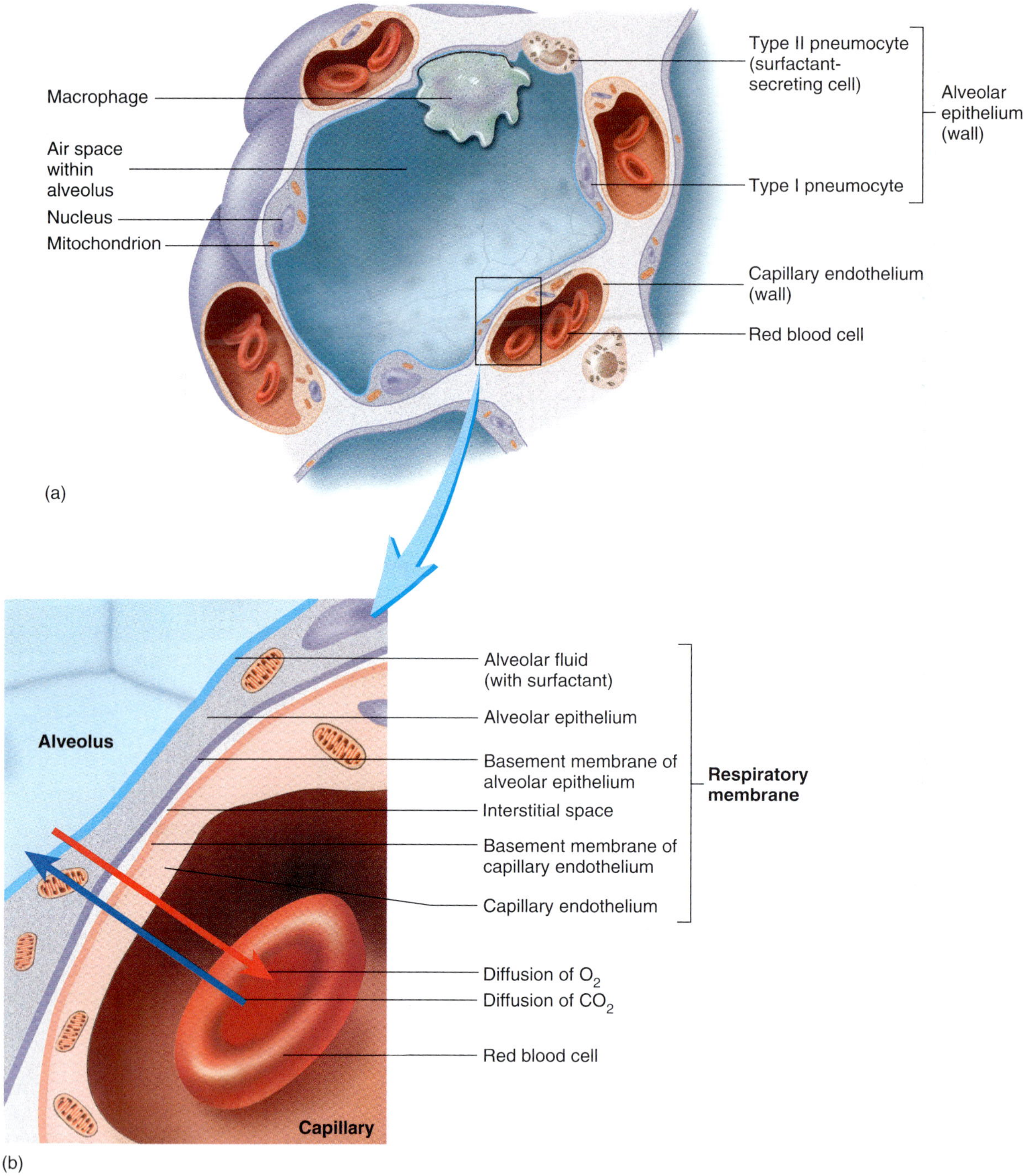

Alveolus and the Respiratory Membrane
Figure 23.8

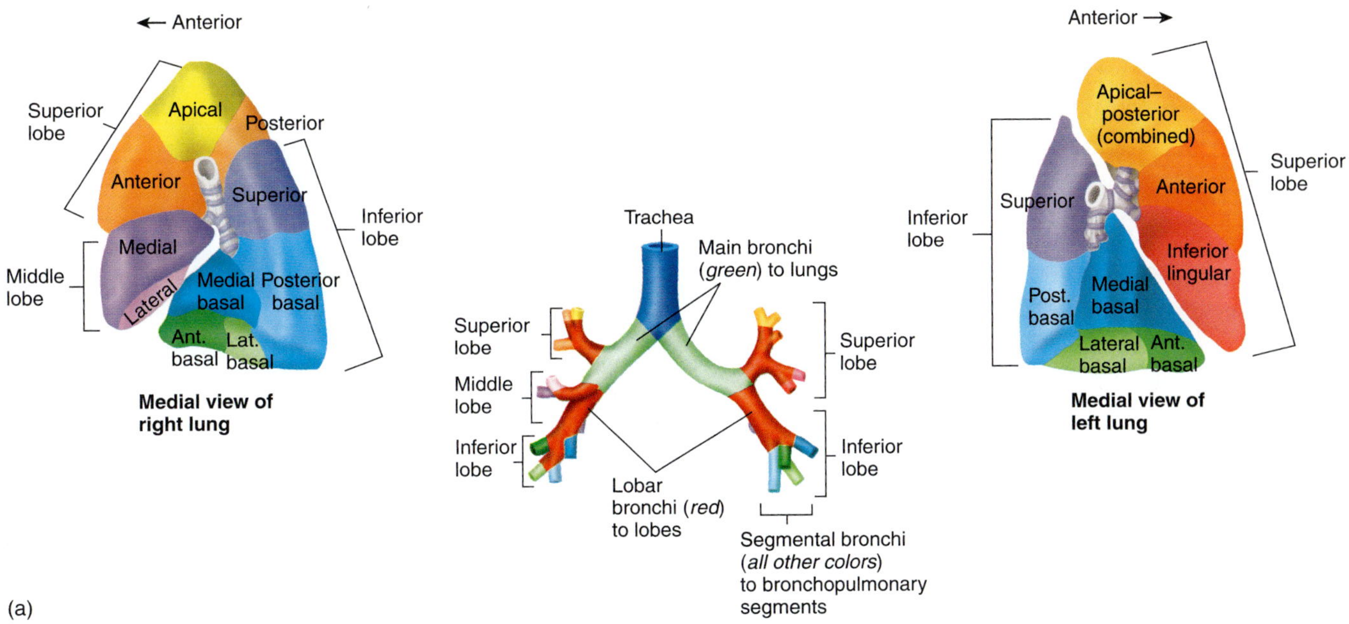

Lobes and Bronchopulmonary Segments of the Lungs
Figure 23.9

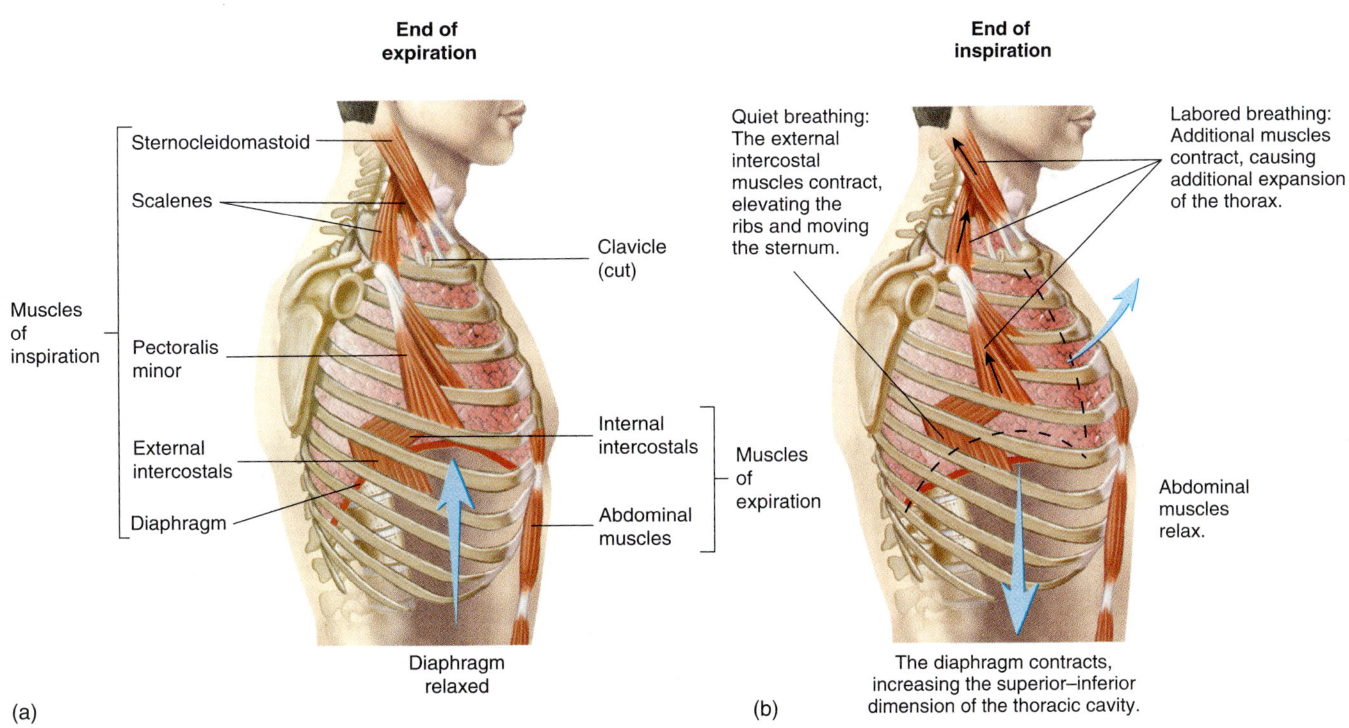

Effect of the Muscles of Respiration on Thoracic Volume
Figure 23.10

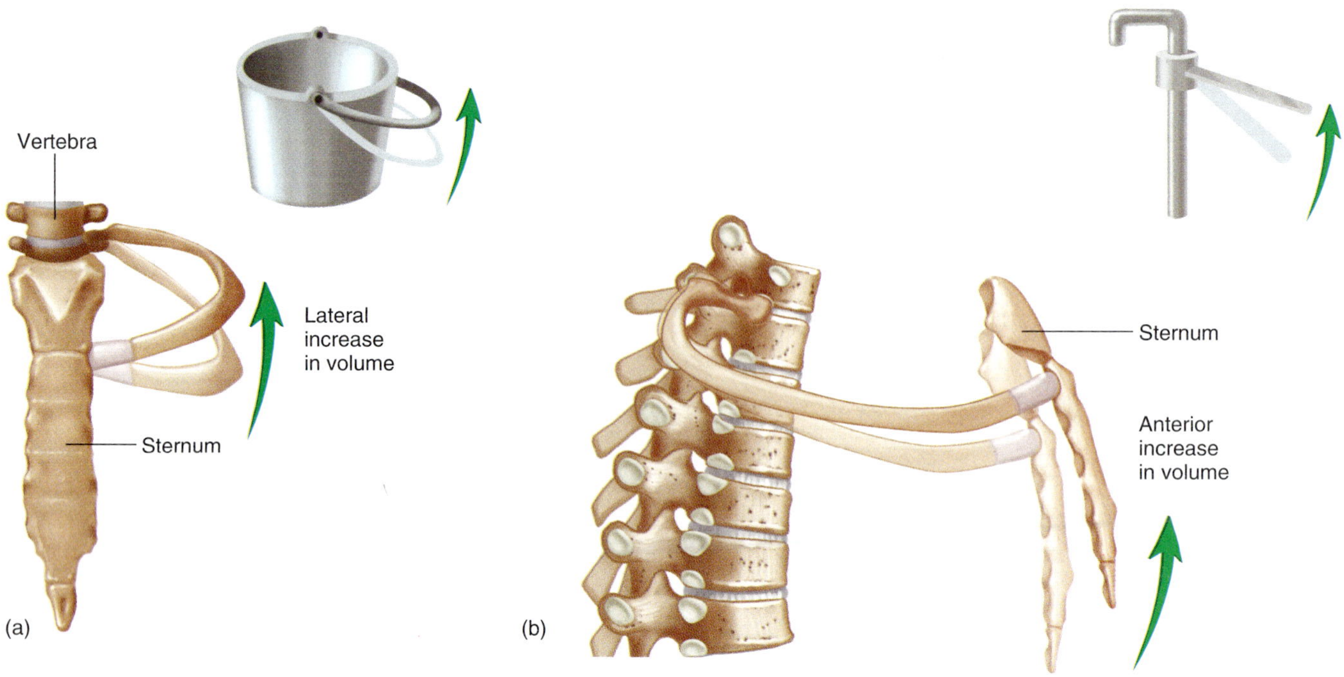

Effect of Rib and Sternum Movement on Thoracic Volume
Figure 23.11

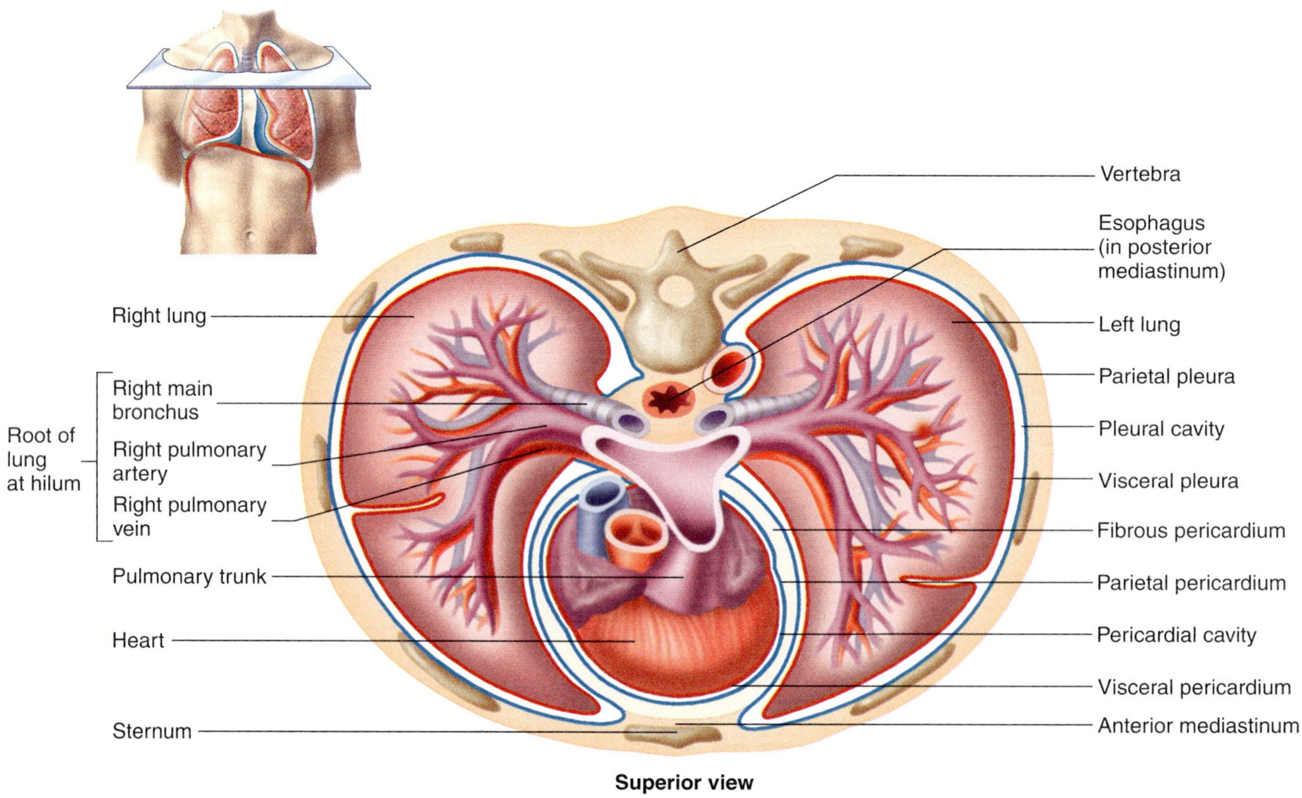

Pleural Cavities and Membranes
Figure 23.12

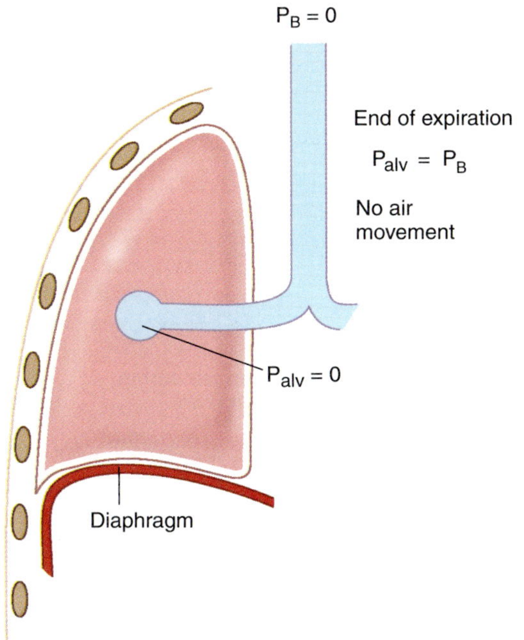

1. At the end of expiration, alveolar pressure (P_{alv}) is equal to barometric air pressure (P_B) and there is no air movement.

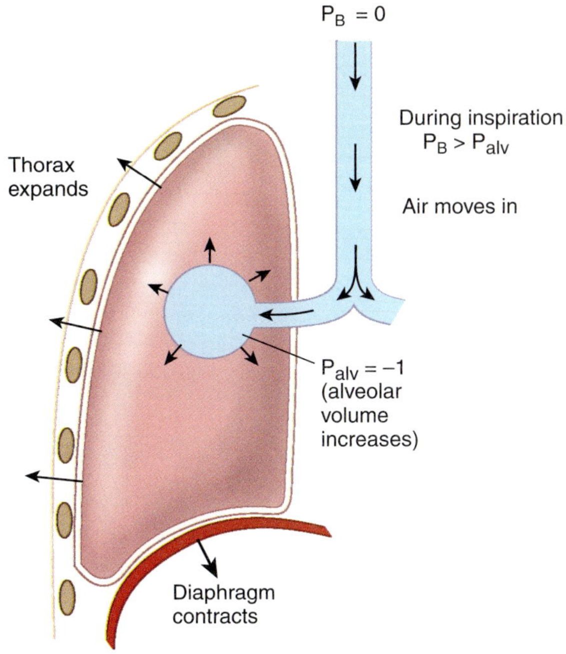

2. During inspiration, increased thoracic volume results in increased alveolar volume and decreased alveolar pressure. Barometric air presure is greater than alveolar pressure, and air moves into the lungs.

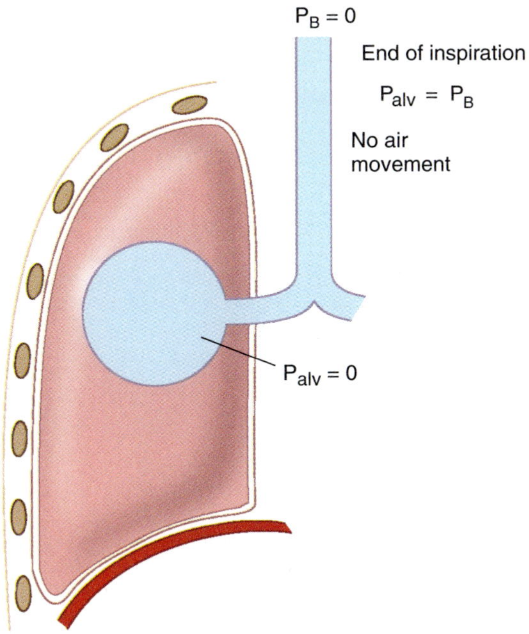

3. At the end of inspiration, alveolar pressure is equal to barometric air pressure and there is no air movement.

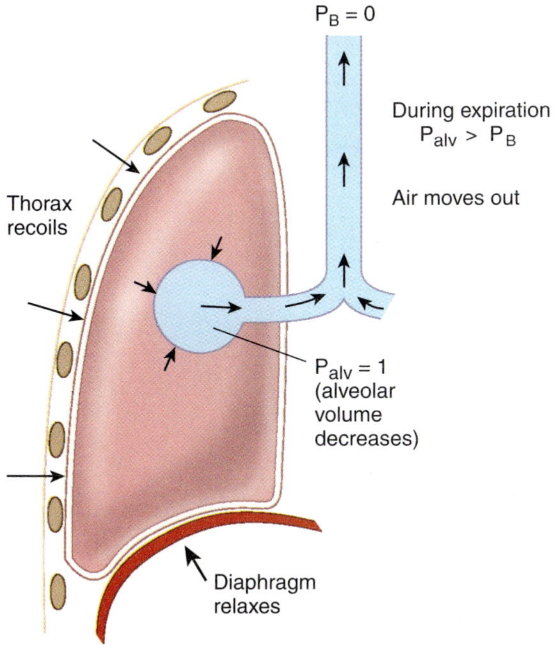

4. During expiration, decreased thoracic volume results in decreased alveolar volume and increased alveolar pressure. Alveolar pressure is greater than barometric air pressure, and air moves out of the lungs.

Alveolar Pressure Changes During Inspiration and Expiration
Process Figure 23.13

Changes during inspiration

1. Pleural pressure decreases because thoracic volume increases.
2. As inspiration begins, alveolar pressure decreases below barometric air pressure (0 on the graph) because the decreased pleural pressure causes alveolar volume to increase. By the end of inspiration, alveolar and barometric air pressure are equal.
3. During inspiration, air flows into the lungs because alveolar pressure is lower than barometric air pressure.

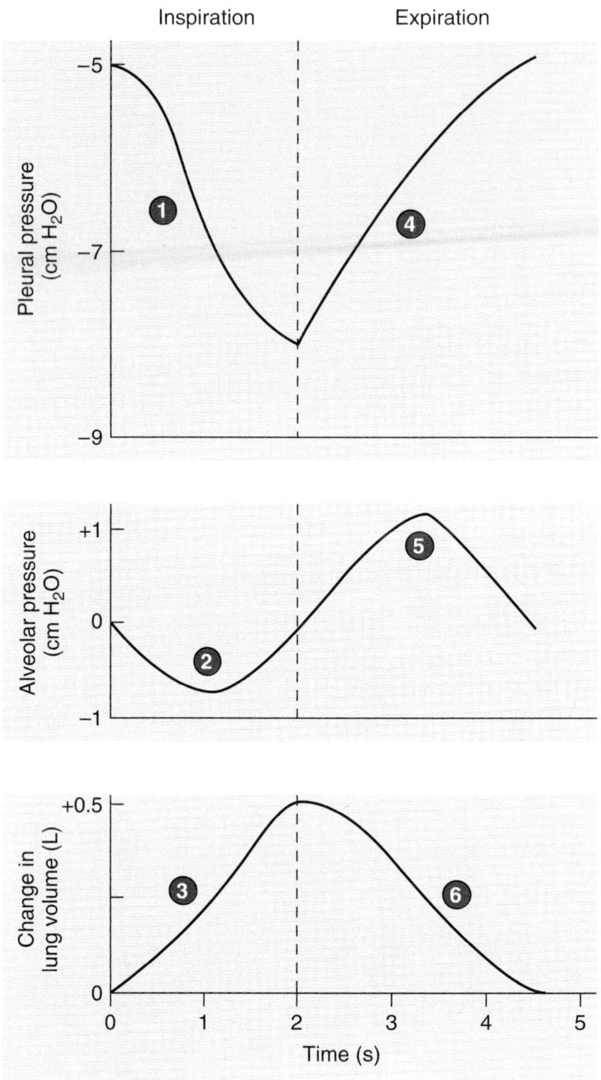

Changes during expiration

4. Pleural pressure increases because thoracic volume decreases.
5. As expiration begins, alveolar pressure increases above barometric air pressure (0 on the graph) because the increased pleural pressure causes alveolar volume to decrease. By the end of expiration, alveolar and barometric air pressure are equal.
6. During expiration, air flows out of the lungs because alveolar pressure is greater than barometric air pressure.

Dynamics of a Normal Breathing Cycle
Process Figure 23.14

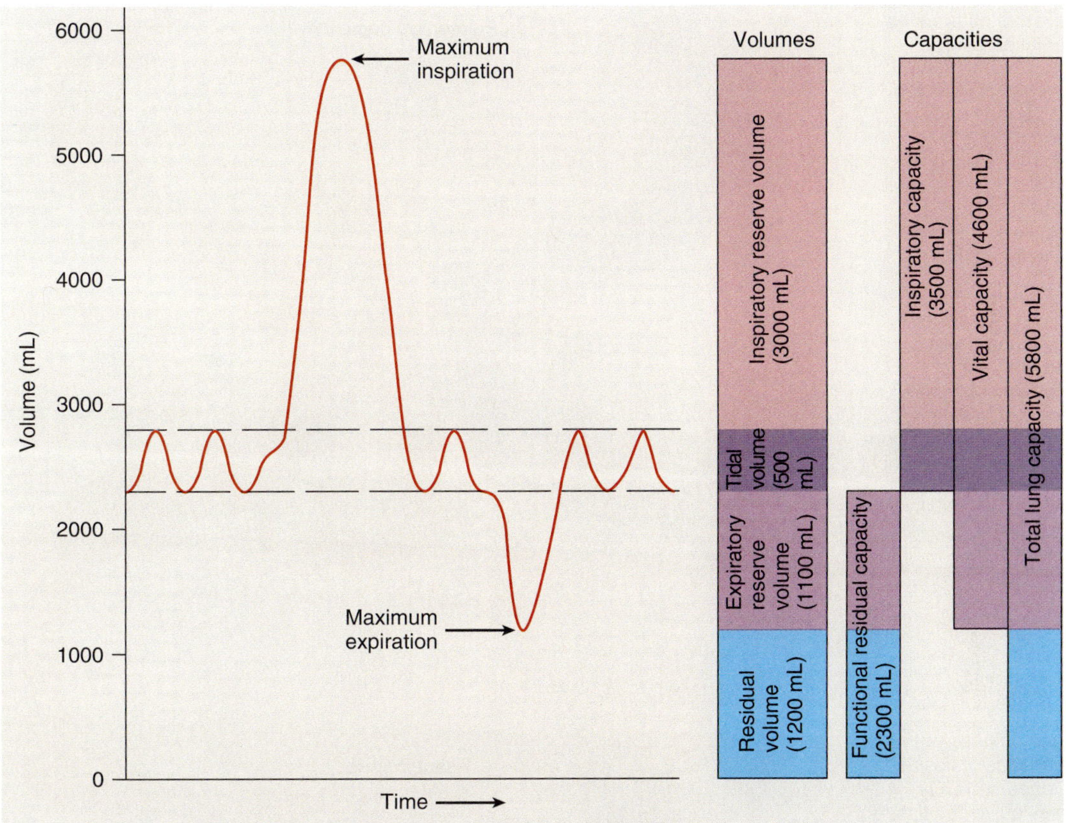

(b)

Spirometer, Lung Volumes, and Lung Capacities
Figure 23.15

1. Oxygen diffuses into the arterial ends of pulmonary capillaries and CO_2 diffuses into the alveoli because of differences in partial pressures.
2. As a result of diffusion at the venous ends of pulmonary capillaries, the Po_2 in the blood is equal to the Po_2 in the alveoli and the Pco_2 in the blood is equal to the Pco_2 in the alveoli.
3. The Po_2 of blood in the pulmonary veins is less than in the pulmonary capillaries because of mixing with deoxygenated blood from veins draining the bronchi and bronchioles.
4. Oxygen diffuses out of the arterial ends of tissue capillaries and CO_2 diffuses out of the tissue because of differences in partial pressures.
5. As a result of diffusion at the venous ends of tissue capillaries, the Po_2 in the blood is equal to the Po_2 in the tissue and the Pco_2 in the blood is equal to the Pco_2 in the tissue. Go back to step 1.

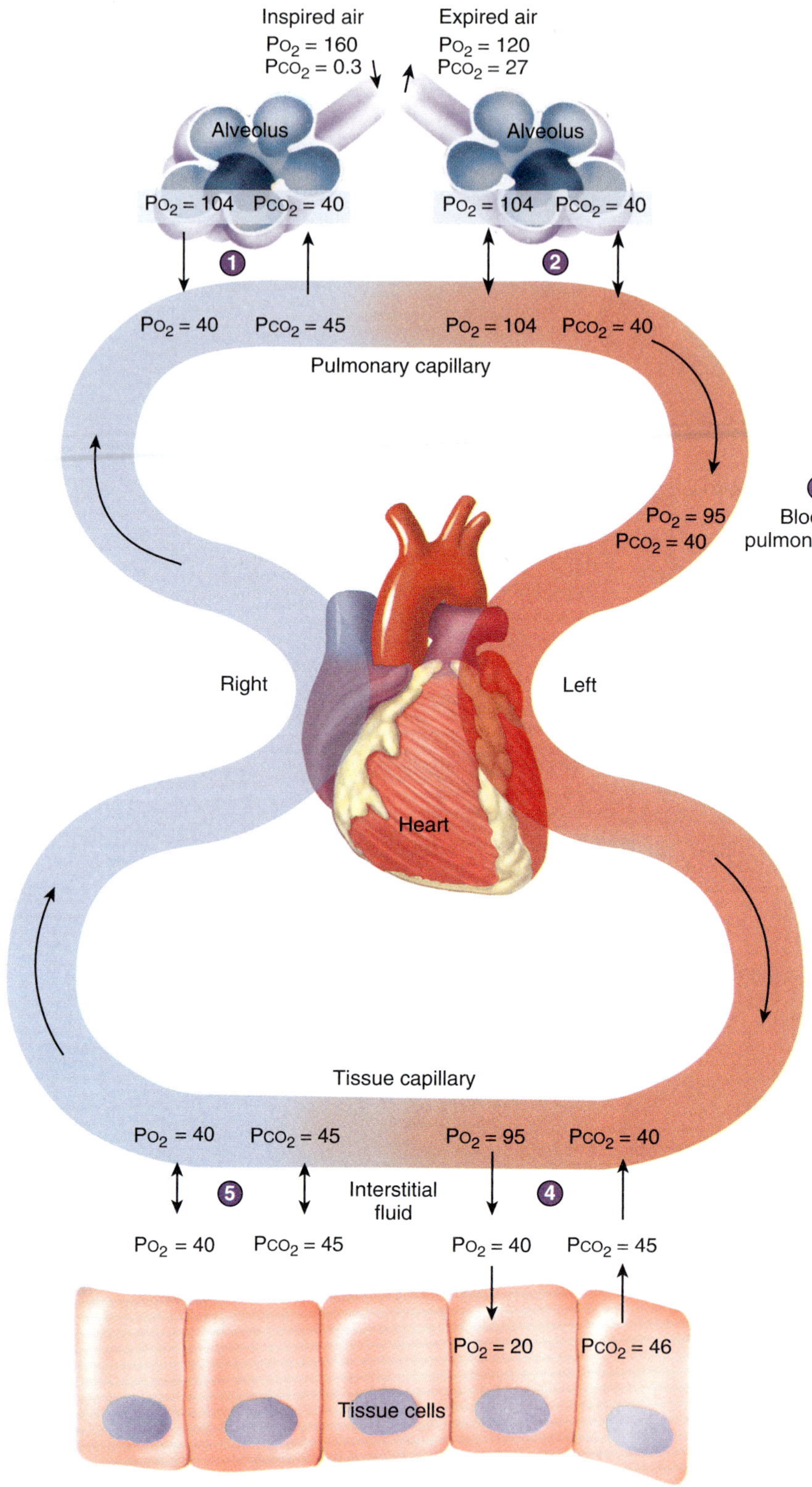

Gas Exchange
Process Figure 23.16

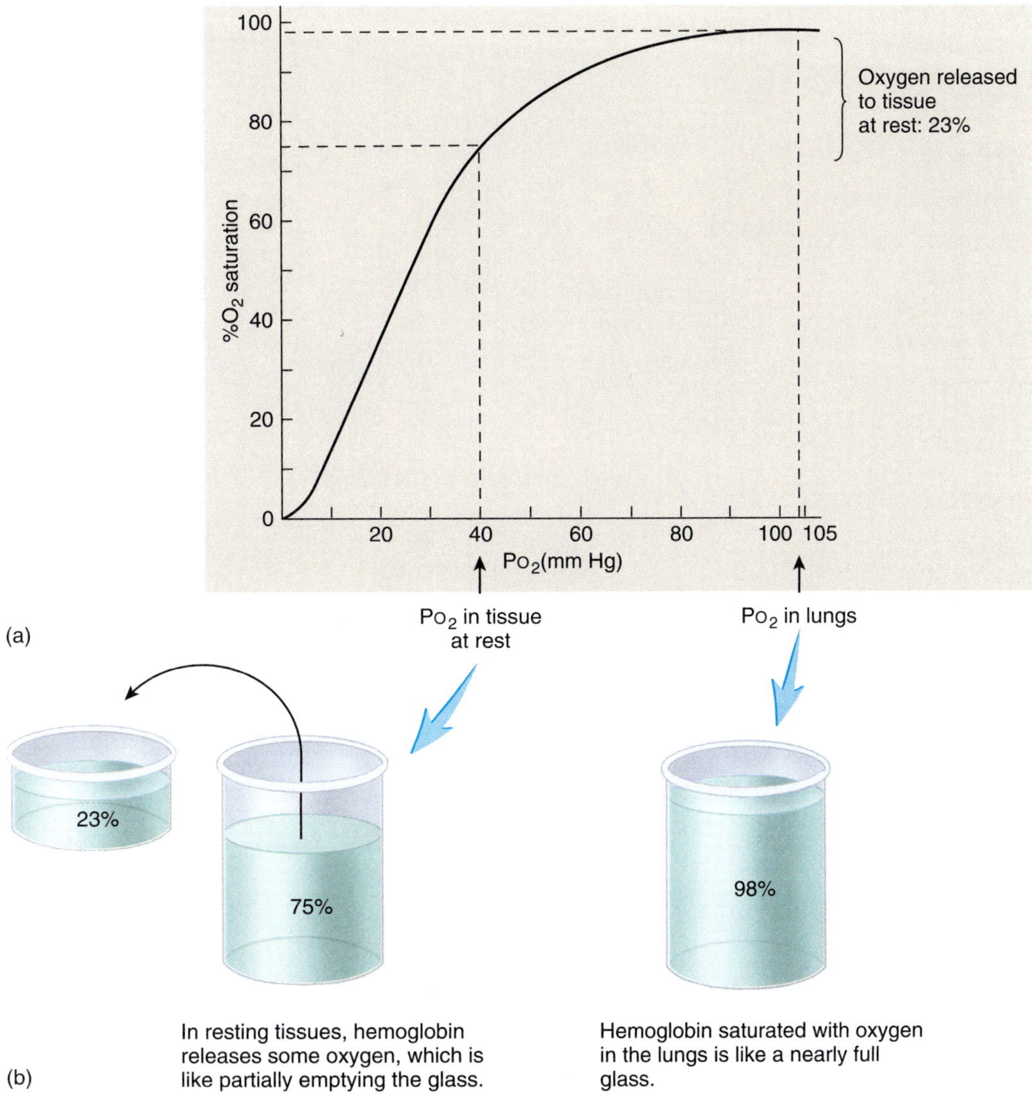

Oxygen–Hemoglobin Dissociation Curve at Rest
Figure 23.17

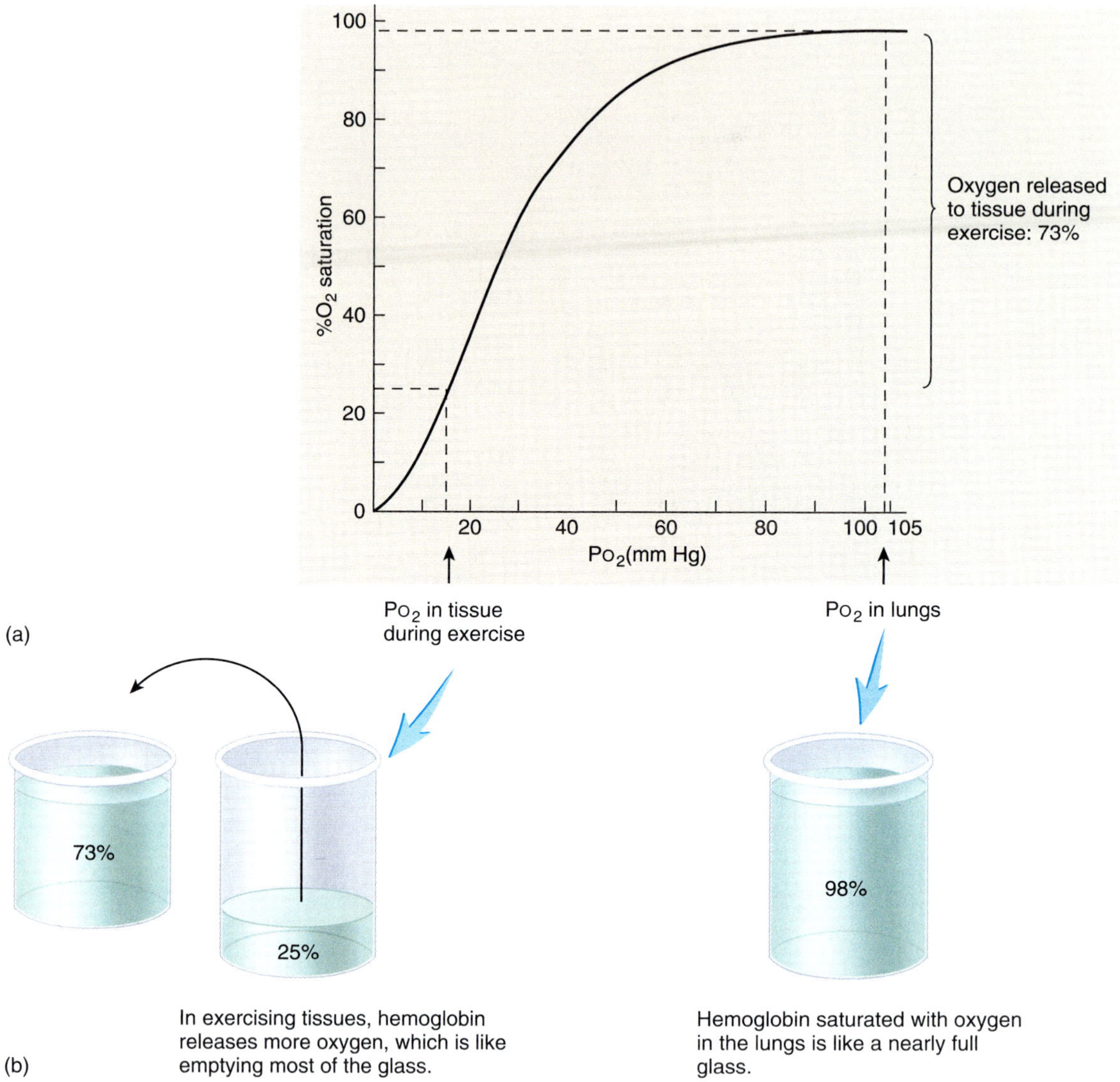

Oxygen–Hemoglobin Dissociation Curve During Exercise
Figure 23.18

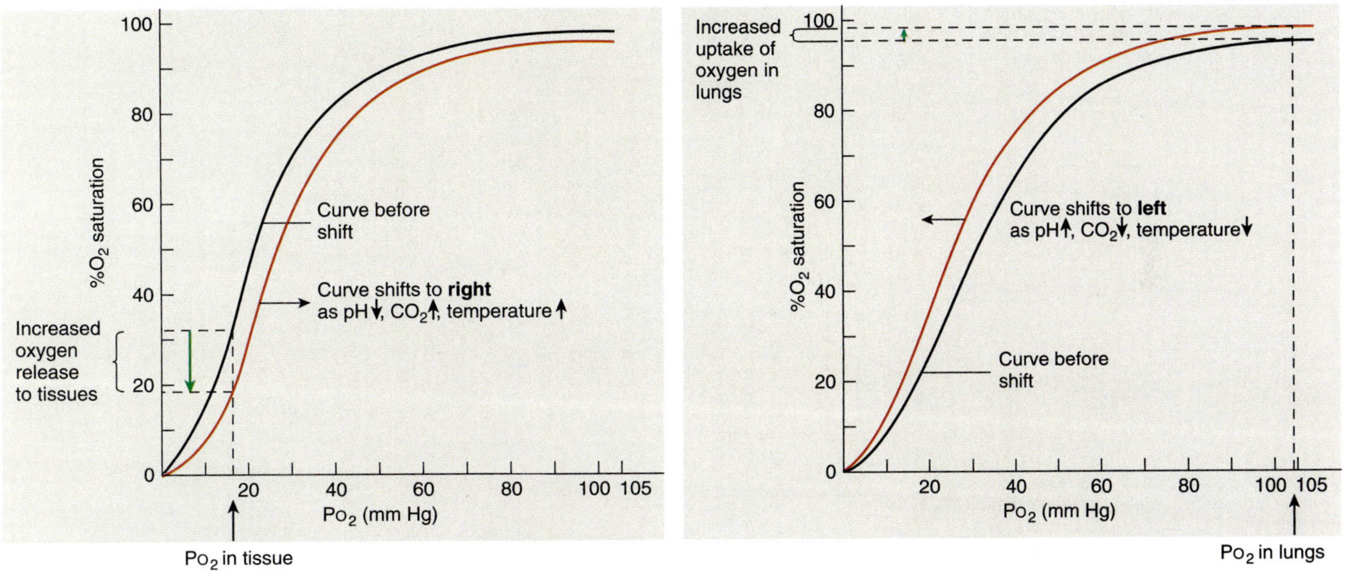

(a) **In the tissues, the oxygen–hemoglobin dissociation curve shifts to the right.** As pH decreases, P_{CO_2} increases, or temperature increases, the curve (*black*) shifts to the right (*red*), resulting in an increased release of oxygen.

(b) **In the lungs, the oxygen–hemoglobin dissociation curve shifts to the left.** As pH increases, P_{CO_2} decreases, or temperature decreases, the curve (*black*) shifts to the left (*red*), resulting in an increased ability of hemoglobin to pick up oxygen.

Effects of Shifting the Oxygen–Hemoglobin Dissociation Curve

Figure 23.19

(a) In the tissue capillaries, CO_2 enters red blood cells and reacts with H_2O to form H_2CO_3, which dissociates to form HCO_3^- and H^+. HCO_3^- are exchanged for Cl^- in the chloride shift. H^+ combine with hemoglobin. Lowering the concentration of HCO_3^- and H^+ inside red blood cells promotes the conversion of CO_2 to HCO_3^-.

(b) In the pulmonary capillaries, CO_2 leaves red blood cells, resulting in the formation of additional CO_2 from H_2CO_3. HCO_3^- and H^+ combine to replace the H_2CO_3. The HCO_3^- are exchanged for Cl^-, and the H^+ are released from hemoglobin.

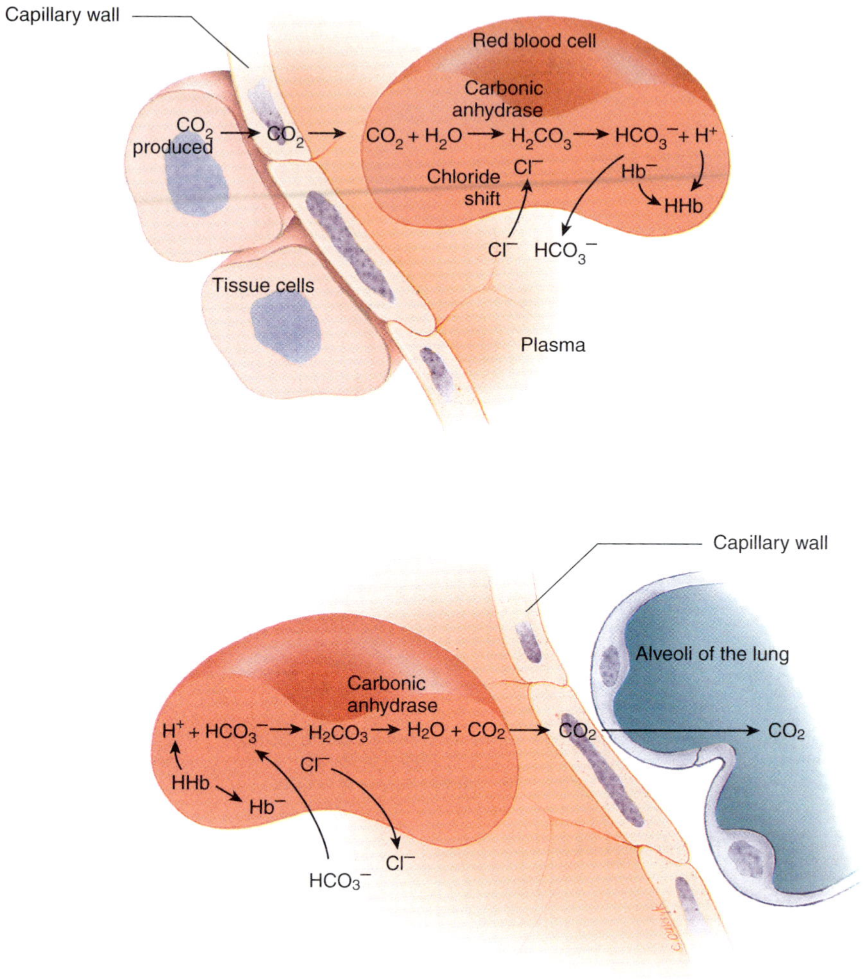

Carbon Dioxide Transport and Chloride Movement
Figure 23.20

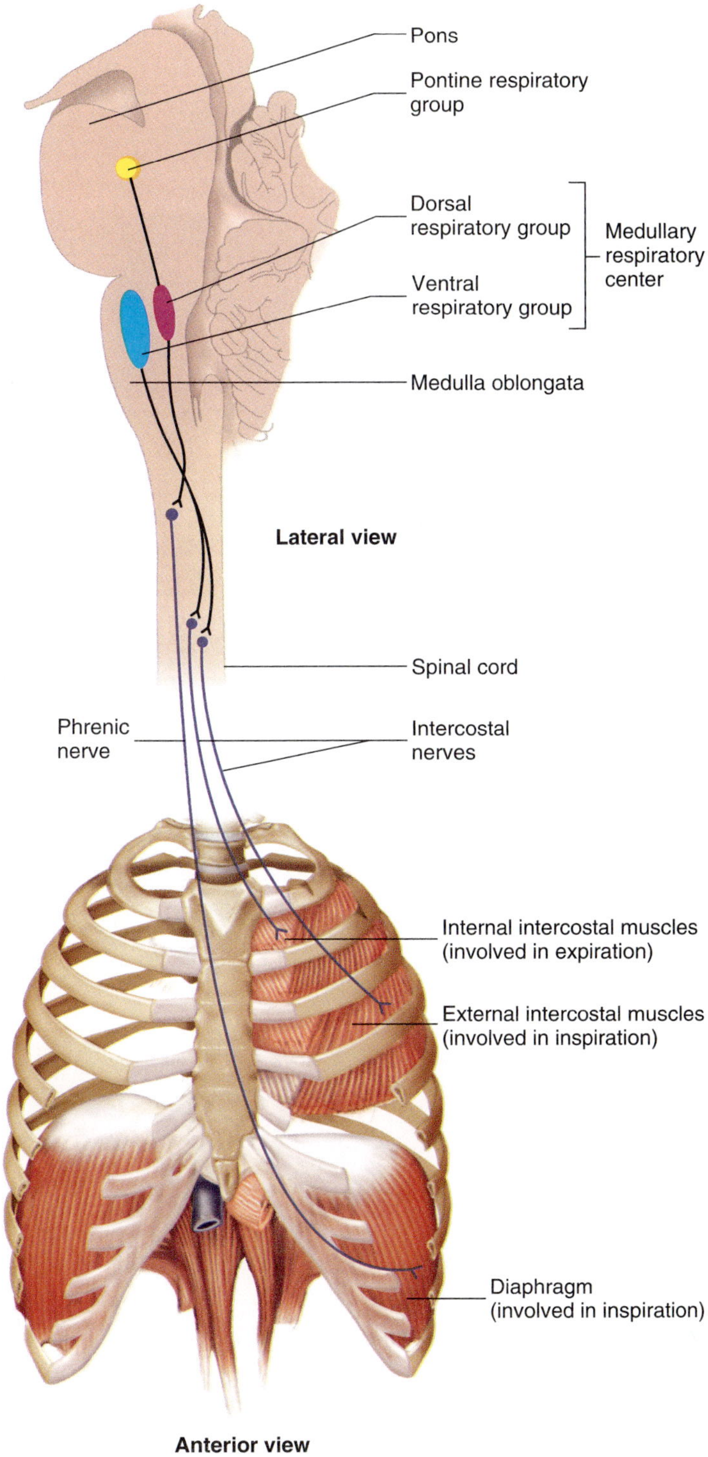

Respiratory Structures in the Brainstem
Figure 23.21

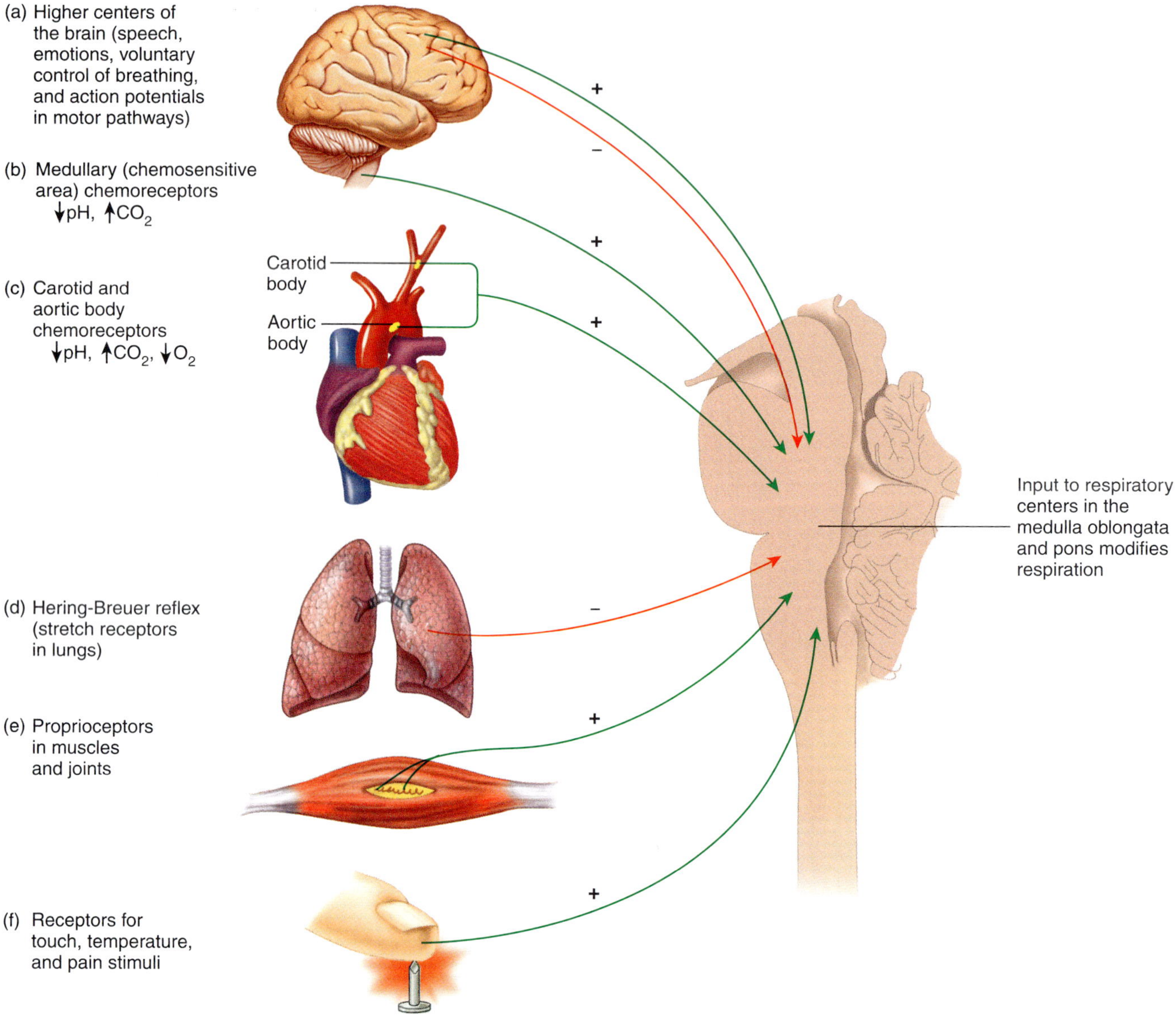

Major Regulatory Mechanisms of Ventilation
Figure 23.22

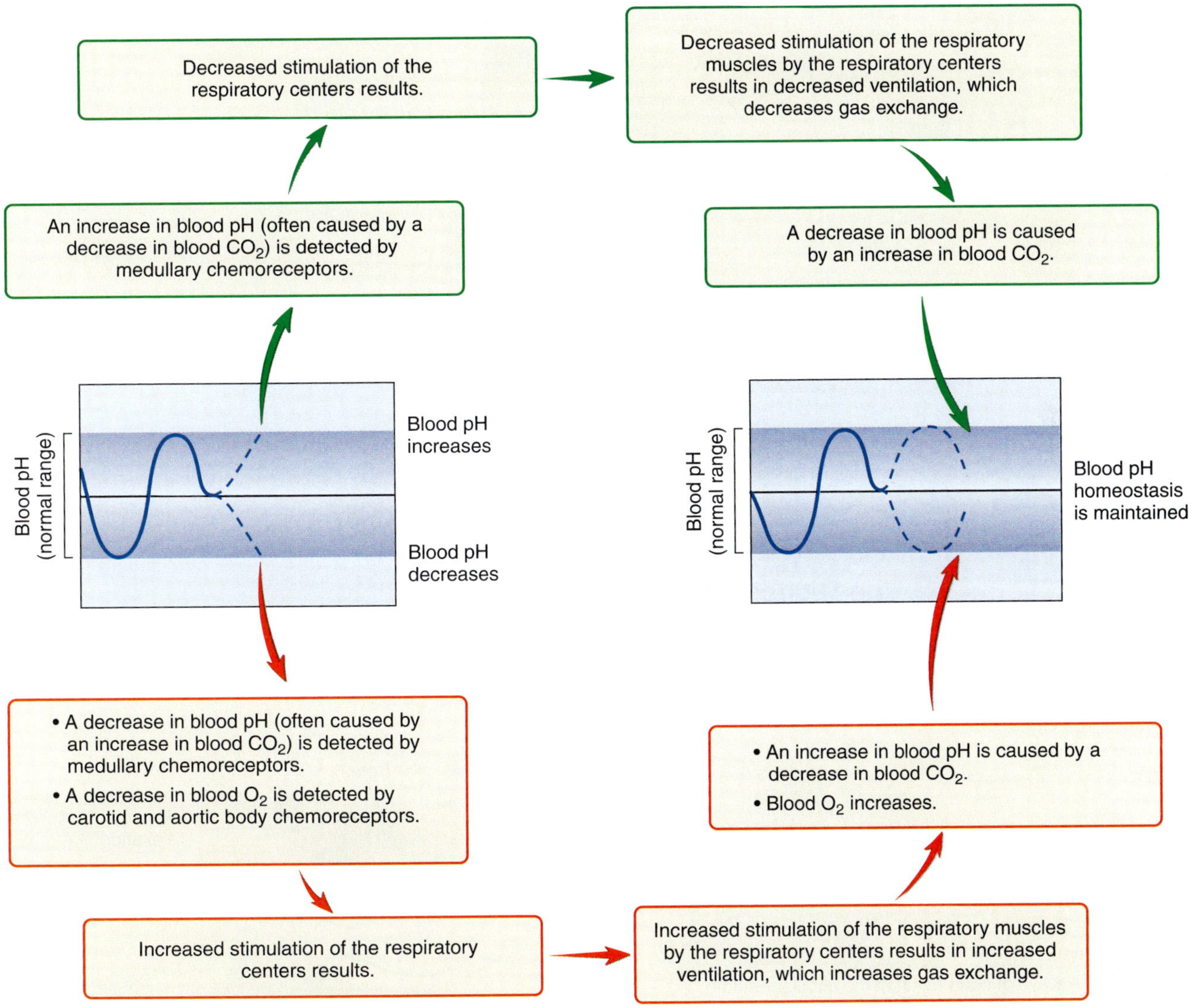

Regulation of Blood pH and Gases
Homeostasis Figure 23.23

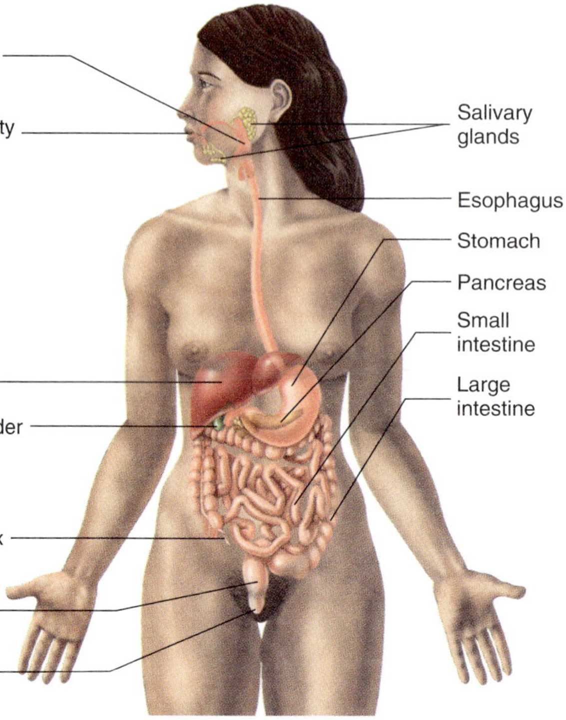

The Digestive System
Figure 24.1

1. A wave of circular smooth muscle relaxation moves ahead of the bolus of food allowing the digestive tract to expand.
2. A wave of contraction of the circular smooth muscles behind the bolus of food propels it through the digestive tract.

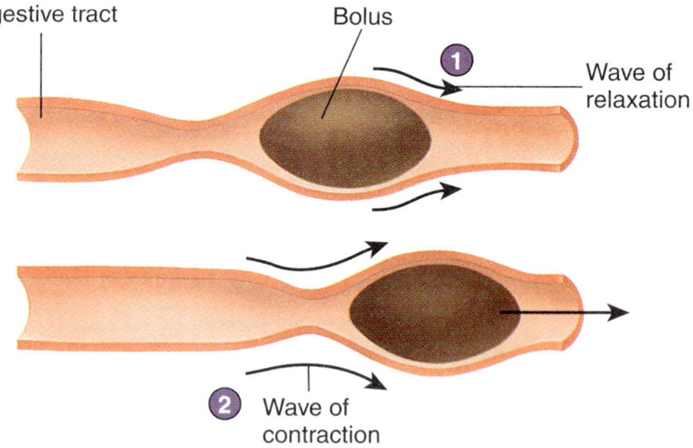

Peristalsis
Process Figure 24.2

1. A secretion introduced into the digestive tract or food within the tract begins in one location.

2. Segments of the digestive tract alternate between contraction and relaxation.

3. Material (*brown*) in the intestine is spread out in both directions from the site of introduction.

4. The secretion or food is spread out in the digestive tract and becomes more diffuse (*lighter color*) through time.

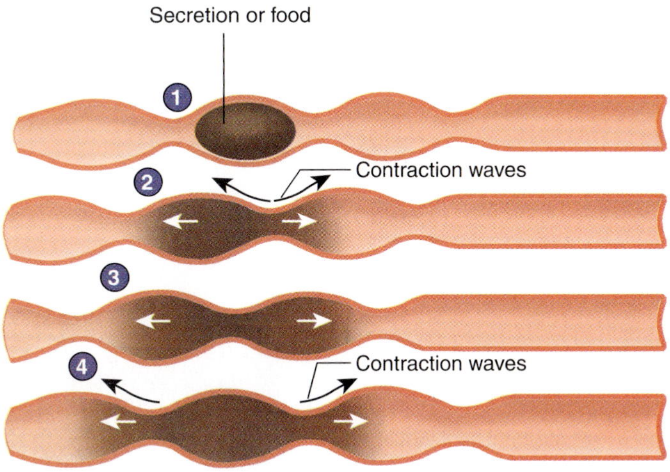

Segmental Contractions
Process Figure 24.3

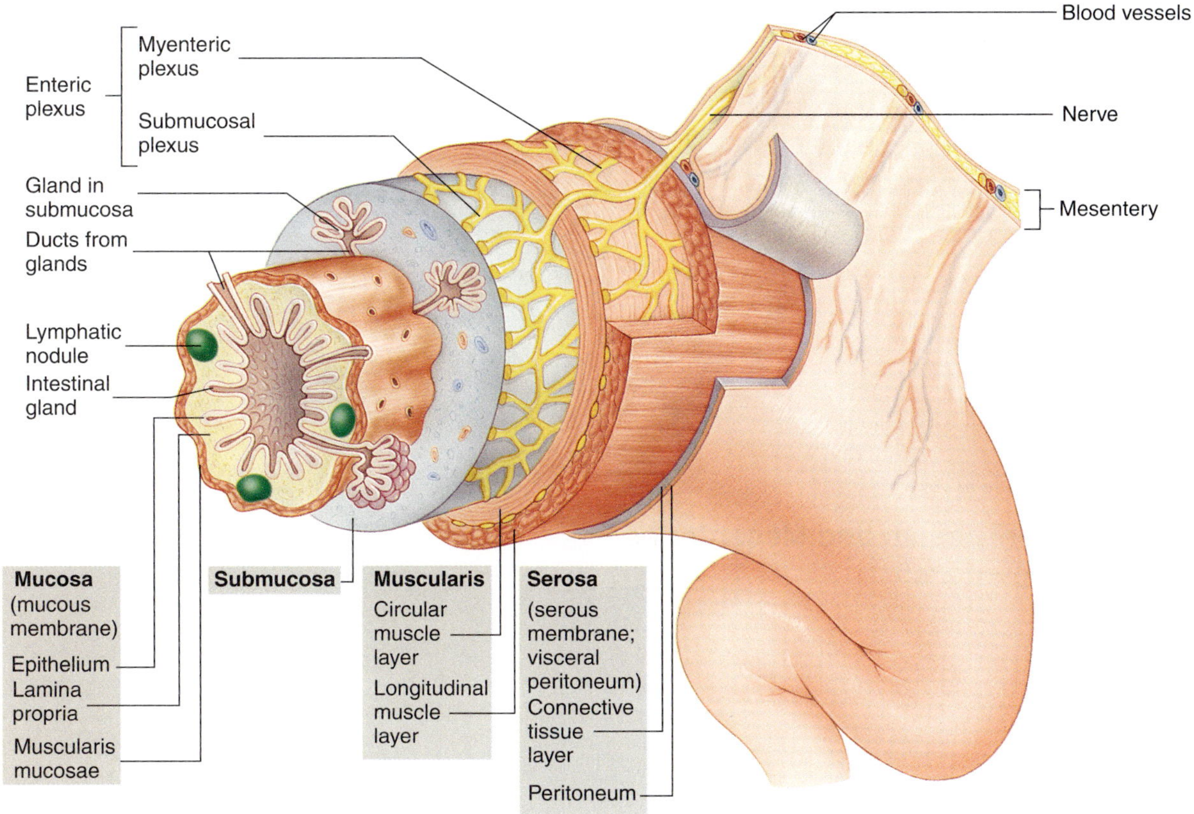

Digestive Tract Histology
Figure 24.4

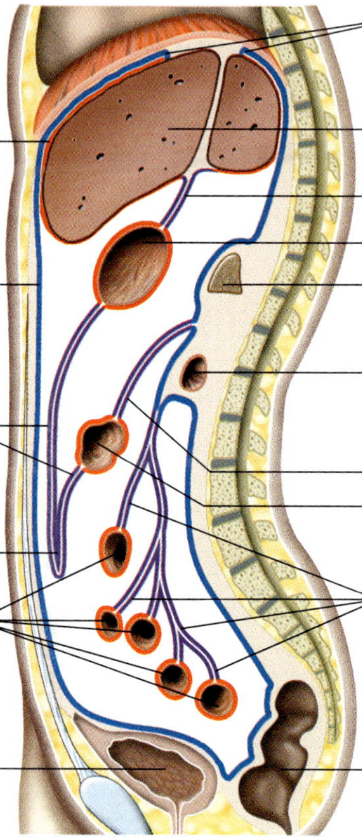

(a) **Medial view**

Peritoneum and Mesenteries

Figure 24.5

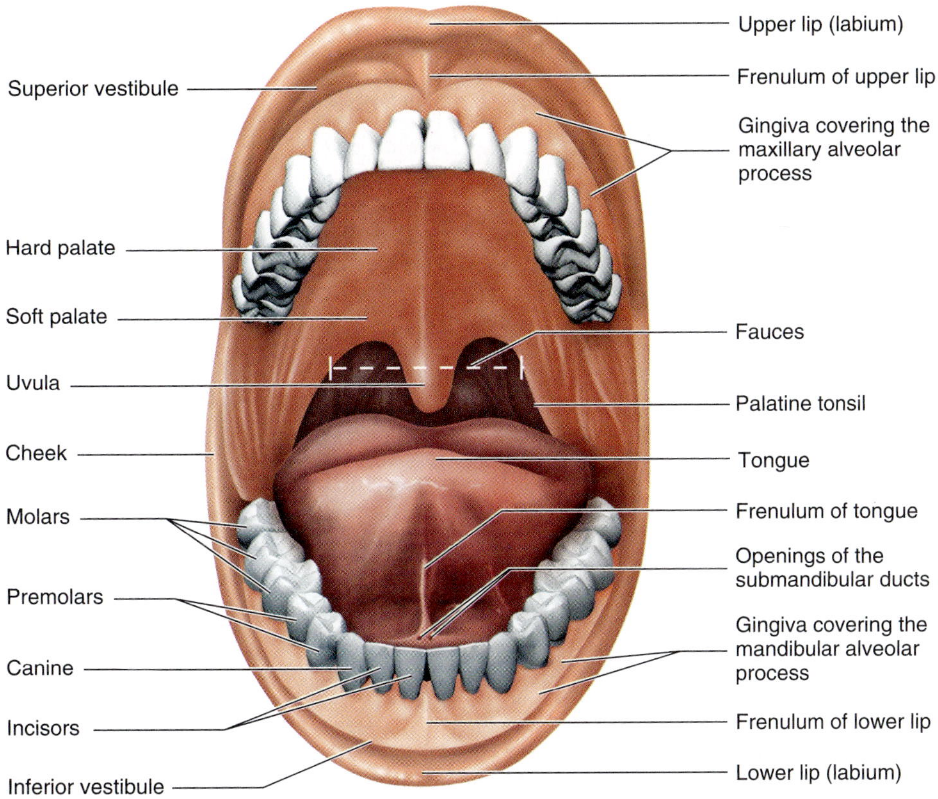
Upper lip (labium)
Superior vestibule
Frenulum of upper lip
Gingiva covering the maxillary alveolar process
Hard palate
Soft palate
Fauces
Uvula
Palatine tonsil
Cheek
Tongue
Molars
Frenulum of tongue
Openings of the submandibular ducts
Premolars
Gingiva covering the mandibular alveolar process
Canine
Incisors
Frenulum of lower lip
Inferior vestibule
Lower lip (labium)

Anterior view

Oral Cavity
Figure 24.6

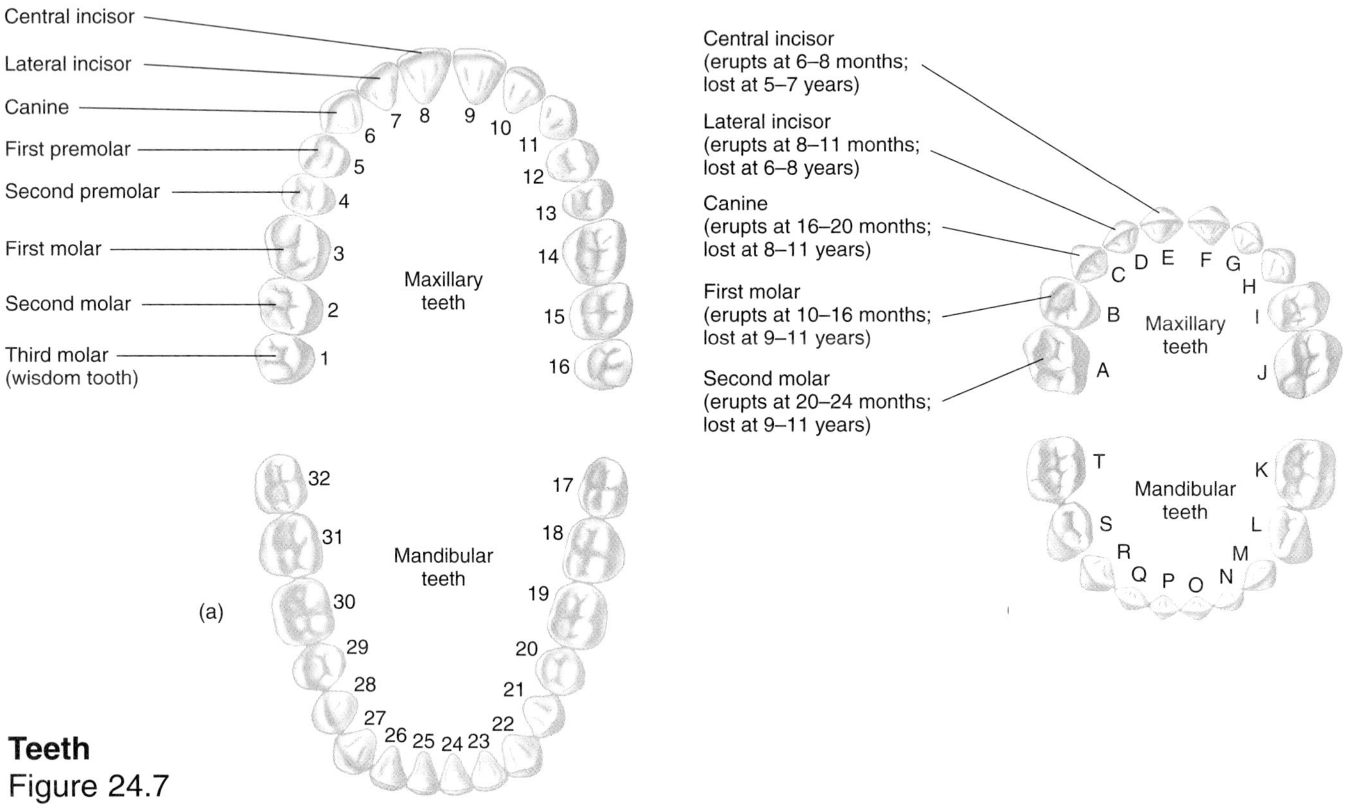
Central incisor
Lateral incisor
Canine
First premolar
Second premolar
First molar
Second molar
Third molar
(wisdom tooth)
Maxillary
teeth
Mandibular
teeth
1
2
3
4
5
6
7
8
9
10
11
12
13
14
15
16
17
18
19
20
21
22
23
24
25
26
27
28
29
30
31
32
(a)
Central incisor
(erupts at 6–8 months;
lost at 5–7 years)
Lateral incisor
(erupts at 8–11 months;
lost at 6–8 years)
Canine
(erupts at 16–20 months;
lost at 8–11 years)
First molar
(erupts at 10–16 months;
lost at 9–11 years)
Second molar
(erupts at 20–24 months;
lost at 9–11 years)
A B C D E F G H I J
K L M N O P Q R S T
Maxillary
teeth
Mandibular
teeth

Teeth
Figure 24.7

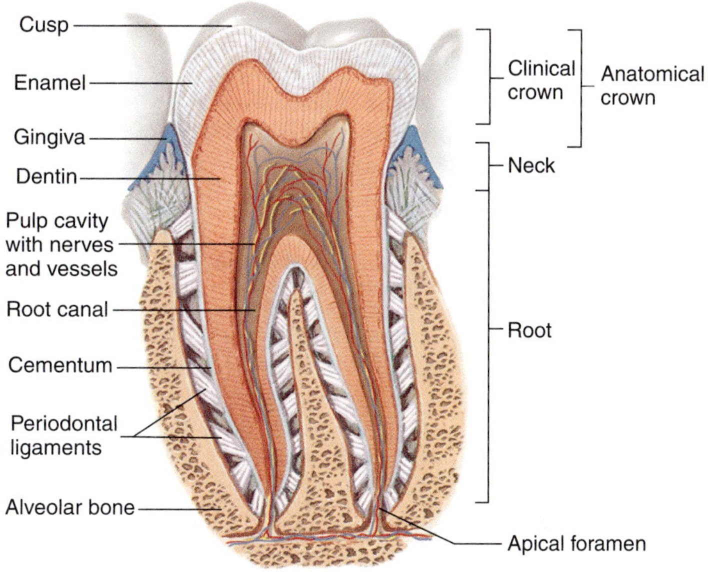

Molar Tooth in Place in the Alveolar Bone
Figure 24.8

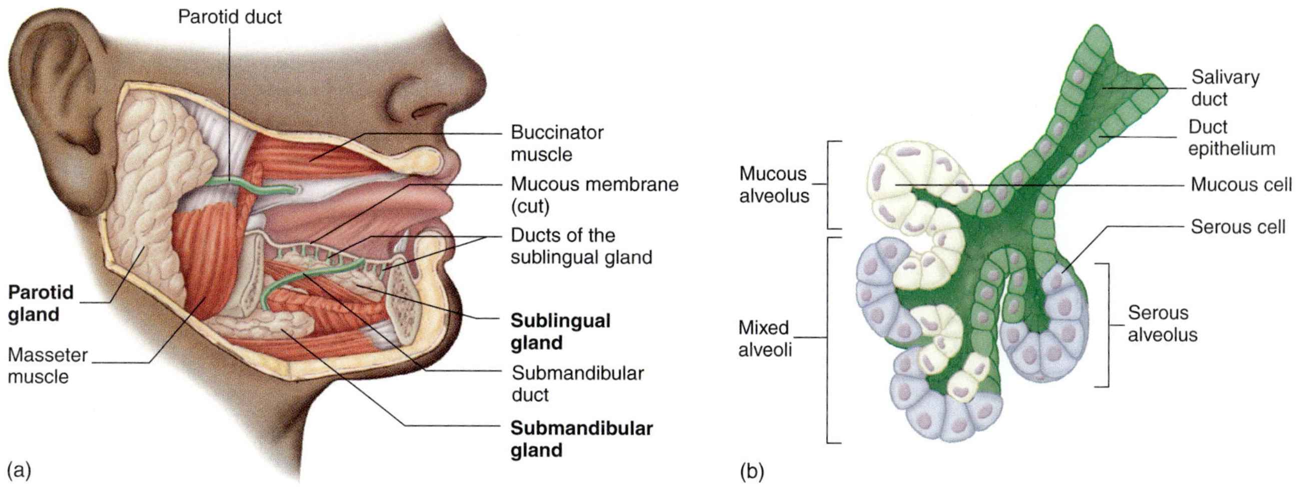

Salivary Glands
Figure 24.9

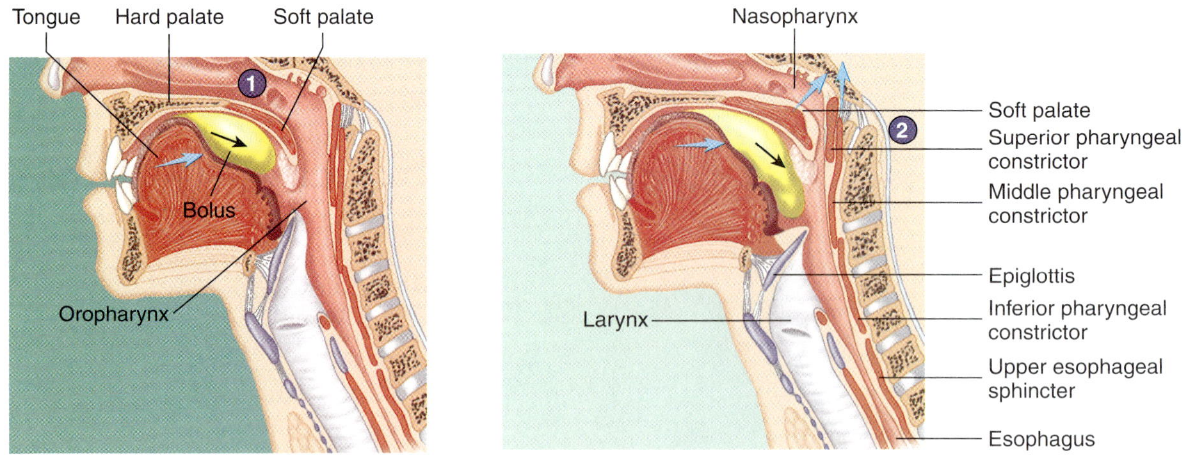

1. During the voluntary phase, a bolus of food (*yellow*) is pushed by the tongue against the hard and soft palates and posteriorly toward the oropharynx (*blue arrow* indicates tongue movement; *black arrow* indicates movement of the bolus). *Tan*: bone, *purple*: cartilage, *red*: muscle.

2. During the pharyngeal phase, the soft palate is elevated, closing off the nasopharynx. The pharynx and larynx are elevated (*blue arrows* indicate muscle movement).

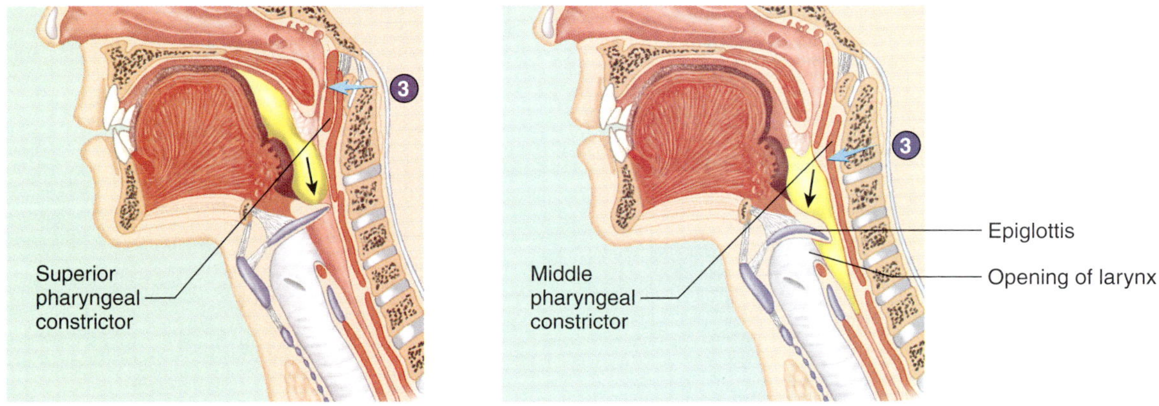

3. Successive constriction of the pharyngeal constrictors from superior to inferior (*blue arrows*) forces the bolus through the pharynx and into the esophagus. As this occurs, the epiglottis is bent down over the opening of the larynx largely by the force of the bolus pressing against it.

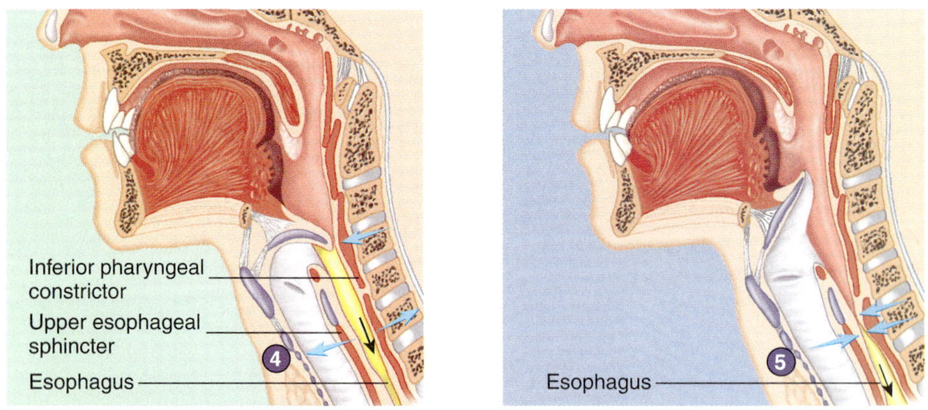

4. As the inferior pharyngeal constrictor contracts, the upper esophageal sphincter relaxes (outwardly directed *blue arrows*), allowing the bolus to enter the esophagus.

5. During the esophageal phase, the bolus is moved by peristaltic contractions of the esophagus toward the stomach (inwardly directed *blue arrows*).

Three Phases of Swallowing (Deglutition)
Process Figure 24.10

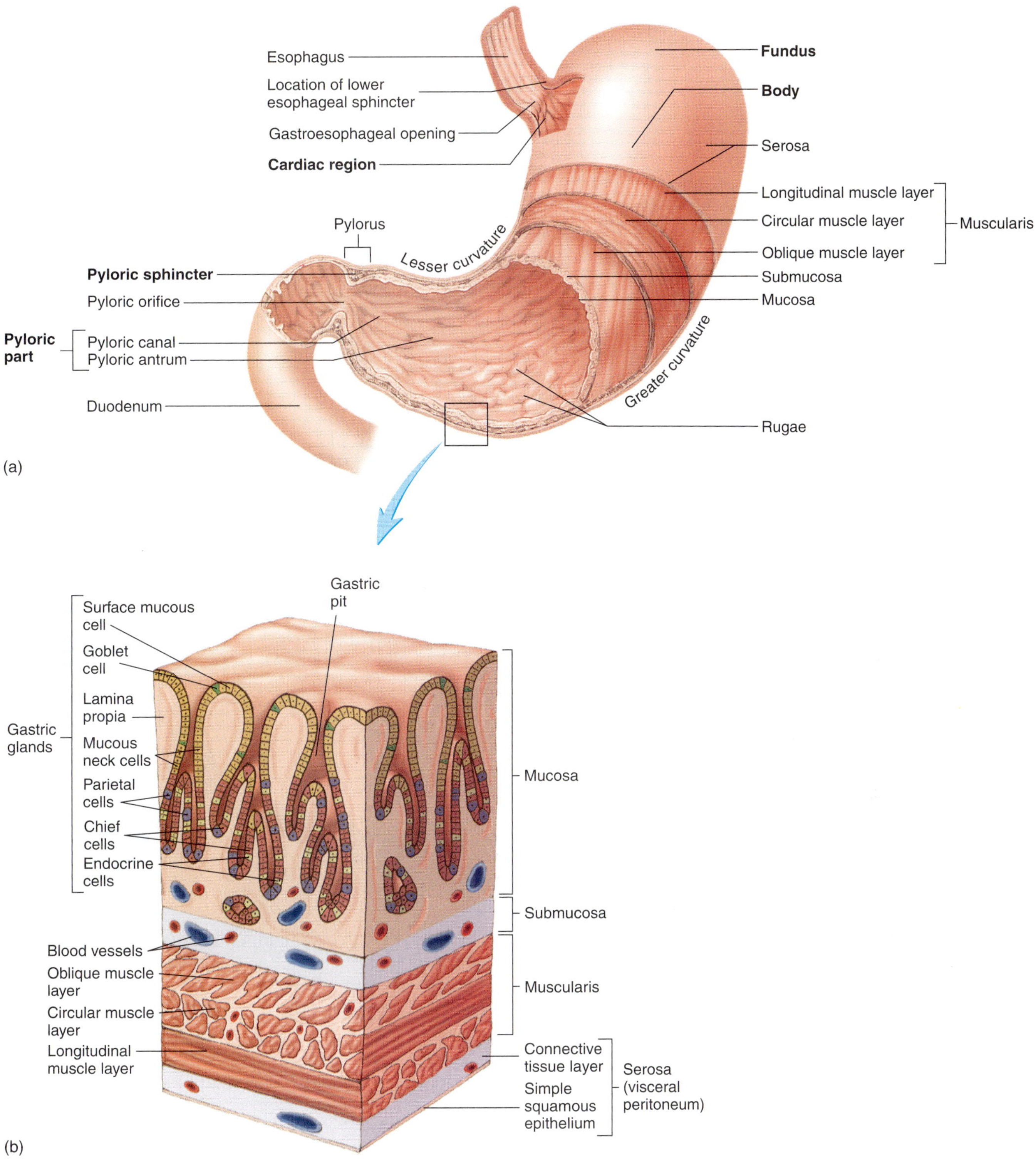

Anatomy and Histology of the Stomach
Figure 24.11

1. Carbon dioxide (CO_2) diffuses into the cell.

2. Carbon dioxide combines with water (H_2O) in an enzymatic reaction that is catalyzed by carbonic anhydrase (CA) to form carbonic acid (H_2CO_3).

3. Carbonic acid dissociates into a bicarbonate ion (HCO_3^-) and a hydrogen ion (H^+).

4. Bicarbonate ions are transported back into the bloodstream. An anion exchange molecule in the plasma membrane exchanges HCO_3^- for a chloride ion (Cl^-) (counter transport).

5. Hydrogen ions are actively transported into the duct of the gastric gland.

6. Chloride ions diffuse with the charged H^+.

7. Some potassium ions (K^+) are counter transported into the cell in exchange for H^+.

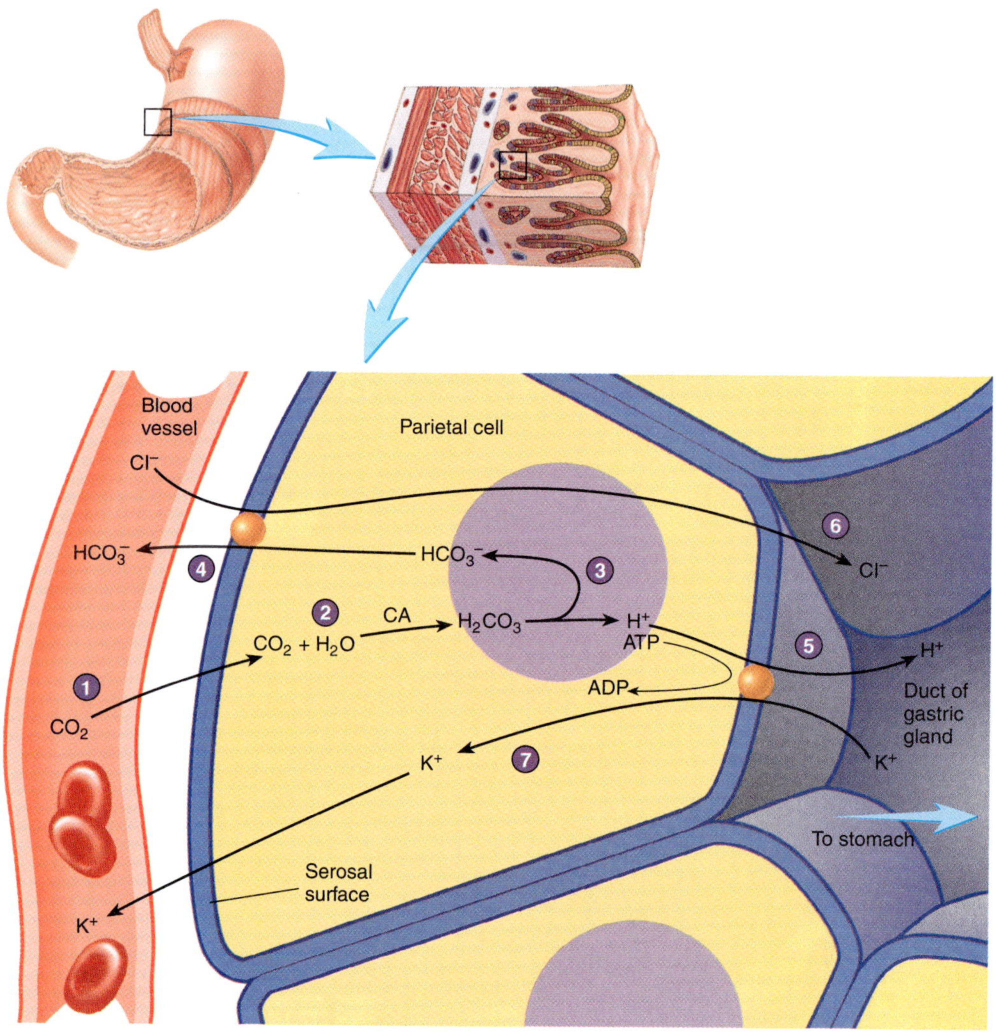

Hydrochloric Acid Production by Parietal Cells in the Gastric Glands of the Stomach
Process Figure 24.12

Cephalic Phase

1. The taste, smell, or thought of food or tactile sensations of food in the mouth stimulate the medulla oblongata (*green arrow*).
2. Parasympathetic action potentials are carried by the vagus nerves to the stomach (*pink arrow*).
3. Preganglionic parasympathetic vagus nerve fibers stimulate postganglionic neurons in the enteric plexus of the stomach.
4. Postganglionic neurons stimulate secretion by parietal and chief cells and stimulate gastrin secretion by endocrine cells.
5. Gastrin is carried through the circulation back to the stomach (*purple arrow*), where it stimulates secretion by parietal and chief cells.

(a)

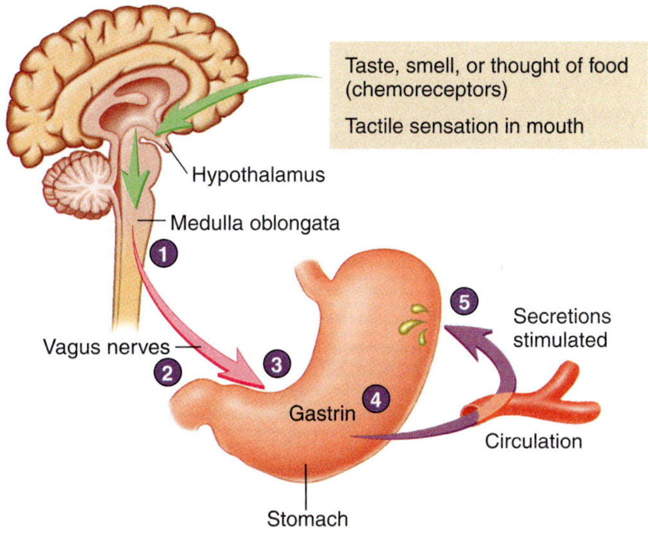

Gastric Phase

1. Distention of the stomach stimulates mechanoreceptors (stretch receptors) and activates a parasympathetic reflex. Action potentials generated by the mechanoreceptors are carried by the vagus nerves to the medulla oblongata (*green arrow*).
2. The medulla oblongata increases action potentials in the vagus nerves that stimulate stomach secretions (*pink arrow*).
3. Distention of the stomach also activates local reflexes that increase stomach secretions (*purple arrow*).

(b)

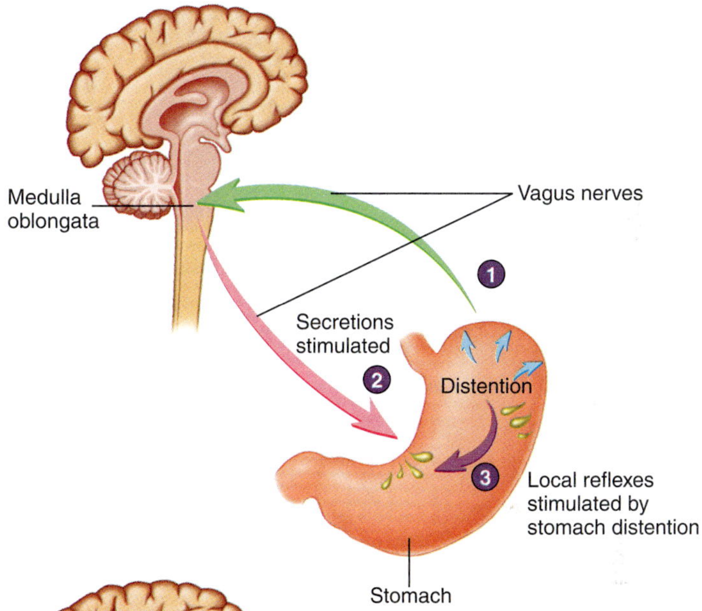

Gastrointestinal Phase

1. Chyme in the duodenum with a pH less than 2 or containing fat digestion products (lipids) inhibits gastric secretions by three mechanisms (2–4).
2. Chemoreceptors in the duodenum are stimulated by H^+ (low pH) or lipids. Action potentials generated by the chemoreceptors are carried by the vagus nerves to the medulla oblongata (*green arrow*), where they inhibit parasympathetic action potentials (*pink arrow*).
3. Local reflexes activated by H^+ or lipids also inhibit gastric secretion (*orange arrows*).
4. Secretin, gastric inhibitory polypeptide, and cholecystokinin produced by the duodenum (*brown arrows*) inhibit gastric secretions in the stomach.

(c)

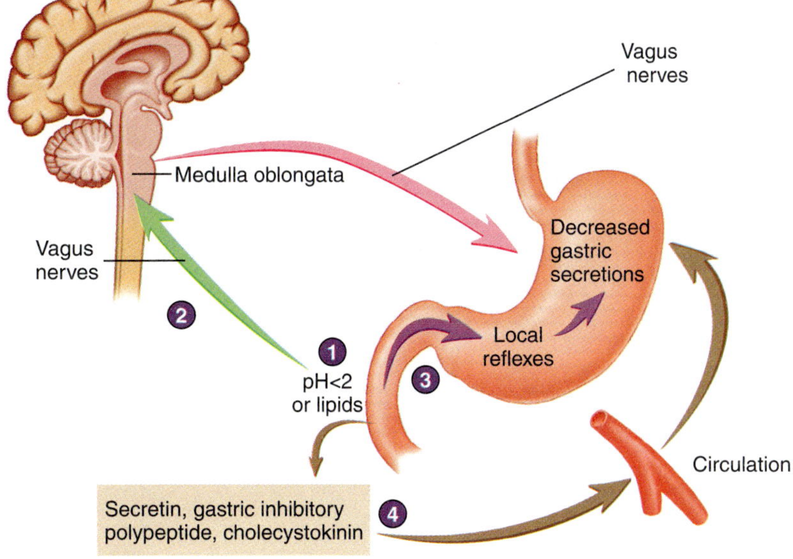

The Three Phases of Stomach Secretion
Process Figure 24.13

1. A mixing wave initiated in the body of the stomach progresses toward the pylorus (*pink arrows directed inward*).
2. The more fluid part of the chyme is pushed toward the pylorus (*blue arrows*) as the mixing wave passes over the more solid center of the chyme. As a result, the more solid center of the chyme is pushed past the constriction back toward the body of the stomach (*orange arrow*).
3. Peristaltic waves (*purple arrows*) move in the same direction and in the same way as the mixing waves but are stronger. As one wave reaches the pyloric part of the stomach, a second wave begins in the body.
4. As with the mixing waves, the more fluid part of the chyme is pushed toward the pylorus (*blue arrows*), whereas the more solid center of the chyme is pushed back toward the body of the stomach (*orange arrow*).
5. Peristaltic contractions force a few milliliters of some of the most fluid chyme through the pyloric orifice into the duodenum (*small blue arrows*). Most of the chyme, including the more solid portion, is forced back toward the body of the stomach for further mixing (*orange arrows*).

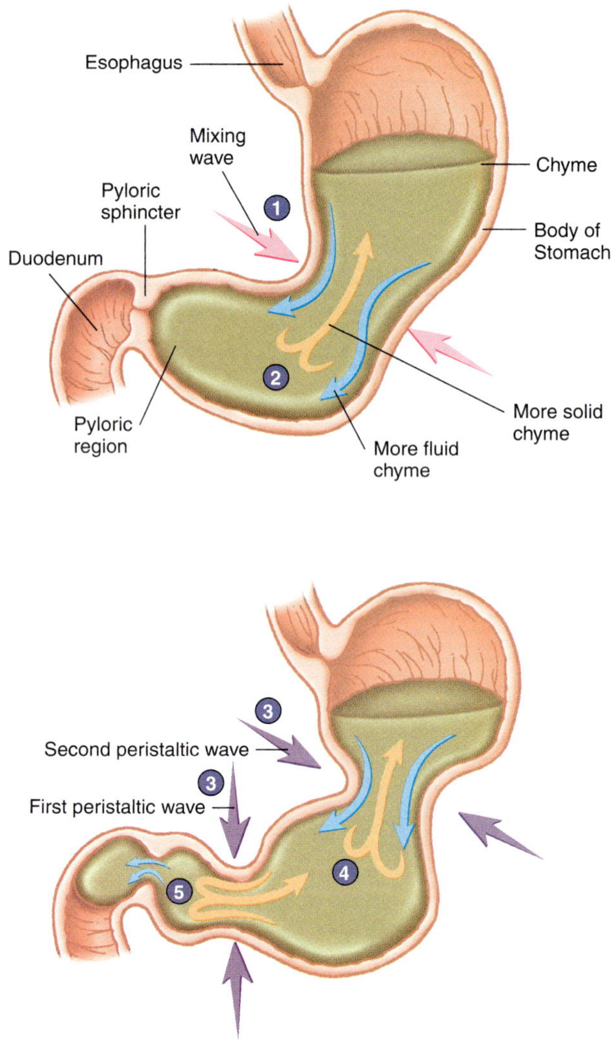

Movements in the Stomach
Process Figure 24.14

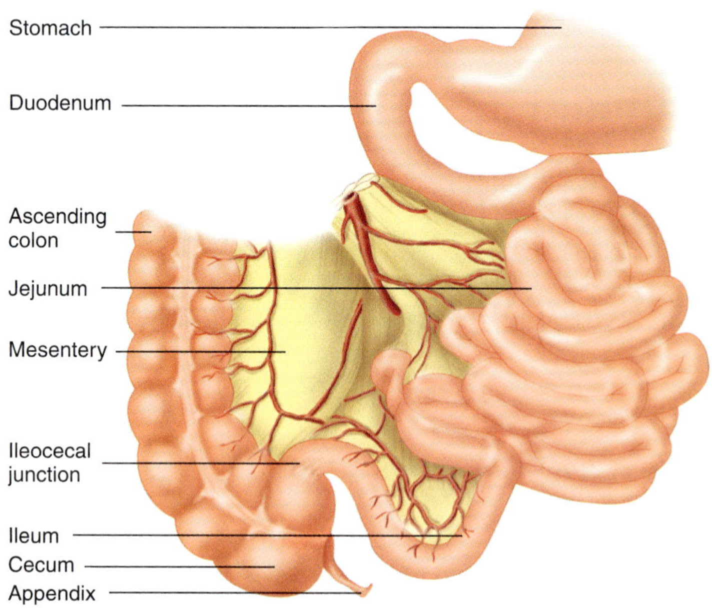

Anterior view

The Small Intestine
Figure 24.15

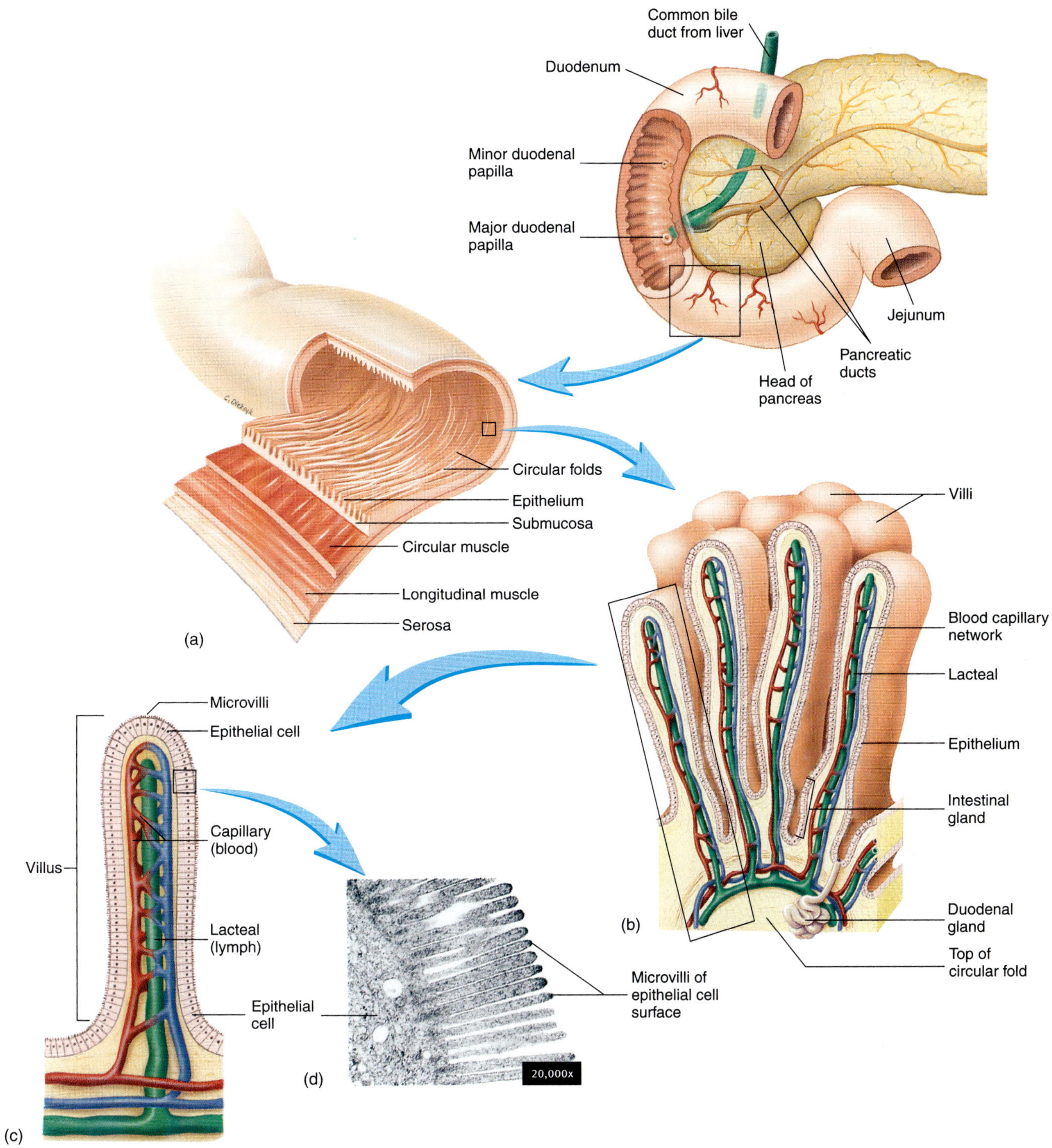

Anatomy and Histology of the Duodenum

Figure 24.16

d: © David M. Phillips/Visuals Unlimited

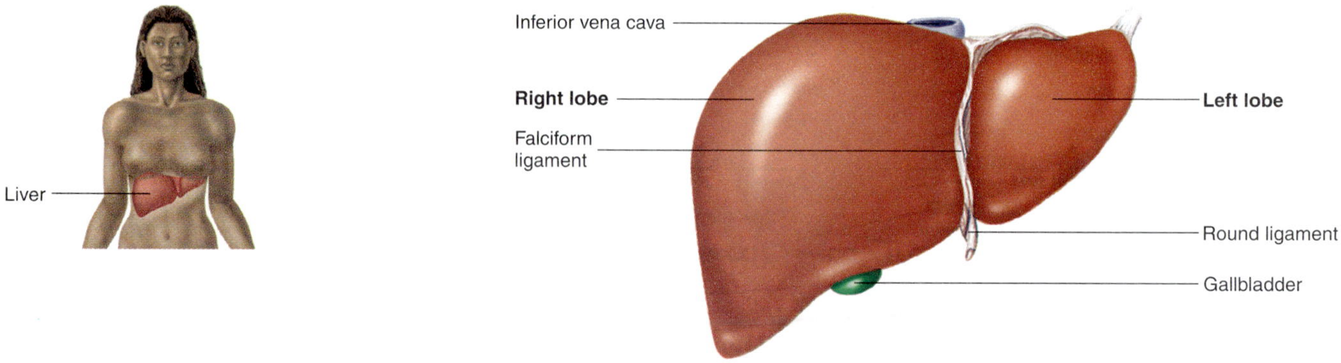

(a) **Anterior view**

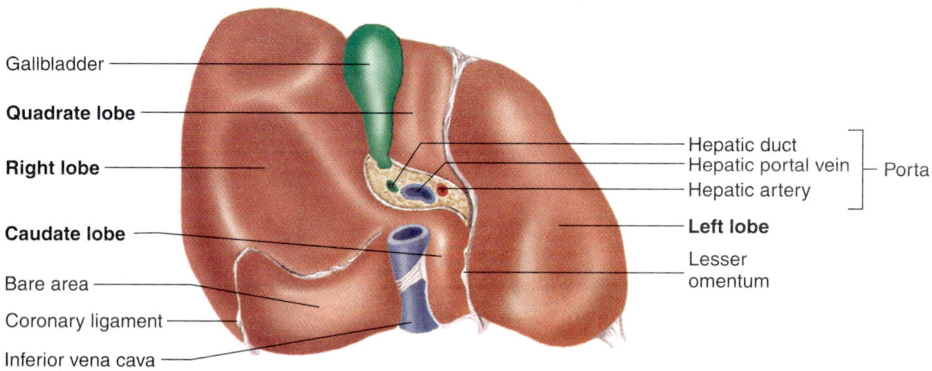

(b) **Inferior view**

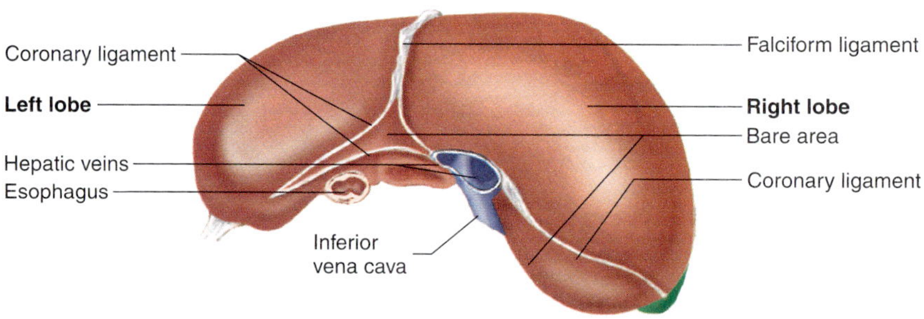

(c) **Superior view**

The Liver
Figure 24.17

1. The hepatic ducts from the liver lobes combine to form the common hepatic duct.

2. The common hepatic duct combines with the cystic duct from the gallbladder to form the common bile duct.

3. The common bile duct and the pancreatic duct combine to form the hepatopancreatic ampulla.

4. The hepatopancreatic ampulla empties into the duodenum at the major duodenal papilla.

5. The accessory pancreatic duct empties into the duodenum at the minor duodenal papilla.

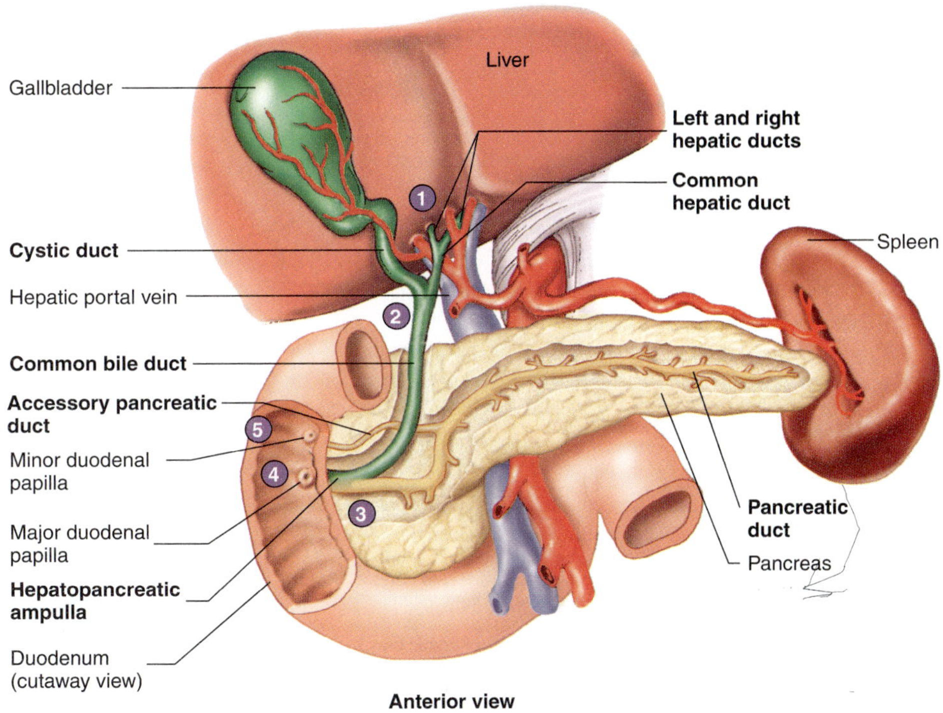

The Liver, Gallbladder, Pancreas, and Duct System
Process Figure 24.18

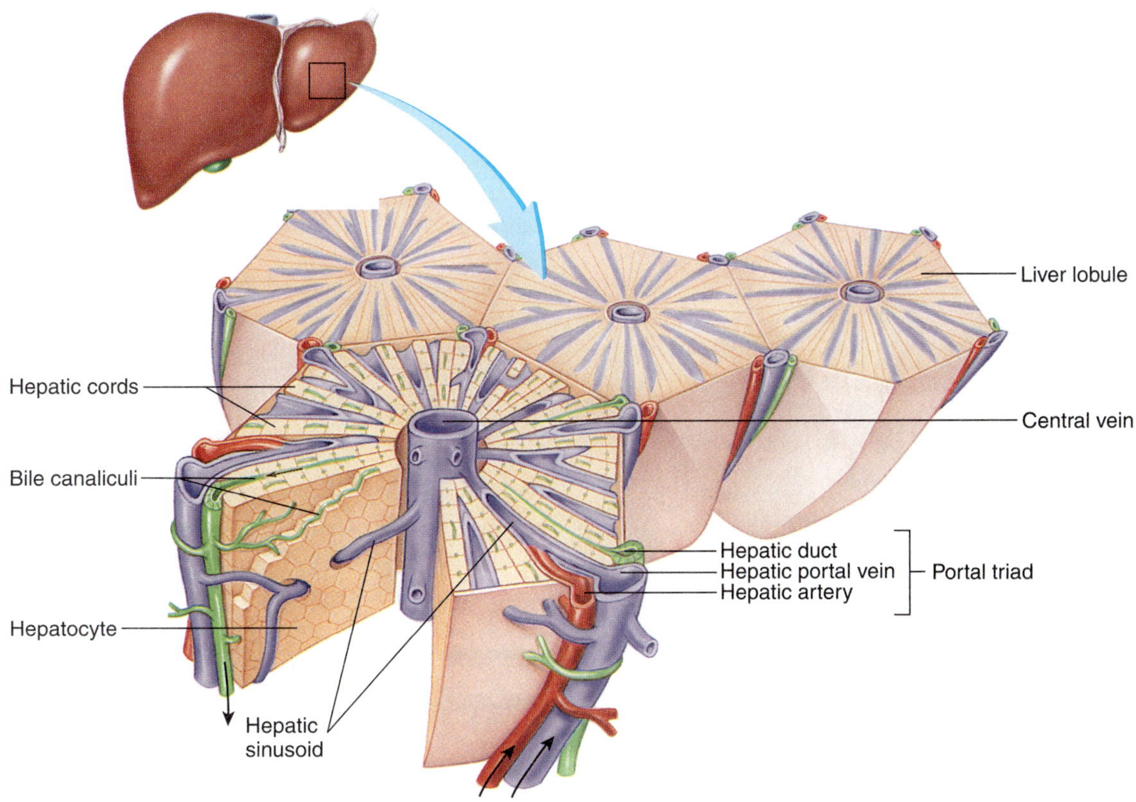

Histology of the Liver
Figure 24.19

1. The hepatic artery carries oxygen-rich blood from the aorta through the porta of the liver. Hepatic artery branches become part of the portal triads. Blood from the hepatic artery branches enters the hepatic sinusoids and supplies hepatocytes in the hepatic cords with oxygen.
2. The hepatic portal vein carries nutrient-rich, oxygen-poor blood from the intestines through the porta of the liver. Hepatic portal vein branches become part of the portal triads. Blood from the hepatic portal vein branches enters the hepatic sinusoids and supplies hepatocytes in the hepatic cords with nutrients.
3. Blood in the hepatic sinusoids that comes from the hepatic artery and hepatic portal vein picks up plasma proteins, processed molecules, and waste products produced by the hepatocytes of the hepatic cords. The hepatic sinusoids empty into central veins. The central veins connect to hepatic veins, which connect to the inferior vena cava.
4. Bile produced by hepatocytes in the hepatic cords enters bile canaliculi, which connect to hepatic duct branches that are part of the portal triads.
5. The hepatic duct branches converge to form the left and right hepatic ducts, which carry bile out the porta of the liver.

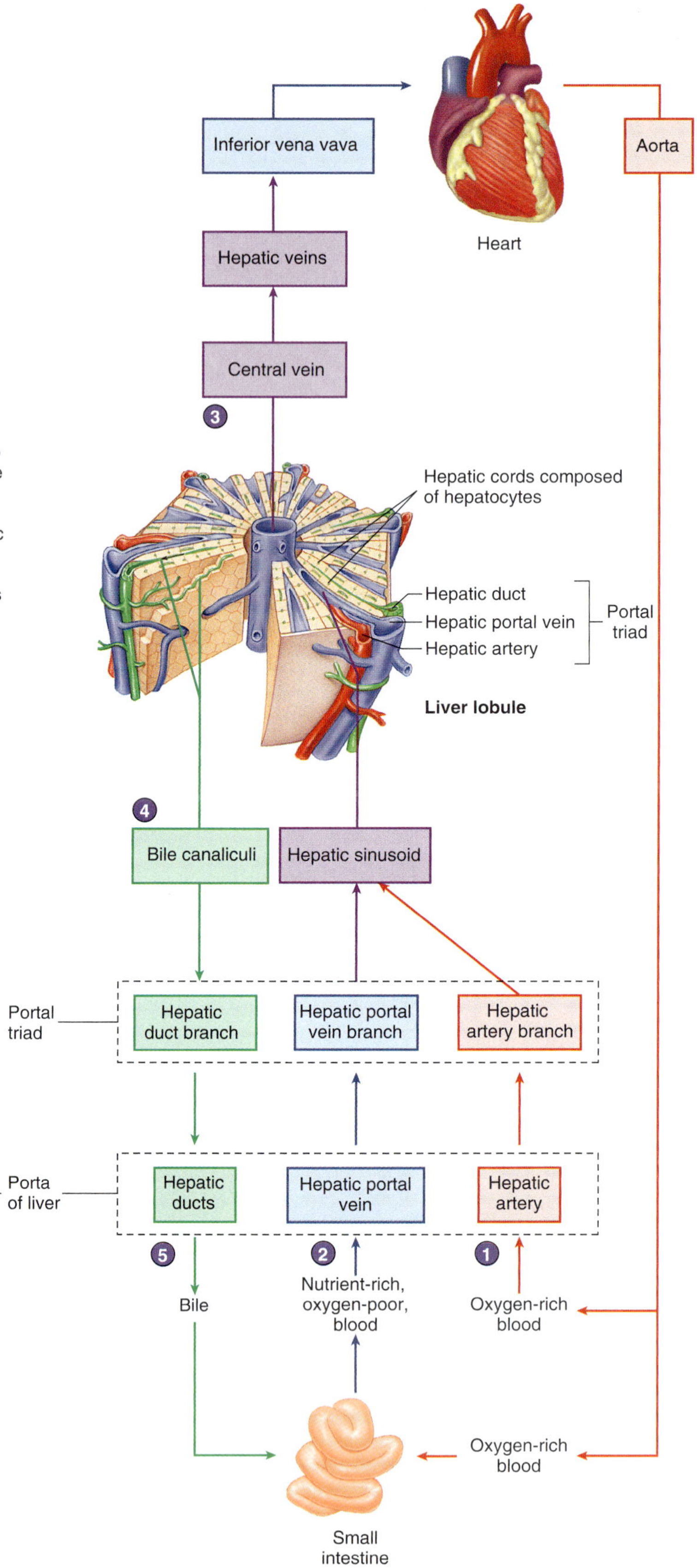

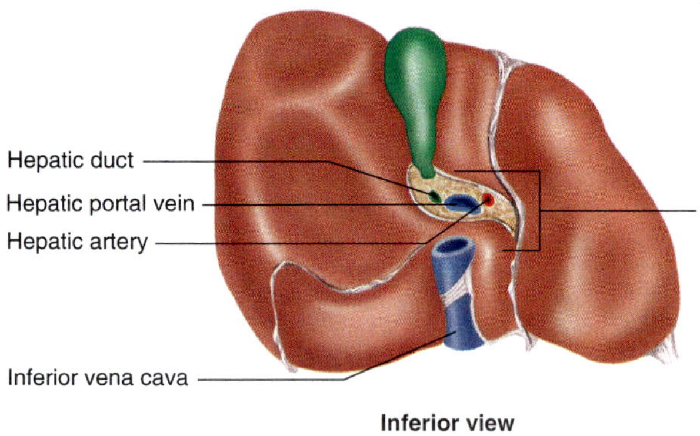

Blood and Bile Flow Through the Liver
Process Figure 24.20

1. Secretin, produced by the duodenum (*purple arrows*) and carried through the circulation to the liver, stimulates bile secretion by the liver (*green arrows inside the liver*).
2. Cholecystokinin, produced by the duodenum (*pink arrows*) and carried through the circulation to the gallbladder, stimulates the gallbladder to contract, thereby releasing bile into the duodenum (*green arrow outside the liver*).
3. Vagus nerve stimulation (*red arrow*) causes the gallbladder to contract, thereby releasing bile into the duodenum.
4. Bile salts also stimulate bile secretion. Over 90% of bile salts are reabsorbed in the ileum and returned to the liver (*blue arrows*), where they stimulate additional secretion of bile salts.

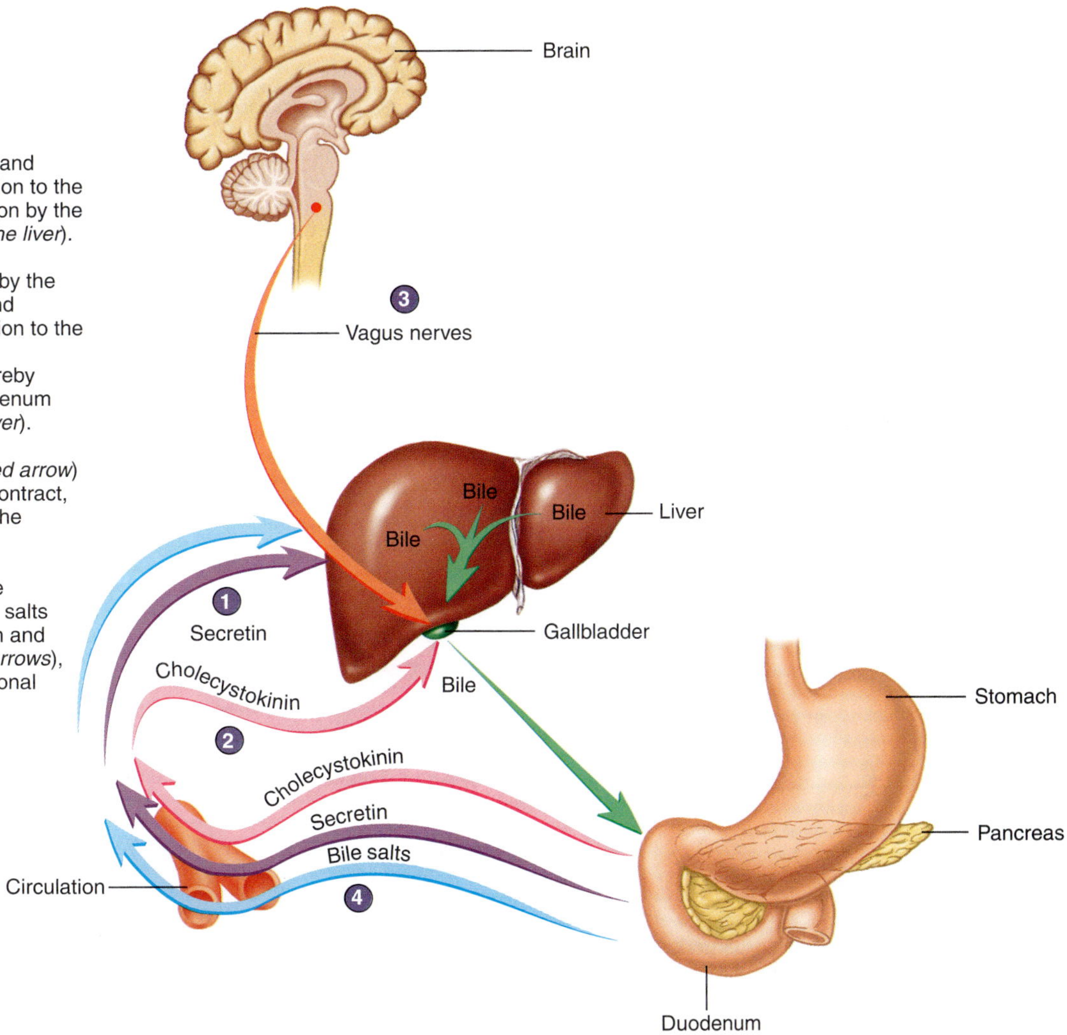

Control of Bile Secretion and Release
Process Figure 24.21

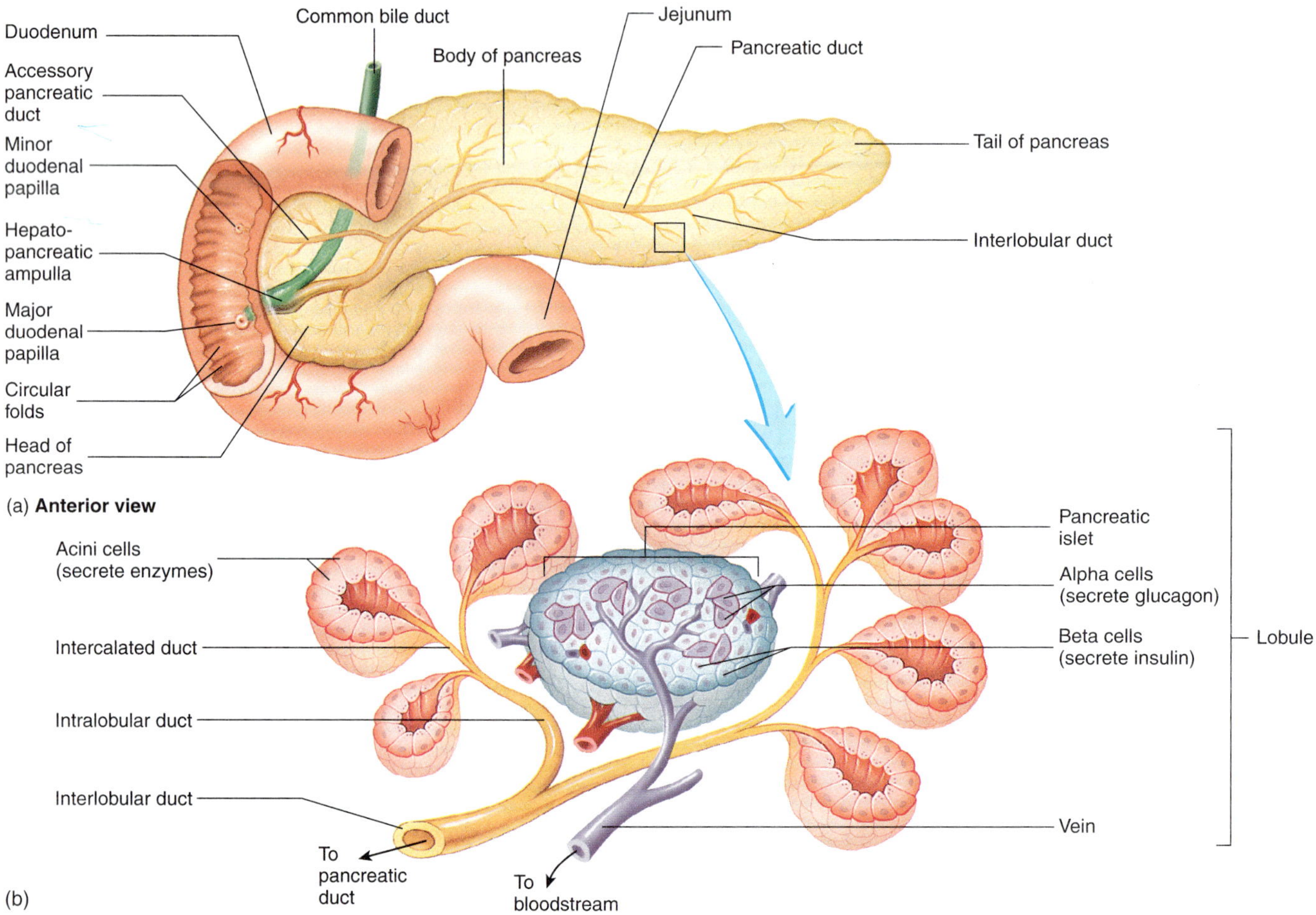

Anatomy and Histology of the Duodenum and Pancreas
Figure 24.22

1. Water (H_2O) and carbon dioxide (CO_2) combine under the influence of carbonic anhydrase (CA) to form carbonic acid (H_2CO_3).
2. Carbonic acid dissociates to form hydrogen ions (H^+) and bicarbonate ions (HCO_3^-).
3. The H^+ are exchanged for sodium ions (Na^+) so H^+ enter and Na^+ are removed from the bloodstream.
4. The HCO_3^- are actively transported into the intercalated ducts. Sodium ions and H_2O follow the HCO_3^- into the ducts.
5. The ions move through the intercalated duct toward the interlobular duct.

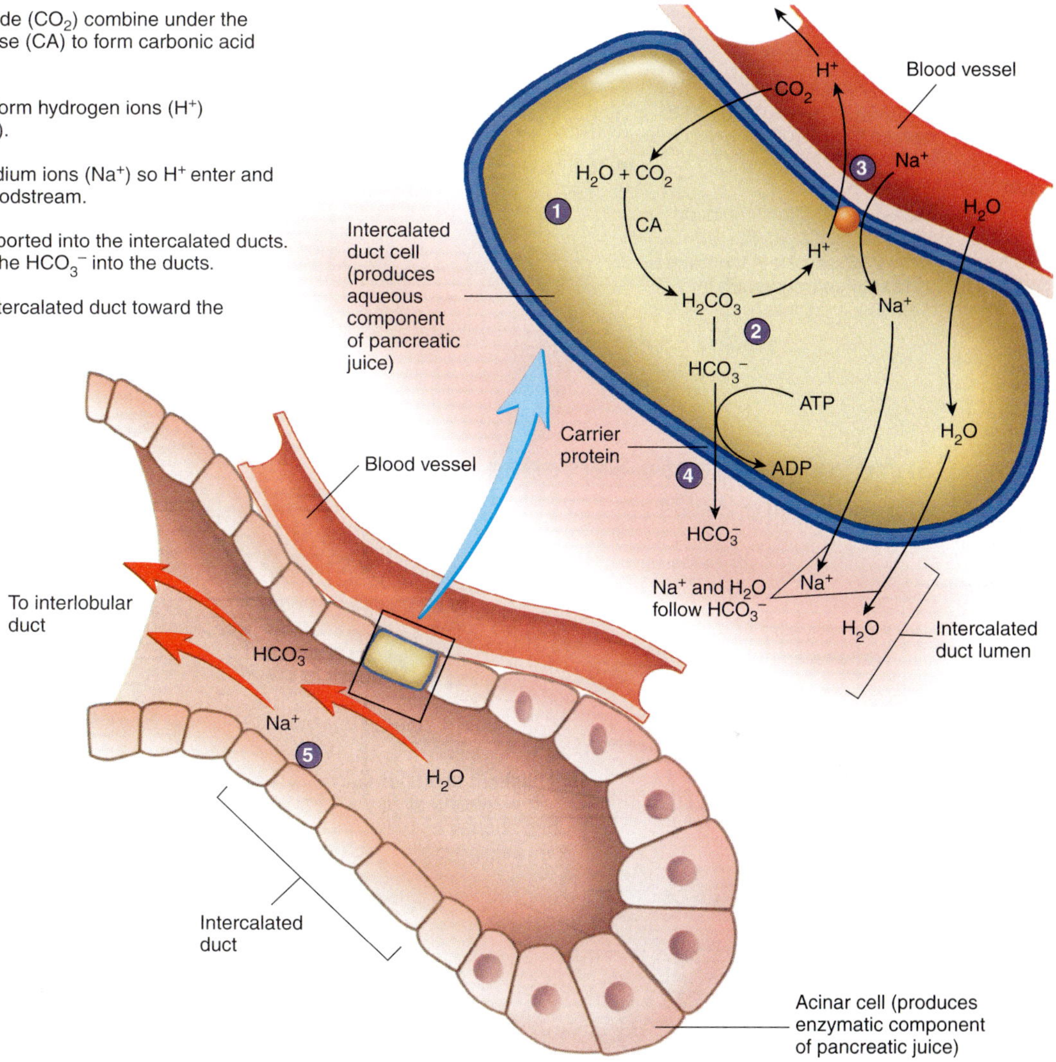

Bicarbonate Ion Production in the Pancreas
Process Figure 24.23

1. Secretin (*purple arrows*) released from the duodenum, stimulates the pancreas to release a watery secretion, rich in bicarbonate ions.
2. Cholecystokinin (*pink arrows*) released from the duodenum, causes the pancreas to release a secretion rich in digestive enzymes.
3. Parasympathetic stimulation from the vagus nerve (*red arrow*) causes the pancreas to release a secretion rich in digestive enzymes.

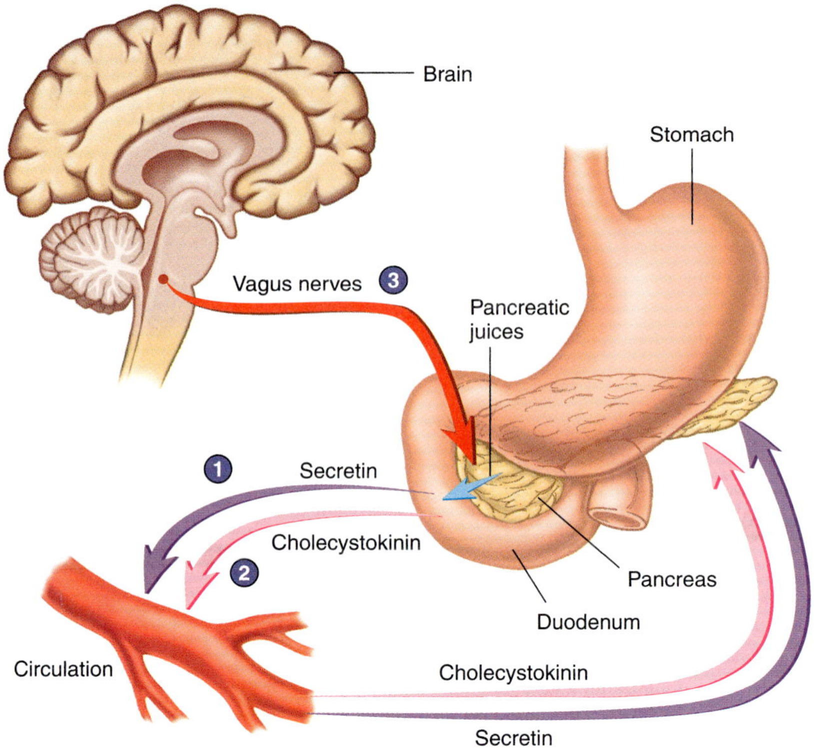

Control of Pancreatic Secretion
Process Figure 24.24

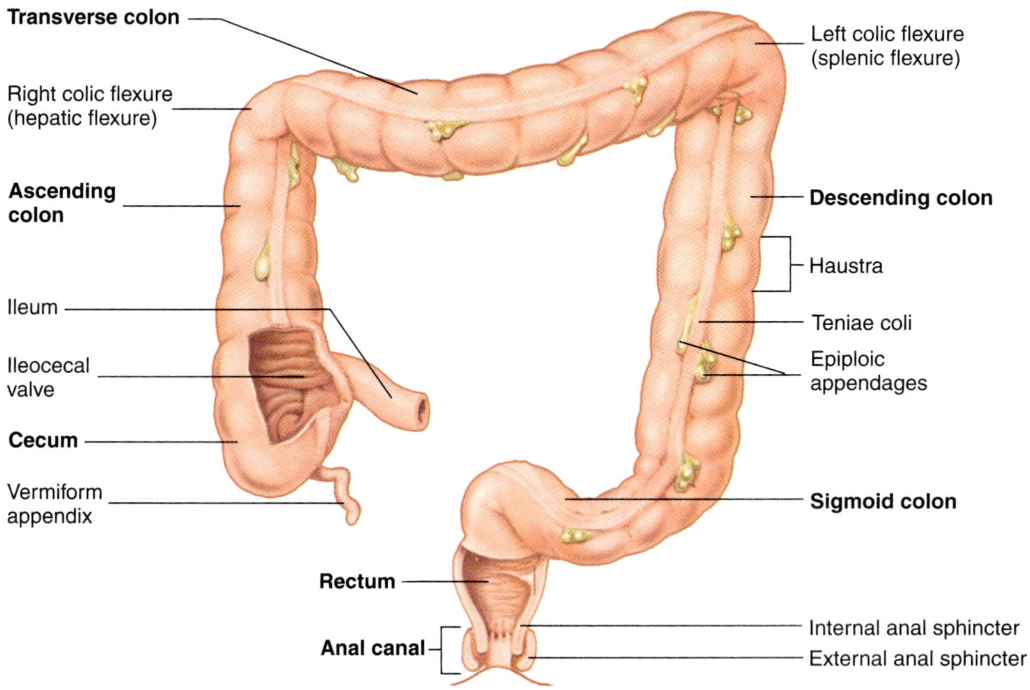

(a) **Anterior view**

Large Intestine
Figure 24.25

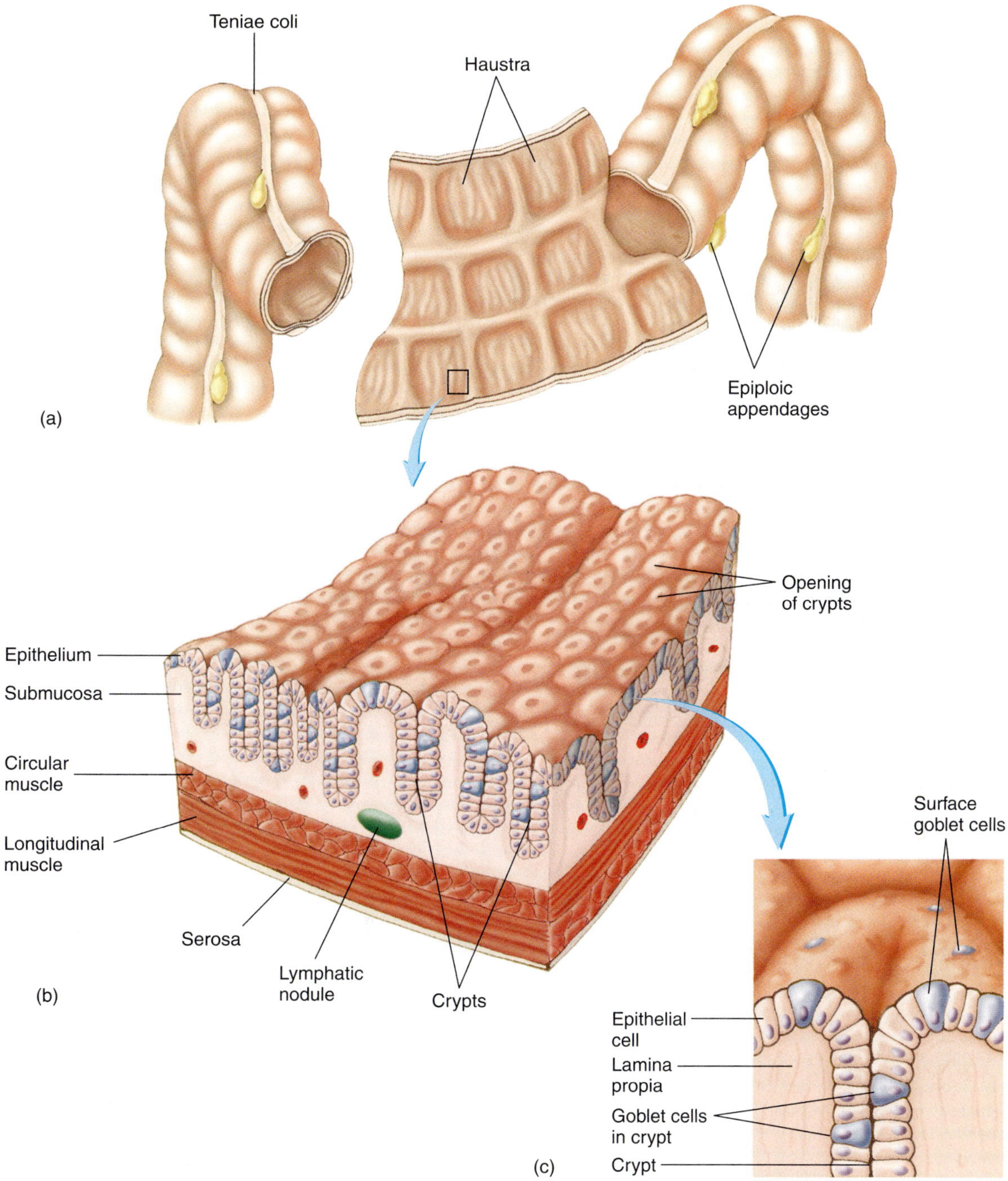
Teniae coli
Haustra
Epiploic appendages
(a)
Opening of crypts
Epithelium
Submucosa
Circular muscle
Longitudinal muscle
Serosa
Lymphatic nodule
Crypts
(b)
Surface goblet cells
Epithelial cell
Lamina propia
Goblet cells in crypt
Crypt
(c)

Histology of the Large Intestine
Figure 24.26

1. The presence of food in the stomach and chyme in the duodenum stimulate mass movement in the colon.

2. Mass movements are integrated by the enteric plexus, located throughout the colon.

3. They propel the contents of the colon toward the rectum.

4. The presence of feces in the rectum stimulates parasympathetic and local reflexes that result in defecation.

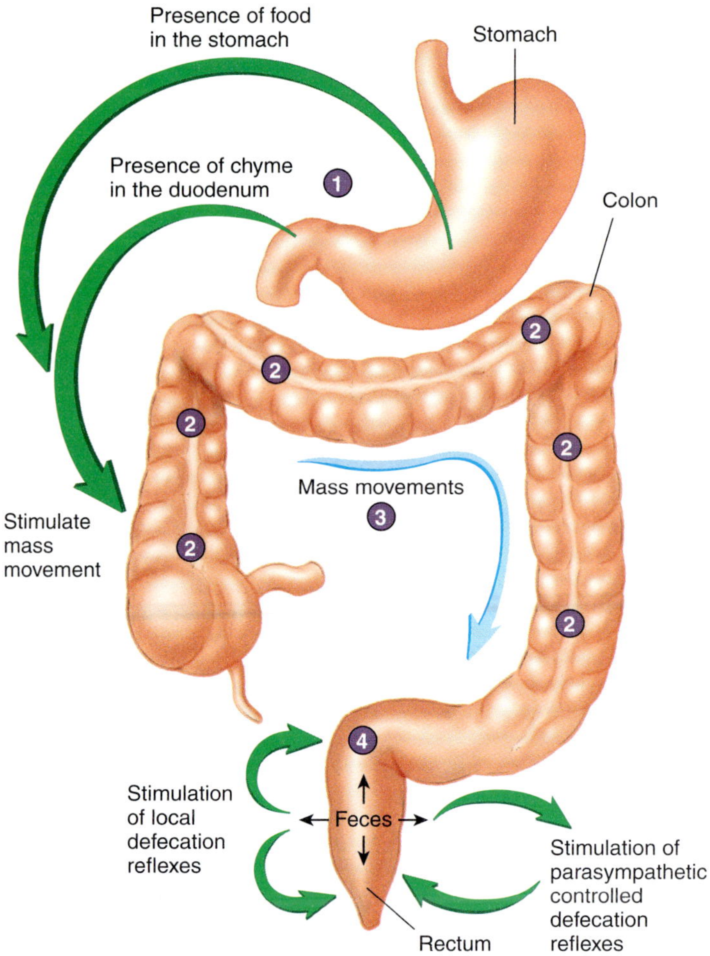

Reflexes in the Colon and Rectum
Process Figure 24.27

Monosaccharide transport

1. Monosaccharides are absorbed by cotransport into intestinal epithelial cells.
2. Monosaccharides move out of intestinal epithelial cells by facilitated diffusion.
3. They enter the capillaries of the intestinal villi and are carried through the hepatic portal vein to the liver.

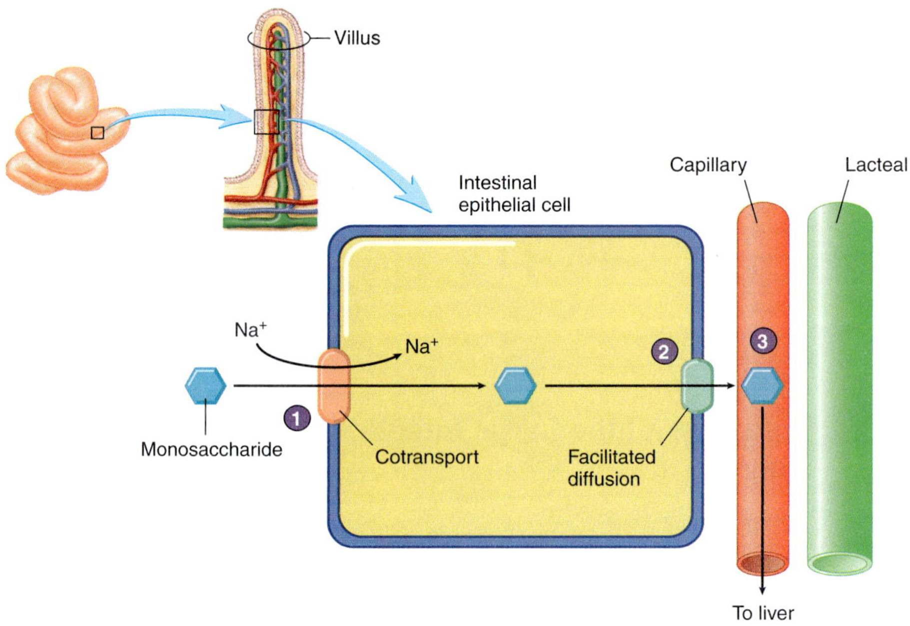

Transport of Monosaccharides Across the Intestinal Epithelium
Process Figure 24.28

Lipid transport

1. Bile salts surround fatty acids and monoglycerides to form micelles.

2. Micelles attach to the cell membranes of intestinal epithelial cells, and the fatty acids and monoglycerides pass by simple diffusion into the intestinal epithelial cells.

3. Within the intestinal epithelial cell, the fatty acids and monoglycerides are converted to triglycerides; proteins coat the triglycerides to form chylomicrons, which move out of the intestinal epithelial cells by exocytosis.

4. The chylomicrons enter the lacteals of the intestinal villi and are carried through the lymphatic system to the general circulation.

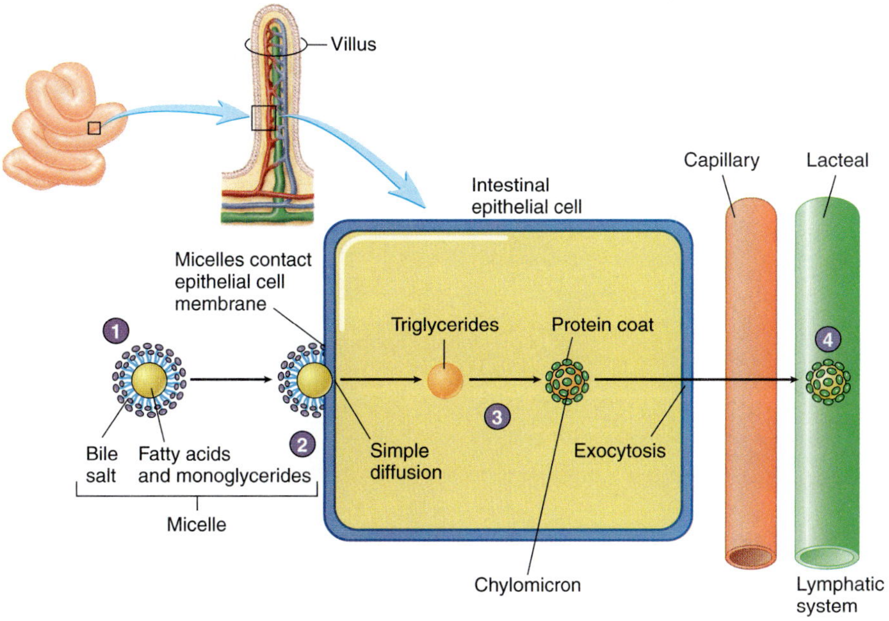

Transport of Lipids Across the Intestinal Epithelium
Process Figure 24.29

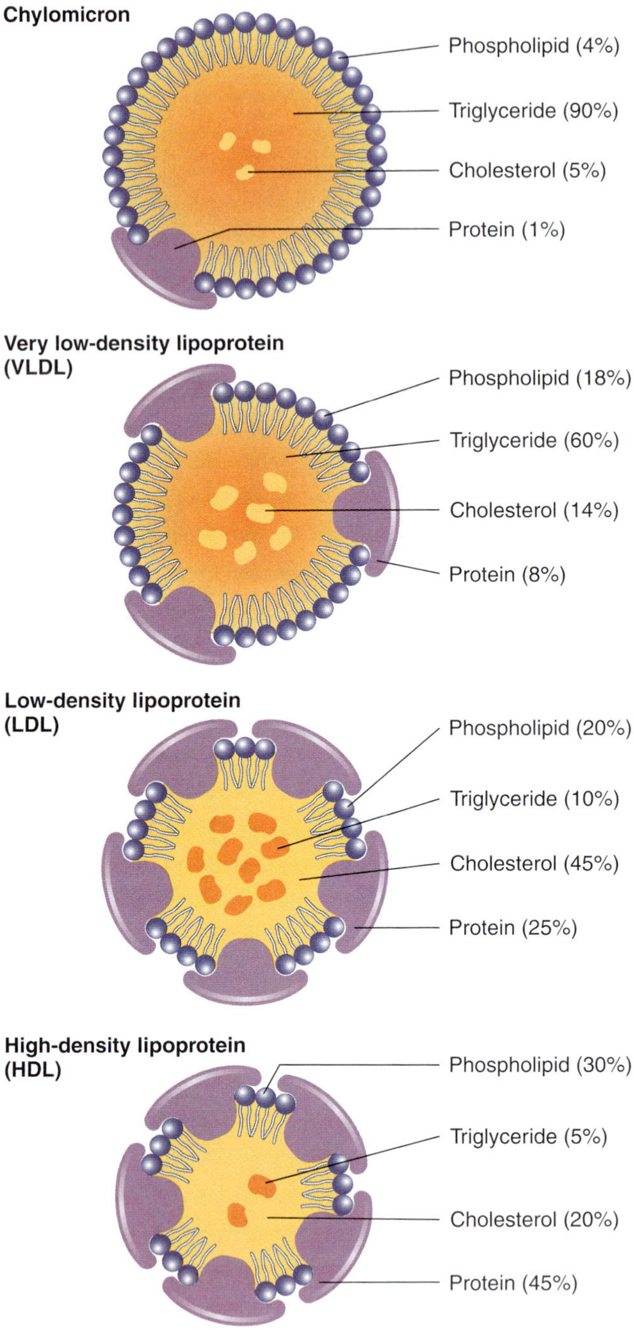
Chylomicron
Phospholipid (4%)
Triglyceride (90%)
Cholesterol (5%)
Protein (1%)
Very low-density lipoprotein
(VLDL)
Phospholipid (18%)
Triglyceride (60%)
Cholesterol (14%)
Protein (8%)
Low-density lipoprotein
(LDL)
Phospholipid (20%)
Triglyceride (10%)
Cholesterol (45%)
Protein (25%)
High-density lipoprotein
(HDL)
Phospholipid (30%)
Triglyceride (5%)
Cholesterol (20%)
Protein (45%)

Lipoproteins
Figure 24.30

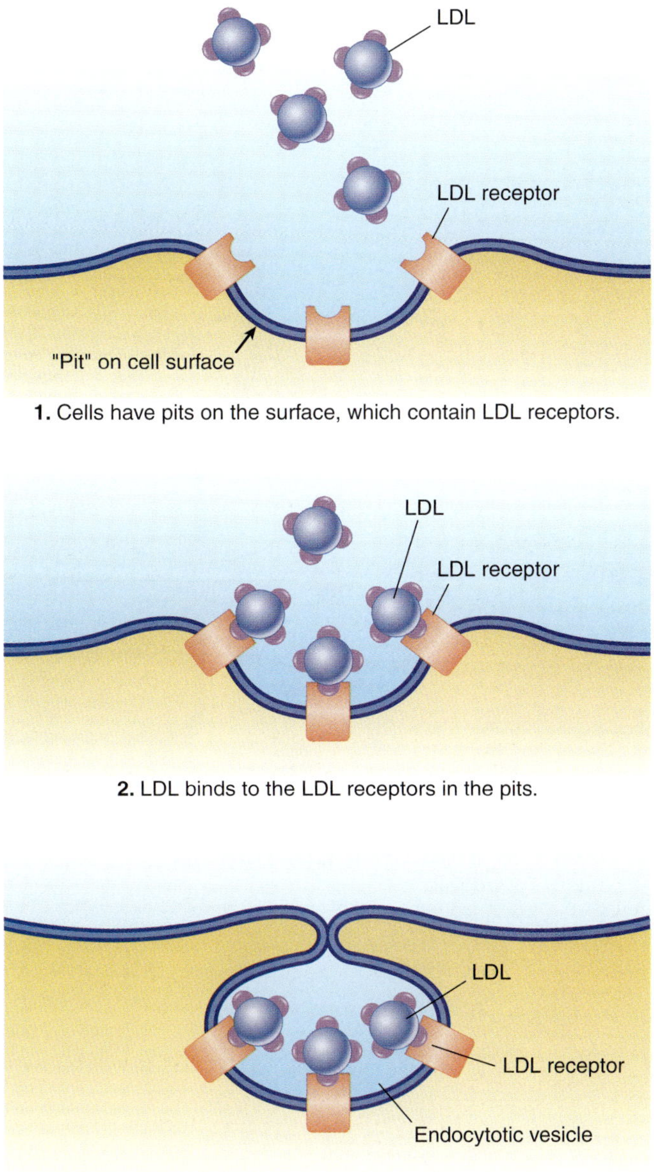

3. The LDL, bound to LDL receptors, is taken into the cell by endocytosis.

Transport of LDL into Cells
Process Figure 24.31

Amino acid transport

1. Tripeptides, dipeptides, as well as acidic and most neutral amino acids, are absorbed by cotransport into intestinal epithelial cells.
2. Amino acids move out of intestinal epithelial cells by active transport.
3. They enter the capillaries of the intestinal villi and are carried through the hepatic portal vein to the liver.

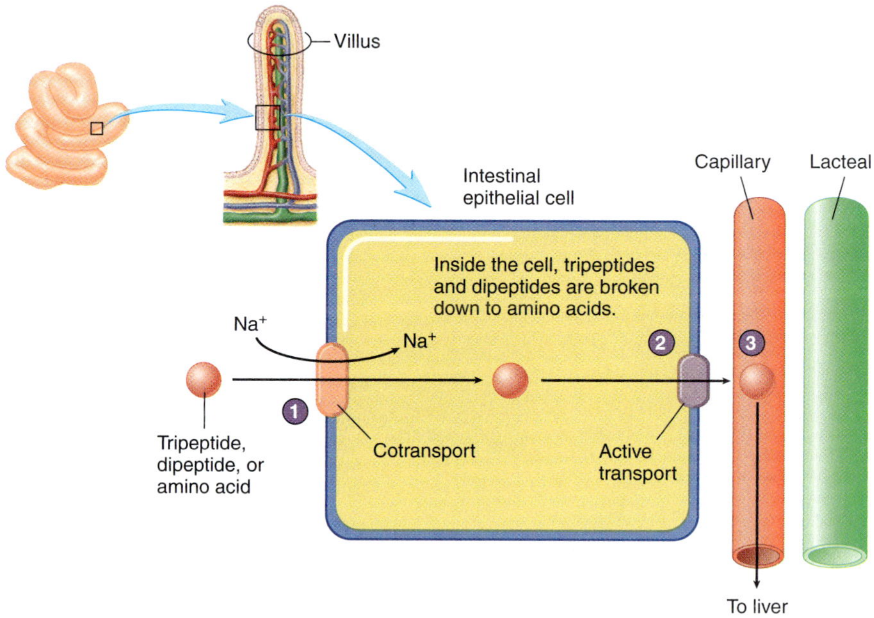

Amino Acid Transport Across the Intestinal Epithelium
Process Figure 24.32

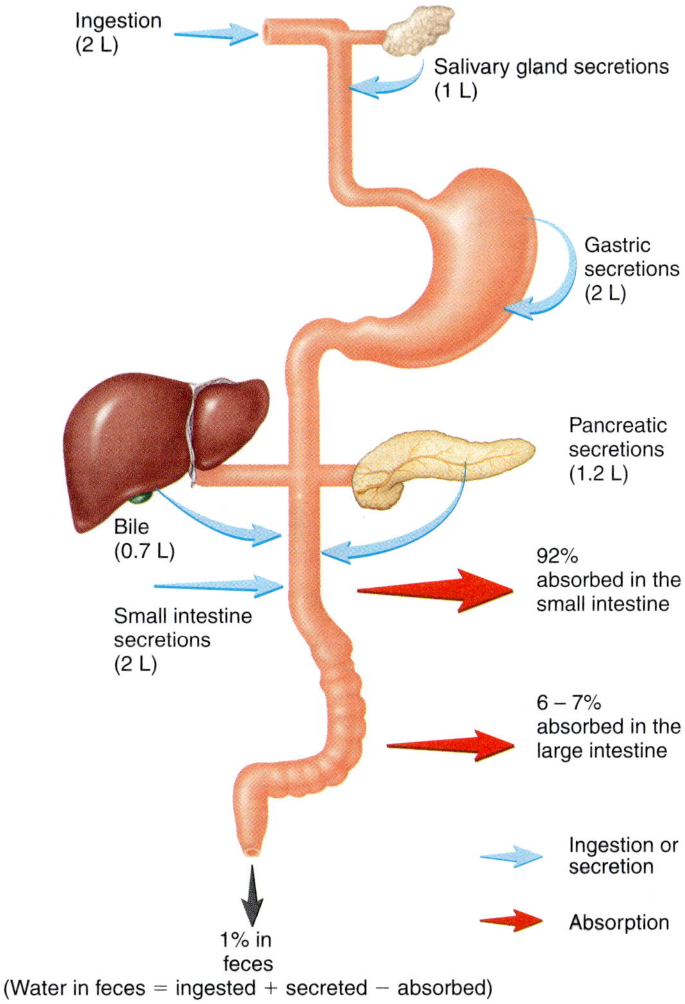

Fluid Volumes in the Digestive Tract
Figure 24.33

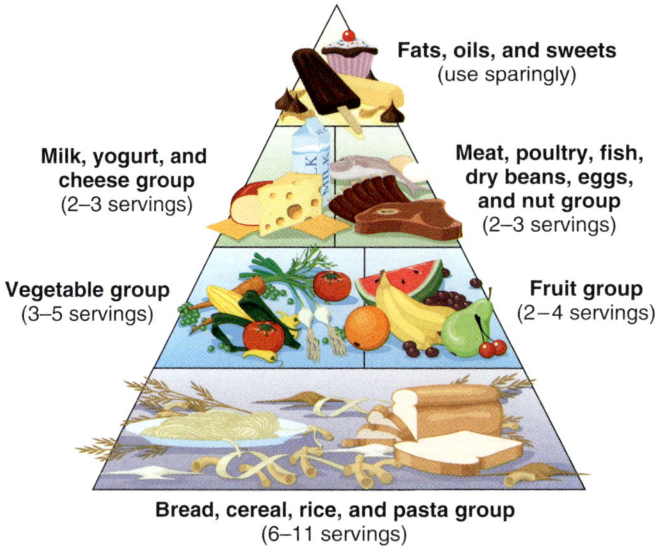

Food Guide Pyramid
Figure 25.1

Nutrition Facts

Serving Size 1 oz. (28g/About 32 chips)
Servings Per Container 2.5

Amount Per Serving

Calories 160	Calories from Fat 90
	% Daily Value*
Total Fat 10g	**16%**
Saturated Fat 1.5g	**7%**
Cholesterol 0mg	**0%**
Sodium 170mg	**7%**
Total Carbohydrate 15g	**5%**
Dietary Fiber 1g	**4%**
Sugars less than 1g	
Protein 2g	

Vitamin A 0% • Vitamin C 0%
Calcium 2% • Iron 0%

* Percent Daily Values are based on a 2,000 calorie diet. Your daily values may be higher or lower depending on your calorie needs:

	Calories:	2,000	2,500
Total Fat	Less than	65g	80g
Sat Fat	Less than	20g	25g
Cholesterol	Less than	300mg	300mg
Sodium	Less than	2,400mg	2,400mg
Total Carbohydrate		300g	375g
Dietary Fiber		25g	30g

Calories per gram:
Fat 9 • Carbohydrate 4 • Protein 4

Food Label
Figure 25.2

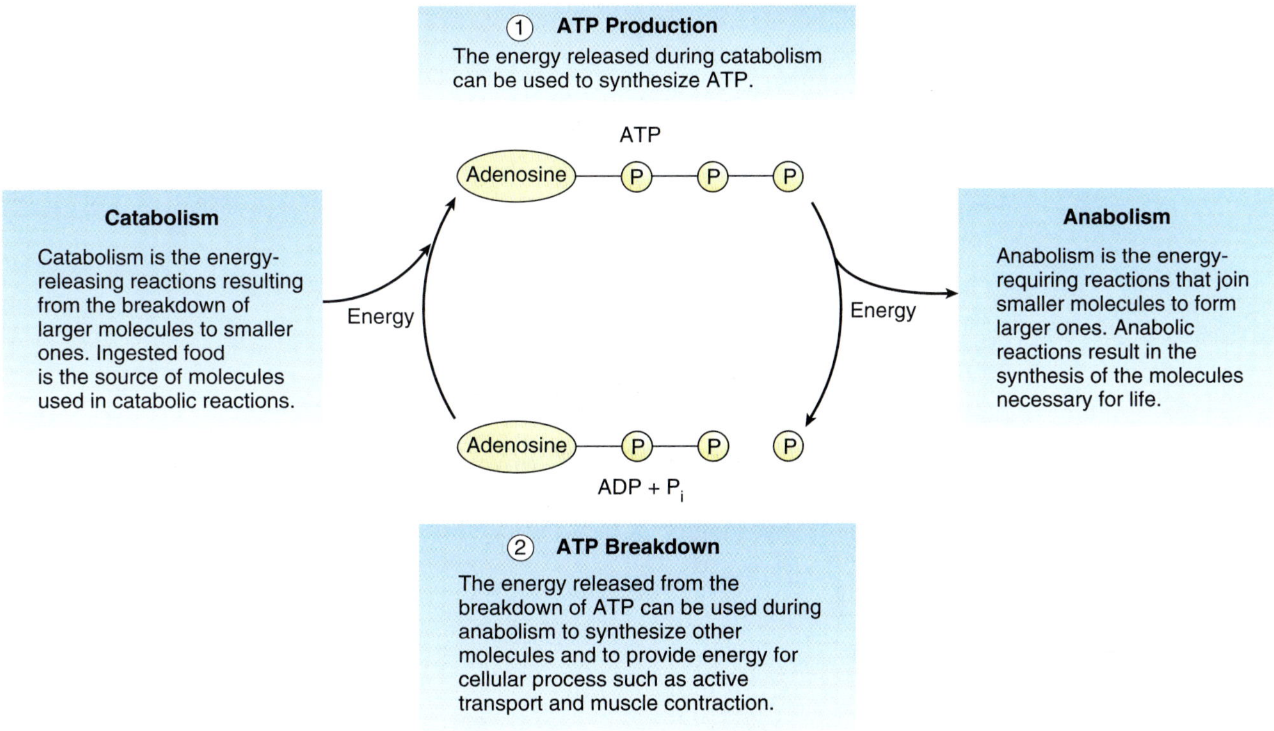

ATP Coupling of Catabolic and Anabolic Reactions

Process Figure 25.3

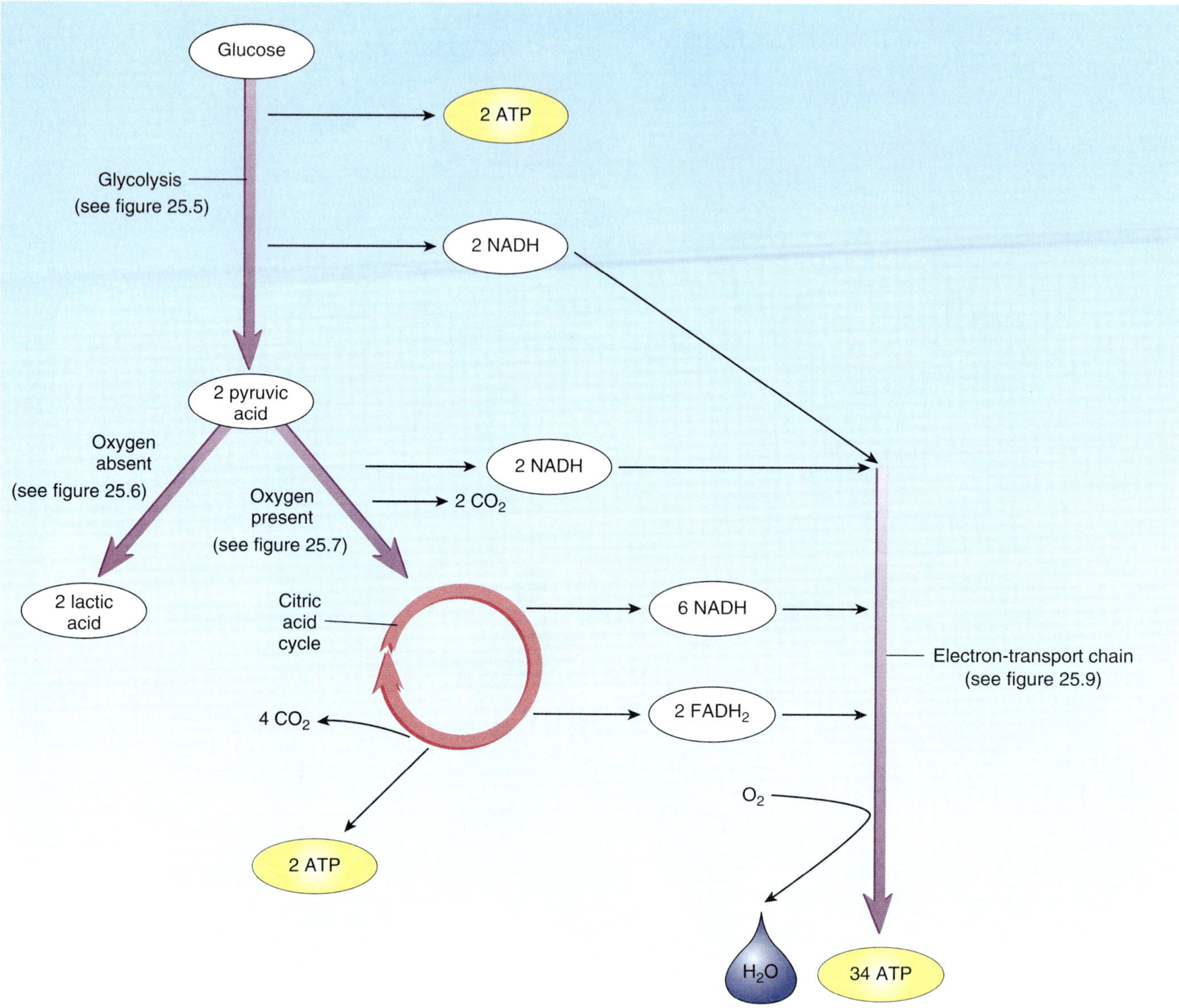
Glucose
2 ATP
Glycolysis
(see figure 25.5)
2 NADH
2 pyruvic acid
Oxygen absent
(see figure 25.6)
Oxygen present
(see figure 25.7)
2 NADH
2 CO_2
2 lactic acid
Citric acid cycle
6 NADH
Electron-transport chain
(see figure 25.9)
4 CO_2
2 $FADH_2$
O_2
2 ATP
H_2O
34 ATP

Cellular Metabolism
Figure 25.4

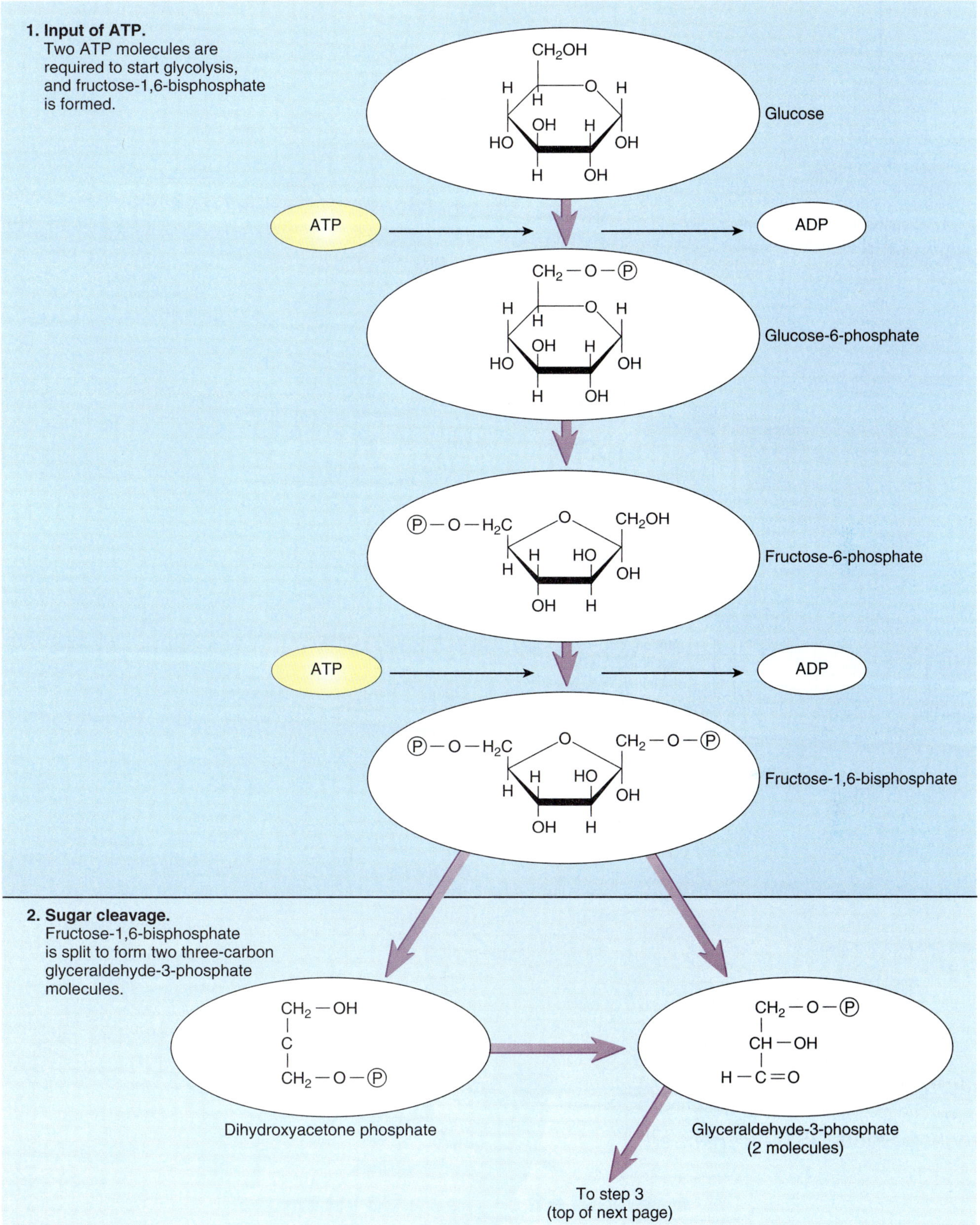
1. Input of ATP.
Two ATP molecules are required to start glycolysis, and fructose-1,6-bisphosphate is formed.
Glucose
ATP
ADP
Glucose-6-phosphate
Fructose-6-phosphate
ATP
ADP
Fructose-1,6-bisphosphate
2. Sugar cleavage.
Fructose-1,6-bisphosphate is split to form two three-carbon glyceraldehyde-3-phosphate molecules.
Dihydroxyacetone phosphate
Glyceraldehyde-3-phosphate (2 molecules)
To step 3 (top of next page)

Glycolysis
Process Figure 25.5

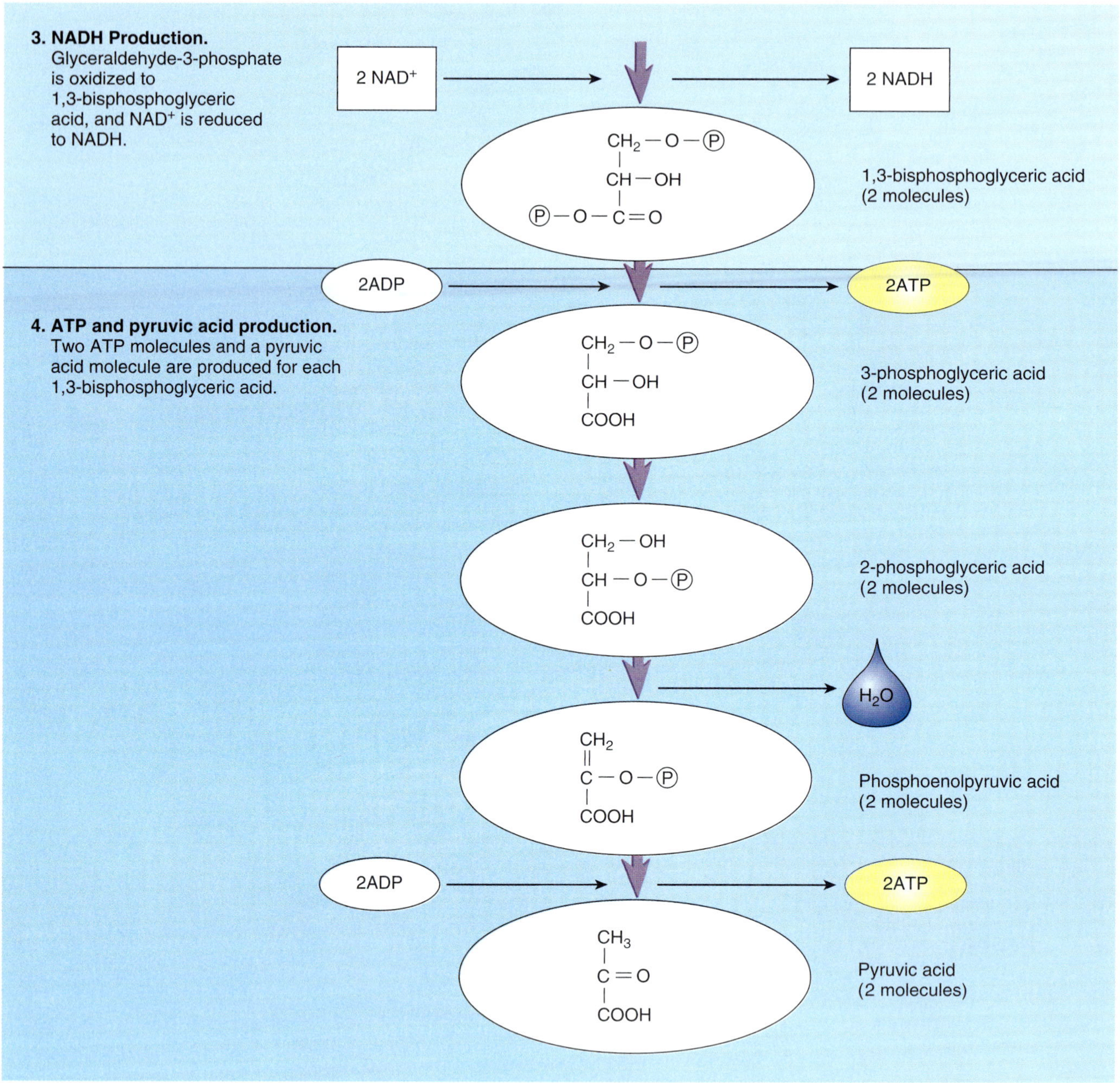
3. NADH Production.
Glyceraldehyde-3-phosphate is oxidized to 1,3-bisphosphoglyceric acid, and NAD+ is reduced to NADH.
2 NAD+
2 NADH
1,3-bisphosphoglyceric acid (2 molecules)
2ADP
2ATP
4. ATP and pyruvic acid production.
Two ATP molecules and a pyruvic acid molecule are produced for each 1,3-bisphosphoglyceric acid.
3-phosphoglyceric acid (2 molecules)
2-phosphoglyceric acid (2 molecules)
H2O
Phosphoenolpyruvic acid (2 molecules)
2ADP
2ATP
Pyruvic acid (2 molecules)

Process Figure 25.5—*continued*

1. Glycolysis converts glucose to two pyruvic acid molecules. The many reactions within the pathway are not shown. There is a net gain of two ATP and two NADH from glycolysis.

2. Anaerobic respiration, which does not require oxygen, includes glycolysis and converts the two pyruvic acid molecules produced by glycolysis to two lactic acid molecules. This conversion requires energy, which is derived from the NADH generated in glycolysis.

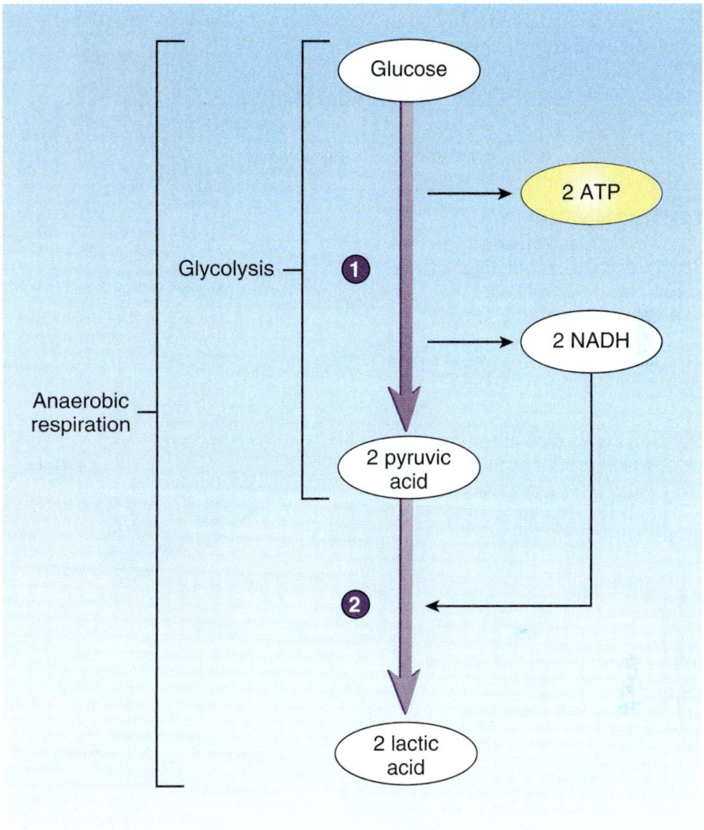

Glycolysis and Anaerobic Respiration
Process Figure 25.6

1. Glycolysis converts glucose to two pyruvic acid molecules and produces two ATP and two NADH. The NADH can go to the electron-transport chain.

2. The two pyruvic acid molecules produced in glycolysis are converted to two acetyl-CoA molecules, producing two CO_2 and two NADH. The NADH can go to the electron-transport chain.

3. The two acetyl-CoA molecules enter the citric acid cycle, which produces four CO_2, six NADH, two $FADH_2$, and two ATP. The NADH and $FADH_2$ can go to the electron-transport chain.

4. The electron-transport chain uses NADH and $FADH_2$ to produce 34 ATP. This process requires O_2, which combines with H^+ to form H_2O.

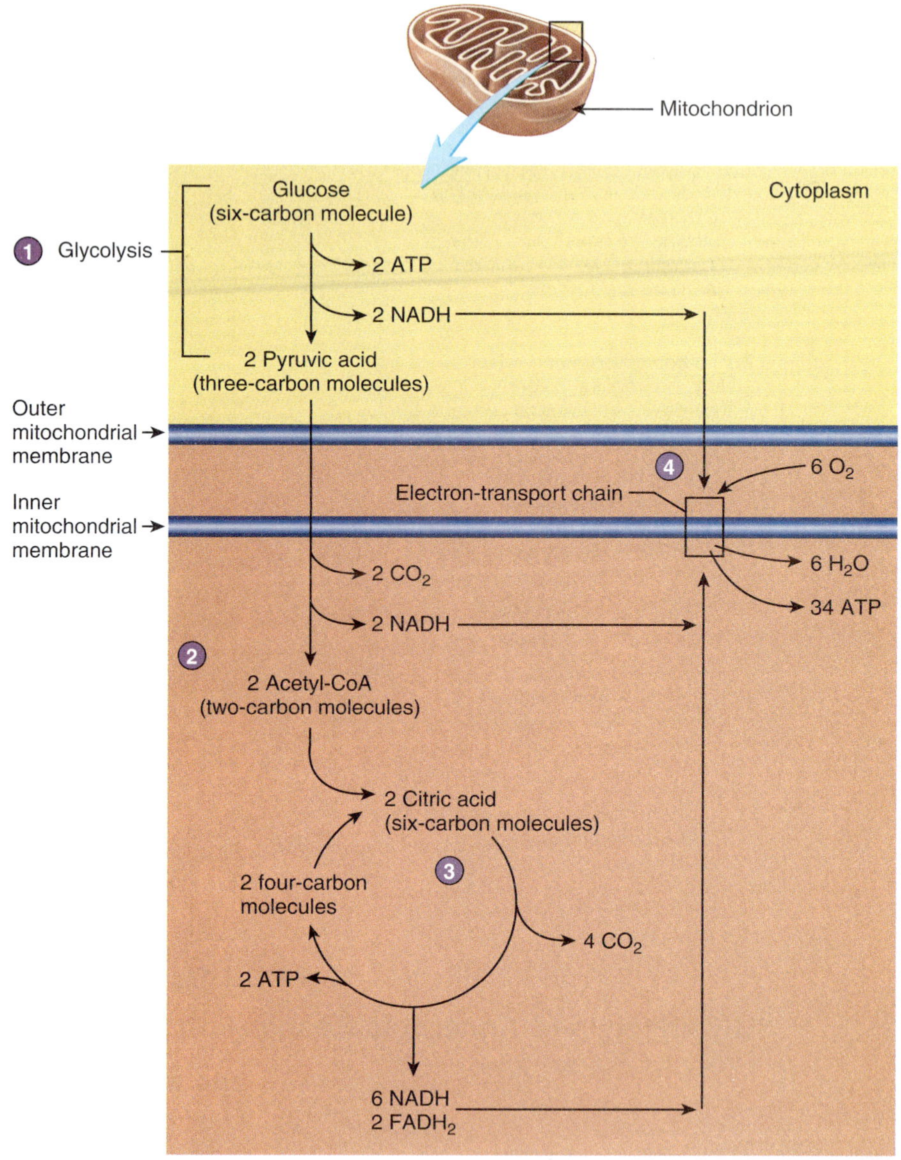

Aerobic Respiration
Process Figure 25.7

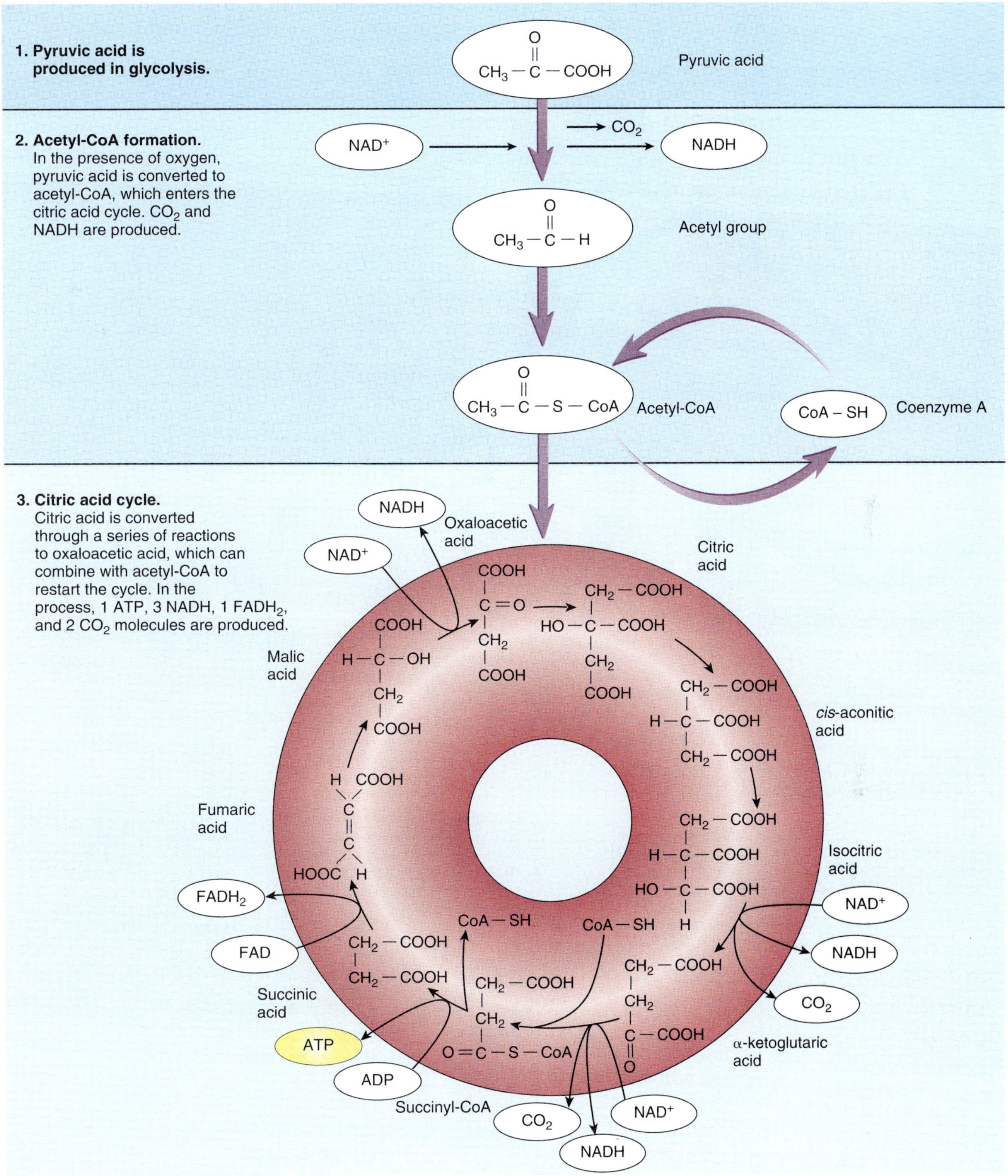

1. Pyruvic acid is produced in glycolysis.
O
CH3 — C — COOH
Pyruvic acid
2. Acetyl-CoA formation. In the presence of oxygen, pyruvic acid is converted to acetyl-CoA, which enters the citric acid cycle. CO2 and NADH are produced.
NAD+
CO2
NADH
O
CH3 — C — H
Acetyl group
O
CH3 — C — S — CoA
Acetyl-CoA
CoA – SH
Coenzyme A
3. Citric acid cycle. Citric acid is converted through a series of reactions to oxaloacetic acid, which can combine with acetyl-CoA to restart the cycle. In the process, 1 ATP, 3 NADH, 1 FADH2, and 2 CO2 molecules are produced.
NADH
NAD+
Oxaloacetic acid
COOH
C=O
CH2
COOH
Citric acid
CH2 — COOH
HO — C — COOH
CH2
COOH
cis-aconitic acid
CH2 — COOH
H — C — COOH
CH2 — COOH
Isocitric acid
CH2 — COOH
H — C — COOH
HO — C — COOH
H
NAD+
NADH
CO2
α-ketoglutaric acid
CH2 — COOH
CH2
C — COOH
O
NAD+
NADH
CO2
CoA — SH
Succinyl-CoA
CH2 — COOH
CH2
O = C — S — CoA
CoA — SH
ADP
ATP
Succinic acid
CH2 — COOH
CH2 — COOH
FAD
FADH2
Fumaric acid
H COOH
C
C
HOOC H
Malic acid
COOH
H — C — OH
CH2
COOH

Aerobic Respiration
Process Figure 25.8

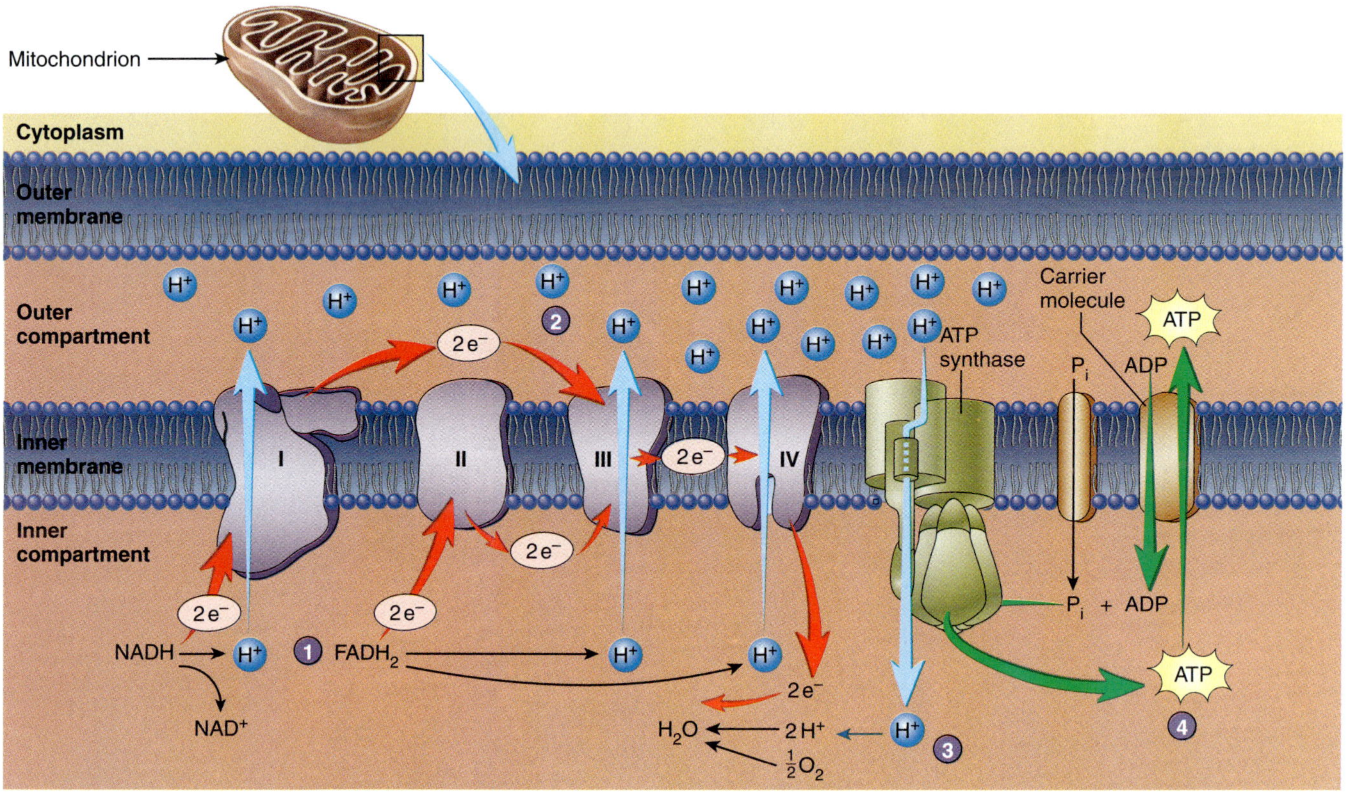

1. NADH or $FADH_2$ transfer their electrons to the electron-transport chain.
2. As the electrons move through the electron-transport chain, some of their energy is used to pump H^+ into the outer compartment, resulting in a higher concentration of H^+ in the outer than in the inner compartment.
3. The H^+ diffuses back into the inner compartment through special channels (ATP synthase) that couple the H^+ movement with the production of ATP. The electrons, H^+, and O_2 combine to form H_2O.
4. ATP is transported out of the inner compartment by a carrier molecule that exchanges ATP for ADP. A different carrier molecule moves phosphate into the inner compartment.

The Electron-Transport Chain
Process Figure 25.9

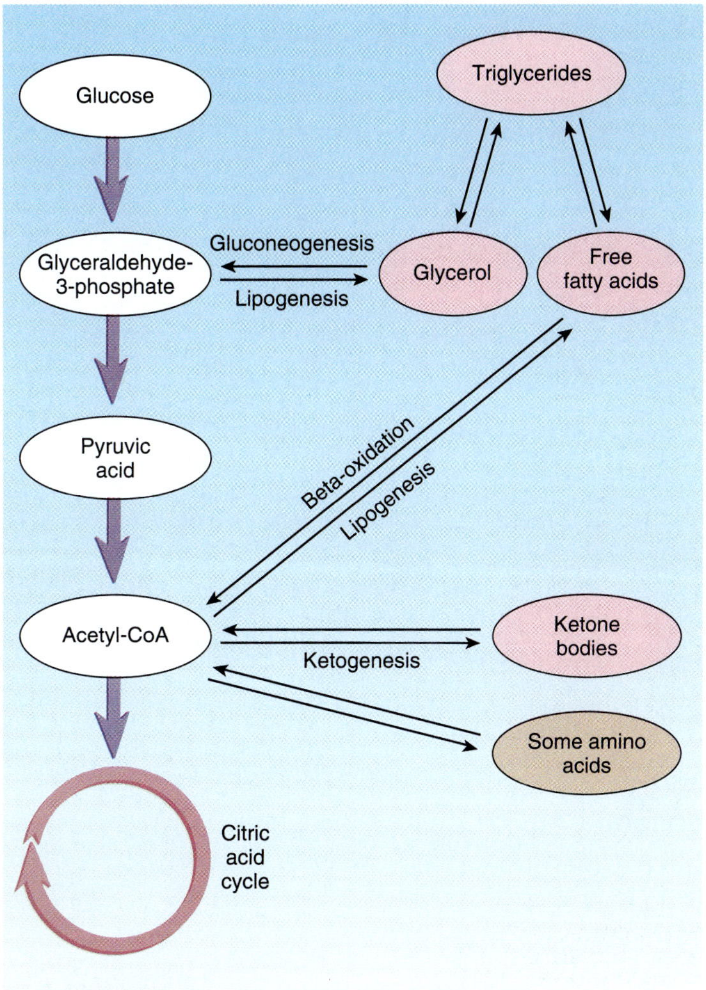
Glucose
Triglycerides
Glyceraldehyde-
3-phosphate
Gluconeogenesis
Lipogenesis
Glycerol
Free
fatty acids
Pyruvic
acid
Beta-oxidation
Lipogenesis
Acetyl-CoA
Ketogenesis
Ketone
bodies
Some amino
acids
Citric
acid
cycle

Lipid Metabolism
Figure 25.10

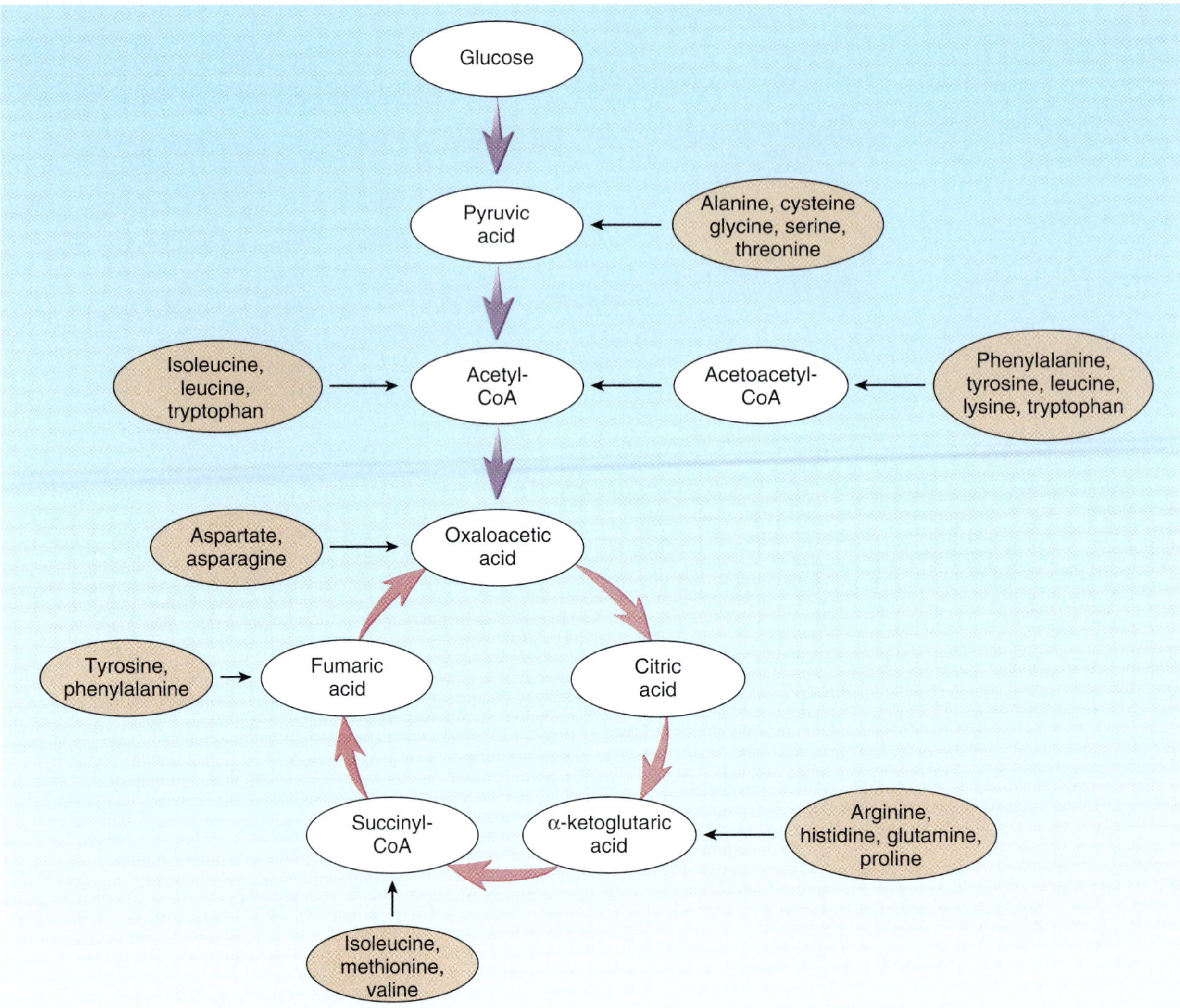

Amino Acid Metabolism
Figure 25.11

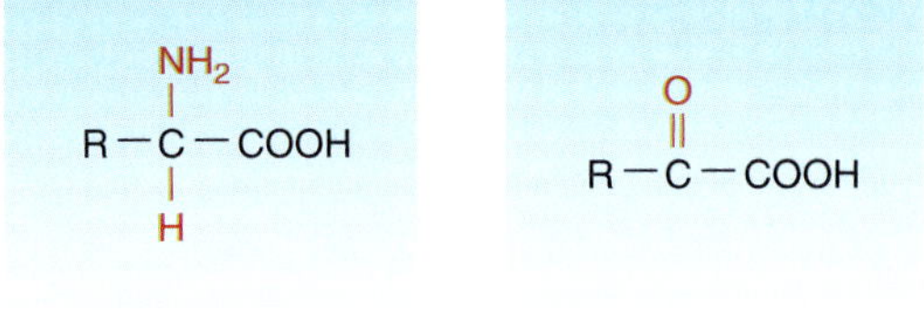

(a) Amino acid (b) Keto acid

General Formulas of an Amino Acid and a Keto Acid
Figure 25.12

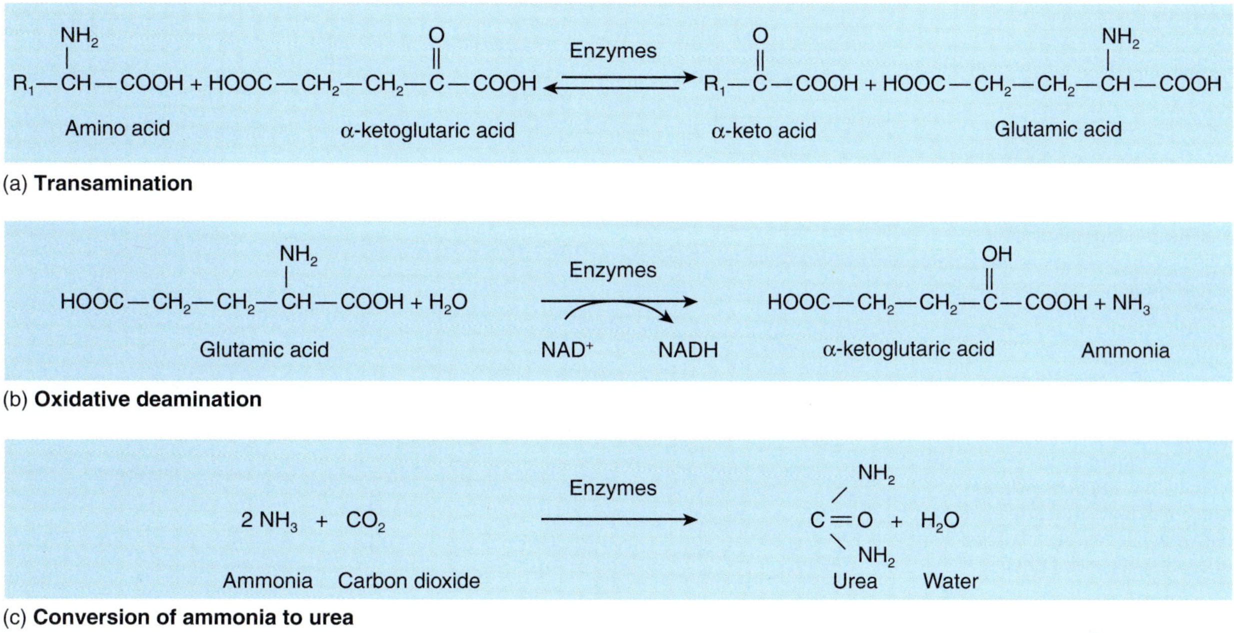

Amino Acid Reactions
Figure 25.13

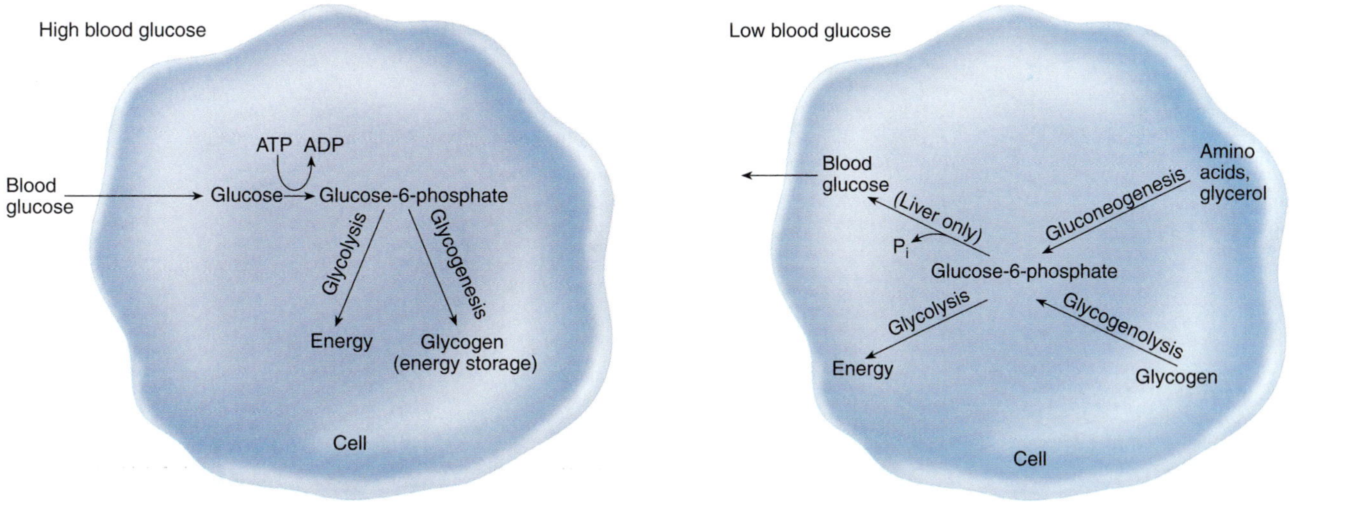

(a) When blood glucose levels are high, glucose enters the cell and is phosphorylated to form glucose-6-phosphate, which can enter glycolysis or glycogenesis.

(b) When blood glucose levels drop, glucose-6-phosphate can be produced through glycogenolysis or gluconeogenesis. Glucose-6-phosphate can enter glycolysis, or the phosphate group can be removed in liver tissue and glucose released into the blood.

Interconversion of Nutrient Molecules
Figure 25.14

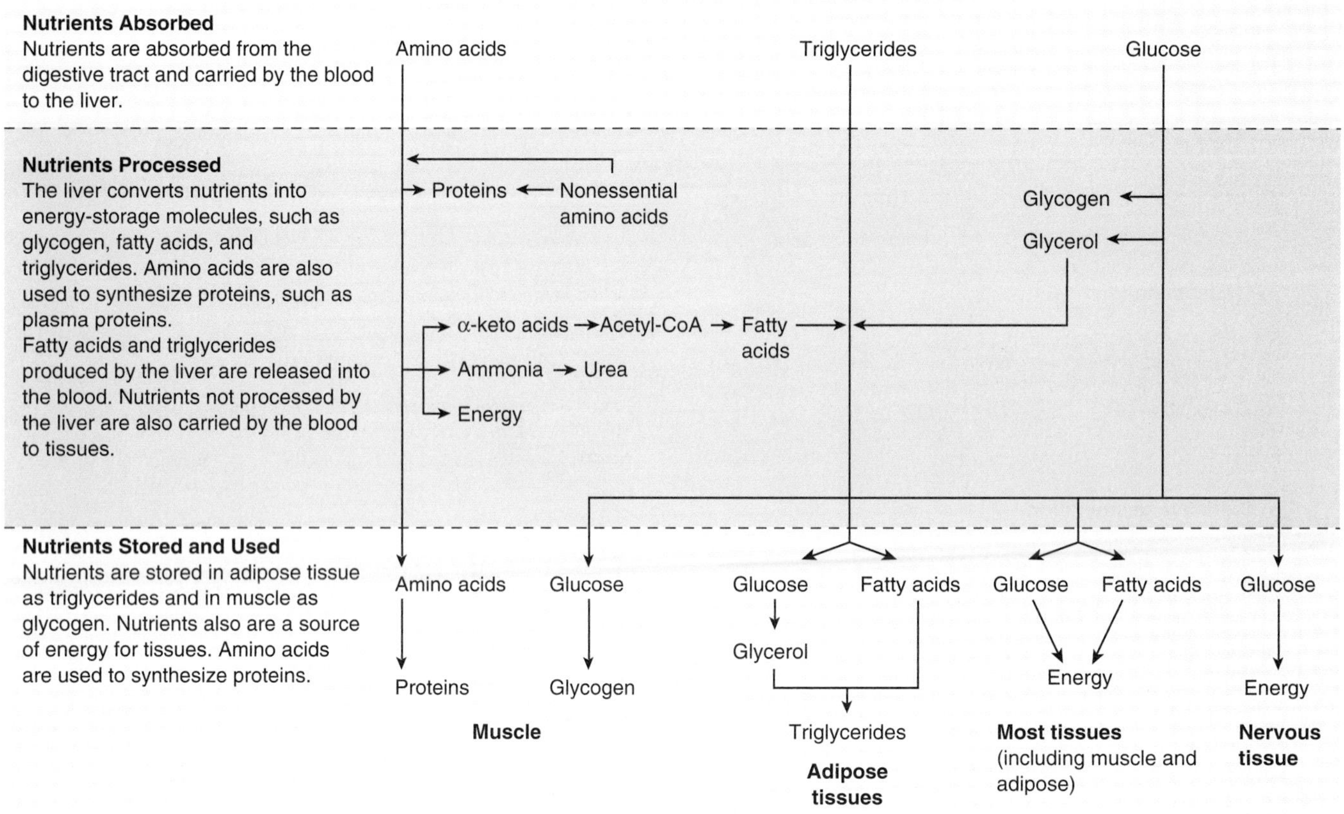

Events of the Absorptive State

Figure 25.15

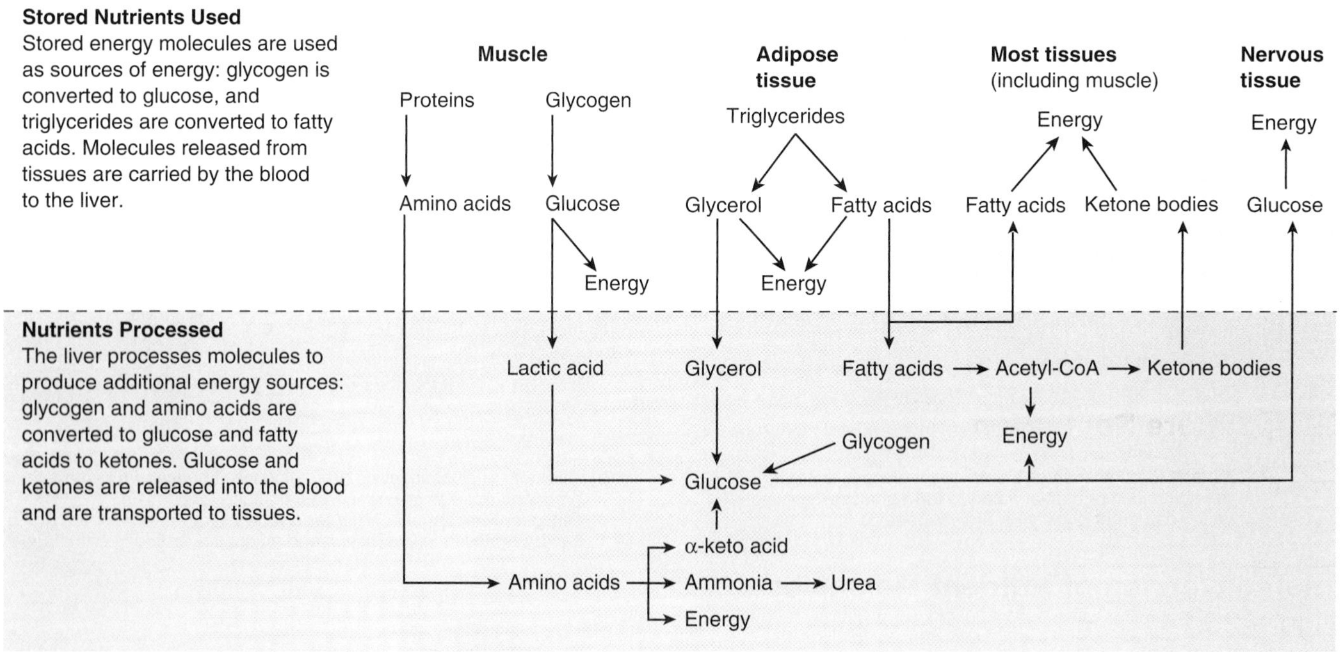

Events of the Postabsorptive State

Figure 25.16

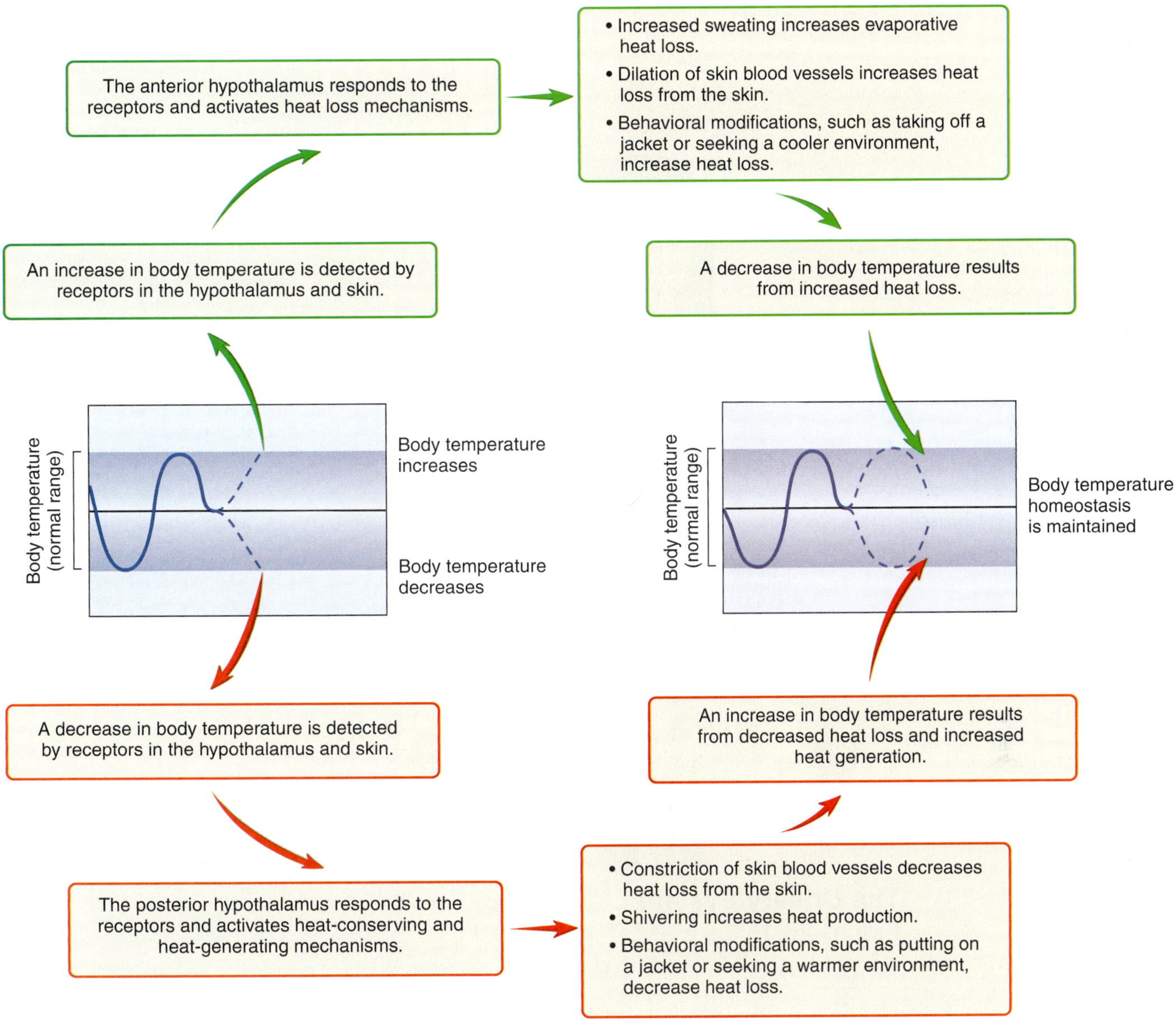

Temperature Regulation
Homeostasis Figure 25.18

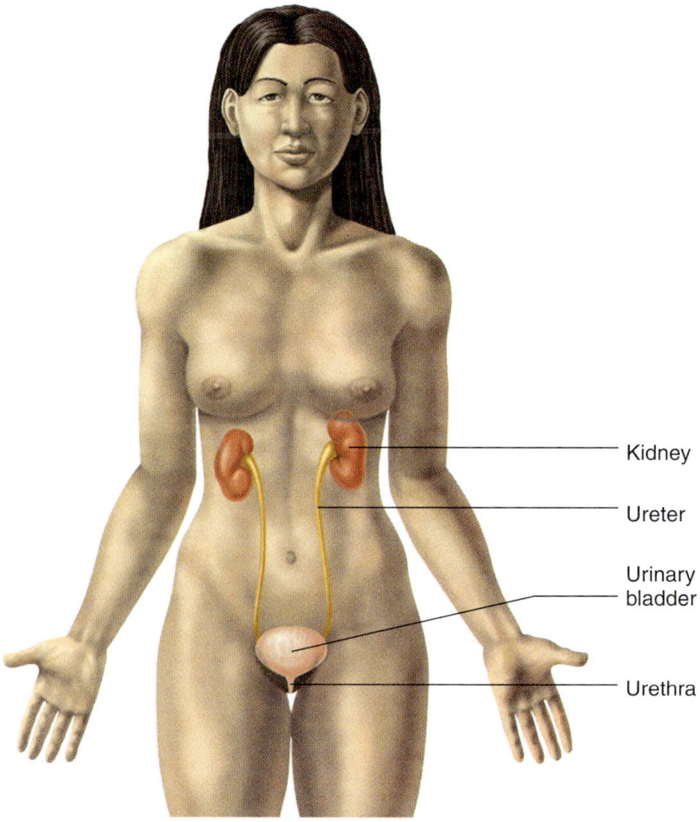
Kidney
Ureter
Urinary
bladder
Urethra

The Urinary System
Figure 26.1

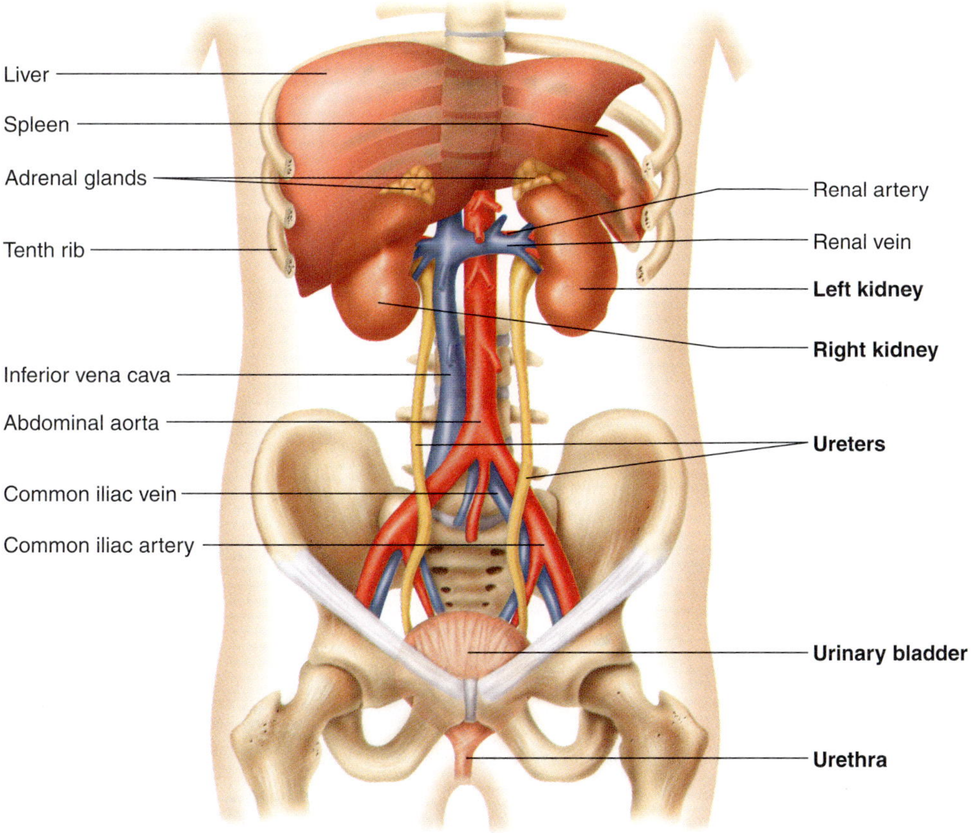

(a) The kidneys are located in the abdominal cavity, with the right kidney just below the liver and the left kidney below the spleen. A ureter extends from each kidney to the urinary bladder within the pelvic cavity. An adrenal gland is located at the superior pole of each kidney.

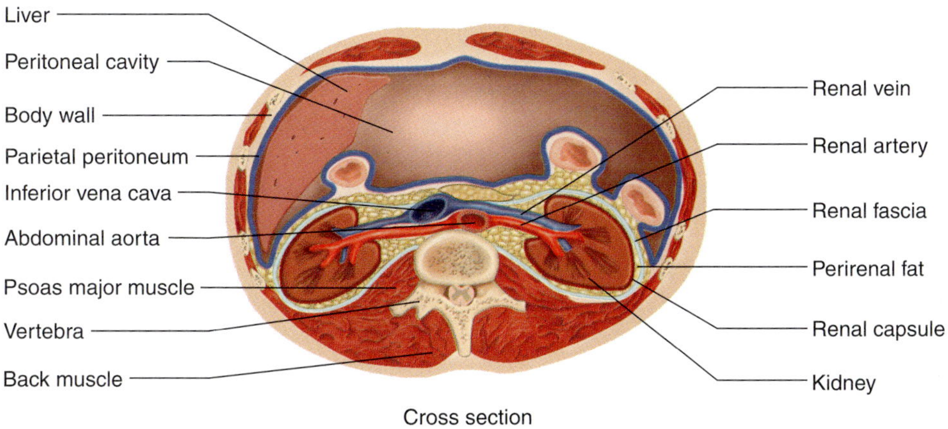

(b) The kidneys are located behind the parietal peritoneum. Surrounding each kidney is the perirenal fat. A connective tissue layer, the renal fascia, anchors the kidney to the abdominal wall. Additional fat is found between the renal fascia and the parietal peritoneum. The renal arteries extend from the abdominal aorta to each kidney, and the renal veins extend from the kidneys to the inferior vena cava.

Anatomy of the Urinary System

Figure 26.2

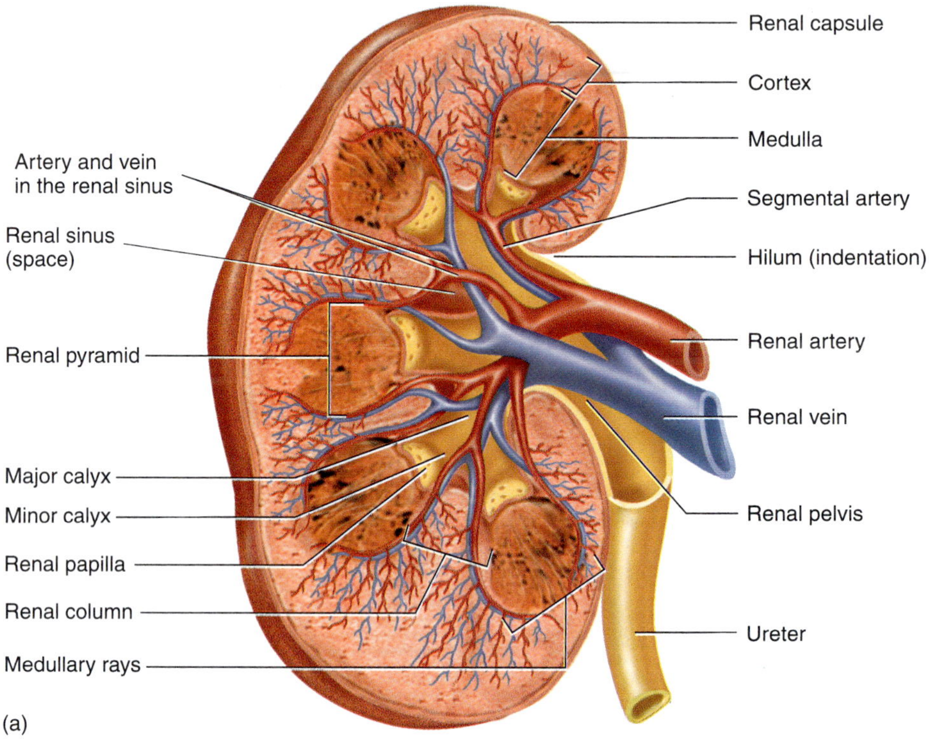

Longitudinal Section of the Kidney and Ureter
Figure 26.3

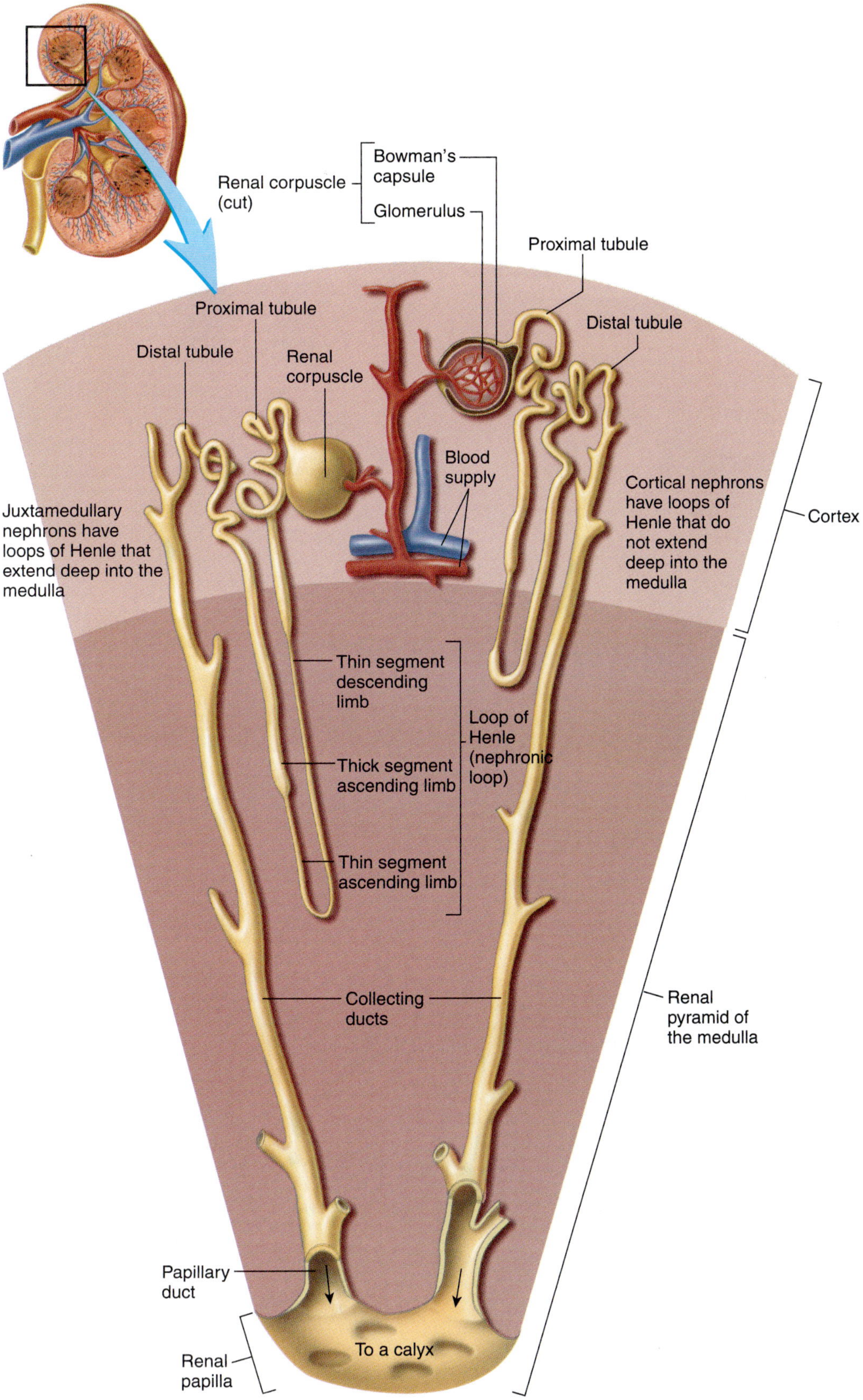

Functional Unit of the Kidney—the Nephron
Figure 26.4

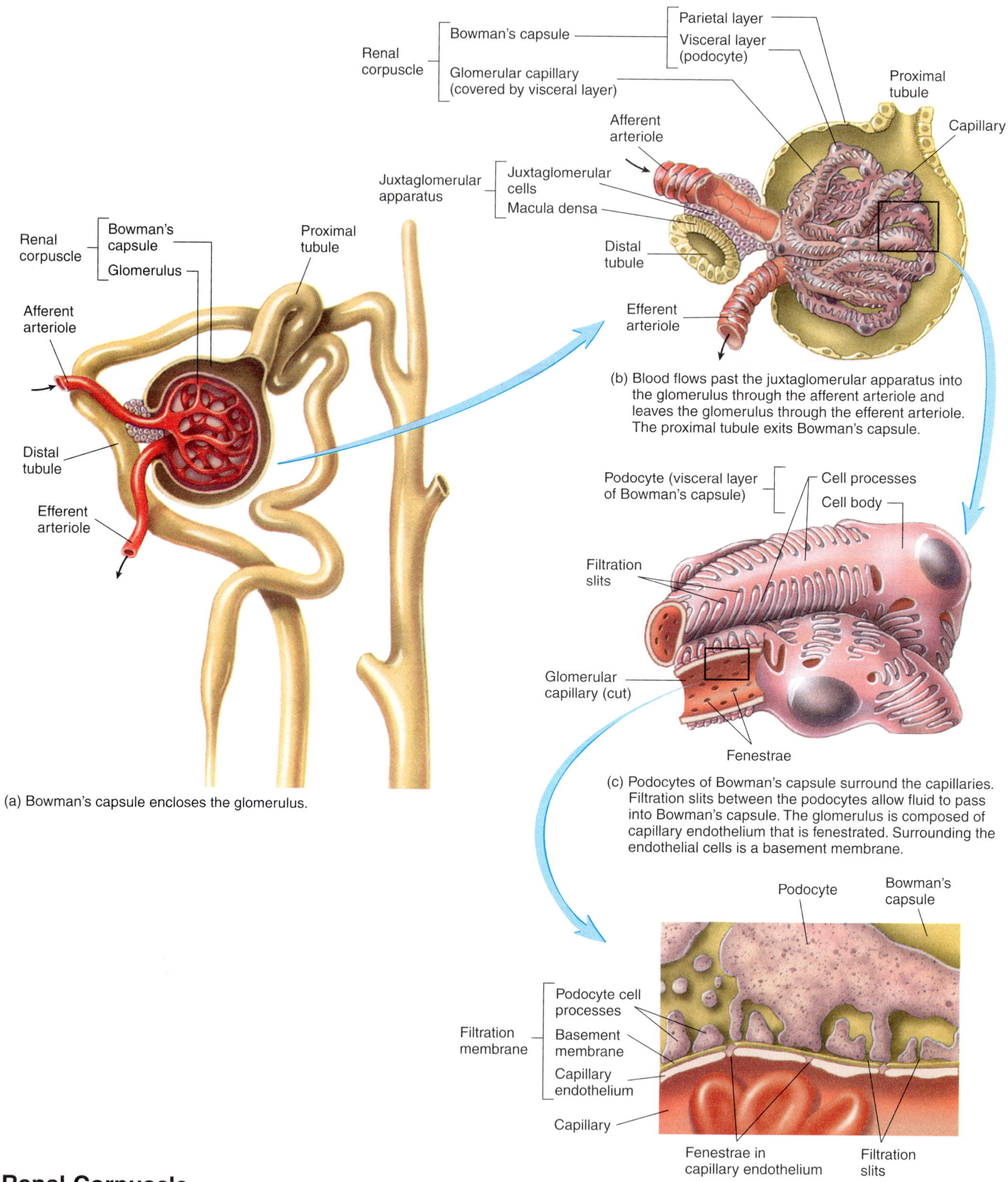

(a) Bowman's capsule encloses the glomerulus.

(b) Blood flows past the juxtaglomerular apparatus into the glomerulus through the afferent arteriole and leaves the glomerulus through the efferent arteriole. The proximal tubule exits Bowman's capsule.

(c) Podocytes of Bowman's capsule surround the capillaries. Filtration slits between the podocytes allow fluid to pass into Bowman's capsule. The glomerulus is composed of capillary endothelium that is fenestrated. Surrounding the endothelial cells is a basement membrane.

(d) Capillary endothelial cells, the basement membrane, and podocytes make up the filtration membrane of the kidney.

Renal Corpuscle
Figure 26.5

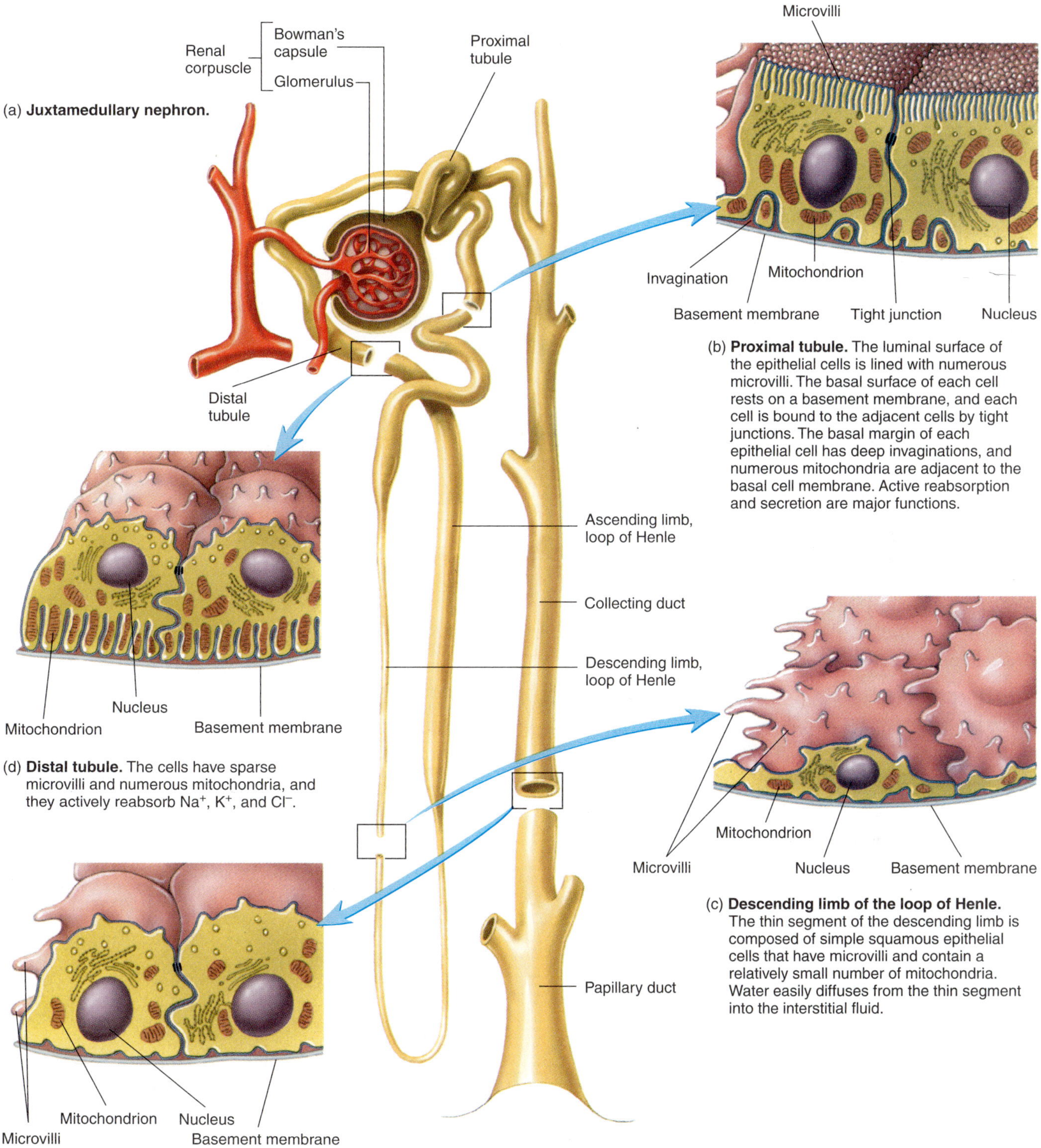

(a) **Juxtamedullary nephron.**

(b) **Proximal tubule.** The luminal surface of the epithelial cells is lined with numerous microvilli. The basal surface of each cell rests on a basement membrane, and each cell is bound to the adjacent cells by tight junctions. The basal margin of each epithelial cell has deep invaginations, and numerous mitochondria are adjacent to the basal cell membrane. Active reabsorption and secretion are major functions.

(c) **Descending limb of the loop of Henle.** The thin segment of the descending limb is composed of simple squamous epithelial cells that have microvilli and contain a relatively small number of mitochondria. Water easily diffuses from the thin segment into the interstitial fluid.

(d) **Distal tubule.** The cells have sparse microvilli and numerous mitochondria, and they actively reabsorb Na^+, K^+, and Cl^-.

(e) **Collecting duct.** The cells have some microvilli and numerous mitochondria, and they actively reabsorb Na^+, K^+, and Cl^-.

Histology of the Nephron
Figure 26.6

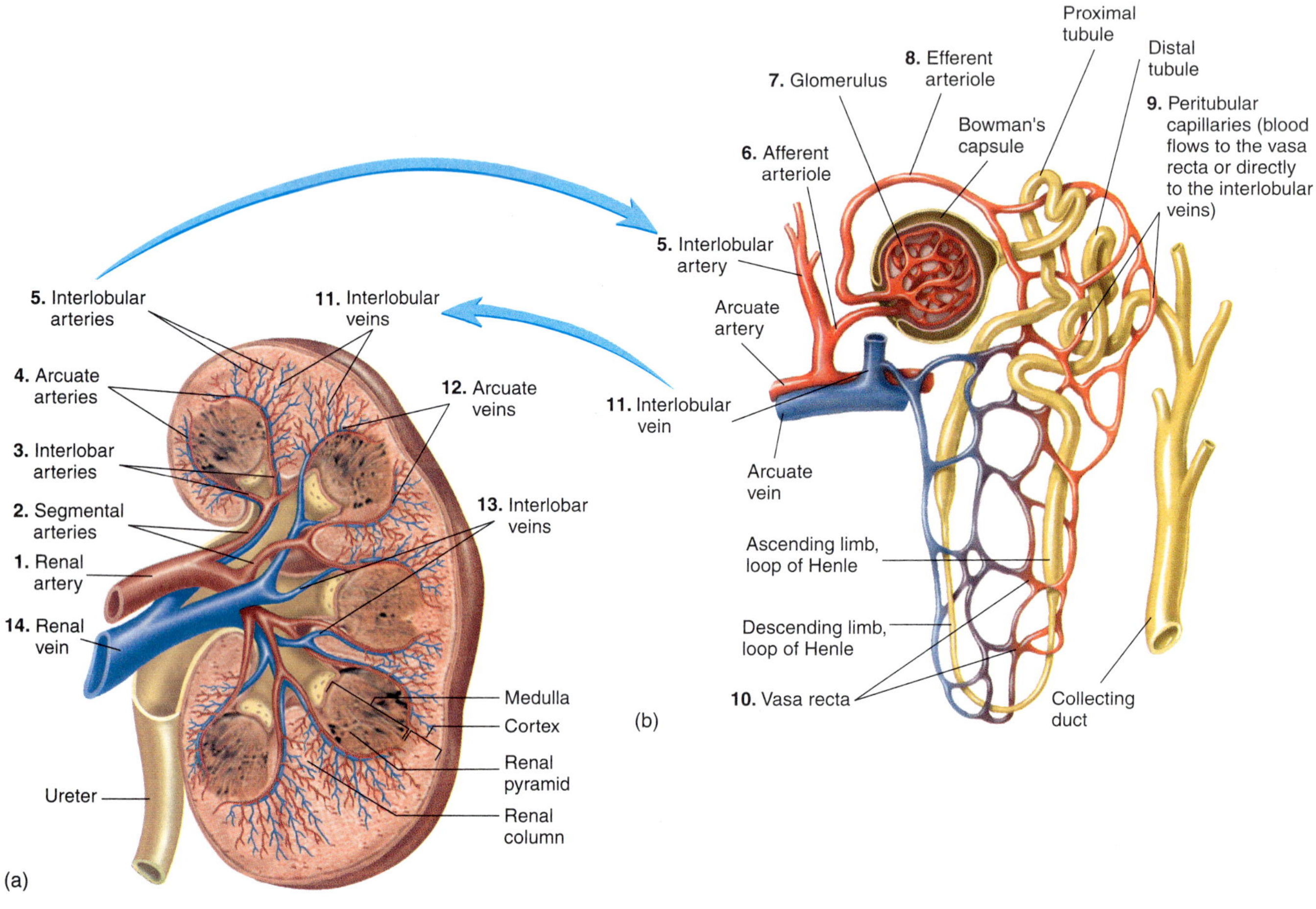
5. Interlobular arteries
11. Interlobular veins
4. Arcuate arteries
12. Arcuate veins
3. Interlobar arteries
2. Segmental arteries
13. Interlobar veins
1. Renal artery
14. Renal vein
Medulla
Cortex
Renal pyramid
Renal column
Ureter
(a)
7. Glomerulus
8. Efferent arteriole
Proximal tubule
Distal tubule
Bowman's capsule
6. Afferent arteriole
9. Peritubular capillaries (blood flows to the vasa recta or directly to the interlobular veins)
5. Interlobular artery
Arcuate artery
11. Interlobular vein
Arcuate vein
Ascending limb, loop of Henle
Descending limb, loop of Henle
10. Vasa recta
Collecting duct
(b)

Blood Flow Through the Kidney
Figure 26.7

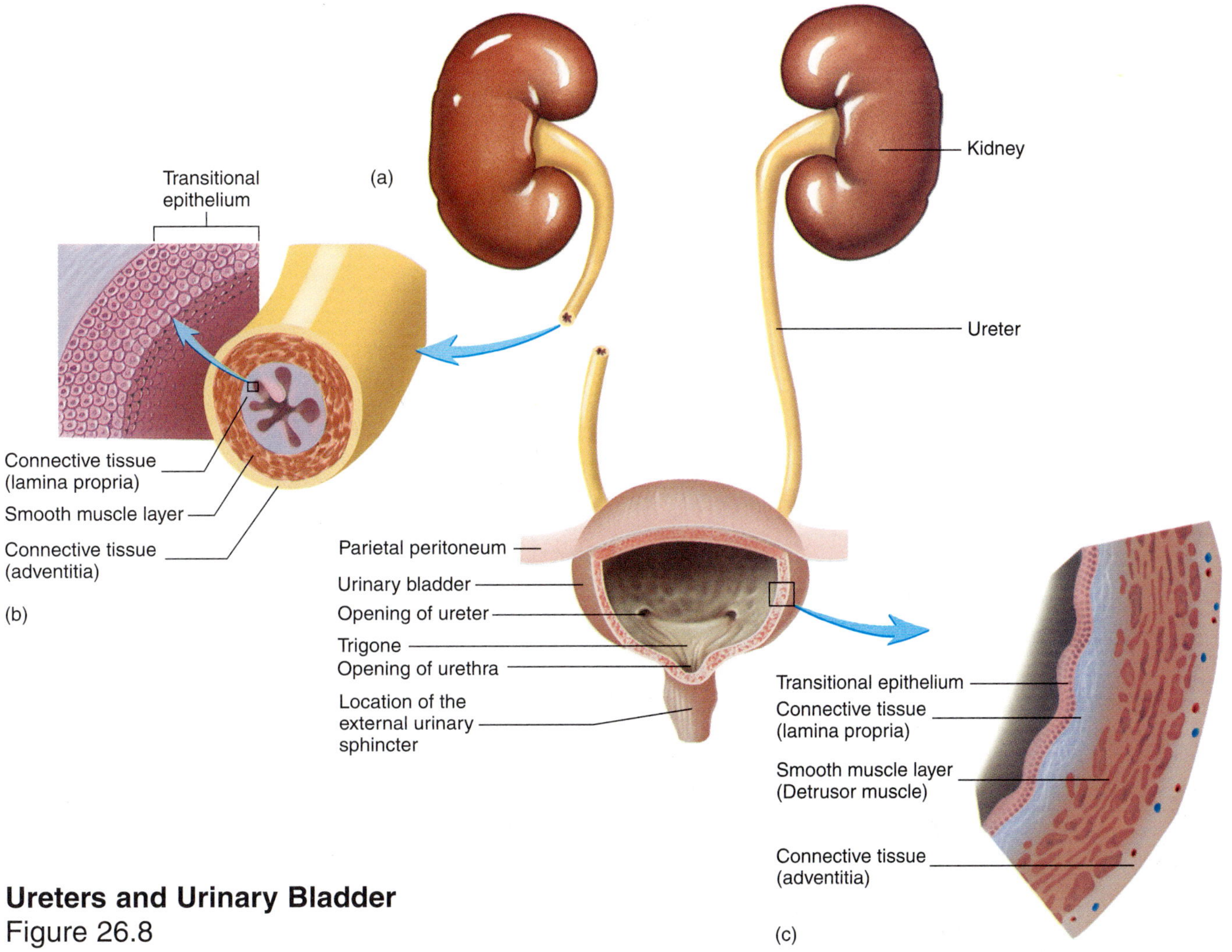

Ureters and Urinary Bladder
Figure 26.8

Urine formation results from the following three processes:

1. **Filtration.** Filtration (*blue arrow*) results in the movement of fluid from glomerular capillaries across the filtration membrane into Bowman's capsule to form filtrate.
2. **Reabsorption.** Solutes are reabsorbed (*purple arrow*) across the wall of the nephron into the interstitial fluid by transport processes, such as active transport and cotransport. The solutes then enter the peritubular capillaries.
   Water is reabsorbed (*green arrow*) across the wall of the nephron into the interstitial fluid by osmosis. The water then enters the peritubular capillaries.
3. **Secretion.** Solutes are secreted (*orange arrow*) across the wall of the nephron into the filtrate.

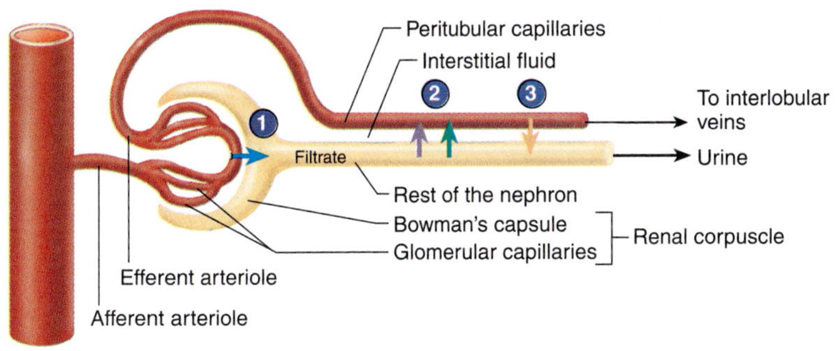

Urine Formation
Process Figure 26.9

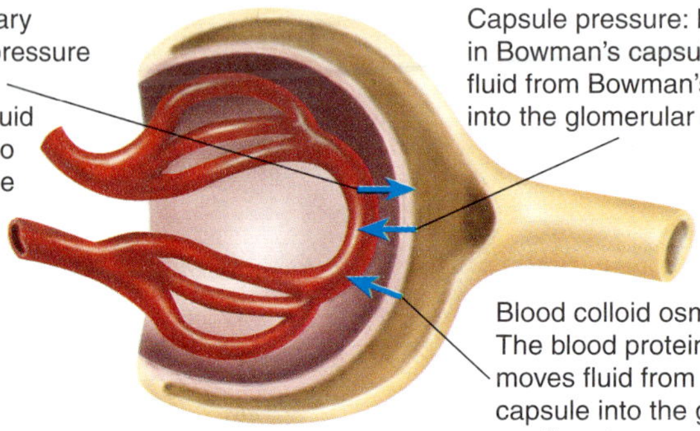

Glomerular capillary pressure
– Capsule pressure
– Colloid osmotic pressure

Filtration pressure

Filtration pressure = GCP – BCOP – CP

50 mm Hg GCP (glomerular capillary pressure)
–30 mm Hg BCOP (blood colloid osmotic pressure)
–10 mm Hg CP (capsule pressure)

10 mm Hg filtration pressure

Filtration Pressure
Figure 26.10

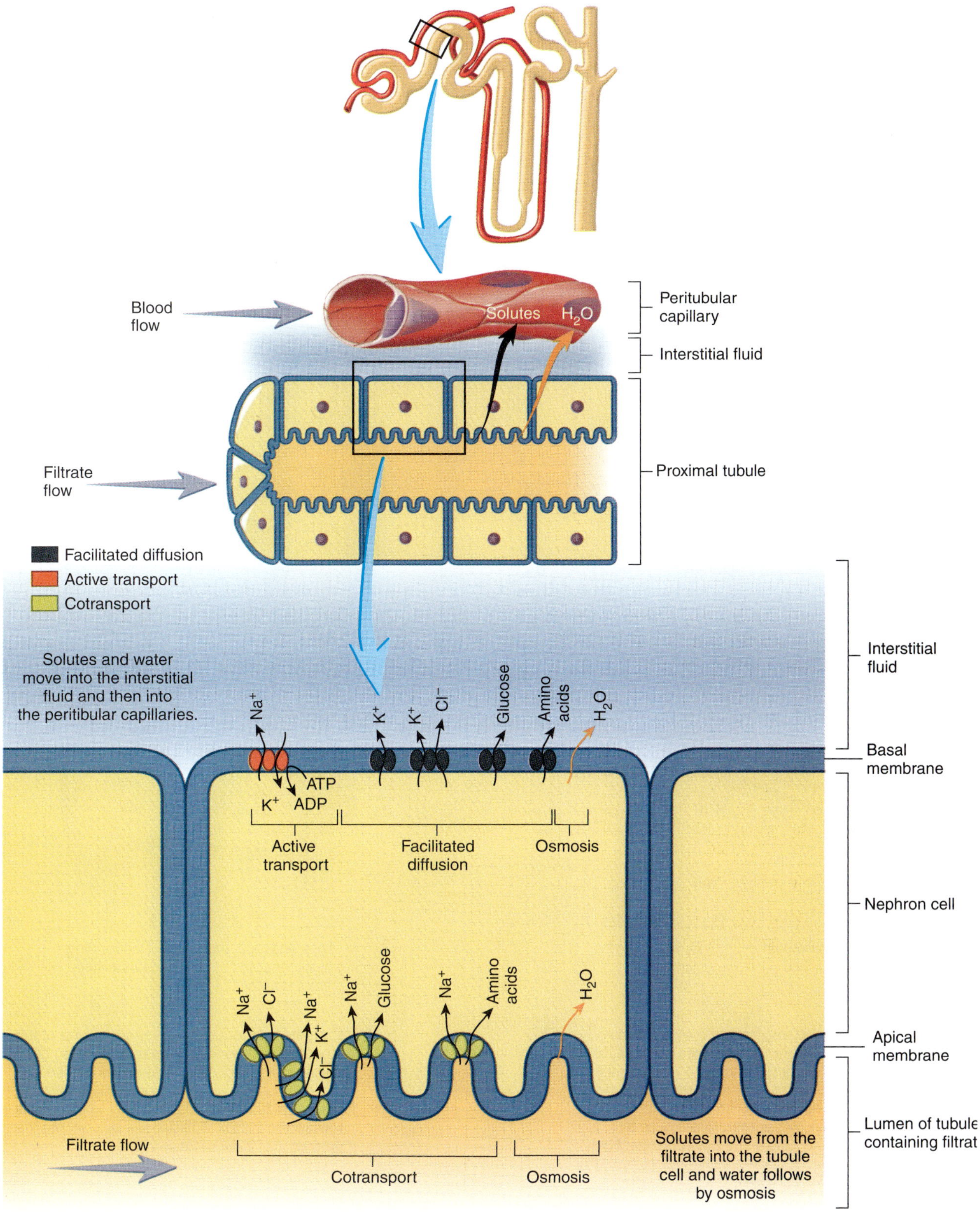

Reabsorption of Solutes in the Proximal Nephron
Figure 26.11

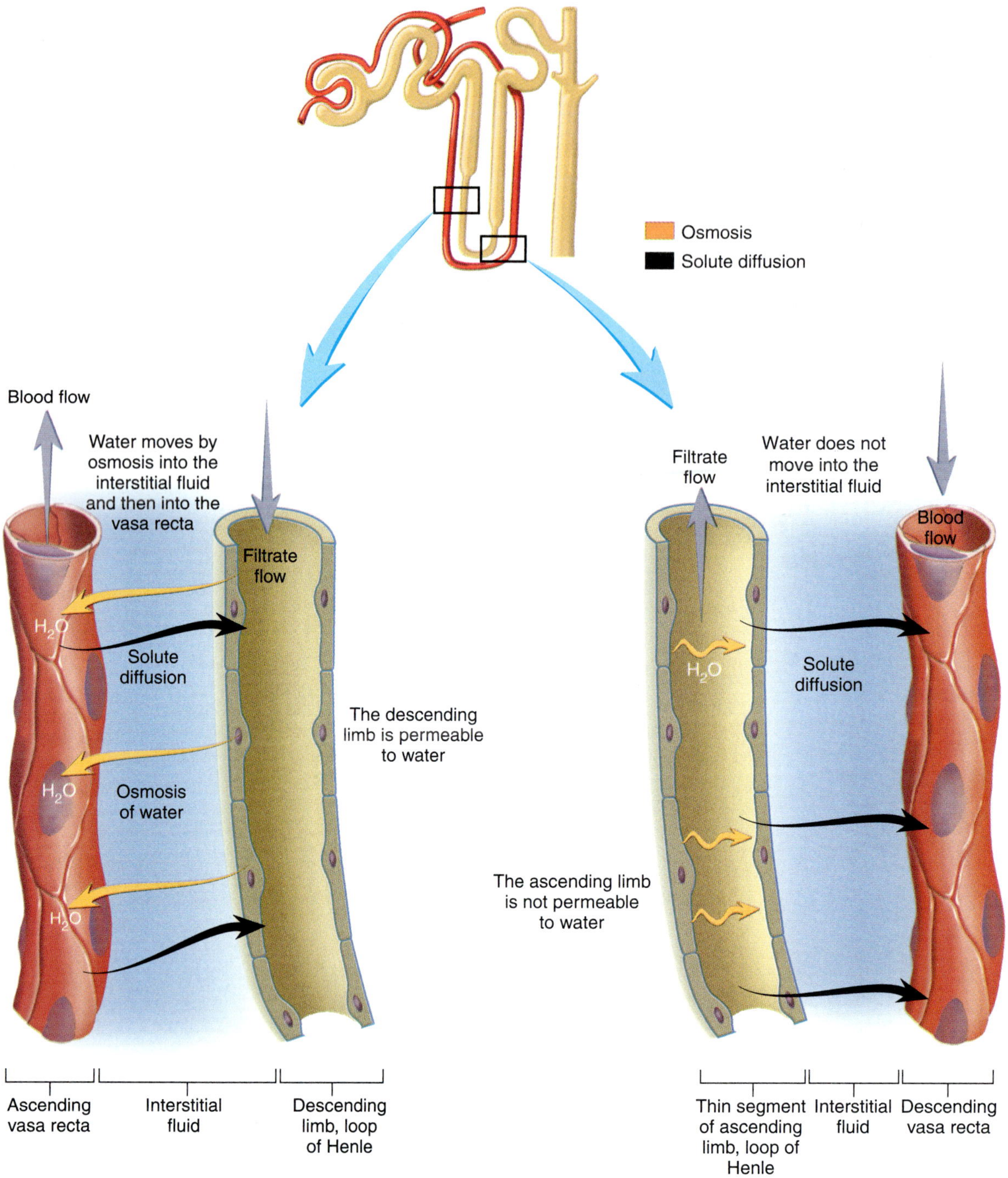

(a) The wall of the thin segment of the descending limb of the loop of Henle is permeable to water and, to a lesser extent, to solutes. The interstitial fluid in the medulla of the kidney and blood in the vasa recta have a high solute concentration (high osmolality). Water therefore moves by osmosis from the tubule into the interstitial fluid and into the vasa recta. An additional 15% of the filtrate volume is reabsorbed.

To a lesser extent, solutes diffuse from the vasa recta and interstitial fluid into the tubule.

(b) The thin segment of the ascending limb of the loop of Henle is not permeable to water, but is permeable to solutes. The solutes diffuse out of the tubule and into the more dilute interstitial fluid as the ascending limb projects toward the cortex. The solutes diffuse into the descending vasa recta.

Reabsorption in the Loop of Henle: The Descending Limb and the Thin Segment of the Ascending Limb
Figure 26.12

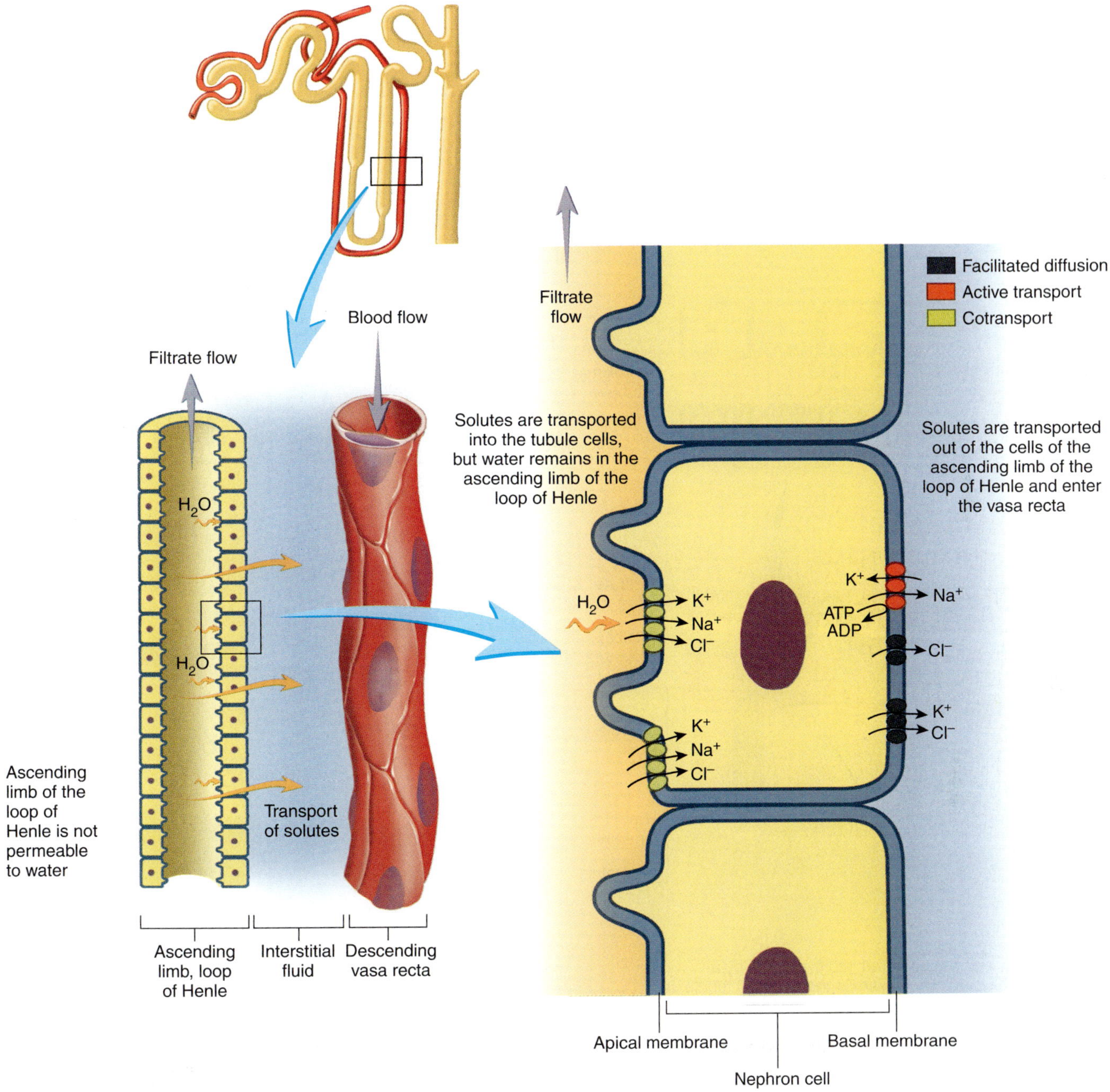

The wall of the ascending limb of the loop of Henle is not permeable to water. Na^+ move across the wall of the basal membrane by active transport, establishing a concentration gradient for Na^+. Potassium ions and Cl^- are cotransported with Na^+ across the apical membrane and ions pass by facilitated diffusion across the basal membrane of the tubule cells.

Reabsorption in the Thick Segment of the Ascending Limb of the Loop of Henle
Figure 26.13

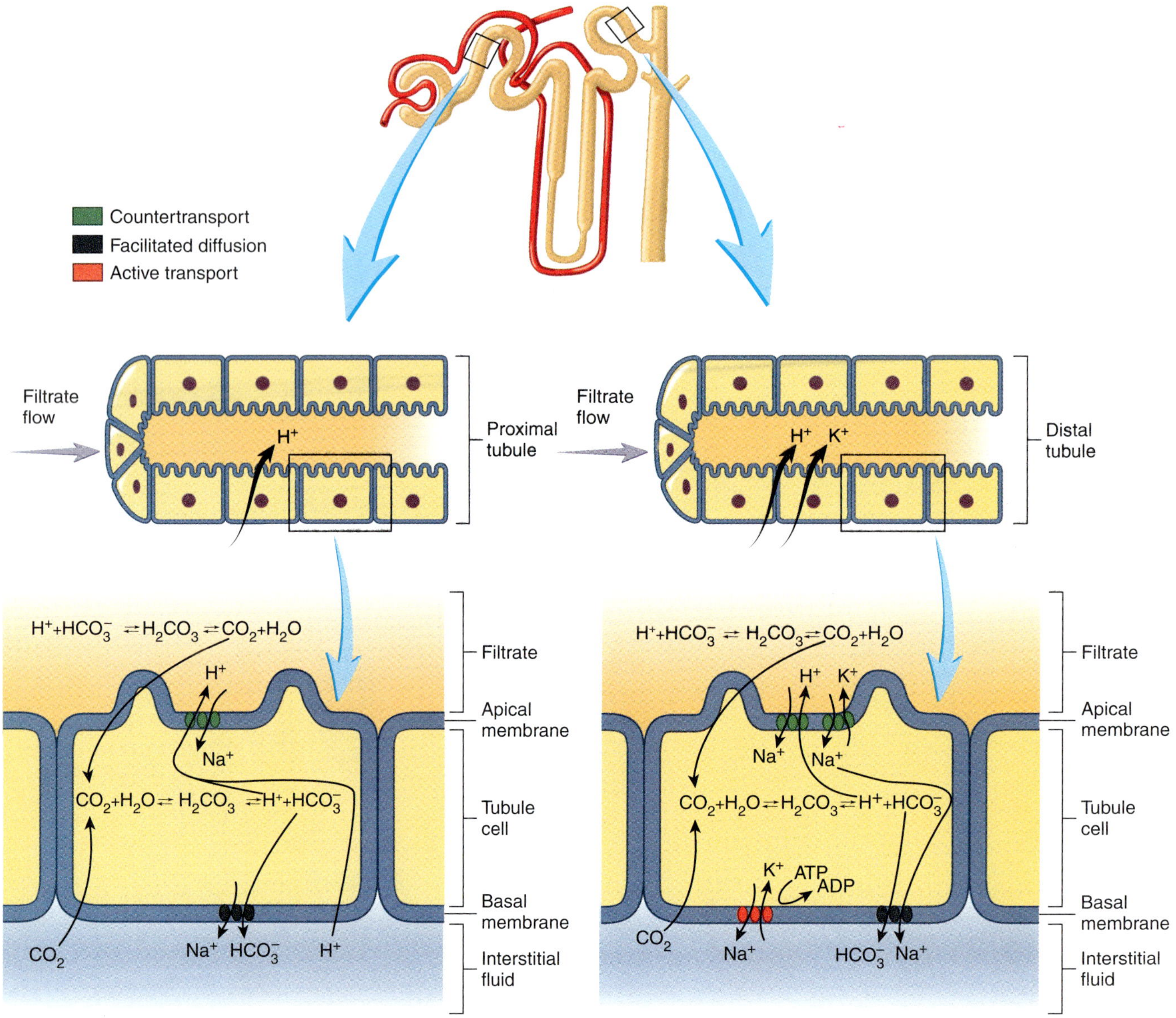

(a) Hydrogen ions are secreted into the filtrate by a countertransport mechanism in the proximal tubule, in which H^+ are exchanged for Na^+. The H^+ are derived from two sources. They diffuse from the peritubular capillaries into the interstitial fluid and then into epithelial cells of the tubule, or they are derived from the reaction between carbon dioxide and water in the cells of the tubule. Sodium ions and HCO_3^- are cotransported across the basal membrane into the interstitial fluid and then diffuse into the peritubular capillaries.

(b) Hydrogen ions and K^+ are secreted into the filtrate by countertransport mechanisms in the distal tubule. Sodium ions and K^+ are moved by active transport across the basal membrane of the tubule cell. Sodium ions and HCO_3^- are cotransported across the basal membrane into the interstitial fluid and then diffuse into the peritubular capillaries.

Secretion of H^+ and K^+ into the Nephron
Figure 26.14

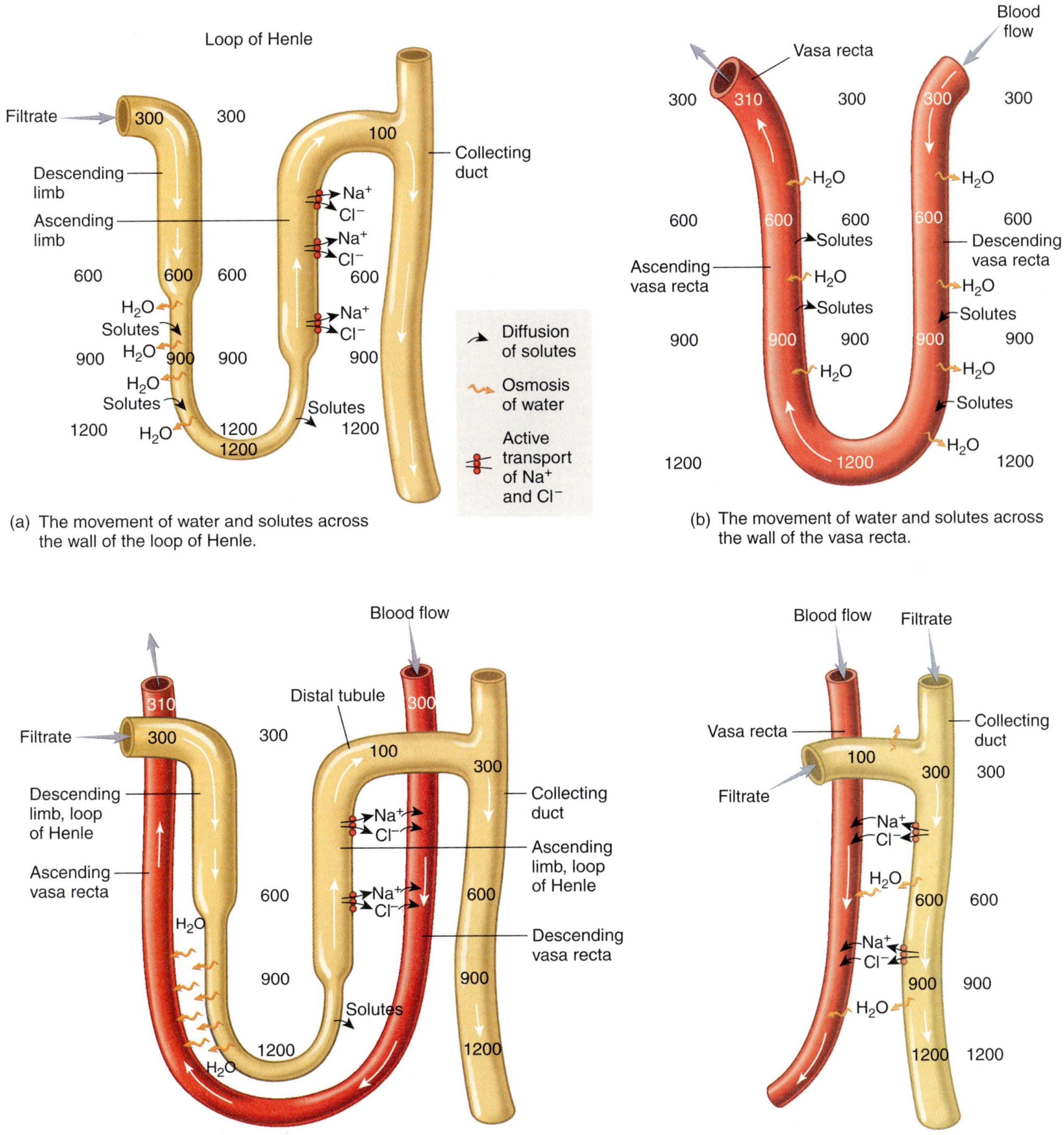

(a) The movement of water and solutes across the wall of the loop of Henle.

(b) The movement of water and solutes across the wall of the vasa recta.

(c) The loop of Henle and the vasa recta function together. Water moves out of the descending limb of the loop of Henle and enters the vasa recta. Solutes diffuse out of the ascending thin segment of the loop of Henle and enter the vasa recta, but water does not. Solutes transported out of the thick segment of the ascending limb of the loop of Henle enter the vasa recta. Excess water and solutes are carried away from the medulla without reducing the high concentration of solutes. The concentration of the filtrate is reduced to 100 mOsm/kg by the time it reaches the distal tubule.

(d) Water and solutes move out of the collecting duct into the vasa recta.

Filtrate Concentration and the Medullary Concentration Gradient
Figure 26.15

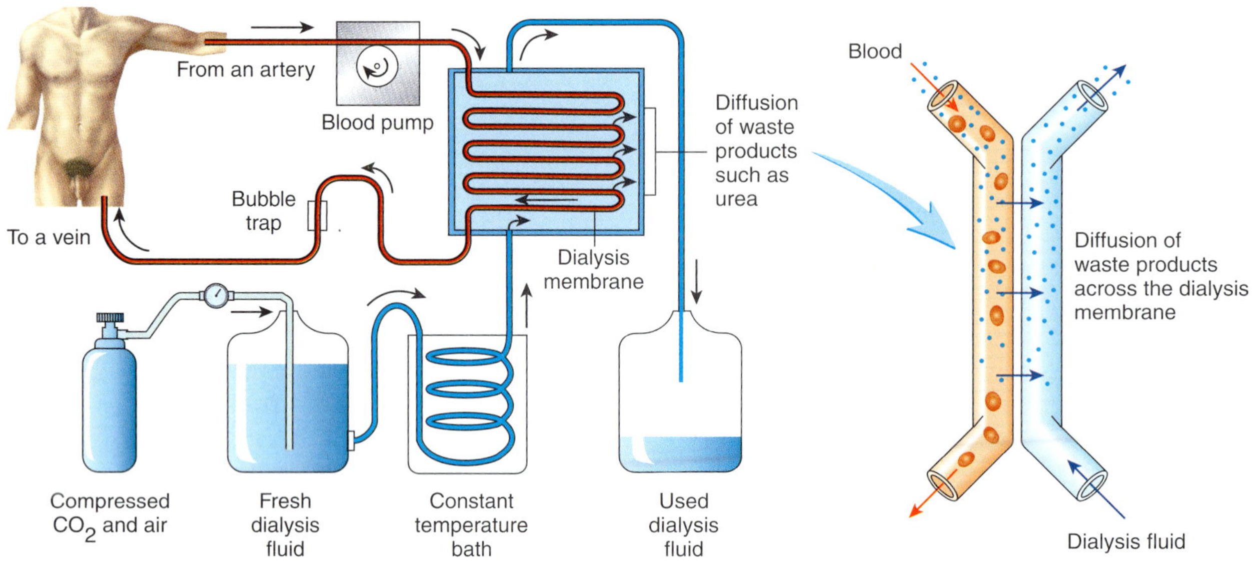

Kidney Dialysis
Figure A

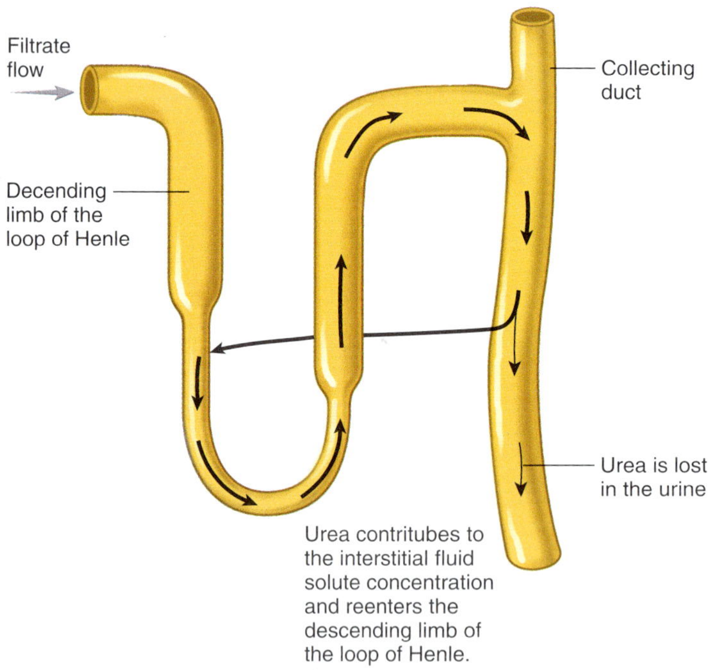

The Medullary Concentration Gradient and Urea Cycling
Figure 26.16

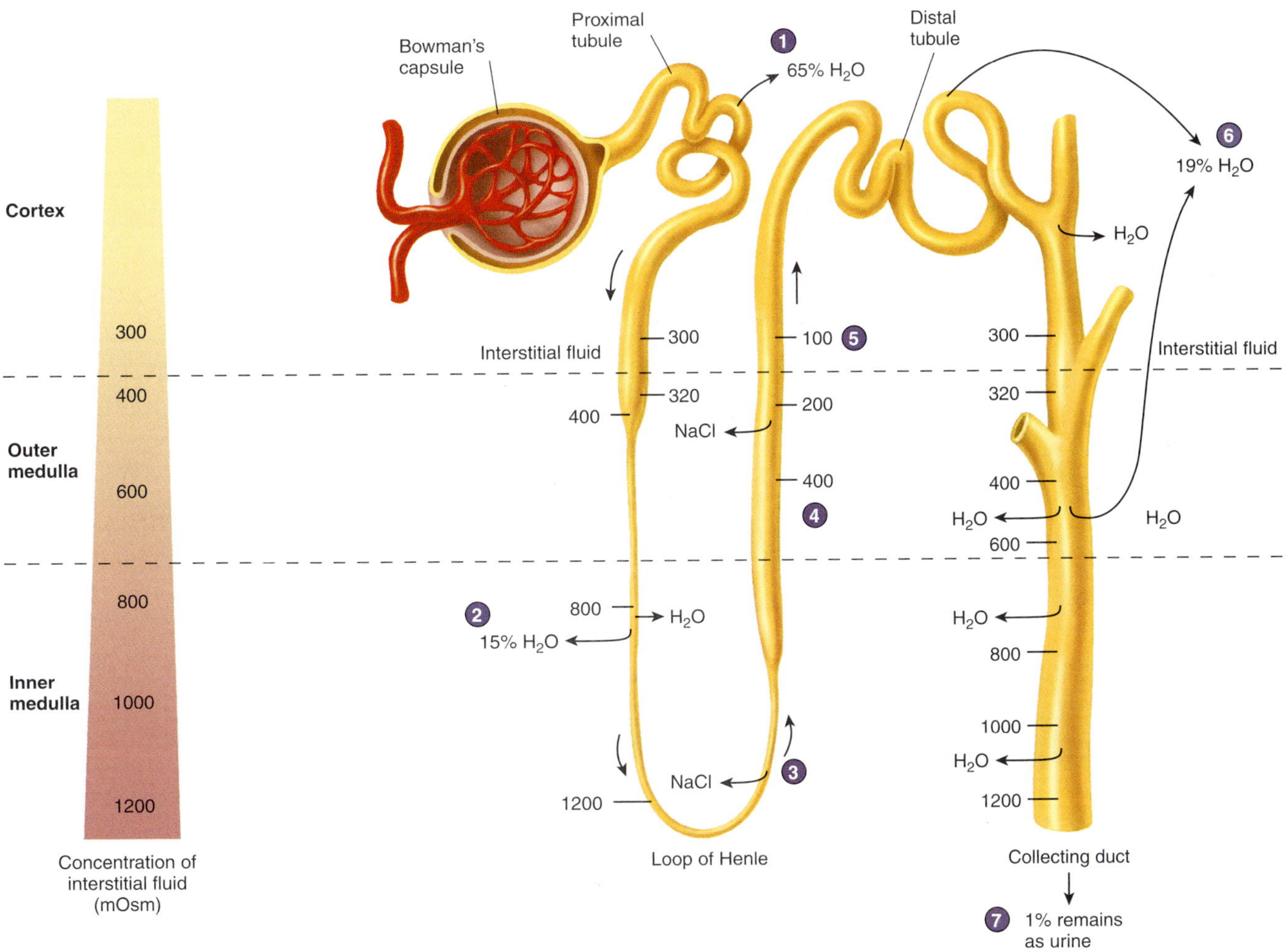

1. Approximately 180 L of filtrate enters the nephrons each day; of that volume 65% is reabsorbed in the proximal tubule. In the proximal tubule, solute molecules move by active transport and cotransport from the lumen of the tubule into the interstitial fluid. Water moves by osmosis because the cells of the tubule wall are permeable to water (see figure 26.11).
2. Approximately 15% of the filtrate volume is reabsorbed in this segment of the descending limb of the loop of Henle. The descending limb passes through the concentrated interstitial fluid of the medulla. Because the wall of the descending limb is permeable to water, water moves by osmosis from the tubule into the more concentrated interstitial fluid (see figure 26.12*a*). By the time the filtrate reaches the tip of the renal pyramid, the concentration of the filtrate is equal to the concentration of the interstitial fluid.
3. The ascending limb of the loop of Henle is not permeable to water. Solutes diffuse out of the thin segment (see figure 26.12*b*).
4. Na^+, K^+, and Cl^- are cotransported across the apical membrane of the thick segment. Na^+ are actively transported and K^+ and Cl^- diffuse across the basal membrane of epithelial cells of the thick segment into the interstitial fluid (see figure 26.13).
5. The volume of the filtrate doesn't change as it passes through the ascending limb, but the concentration is greatly reduced (see figure 26.13). By the time the filtrate reaches the cortex of the kidney, the concentration is approximately 100 mOsm/L, which is less concentrated than the interstitial fluid of the cortex (300 mOsm/L).
6. The distal tubule and collecting duct are permeable to water if ADH is present. If ADH is present, water moves by osmosis from the less concentrated filtrate into the more concentrated interstitial fluid. By the time the filtrate reaches the tip of the renal pyramid an additional 19% of the filtrate is reabsorbed.
7. One percent or less remains as urine.

Urine Concentrating Mechanism
Process Figure 26.17

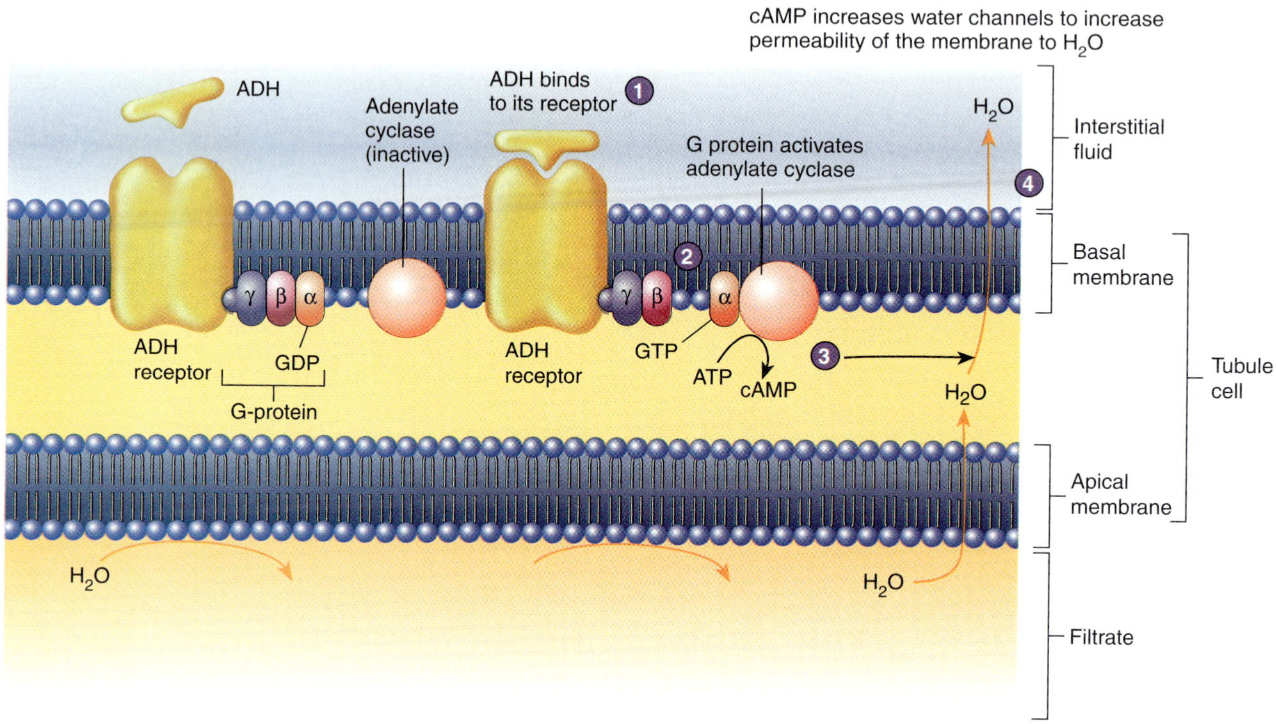

1. ADH binds to ADH receptors in the plasma membranes of the distal tubule cells and the collecting duct cells.
2. When ADH is bound to its receptor, a G protein mechanism is activated, which, in turn, activates adenylate cyclase.
3. The increase in the rate of cyclic adenosine monophosphate (cAMP) synthesis increases the permeability of the epithelial cells to water by increasing the number of water channels in their plasma membranes.
4. Water then moves out of the distal tubule and collecting duct into the interstitial fluid by osmosis, both decreasing the volume of the filtrate and increasing its concentration (osmolality).

The Effect of Antidiuretic Hormone (ADH) on the Nephron
Process Figure 26.18

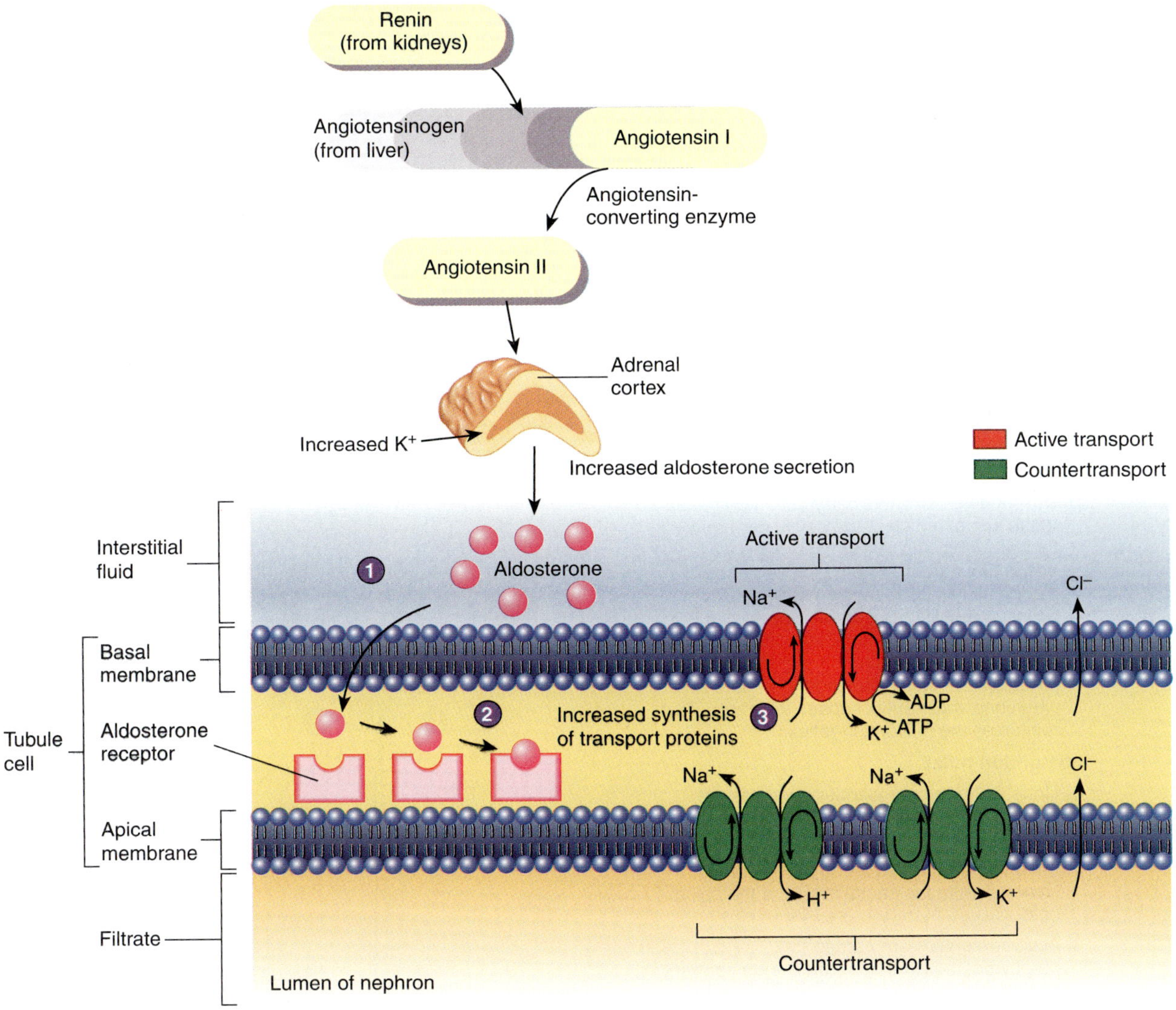

1. Aldosterone secreted from the adrenal cortex enters cells of the distal tubule.
2. Aldosterone binds to intracellular receptors and increases the synthesis of transport proteins of the apical and basal membranes.
3. Newly synthesized transport proteins increase the rate at which Na^+ are absorbed and K^+ and H^+ are secreted. Chloride ions move with the Na^+ because they are attracted to the positive charge of Na^+.

Effect of Aldosterone on the Distal Tubule
Process Figure 26.19

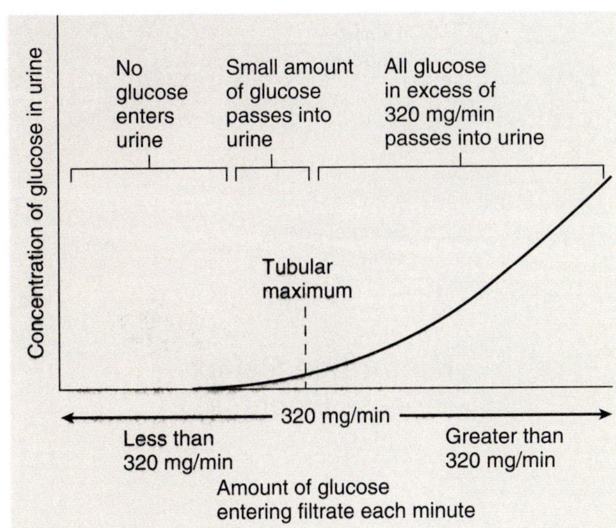

Tubular Maximum for Glucose
Figure 26.20

Control of the micturition reflex by higher brain centers

A. Ascending pathways carry an increased frequency of action potentials up the spinal cord to the pons and cerebrum when the urinary bladder becomes stretched. This increases the conscious desire to urinate.

B. Descending pathways carry action potentials to the sacral region of the spinal cord to tonically inhibit the micturition reflex, preventing automatic urination when the bladder is full. Descending pathways carry action potentials from the cerebrum to the sacral region of the spinal cord to stimulate the reflex when stretch of the urinary bladder produces the conscious urge to urinate and when one voluntarily chooses to urinate. This reinforces the micturition reflex.

Micturition reflex

1. Urine in the urinary bladder stretches the bladder wall.
2. Action potentials produced by stretch receptors are carried along pelvic nerves (*green line*) to the sacral region of the spinal cord.
3. Action potentials are carried by parasympathetic nerves (*red line*) to contract the smooth muscles of the urinary bladder. Decreased action potentials carried by somatic motor nerves (*purple line*) cause the external urinary sphincter to relax.

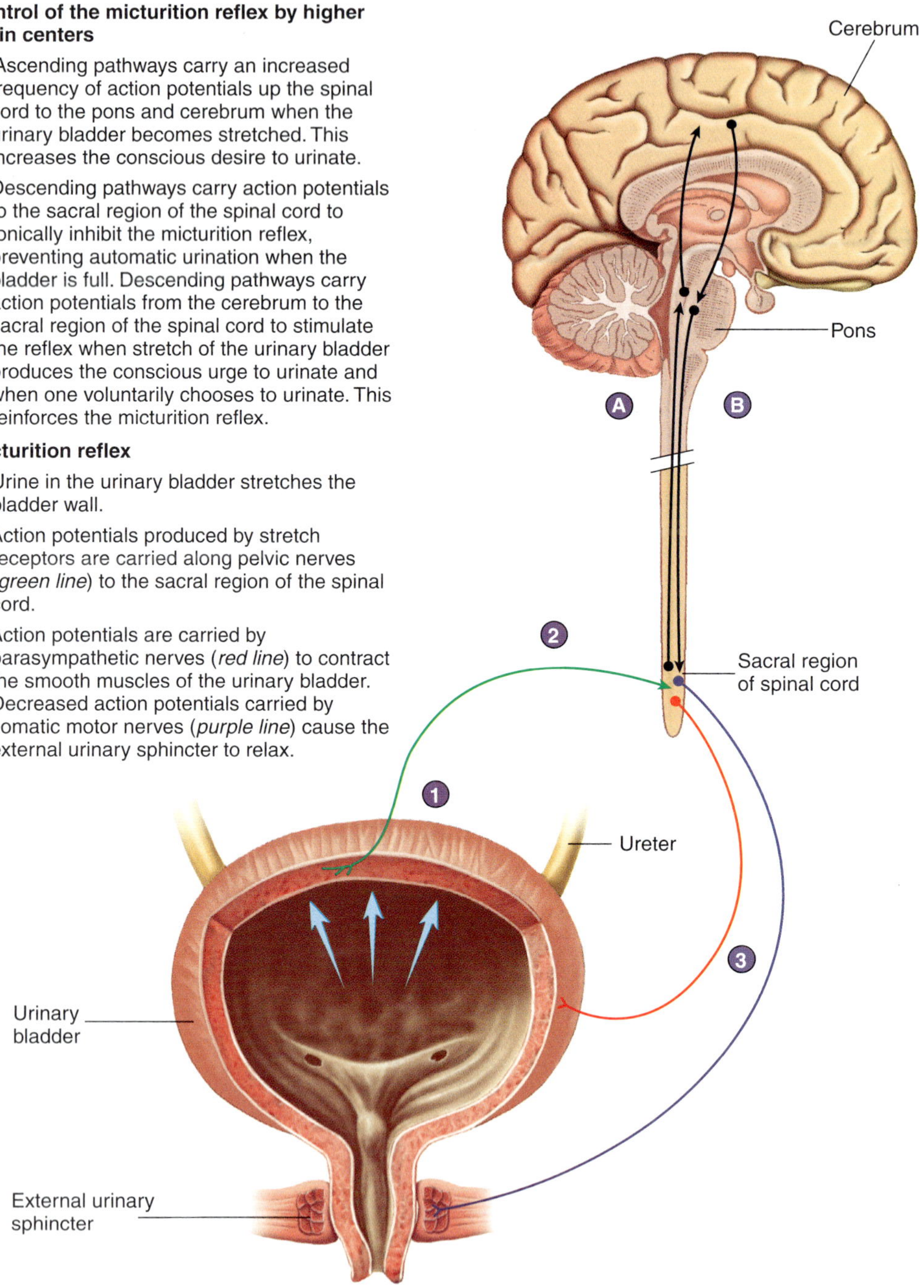

Micturition Reflex
Process Figure 26.21

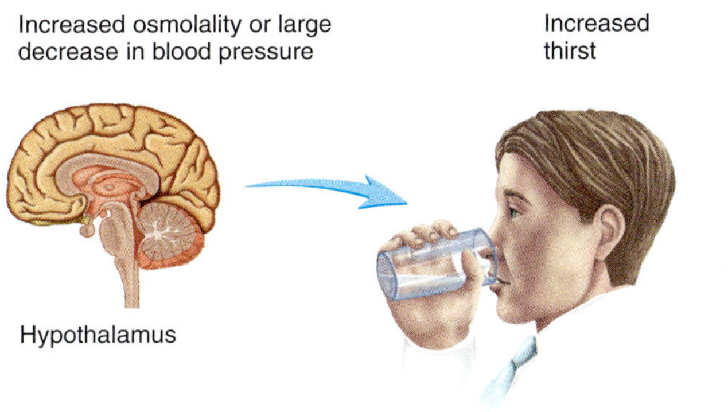

Effect of Blood Osmolality and Pressure on Water Intake
Figure 27.1

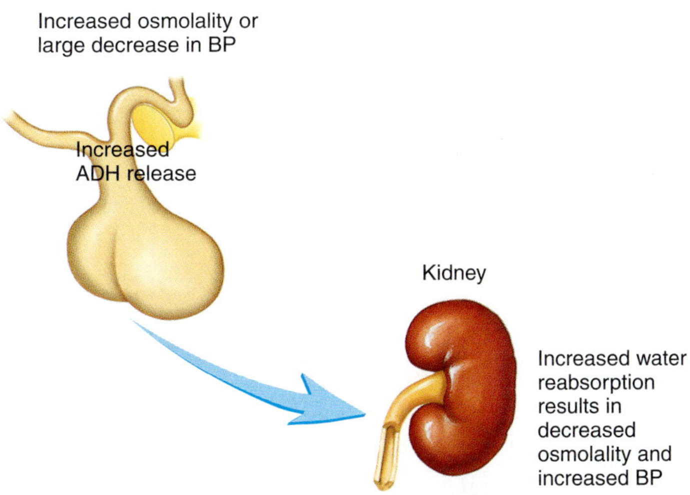

Effect of Blood Osmolality and Blood Pressure on Water Reabsorption in the Kidneys
Figure 27.2

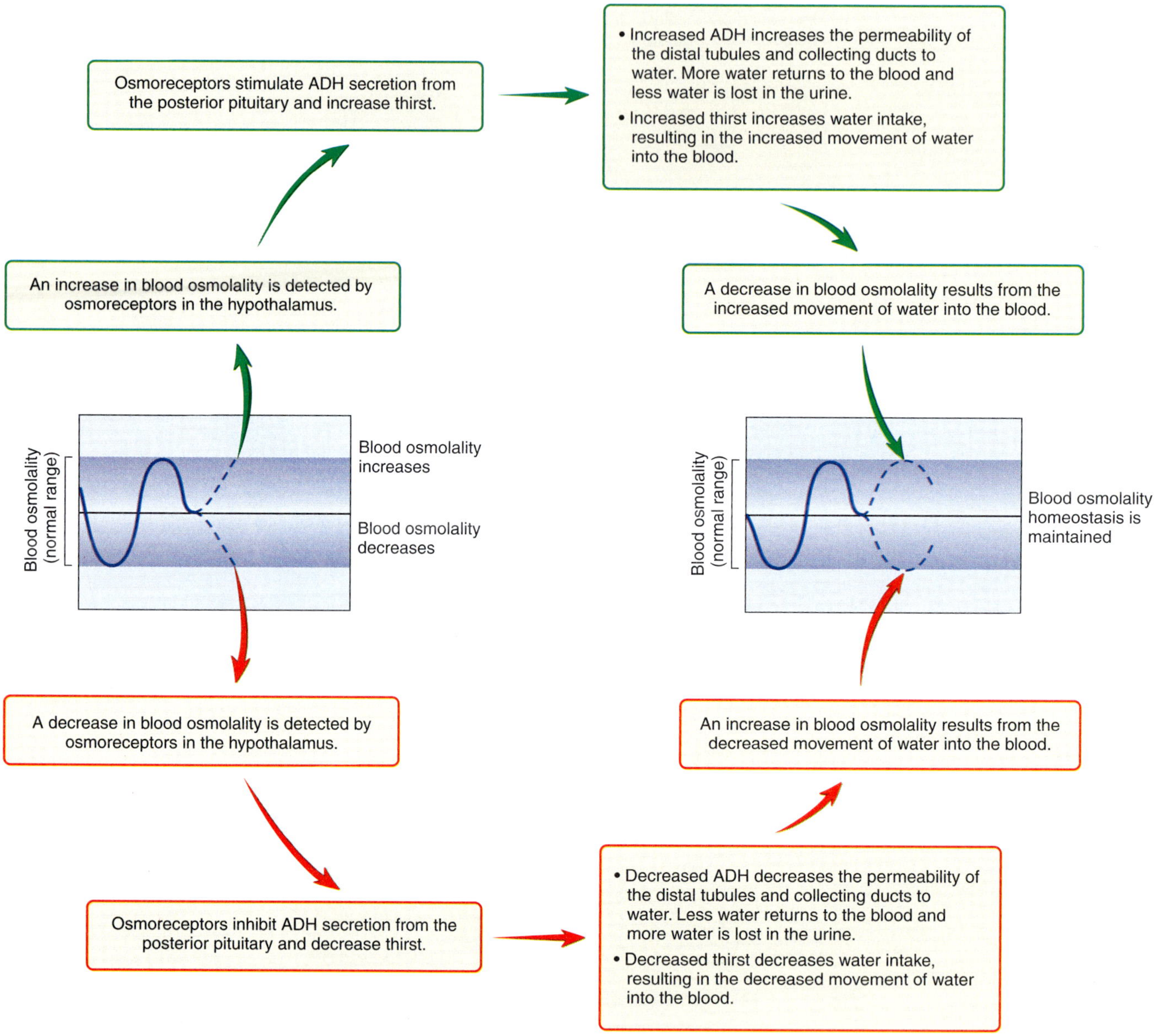

Hormonal Regulation of Blood Osmolality
Homeostasis Figure 27.3

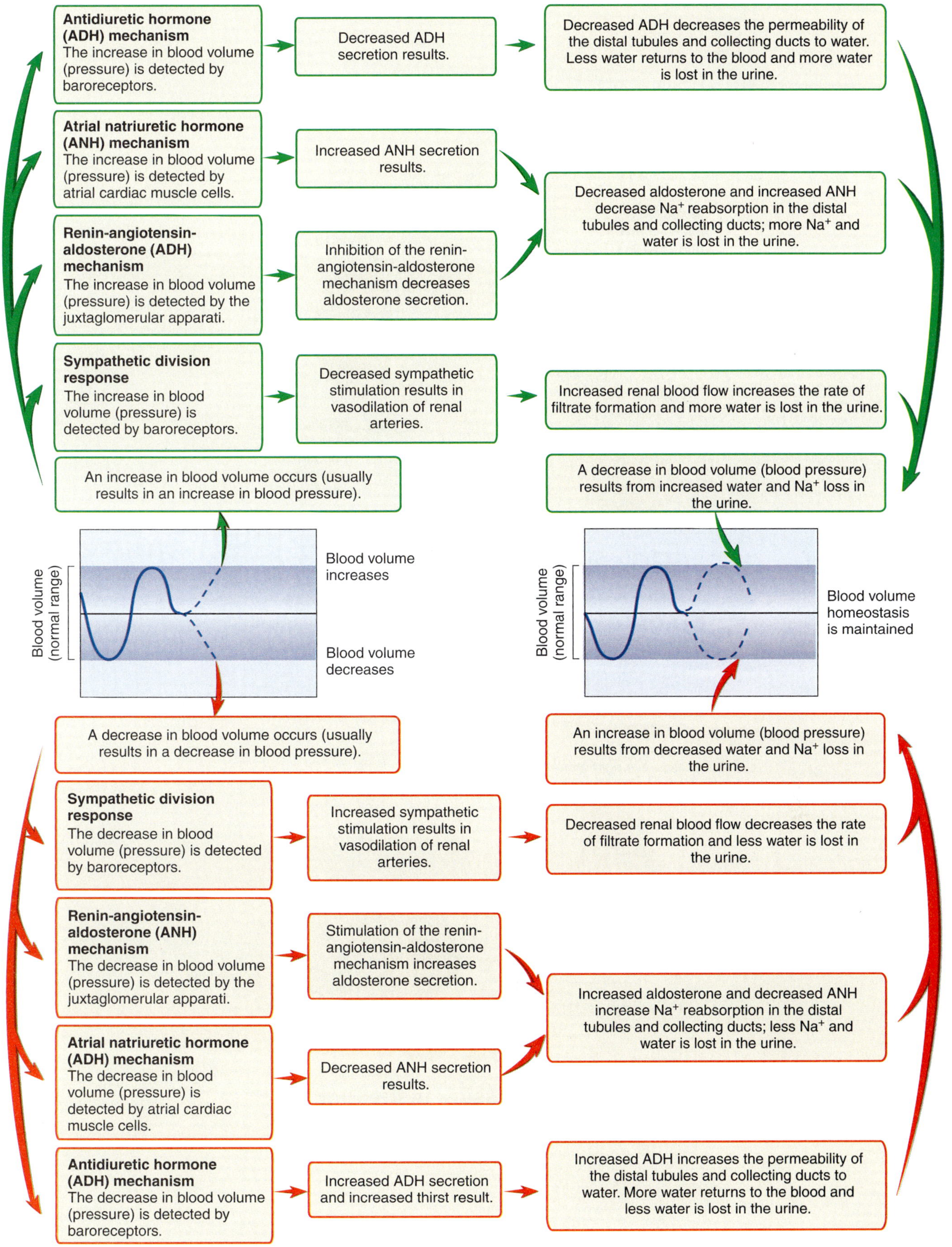

Regulation of Blood Volume
Homeostasis Figure 27.4

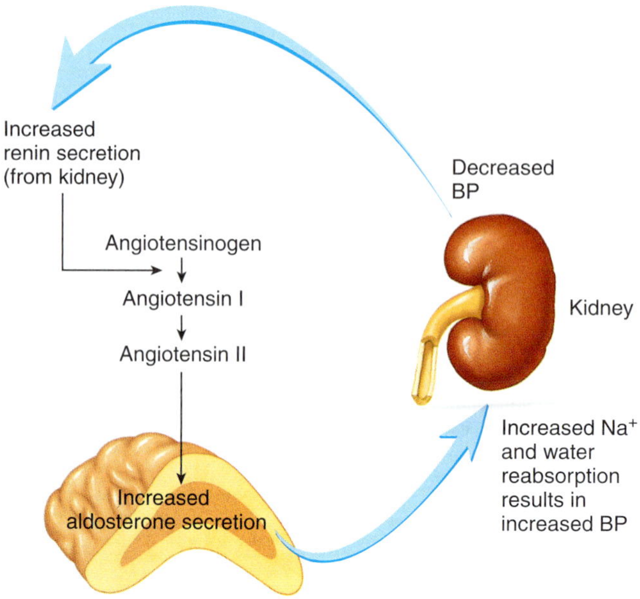

Effect of Blood Pressure on Na^+ and Water Reabsorption in the Kidneys
Figure 27.5

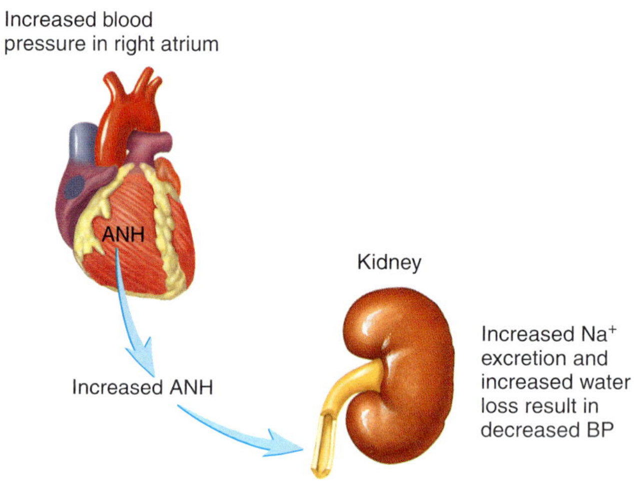

Effect of Blood Pressure in the Right Atrium on Na^+ and Water Excretion
Figure 27.6

1. Large organic molecules such as proteins, which cannot cross the plasma membrane, are synthesized inside cells and influence the concentration of solutes inside the cells.

2. The transport of ions across the plasma membrane, such as Na^+, K^+, and Ca^{2+}, influences the concentration of ions inside and outside the cell.

3. An electric charge difference across the plasma membrane influences the distribution of ions inside and outside the cell.

4. The distribution of water inside and outside the cell is determined by osmosis.

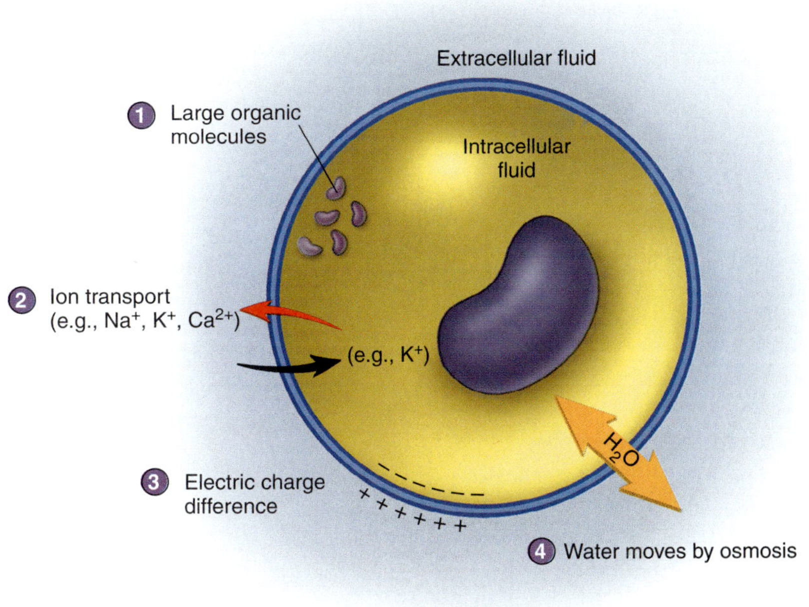

Regulation of Intracellular and Extracellular Distribution of Water and Solutes
Process Figure 27.7

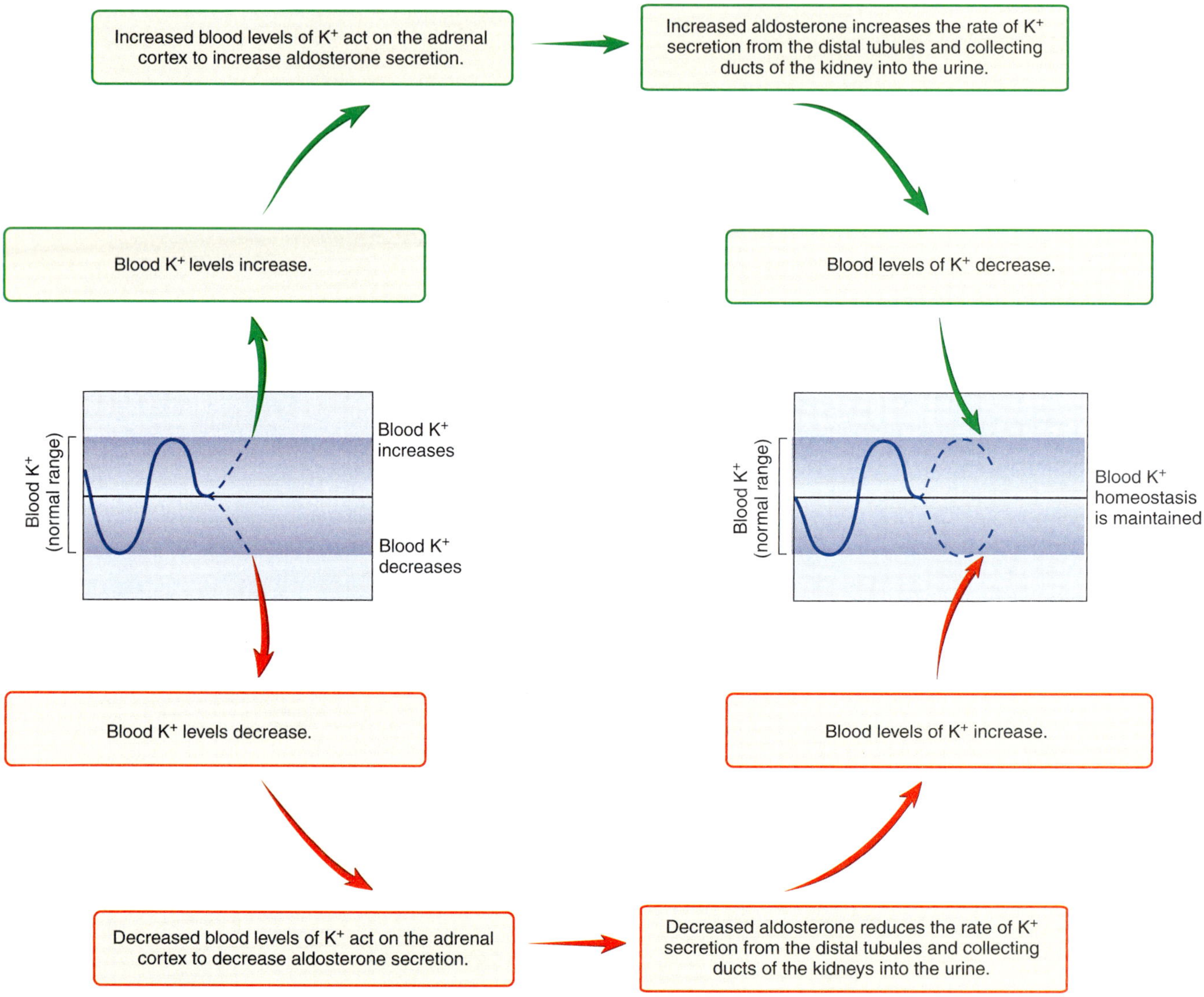

Regulation of Potassium Ions in the Extracellular Fluid
Homeostasis Figure 27.8

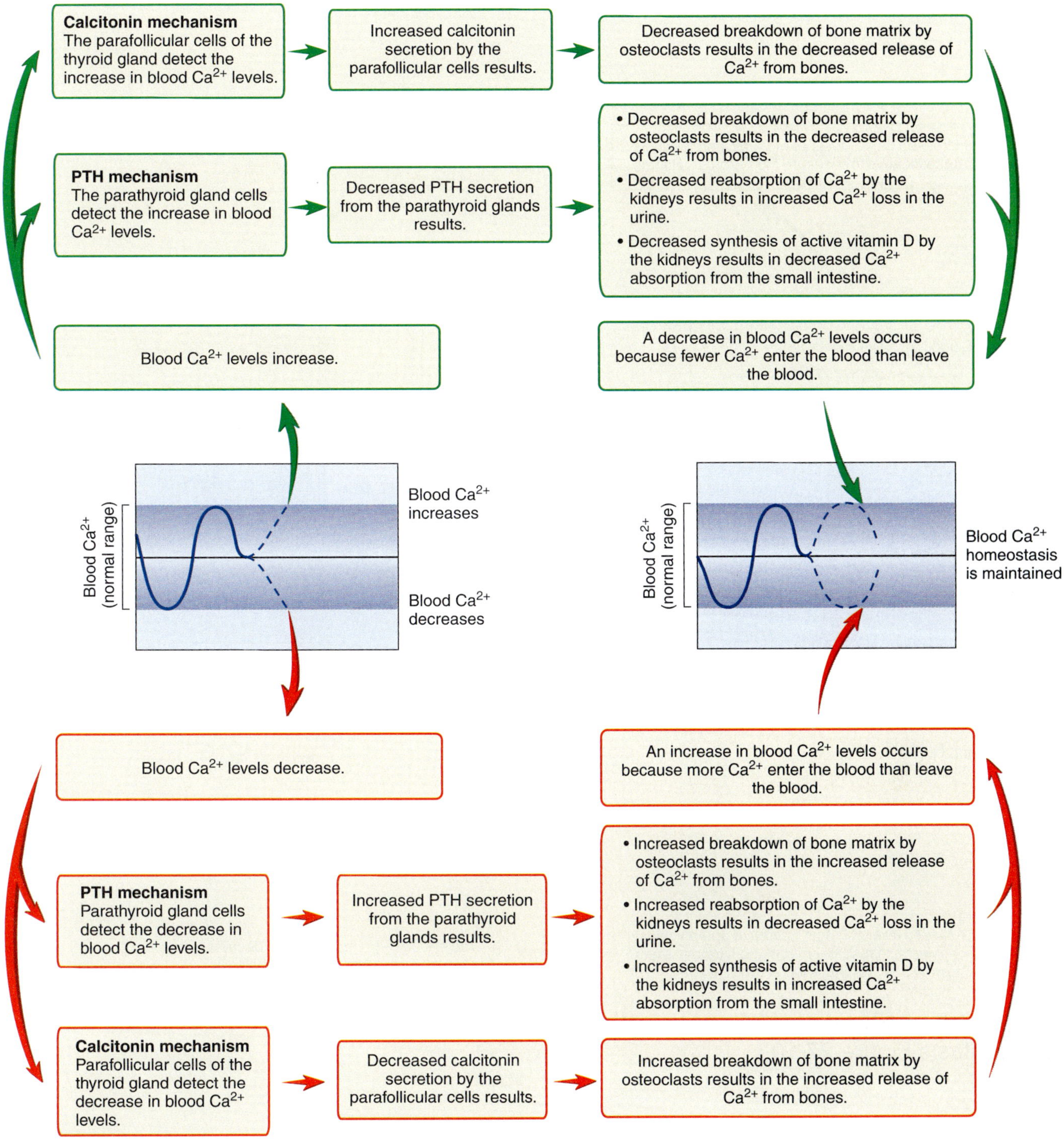

Regulation of Ca^{2+} in the Extracellular Fluid
Homeostasis Figure 27.9

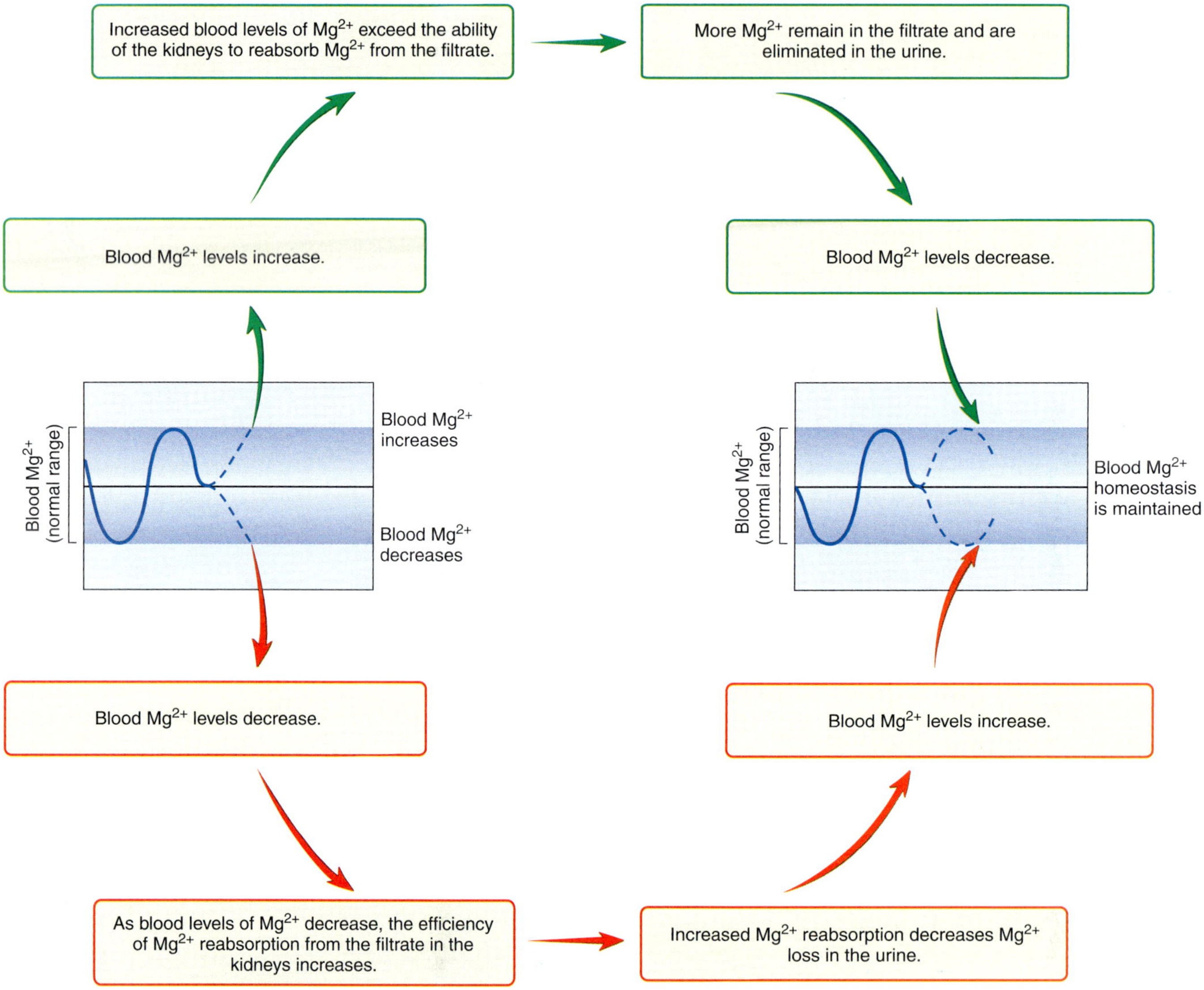

Regulation of Blood Mg^{2+} Concentration in the Extracellular Fluid
Homeostasis Figure 27.10

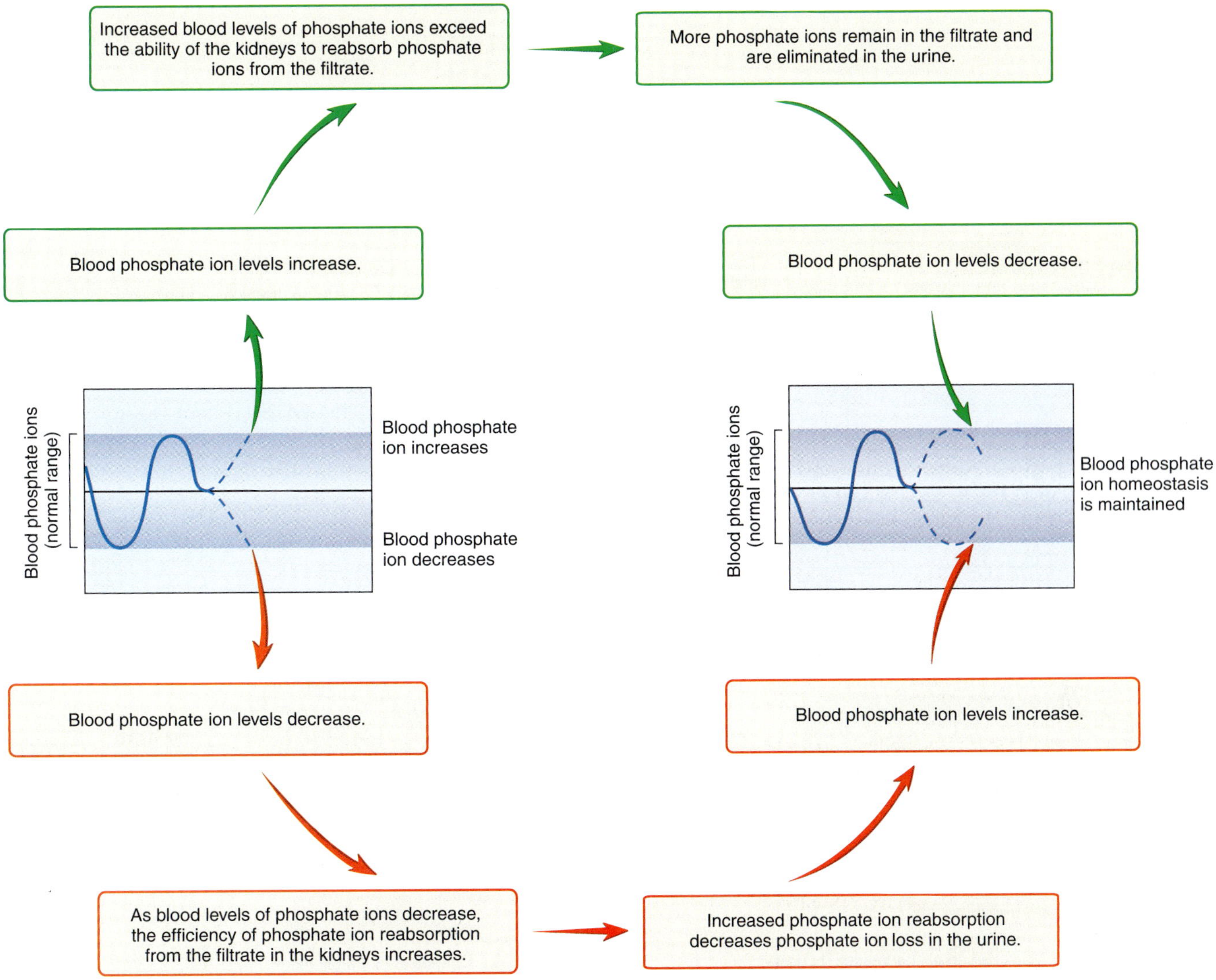

Regulation of Blood Phosphate Concentration in the Extracellular Fluid
Homeostasis Figure 27.11

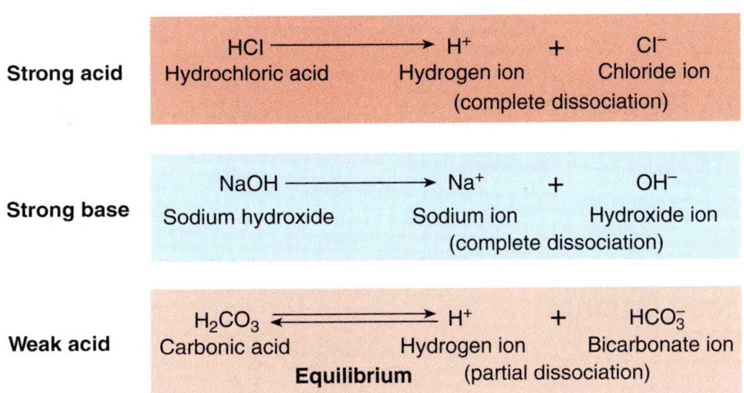

Comparison of Strong and Weak Acids
Figure 27.12

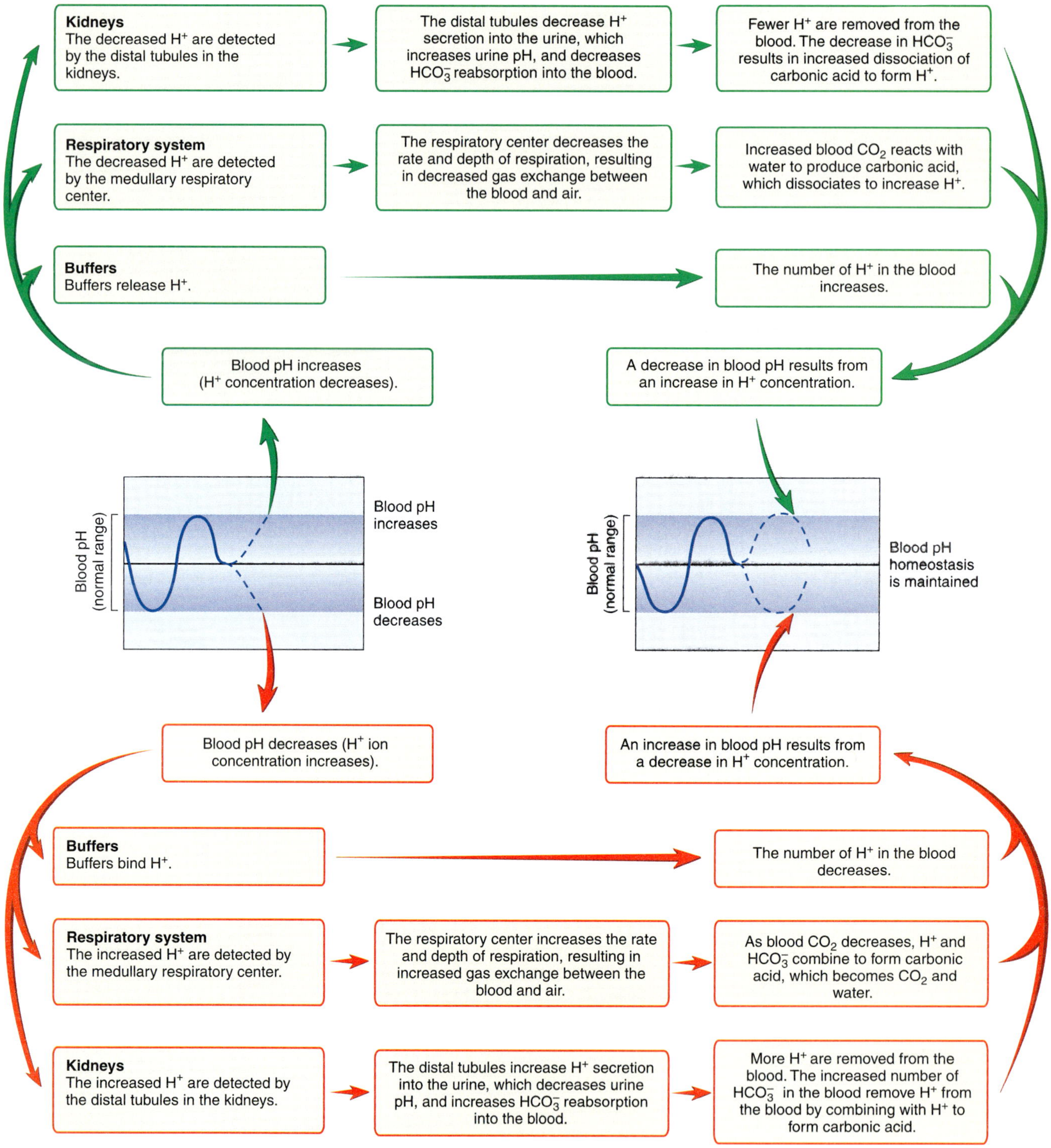

Regulation of Acid–Base Balance
Homeostasis Figure 27.13

1. CO_2 reacts with H_2O to form H_2CO_3. An enzyme, carbonic anhydrase, found in red blood cells and on the surface of blood vessel epithelium, catalyzed the reaction. H_2CO_3 dissociates to form H^+ and HCO_3^-. An equilibrium is quickly established.

2. A decreased pH of the extracellular fluid stimulates the respiratory center and causes an increased rate and depth of respiration.

3. An increased rate and depth of respiration causes CO_2 to be expelled from the lungs, thus reducing the extracellular CO_2 levels. As CO_2 levels decrease, the extracellular concentration of H^+ decreases, and the extracellular fluid pH increases.

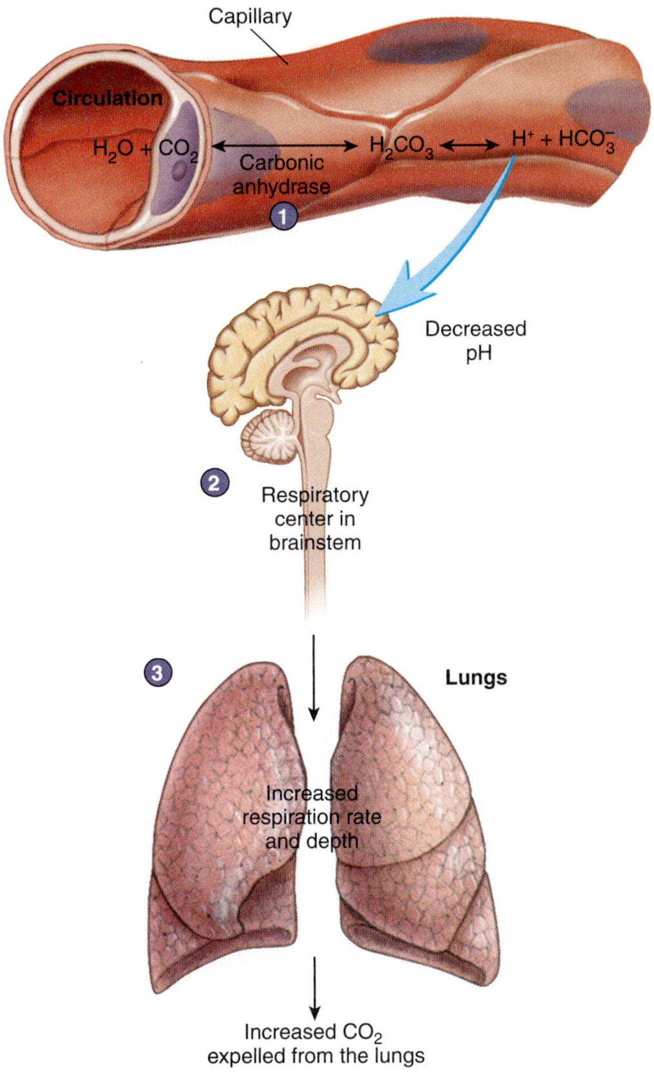

Respiratory Regulation of Body Fluid Acid–Base Balance
Process Figure 27.14

1. When the filtrate or blood pH decreases, H^+ combine with HCO_3^- to form carbonic acid that is converted into CO_2 and H_2O. The CO_2 diffuses into tubule cells.
2. In the tubule cells, CO_2 combines with H_2O to form H_2CO_3 that dissociates to form H^+ and HCO_3^-.
3. A countertransport mechanism secretes H^+ into the filtrate in exchange for Na^+ from the filtrate. As a result, filtrate pH decreases.
4. Bicarbonate ions are cotransported with Na^+ into the interstitial fluid. They then diffuse into capillaries.
5. In capillaries, HCO_3^- combine with H^+. This decreases the H^+ concentration and increases blood pH.

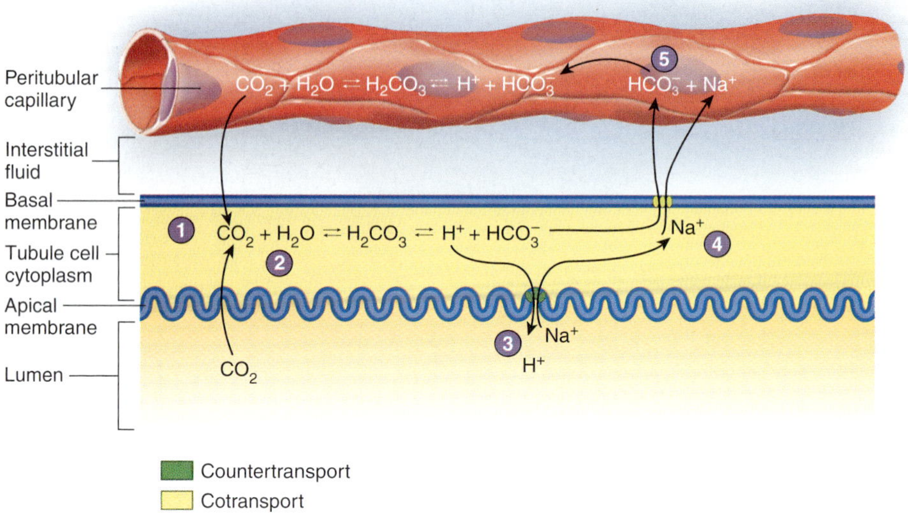

Kidney Regulation of Body Fluid Acid–Base Balance

Process Figure 27.15

1. Hydrogen ions secreted into the filtrate are buffered.
2. Hydrogen ions can react with HCO_3^- that enters the filtrate to form H_2CO_3 which is in equilibrium with H_2O and CO_2.
3. Hydrogen ions can react with HPO_4^{2-} that enters the filtrate to form $H_2PO_4^-$.
4. Hydrogen ions can react with NH_3 formed by amino acid deamination and secreted into the nephron to form NH_4^+.

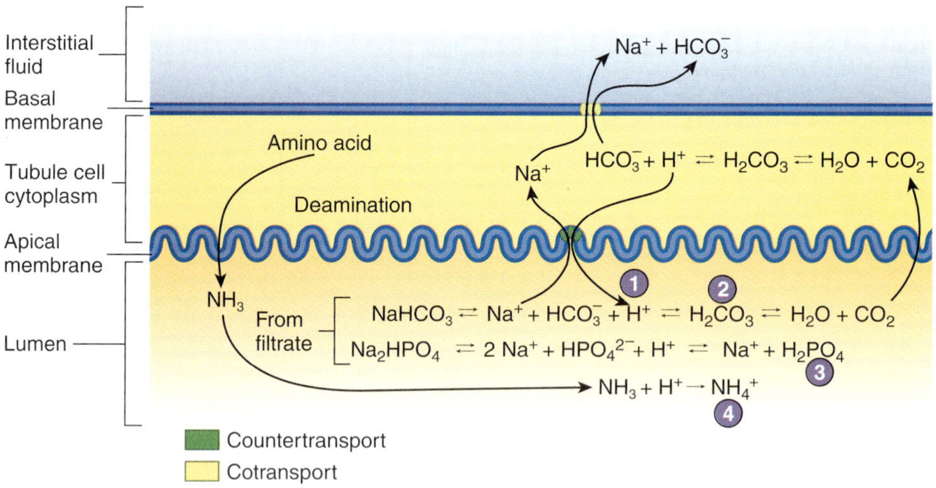

Hydrogen Ion Buffering in the Filtrate

Process Figure 27.16

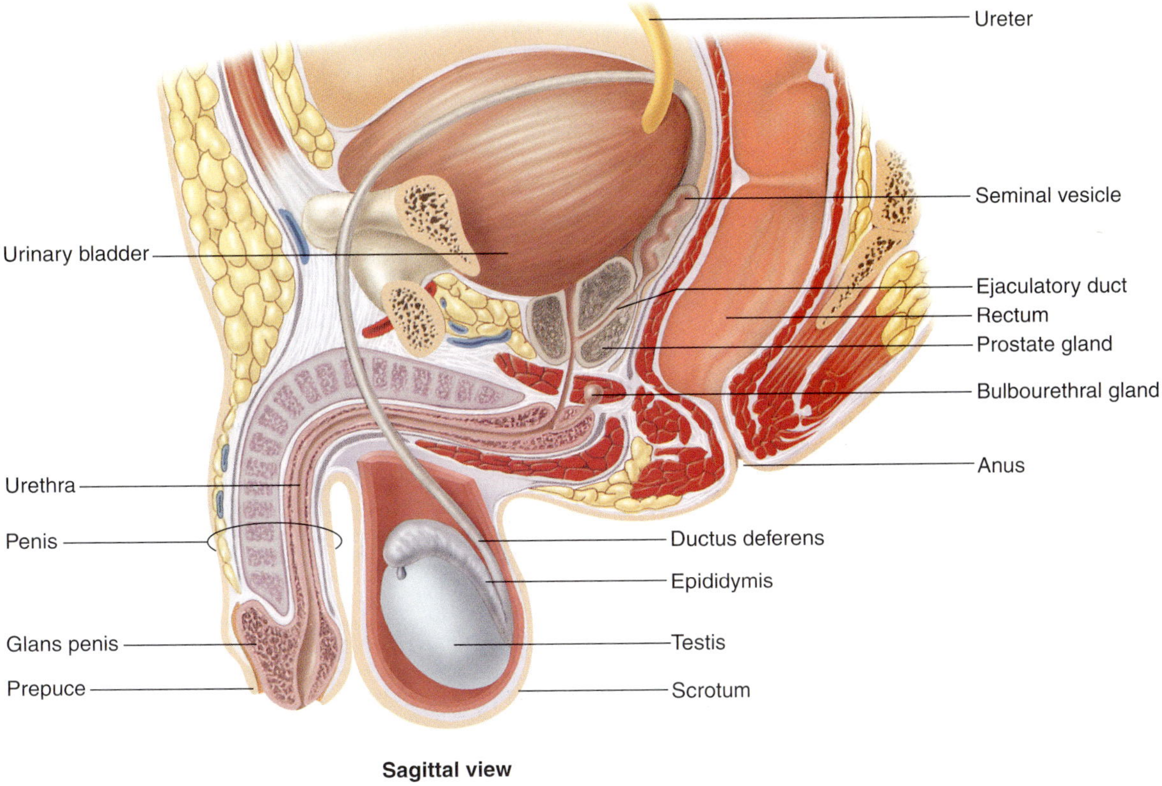

Male Reproductive Structures
Figure 28.1

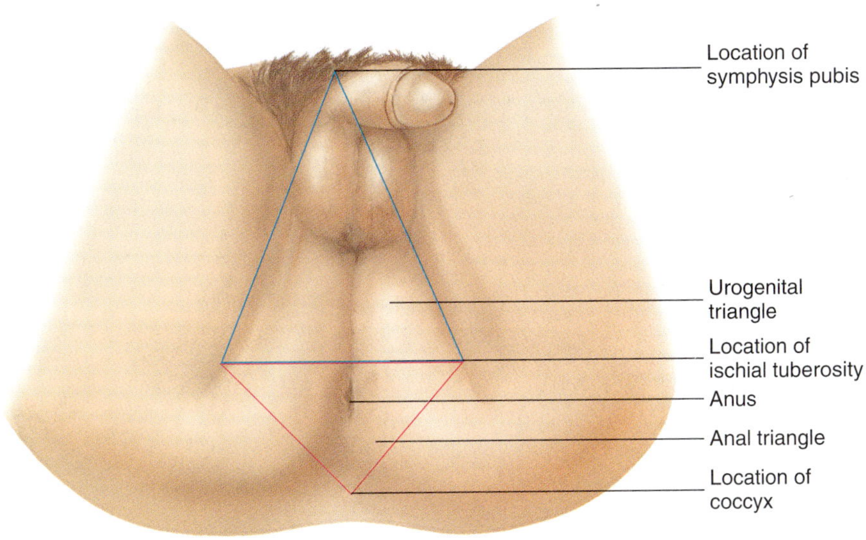

Male Perineum
Figure 28.2

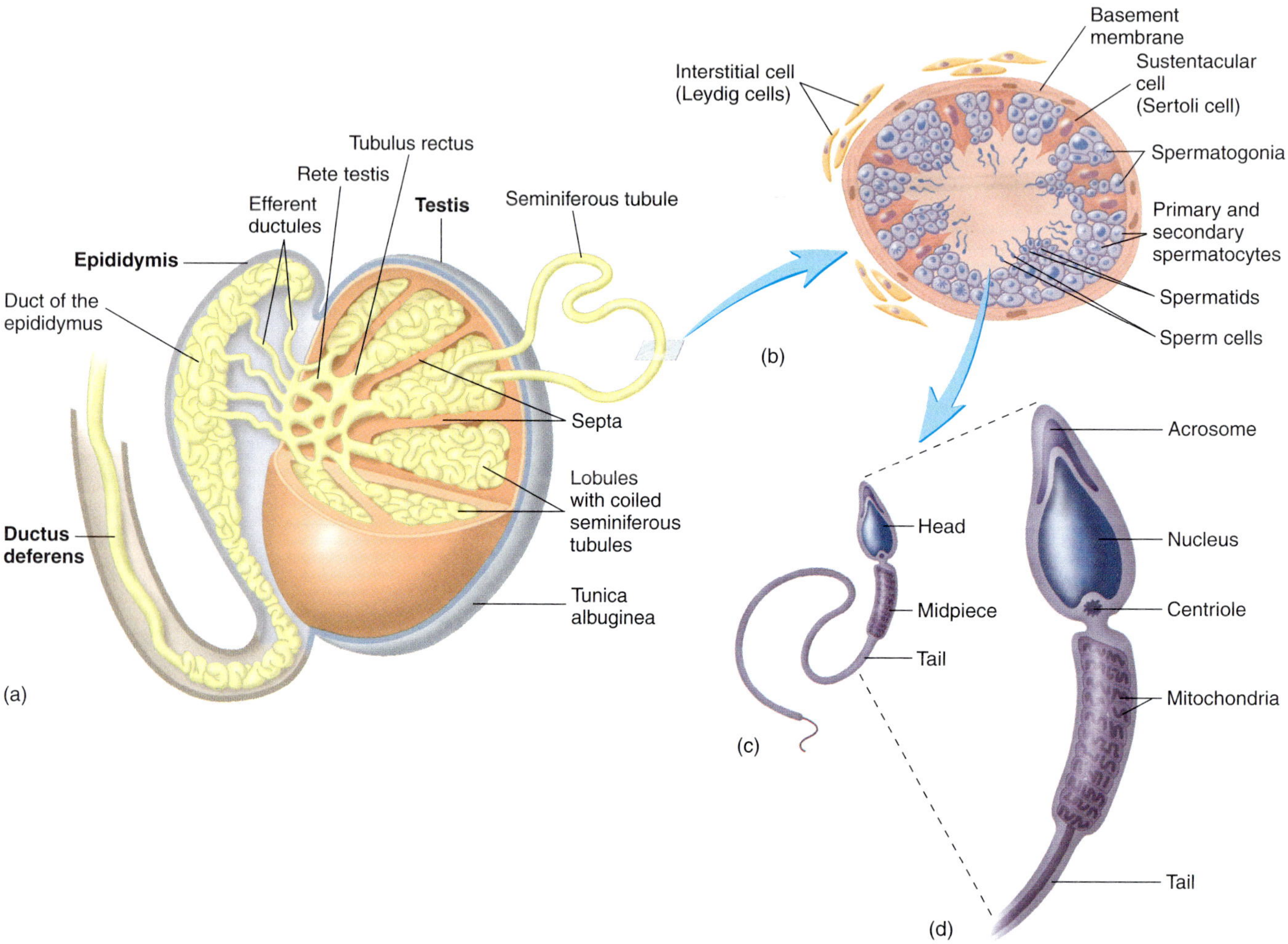

Tubulus rectus
Rete testis
Efferent ductules
Testis
Seminiferous tubule
Epididymis
Duct of the epididymus
Septa
Lobules with coiled seminiferous tubules
Ductus deferens
Tunica albuginea
(a)
Basement membrane
Interstitial cell (Leydig cells)
Sustentacular cell (Sertoli cell)
Spermatogonia
Primary and secondary spermatocytes
Spermatids
Sperm cells
(b)
Acrosome
Head
Nucleus
Midpiece
Centriole
Tail
Mitochondria
(c)
Tail
(d)

Histology of the Testis
Figure 28.3

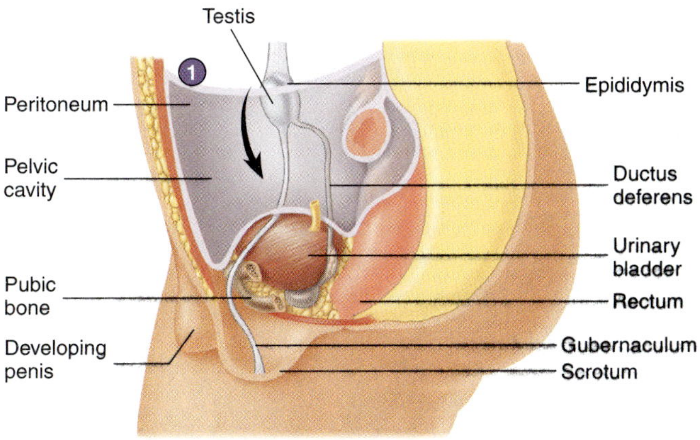

Approximately 8 weeks

(1) Each testis forms as a retroperitoneal structure near the level of the kidney and begins to descend.

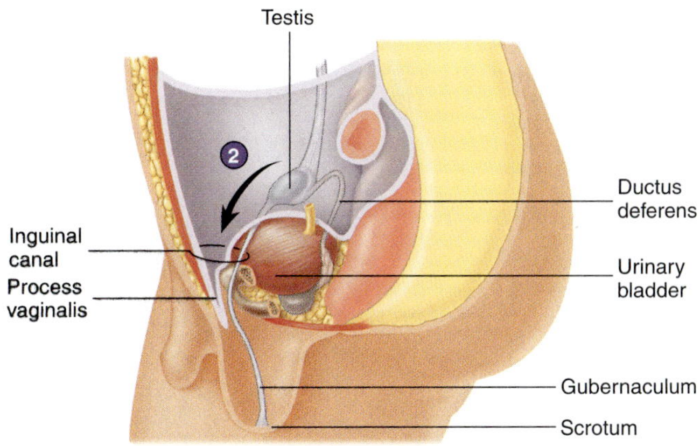

Approximately 12 weeks

(2) The testis beneath the parietal peritoneum descends toward the inguinal canal.

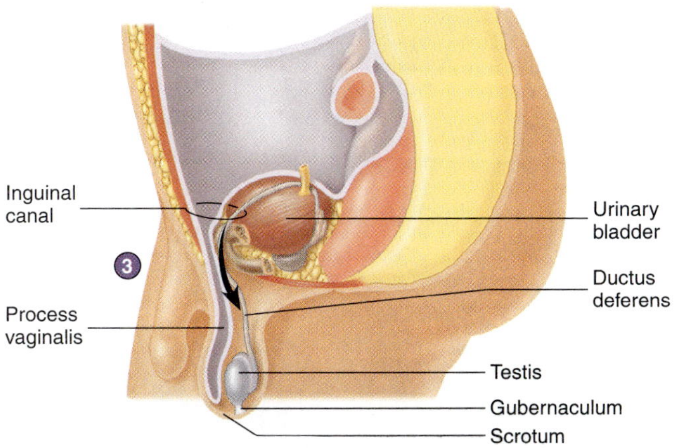

Approximately birth

(3) The testis descends through the inguinal canal into the scrotum.

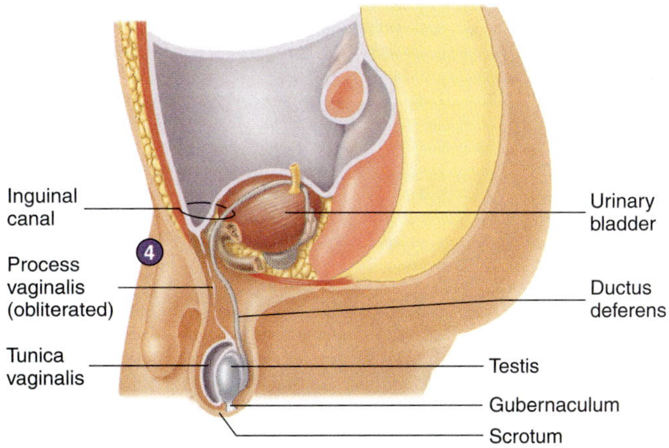

Adult

(4) Between birth and adulthood the process vaginalis is obliterated and its inferior portion becomes the tunica vaginalis.

Descent of the Testes
Process Figure 28.4

1. Spermatogonia are the cells from which sperm cells arise. The spermatogonia divide by mitosis. One daughter cell remains a spermatogonium that can divide again by mitosis. One daughter cell becomes a primary spermatocyte.
2. The primary spermatocyte divides by meiosis to form secondary spermatocytes.
3. The secondary spermatocytes divide by meiosis to form spermatids.
4. The spermatids differentiate to form sperm cells.

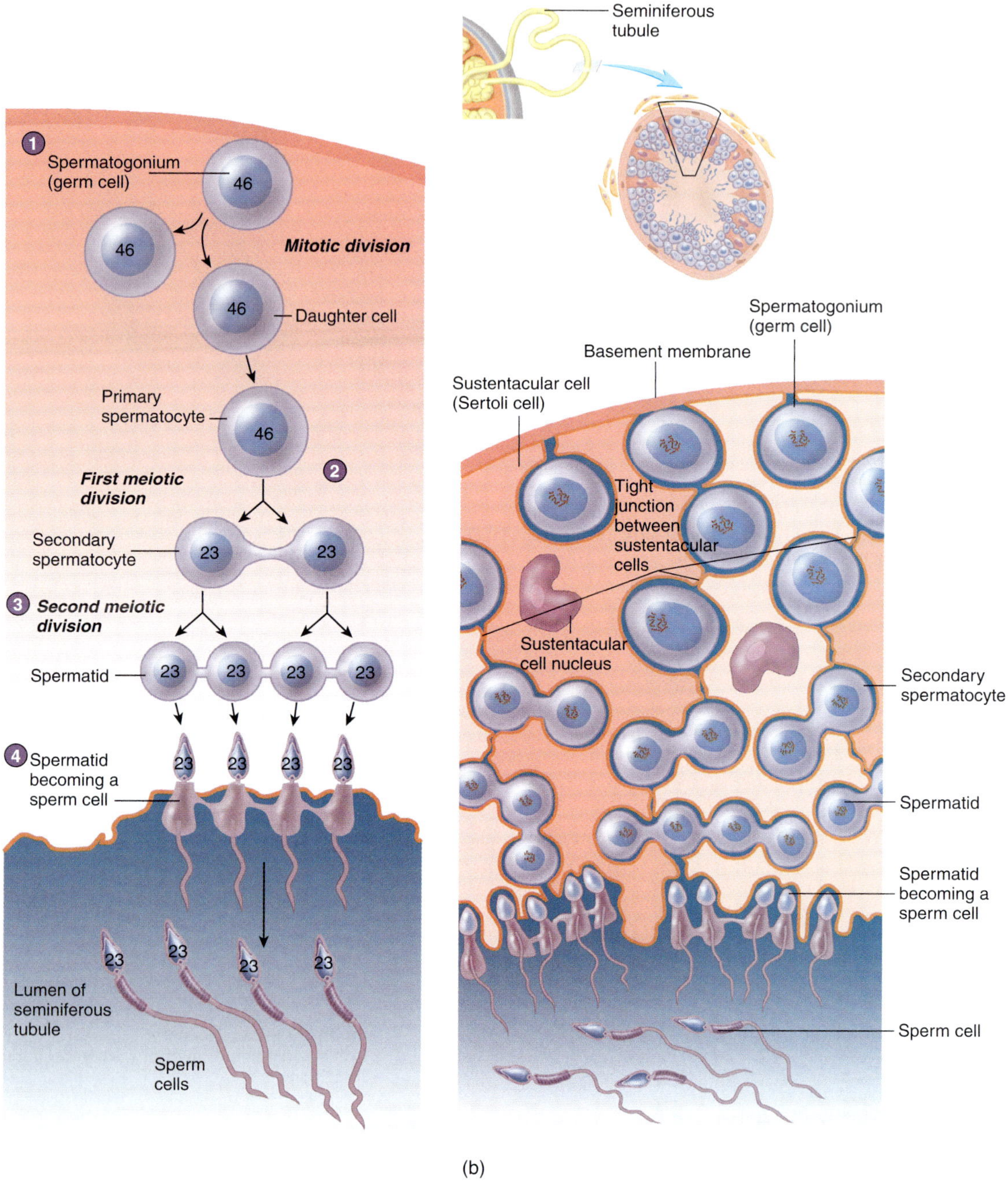

Spermatogenesis
Process Figure 28.5

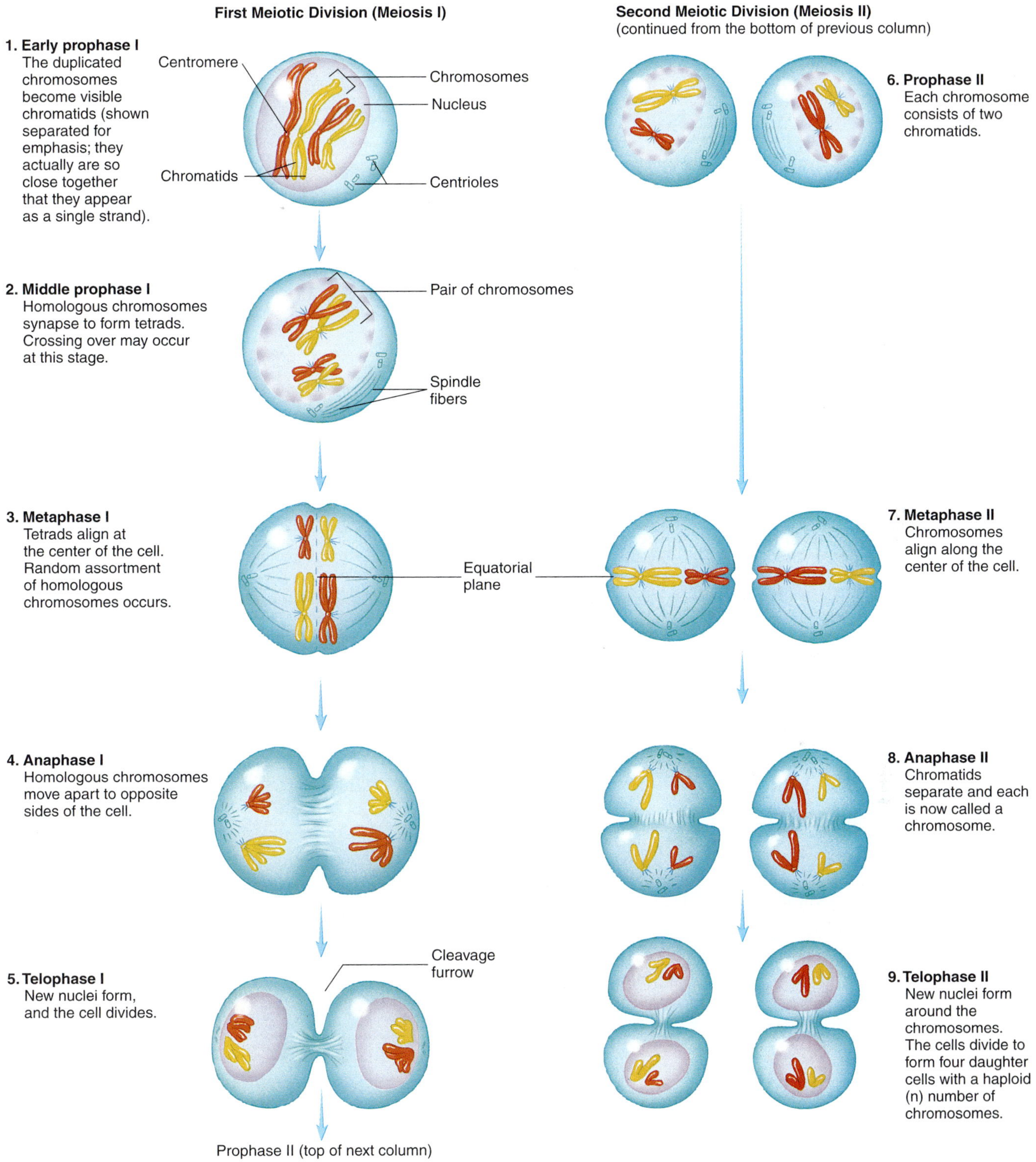

First Meiotic Division (Meiosis I)
1. Early prophase I
The duplicated chromosomes become visible chromatids (shown separated for emphasis; they actually are so close together that they appear as a single strand).
Centromere
Chromosomes
Nucleus
Chromatids
Centrioles
2. Middle prophase I
Homologous chromosomes synapse to form tetrads. Crossing over may occur at this stage.
Pair of chromosomes
Spindle fibers
3. Metaphase I
Tetrads align at the center of the cell. Random assortment of homologous chromosomes occurs.
Equatorial plane
4. Anaphase I
Homologous chromosomes move apart to opposite sides of the cell.
5. Telophase I
New nuclei form, and the cell divides.
Cleavage furrow
Prophase II (top of next column)
Second Meiotic Division (Meiosis II)
(continued from the bottom of previous column)
6. Prophase II
Each chromosome consists of two chromatids.
7. Metaphase II
Chromosomes align along the center of the cell.
8. Anaphase II
Chromatids separate and each is now called a chromosome.
9. Telophase II
New nuclei form around the chromosomes. The cells divide to form four daughter cells with a haploid (n) number of chromosomes.

Meiosis
Process Figure A

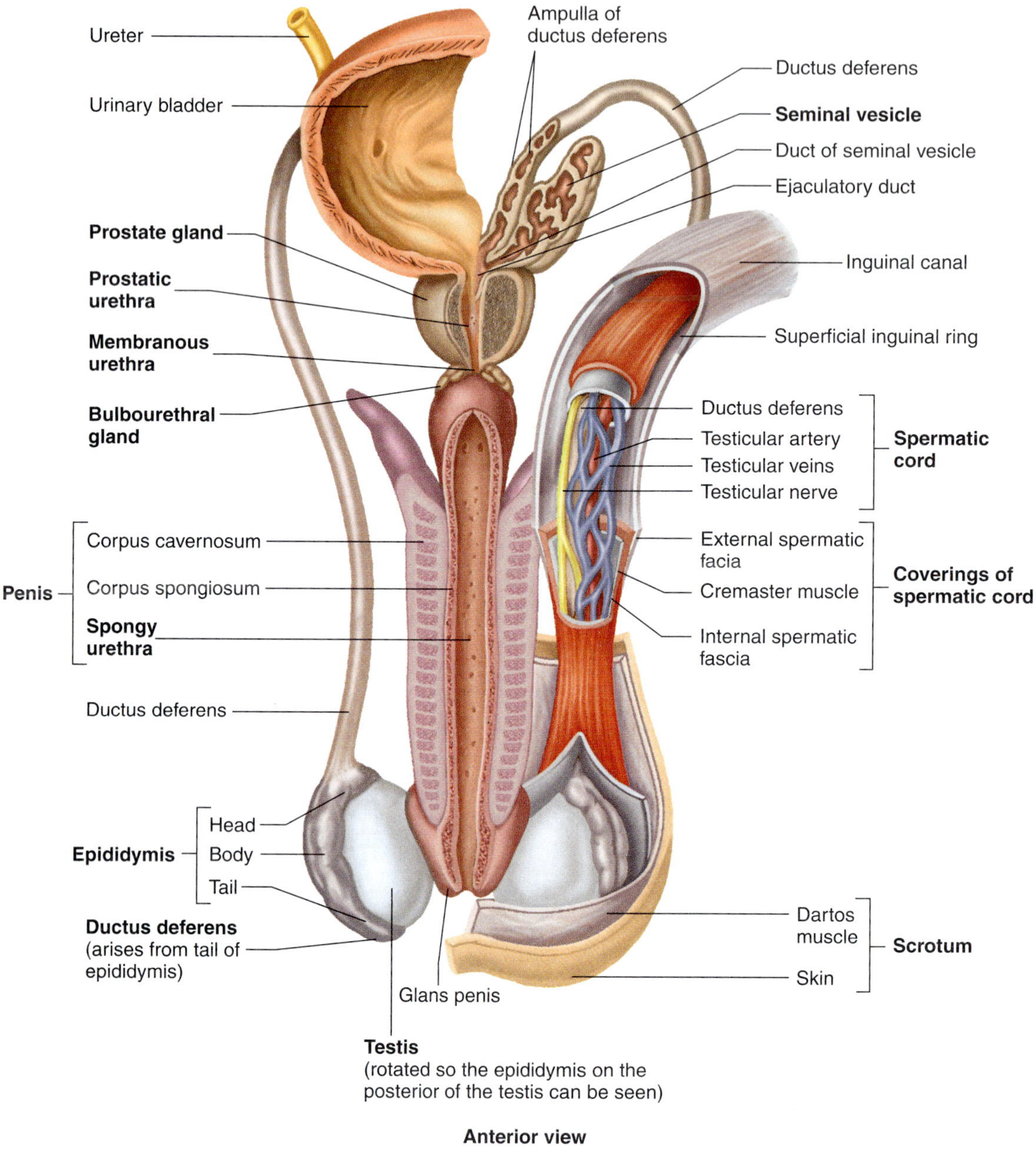

Ureter
Urinary bladder
Ampulla of ductus deferens
Ductus deferens
Seminal vesicle
Duct of seminal vesicle
Ejaculatory duct
Prostate gland
Prostatic urethra
Membranous urethra
Bulbourethral gland
Inguinal canal
Superficial inguinal ring
Ductus deferens
Testicular artery
Testicular veins
Testicular nerve
Spermatic cord
External spermatic facia
Cremaster muscle
Internal spermatic fascia
Coverings of spermatic cord
Penis
Corpus cavernosum
Corpus spongiosum
Spongy urethra
Ductus deferens
Epididymis
Head
Body
Tail
Ductus deferens (arises from tail of epididymis)
Glans penis
Testis (rotated so the epididymis on the posterior of the testis can be seen)
Dartos muscle
Skin
Scrotum
Anterior view

Male Reproductive Structures
Figure 28.6

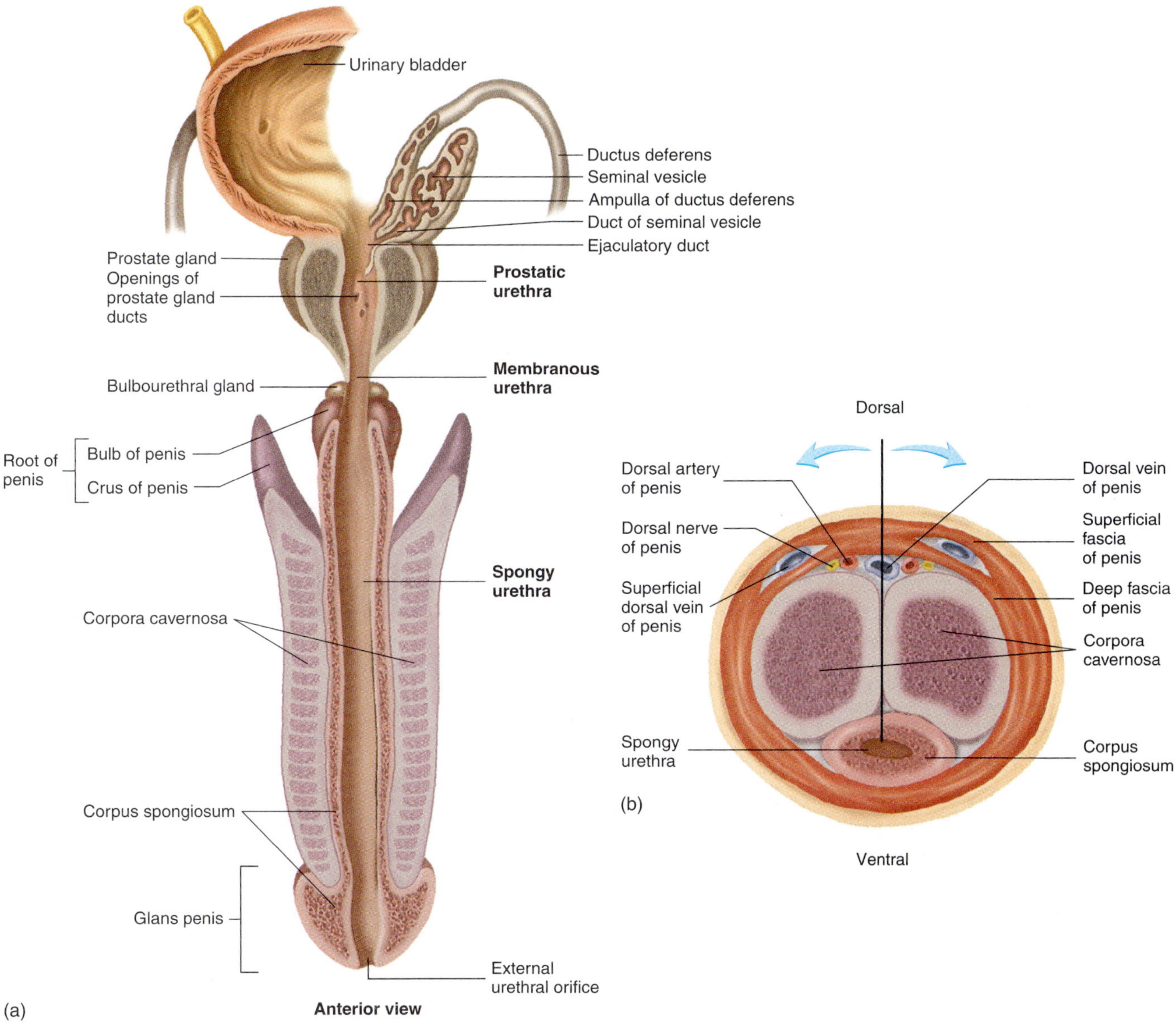

Urinary bladder
Ductus deferens
Seminal vesicle
Ampulla of ductus deferens
Duct of seminal vesicle
Ejaculatory duct
Prostate gland
Openings of prostate gland ducts
Prostatic urethra
Bulbourethral gland
Membranous urethra
Root of penis
Bulb of penis
Crus of penis
Spongy urethra
Corpora cavernosa
Corpus spongiosum
Glans penis
External urethral orifice
(a)
Anterior view
Dorsal
Dorsal artery of penis
Dorsal vein of penis
Dorsal nerve of penis
Superficial fascia of penis
Superficial dorsal vein of penis
Deep fascia of penis
Corpora cavernosa
Spongy urethra
Corpus spongiosum
(b)
Ventral

Penis
Figure 28.7

1. GnRH from the hypothalamus stimulates the secretion of LH and FSH from the anterior pituitary.
2. LH stimulates testosterone secretion from the interstitial cells.
3. FSH stimulates sustentacular cells of the seminiferous tubules to increase spermatogenesis and to secrete inhibin.
4. Testosterone has a stimulatory effect on the sustentacular cells of the seminiferous tubules, and it has a stimulatory effect on the development of sex organs and secondary sex characteristics.
5. Testosterone has a negative-feedback effect on the hypothalamus and pituitary to reduce LH and FSH secretion.
6. Inhibin has a negative-feedback effect on the anterior pituitary to reduce FSH secretion.

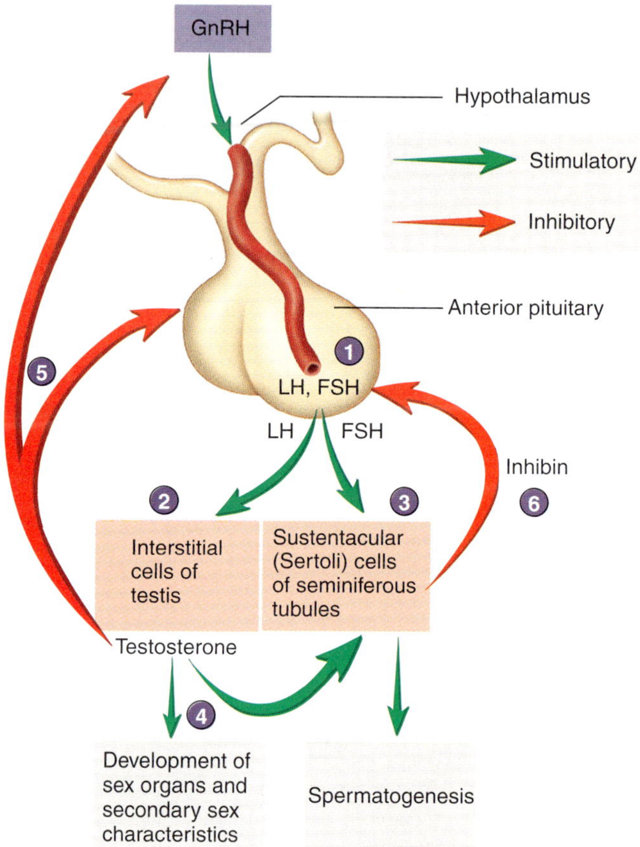

Regulation of Reproductive Hormone Secretion in Males
Process Figure 28.8

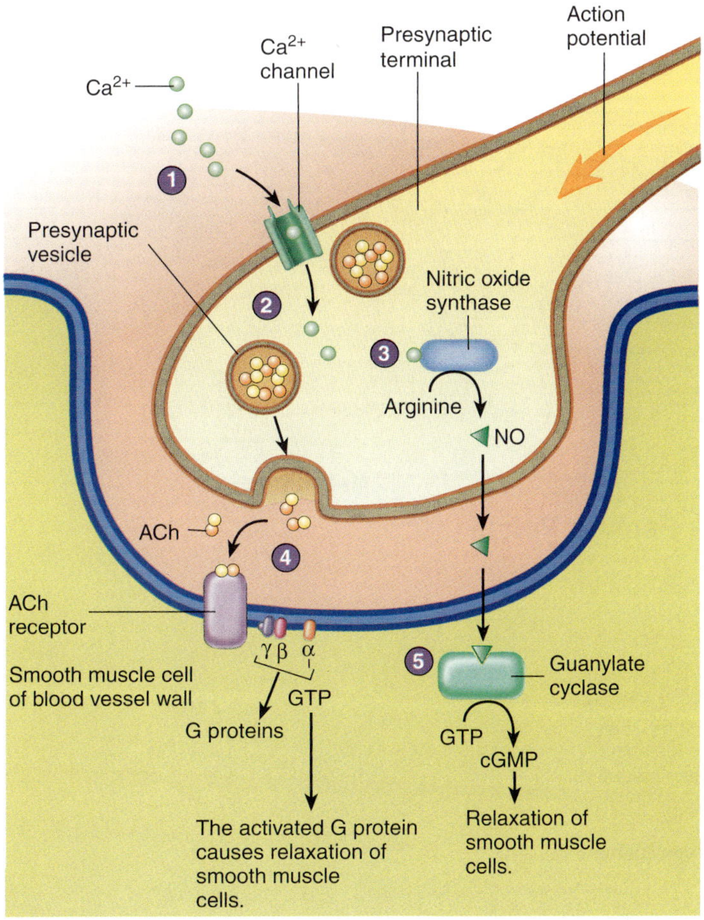

1. Action potentials in parasympathetic neurons cause voltage-gated Ca^{2+} channels to open and Ca^{2+} diffuse into the presynaptic terminals.
2. Calcium ions initiate the release of acetylcholine (ACh) from presynaptic vesicles.
3. Calcium ions also activate nitric oxide synthase, which promotes the synthesis of nitric oxide (NO) from arginine.
4. ACh binds to ACh receptors on the smooth muscle cells and activates a G protein mechanism. The activated G protein causes the relaxation of smooth muscle cells and erection of the penis.
5. NO binds to guanylate cyclase enzymes and activates them. The activated enzymes convert GTP to cGMP, which causes relaxation of the smooth muscle cells and erection of the penis.

Neural Control of Erection
Process Figure 28.9

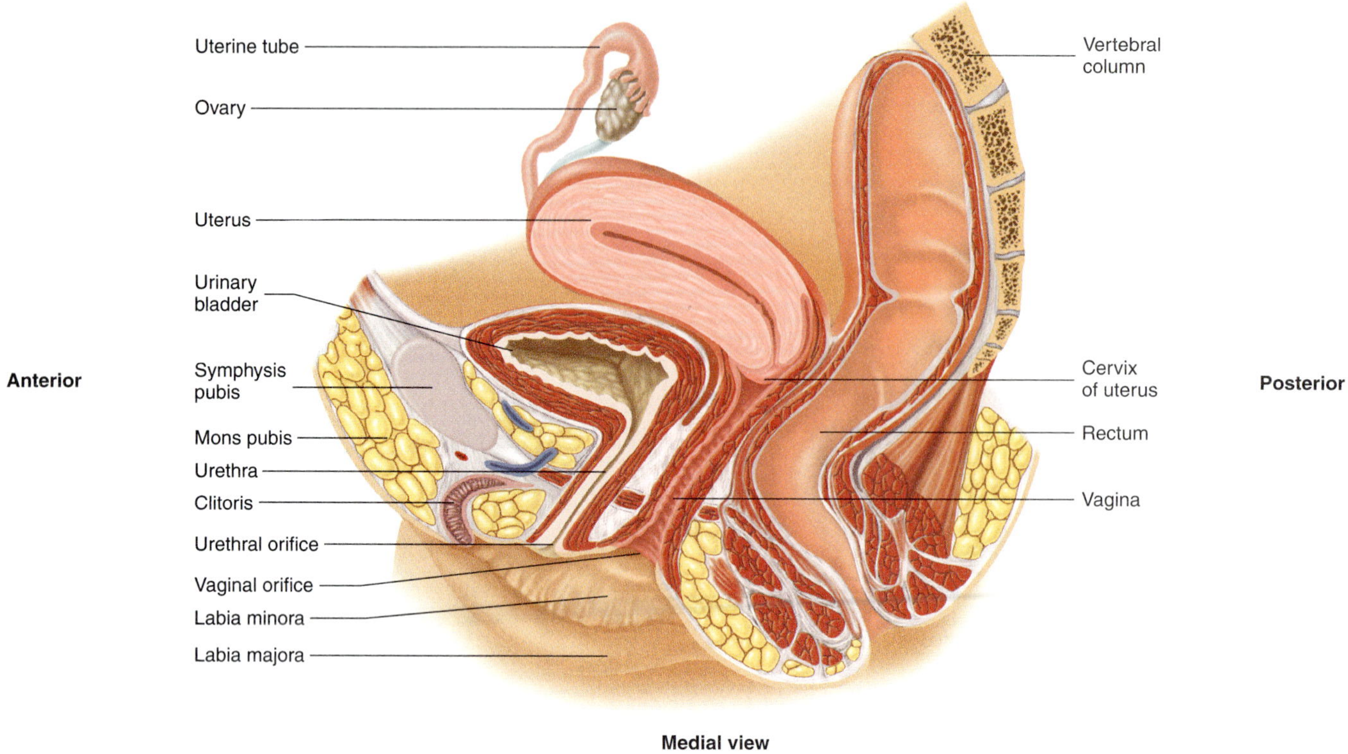

Sagittal Section of the Female Pelvis
Figure 28.10

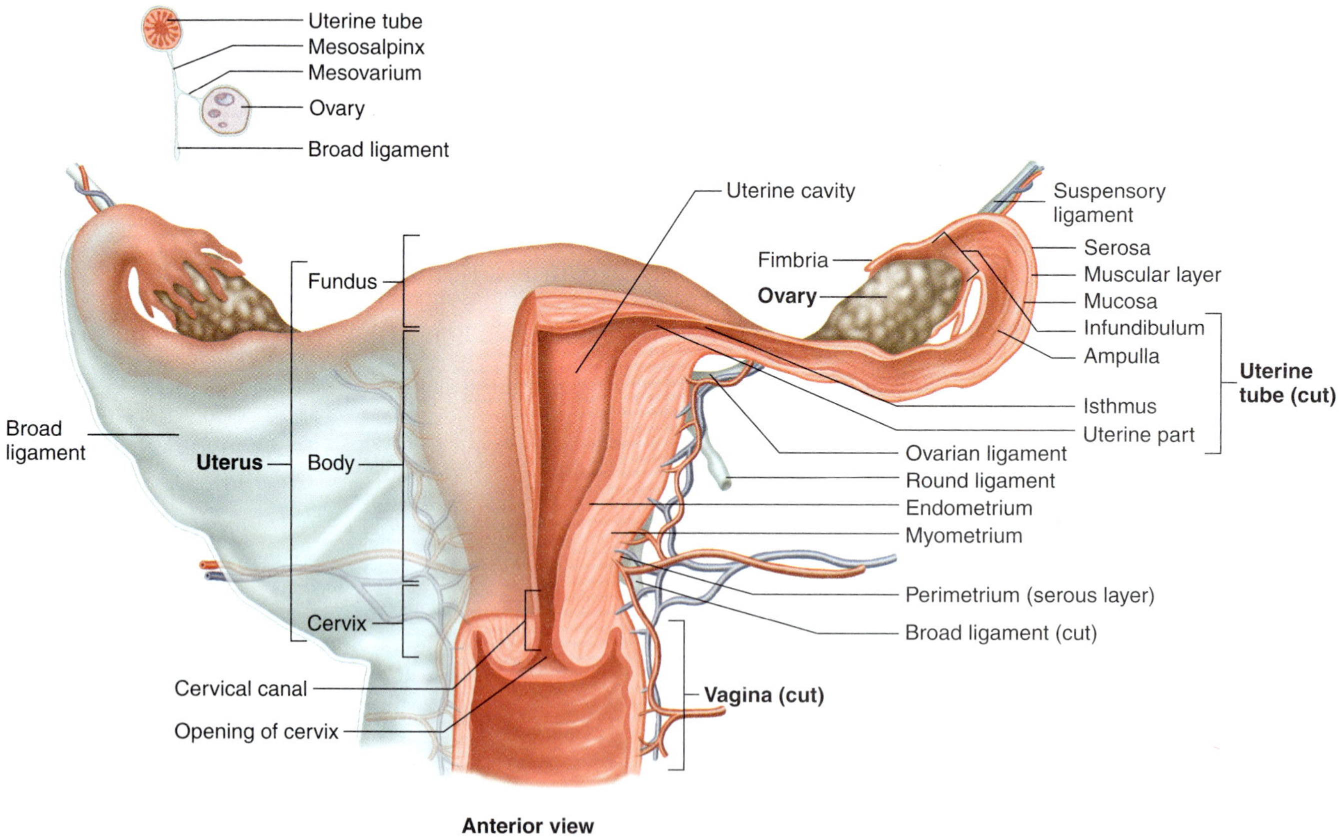

Uterus, Vagina, Uterine Tubes, Ovaries, and Supporting Ligaments
Figure 28.11

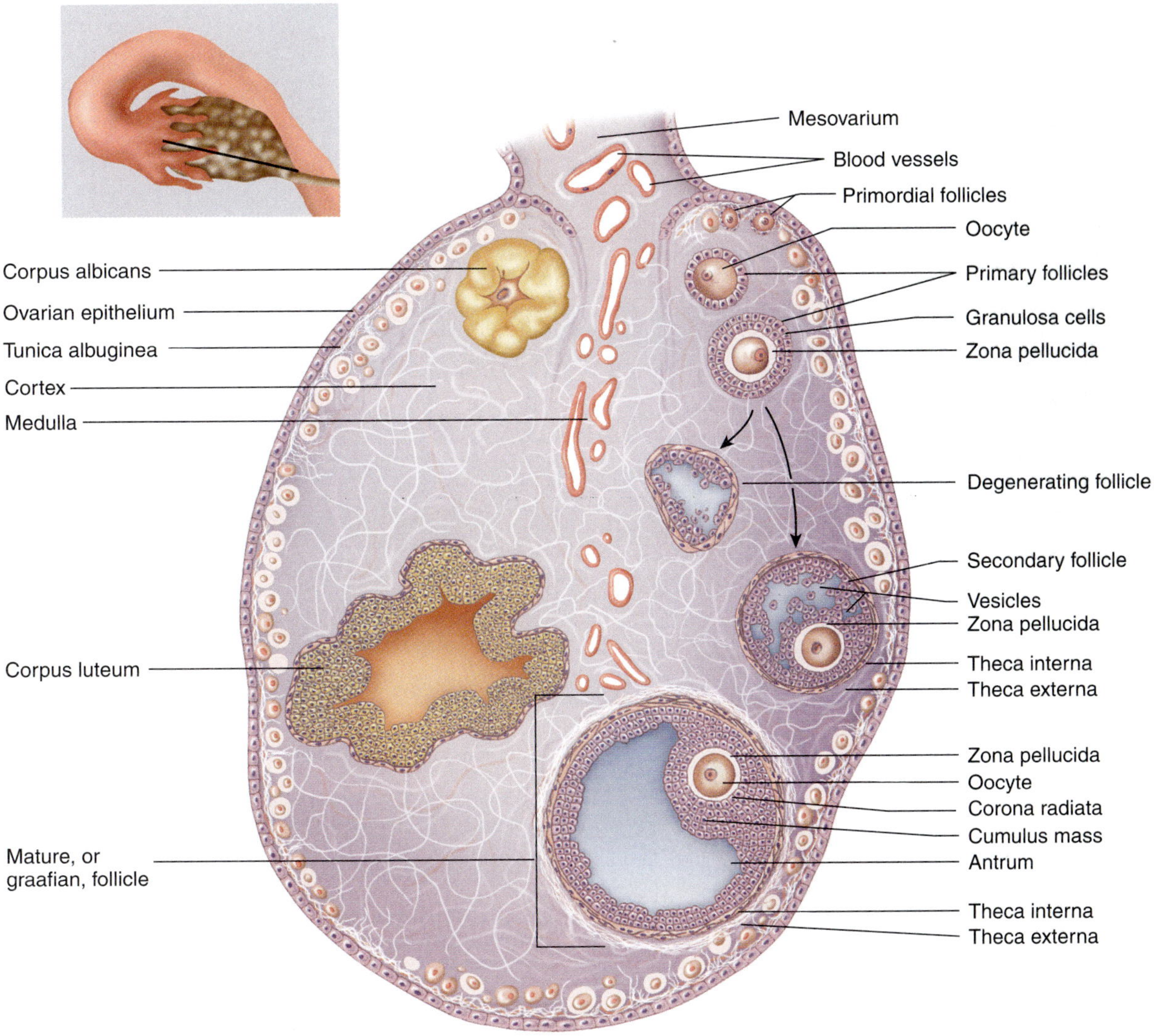

Mesovarium
Blood vessels
Primordial follicles
Oocyte
Primary follicles
Granulosa cells
Zona pellucida
Corpus albicans
Ovarian epithelium
Tunica albuginea
Cortex
Medulla
Degenerating follicle
Secondary follicle
Vesicles
Zona pellucida
Theca interna
Theca externa
Corpus luteum
Zona pellucida
Oocyte
Corona radiata
Cumulus mass
Antrum
Mature, or graafian, follicle
Theca interna
Theca externa

Histology of the Ovary
Figure 28.12

1. Oogonia are the cells from which oocytes arise. The oogonia divide by mitosis to produce other oogonia and primary oocytes.

2. Five million oocytes may be produced by the fourth month of prenatal life. Primary oocytes begin the first meiotic division but stop at prophase I. All of the primary oocytes remain in this state until puberty.

3. The first meiotic division is completed in a single mature follicle just before ovulation during each menstrual cycle. A secondary oocyte and the first polar body result from the unequal division of the cytoplasm.

4. The secondary oocyte begins the second meiotic division but stops at metaphase II.

5. The second meiotic division is completed after ovulation and after a sperm cell unites with the secondary oocyte. A secondary oocyte and a second polar body are formed.

6. Fertilization is completed after the nuclei of the secondary oocyte and the sperm cell unite. The resulting cell is called a zygote.

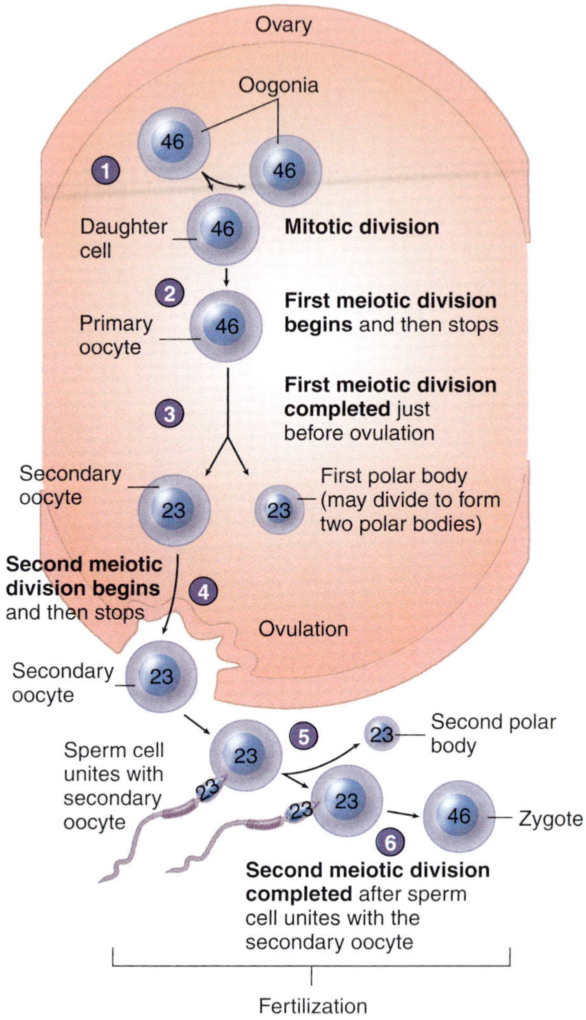

Maturation and Fertilization of the Oocyte
Process Figure 28.13

1. The primordial follicle consists of an oocyte surrounded by a single layer of squamous granulosa cells.

2. A primordial follicle becomes a primary follicle as the granulosa cells become enlarged and cuboidal.

3. The primary follicle enlarges. Granulosa cells form more than one layer of cells. A zona pellucida forms around the oocyte.

4. A secondary follicle forms when fluid-filled vesicles (spaces) develop among the granulosa cells and a well-developed theca becomes apparent around the granulosa cells.

5. A mature follicle forms when the fluid-filled vesicles form a single antrum. When a follicle becomes fully mature, it is enlarged to its maximum size, a large antrum is present, and the oocyte is located in the cumulus mass. Cells of the cumulus mass called the corona radiata surround the oocyte.

6. During ovulation the oocyte is released from the follicle, along with cells of the corona radiata.

7. Following ovulation, the granulosa cells divide rapidly and enlarge to form the corpus luteum.

8. When the corpus luteum degenerates, it forms the corpus albicans.

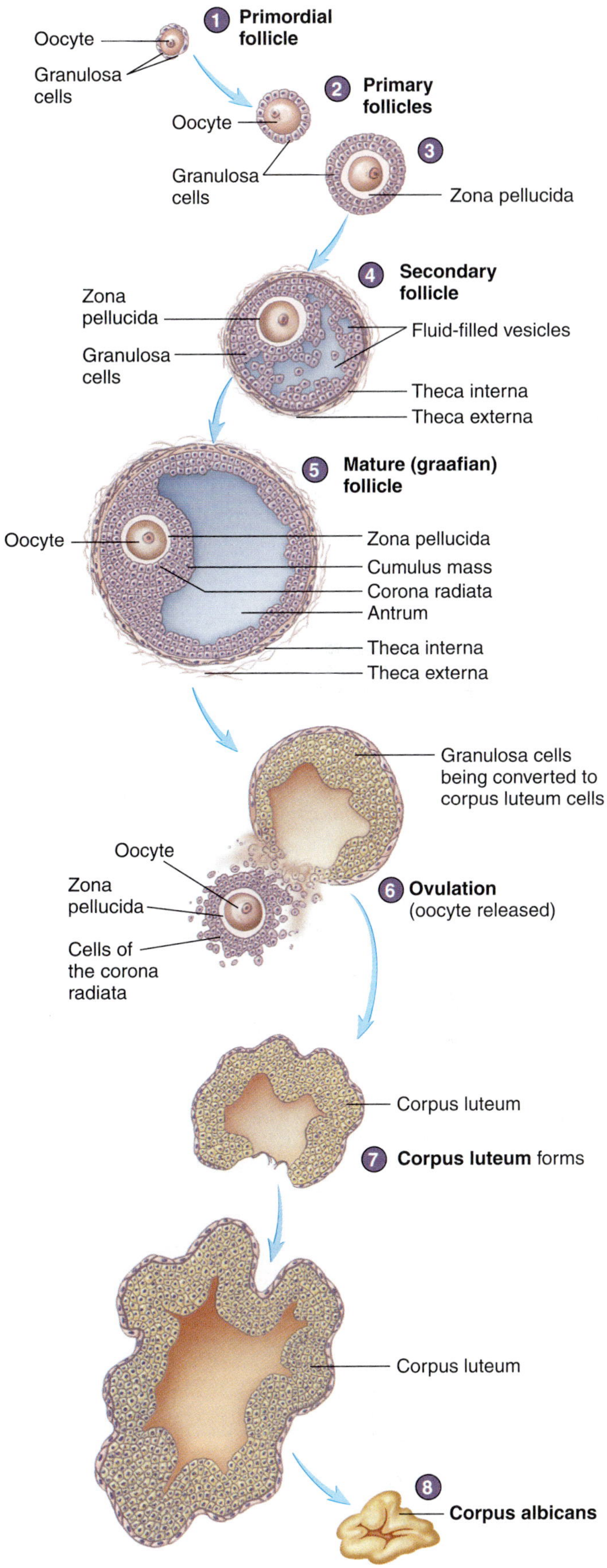

Maturation of the Follicle and Oocyte
Process Figure 28.14

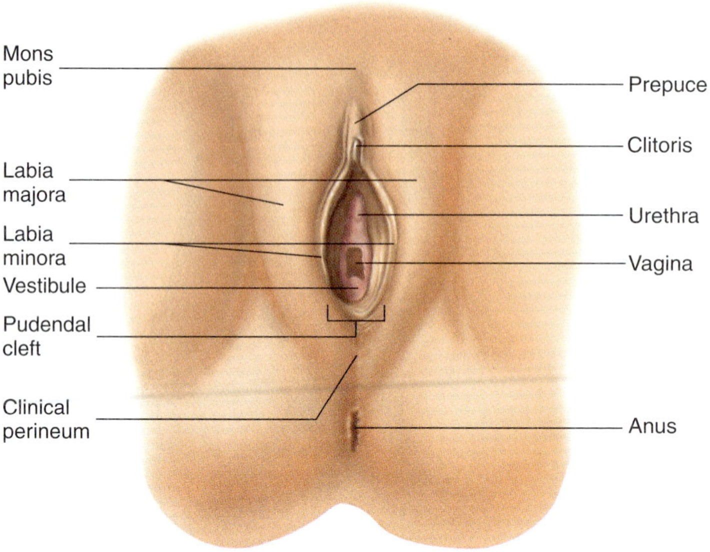

Female External Genitalia
Figure 28.15

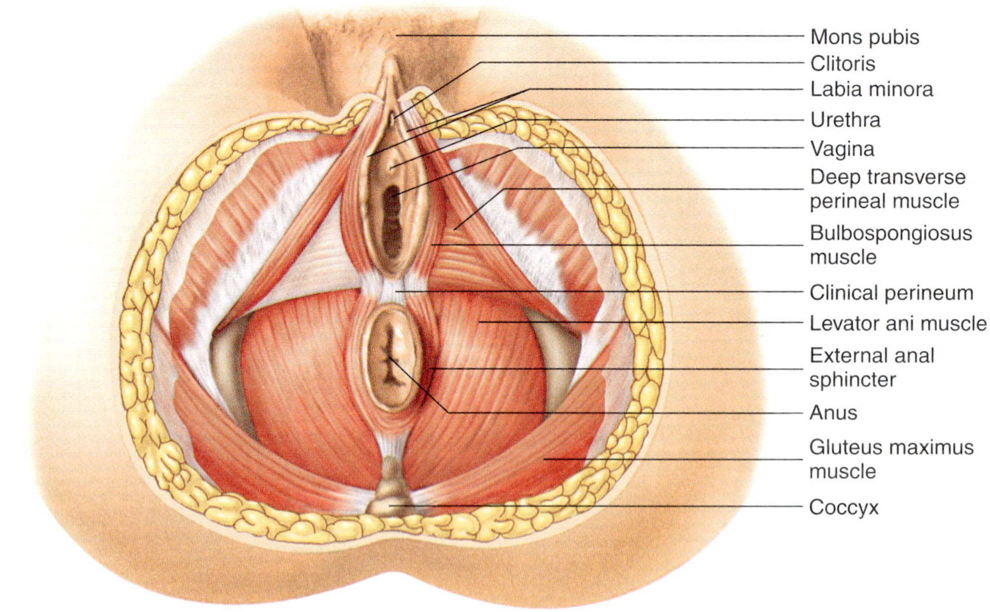

Inferior View of the Female Perineum
Figure 28.16

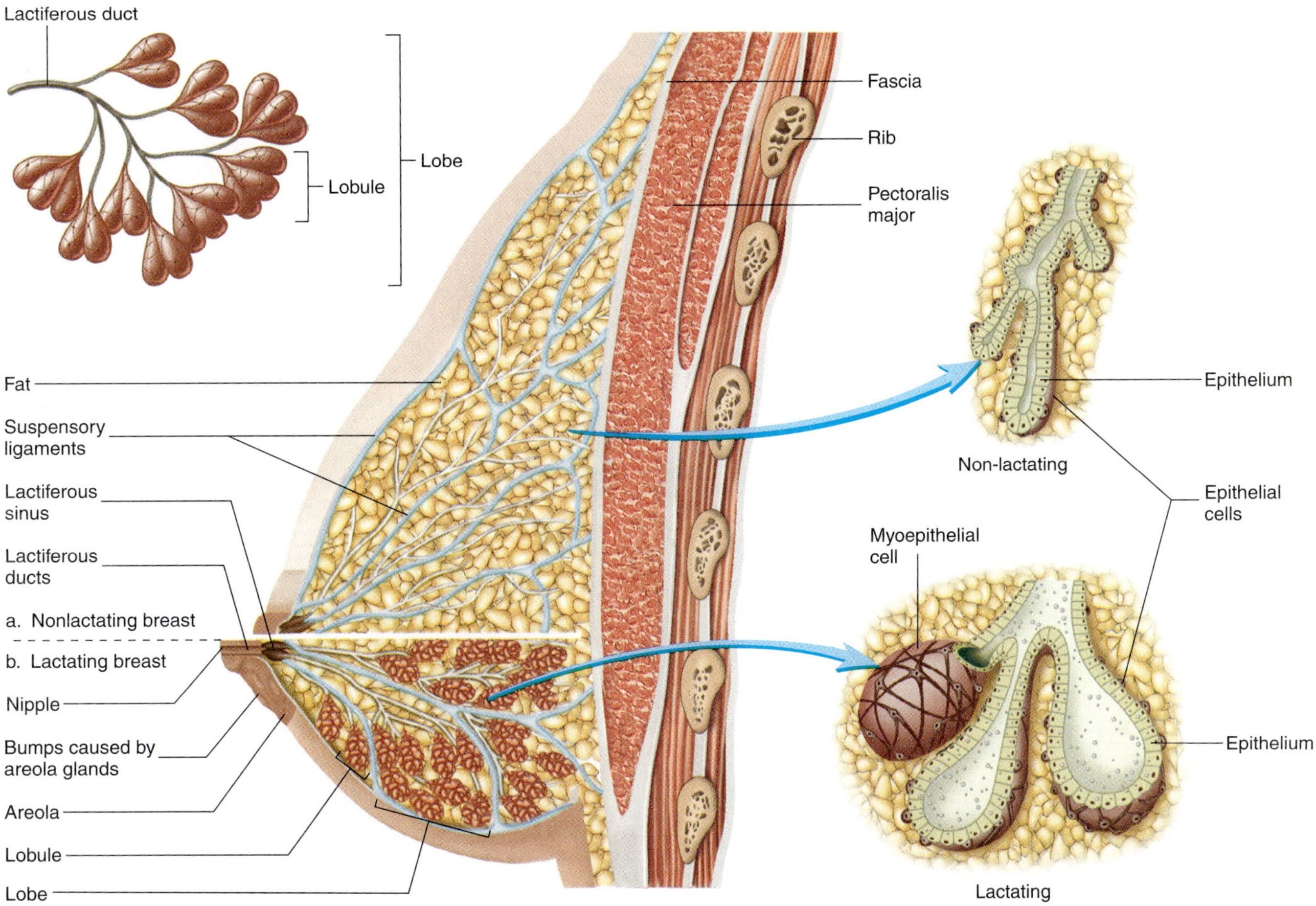
Lactiferous duct
Lobule
Lobe
Fascia
Rib
Pectoralis major
Fat
Suspensory ligaments
Lactiferous sinus
Lactiferous ducts
a. Nonlactating breast
b. Lactating breast
Nipple
Bumps caused by areola glands
Areola
Lobule
Lobe
Epithelium
Non-lactating
Epithelial cells
Myoepithelial cell
Epithelium
Lactating

Anatomy of the Breast
Figure 28.17

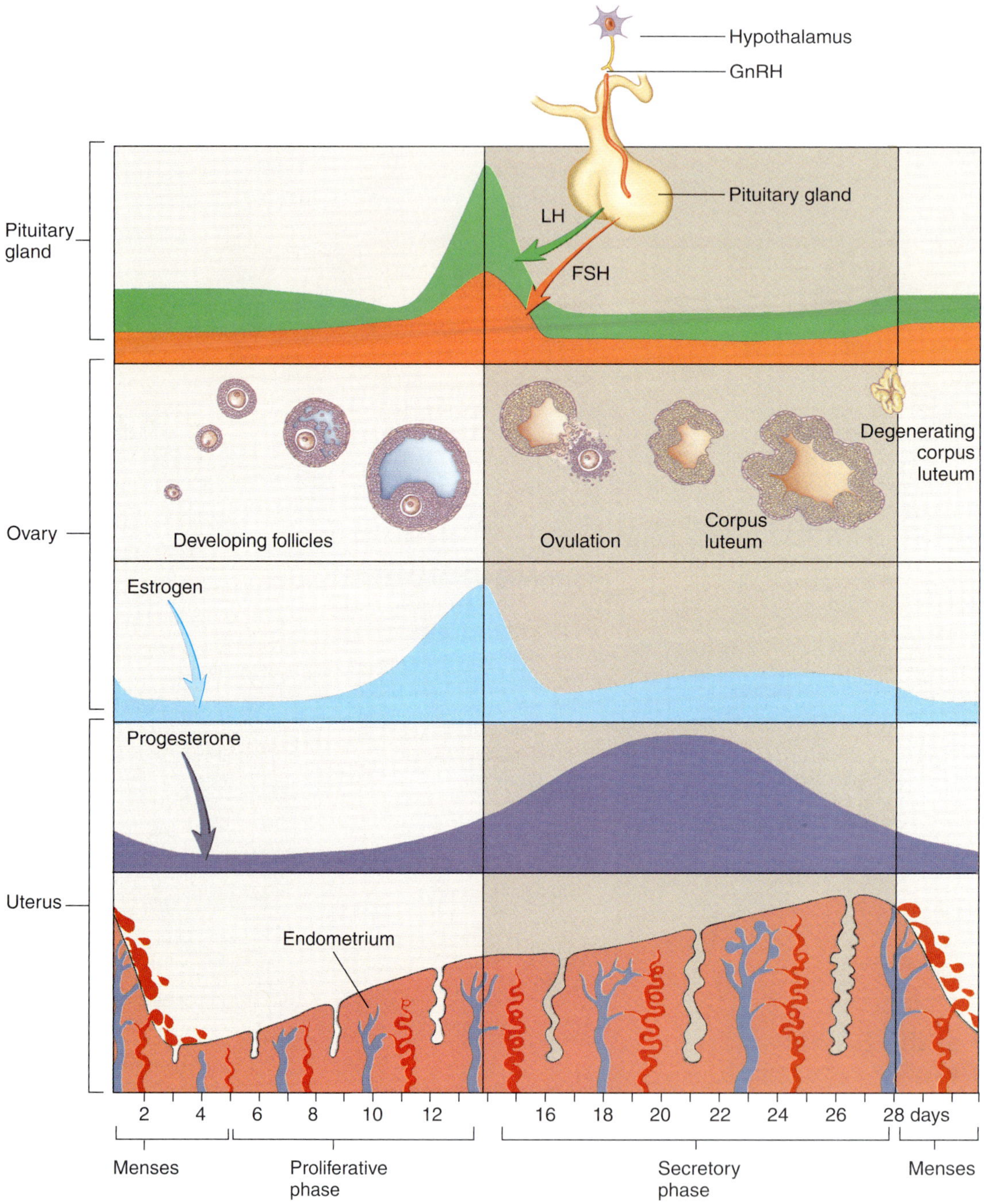

Hypothalamus
GnRH
Pituitary gland
Pituitary gland
LH
FSH
Ovary
Developing follicles
Ovulation
Corpus luteum
Degenerating corpus luteum
Estrogen
Progesterone
Uterus
Endometrium
2
4
6
8
10
12
16
18
20
22
24
26
28 days
Menses
Proliferative phase
Secretory phase
Menses

The Menstrual Cycle
Figure 28.18

1. Before ovulation there is an increase in FSH secretion which stimulates follicles to develop and secrete estrogen.

2. Estrogen causes the endometrium to proliferate and the hypothalamus to increase LH secretion, which results in the LH and FSH surges prior to ovulation.

3. The LH surge causes a follicle to mature and ovulate. The corpus luteum develops and secretes progesterone and some estrogen.

4. The progesterone causes hypertrophy of the endometrium and has a negative-feedback effect on LH and FSH secretion. The corpus luteum continues to secrete progesterone for approximately 12 days after ovulation.

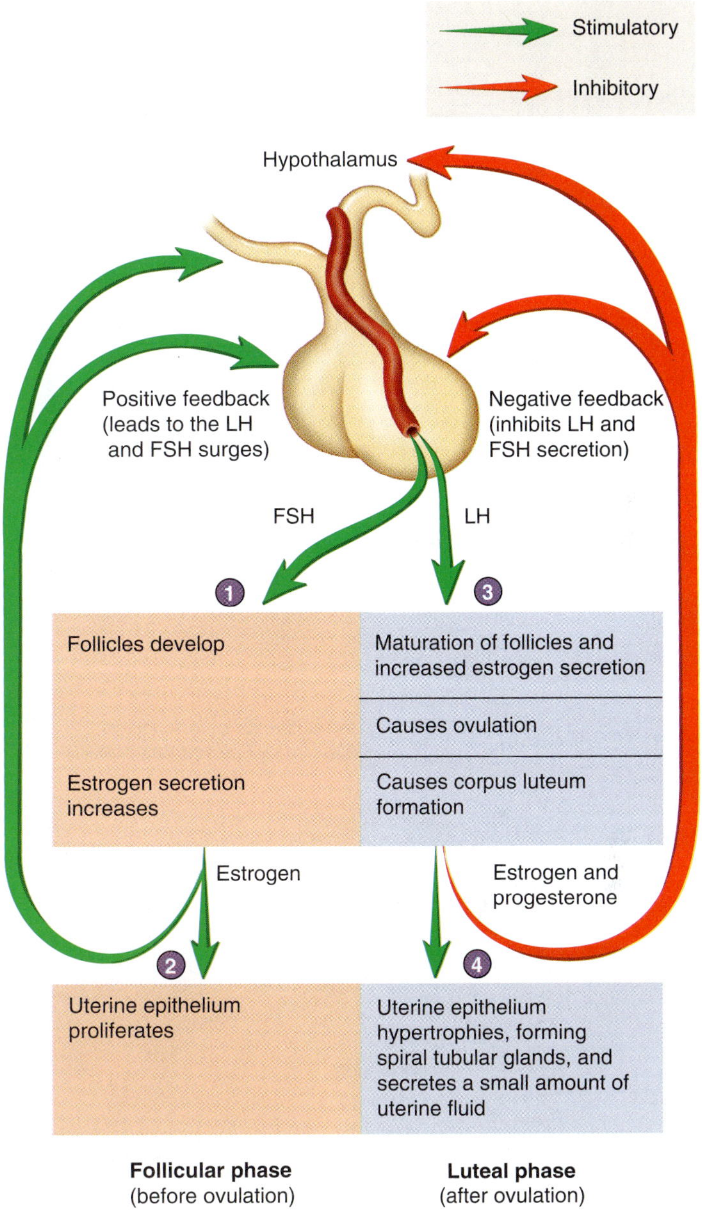

Regulation of Hormone Secretion During the Menstrual Cycle
Process Figure 28.19

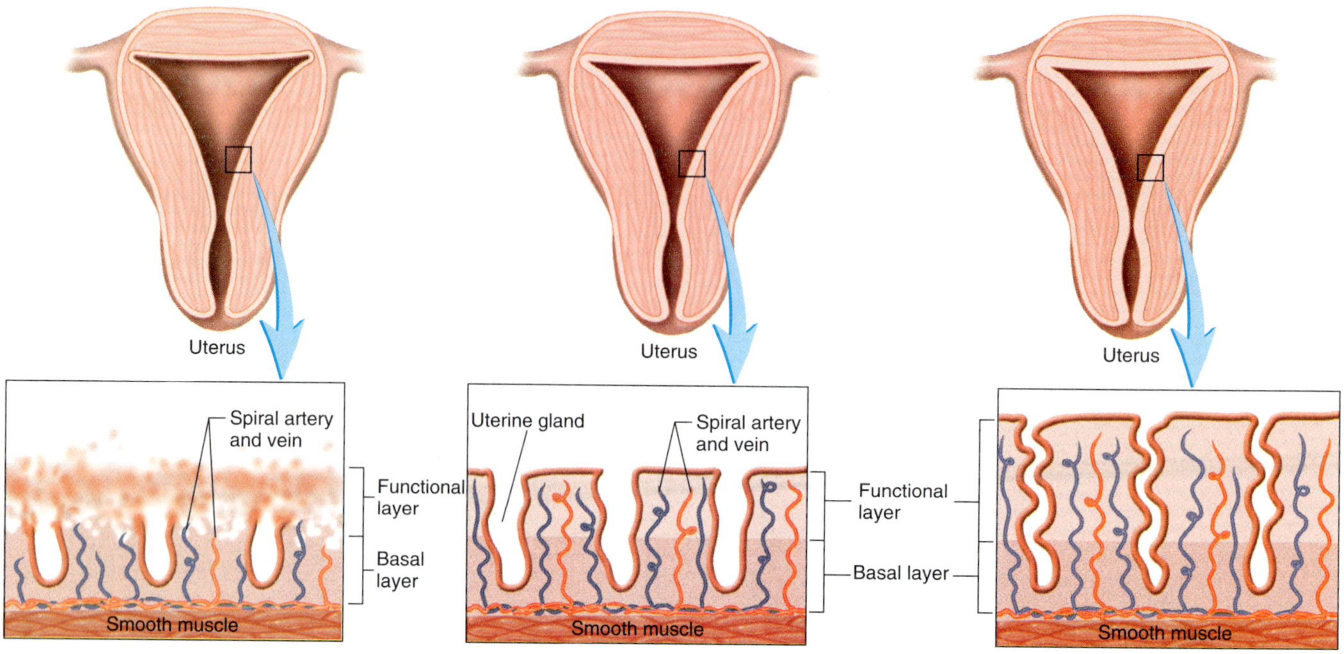

The Uterine Cycle
Process Figure 28.20

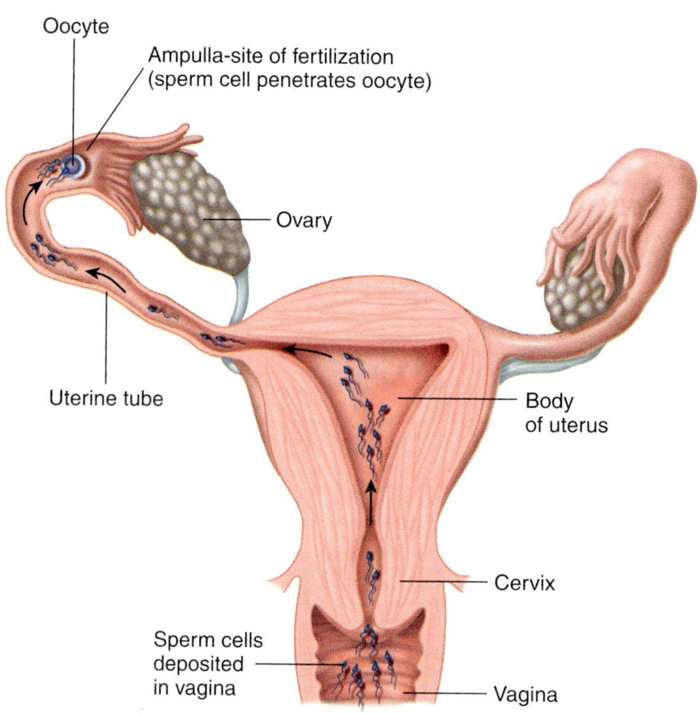

Sperm Cell Movement
Figure 28.21

1. Human chorionic gonadotropin (HCG) increases until it reaches a maximum concentration near the end of the first 3 months of pregnancy and then decreases to a low level thereafter.
2. Progesterone continues to increase until it levels off near the end of pregnancy. Early in pregnancy, progesterone is produced by the corpus luteum in the ovary, later production shifts to the placenta.
3. Estrogen levels increase slowly throughout pregnancy, but they increase more rapidly as the end of pregnancy approaches. Early in pregnancy, estrogen is produced only in the ovary; later production shifts to the placenta.

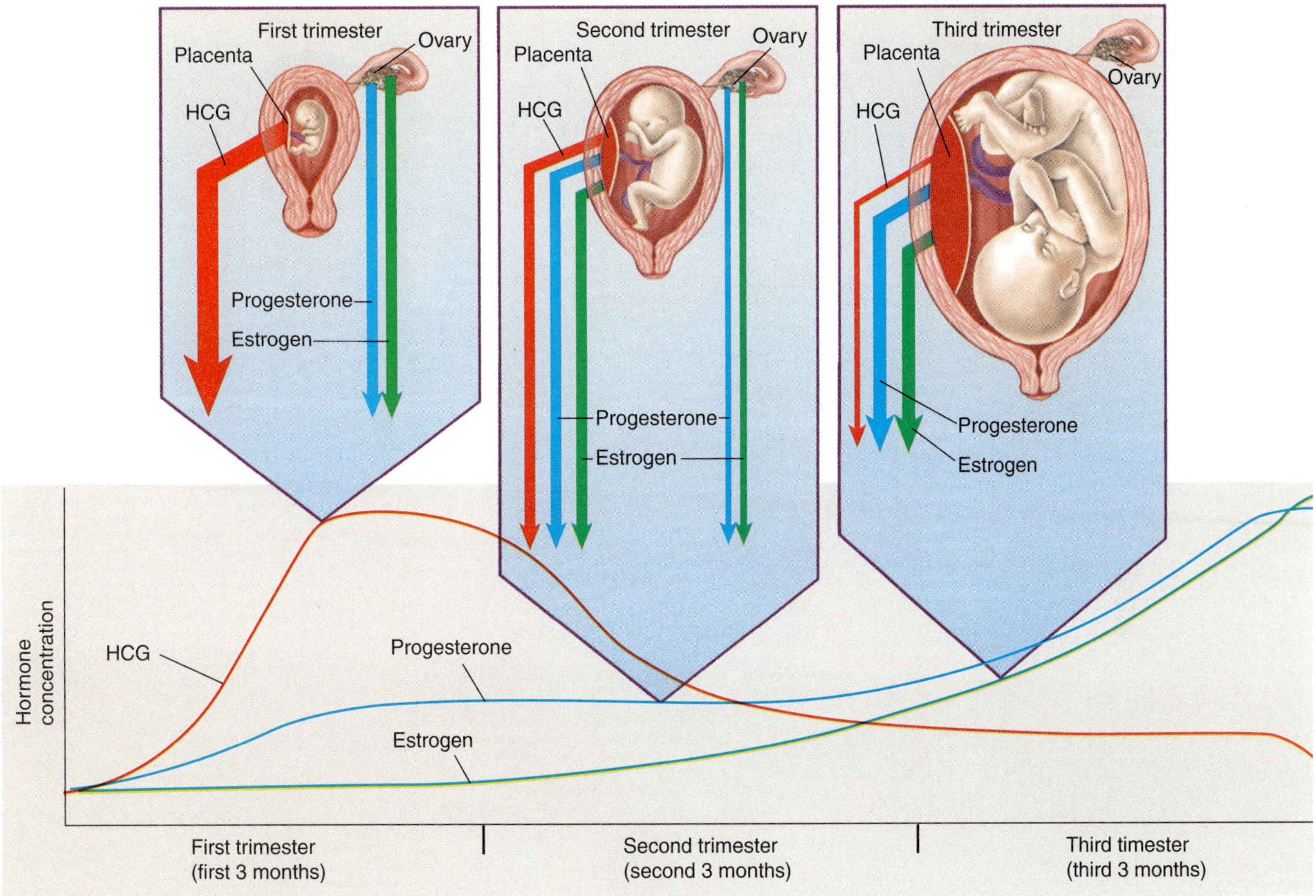

Changes in Hormone Concentration During Pregnancy
Process Figure 28.22

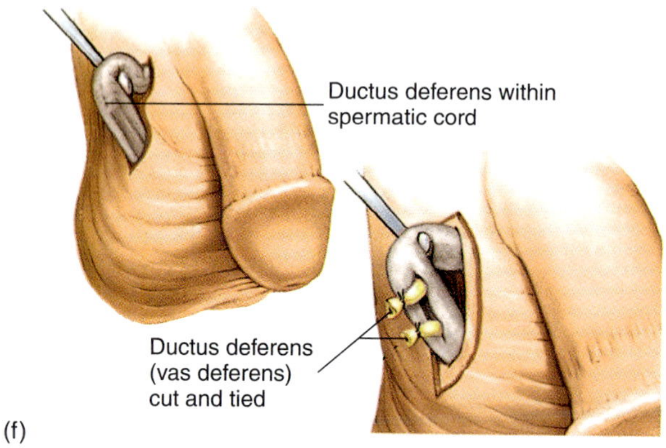

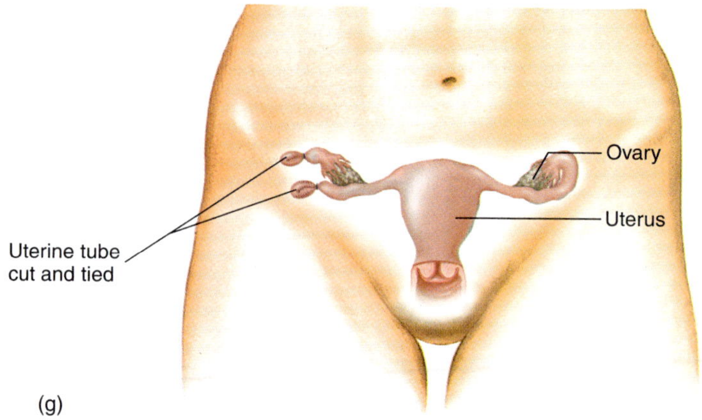

Contraceptive Devices and Techniques
Figure B

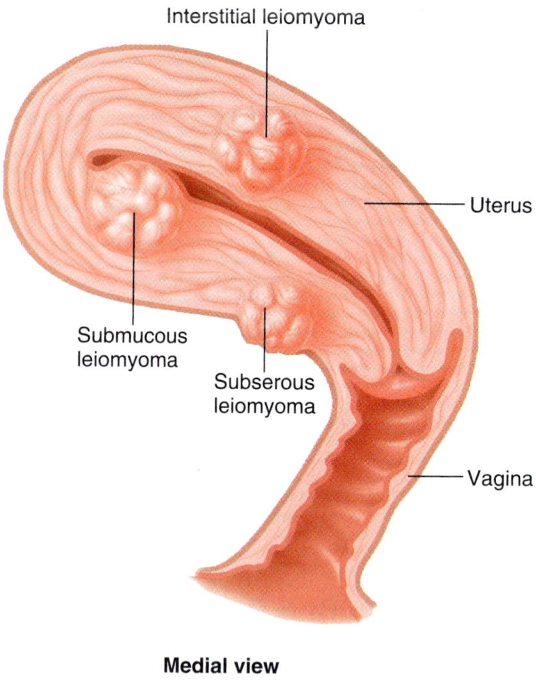

Leiomyomas or Fibroid Tumors
Figure C

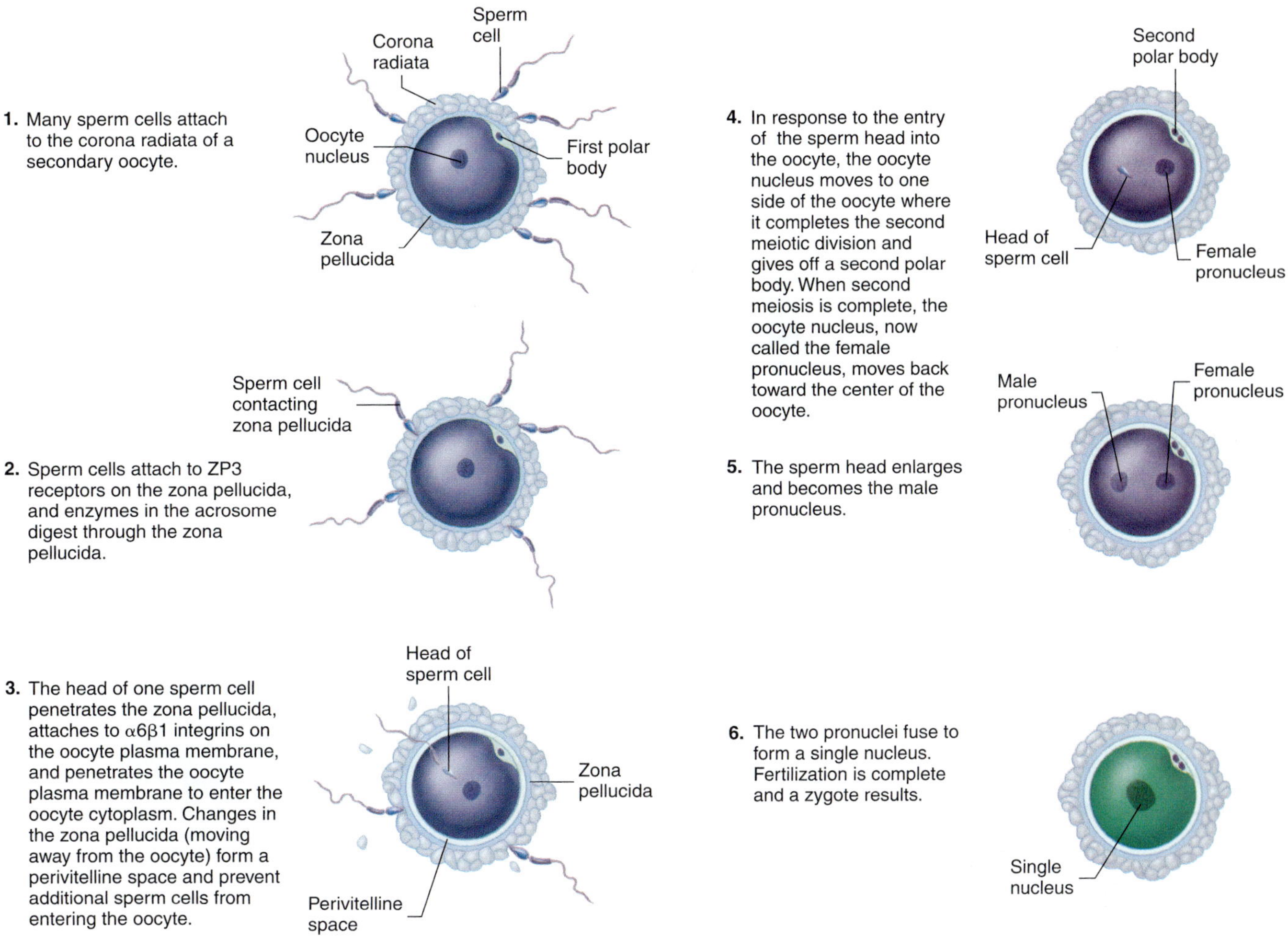
1. Many sperm cells attach to the corona radiata of a secondary oocyte.
Sperm cell
Corona radiata
Oocyte nucleus
First polar body
Zona pellucida
2. Sperm cells attach to ZP3 receptors on the zona pellucida, and enzymes in the acrosome digest through the zona pellucida.
Sperm cell contacting zona pellucida
3. The head of one sperm cell penetrates the zona pellucida, attaches to α6β1 integrins on the oocyte plasma membrane, and penetrates the oocyte plasma membrane to enter the oocyte cytoplasm. Changes in the zona pellucida (moving away from the oocyte) form a perivitelline space and prevent additional sperm cells from entering the oocyte.
Head of sperm cell
Zona pellucida
Perivitelline space
4. In response to the entry of the sperm head into the oocyte, the oocyte nucleus moves to one side of the oocyte where it completes the second meiotic division and gives off a second polar body. When second meiosis is complete, the oocyte nucleus, now called the female pronucleus, moves back toward the center of the oocyte.
Second polar body
Head of sperm cell
Female pronucleus
5. The sperm head enlarges and becomes the male pronucleus.
Male pronucleus
Female pronucleus
6. The two pronuclei fuse to form a single nucleus. Fertilization is complete and a zygote results.
Single nucleus

Fertilization
Process Figure 29.1

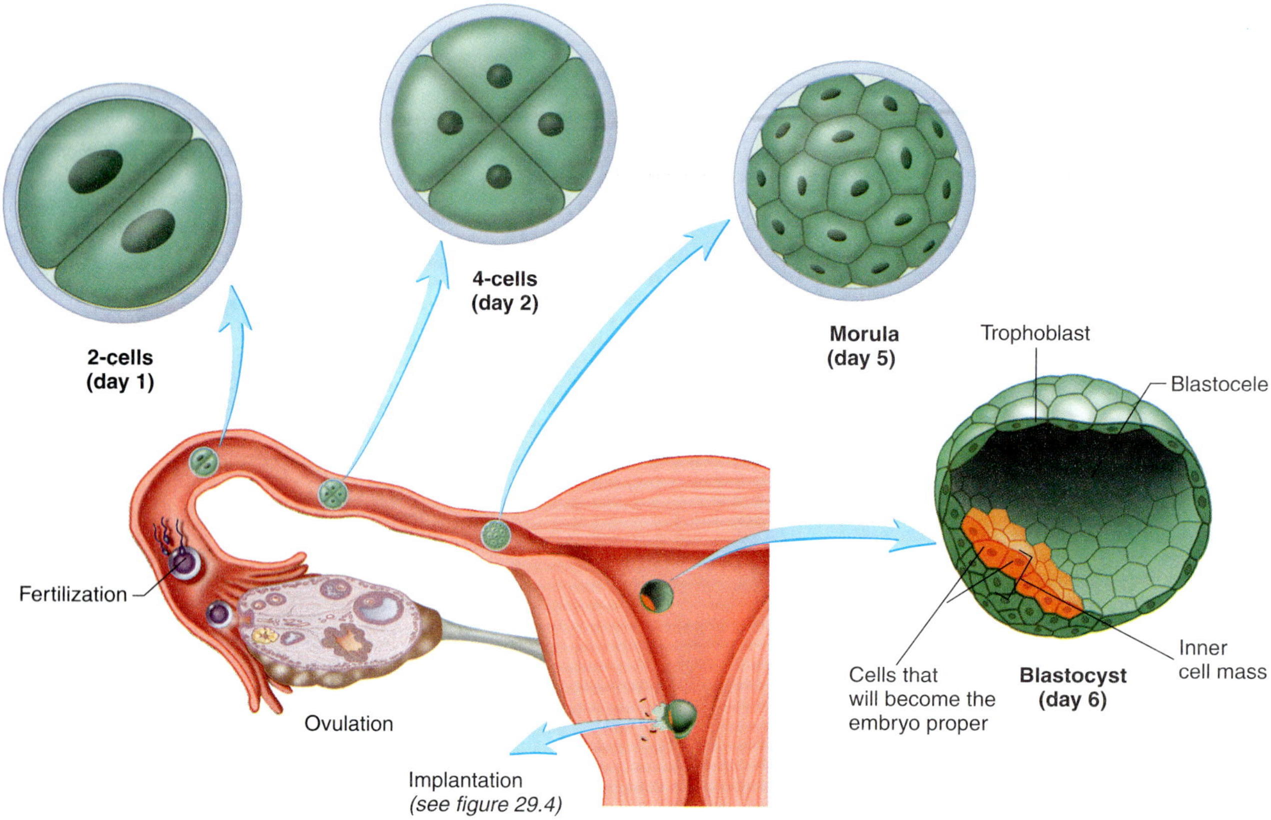

2-cells
(day 1)
4-cells
(day 2)
Morula
(day 5)
Trophoblast
Blastocele
Fertilization
Ovulation
Implantation
(see figure 29.4)
Cells that
will become the
embryo proper
Blastocyst
(day 6)
Inner
cell mass

Blastocyst
Figure 29.3

1. Frontal section of the uterus and uterine tube showing development 7 days after fertilization.

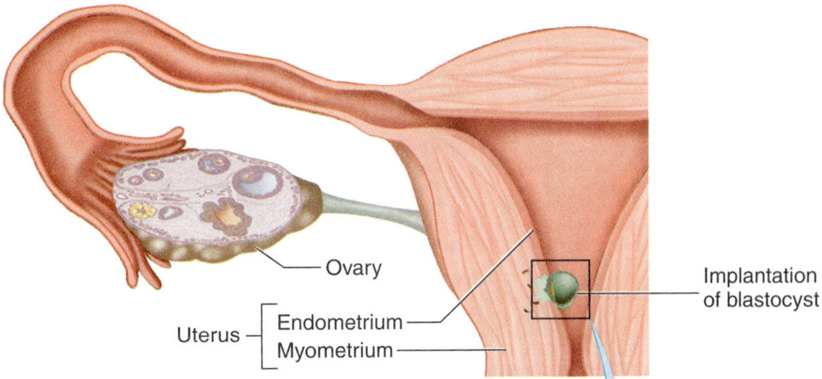

2. Implantation of the blastocyst with syncytiotrophoblast beginning to invade the uterine wall (at about 8–12 days).

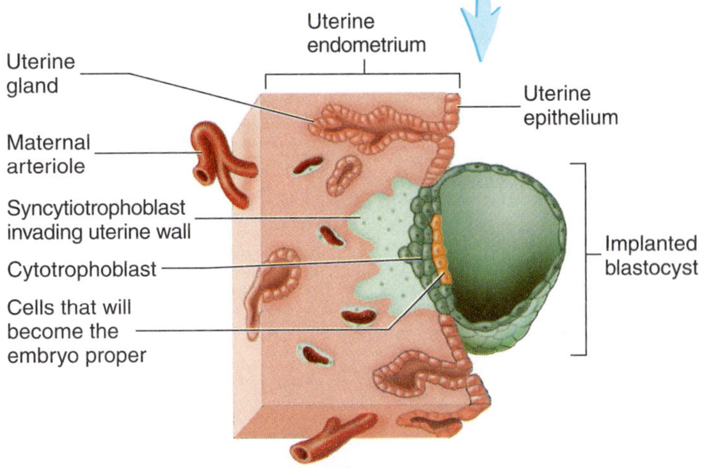

3. Intermediate stage of placental formation (at about 14–20 days). As maternal blood vessels are encountered by the syncytiotrophoblast, lacunae are formed and filled with maternal blood.

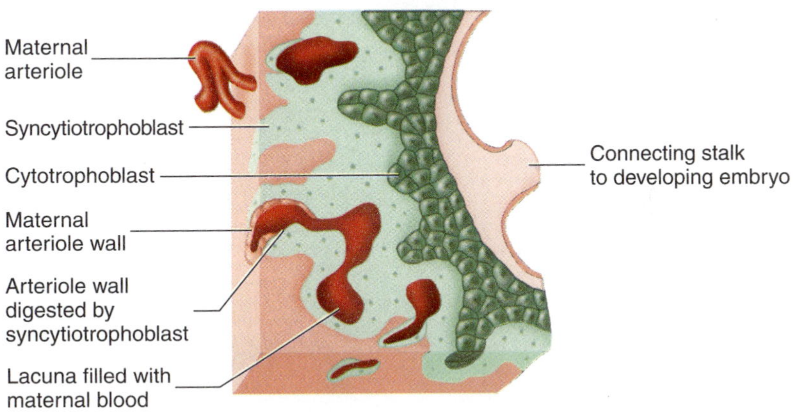

4. Cytotrophoblast cords surround the syncytiotrophoblast and lacunae, and embryonic blood vessels enter the cord (at about 1 month).

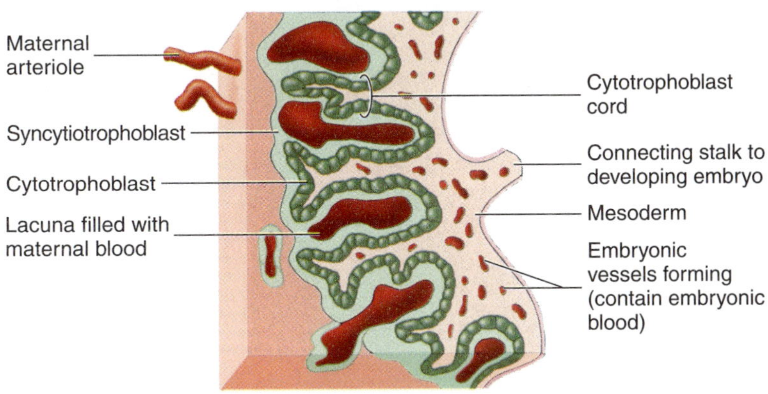

Formation of the Placenta
Process Figure 29.4

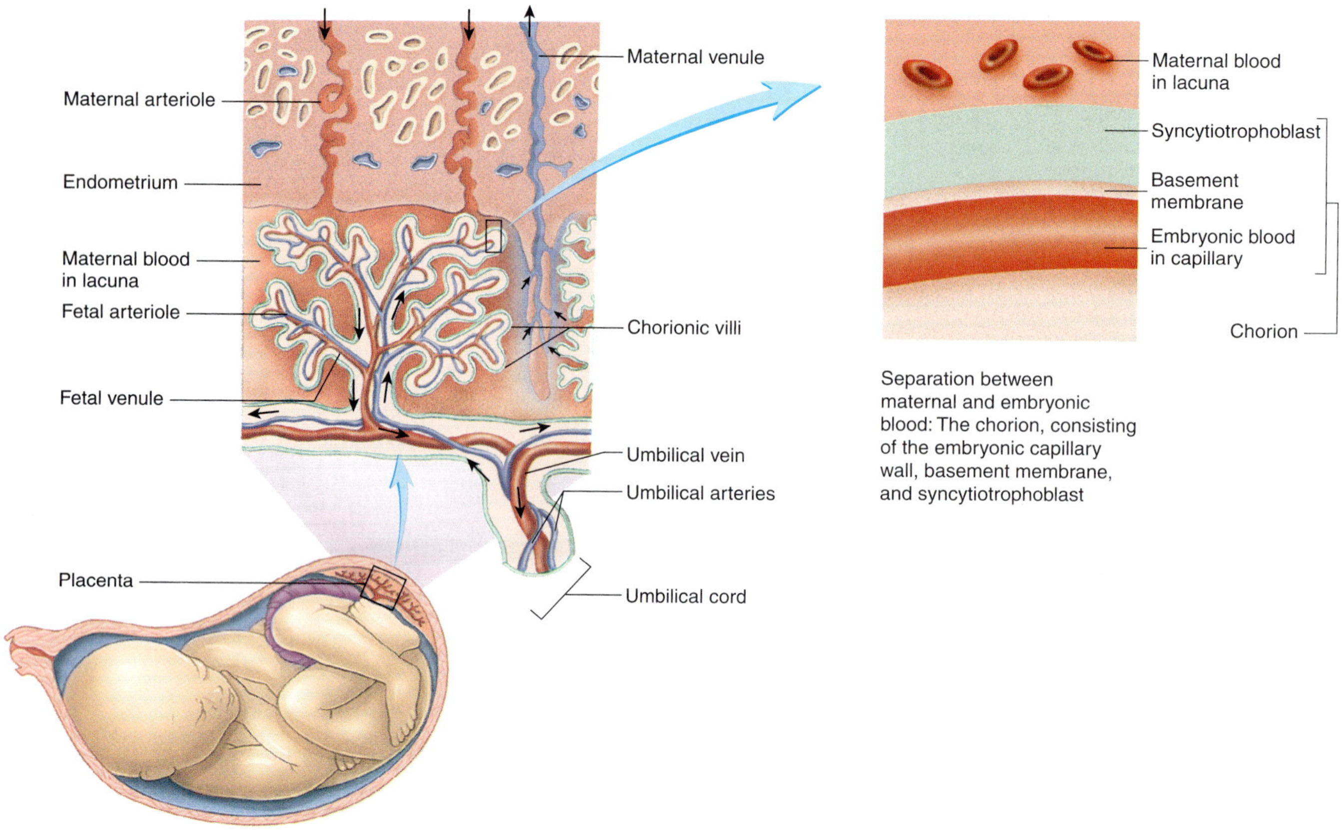

Mature Placenta and Fetus
Figure 29.5

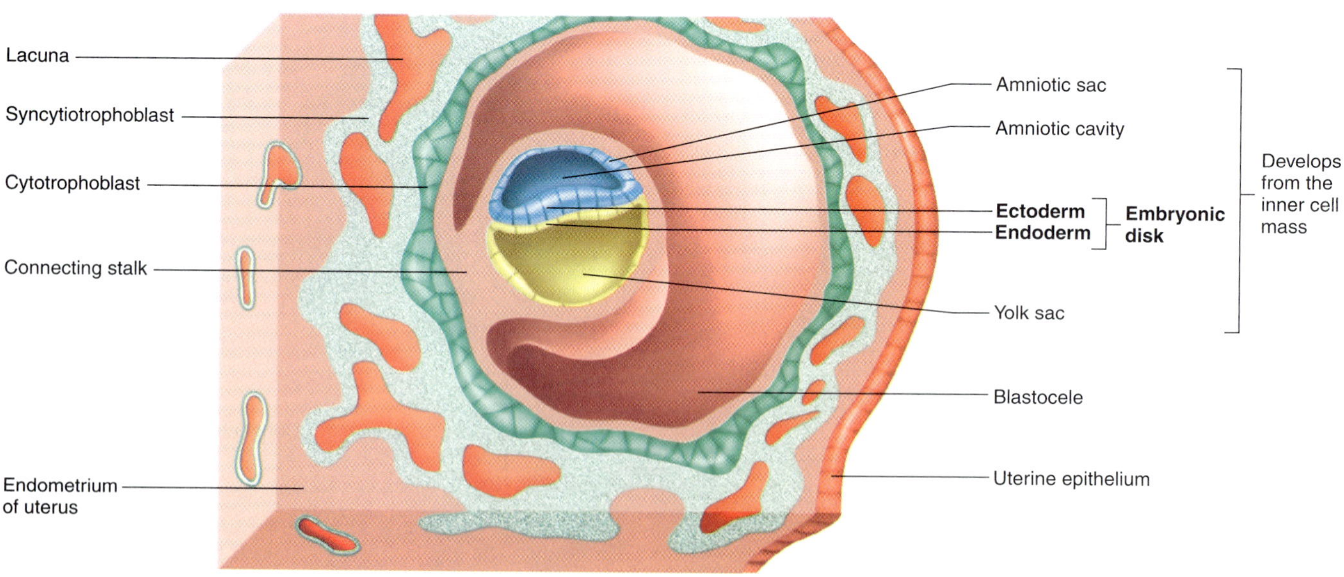

Embryonic Disk
Figure 29.6

1. Cells in the surface ectoderm move toward the primitive streak and migrate through the streak (*blue arrow tails*).

2. Cells of the ectoderm that migrate through the primitive streak become mesodermal cells (*red arrows*).

3. The mesoderm (*red*) lies between the ectoderm (*blue*) and endoderm (*yellow*).

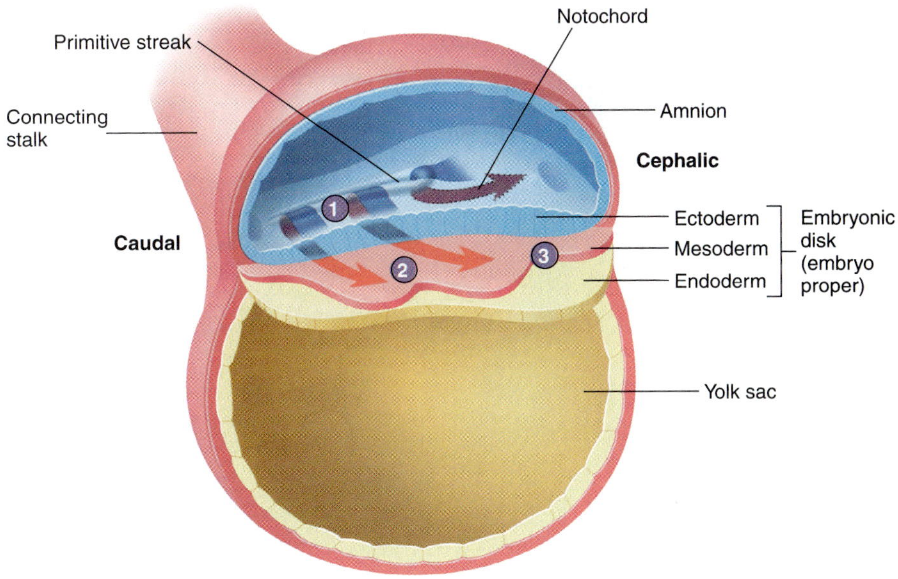

Primitive Streak
Process Figure 29.7

1. The neural plate is formed from ectoderm.

2. Neural folds form as parallel ridges along the embryo.

3. Neural crest cells begin to form from the crest of the neural folds.

4. The neural folds meet at the midline to form the neural tube and neural crest cells separate from the neural folds.

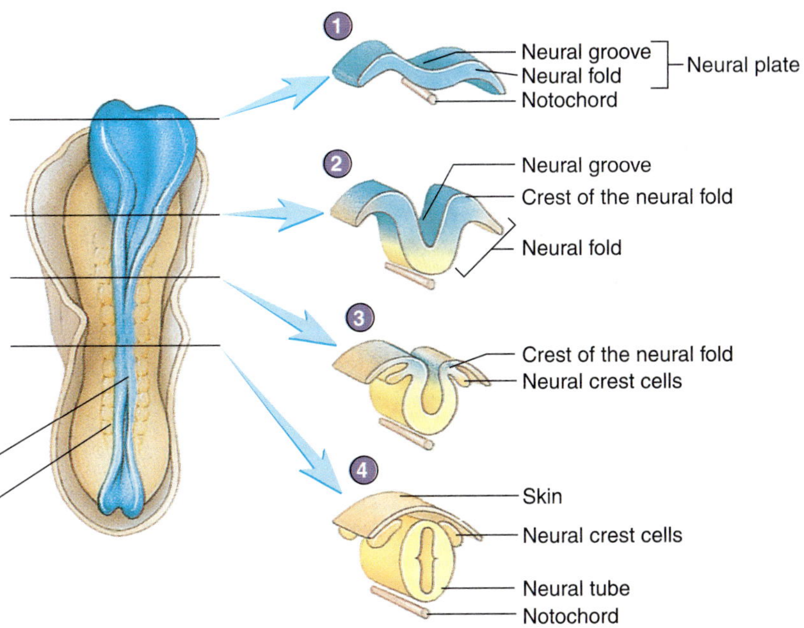

Formation of the Neural Tube
Process Figure 29.8

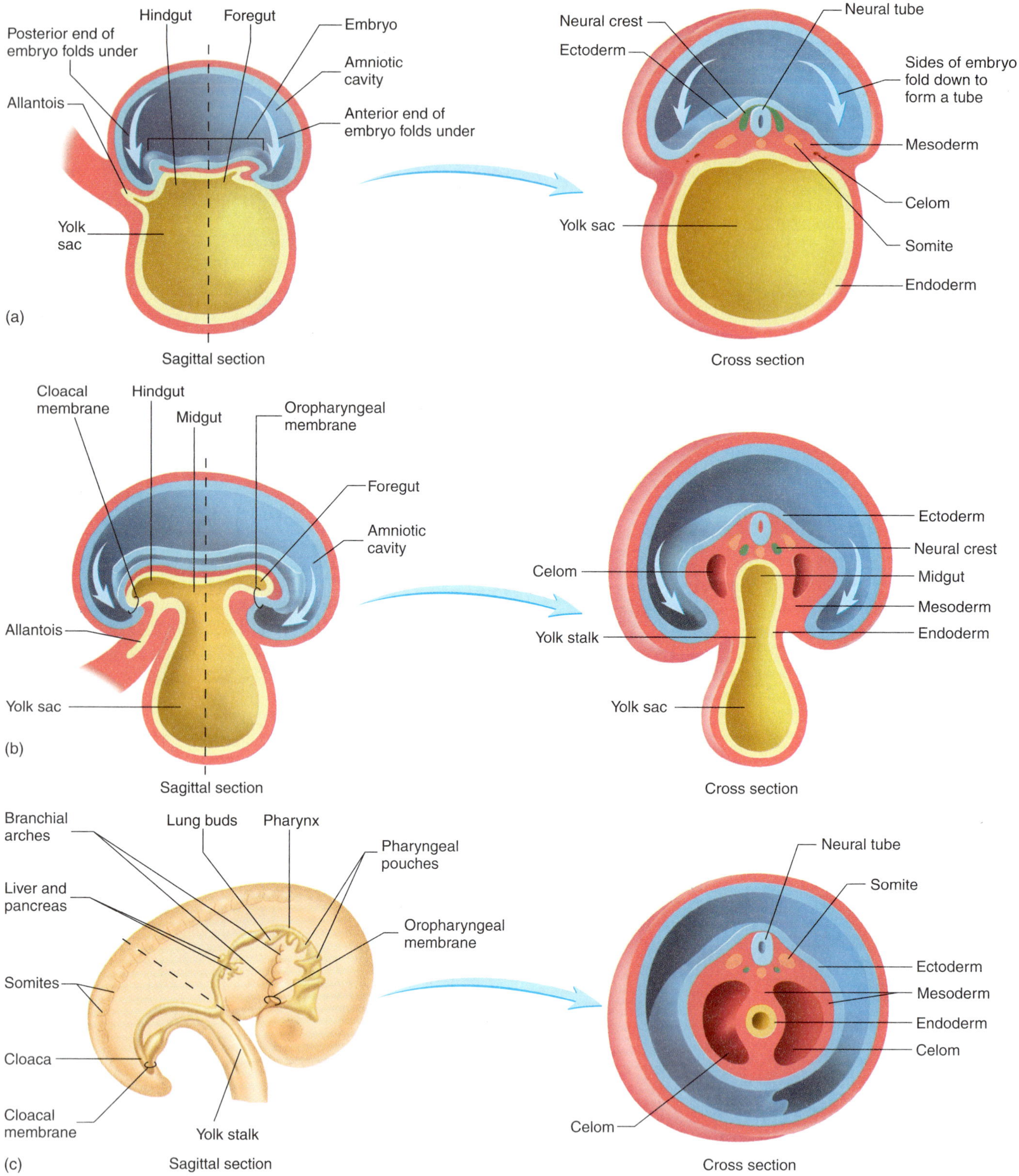

Hindgut
Foregut
Posterior end of embryo folds under
Embryo
Amniotic cavity
Allantois
Anterior end of embryo folds under
Yolk sac
(a)
Sagittal section
Neural crest
Ectoderm
Neural tube
Sides of embryo fold down to form a tube
Mesoderm
Celom
Yolk sac
Somite
Endoderm
Cross section
Cloacal membrane
Hindgut
Midgut
Oropharyngeal membrane
Foregut
Amniotic cavity
Allantois
Yolk sac
(b)
Sagittal section
Ectoderm
Neural crest
Celom
Midgut
Mesoderm
Yolk stalk
Endoderm
Yolk sac
Cross section
Branchial arches
Lung buds
Pharynx
Pharyngeal pouches
Liver and pancreas
Oropharyngeal membrane
Somites
Cloaca
Cloacal membrane
Yolk stalk
(c)
Sagittal section
Neural tube
Somite
Ectoderm
Mesoderm
Endoderm
Celom
Celom
Cross section

Formation of the Digestive Tract
Figure 29.9

1. 28 days after fertilization
The face develops from five processes: frontonasal (*blue*), two maxillary (*yellow*), and two mandibular (*orange*; already fused).

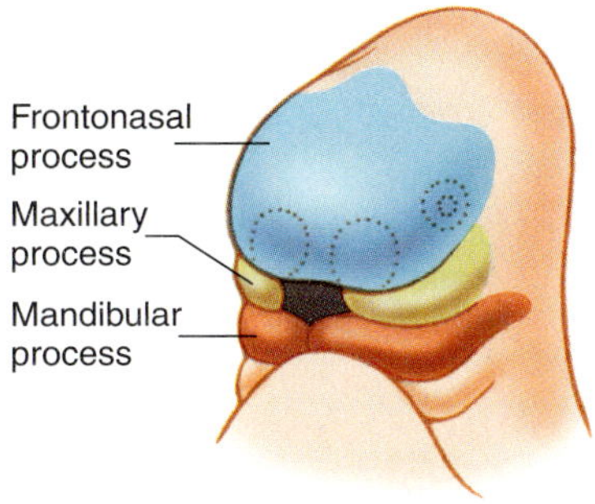

2. 33 days after fertilization
Nasal placodes appear in the frontonasal process.

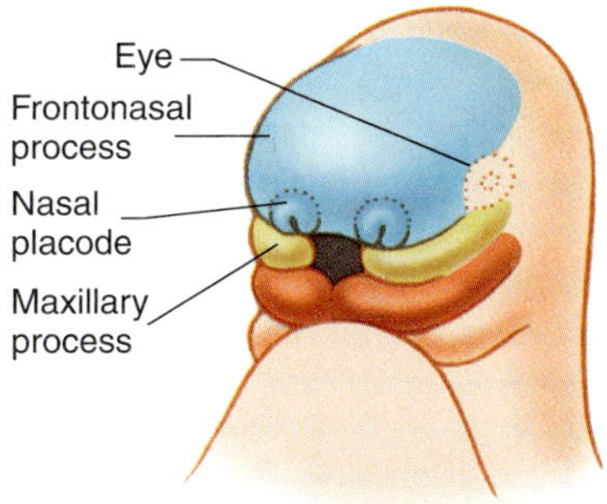

3. 40 days after fertilization
Maxillary processes extend toward the midline. The nasal placodes also move toward the midline and fuse with the maxillary processes to form the jaw and lip.

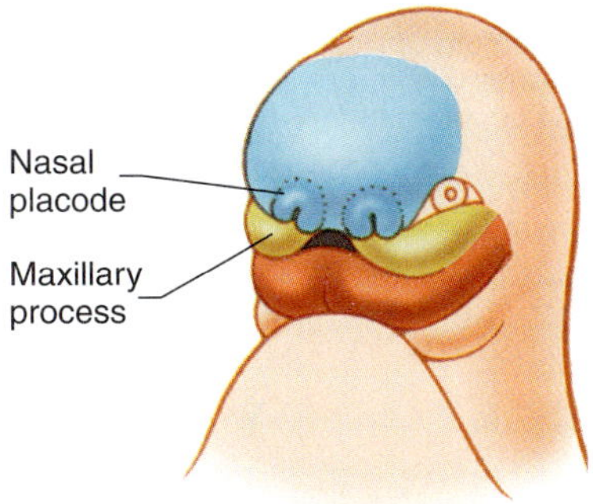

4. 48 days after fertilization
Continued growth brings structures more toward the midline.

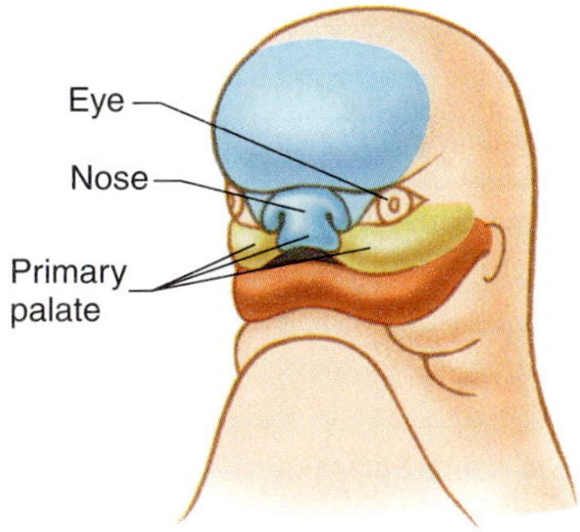

5. 14 weeks after fertilization
Colors show the contributions of each process to the adult face.

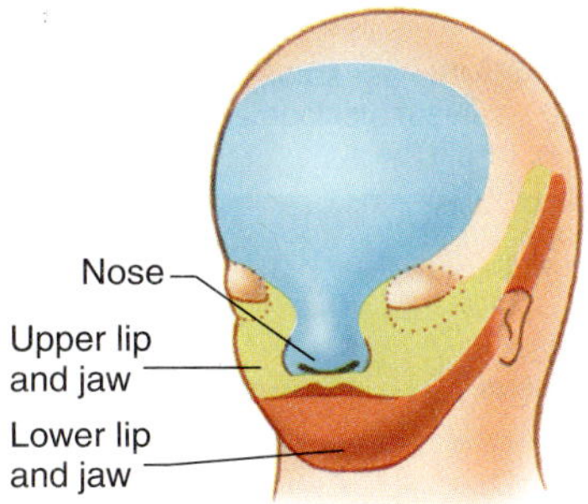

Development of the Face
Process Figure 29.11

1. **20 days after fertilization**
   At this age, the heart consists of two parallel tubes.

2. **22 days after fertilization**
   The two parallel tubes have fused to form one tube. This tube bends as it elongates (*blue arrows* suggest the direction of bending) within the confined space of the pericardium.

3. **31 days after fertilization**
   The interatrial septum (septum primum, *green*) and the interventricular septum grow toward the center of the heart.

4. **35 days after fertilization**
   The interventricular septum is nearly complete. A foramen opens in the septum primum (*green*) as the septum secundum begins to form (*blue*).

5. **The final embryonic condition of the interatrial septum**
   Blood from the right atrium can flow through the foramen ovale into the left atrium. After birth, as blood begins to flow in the other direction, the left side of the interatrial septum is forced against the right side, closing the foramen ovale.

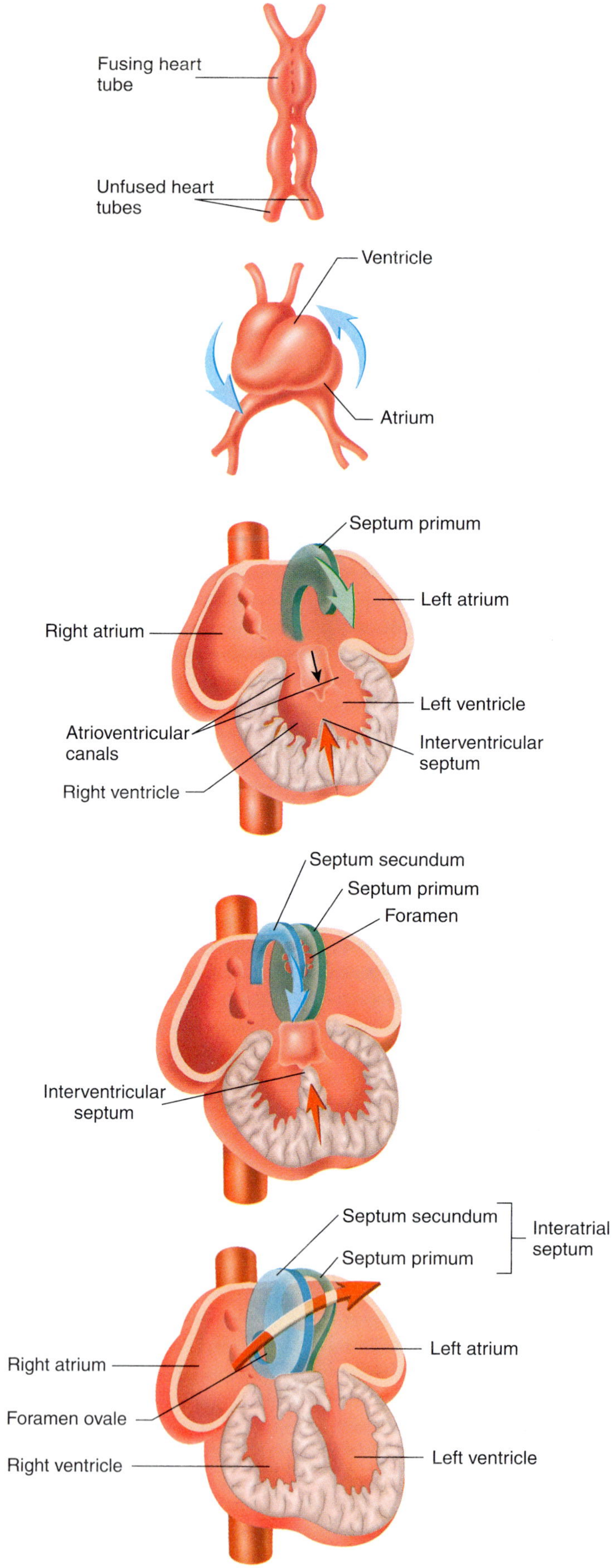

Development of the Heart
Process Figure 29.12

1. 28 days after fertilization
A single bud forms and divides into two buds, which will become the lungs and primary bronchi.

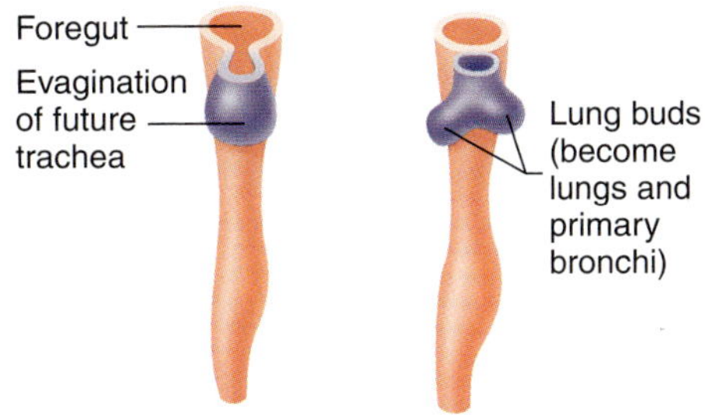

2. 32 days after fertilization
Primary bronchi branch to form secondary bronchi, which supply the lung lobes.

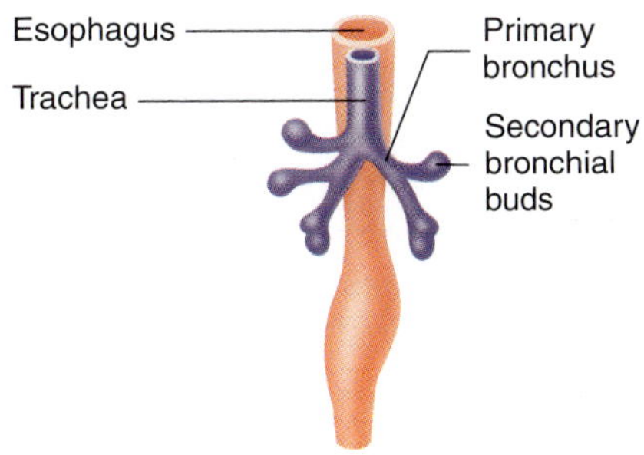

3. 35 days after fertilization
Secondary bronchi branch to form tertiary bronchi, which supply the bronchopulmonary segments.

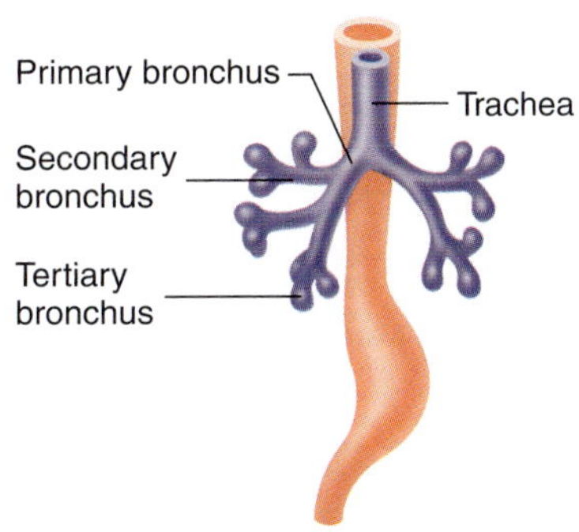

4. 50 days after fertilization
The branching will continue to eventually form the extensive respiratory passages in the lungs.

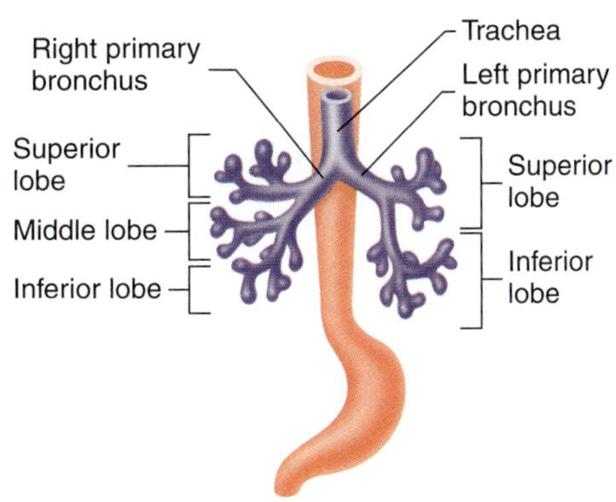

Development of the Lung
Process Figure 29.13

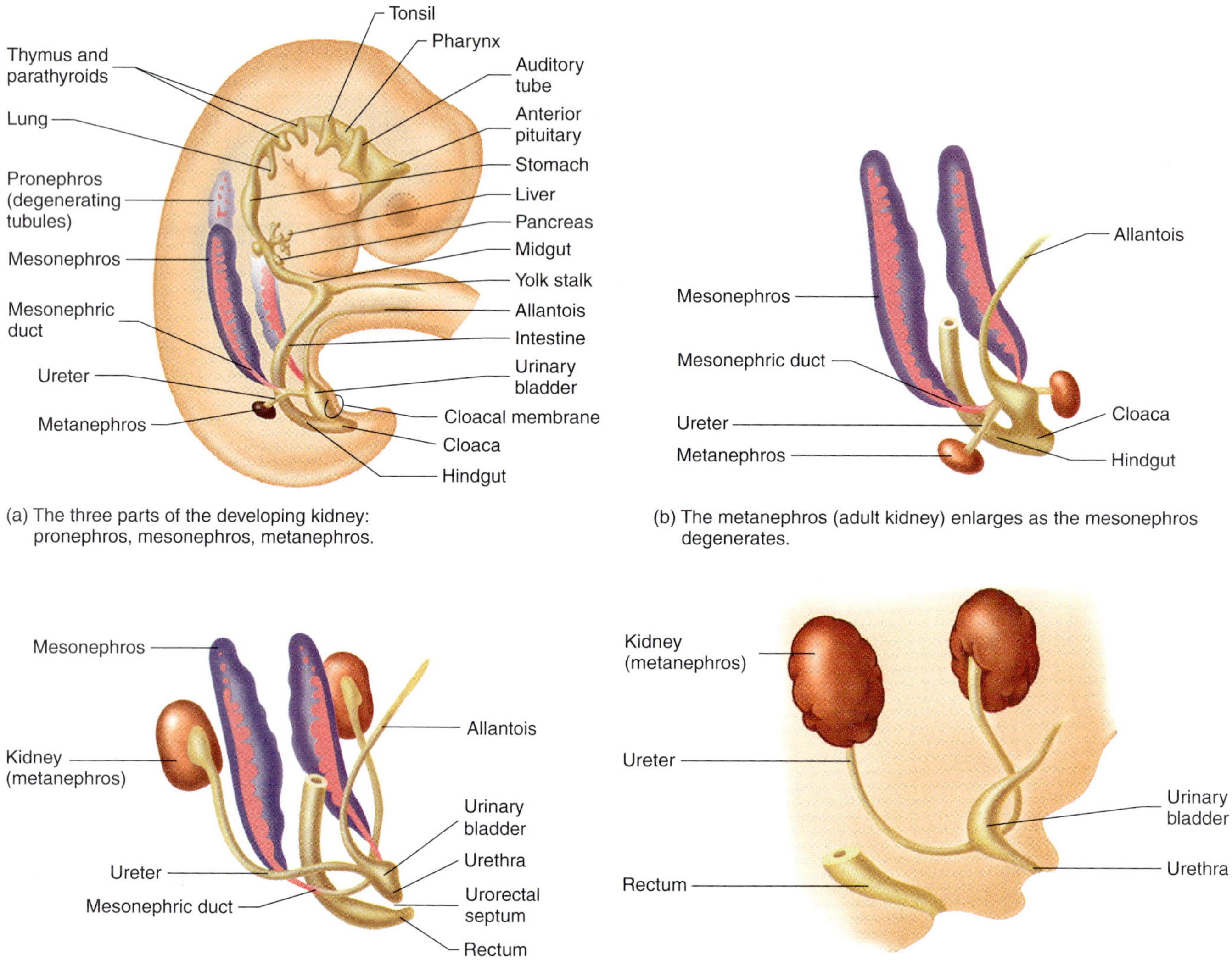

(a) The three parts of the developing kidney: pronephros, mesonephros, metanephros.

(b) The metanephros (adult kidney) enlarges as the mesonephros degenerates.

(c) The kidney continues to grow and develop.

(d) Final duct relationships.

Development of the Kidney and Urinary Bladder
Figure 29.14

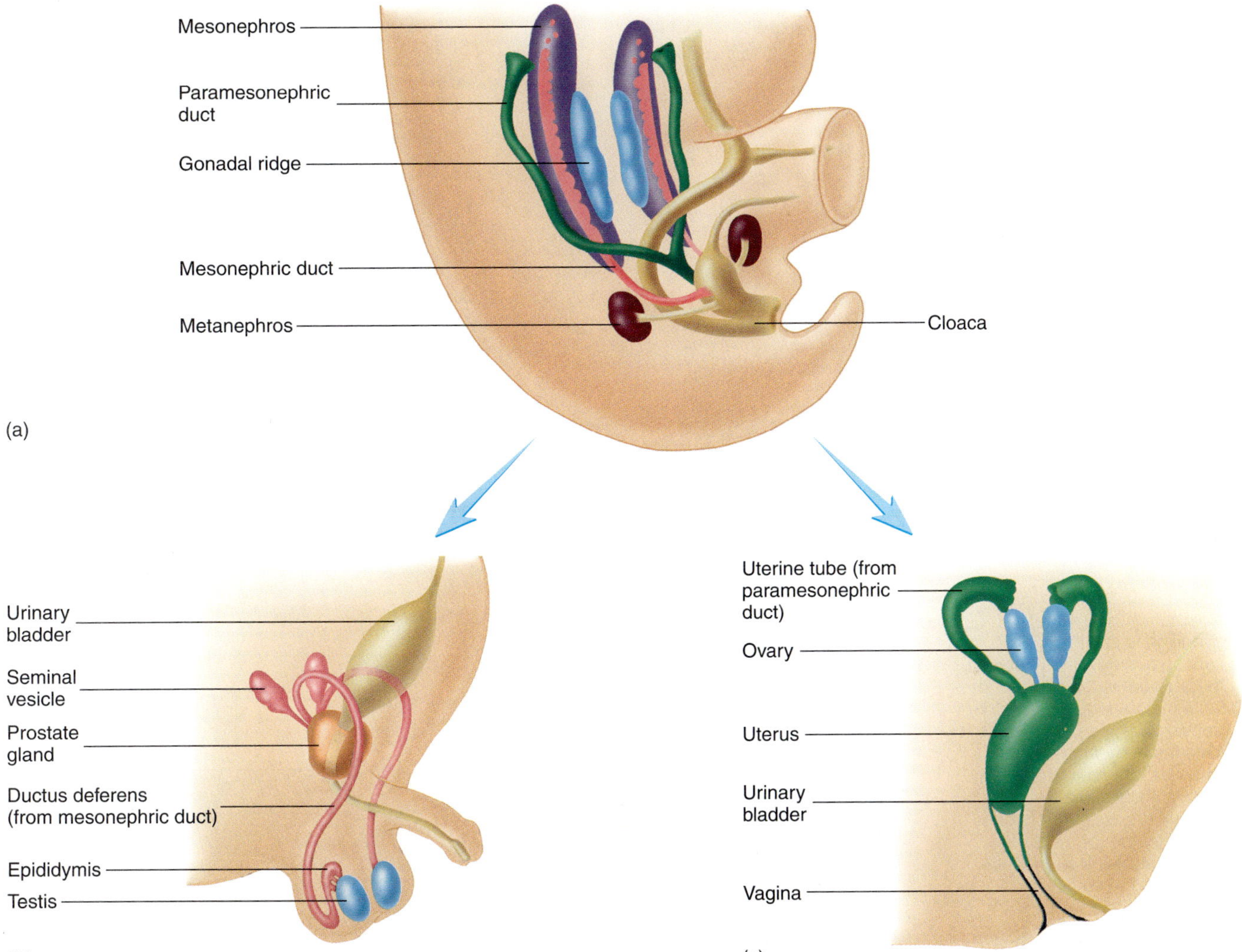

Development of the Reproductive System
Figure 29.15

1. At approximately 5 weeks after fertilization, the genital tubercle, genital folds, and labioscrotal swellings are the same for male and female.

2. By 10 weeks of development, the male penile structures are somewhat larger than those of the female clitoris. The urethral groove is still open in male and female.

3. Near term, the general adult condition is achieved. The labioscrotal swellings become the scrotum in the male, and labia majora in the female. The genital folds fuse in the male to form the body of the penis, whereas they form the labia minora in the female. The urethral groove in the male is closed by the raphe of the penis. It remains open in the female as the vestibule.

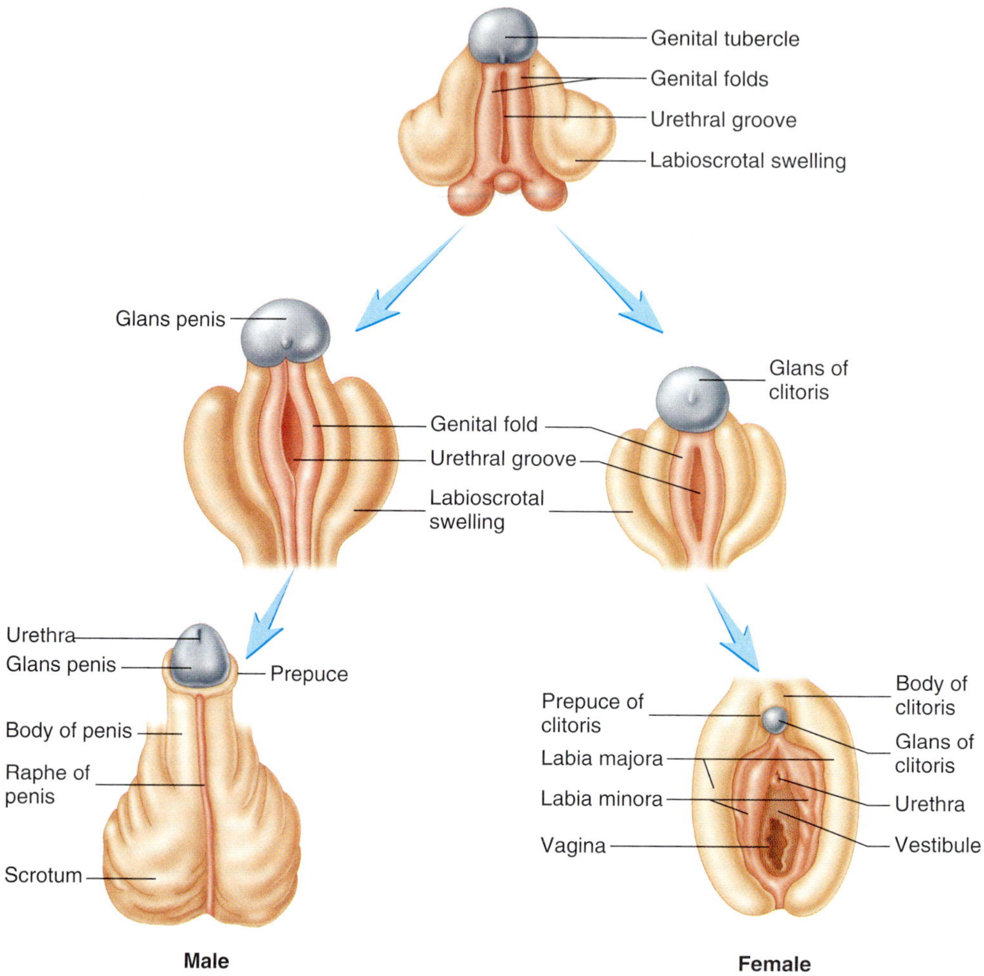

Development of the External Genitalia
Process Figure 29.16

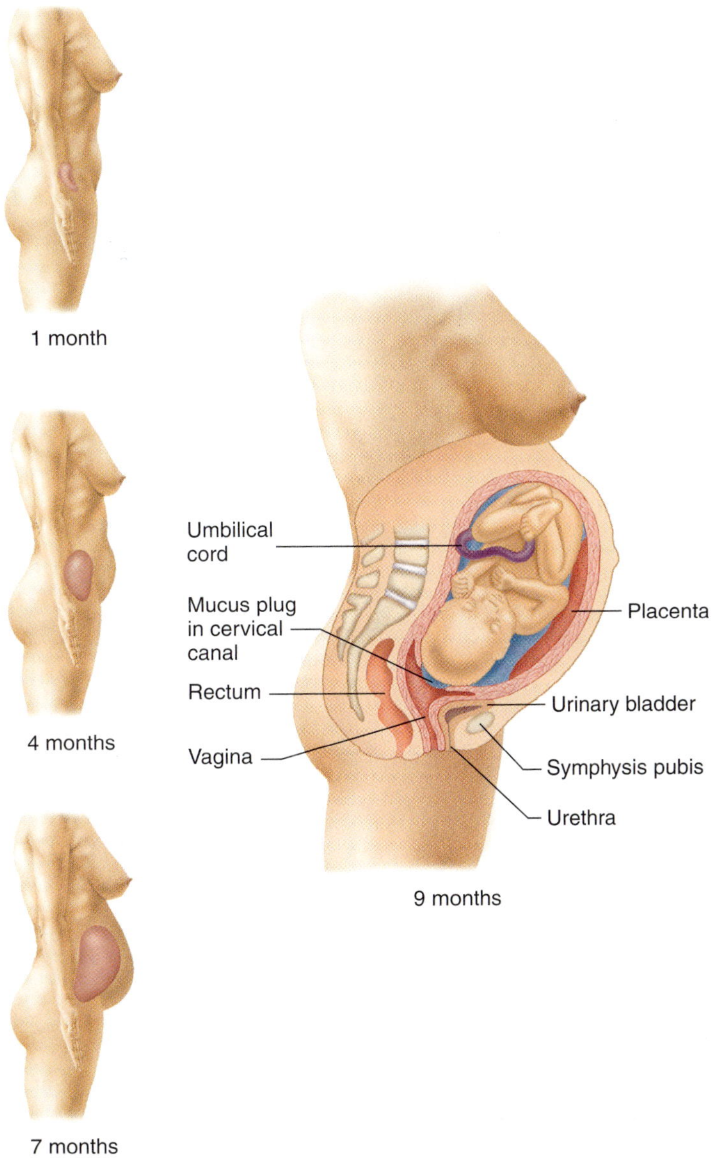

Enlargement of the Uterus During Fetal Development
Figure 29.18

Removal of Amniotic Fluid for Amniocentesis
Figure B

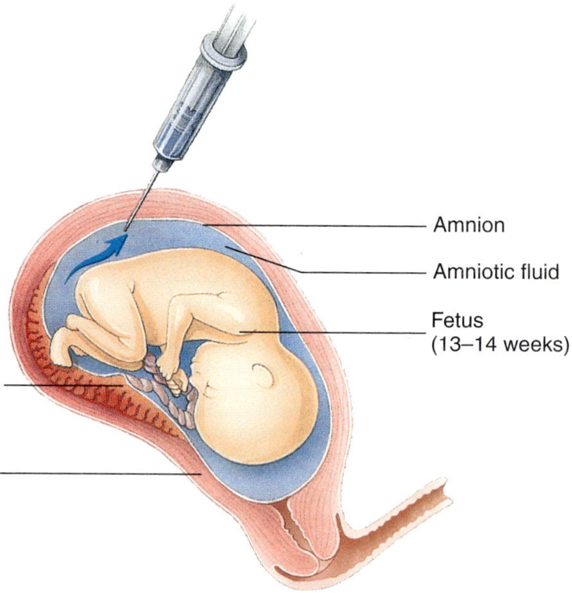

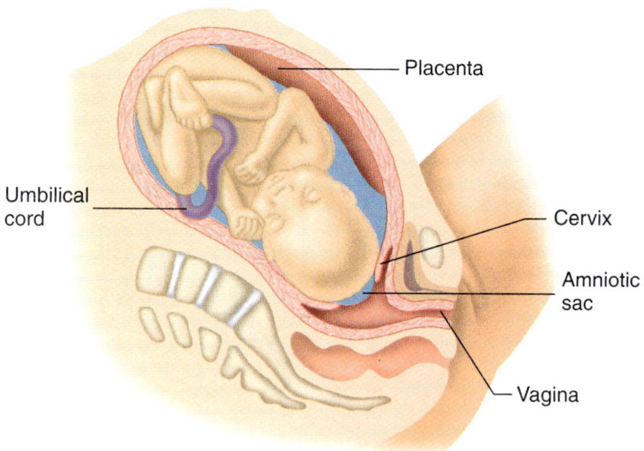

1. First stage: the cervix begins to dilate.

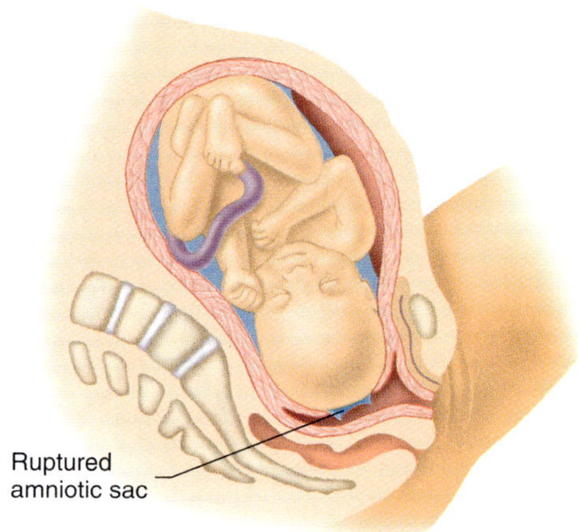

2. Second stage: further dilation of the cervix and rupture of the amniotic sac occur.

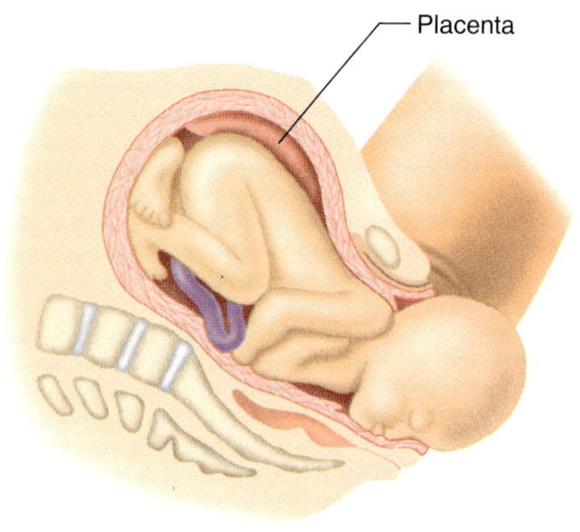

3. Third stage: the fetus is expelled from the uterus.

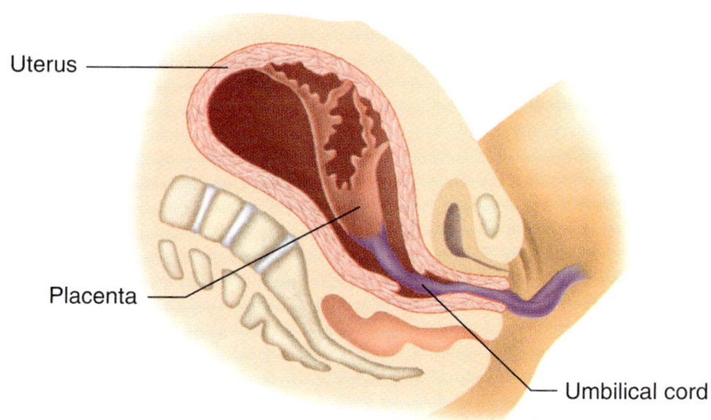

4. Fourth stage: the placenta is expelled.

Parturition
Process Figure 29.19

1. The fetal hypothalamus secretes a releasing hormone, corticotropin-releasing hormone (CRH), which stimulates adrenocorticotropic hormone (ACTH) secretion from the pituitary. The fetal pituitary secretes ACTH in greater amounts near parturition.

2. ACTH causes the fetal adrenal gland to secrete greater quantities of adrenal cortical steroids.

3. Adrenal cortical steroids travel in the umbilical blood to the placenta.

4. In the placenta the adrenal cortical steroids cause progesterone synthesis to level off and estrogen and prostaglandin synthesis to increase, making the uterus more irritable.

5. The stretching of the uterus produces action potentials that are transmitted to the brain through ascending pathways.

6. Action potentials stimulate the secretion of oxytocin from the posterior pituitary.

7. Oxytocin causes the uterine smooth muscle to contract.

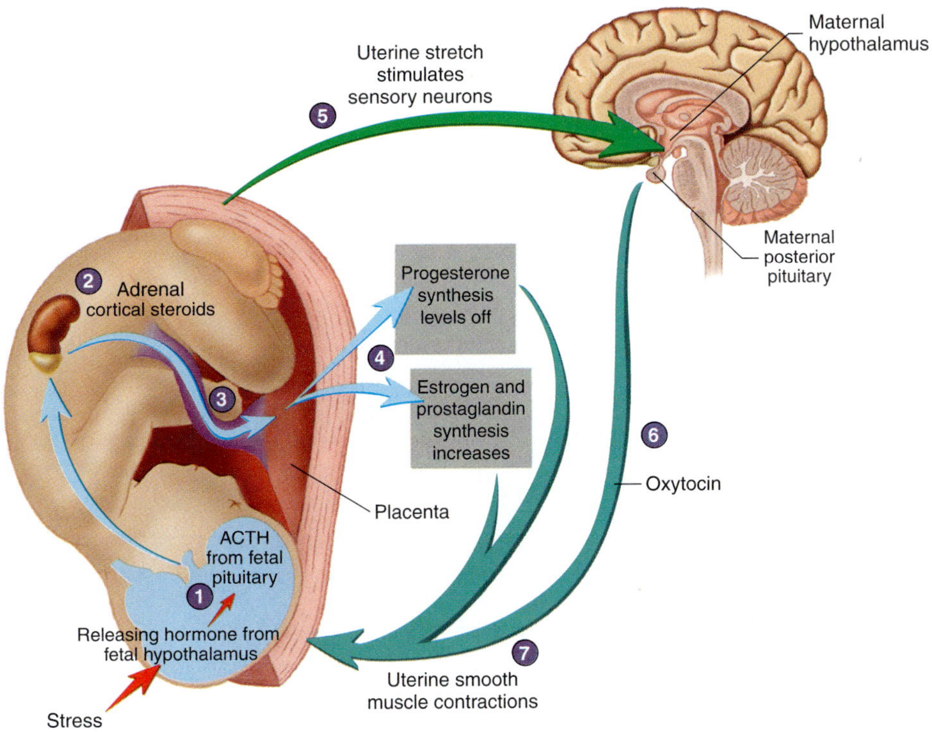

Factors That Influence the Process of Parturition
Process Figure 29.20

1. Blood bypasses the lungs by flowing from the pulmonary trunk through the ductus arteriosus to the aorta.

2. Blood also bypasses the lungs by flowing from the right to the left atrium through the foramen ovale.

3. Blood bypasses the liver sinusoids by flowing through the ductus venosus.

4. Oxygen-rich blood is returned to the fetus from the placenta by the umbilical vein.

5. Oxygen-poor blood is carried from the fetus to the placenta through the umbilical arteries.

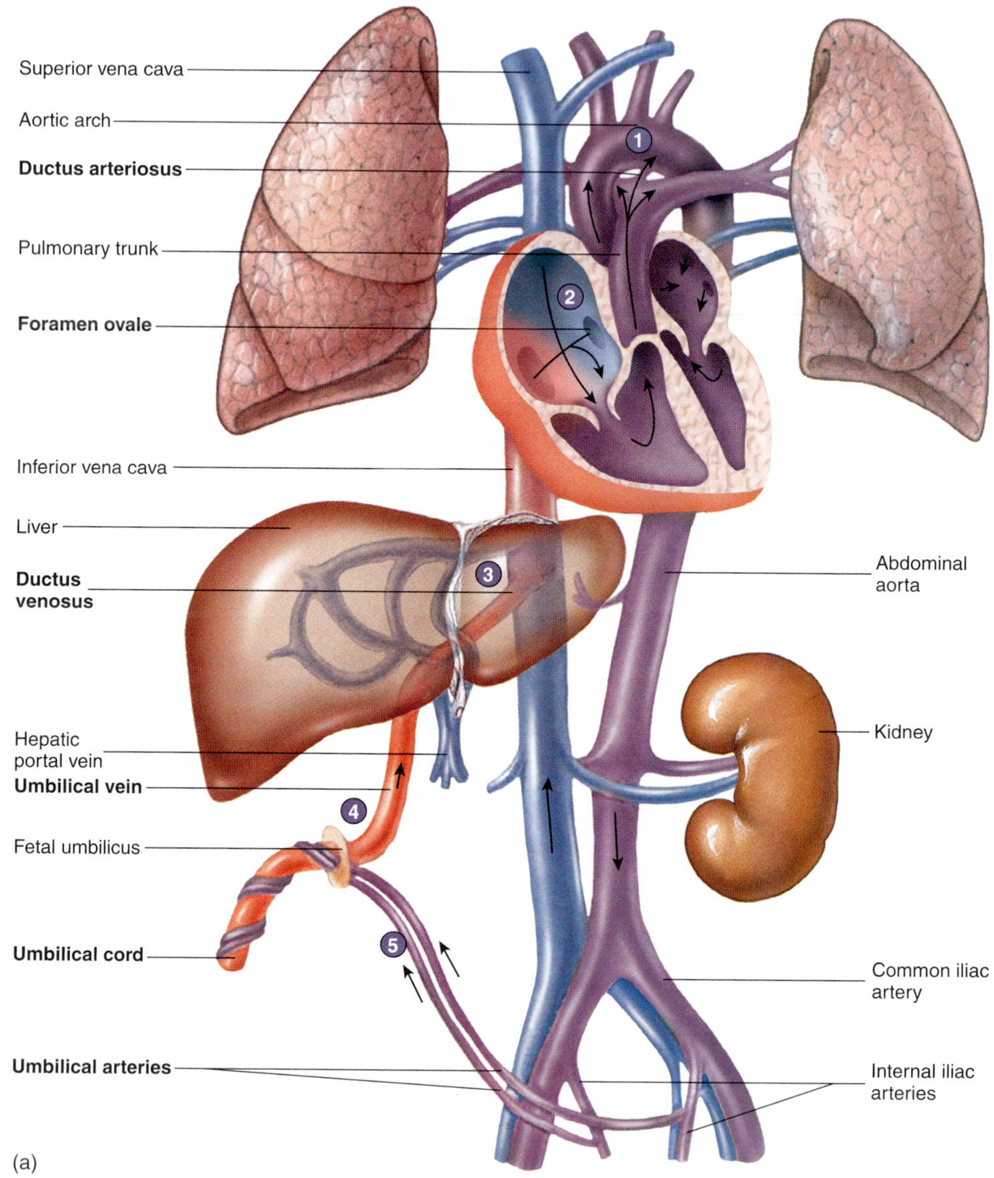

(a)

Circulatory Changes at Birth

Process Figure 29.21

1. When air enters the lungs, blood is forced through the pulmonary arteries to the lungs. The ductus arteriosus closes and becomes the ligamentum arteriosum (*gray*).

2. The foramen ovale closes and becomes the fossa ovalis. Blood can no longer flow from the right to the left atrium.

3. The ductus venosus degenerates and becomes the ligamentum venosum (*gray*).

4. The umbilical arteries and vein are cut. The umbilical vein becomes the round ligament of the liver (*gray*).

5. The umbilical arteries also degenerate and become the cords of the umbilical arteries (*gray*).

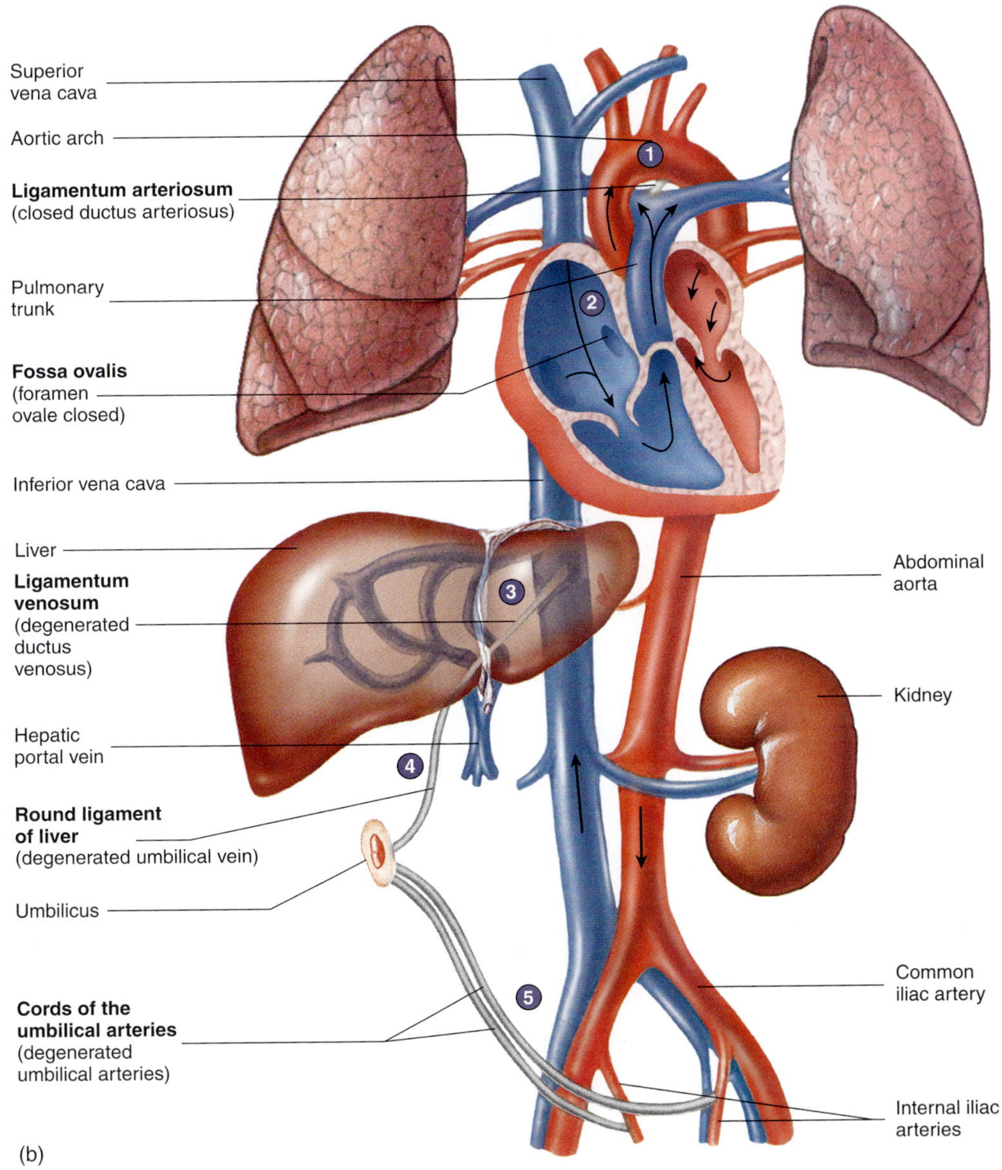

(b)

Circulatory Changes at Birth
Process Figure 29.21 (continued)

1. Stimulation of the nipple by the baby's suckling initiates action potentials in sensory neurons that connect with the hypothalamus.

2. In response, the hypothalamus stimulates the posterior pituitary to release oxytocin and the anterior pituitary to release prolactin.

3. Oxytocin stimulates milk release from the breast. Prolactin stimulates additional milk production.

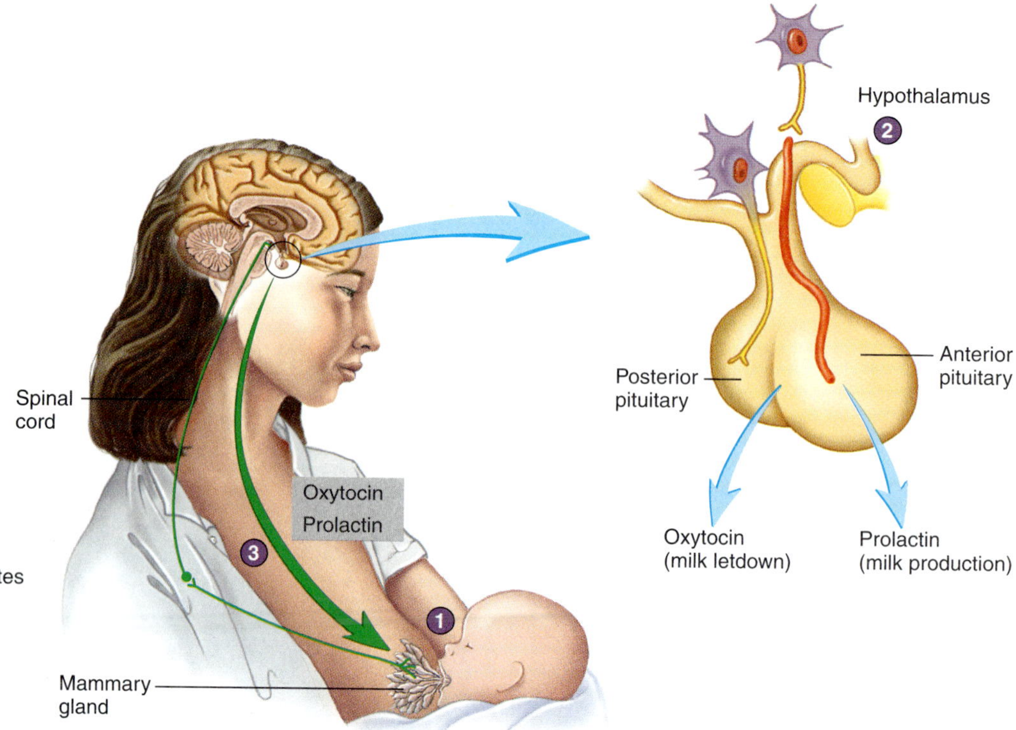

Hormonal Control of Lactation
Process Figure 29.22

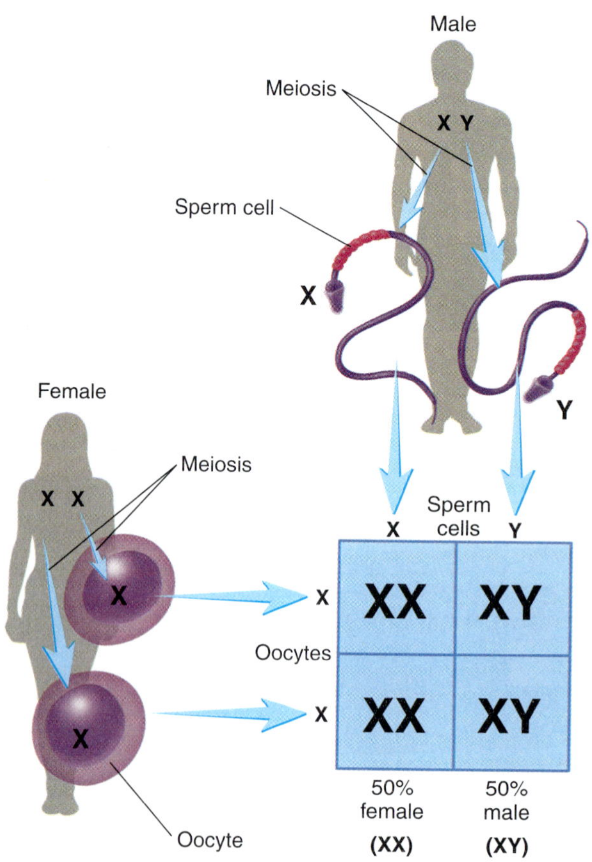

Inheritance of Sex
Figure 29.25

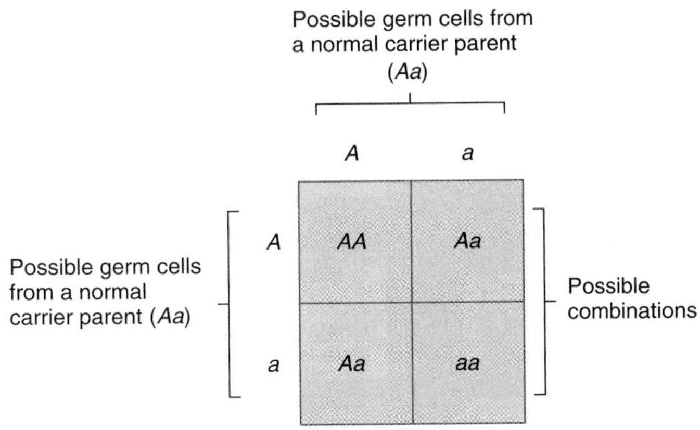

Possible outcome in children:
$\frac{1}{4}$ *AA* (normal) : $\frac{1}{2}$ *Aa* (normal carrier) : $\frac{1}{4}$ *aa* (albino)

Inheritance of a Recessive Trait: Albinism
Figure 29.26

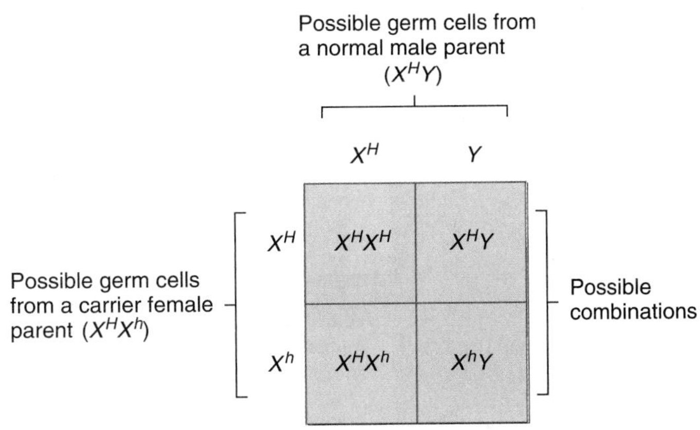

Possible outcome in children:
$\frac{1}{4}$ X^HX^H (normal female) : $\frac{1}{4}X^HX^h$ (carrier female) :
$\frac{1}{4}$ X^HY (normal male) : $\frac{1}{4}X^hY$ (male with hemophilia)

Inheritance of an X-Linked Trait: Hemophilia
Figure 29.27

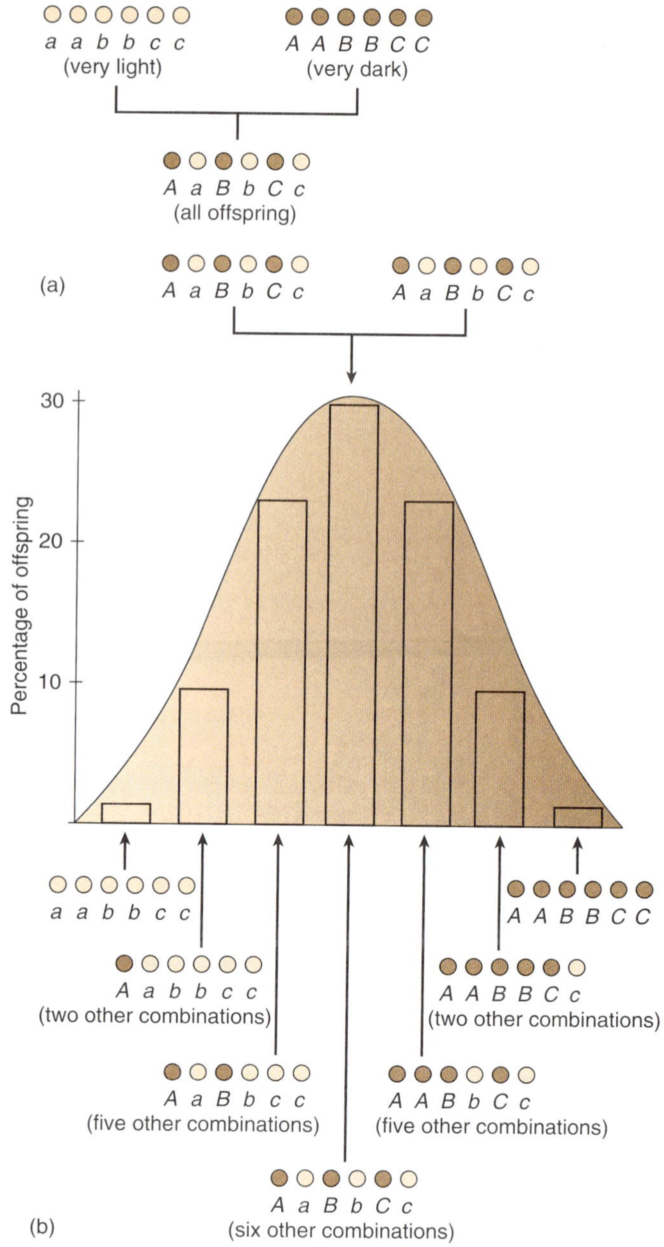

Inheritance of a Polygenic Trait: Skin Color
Figure 29.28

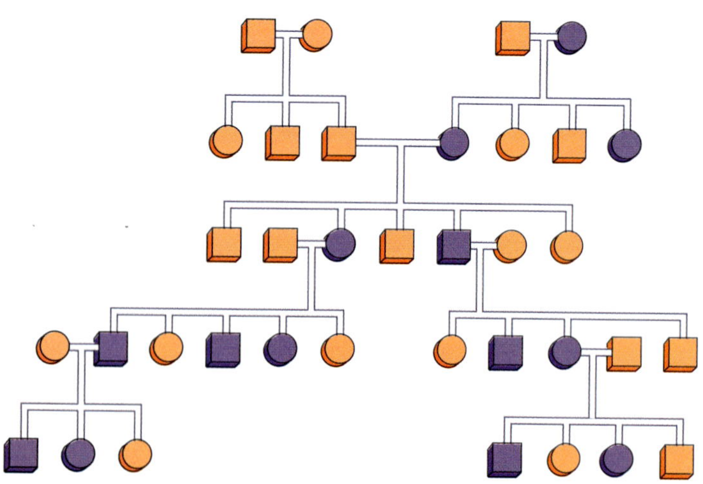

Pedigree of a Simple Dominant Trait
Figure 29.29

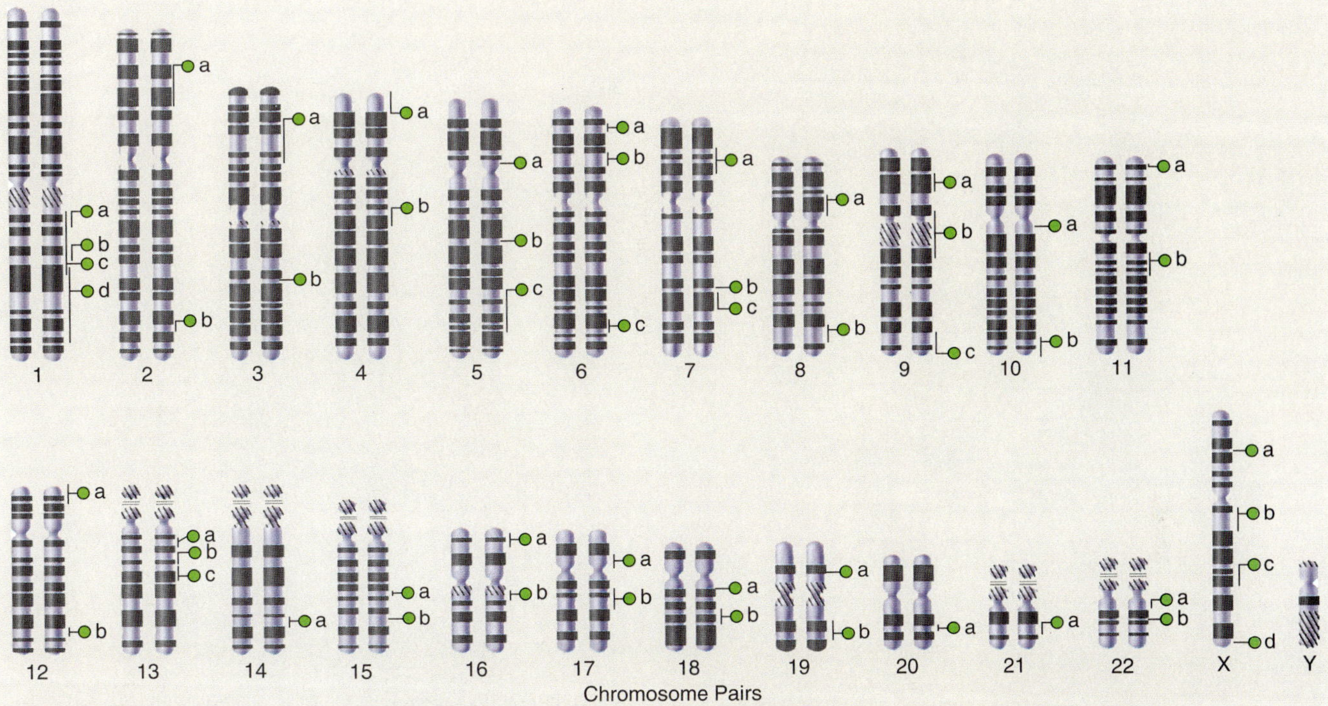

1. a. Gaucher's disease
   b. Prostate cancer
   c. Glaucoma
   d. Alzheimer's disease*
2. a. Familial colon cancer*
   b. Waardenburg syndrome
3. a. Lung cancer
   b. Retinitis pigmentosa*
4. a. Huntington's chorea
   b. Parkinson's disease
5. a. Cockayne's syndrome
   b. Familial polyposis of the colon
   c. Asthma
6. a. Spinocerebellar ataxia
   b. Diabetes*
   c. Epilepsy*
7. a. Diabetes*
   b. Cystic fibrosis
   c. Obesity*
8. a. Werner's syndrome
   b. Burkitt's lymphoma
9. a. Malignant melanoma
   b. Friedreich ataxia
   c. Tuberous sclerosis
10. a. Multiple endocrine neoplasia, type 2
    b. Gyrate atrophy
11. a. Sickle-cell disease
    b. Multiple endocrine neoplasia
12. a. Zellweger syndrome
    b. Phenylketonuria (PKU)
13. a. Breast cancer*
    b. Retinoblastoma
    c. Wilson's disease
14. a. Alzheimer's disease*
15. a. Marfan's syndrome
    b. Tay-Sachs disease
16. a. Polycystic kidney disease
    b. Crohn's disease*
17. a. Peroneal muscular atrophy
    b. Breast cancer*
18. a. Amyloidosis
    b. Pancreatic cancer*
19. a. Familial hypercholesterolemia
    b. Myotonic dystrophy
20. a. Severe combined immunodeficiency
21. a. Amyotrophic lateral sclerosis*
22. a. DiGeorge's syndrome
    b. Neurofibromatosis, type 2

X a. Duchenne's muscular dystrophy
  b. Menkes's syndrome
  c. X-linked severe combined immunodeficiency
  d. Factor VIII deficiency (hemophilia A)

*Gene responsible for only some cases.

The Human Genomic Map
Figure C